DATE DUE

Dec 15 '76			
Dec 7 78			
Jul 12 79			
Dec 14 79			
Dec 3 81			
Nov 10 '81			
Dec 2 '82			
GAYLORD M-2			PRINTED IN U.S.A.

CYTOLOGY AND EVOLUTION

2nd EDITION

CYTOLOGY AND EVOLUTION

2nd EDITION

E. N. WILLMER

FELLOW OF CLARE COLLEGE, CAMBRIDGE
EMERITUS PROFESSOR OF HISTOLOGY
UNIVERSITY OF CAMBRIDGE
CAMBRIDGE, ENGLAND

1970

ACADEMIC PRESS New York and London

ACADEMIC PRESS, INC.
111 Fifth Avenue, New York, New York 10003

United Kingdom Edition published by
ACADEMIC PRESS, INC. (LONDON) LTD.
Berkeley Square House, London W1X 6BA

LIBRARY OF CONGRESS CATALOG CARD NUMBER: 68-26637

PRINTED IN GREAT BRITAIN BY
WILLMER BROTHERS LIMITED, BIRKENHEAD

PREFACE TO THE SECOND EDITION

On page 334 of the first edition of "Cytology and Evolution" reference was made to "the nemertine as a prototype". This was done somewhat tentatively, since the idea had been more or less ridiculed by zoologists from the time it was first put forward by Hubrecht in 1883.

Shortly after the publication of the first edition, I was fortunate enough to spend some months at the University of Chicago where I found the excellent library facilities and the stimulus of a very lively seminar group conducive to further research on nemertines and their potentialities. The results of these further investigations have, after the lapse of some years, culminated in this second edition of "Cytology and Evolution", in which the basic cell types as observed in tissue cultures are first described very much as they were in the first edition. In the second half of the book a discussion is presented on how the various tissues and organs, which we accept as characteristic of vertebrates, could first have come into being, and on how they could have then evolved into their present pattern of form and function. The nemertines have played a large part in this.

The second edition will, I believe, be found to be as unorthodox and unusual as the first, but if it provokes interest and investigation into topics which have been neglected or misinterpreted in the past and suggests ideas for further research its purpose will have been served. It is not intended to be a text-book on cytology; rather, it is meant to be a thought-provoking essay on how cellular mechanisms may possibly have evolved and how they might be reassessed.

I am all too conscious of the vast fields which I have left unexplored and of the many papers which must contain relevant and indeed important material which I have not quoted and have not even seen. I apologise to all who feel that their contributions should have been included. They may, however, contemplate the size of a volume which would have included all relevant information! If, with this essay, I have stimulated research workers to think more inquisitively along evolutionary lines, involving both structure and function, perhaps a lively future of cytological and histological investigation is in store for them.

I would like to express my thanks to all those who have helped to make this volume what it is. They include my hosts and friends at Chicago University; all those who have contributed or made accessible photographs or other material for illustrations; my colleagues in Cambridge who have generously read and commented on some or all of the text; Miss Sylvia Elton who has typed much of this second edition as patiently and competently as she did the first; Mrs. Hood for other excellent typing; Mr. Starling for his photographic skill in preparing a large number of new illustrations; and finally, all those with whom I have discussed the hypotheses, interpretations, and observations which form the nucleus of the book.

Cambridge, England
June, 1969

E. N. WILLMER

CONTENTS

Preface v

Introduction 1

1. **Tissue Culture and the Study of Living Cells**

The Method of Tissue Culture 7
The Behaviour of Cells in Culture 13

2. **Cells in Tissue Culture: Mechanocytes**

Cultures from the Chick Heart 23
The Nature of Mechanocytes 27
The Growth of Mechanocytes 36
Pure Strains of Mechanocytes 46
Synthetic Media 47
The Races of Mechanocytes 48
Myxoblasts and Myoblasts 52

3. **The Growth of Epitheliocytes**

The Products of Epitheliocytes 68
Requirements for Growth 72
Differentiation of Epithelia 73

4. **The Growth of Amoebocytes** 87

5. **Nerve Cells, Neuroglia, and Schwann Cells**

Schwann Cells and Neuroglia 114

6. **The Basic Cell Types**

The Nature of Differentiation 123
Cell Types in Relation to the Germ Layers 134

7. **Cell Types in Sponges and Their Origin**

Cell Lineage in Sponges 150

8. **The Naeglerioid Stage**

The Physiology of *Naegleria* 166

9. **The Blastuloid Stage**

The Use and Storage of Genetic Information 178
From the Unicellular to the Multicellular 184
The Properties of a Blastuloid 187
Gradients 192
The Stability of the Blastuloid and the Internal Environment 198

10. **The Planuloid and Acoeloid Stages**

The Planuloid Stage 206
On the Nature of Mechanocytes 213
Planuloid Organization 224
The Acoeloid Stage 229

11. **The Rhabdocoeloid Stage**

Epidermis 239
Neural Tissues 242
Muscles 249
Alimentary System 250
Urinogenital Systems 251
Pigmentation 255

12. **The Nemerteoid Stage** 260

13. From Nemerteoids to Vertebrates: 1. Neuromuscular System

Divergent Paths from the Nemerteoid 281
Protochordates 285
Other Chordates 294
The Notochord and the Somites 299
The Heterogeneity of Muscle 315
Nervous Control and the Tubular Nervous System 325

14. From Nemerteoids to Vertebrates: 2. The Pituitary and the Pharyngeal Complex

Pituitary 338
Pharyngeal Derivatives 349

15. From Nemerteoids to Vertebrates: 3. Sense Organs

Frontal Organ 363
Ciliated Grooves 365
The Cephalic Organ 370
Eye-Spots, Eyes, and Associated Tissues 379
Eyes of Cephalopod Molluscs 385
The Cephalic Organ and the Hypothalamus 386
Eyes and Pineal Eyes 390
Rods, Cones, and Related Structures 400

16. From Nemerteoids to Vertebrates: 4. The Vascular System, Coelom, and Urinogenital System

The Vascular System 425
The Nature of Body Cavities 432
Urinogenital Tubules in the Invertebrates 436
Urinogenital Tubules: Transition to the Vertebrate Pattern 444
Origin of the Coelom of Vertebrates 453

17. Within the Vertebrates: 1. Urinogenital System

The Divergence of Cell Types 464
Urinogenital Tubes and Sex Cords 467
The Differentiation of Germ Cells 475
The Adrenal Cortex 482
The Problem of the Steroids 487
The Mode of Action of Steroids 488
Ecballism, Emballism, and Renal Function 498
The Pituitary and Control Systems 505

18. Within the Vertebrates: 2. The Alimentary Canal, Blood Cells, Bones, and Teeth

The Alimentary Canal 516
The Cells of the Blood 525
Bones and Teeth 544

19. The Concept of Evolutionary Cytology

The Principle of Perpetual Adaptation 564
The Principle of the Constancy of the Specific Environment: The Agoranome 566
The Principle of Balanced Activity 569
The Principle of the Permanence of the Genome 573

Author Index 577

Index of Animal and Plant Names 587

Subject Index 591

INTRODUCTION

The transition from the Victorian carriage to the racing car of today or from the Wright brothers' flying machine to the "Concorde" are object lessons in evolution. Step by step, some features have been modified and adapted; others have been discarded. Some entirely new characters have appeared, while certain basic structures have been maintained relatively unchanged throughout.

A knowledge of this history is not perhaps vital to the automobile or aeronautical engineers of today, but it certainly helps the ordinary man to explain and understand the particular pattern which has evolved, and it demonstrates the extreme changes which may occur by constant small "adaptations". Similarly a study of *human* embryonic development alone may be sufficient for the practising gynaecologist, but "what should they know of England, who only England know"? A deeper understanding of the embryology of other animals may lead to better practice.

Step by step man has evolved from more primitive forms which are now forever lost. So also have other animals. Nevertheless, by combining a study of the evolution of a wide variety of animals, including man, with a study of their embryological development, certain basic patterns of the evolutionary process can be discerned; hypothetical, if not actual, "common ancestors" begin to appear; and evolutionary history becomes more nearly a reality.

The basic units of the animal body are its cells, and, like the organs which they constitute and the animals to which they belong, they, too, have evolved. The evolution of animals is accepted and "the outline of history" has been written; the evolution of the organs of the animal body is the basis of much zoological training; but the story of the evolution of the constituent cells and tissues is only beginning to be told. Cells are minute, and the methods of biochemistry and of molecular biology, refined as many of them now appear to be, are, by contrast, gross or blundering. Fossils, which are so valuable in tracing the evolution of bones and organs, cannot help with cells. In recent years, the advent of new microscopical methods and the development of better physical and chemical techniques for studying cells in the living condition, and in their interactions with one another, have made the study of comparative cellular physiology a much more practicable and profitable pursuit and with its help some insight can be gained into the evolution of cells and tissues. Even so, by comparison with the refinement of the cells and their physiological processes, the methods are crude.

In the animal body each organ is well known to be specially adapted for carrying out certain definite functions, and yet the whole animal is built to a precise plan characteristic of the species. In many ways, therefore, each species of animal must be considered as a somewhat special case and this specialization naturally entails particular problems for the physiologist. On the other hand, it is also well recognized that throughout the whole kingdom of vertebrates, for example, there is a common plan of organization, and that many of the general functions carried out by the different organs may remain much the same in all species; within this common plan, however, there may also be conspicuous differences in detail, from species to species, according to the special adaptations of the animal to its particular environment. These differences, unfortunately, sometimes tend to obscure the main pattern, for often, because of local conditions, human interest or some other essentially trivial cause, there is attached to them a much greater significance than they deserve in relation to the pattern as a whole.

The student of evolutionary cytology is therefore always up against the difficulty of deciding whether any particular structure or function has general significance, or whether it should be regarded as some aberrant specialization, produced by the organism in question to meet the peculiar demands of the environment, this last term being used in its widest sense. It need hardly be emphasized that the various tissues and structures in animals are often extremely complex and intricately organized, so that when the pattern of organization is only slightly modified there may be far-reaching secondary effects, and it is thus a matter of some difficulty to comprehend the nature of the primary modification and to separate it from the secondary results produced. At first sight, the adrenal medulla and a sympathetic ganglion do not appear as very similar structures, but, on further investigation of their embryological and phylogenetic development, the evolution of both from similar cells becomes obvious and the functions of both can be, at least partly, interpreted as variations on a common theme. The same might be said about the inner ear of man and the lateral-line organs of fishes, about his thyroid gland and the endostyle of *Amphioxus*, and many other similar examples.

On the anatomical scale, such homologues among the vertebrates as pectoral fins, forelegs, wings, and arms are all, of course, now universally accepted; but on the cellular level there is still much that is mysterious.

While the classical methods of histology and cytology, with their accent on cell structure and histological organization, can contribute much to the picture of the evolution of many of the various organs of the vertebrate body, they cannot reach beyond a certain point, because functional differences do not always involve parallel and visible structural or morphological differences. It is therefore obvious that the newer techniques which give information about histo- and cyto-chemistry and, especially, about cell behaviour will, in the future, have an ever-increasing contribution to make.

In the chapters which follow, the main accent will therefore be placed on those techniques and their results which allow cells to be investigated during activity, and a search will be made for those features of cell behaviour or their chemical activity which enable cells to be classified together as being similar on the one hand, or differentiated from each other in a fundamental manner on the other. Naturally, there is always the inherent difficulty of deciding what constitutes a fundamental manner, and it is in elucidating this point that many of the real problems are raised. A few examples may illustrate the general standpoint which will be adopted.

Ciliated or flagellated cells occur in an extraordinary variety of situations in the bodies of animals belonging to nearly all the phyla. Yet, in all those animals which have been investigated except the Crustacea, the cilia and flagella, as seen by the electron-microscope and other optical techniques, seem to be built on a common plan, with variations of a rather secondary character (Fig. 1). All have a core with nine peripheral fibres, which are double, and most have two central fibres. Is the conclusion to be drawn from this that there is only one possible type of flagellum, or is it that the essential flagellar pattern was developed early in evolution and has continued to be inherited as a useful potentiality by certain groups of cells, even if it is not actually called into play? Surely the latter is the more likely explanation, even if it does raise difficulties concerning the distribution of ciliated cells in different regions or parts of animals and with regard to the factors which cause their appearance or differentiation.

Similarly, muscular contraction and the presence in the cell of the proteins actin and myosin seem to be closely linked; are these proteins the only possible contractile elements? Or is their production again a primitive basic trait which has been handed on from one generation to another and from one animal to another to be produced, or not, according to circumstances and, when it is used, to be incorporated into special mechanisms adapted to serve particular ends?

The study of any one animal in isolation, whether it be man or amoeba, does not allow the wood to be seen for the trees. The specialization cannot be differentiated from the fundamental and common plan. Consequently it is necessary to try to extract the skeleton-key for the evolution of an organ or structure by studying features common to large numbers of organisms, and in studying the cells of that organ it may often be advantageous to rob them, if possible, of some of their higher flights of specialization so that their more fundamental basic pattern may emerge more clearly.

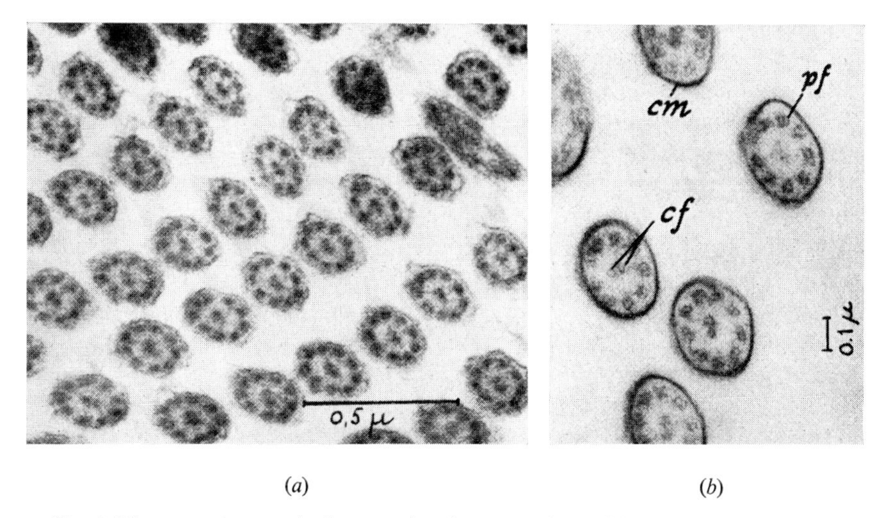

FIG. 1. Electron-microscopic photographs of cross-sections of (*a*) cilia from a mammalian renal tumour and (*b*) cross-sections of cilia from *Paramecium*, showing the typical pattern and arrangement of internal fibres. Note the similar arrangement of two central (cf) and nine pairs of peripheral (pf) fibres enclosed within the cell membrane (cm) in cilia from these very different sources. (Mannweiler and Bernhard, 1957; Sedar and Porter, 1955.)

Fortunately it seems likely, when tissues are grown by certain of the "tissue-culture" techniques, that the constituent cells almost immediately lose some of their morphological and physiological specializations, either temporarily or permanently, and such "simplified" cells may then be able to point to the road along which they have evolved. Admittedly, tissue-culture techniques are not without their specializing actions on cells, for cells have an extraordinary capacity for adapting themselves to their surroundings and, within limits, of adapting their surroundings to themselves. Indeed, evidence is accumulating from the study of "pure strains" of cells that prolonged life *in vitro* leads to many fundamental changes in cell activity, which can be profitably regarded as adaptations to the particular conditions of cell culture. Nevertheless, judicious comparison between the behaviour of cells in tissue culture, in embryological material, in pathological states, and their behaviour in corresponding or homologous tissues in a wide variety of organisms may often throw the main patterns of cell behaviour into relief. The actual meaning to be applied to the term "homologous" will have to be considered rather more closely at a later stage.

In bacteriological studies, it has become increasingly obvious in recent years that bacteria can often become adapted very quickly to life in surroundings which were at first unfavourable to them, even to the extent of producing new enzymes to deal with unfamiliar substrates. Whether this occurs by selection of favourable mutants or by some more directly adaptive mechanism is not always clear, but the facts are undisputed and there is very good evidence that enzymes can actually be produced *de novo* to deal with new substrates. How far similar adaptability is possible among the cells of metazoa is not yet so clear, but it must certainly be kept in the foreground as a possibility, and even a probability, when we are considering how cells could behave when the conditions of their life are altered. For example, there is now evidence that two clones of human cells, both originally derived from the same single cell, have developed quite different properties, one strain being highly malignant, the other not malignant at all (Sanford *et al.*, 1954).

The growth of cells *in vitro* by tissue culture methods has thrown into prominence, for the author at least, many important features of cell physiology. For this reason, some account will first be given of cells as they are seen under these conditions. This will be followed by a discussion of a possible course of evolution from the "unicellular" or "acellular" protozoa to the simple metazoa and hence to organisms containing several basic cell types. From these early beginnings of "phylogenetic differentiation" the evolutionary process will be followed through several possible stages till an organism not unlike some of the existing nemertine worms may have arisen. The cytology and cellular physiology of such organisms will then be investigated in relation to subsequent evolutionary processes which eventually culminate in the production of vertebrates in general and of primates in particular. Throughout the whole argument, the evolution of cells and of cellular behaviour-patterns will dominate the reasoning, and the sophisticated organs and tissues of man will then be seen as the results of perpetual modification and selection of cellular form and function.

References

Mannweiler, K. I., and Bernhard, W. (1957). Recherches ultrastructurales sur une tumeur rénale expérimentale du Hamster. *J. Ultrastruct. Res.* **1**, 158.

Sanford, K. K., Likely, G. D., and Earle, W. R. (1954). The development of variations in transplantability and morphology within a clone of mouse fibroblasts transformed to sarcoma-producing cells *in vitro*. *J. Natl. Cancer. Inst* **15,** 215.

Sedar, A. W., and Porter, K. R. (1955). The fine structure of cortical components of *Paramecium multimicronucleatum*. *J. Biophys. Biochem. Cytol.* **1,** 583.

CHAPTER 1

TISSUE CULTURE AND THE STUDY OF LIVING CELLS

The critical study of cells in tissue culture can, as indicated, provide a first approach to the study of the cells of the higher animals, and this approach is both illuminating and provocative. Some of the various methods used and the results obtainable with them must therefore be first reviewed in outline, for the behaviour of cells in culture depends very greatly on the method of culture employed. Each method imposes different conditions on the cells, and it is necessary to appreciate these effects in translating results from tissue-culture experiments to the behaviour of cells in the animal. Tissue culture, even if it does nothing else, succeeds admirably in emphasizing the plasticity and vital qualities of cells on the one hand, and in pointing towards the existence of a limited number of extremely characteristic and probably fundamental modes of cell behaviour on the other. It is this latter characteristic which is of peculiar interest in the study of evolutionary cytology.

The Method of Tissue Culture

The development of the tissue-culture method might have occurred long ago had it not had to wait for two main developments of technique and of knowledge in other fields. First, for all tissues from warm-blooded animals a satisfactory aseptic technique had to be evolved. Tissues isolated in culture have only very limited defence against bacteria and moulds, so that their strict asepsis or the use of antibiotics is necessary, and both are nowadays commonly employed. Second, the tissues of higher animals only survive and remain active in rather special media, so that much had first to be learned about balanced salt solutions, nutrients, growth stimulants, etc., before any real progress could be made in the way of keeping tissues alive outside the body for any length of time. Although there had been earlier beginnings (Loeb, 1897, 1898), Harrison (1907, 1910) is credited with the first successful tissue culture when he obtained growth of amphibian nerve fibres, *in vitro*, in a medium of clotted lymph. From these simple beginnings, progress was at first rapid, and numerous techniques, each with its own special application, were quickly developed. To the pioneering work of Carrel and his school at the Rockefeller Institute from about 1912 onwards are due many of the main methods of tissue culture (Carrel, 1912, 1913, 1923). Indeed, this school, together with that of Fischer (1930) in Copenhagen, produced much of the basic description of types of cell behaviour *in vitro*. In England, the division of cells *in vitro* was first described as a continuous process by Strangeways (1922) who, together with Canti (Strangeways and Canti, 1927; Canti, 1928), produced in 1927 a revolutionary, and now classical, cine-film which brought this process, and indeed many other essentials of cell behaviour, vividly before the eyes of the world. Much of the early work was carried out on embryo-chick tissues because these could be readily obtained in a very actively growing state, uncontaminated with bacteria, etc. The medium used for the growth of many of these pioneering cultures was a mixture of fowl blood plasma and an extract of crushed chick embryo in a physiologically balanced saline solution containing Na^+, K^+, Ca^{2+}, Mg^{2+}, Cl^-, PO_4^{3-}, HCO_3^-, and glucose. Fowl plasma was used because, unlike that of mammals, it does not clot readily on standing, but produces a firm coagulum in the presence of tissue juices. The "embryo-extract" contains both blood-coagulating and growth-promoting factors and can be readily obtained in a sterile condition. The strictest asepsis was preserved throughout the whole technique.

This was many years ago and the position has now greatly changed. Since 1950, various synthetic media, in which all the constituents are of known chemical composition, have been successfully developed for certain strains of "normal" cells (Morgan *et al.*, 1950; Evans *et al.*, 1953; Healy *et al.*, 1954; Waymouth, 1955; Morgan, 1958); antibiotics are also extensively used nowadays to simplify the complicated techniques of asepsis, though there are certain limitations and possible dangers in their use. For example, the sulphonamide drugs, while relatively harmless to cells *in vitro* in comparison with their action on bacteria, do produce a reversible inhibition of cell division when used at concentrations of more than about 1:1,000 (Jacoby *et al.*, 1941). Sulphanilamide itself is also known to poison the enzyme carbonic anhydrase. Other antibiotics, e.g. actinomycin and puromycin, inhibit protein synthesis.

Tissue culture today thus promises to become much simpler and to have more of its initially numerous variables under control. In consequence, more and more problems seem to be capable of solution by its use.

The primary aim of tissue culture is to maintain cells and tissues alive, active, differentiating, functioning, or growing under conditions in which the various processes can be observed directly or indirectly, measured and analysed in a situation uncomplicated by the presence of the interfering influences of other tissues in the body. Naturally the method can be applied to cytological, histological, embryological, biochemical, and pathological problems, and appropriate techniques have consequently been devised for particular purposes.

It is not necessary to describe the details of all the numerous techniques available; but some of the more important conditions of culture must be mentioned, so that the results obtained with them may be evaluated. The simplest technique is that known as the hanging-drop. In this, a small fragment of tissue (not more than few mm^3) is placed in a drop of a physiologically balanced salt solution or nutrient medium, very often mixed with blood plasma, on a coverslip, and the whole is inverted over a hollow-ground slide and ringed with paraffin wax in order to maintain an air-tight seal. This, in essentials, was the original technique used by Harrison for the express purpose of finding out whether nerve processes grew out from nerve cells in the spinal cord of the frog, and by which he produced a major piece of evidence in support of the neurone theory of the central nervous system. By this method he was able to observe directly under the microscope how, after a few hours, the processes of the cells thrust themselves outwards as nerve fibres, by a sort of amoeboid movement, into the coagulated lymph which he used as medium (Fig. 1.1). Incidentally, the essentially amoeboid character of the terminals of the nerve processes which he then described is something which is perhaps insufficiently appreciated by neurologists and neurophysiologists of the present day as a potential property of nerve cells within the living body. Conditions *in vivo* and *in vitro* are, of course, very different, but it is unlikely that complete stability of the processes of nerve cells reigns in the body.

The hanging-drop method, with modifications, is eminently suited to cytological studies of short duration, and can be applied to a wide variety of tissues. Unless, however, the medium is renewed at frequent intervals or the tissues are transplanted to fresh medium, a limit of a few days only is set to their activity, by the scarcity of oxygen, high CO_2, lack of food, or the accumulation of toxic metabolites. In the early stages of such cultures, cells of various types may be persuaded by suitable media, temperature, etc., to emerge from the original piece of tissue (Fig. 1.2), especially if this is embryonic in character, or if more adult tissue has been previously exposed to the action of a dilute solution of trypsin (Simms and Stillman, 1937). On the whole, the younger the tissue the more actively do the cells emigrate; for this reason, embryonic tissues have been most studied. The outgrowing cells mainly creep upon the surface of the glass or on the interface between the

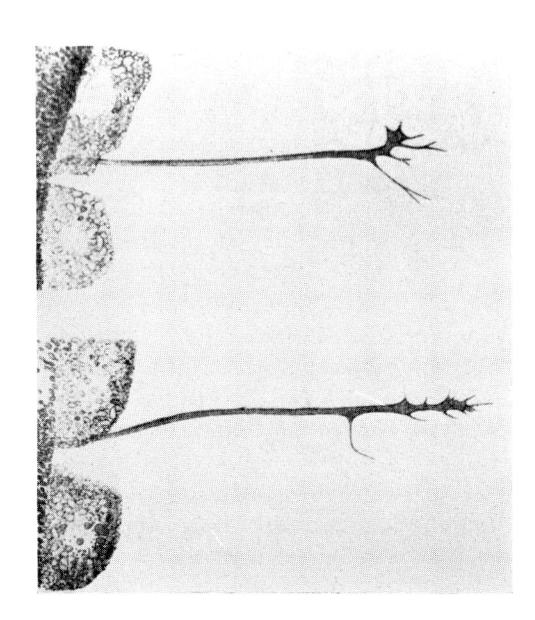

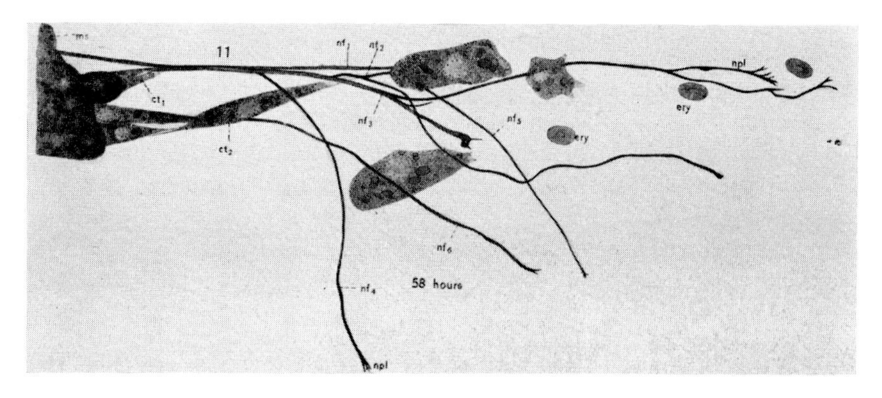

FIG. 1.1. The outgrowth of nerve fibres from the spinal cord of a frog in a medium of clotted lymph. In the upper figure the amoeboid processes at the ends of the fibres are clearly seen. ms, central mass; ery, erythrocyte; nf., nerve fibre; npl, protoplasmic end of fibre; ct, connective tissue cell. (Harrison, 1907, 1910.)

medium and the air. Cells which become loose in the fluid tend to float away and get lost. If the medium is in the form of a gel, e.g. a plasma coagulum, then cells may also be able to invade the substance of the gel. The thinner the gel the more extensively is it invaded by the growing cells. Since, however, some cells liquefy the plasma coagulum, gels which are too thin sometimes break down, and thus there is a limit to the extent to which the blood plasma can be diluted for use as a supporting framework for the cells of each tissue.

In hanging-drop cultures the cells can be directly observed under the highest powers of the microscope, either with direct illumination or with such devices as polarized light, dark-field illumination, phase contrast, or other similar methods. They are also immediately available for experimental study, by microdissection, microinjection, electrical stimulation, and most of the cytochemical techniques. Even more important, perhaps, is the fact

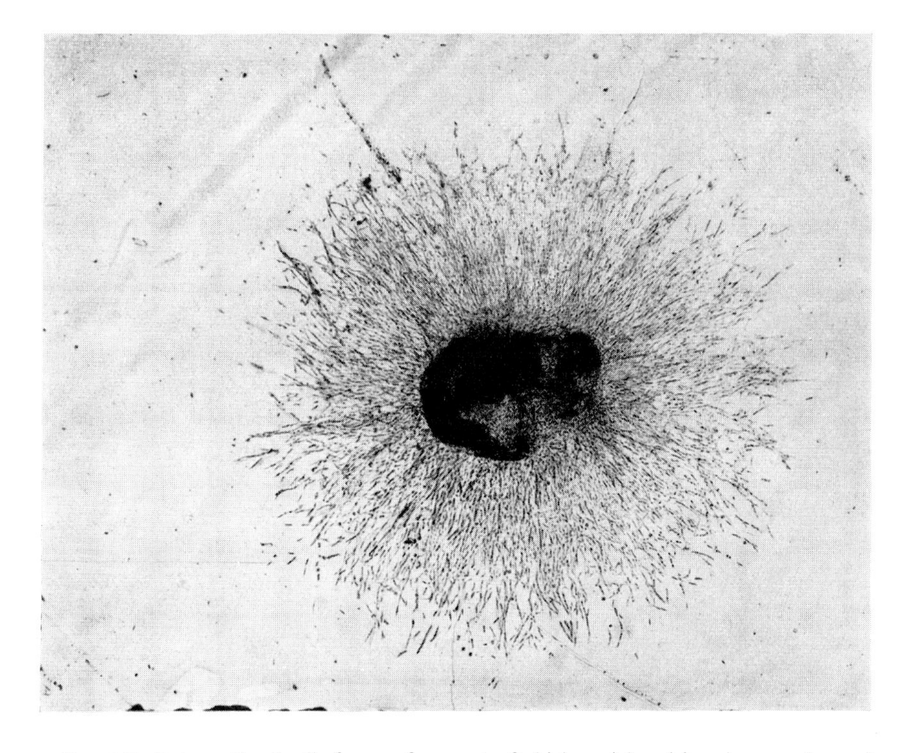

FIG. 1.2. Outgrowth of cells from a fragment of chick perichondrium in a medium of plasma and embryo extract. (Photograph by F. Jacoby.)

that under these conditions the activities of the cells may be photographically recorded on cine-films, or directly on recording paper, either at normal speed, or at some higher speed when the processes to be observed are slow; the photography can also be combined with a stroboscopic device to record any rhythmic movements which may be occurring at too high a frequency for direct observation, as in the beating of cilia.

Although the cells under these conditions are ideally situated for study from the optical point of view, there is one major difficulty. The cells which migrate outwards from the central explant are not often immediately recognizable in appearance as normal constituents of the tissue explanted: in other words, they almost immediately change their shape and pattern as they leave their normal position. Thus, it must always be remembered that the cells which emerge are, in general, not physiologically, and certainly not morphologically, identical with the original cells of the tissue. Not only do the cells themselves change, but the histological organization of the original tissue ceases to be maintained in the zone of outgrowing cells; it may, however, be so retained in the main mass of tissue, to a greater or lesser extent, depending on the conditions of culture.

The change in the outgrowing cells nearly always involves a simplification of their morphology. Moreover, the character of the change in the cells depends on the mechanical and physico-chemical properties of the medium into which they migrate; it depends also on whether the cells find a suitable surface upon which to cling, for when they become detached from the tissue into a fluid medium they immediately round up and sometimes remain relatively inert at the bottom of the drop, just as the white cells in the blood are spherical while they are in the circulation, but become more or less amoeboid when they make contact with a surface which they "wet" and to which they can adhere. This inertness of isolated cells may be only relative or temporary because it has now been found possible to maintain active growth in cells kept in a constantly agitated fluid medium where the cells are freely suspended (Earle *et al.*, 1954, 1955; Owens *et al.*, 1954).

Cells in culture, to some extent following the lines of least resistance, tend to flatten on to surfaces, to become spindly when enclosed in the substance of the coagulum, and to follow the pattern of any fibrous supporting framework which may be provided for them. In this connection it is worth while to contrast the local conditions of a columnar epithelial cell *in situ* with those which it meets when "growing" on the surface of a glass coverslip. *In situ* it has one surface resting on a "basement membrane" and bathed with tissue fluid. The opposite surface of the cell may be in contact with the fluid in the lumen of some hollow viscus. All other surfaces of the cell are adjacent to the surfaces of other similar cells. *In vitro* nearly half the cell surface is in contact with glass, nearly half is bathed in some nutrient, but certainly abnormal medium, and the small remainder is in contact with adjacent but similarly abnormal cells. It is clear therefore that epithelial cells at least cannot be expected to function exactly as they do in the body when placed under these artificial conditions; deductions from cell behaviour in tissue cultures have therefore to be applied with the greatest caution to *in vivo* situations.

The Behaviour of Cells in Culture

The simple methods of tissue culture, as just described, immediately and dramatically emphasize several points which are essential to the appreciation of cell behaviour as a whole. The cells of a tissue react almost immediately to changed conditions; many respond not only biochemically but also by increased movement and by changes of shape and form, which in tissue cultures are often in the direction of greater simplification with loss of morphological characteristics. The movement of the cells, in most cases, eventually leads to a thinning out of the explant and to a more uniform spacing of the cells on the available surface, so that each cell gets a suitable living-space with sufficient supplies of oxygen, etc. Indeed, in cultures obtained either by first dissociating tissues with Ca^{2+}- and Mg^{2+}-free trypsin solutions and then making suspensions of the separated cells and allowing them to settle on the glass surface of the culture vessel (Moscona, 1952; Rinaldini, 1954) or by some similar technique, it has been shown that a given size of vessel, as with bacteria, can support only a limited number of cells, irrespective of the amount of nutrient medium in the vessel (Earle *et al.*, 1951). Necrosis sets in when this number is exceeded.

The rate and direction of movement of cells is always partly determined by the presence of neighbouring cells which may be of similar or of different types, and there are many other factors, some intrinsic in the actual tissue itself and others dependent on the environment, which determine the activity of the cells in a culture. Among external factors, the orientation of any micellar particles in the medium is of great importance. For example, if a fragment of embryonic heart muscle is placed at the centre of a plasma droplet suspended within the space of a minute triangle of glass rod, the tissue first of all brings about the coagulation of the plasma and then its cells begin to emerge into the plasma. They do not, however, emerge evenly, but come out in greater numbers and proceed more quickly in directions at right angles to the glass sides of the triangle (Figs. 1.3, 1.4). The reason for this is that, during and after the formation of the clot, there is some contraction of the gel, and this produces uneven stresses in the whole coagulum (Weiss, 1929). It is these lines of force which appear to guide the movement of the cells, probably not directly but because the fibrous protein particles in the gel have become orientated during the contraction. In some way these particles induce the cells to elongate in a direction parallel to their own long axes. A similar result is obtained by culturing cells on the scales of certain fishes when it is found that the cells align themselves along the grooves in the scales (Figs. 1.5, 1.6) (Weiss, 1959).

Not only can the cells be orientated in this way, but their shapes may be similarly influenced; cells tend to be more spindle-shaped when they are

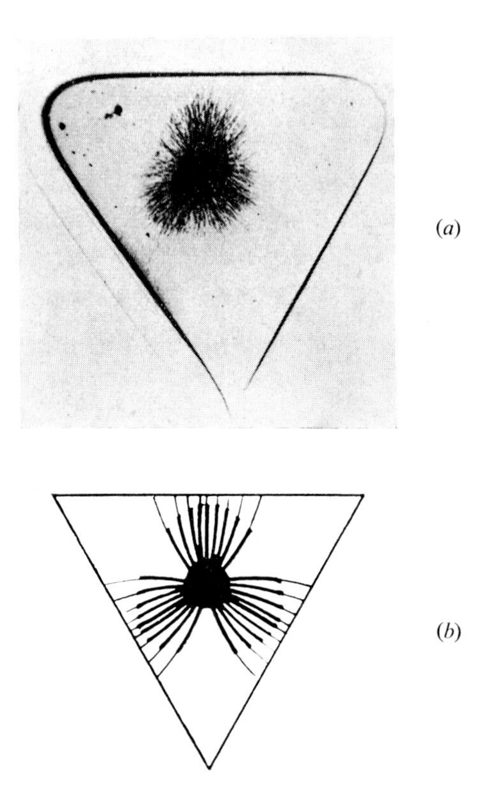

FIG. 1.3. (*a*) Cells emerging from a tissue culture suspended in a plasma clot contained within a triangular frame. Note the tendency for the cells to orientate themselves towards the sides, rather than towards the angles, of the triangle. (*b*) Diagram indicating lines of tension. (Weiss, 1929.)

under tension or lying on an orientated surface. It is notable that cells in a coagulum are always more truly spindle-shaped (i.e. in three dimensions) than when growing on a surface. The microscope, which gives a sharp image only of a single optical plane, thus easily leads to false interpretations of the size and shape of cells unless the conditions are fully appreciated "in the solid". The spindle-shaped cells often appear much smaller than those flattened on the surface, though in fact their volumes are presumably similar.

The first and vitally important point, therefore, which tissue culture in this simple form emphasizes, is the lability of cell morphology in response to changed conditions, and this property of cells is seldom fully appreciated by histologists and cytologists accustomed only to the study of fixed and stained material. It may perhaps be argued that conditions in tissue culture are very different from those in the body and, consequently, that they are in

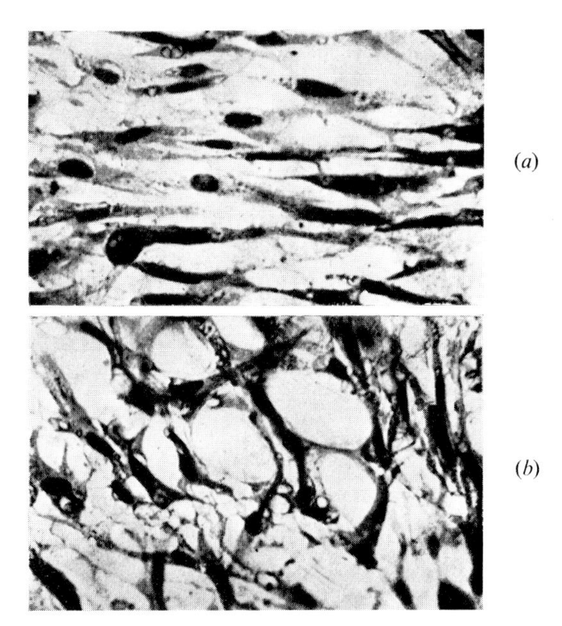

FIG. 1.4. High-power photographs of the cells in the culture shown in Fig. 1.3a. (*a*) Cells facing the sides of the triangle; (*b*) cells facing the angles of the triangle. (Weiss, 1929.)

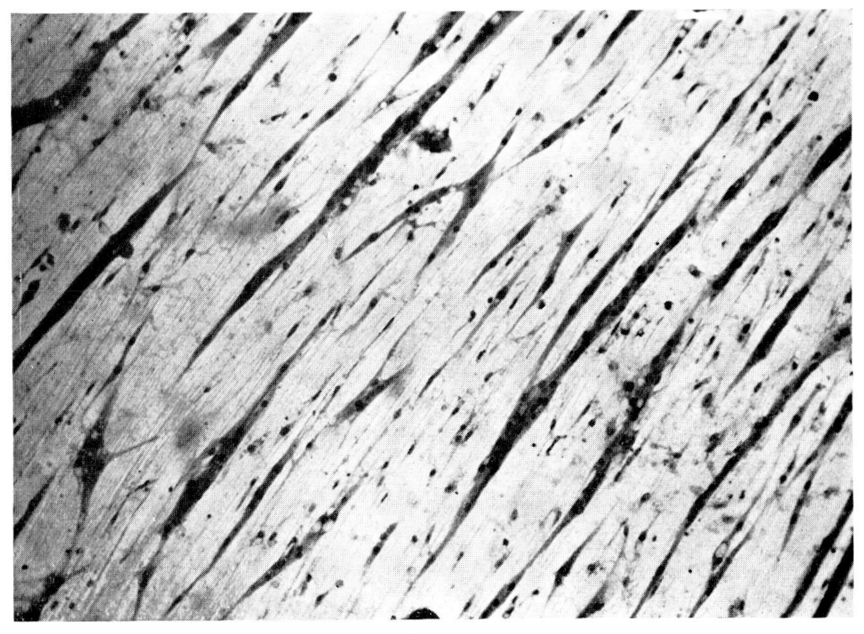

FIG. 1.5. Cells of a tissue culture orientated along the grooves of a fish-scale. (Photograph by P. Weiss.)

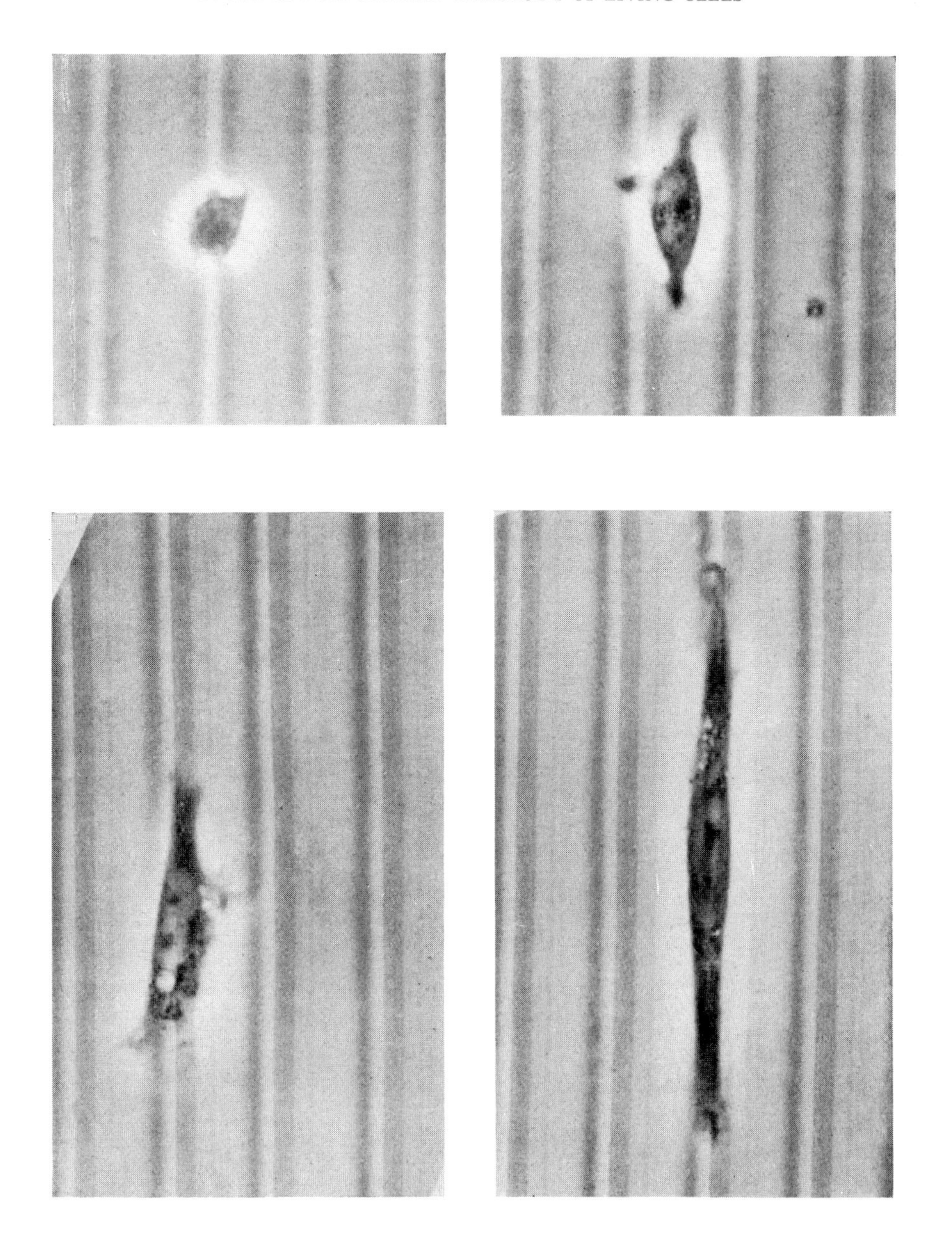

FIG. 1.6. Four stages in the orientation of a cell along the groove of scored glass. (Photographs by P. Weiss.)

some ways irrelevant to normal cell physiology: in general, this is certainly true. Nevertheless, the conditions in a wound, for example, or in the culture chambers which have been used for the observation of cells *in situ* in different situations in the animal body, as in the anterior chamber of the eye, in the "windows" sometimes used in the ears of rabbits (Sandison, 1928; Clark and Clark, 1932; Ebert *et al.*, 1940), or in the skin-flaps on the backs of mice (Algire, 1943; Algire and Legallais, 1949) are not actually very different from those in a tissue culture, and it is generally assumed that the cells in these situations are doing very much what they normally do, or can do, under conditions which can be truthfully described as "physiological". The main and most constant difference between the environment of cells in culture and that of cells in the body probably lies in the presence or absence of a circulating blood supply and of the cells which come with it. Other factors may, of course, also arise to modify the immediate surroundings of the cells, but these depend very much on the method of culture and are sometimes important and sometimes not.

The second point which emerges from the study of cultures involving the use of tissue explants concerns the relationship between the cells which emigrate from the explant and those which were present in the original tissue when it was explanted. What are the changes, other than those of purely morphological character and those dependent on the mechanical conditions reigning in the culture, which occur in the emigrating cells, and how far are the cells of the outgrowth different from those in their original situations? This is of such fundamental importance that it will form the subject matter of the succeeding chapters and provide a starting point for a study of evolutionary cytology.

Nearly everything which has been said in this chapter about tissue culture has had direct reference to the hanging-drop method. In cultures in Carrel flasks, in roller tubes, and made by all those methods whereby the cells are encouraged to emigrate into the medium (Parker, 1958; Willmer, 1958, 1965), the same considerations concerning cell morphology and function apply, and they apply with even greater force to the more recent techniques in which cells are first obtained in suspension and are then grown under conditions more akin to those in which bacteria are usually cultured. On the other hand, in the methods of "organ culture" in which pieces of tissue are cultured in such a way as to restrict outgrowth, and in which attention is concentrated on preserving functional activity within the explant, the cells may remain very much as they appear *in situ* in the body and their functional activities are probably much more akin to those in which they are normally involved. For many purposes, this type of culture is ideal, and observations dependent upon its use will be freely quoted, but for continuous studies of cell behaviour it is not so convenient as those other methods whereby the comparatively isolated cells can be kept under observation almost continuously. Broadly speaking, "organ cultures" are used for investigations of tissue function and for studies of organized growth. For unorganized growth and cytological studies, other methods are at present more suitable, but the results obtained by any method always require interpretation on the lines which are under discussion.

References

Algire, G. H. (1943). An adaptation of the transparent chamber technique to the mouse. *J. Natl. Cancer Inst.* **4,** 1.

Algire, G. H., and Legallais, F. Y. (1949). Recent developments in the transparent chamber technique as adapted to the mouse. *J. Natl. Cancer Inst.* **10,** 225.

Canti, R. G. (1928). Cinematograph demonstration of living tissue cells growing *in vitro. Arch. Exptl. Zellforsch. Gewebezücht.* **6,** 86.

Carrel, A. (1912). On the permanent life of cells outside the organism. *J. Exptl. Med.* **15,** 516.

Carrel, A. (1913). Artificial activation of the growth *in vitro* of connective tissue. *J. Exptl. Med.* **17,** 14.

Carrel, A. (1923). A method for the physiological study of tissues *in vitro. J. Exptl. Med.* **38,** 407.

Clark, E. R., and Clark, E. L. (1932). Observations on living perfused blood vessels as seen in a transparent chamber inserted into the rabbit's ear. *Am. J. Anat.* **49,** 441.

Earle, W. R., Sanford, K. K., Evans, V. J., Waltz, H. K., and Shannon, J. E. (1951). The influence of inoculum size on proliferation in tissue cultures. *J. Natl. Cancer Inst.* **12,** 133.

Earle, W. R., Schilling, E. L., Bryant, J. C., and Evans, V. J. (1954). The growth of pure strain-L cells in fluid-suspension cultures. *J. Natl. Cancer Inst.* **14,** 1159.

Earle, W. R., Bryant, J. C., Schilling, E. L., and Evans, V. J. (1955). Growth of cell suspensions in tissue culture. *Ann. N. Y. Acad. Sci.* **63,** 666.

Ebert, R. H., Sanders, A. G., and Florey, H. W. (1940). Observations on lymphocytes in chambers in the rabbit's ear. *Brit. J. Exptl. Pathol.* **21,** 212.

Evans, V. J., Shannon, J. E., Bryant, J. C., Waltz, H. K., Earle, W. R., and Sanford, K. K. (1953). A quantitative study of the effect of certain chemically defined media on the proliferation *in vitro* of strain-L cells from the mouse. *J. Natl. Cancer Inst.* **13,** 773.

Fischer, A. (1930). "Gewebezüchtung". Müller & Steinicke, Munich.

Harrison, R. G. (1907). Observations on the living developing nerve fibre. *Proc. Soc. Exptl. Biol. Med.* **4,** 140.

Harrison, R. G. (1910). The outgrowth of the nerve fibre as a mode of protoplasmic movement. *J. Exptl. Zool.* **9,** 787.

Healy, G. M., Fisher, D. C., and Parker, R. C. (1954). Nutrition of animal cells in tissue culture. (IX). Synthetic medium No. 703. *Can. J. Biochem. Physiol.* **32,** 327.

Jacoby, F., Medawar, P. B., and Willmer, E. N. (1941). The toxicity of sulphonamide drugs to cells *in vitro. Brit. Med. J.* **ii,** 149.

Loeb, L. (1897). "Über die Entstehung von Bindegewebe, Leucocyten und roten Blutkörperchen aus Epithel und über eine Methode isolierte Gewebezelle zu züchten". Chicago, Illinois.

Loeb, L. (1898). "Über Regeneration des Epithels. *Arch. Entwicklungsmech. Organ.* **6,** 297.

Morgan, J. F. (1958). Tissue culture nutrition. *Bacteriol. Rev.* **22,** 20.

Morgan, J. F., Morton, H. J., and Parker, R. C. (1950). Nutrition of animal cells in tissue culture. 1. Initial studies on a synthetic medium. *Proc. Soc. Exptl. Biol. Med.* **73,** 7.

Moscona, A. (1952). Cell suspensions from organ rudiments of the early chick embryo. *Exptl. Cell Res.* **3,** 535.

Owens, O. von H., Gey, M. K., and Gey, G. O. (1954). Growth of cells in agitated fluid medium. *Ann. N. Y. Acad. Sci.* **58,** 1039.

Parker, R. C. (1958). "Methods of Tissue Culture". Hamish Hamilton, London.

Rinaldini, L. M. (1954). A quantitative method for growing animal cells *in vitro*. *Nature* **173,** 1134.

Sandison, J. C. (1928). The transparent chamber of the rabbit's ear, giving a complete description of improved technique, of the structure and introduction and general account of growth and behaviour of living cells and tissues as seen with the microscope. *Am. J. Anat.* **41,** 447.

Simms, H. S., and Stillman, N. P. (1937). Substances affecting adult tissue *in vitro*. 1. The stimulating action of trypsin on fresh adult tissues. *J. Gen. Physiol.* **20,** 603.

Strangeways, T. S. P. (1922). Observations on the changes seen in living cells during growth and division. *Proc. Roy. Soc.* (*London*) **B94,** 137.

Strangeways, T. S. P., and Canti, R. G. (1927). The living cell *in vitro* as shown by dark field illumination and the changes induced in such cells by fixing agents. *Quart. J. Microscop. Sci.*, **71,** 1.

Waymouth, C. (1955). Simple nutrient solutions for animal cells. *Texas. Rept. Biol. Med.* **13,** 522.

Weiss, P. (1929). Erzwingung elementarer Strukturverschiedenheiten am *in vitro* wachsenden Gewebe. *Arch. Entwicklungsmech. Organ.* **116,** 438.

Weiss, P. (1959). Cellular dynamics. *Rev. Mod. Phys.* **31,** 11.

Willmer, E. N. (1958). "Tissue Culture", 3rd Ed. Methuen, London.

Willmer, E. N. ed. (1965). "Cells and Tissues in Culture", Vol. 1. Academic Press, New York.

CHAPTER 2

CELLS IN TISSUE CULTURE: MECHANOCYTES

The study of cells emigrating during the early stages of growth from cultures of embryonic or adult tissues from a wide variety of vertebrates including, for example, fish, amphibia, birds, mammals, and man, reveals the interesting and perhaps surprising phenomenon that the resulting morphological appearance of the cells, although not always identical, generally conforms to one of a very limited number of patterns. In other words, the cells which emerge are neither all immediately reduced to one tissue-culture type nor do they maintain the great diversity of types to be found in the original explants. The emergent cells can, in general, be grouped into three or four main categories according to their appearance, type of movement, staining reactions, and general growth-habits and requirements. These main categories are: epitheliocytes, mechanocytes (fibroblasts), and amoebocytes (wandering cells). Nerve cells form a further category with rather different characteristics and will need to be considered separately, and so too will one or two other smaller groups of cells.

Each tissue or organ, from whatever animal it is obtained, produces a pattern of growth which, while not specific to that organ or tissue alone, is characteristic of it. The significance of these major patterns is obviously, in itself, a point of some interest, but the interest must inevitably deepen when it is realized that the patterns are not only skin-deep and morphological but also reflect the existence of underlying differences in physiology and behaviour among the cells present.

Although, on the basis of what we have already discussed in relation to the lability of cell form, too much weight must obviously not be placed upon the outward appearance of cells for purposes of diagnosis, the existence of only a few varieties of cell form as observed in hanging-drop cultures will be seen to argue in favour of the existence of a correspondingly limited number of standard types of cells which do not readily transform from one into the other; it will be shown also that each of these main structural patterns reflects a different pattern of behaviour. Moreover, these differences in behaviour are not generally connected in any causal manner with the different conditions in which the cells find themselves in the culture. It is as though the cells which emerge in hanging-drop cultures either belong to, or revert to, only a few types which then continue to reproduce their own kind. It is our immediate problem, therefore, to enquire into the nature and properties of these types, for they cannot be accidental, nor can they only be the outcome of the environment in the tissue culture, for it frequently happens that two or more types of cell exist simultaneously in one and the same culture (Figs. 2.1, 2.2). Some features in the morphology of the cells in tissue culture may

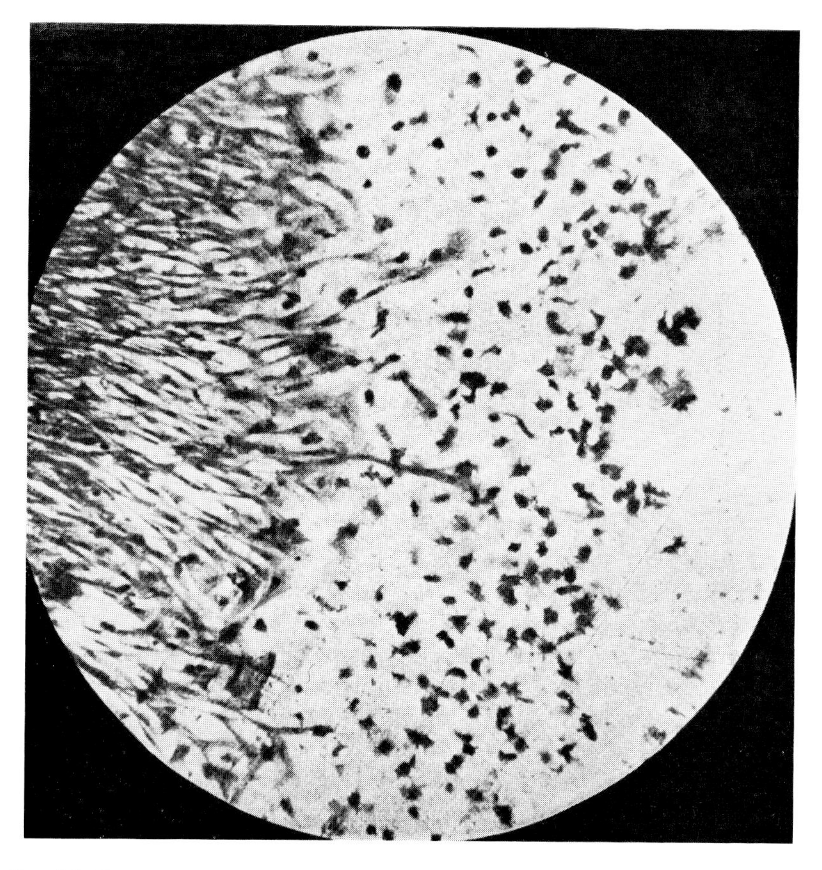

FIG. 2.1. The outgrowth from a culture of chick heart in which mechanocytes (fibroblasts) are seen on the left and amoebocytes (wandering cells) appear on the right. (Fischer, 1930.)

has been observed to be stationary and the cytoplasm appears to flow through it (de Bruyn, 1946; Lewis, 1931). Whether this is also true of these cells emigrating from the heart is not yet certain. While the cells from the heart move individually and separately, they are influenced by others in their neighbourhood and are slowed down by contact with each other (Abercrombie and Heaysman, 1952). The whole pattern of growth assumes the form of a network of clear cells, each somewhat fan-shaped in the periphery of the outgrowth and inclining more to be spindle-shaped nearer the original tissue,

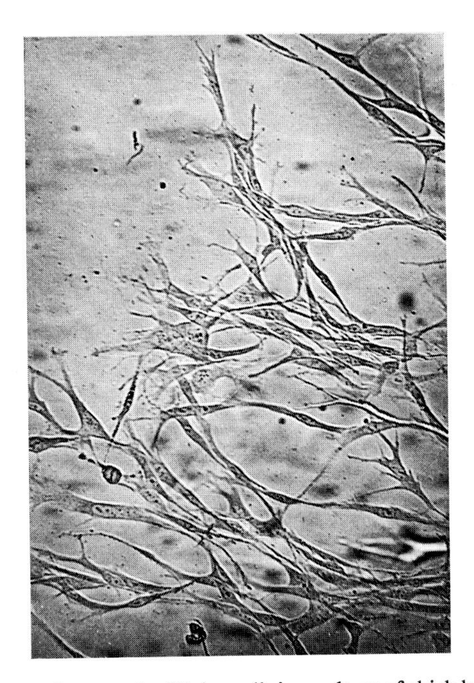

FIG. 2.3. Photograph of living cells in a culture of chick heart.

where they are densely packed together. Most of the cells have a single nucleus with two nucleoli, though the number of nucleoli seems to vary, from one to three or four, in a manner which is as yet unexplained. The cytoplasm contains numerous rod-shaped or filamentous mitochondria which move about independently within the cell and are constantly changing shape. They can sometimes be seen to fragment. The cytoplasm may also contain fat droplets, a few granules, and vacuoles which increase in number when the cells are short of sugar and which stain with neutral red (M. R. Lewis, 1922). When examined in the electron-microscope these cells generally have a fairly well-developed endoplasmic reticulum with a characteristic arrangement of ribosomes (see Fig. 2.25). If presented with acid vital dyes, like trypan blue, these cells, by comparison with certain others, e.g. macrophages, only take up a very small quantity, and that rather slowly. These cells are generally known as "fibroblasts" though, for reasons which will be discussed later, the term "mechanocyte" is preferable. The term "fibroblast" already has a more restricted use in connexion with those cells in the body which actually form the fibres of connective tissue.

Occasionally, in parts of a culture or even throughout the whole outgrowth of a culture from chick heart, the migrating cells may be of a different form. They may appear in the form of a sheet of very large flattened cells connected together on all their surfaces of contact in much the same way as the cells of a pavement epithelium are arranged. These cells often show a modified region of the cytoplasm near the nucleus—called the centrosphere. Lewis (1923a, b) has described all stages of transition between the two forms of cell, i.e. a mechanocyte may transform into a flat "epithelioid" cell, and, *vice versa*, epithelioid membranes have been observed to break up into a network of cells of the "mechanocyte" type. Furthermore, cells of the mechanocyte type, though with somewhat different cell processes, have been observed to contract rhythmically like normal cardiac muscles: this they may do either as isolated units or in groups (Lewis and Lewis, 1924; Rinaldini, 1958). Sometimes clear cross-striations are visible in the cytoplasm, but they do not appear to be necessary for normal rhythmic contraction (Fig. 2.4). At the ultrastructural level, however, striated myofibrils are generally present as isolated units or in small groups (see Fig. 13.31). These observations suggest that the cells which migrate out from the heart include not only true cardiac muscle cells, hence the beating cells, but also endothelial or mesothelial cells, for this would be the simplest explanation for the "pavement" behaviour. Moreover, when heart cultures in which the cells are growing as a network are allowed to grow for a few days, particularly in a medium containing blood plasma into which the cells penetrate, and they are then treated by silver methods for the demonstration of reticulin or argyrophil fibres, it will be found that such fibres, which are normally regarded as the products of connective tissue cells (true fibroblasts and fibrocytes), are present in the interstices between the cells of the "mechanocyte" network. Under some conditions these fibres are transformed into typical collagen fibres. Elastic fibres also may be found to have developed (Bloom, 1930). True fibroblasts may thus be present in the outgrowth, as they are certainly present in the original tissue. Two explanations of the behaviour of the cells in the outgrowth seem therefore to be possible. Since, apart from macrophages, the outgrowth of cells may actually include cells of all the main classes present in the original tissue, heart muscle, connective tissue, endothelium, and perhaps mesothelium, it is possible that the apparent uniformity arises because some or all of them can change their

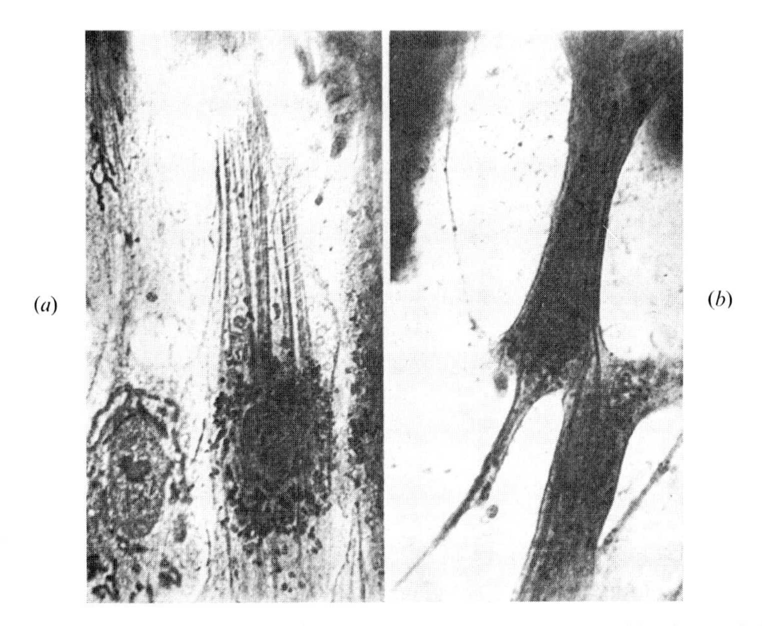

FIG. 2.4. (*a*) Heart muscle cell, showing striations in a seven-day-old culture of chick heart; (*b*) two heart muscle cells in a two-day-old culture of chick heart. Each was pulsating with its own rhythm. Striations not apparent. (Lewis and Lewis, 1924.)

FIG. 2.5. Heart muscle (left) and other cells (right) in a five-day-old culture of chick heart. (Lewis and Lewis, 1924.)

behaviour, e.g. cardiac muscle cells may actually become fibroblasts and so be capable of forming fibres: in other words, heart muscle, fibroblasts, and endothelium may all grow *in vitro* as "mechanocytes". Alternatively, the cells which emigrate and survive may belong to one category only, i.e. only the fibroblasts flourish under tissue culture conditions. This, however, is inconsistent with the presence of contractile cells and cells showing an epithelioid type of behaviour, for both of these may persist in the outgrowth for many days (Fig. 2.5), and show no obvious signs of necrosis or of being overgrown by the "fibroblasts". It is therefore probable that the three original types of cells, with their many different physiological characteristics, all become morphologically extremely similar, if not actually indistinguishable, under tissue-culture conditions. As already indicated, this morphological "dedifferentiation" of the cells of the heart raises far-reaching problems as to its significance and as to the bearing which it may have on physiological dedifferentiation.

When fresh heart muscle from a developing chick is treated with a dilute solution of crude trypsin or of elastase in a calcium- and magnesium-free physiological salt solution (e.g. Tyrode's solution containing sodium citrate in place of $CaCl_2$ and $MgCl_2$), the living cells can be made to fall apart relatively undamaged (Rinaldini, 1958). Suspensions can be made from these cells in a plasma medium which is then allowed to clot, so that the cells, completely isolated from each other or in only very small aggregates, are scattered and suspended throughout the coagulum. Under such conditions and with proper precautions against damage to the cells, most of the cells survive and, though they appear round and formless at first, they quickly spread out. Several types of cell can then be recognized; and some cells can be observed to beat while still in complete isolation. Eventually, however, most assume essentially mechanocytic characteristics. In the early stages of such cultures, therefore, some undoubted heart-muscle cells are visibly separable from other cells which might originate from the connective tissue elements of the original tissue; the contractile cells stain rapidly with the intravital dye, neutral red, but after a few days such differences disappear and the whole culture appears to consist of mechanocytes; contractility may, however, still be displayed by many of the clusters of cells. It is not clear yet what happens to the cells between the stage of their being recognizable as heart muscle and the stage when the cells all appear to be mechanocytes. Some death and autolysis of cells may occur, but it is unlikely that this explains events better than does a transformation of muscle cells to the morphology of mechanocytes, especially in view of the frequent persistence of contractility in the cell groups.

The Nature of Mechanocytes

Not only may such cells, i.e., mechanocytes, arise from fragments of chick heart, but, as already indicated, cells which habitually appear the same in tissue culture and which also behave in a very similar way emigrate from fragments of many other tissues, e.g. from the walls of blood vessels, from loose connective tissue, from developing or adult bone, from perichondrium and periosteum, from skeletal muscle and the muscular coats of the intestine, and from many other sources. Since nearly all these sources are from tissues which are concerned with the structural and hydraulic systems of the body, the term "mechanocyte" seems to be appropriate (Fig. 2.6).

The view that such "mechanocytes" are derived from only one type of cell (e.g. the fibroblast, or fibrocyte, of the connective tissue present in the

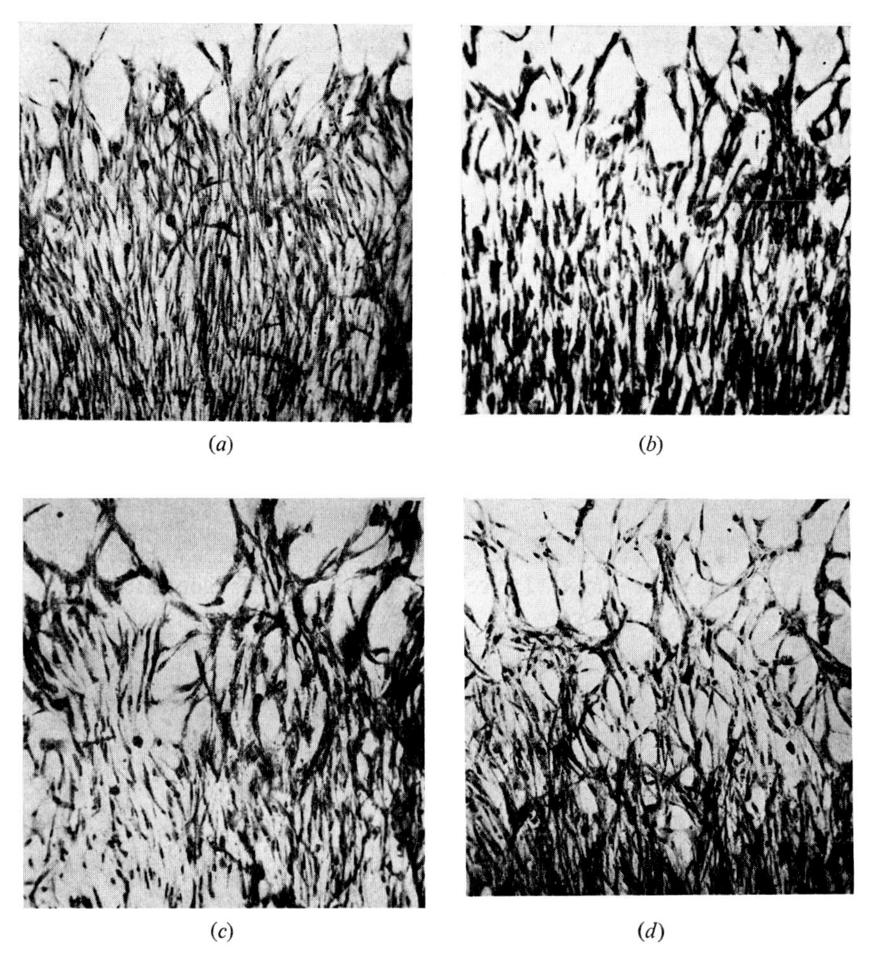

(*a*) (*b*)

(*c*) (*d*)

FIG. 2.6. Growth of mechanocytes derived from four different sources. (*a*) Muscle; (*b*) chondroblasts; (*c*) heart muscle; (*d*) osteoblasts. (Fischer, 1930.)

original explant of tissue), while attractive in its simplicity, has already been shown to be improbable in the case of cultures from the heart. In cultures of skeletal or striated muscle, the muscle fibres can be observed to split up into individual myoblasts (de Rényi and Hogue, 1934) (Figs. 2.7, 2.8),

FIG. 2.7. (*a*) A 'muscle-bud' in the outgrowth of a tissue culture of limb muscle of a chick embryo; (*b*) the same 'muscle-bud' splitting into myoblasts 24 hours later. (de Rényi and Hogue, 1934.)

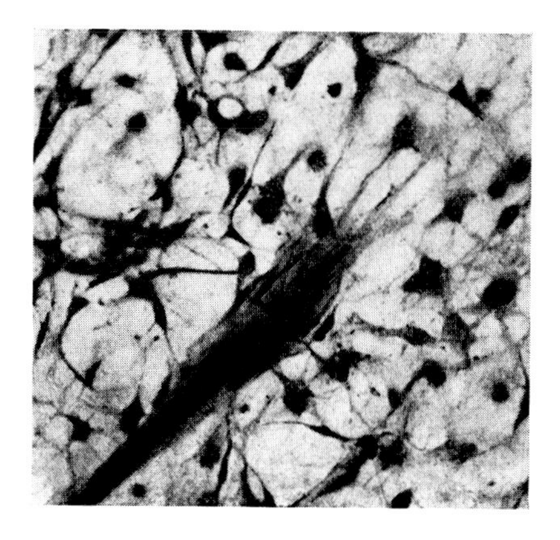

FIG. 2.8. A 'muscle-bud' splitting into myoblasts. (Lewis and Lewis, 1917.)

eventually to become, morphologically, mechanocytes. This splitting up of the muscle fibres is often preceded or accompanied by the presence of peculiar multinucleate muscle-buds probably formed by the migration of the nuclei into the ends of the fibres (Fig. 2.9). It should, however, be mentioned that certain other types of cell (see p. 321) may also emerge from cultures of muscle, and it is possible that muscle is not the uniform tissue which it appears to be at first sight. Indeed, the formation of multinucleate cells is a characteristic of another group of cells (see p. 95), and it is not every muscle fibre in a culture which shows this curious form of behaviour.

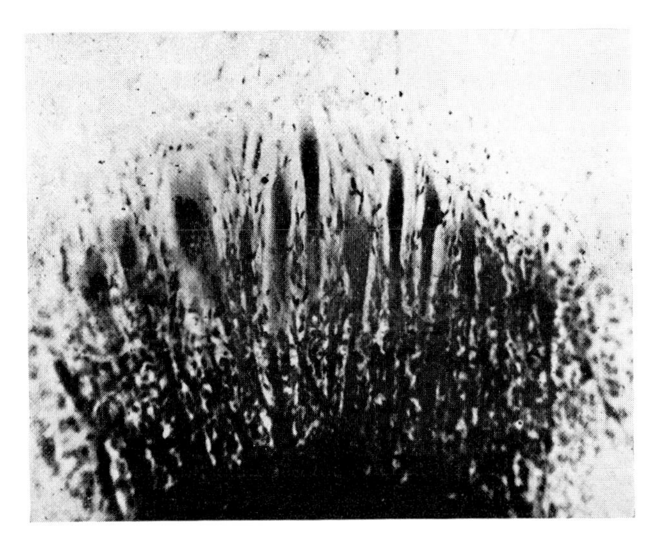

FIG. 2.9. Multinucleate cell masses developing from muscle fibres in a two-day-old culture. (Lewis and Lewis, 1917.)

In addition to the evidence of this sort against the uniformity of origin of the mechanocyte population in a tissue culture, there are also further arguments which can be raised against it. There is good evidence, for example, that mechanocytes in tissue cultures, though morphologically similar, are not all physiologically identical. Mechanocytes of different origin do not all grow at the same rate, but in any given medium each has its own characteristic rate of growth and frequency of cell division (Parker, 1933b; Willmer and Jacoby, 1936). Moreover, the study of the ultimate fate of such cells, when their cultivation *in vitro* is successfully prolonged for several weeks or months, especially in conditions which favour their subsequent differentiation and functional activity, emphasizes their inherent differences. To encourage differentiation rather than growth, tissues may be planted in a serum medium on the flat bottom of special flasks (Carrel flasks) or they may be embedded in a plasma clot on the bottom of such a flask. At frequent intervals the cultures are washed with fresh serum, or the surface of the plasma clot is irrigated with fresh plasma to which sufficient heparin has been added to prevent its coagulation, so that more food substances are added and metabolites are to some extent removed without increasing the depth of the clot (Fischer and Parker, 1929). Alternatively, they may be placed on the surface of a solid nutrient medium or supported by a raft of cigarette paper or rayon (cellulose acetate) on a fluid medium in an embryological watch-glass (Fell and Robison, 1929; Shaffer, 1956) or supported on a tantalum grid at the surface of a fluid medium (Trowell, 1954). Under such conditions the cells do not continue to emigrate indefinitely into the medium. After a brief period, there is very little more, if any, outgrowth, and the cells in the central explant begin to differentiate. The new tissues which are then formed are recognizably similar in character to those from which the cultures were derived; they are quite stable and remain alive for long periods. For example, cultures of muscle have remained alive for about a year and could then still be activated to grow as free mechanocytes by transferring them to hanging-drop

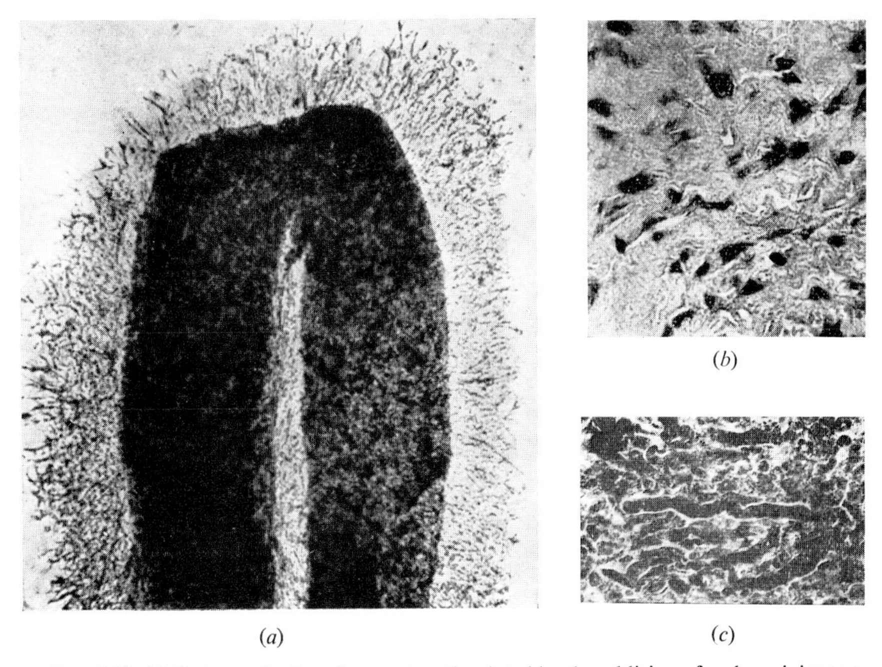

FIG. 2.10. (*a*) Outgrowth of mechanocytes stimulated by the addition of embryo juice to a colony of muscle tissue kept in "latent life" for a year; (*b*) and (*c*) sections through the central explant of a muscle culture kept in "latent life" for a year, (*b*) showing collagen fibres and (*c*) showing a few muscle fibres. (Parker, 1936.)

cultures in a growth-promoting medium. Such cultures, when sectioned and examined histologically, show the development of large amounts of connective tissue fibre and of occasional muscle fibres (Parker, 1936) (Fig. 2.10).

Admittedly, the relative increase in connective tissue and the decrease in amount of muscle is a little disturbing, but possibly the conditions under which muscle develops *in vitro* are rather more complex than those required for the formation of collagen fibres, for these seem to be produced rather easily, first as argyrophil fibres and then as true collagen. Moreover, muscle is a tissue which does not readily regenerate *in vivo* and also tends to atrophy with disuse, so it would be surprising if it did better *in vitro*.

The cells which remain in the periphery of these relatively dormant cultures often begin to differentiate also and to assume forms quite unlike the usual mechanocytes seen in cultures treated with the growth-promoting embryo juice (Parker, 1933a) (Fig. 2.11); the precise identification of such cells is, however, uncertain.

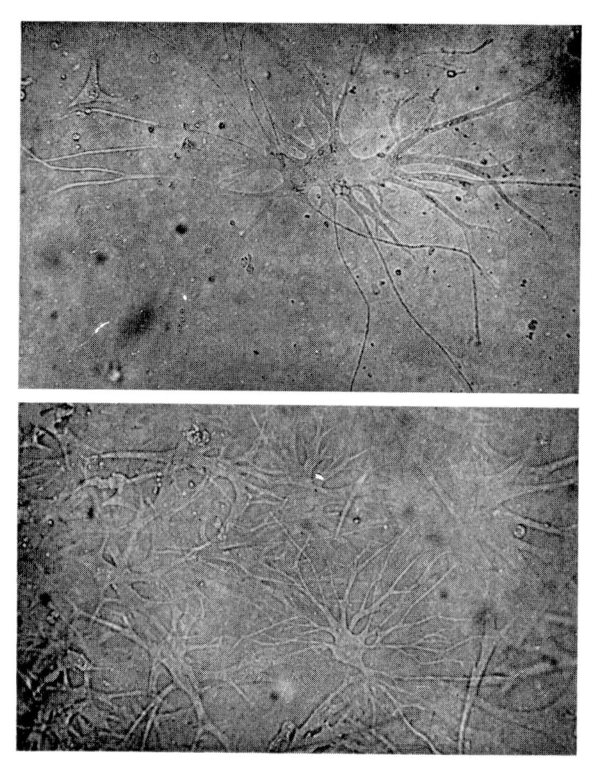

FIG. 2.11. "Fibroblasts" in the outgrowth of colonies of muscle tissue kept in "latent life" for a year. (Parker, 1933a.)

Another example in which the true nature of the cells in culture can be determined by their capacity to re-differentiate is very illuminating. When cultures of mechanocytes are obtained from fragments of bone taken from the shaft of a tibia previously deprived of all adherent cells (periosteal, endosteal, and those from the bone marrow), presumably the cells emerging in such cultures are likely to have come only from bone cells (osteoblasts, osteocytes, or osteoclasts), or from the cells associated with the developing Haversian systems and blood vessels (Fig. 2.12). The cells in the outgrowth

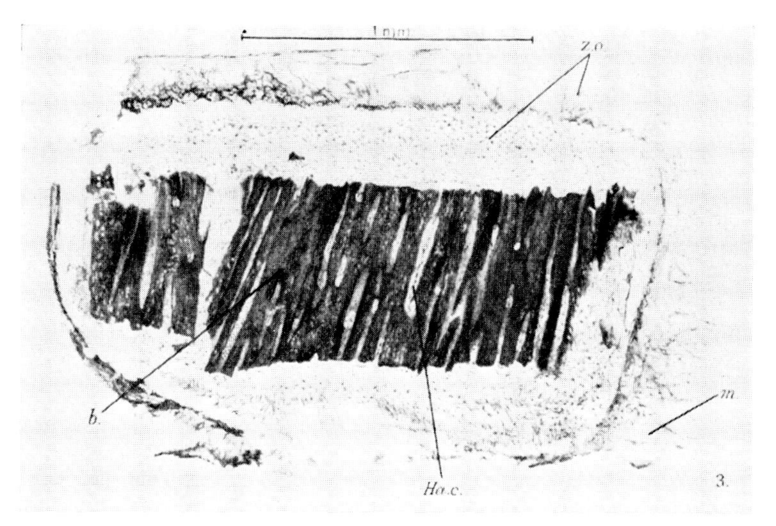

FIG. 2.12. Culture of endosteal bone from the tibia of a chick, showing the original explant, b., the zone of outgrowing cells, z.o., in the medium, m., and Haversian canals, Ha.c. (Fell, 1933.)

of these cultures mostly behave like the cells from the heart, i.e. like mechanocytes: they migrate out copiously with the same sort of movement and form a network of spindle-shaped or fan-shaped cells. After a time, the original explant of bone can be removed, and the space which it occupied can be filled with fresh plasma coagulum. The surrounding cells immediately begin to grow back into this vacant space, especially if the medium as a whole is growth-promoting and composed of a mixture of plasma and of extract of embryo tissues in physiological salt solution (i.e. embryo extract). When the space has become entirely filled with cells, the concentration of the growth-promoting substances in the medium is reduced by decreasing the concentration of embryo extract. Under these conditions the cells in the centre, which are, of course, derived from those that originally migrated away from the bone, now begin to differentiate. In most cases, the cytoplasm of the cells becomes hazy in appearance, a ground substance or matrix is laid down between the cells and this matrix then calcifies; histological sections of the differentiated tissue show that osteoid tissue or even true bone has been formed (Fell, 1932) (Fig. 2.13). The cells therefore which had emigrated and were behaving as generalized mechanocytes in the original outgrowth from the bone fragment must have contained among their numbers at least some

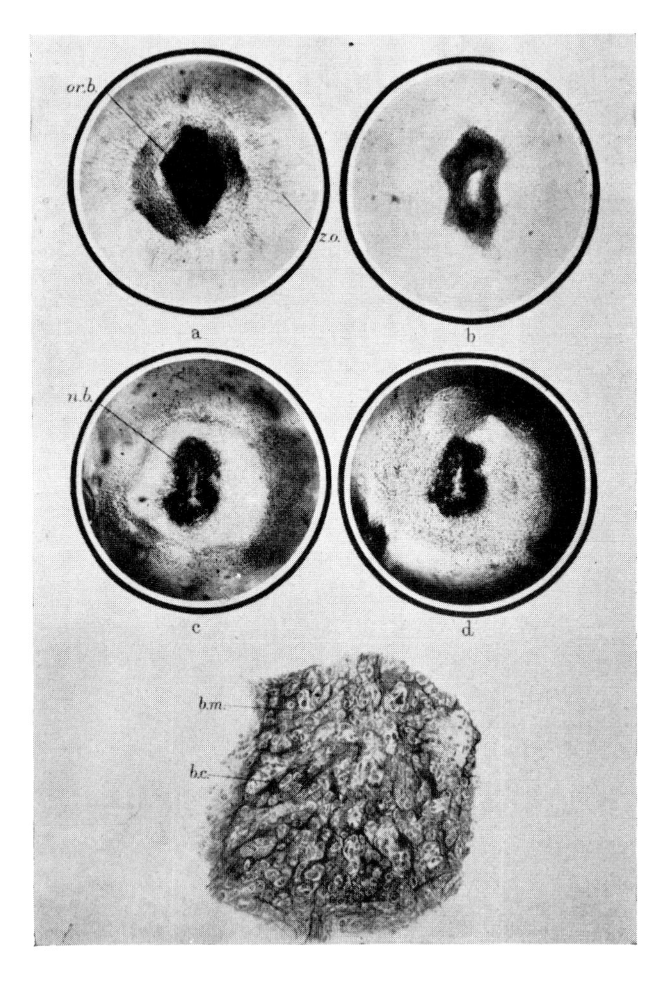

FIG. 2.13. Development of bone from mechanocytes. (*a*) Original culture showing outgrowth of mechanocytes; (*b*) original explant removed, with retraction of the zone of outgrowth; (*c*) and (*d*) new bone developing in the centre of the colony; (*e*) section of the newly formed "bone". b.c. Bone cell; b.m. bone matrix; n.b. new bone; or.b. original bone; z.o. zone of outgrowth. (Fell, 1932.)

potential bone cells (osteoblasts and osteocytes); moreover, the frequency with which the differentiation of bone occurs in such cultures suggests that bone cells are normally numerous in the original outgrowth and must therefore constitute a large proportion of the mechanocytes present. The method by which the osteocytes emerge from the bone has not been determined, but it clearly poses some interesting questions concerning the relationship between the cells and the matrix which they have produced.

Although, as we have seen, the mechanocytes in cultures from heart or muscle tissue, and, in fact, in cultures from many tissues other than bone, may differentiate under conditions of decreased growth-stimulation and form fibrous tissue, and although precipitates of calcium salts may occur in some of these cultures, e.g. as in cultures of heart tissue under some conditions, yet true bone with its characteristic fibrous pattern and crystalline structure only differentiates from mechanocytes derived from tissues originally possessing osteogenic potentialities. On these grounds, therefore, the osteogenic mechanocytes must be considered as a special breed and physiologically different from the muscle mechanocytes.

Since bone cells, heart cells, and cells from the other sources indicated (see p. 28) all seem to change their morphological characteristics when they emigrate in tissue culture, and since, in the case of the heart muscle treated with trypsin, various other morphological forms appear in the early stages, it may be asked whether there is any real significance to be attached to the mechanocyte form, other than that it is the characteristic form which cells take up under certain conditions of tissue culture, particularly those which favour the migration of cells and their growth by cell division. In later pages it will become clear that this form almost certainly has a biological significance and indeed a fundamental one, but, at this stage, it must suffice to say that many tissues, such as those already listed, habitually and regularly produce mechanocytes *in vitro*. By way of contrast, however, it must be emphasized that cells from certain other sources of tissue, just as habitually and just as consistently, behave in other equally characteristic ways and do not normally, if ever, assume the mechanocyte form. They may perhaps be able to do so under certain special conditions, but the evidence for such transformations is often equivocal and in one or two cases will be the subject of later discussions.

Although tissue culture teaches, above all things, that too much accent must not be placed on mere outward morphology, the mechanocyte form is nevertheless a definite and recognizable form of cell appearance and characterizes the behaviour of cells derived *in vitro* from certain well-defined sources. Furthermore, it will be shown that this cell form is connected with a particular pattern of behaviour, and that the mechanocyte has a definite series of potentialities for further differentiation, when suitable conditions are provided to evoke them.

The Growth of Mechanocytes

The general morphology, the orientated gliding movement, the capacity to form argyrophil and collagen fibres (for which reason the original term "fibroblast" is not altogether inappropriate), and the property of linking up with its neighbours to form a cellular network have already been stressed as diagnostic characters of the mechanocyte. So too are the special conditions which favour active growth by cell division among these cells.

When embryo tissues from a wide variety of sources—particularly those already mentioned, e.g. heart, periosteum, bone, connective tissue, etc.—are cultured in media containing plasma and embryo extract and the medium is frequently renewed, exuberant colonies of mechanocytes are obtained. The embryo-tissue juice (or extract) is an essential factor in this growth: it both stimulates the movement of the cells, so that they migrate more quickly into the medium, and at the same time it fosters and maintains cell division. Cells divide more frequently and more quickly in its presence. In fact, with suitable renewal of such media, cells may be kept alive and growing for an apparently indefinite time *in vitro*, and in the outgrowth of such cultures it is not unusual to record mitotic rates of more than 5%: that is to say, that more than 5% of the cells may be seen in mitosis at any one time. Since mitosis generally lasts for a little under an hour; this means that approximately 5% of the cells divide per hour so that the population, as a whole, may double itself within a day or so. These growth-rates can be conveniently recorded by time-lapse cine-microphotography directly onto sensitive paper. Unfortunately, only one particular zone of the explant can be examined in this way at any one time, but repeatable results can be readily obtained. The frequency of the exposures (e.g. 6-minute intervals) is arranged so that there is little chance of missing any of the cell divisions occurring in a population of several hundred cells. The method, of course, gives no information concerning the cells in the centre of the explant, and can only be applied when the cells are spreading in a single layer (Willmer, 1933a), and are thus under distinctly unnatural conditions.

If the embryo juice is omitted from the medium, then this exuberant growth does not continue after the first few hours and photographic records show that mitosis practically ceases in the outgrowth after approximately 48 hours, and that cell movement only continues thereafter at a very slow rate (Willmer, 1933b) (Fig. 2.14).

Recent work on clones and pure strains of cells, particularly when these are grown on more or less synthetic media, has indicated that after prolonged culture *in vitro* mechanocytes, and indeed other cells also, may acquire new properties (Sanford *et al.*, 1954; Puck and Fisher, 1956). For example, cells which were originally mechanocytes, and as such would need embryo extract,

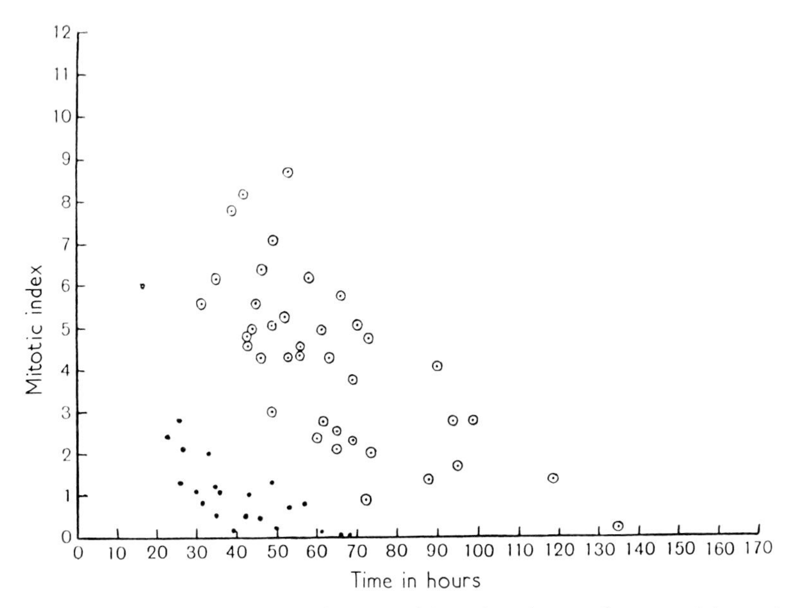

FIG. 2.14. Mitotic index plotted against age of the culture in a medium containing embryo extract (⊙) as compared with that in a medium of plasma only (·). (Willmer, 1933b.)

have now been kept actively multiplying for many months on a synthetic medium without any embryo juice (Healy *et al.*, 1954b; White, 1949). It is probable, however, that in such cases the cell colonies actually have acquired new properties in response to changed conditions in much the same way as bacteria "adapt" to new media, etc. Figure 2.15 shows cell cultures derived by cloning from cells which were originally fibroblasts, and it is clear that the morphology of these cells is very different from that of fibroblasts derived directly from fresh tissues. The picture shows also that not only does the morphology change with the density of the cells in the culture, but that the behaviour and form of the cells in any given field of the culture is very variable.

The type of behaviour of cells now under discussion, however, concerns cells recently derived from the animal and with as little "adaptation" to tissue-culture conditions as possible. Even though migration and mitosis almost cease in the plasma medium after a few days, the cells are still perfectly healthy; indeed, if washed periodically with serum, they may remain alive and functional for years. From the point of view of growth, however, they are quiescent, and are now in a condition very suitable for the investigation of the effects of potential growth-promoting media, which may be added at this time, i.e. say 60 hours after explantation into a medium consisting of plasma only. These quiescent mechanocytes respond particularly favourably

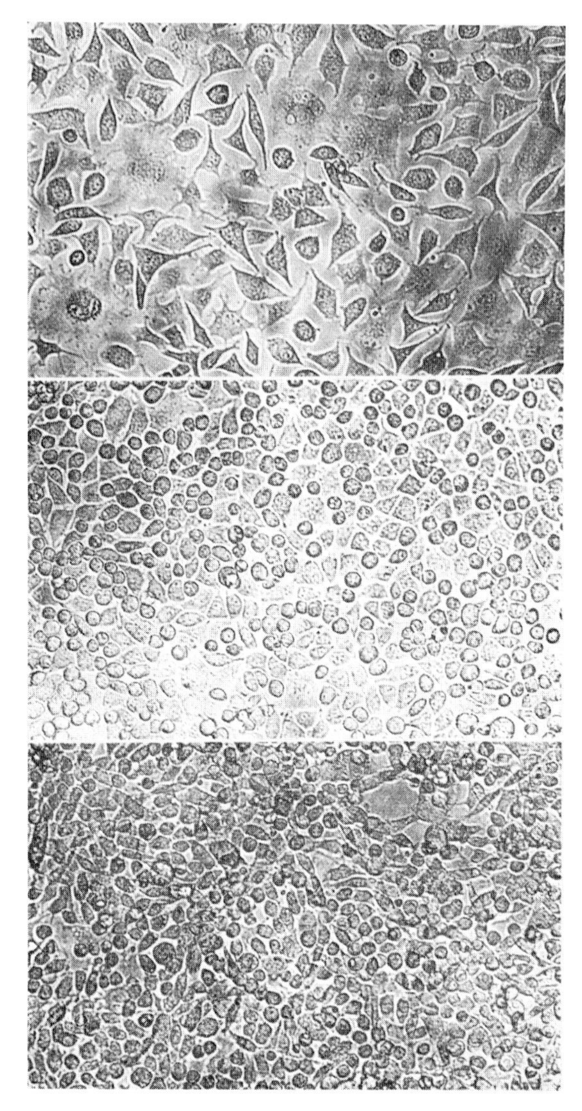

FIG. 2.15. Variations in the form of cells in a clone of mouse "fibroblasts." Note the changes in form with increasing density of population. These cells were grown in a protein-free, chemically-defined medium. (Earle, 1958.)

to treatment with embryo juice; with it they begin to grow again: without it they normally do not grow—though they may, as indicated above, survive for very long periods and though they may, under special conditions, "adapt" and then grow spontaneously. Within 2 hours of the addition of the extract, cell movement is measurably accelerated and a velocity of 12 μ per hour may rapidly climb to 40 μ or even 50 μ per hour (Willmer and Jacoby, 1936) (Fig. 2.16). There is a lag period of about 10 hours before cell division is resumed. This lag period always occurs and may be connected with the building up of the nuclear deoxyribonucleoprotein (see p. 41) indicating that the cells stop growing in the first (G_1) period of the cell cycle, i.e. in the period before the synthesis of the DNA which would be required before further division could occur (Howard and Pelc, 1953; Firket, 1965).

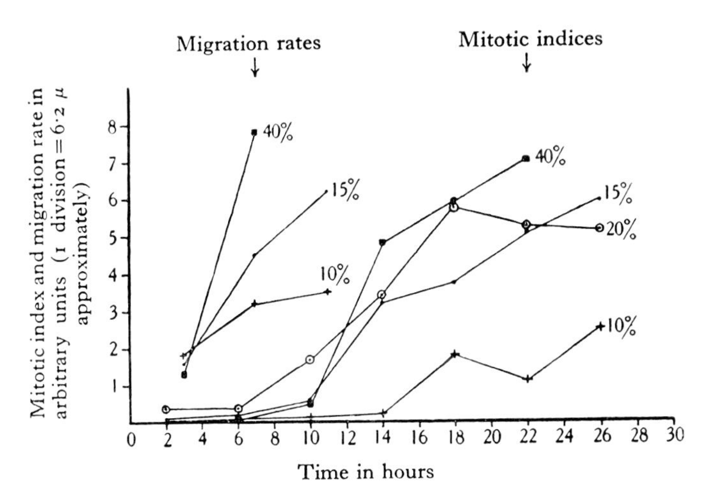

FIG. 2.16. Diagram showing the almost immediate increase in migration rate, and the delayed increase in the mitotic rate when different concentrations of embryo juice are added to cultures of "osteoblasts" reduced to quiescence by previous culture in plasma only. (Willmer and Jacoby, 1936.)

It is almost immaterial from which of the various sources mentioned above the tissue is derived; any growth of mechanocytes of which the tissue is capable always seems to be activated to a greater or lesser extent by the presence of embryo extract, and mixed colonies of cells from complex tissues often end up, after continued cultivation and sub-cultivation in its presence, as apparently pure colonies of mechanocytes. On the other hand, in spite of this general sensitivity to embryo extract, there is evidence that mechanocytes from different sources respond quantitatively differently (Parker, 1933b). Embryo juice can be conveniently prepared by washing 7- to 10-day chick embryos free from blood, yolk, etc., crushing them to a pulp, and then centrifuging away the tissue debris. The opalescent supernatant fluid may then be regarded as 100% embryo juice which can be diluted to any desired concentration with Tyrode's solution (or other suitable medium). The age of the chick from which the juice is made is of some importance and should be standardized as far as possible in any given series of experiments. It has then been found that, when such dilutions are applied to quiescent cultures of races of mechanocytes from different sources, the optimal concentrations are quite different. For mechanocytes from the frontal bone of the chick, 40% juice produces decidedly greater rates of growth than 15% juice and full growth-stimulation does not occur till the concentration is raised to this upper limit. On the other hand, in mechanocytes derived from the heart of the chick the maximum mitotic rates are produced by 15% juice and these are generally rather lower than those occurring in the tissue cultures derived from periosteum under the same conditions (Willmer and Jacoby, 1936). Moreover, when the concentration is raised above 15% there is an increasing tendency for some races of mechanocytes, e.g. those from the periosteum, to digest or liquefy the supporting plasma coagulum.

These experiments again emphasize certain common properties shared by mechanocytes, but at the same time they show that, within the mechanocyte "family", there may be many races or genera.

Some interesting sidelights are thrown on the behaviour of mechanocytes by experiments in which the embryo juice only remains in contact with the tissue for a limited time (e.g. 1 hour) and is then removed and replaced by Tyrode's solution (Jacoby *et al.*, 1937). Under these conditions, mitoses occur after the usual latent period (10–12 hours), and their number depends on the concentration of the embryo juice. They now, however, cease to occur after a further period of about 12 hours, instead of continuing for several days as they would if the juice were left in contact with the cells (Fig. 2.17). Repetition of the treatment with the juice can again produce mitoses after 10 hours, but this only occurs when the second dose is given after the first dose has already become effective, i.e. if given more than 10 hours after the first dose. When it is applied earlier than that, i.e. during the "latent period" of the first dose, then the second dose is ineffective. The interpretation would seem to be that the active agents in the extract penetrate the cell quickly, i.e. within the hour, and alter its metabolism in such a way that the movement of the cell is visibly more active within the first 2 hours, and metabolic processes are inaugurated, which finally lead up to cell division after a period of 10 or more hours. Those cells which have been activated are not susceptible to further treatment with embryo juice until the first division has occurred, but then they may be re-charged, so to speak, and produce a second outcrop of mitoses after another lag period of about 10 hours. Such evidence as there is encourages the belief that the cells of the second crop, as implied above, are actually the daughter cells from the first crop. Since high concentrations of embryo juice produce numerous mitoses in these experiments, while the numbers of divisions are much less in low concentrations, it is probable that there are, in a colony of mechanocytes, cells whose sensitivities to the stimulating factors

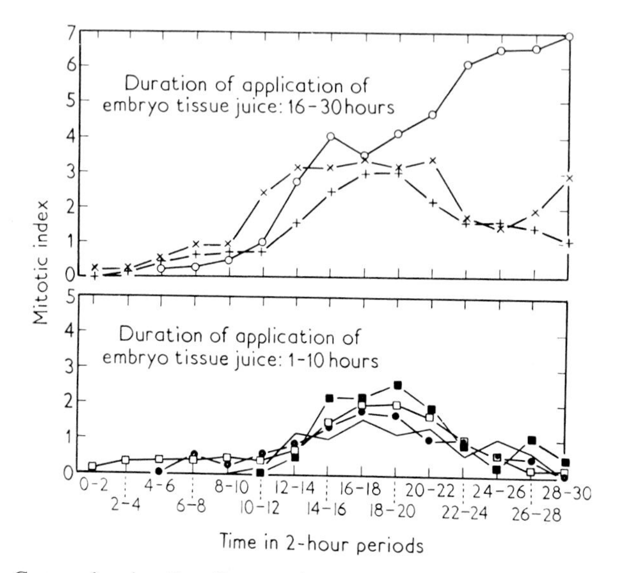

FIG. 2.17. Curves showing the effect on the mitotic index of applying embryo juice to cultures of periosteal fibroblasts for different lengths of time, from 1 to 30 hours. O—O 30 hours; ×—× 20 hours; +—+ 16 hours; ■—■ 10 hours; ●—● 6 hours; □—□ 3 hours;—— 1 hour. (Jacoby *et al.*, 1937.)

of embryo juice differ. There are some which are, as it were, nearly ready to divide and only need about 5% juice in the medium to send them into division, while others are much further from the division state and require 15% or even 40% juice before they are raised to the pitch at which they can respond. From the point of view of growth, therefore, an apparently uniform population of cells may, in fact, be anything but uniform. Moreover, the cells which respond by division to weak concentrations of juice are scattered in an apparently random manner throughout the outgrowth, so that it is not simply a question of the availability of limited quantities of some activating substance.

Exactly how the embryo juice activates the cells to divide is still uncertain. There is evidence that RNA (or ribonucleoproteins) increases within the cells rather quickly after the addition of the juice (Willmer, 1942; Davidson *et al.*, 1949), but whether this coincides with the S period of DNA synthesis is not known. The time sequence and analogy with the behaviour of cells of other sorts under somewhat different conditions (Seed, 1965) suggest that the cells before activation are in the G_1 period and that the embryo juice sooner or later induces the S period, but there does not seem to be any direct evidence, on cultures of this kind, that this is so. Studies of the deoxyribonucleic acid content (DNA) of tissue-culture cells indicate that the amount

present at division is constant, but that between divisions it is significantly less, presumably because at division the amount is halved between the daughter cells and these then build it up over a limited period, the S period, to the "tetraploid" amount (Walker and Yates, 1952). The relationship between the DNA content of the nucleus and the type of cell is obviously very important in connexion with the nature of cell differentiation and the ability of races of cells to breed true (Healy *et al.*, 1954a). Fibroblasts under the influence of embryo juice and plasma normally breed true and do not give rise to other types of cell at all readily. It is not yet known whether this capacity to breed true is a nuclear or a cytoplasmic phenomenon. If it is nuclear, how is the DNA content of the different races affected, or is the difference between races dependent on some difference in the protein part of the chromosomes, e.g. the histones, or in the RNA of the cells? The DNA in the mitochondria could perhaps be responsible for these essentially cytoplasmic differences which characterize the different races of cells (Chèvremont and Frédéric, 1968). It is unlikely to be anything so drastic as a major change in chromosome number or a fundamental alteration in the DNA because, although the races of cells in an organism may differ from each other and breed true as such in culture, they still remain species specific. It is true that the chromosomes in somatic cells of the chick are peculiar in that they show a tendency to fragment, but it may be of interest to observe that the number of fragments visible in the mitoses of chick mechanocytes in tissue culture is in general greater than the number of fragments in the macrophages of the same species under similar conditions, thus perhaps indicating some difference in the constitution of the chromosomes in the different somatic cells. In any one somatic cell it is probable that only a part of the total DNA coding system is being used for building the necessary proteins, so that differentiation must in some way suppress the potential synthetic activities of the rest of the DNA. Thus cells like mechanocytes, which breed true in tissue culture, must in some way duplicate this suppressing system when they divide.

In the presence of embryo juice, even in very small amounts, the growth of the mechanocytes is greatly accelerated by the presence of proteoses, especially when these are derived from fibrin by partial digestion with pepsin (Carrel and Baker, 1926; Baker and Carrel, 1928). The fact that the digest of fibrin is better than that of other proteins is perhaps a point of physiological importance. In the first place it indicates that certain groupings of amino acids are probably more suitable than others; but it also suggests that the fibrin deposited in a wound in the body may not be merely an inert scaffolding in relation to the subsequent repair process in which mechanocytes, as potential tissue cells, play a large part. Furthermore, the fibrinogen circulating in the blood and lymph perhaps has potentialities as a normal food substance for

the mechanocytes and their derivatives in the body and is not only concerned with the clotting of the blood or lymph. This appropriateness of fibrinogen as a nutrient for fibroblasts is worthy of further study and should be borne in mind in relation to theories of the origin of "fibroblasts" (see pp. 99 and 210).

In the absence of embryo juice the proteoses are apparently much less effective (Willmer and Kendal, 1932). It is as though the embryo juice increased the protein metabolism, and the presence of proteoses in the medium provided a raw material or substrate which is far more effective than either whole proteins or most amino acids. However, the embryo extract also accelerates the uptake of glucose from the medium (Willmer, 1942), so the mode of its action is anything but clear; nor is it yet possible to make any concentrate from it which is more active than the extract as a whole. Probably there are several factors concerned, some at least of which are thermolabile and non-dialysable (perhaps lipoproteins or ribonucleoproteins); others are thermostable and dialysable and probably include certain amino acids (Baker and Carrel, 1926a, b; Fischer, 1941; Harris and Kutsky, 1954; Hueper *et al.*, 1933; Jacoby, 1937; Lasnitski, 1937).

The whole question of growth-stimulation by embryo extract, proteoses, etc., has been treated at some length here, because it is particularly relevant to the behaviour of mechanocytes, which, unlike certain other cells (see p. 90), will normally not continue to divide after the first few hours in a plasma medium devoid of embryo extract; they can, however, be brought once again into a state of division, even after they may have been dormant for very long periods *in vitro* in a medium of plasma or serum, if they are treated with embryo juice. Even in the tissues from an adult organism, where the fibroblasts must have been relatively dormant for years, they again become active when the tissue is treated for some time with embryo juice in tissue culture conditions. The effects can be accelerated by previous treatment of the tissue with solutions of trypsin (Simms and Stillman, 1937). The discovery of the effects of embryo juice was made by Carrel in 1913 and there is still nothing more effective in causing cell divisions among mechanocytes, though saline extracts of some adult tissues, e.g. brain, thymus (Trowell and Willmer, 1939), cardiac muscle (Doljanski and Hoffman, 1939), and cartilage (Davidson and Waymouth, 1943), show some similar but weaker activity. It is very difficult to correlate the capacity of a tissue to provide saline extracts of high growth-promoting quality with any other metabolic features possessed by that tissue (Pomerat and Willmer, 1939). High ribonucleoprotein content and high glycolytic activity, especially under aerobic conditions, seem to be features common to many of the active tissues, but whether these features are more than incidental is doubtful. The high mucoprotein content of some of the active extracts is possibly significant. Growth of fibroblasts occurs more freely when the cells are in motion (Abercrombie *et al.* (1968) have pointed out

that contact inhibition among fibroblasts involves inhibition not only of movement but of mitosis also.) And there is some evidence that mesenchymal (fibroblastic) cells put out their pseudopodia more easily when sulphated polysaccharides are present in their surroundings (Immers, 1961).

It should be remembered that, in order to test the growth-promoting activity of tissue extracts, these are usually left in contact with the cells for a considerable time, and any toxic properties which they may possess are thus also given every opportunity to show themselves. Moreover, saline extracts of some tissues must inevitably contain substances which are inhibitory to cell activity, so that the total effect of each tissue-extract must be the resultant of the growth-stimulating and the growth-inhibiting actions of the substances extracted. For example, in extracts of both liver and kidney either the inhibitors preponderate or there are no growth-promoting agents present, for these extracts are usually ineffective in stimulating the growth of fibroblasts. Only a fine line of distinction can be drawn between inhibitors and toxic agents. Reversibility of the effect is perhaps the simplest criterion. There are, however, several substances which more or less reversibly suppress the activity of mechanocytes, and some of these (though not all, of course) act specifically on mechanocytes, leaving other tissues active and apparently unaffected. For example, when certain aldehydes (0.002 M) are added to the culture medium of mechanocyte colonies, the peripheral cells rather quickly withdraw their processes and round up. They may remain apparently inert in this condition for days, but as soon as the aldehyde is removed they spread out again and go on growing as before without any permanent ill-effects. Glyceraldehyde and propylaldehyde are particularly effective (Pomerat and Willmer, 1939). The lowest aliphatic aldehydes, formaldehyde and acetaldehyde, are, as might be expected, toxic. Many of the sulphonamide drugs, in relatively high concentrations, act in much the same way, and, again as might be expected, the more soluble they are the more effective they are (Jacoby *et al.*, 1941). A particularly interesting inhibitory substance in relation to the present discussion is found in malt extracts (Heaton, 1926; Medawar, 1937), and has been identified as a hexenolactone (Medawar *et al.*, 1943). This substance may entirely suppress the activity of mechanocytes while leaving cells of some other types (e.g. epithelium) in similar cultures apparently untouched (Fig. 2.18). When applied to the whole animal the effects are, as might be expected, more complex (Briggs, 1946; Hauschka, 1946), and even in tissue cultures the distinction between mechanocytes and epithelial cells in their sensitivity to hexenolactone has been challenged (Royle, 1945). Other factors like cysteine, which reduces the toxicity, and glutamic acid, which increases it (Hauschka, 1946), are probably involved; serum has a protective action on the cells.

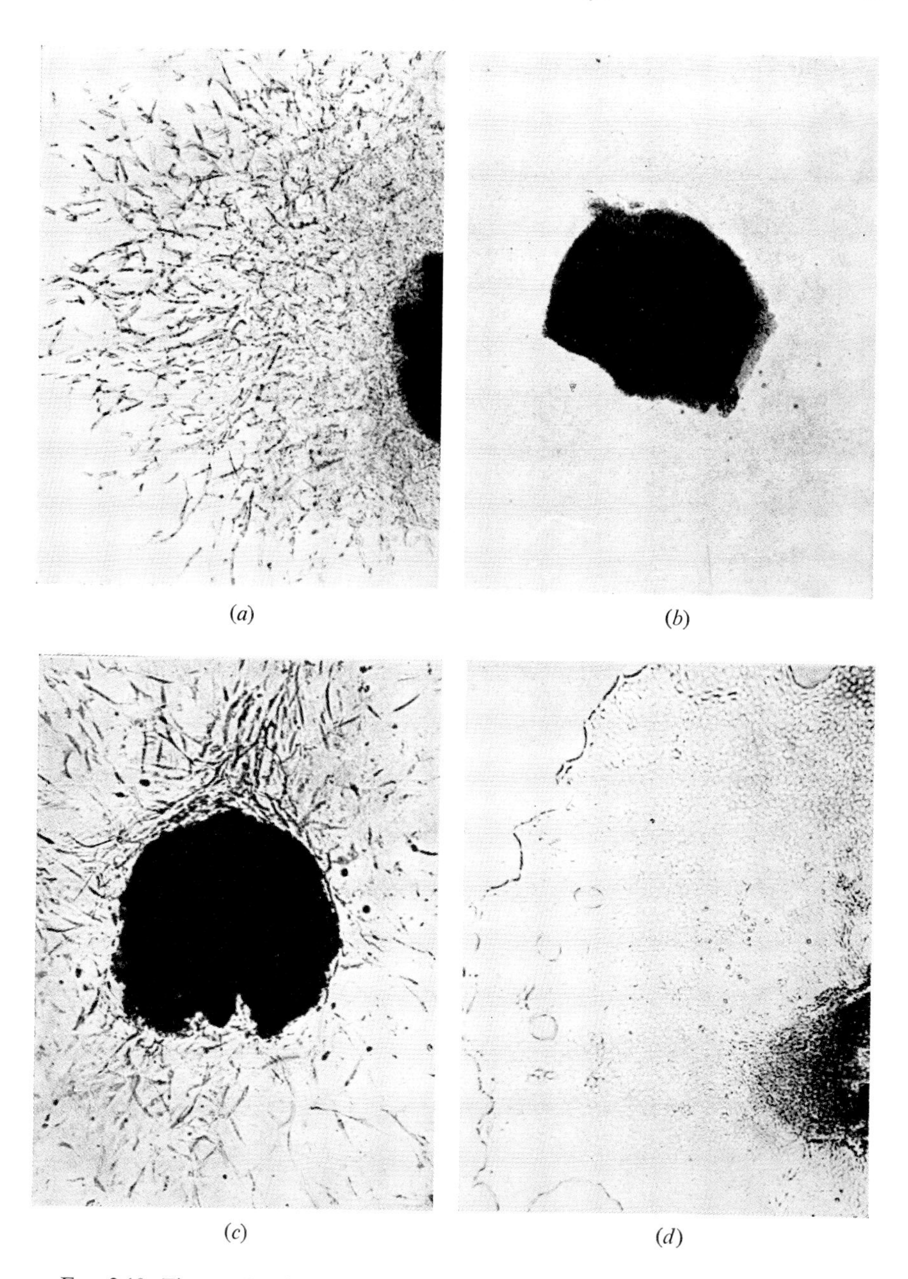

FIG. 2.18. Figures showing the inhibition of mechanocytes by a substance present in malt and maize extracts. (*a*) Normal growth of mechanocytes; (*b*) comparable culture with inhibitor; (*c*) growth of a culture like that in (*b*) after removal of inhibitor; (*d*) growth of epithelium in medium containing inhibitor. (Medawar, 1937.)

Pure Strains of Mechanocytes

Since it appears that there are different races of mechanocytes, and since most cultures must contain mixtures of these different kinds of cells, or of cells in different physiological states, it is, at first sight, surprising that the making of pure cultures of the different breeds of mechanocytes from single cells has been so long delayed. The main reason which delayed this approach emphasizes a peculiar property of mechanocytes, namely that they survive and grow only with great difficulty in isolation. The typical network-like growth of these cells has already been emphasized and it is comparatively rarely that a cell leaves this network and survives to divide as an independent unit. Successful pure colonies from single cells of mechanocyte-type have been made by Earle and his school (Sanford *et al.*, 1948; Likely *et al.*, 1952) by explanting single cells in media which have already supported colonies of cells, and also by Puck *et al.* (1956, 1957) who sometimes used "feeder" cells, i.e. cells whose growth had been stopped by irradiation, to initiate the growth of normal cells isolated on a layer of such cells. In the ordinary media, isolated mechanocytes, unlike one or two other types of cells, do not readily survive unless the cells are enclosed within a very limited quantity of medium as, for example, in a small capillary tube, or are given some other special treatment. It appears that the plasma-embryo juice medium, as normally used, is not itself entirely satisfactory for the rather fastidious mechanocyte, when it is completely isolated, but can be made so by the combined action of numerous cells, even when these are all of the same type, i.e. mechanocytes. The situation is somewhat parallel to the common failure of single fertilized echinoderm eggs to form blastulae when isolated on coverslips in hanging drops of sea-water, but their complete success when placed in hanging drops containing more than about 10 eggs. In this case, one of the factors is almost certainly the pH of the sea-water, which tends to become too high owing to solution of alkali from the glass; the CO_2 production by numerous cells can presumably counteract this, while that from a single egg may be insufficient. This is obviously a very simple consequence of the technique employed and other more complex factors are certainly involved in the tissue culture problem, though what they are is not yet known. Fresh mechanocytes which are growing on a coverslip in a completely fluid medium of embryo juice are very easily damaged by washing them, even with fresh embryo juice, and one of the actions of a plasma coagulum in encouraging mechanocyte growth may well be the provision of a more constant local environment for the cells. On the other hand, the pure clones of cells originally derived from mechanocytes, which are now being successfully cultured, can be kept in large volumes of agitated fluid media and the cells multiply freely under these conditions (Earle *et al.*, 1954). As discussed earlier, however, such cells should probably not any longer be considered as typical mechanocytes, but as "adapted tissue-culture cells". There is some evidence that myoinositol and perhaps other constituents of serum have a stabilizing action on these suspended cells (Eagle *et al.*, 1957b).

If the cell membrane is relatively permeable to amino acids and other small molecules, necessary metabolites may become so diluted in the cells by escaping into the medium that adequate metabolic pools can no longer be maintained and the cells die of overwork or inanition (Eagle and Levintow, 1965).

Synthetic Media

Not only has it been found difficult to establish cultures from single mechanocytes from normal tissues, but this family of cells is equally difficult to nourish with a purely synthetic medium. Such media have been devised, and many of them are very complex. One of them initially contains 58 constituents in addition to water, though how much these interact and how many remain distinct in the final medium is another question. Nevertheless, they still do not support the growth of mechanocytes indefinitely. Probably not all these ingredients are essential but investigators in this field (Eagle, 1954; Eagle *et al.*, 1957a; Healy *et al.*, 1954b; Waymouth, 1955; White, 1949) have considered it advantageous to incorporate in the basic medium all those substances which might be important to the cells, and then to discard them one by one in order to find out which are essential, rather than to build up the medium from the simplest necessities. The reason for this is that so often a combination of substances is more efficacious than any one of them alone, and it is easier to spot these combinations subtractively than additively. While "organ cultures", i.e. those in which a complex tissue is encouraged to differentiate or function under tissue-culture conditions, can be successfully provided with synthetic media and often develop well on them for many weeks, the indefinite survival of pure strains of normal mechanocytes in the complete absence of traces of serum or plasma is more doubtful. So far, the survival time of cultures of normal cells on purely synthetic media does not seem to be more than a few months, while in serum and embryo juice there appears to be no limit. The synthetic media have proved more successful with malignant cells or with cells which have become in some way adapted to tissue culture conditions. In any case, it is necessary to distinguish between survival, with or without differentiation, and growth.

These purely synthetic media naturally contain a large number of amino acids, either in more or less empirical amounts or in amounts based on the amino acids essential for blood formation or on the composition of digests of fibrin, etc. Mixtures of the L-amino acids are, on the whole, more effective than those of the racemic or D-forms. The effects of a few amino acids have been tried alone or in simple combinations and something positive is now beginning to be known about the requirements of mechanocytes in this direction. For a strain of cells originally derived from a colony of mouse fibroblasts (strain L) thirteen amino acids have been found to be essential (Eagle, 1954). However, the "normality" of these cells is suspect, owing to their tendency to become malignant, and further work is necessary on cells still in the normal mechanocyte state and less "adapted" to tissue culture conditions. Nevertheless it may be found to be very significant in relation to a later discussion (see p. 220) on the nature of mechanocytes that among the amino acids which are not essential for the growth of this strain are glycine, proline, hydroxyproline, alanine, glutamic acid, aspartic acid, and serine, which together account for over 60% of the total protein nitrogen of collagen, the main product of the mechanocyte (see p. 214). Cystine has been found to be essential for the growth of mechanocytes from bone (osteoblasts) and from muscle (myoblasts), and it could not be replaced by methionine (Fischer, 1948) as it can for some purposes in the body, though its action is assisted by methionine (Morgan and Morton, 1955). In tissue cultures, lysine is not necessary for the growth of myoblasts (mechanocytes from muscle), but does benefit the growth of osteoblasts (mechanocytes from bone) (Fischer, 1948): this is interesting because lysine is, of course, essential for the growth of the whole animal. Once again, this is further evidence that the cells which masquerade in tissue cultures as mechanocytes are not all identical, but that there are metabolic differences between them, and that they exist in certain fairly well-defined races. Glutamine does not appear to be essential for chick heart fibroblasts (Fischer, 1948; Pasieka *et al.*, 1956) but is required by periosteal fibroblasts, and is required by the L-strain of mouse fibroblasts (Eagle *et al.*, 1955).

The Races of Mechanocytes

It is thus probably true to say that although by prolonged cultivation *in vitro* one may obtain a pure culture of mechanocytes to all visible appearances, yet such a so-called pure strain may well be a mixture of mechanocytes of several kinds. Moreover, as pointed out earlier, it may have acquired certain new characteristics by adaptation to the conditions of tissue culture. This idea of a mixture of types of mechanocytes can be verified to some extent by studying the nature of the tissue developed when cultures are allowed to redifferentiate once more after a period of rapid proliferation, i.e. of unorganized growth. The results with osteoblast cultures and muscle cultures have already been partly described, and it was observed in the former case that such mechanocytes have the power to differentiate, as one would expect them to do, into a bone-forming tissue. Nevertheless, in a small number of cases, the differentiating cells from pure cultures of endosteal tissue produce cartilage and not bone (Fell, 1933) (Fig. 2.19). Either one must assume that cartilage cells (chondroblasts) or undifferentiated "reserve" cells were somehow enclosed in the original explant, though since endosteal bone from the centre of the diaphysis was used, this does not seem very likely; alternatively, mechanocytes derived from bone (osteoblasts) still have the power, should the conditions evoke it, to produce cartilage instead of bone, i.e. to become chondroblasts.

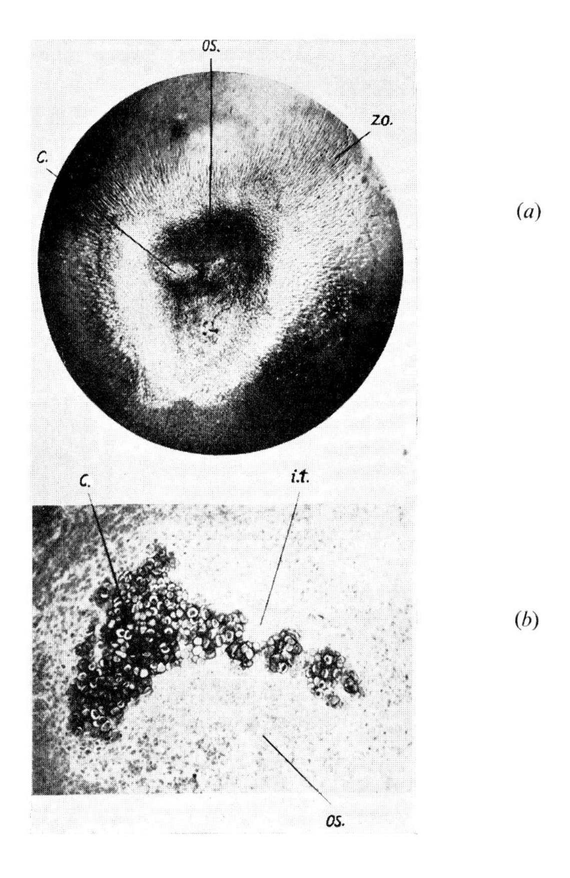

FIG. 2.19. Culture, originally derived from endosteal bone, showing the development of cartilage, *c.*, as well as of bone, *os.*, when encouraged to redifferentiate after a period of unorganized growth. (*a*) Living culture showing cartilage nodule; *zo.* zone of outgrowth. (*b*) Fixed and stained section through the culture; *i.t.*, intermediate tissue (Fell, 1933.)

As a visual example of the sort of differentiation involved in these cultures, attention may be called to the fact that if the lateral surface of the epiphysis of a developing long bone is examined in histological section, there generally will be found one region where the cells are histologically more or less undifferentiated (Fig. 2.20). A small movement of the section in one direction shows these cells developing into indubitable cartilage cells, or chondroblasts busy with the formation of hyaline cartilage. A movement in the opposite direction shows a series of almost imperceptible changes, cell by cell, from the

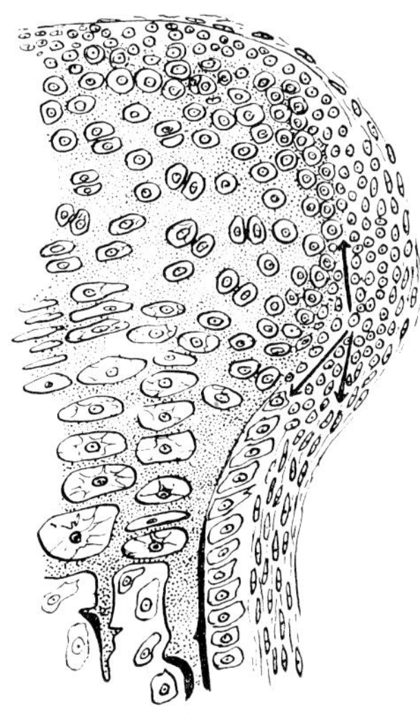

FIG. 2.20. A diagrammatic section of the margin of the epiphysis of a long bone showing the differentiation of primitive mesenchyme cells into chondroblasts, osteoblasts, or fibroblasts.

undifferentiated cells at the starting point to well-defined periosteal osteoblasts, which lie along the surface of the hypertrophic cartilage and are responsible for the formation of the developing bony diaphysis. If the field of view is now shifted once more from the osteoblasts of the periosteum to the fibrous layer of this membrane it will be found that, if this layer is followed towards the epiphysis and the point of origin, the fibrocytes of the periosteum grade steadily into the indifferent or undifferentiated cells once more. This particular zone on the surface of the epiphysis thus illustrates very clearly three potentialities of morphological development for the primitive mesenchymal cell which is probably in this case very much the equivalent of the tissue-culture mechanocyte, namely towards chondroblast, osteoblast or fibroblast. Cells from the synovial membranes (synovioblasts) can probably also be included in the same group, as they behave very similarly *in vitro*

(Vaubel, 1933a, b). Odontoblasts, the dentine-forming cells of teeth, are also probably similar, since they closely resemble osteoblasts *in vitro*.

These four types of behaviour (i.e. bone formation, cartilage formation, fibre formation, and synovial membrane formation) are also four of the most easily demonstrated potentialities of one class of mechanocytes as seen in tissue culture, and as we shall see, other mechanocytes have other potentialities. The particular type of differentiation which is followed thus appears to be determined, at least partly, by conditions external to the cell. When cultures of osteoblasts are treated with embryo juice, they grow rapidly, and the only sign of differentiation is that they produce argyrophil fibres, but in less growth-promoting media such cultures produce true collagen fibres and osteoid tissue, and, in a few cases, cartilage. This little subfamily of cells, including true "fibroblasts," chondroblasts, osteoblasts, synovioblasts, and probably a few others, e.g. odontoblasts, is thus characterized by its capacity to produce and to lay down between the cells varying proportions of collagen-type proteins and also mucoproteins, mostly containing hyaluronic acid or chondroitin sulphuric acid, and sometimes to produce the enzyme alkaline phosphatase, which may then, in a suitable medium, determine the deposition of calcium salts.

How far the different types of activity are mutually interchangeable is uncertain; probably there may be steps in the process of differentiation which demand rather special conditions. For example, if the limb bud from a 3-day-old chick embryo is cultured on the surface of a plasma coagulum, the limb develops for a time as a whole and a cartilaginous skeleton appears within it, much as it might do in the intact animal. Such a skeleton, however, reaches a certain stage of development but does not develop further; no phosphatase appears in the cells; no hypertrophy of the cartilage cells takes place; and there is no true ossification. On the other hand, if the original tissue explanted is the limb bud of the 6-day embryo, then it continues to develop more completely; in it the cartilage hypertrophies and calcified bone appears in the usual places (Fell and Robison, 1929). Obviously, some conditions are established within the tissue, between the third and sixth days, which eventually allow the development of phosphatase by the relevant cells, and the full differentiation of the osteoblast.

A somewhat similar situation arises when two developing ribs are planted in a tissue culture at right angles to each other in the form of a T (Glücksmann, 1939, 1942). As the ribs continue their growth and development, one presses against the other and bends it. Under these conditions the morphology of the bent bone changes so that bone appears mostly where tension is highest, namely on the outer surface of the curved bone, while cartilage tends to appear from the perichondrium, and even the periosteum, on the lesser curvature where the cells are under compression.

Another particularly striking example of the effects of external conditions on the activities of mechanocytes is seen in the results obtained by Fell and Mellanby (1950, 1952) on the growth of developing long-bones in media containing a large amount of vitamin A in the form of the alcohol or acetate. In these cultures the cartilage cells not only laid down no matrix but they actually resolved that which they had already produced, and gradually the cartilage disappeared as such and the still healthy cells became lost among, and indistinguishable from, the normal connective tissue cells. This change in the matrix probably involved the mucoprotein part rather than the collagenous part, for the basophilia and metachromasia disappeared while the affinity for Van Gieson's stain was enhanced. It is probable that the protein part of the mucoprotein is digested by enzymes liberated from the lysosomes of the cells as these are made more permeable by the action of the vitamin (Dingle and Lucy, 1965). In similar experiments on the ossifying bones from the mouse, there was also destruction of bone matrix caused by the excess vitamin A, but in the chick this was not so evident.

Myxoblasts and Myoblasts

All these observations point to the close functional relationship between those mechanocytes which can be classed as osteoblasts, chondroblasts, fibroblasts, and synovioblasts. On the other hand, there is no suggestion that osteoblasts or their relatives ever become heart muscle cells, plain muscle cells, or skeletal muscle cells which are all cells which can also behave *in vitro* as mechanocytes. Nevertheless, there are reasons for believing that these three types of muscle are themselves not unrelated. There are records of plain muscle, as in the bladder of the dog, becoming striated when it is made to contract more frequently than normal. In tissue cultures of skeletal muscle the myoblasts which separate from the muscle behave for a time very similarly to the separate cardiac muscle fibres and may develop very similar patterns of longitudinal striations or "tension striae." The cross-striations are not an essential part of the basic contractile process and all three types of muscle may contract without any visible striations (Lewis and Lewis, 1917, 1924; de Rényi and Hogue, 1934). The electron-microscope, however, reveals that myofibrils with typical striation are generally present in cells of cardiac and skeletal muscle that are contractile (see p. 324). Moreover, under conditions of tension all may develop "tension-striae" within their cytoplasm. Biochemically, their metabolisms do not seem to be very dissimilar, though the organization of the tissues and the time relationships of contraction may be very different.

It would seem justifiable therefore to suggest that the main family of mechanocytes, as classified by their type of growth and general behaviour in tissue culture, may be subdivided into at least two subfamilies which could be conveniently called myoblasts or potential muscle cells, and myxoblasts or cells potentially concerned with the formation of skeletal and supporting elements. This term "myxoblast" refers to the outstanding property shared by fibroblasts, chondroblasts, synovial cells, osteoblasts, and odontoblasts of producing and liberating into their surroundings varying amounts of mucoprotein [μύξα (Gr.) = mucus]. From the point of view of the actual mucoprotein that these various classes of cells produce in the body, it is interesting to note that synovioblasts and the cells of the vitreous body in the eye produce mainly hyaluronic acid; dermal fibroblasts produce hyaluronic acid, chondroitin sulphate C, and chondroitin sulphate B; chondroblasts produce mostly chondroitin sulphate C and, as they age, some chondroitin sulphate A is formed; osteoblasts produce chondroitin sulphate A and some keratosulphate (Mathews, 1967). This property of mucoprotein production may be contrasted with the similarly well-developed property of the myoblasts which is to form within themselves the contractile protein system of actomyosin. This does not mean that myoblasts cannot produce muco-proteins nor that myxoblasts are unable to form actomyosin, but simply that in each case these properties are not well developed. Collagen production, one of the primary and most typical characters of mechanocytes, is on the whole better developed in the myxoblasts than in the myoblasts, but it is certainly not their exclusive property. Prolonged cultivation of muscle in conditions not favouring growth has produced, as noted above, cultures containing little muscle and much collagen; and cardiac tissue has produced argyrophil fibres *in vitro* (Bloom, 1930). The origin of this collagen is, of course, not certain—it may have been produced by the interstitial connective tissue cells of the original explant, or it may genuinely have been developed by the dedifferentiated muscle cells. This, incidentally, illustrates one of the main difficulties of the tissue culture method, namely the correct identification of cell types and the origin of the various cells and structures which develop. Tissues as such are seldom, if ever, pure, and it is all too easy for an apparently uniform culture of cells to contain small numbers of cells of other types and for these to escape notice.

Collagen-type proteins, which are essentially chains containing a high proportion of glycine, proline, and hydroxyproline molecules having a distinctive X-ray diffraction pattern, are characteristic of many of the tissues from which mechanocytes emerge in tissue culture. These tissues are all derivatives of the loose mesenchyme, the majority of whose cells are virtually synonymous with the myxoblasts. With one or two interesting exceptions, collagens are not normally produced in any appreciable amount by any of the essentially epithelial tissues, glands, or strictly neural elements in the body. From the work of Gross and others (Gross, 1956; Gross *et al.*, 1954) it appears that collagen results from the building up of "tropocollagen" units in a rather special manner into fibres (Fig. 2.21). Tropocollagen units are stiff rods, about 2900 Å long and 14 Å wide, each consisting of three intertwined helical polypeptide chains, and having a "molecular" weight of about 340,000. The

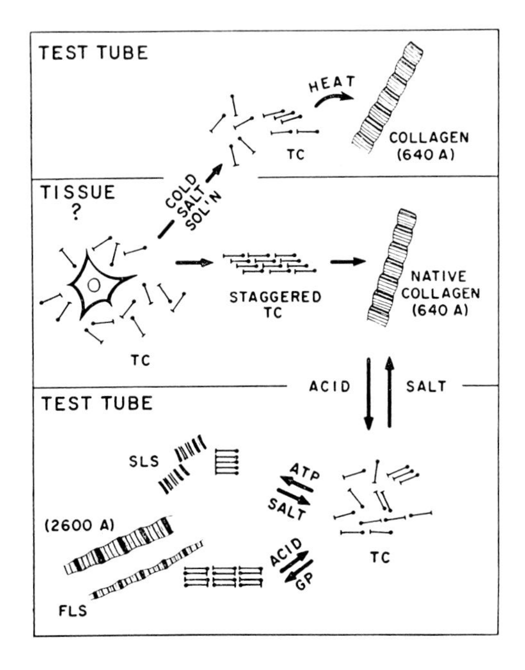

FIG. 2.21. Scheme for the building of collagen fibres from tropocollagen units, both artificially and naturally. ATP, Adenosine triphosphoric acid; FLS, fibrous long spacing; GP, glycoprotein; SLS, segment long spacing; TC, tropocollagen. (Gross, 1956.)

tropocollagen units, or more probably their polypeptide precursors, are liberated from the cells more or less in proportion to the growth rate of the tissue and appear in the surrounding medium where, given the right conditions, they line up in a staggered pattern to produce cross-banded fibres with a spacing of 640 Å (Fig. 2.22). Although it is possible in the test tube to produce the 640 Å banding of natural collagen fibres from tropocollagen units, the actual conditions for their formation in the body are not yet known. Mucopolysaccharides and vitamin C appear to be involved *in vivo* in argyrophil fibre formation, but, so far, it has not been shown how they contribute biochemically.

Apart from collagen formation on the one hand and contractility on the other as distinguishing features, the importance of lysine for osteoblasts and

FIG. 2.22. Collagen fibres showing the typical cross-banding. Electron micrograph of leg-tendon of fowl. (× 31,000.) (Randall, 1953.)

not for myoblasts, as mentioned earlier, may perhaps point to a further biochemical difference between the two subfamilies, myxoblasts and myoblasts.

Although the formation of elastic fibres has been less investigated in tissue cultures, it is probable that they too are the products of mechanocyte activity and it is interesting to note that lysine plays a dominant role in cross-linking the polypeptide chains and thus imparting elasticity (Partridge *et al.*, 1965).

For various reasons, neither osteoblasts nor chondroblasts tolerate high concentrations (i.e. more than 15%) of embryo juice for long, though the mitotic rate in osteoblasts may be increased temporarily by concentrations up to 40%: heart and muscle mechanocytes, on the other hand, are not so adversely affected by high concentrations of juice nor are their mitotic rates much increased by concentrations between 15% and 40%. All these observations taken together support the conception of a division of mechanocytes into two main families, in each of which there may prove to be many "genera" and "species."

Before leaving mechanocytes to consider other families of cells, something more may perhaps be said about the endothelium of blood vessels, especially as it emerges from cultures of the heart and aorta. From cultures of fresh heart tissue the endothelial cells often emerge as a continuous sheet of polygonal cells. Such cells from the heart often show tension-striae *in vitro*

(Lewis and Lewis, 1924), and endothelial cells *in vivo* are mildly contractile. Their typically membranous growth of extended and flattened cells easily breaks up and gives rise to separate cells not unlike fibroblasts; in fact, Lewis (1922, 1923 a, b) has described all stages in the transformation. The simplest hypothesis, admittedly based on purely morphological criteria, would therefore be that these endothelial membranes represent the reaction of myoblasts, which are normally present in the mesenchyme, to growth on a surface, though the actual conditions which evoke the epithelioid type of behaviour have not yet been worked out. Such endothelia (and they are the majority of those in the body) are not phagocytic and are probably to be distinguished from a few special endothelia derived from such situations as bone marrow, spleen, liver sinusoids, and adrenal cortex, where cells of a different type may be involved (see p. 95). These are distinguishable by their well developed phagocytic properties. If, as tentatively suggested, this flattening of the typical endothelial cells is the reaction of myoblasts to contact with a surface, then the synovial membranes (King, 1935) (Fig. 2.23) and

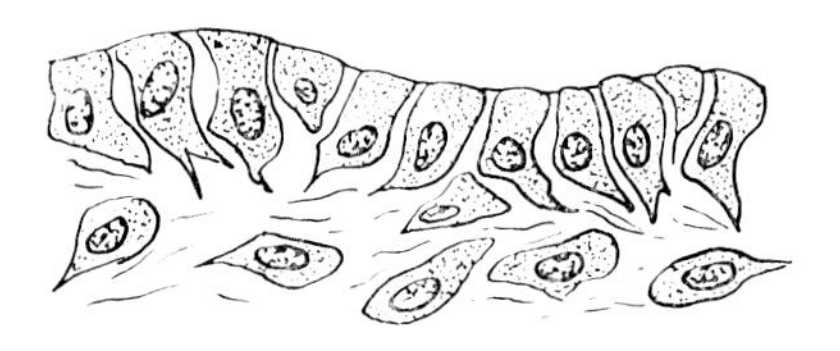

FIG. 2.23. Synovioblasts showing the semicolumnar arrangement which these cells sometimes assume. (Redrawn from King, 1935.)

the layer of odontoblasts (Fig. 2.24) in teeth and, to a lesser extent, the osteoblasts of the periosteum, may well represent the manner of behaviour of myxoblasts under similar conditions. In these two cases the cells tend to take up a much more columnar shape and the whole membrane resembles a somewhat loose columnar epithelium rather more than a pavement. It should, however, be stated that on the surfaces provided in tissue cultures, synovioblasts may also produce a membrane-like growth of flattened cells (Vaubel, 1933a).

From all that has been said so far it seems possible to conclude that there is a large group of tissues in the animal body which, when explanted *in vitro*, give rise to cells which display mechanocytic form and behaviour (see Figs. 2.3 and 2.25); this behaviour is undoubtedly a simplification of normal cell function, but is only partly a dedifferentiation. Individual potentialities may still persist more or less reversibly and can be evoked again by suitable conditions. Within the main family of mechanocytes there is the suggestion of at least two subfamilies, myoblasts and myxoblasts, and, pushing this idea of classification further, the myoblasts could be divided into genera of cardio-myoblasts, skeleto-myoblasts, viscero-myoblasts, and perhaps others like "endothelio-myoblasts," while the myxoblasts would fall mainly into three rather closely related genera of osteoblasts, chondroblasts, and fibroblasts and the more aberrant synovioblasts and odontoblasts. It must be remembered,

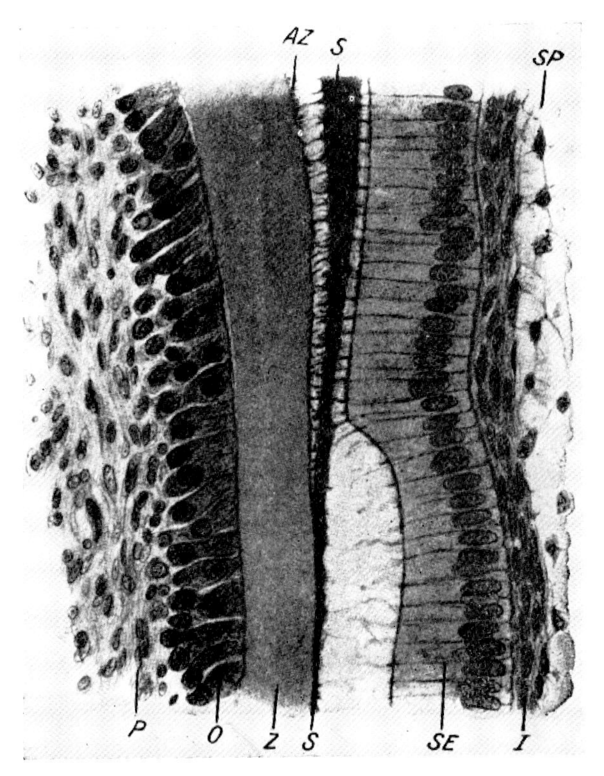

FIG. 2.24. Diagram showing the pseudocolumnar arrangement of odontoblasts O. AZ, dentino-enamel junction; I, flattened cells of enamel organ; P, Pulp; S, enamel; SE, ameloblasts; SP, stellate cells of enamel organ: Z, dentine. (Maximow and Bloom, 1930.)

in using this analogy with the classification of species, that unlike the latter it is not based on quite such an irreversible system. In other words, while species cannot change from one into the other, though the boundaries between them may sometimes be ill-defined, cells, by reversing their development, appear to be able, to some extent, to start again on another line; this at least follows if the appearance of cartilage in pure cultures from endosteal bone as described above has been correctly interpreted. However this may be, a question

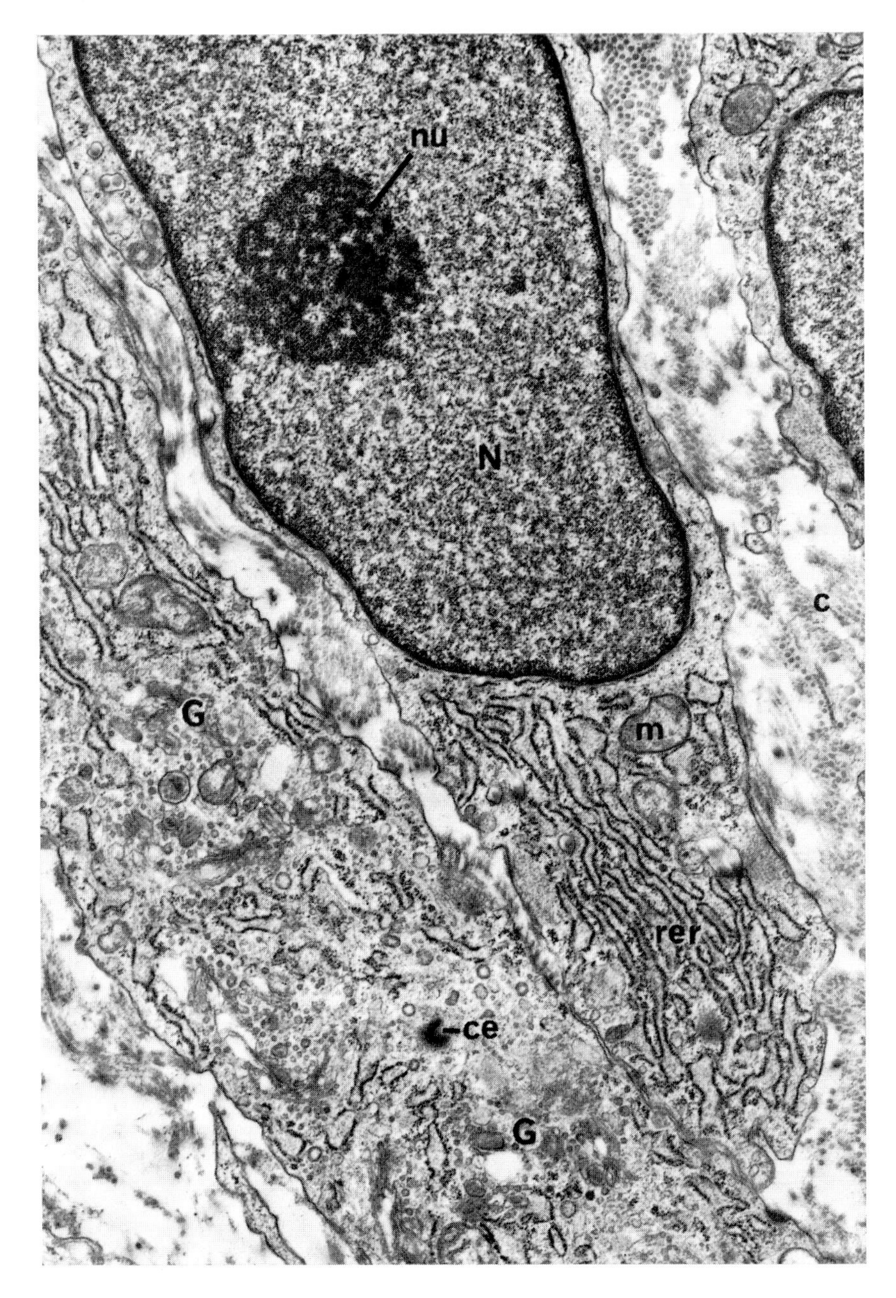

FIG. 2.25. Fibroblasts from a healing wound (× 14,000). c, Collagen; ce, centriole; G, Golgi complex; m, mitochondria; N, nucleus; nu, nucleolus; rer, rough endoplasmic reticulum. (Photo by R. Ross.)

which must be asked is: "How do myxoblasts in tissue culture continue to produce myxoblasts, and myoblasts to reproduce myoblasts?" As we shall see, mechanocytes are recognizably different from other cell types *in vitro*, but after prolonged periods of continuous subcultivation, in a medium of of plasma and embryo-tissue juice, cultures of mechanocytes *in vitro* have remained essentially the same as they were at the end of the first passage. They have bred true as mechanocytes. It should be stressed, however, that under other conditions of cultivation, generally involving more complete isolation, mechanocytes have acquired new characteristics *in vitro* and have adapted themselves to the new medium and conditions, and become "tissue-culture cells."

References

Abercrombie, M., and Heaysman, J. E. M. (1952). Observations on the social behaviour of cells in tissue culture. 1. Speed of movement of chick heart fibroblasts in relation to their mutual contacts. *Exptl. Cell Res.* **5,** 112.

Abercrombie, M, Lamont, D. M., and Stephenson, E. M. (1968). The monolayering in tissue culture of fibroblasts from different sources. *Proc. Roy. Soc. London* **B170,** 349.

Ambrose, E. J. (1961). The movements of fibrocytes. *Exptl. Cell Res. Suppl.* **8,** 54.

Baker, L. E., and Carrel, A. (1926a). The action on fibroblasts of the protein fraction of embryo tissue extract. *J. Exptl. Med.* **44,** 387.

Baker, L. E., and Carrel, A. (1926b). Effects of the amino acids and dialysable constituents of embryo tissue juice on the growth of fibroblasts. *J. Exptl. Med.* **44,** 397.

Baker, L. E., and Carrel, A. (1928). The effects of digests of pure proteins on cell proliferation. *J. Exptl. Med.* **47,** 353.

Bloom, W. (1930). Studies on fibres in tissue culture. (ii). The development of elastic fibres in cultures of embryonic heart and aorta. *Arch. Exptl. Zellforsch. Gewebezücht.* **9,** 6.

Briggs, R. (1946). Effects of growth inhibitor, hexenolactone, on frog embryos. *Growth* **10,** 45.

Carrel, A. (1913). Artificial activation of the growth *in vitro* of connective tissue. *J. Exptl. Med.* **17,** 14.

Carrel, A., and Baker, L. E. (1926). The chemical nature of substances required for cell multiplication. *J. Exptl. Med.* **44,** 503.

Chèvremont, M., and Frédéric, J. (1968). Les ADN cytoplasmiques, en particulier les ADN mitochondriaux. *Excerpta Med. I.C.S.* **166,** 7.

Davidson, J. N., and Waymouth, C. (1943). Factors influencing the nucleoprotein content of fibroblasts growing *in vitro*. *Biochem. J.* **37,** 271.

Davidson, J. N., Leslie, I., and Waymouth, C. (1949). Nucleoprotein content of fibroblasts growing *in vitro*. 4. Changes in the ribonucleic acid phosphorus and deoxyribonucleic acid phosphorus content. *Biochem. J.* **44,** 5.

de Bruyn, P. P. H. (1946) The amoeboid movement of the mammalian leucocytes in tissue culture. *Anat. Record* **95,** 177.

de Rényi, G. S., and Hogue, M. J. (1934). Studies on skeletal muscle grown in tissue cultures. *Arch. Exptl. Zellforsch. Gewebezücht.* **16,** 167.

Dingle, J. T., and Lucy, J. A. (1965). Vitamin A, carotenoids and cell function. *Biol. Rev. Cambridge Phil. Soc.* **40,** 422.

Doljanski, L., and Hoffman, R. S. (1939). Stimulation de la croissance de colonies de fibroblastes *in vitro* par des extraits de tissu adulte. L'action d'extraits de muscle cardiaque. *Compt. Rend. Soc. Biol.* **130,** 1246.

Eagle, H. (1954). The specific amino-acid requirements of a mammalian cell (strain L) in tissue culture. *J. Biol. Chem.* **214,** 839.

Eagle, H., and Levintow, L. (1965). Amino acid and protein metabolism. 1. The metabolic characteristics of serially propagated cells. *In* "Cells and Tissues in Culture" (E. N. Willmer, ed.), Vol. 1, p. 277. Academic Press, New York.

Eagle, H., Oyama, V. I., Levy, M., Horton, C. L., and Fleischman, R. (1955). The growth response of mammalian cells in tissue culture to L-glutamine and L-glutamic acid. *J. Biol. Chem.* **218,** 607.

Eagle, H., Oyama, V. I., and Levy, M. (1957a). Amino acid requirements of normal and malignant human cells in tissue culture. *Arch. Biochem. Biophys.* **67,** 432.

Eagle, H., Oyama, V. I., Levy, M., and Freeman, A. E. (1957b). Myoinositol as an essential growth factor for normal and malignant human cells in tissue culture. *J. Biol. Chem.* **226,** 191.

Earle, W. R. (1958). Long-term cultivation of animal tissue cells in large cultures. *Federation Proc.* **17,** 967.

Earle, W. R., Schilling, E. L., Bryant, J. C., and Evans, V. J. (1954). The growth of pure strain-L cells in fluid suspension cultures. *J. Natl. Cancer Inst.* **14,** 1159.

Fell, H. B. (1932). The osteogenic capacity *in vitro* of periosteum and endosteum isolated from the limb skeleton of fowl embryos and young chicks. *J. Anat.* **66,** 157.

Fell, H. B., (1933). Chrondrogenesis in cultures of endosteum. *Proc. Roy. Soc. London* **B112,** 417.

Fell, H. B., and Mellanby, E. (1950). Effects of hypervitaminosis A on foetal mouse bones cultivated *in vitro*. *Brit. Med. J.* **ii,** 535.

Fell, H. B., and Mellanby, E. (1952). The effect of hypervitaminosis A on embryonic limb bones cultivated *in vitro*. *J. Physiol. (London)* **116,** 320.

Fell, H. B., and Robison, R. (1929). The growth, development and phosphatase activity of embryonic avian femora and limb-buds cultivated *in vitro*. *Biochem. J.* **23,** 767.

Firket, H. (1965). Cell division. *In* "Cells and Tissues in Culture" (E. N. Willmer, ed.), Vol. 1., p. 203. Academic Press, New York.

Fischer, A. (1930). "Gewebezüchtung". Müller & Steinicke, Munich.

Fischer, A. (1941). Die Bedeutung der Aminosäuren für die Gewebezellen *in vitro*. *Acta Physiol. Scand.* **2,** 143.

Fischer, A. (1948). Amino acid metabolism of tissue cells *in vitro*. *Biochem. J.* **43,** 491.

Fischer, A., and Parker, R. C. (1929). A new technique for the study of tissues in a state of latent life *in vitro*. *Proc. Soc. Exptl. Biol. Med.* **26,** 585.

Glücksmann, A. (1939). Studies on bone mechanics *in vitro*. (ii). The role of tension and pressure in chondrogenesis. *Anat. Record* **73,** 39.

Glücksmann, A. (1942). The role of mechanical stresses in bone formation *in vitro*. *J. Anat.* **76,** 231.

Gross, J. (1956). The behaviour of collagen units as a model in morphogenesis. *J. Biophys. Biochem. Cytol.* **2,** 261.

Gross, J., Highberger, J. H., and Schmitt, F. O. (1954). Collagen structures considered as states of aggregation of a kinetic unit. The tropocollagen particle. *Proc. Natl. Acad. Sci. U.S.* **40,** 679.

Harris, M., and Kutsky, R. J. (1954). Synergism of nucleoprotein and dialysate growth factors in chick embryo extract. *Exptl. Cell Res.* **6,** 327.

Hauschka, T. (1946). Effects of the growth-inhibitor, hexenolactone, on flatworms. *Growth* **10,** 193.
Healy, G. M., Fisher, D. C., and Parker, R. C. (1954a). Nutrition of animal cells in tissue culture. (viii). Desoxyribonucleic acid phosphorus as a measure of cell multiplication in replicate culture. *Can. J. Biochem. Physiol.* **32,** 319.
Healy, G. M., Fisher, D. C., and Parker, R. C. (1954b). Nutrition of animal cells in tissue culture. (ix). Synthetic medium 703. *Can. J. Biochem. Physiol.* **32,** 327.
Heaton, J. B. (1926). The nutritive requirements of growing cells. *J. Pathol. Bacteriol.* **29,** 293.
Howard, A., and Pelc, S. R. (1953). Synthesis of deoxyribonucleic acid in normal and irradiated cells and its relation to chromosome breakage. *Heredity* **6,** Suppl. 261.
Hueper, W. C., Allen, A., Russell, M., Woodward, G., and Platt, M. (1933). Studies on the growth factors of embryo extract. *Am. J. Cancer* **17,** 74.
Immers, J. (1961). Comparative study of the localization of incorporated ^{14}C-labeled amino acids and $^{35}SO_4$ in the sea urchin ovary, egg and embryo. *Exptl. Cell Res.* **24,** 356.
Jacoby, F. (1937). Migratory and mitotic activity of chick fibroblasts under the influence of dialysate of embryo juice. *Arch. Exptl. Zellforsch. Gewebezücht.* **19,** 241.
Jacoby, F., Trowell, O. A., and Willmer, E. N. (1937). Further observations on the manner in which cell division of chick fibroblasts is affected by embryo tissue juice. *J. Exptl. Biol.* **14,** 255.
Jacoby, F., Medawar, P. B., and Willmer, E. N. (1941). The toxicity of sulphonamide drugs to cells *in vitro*. *Brit. Med. J.* **ii,** 149.
King, E. S. J. (1935). The Golgi apparatus of synovial cells under normal and pathological conditions and with reference to the formation of synovial fluid. *J. Pathol. Bacteriol.* **41,** 117.
Lasnitski, I. (1937). The action of heated embryo extract upon the growth of fibroblast cultures. *Skand. Arch. Physiol.* **76,** 303.
Lewis, M. R. (1922). The importance of dextrose in the medium of tissue cultures. *J. Exptl. Med.* **35,** 317.
Lewis, M. R., and Lewis, W. H. (1917). Behaviour of cross-striated muscle in tissue cultures. *Am. J. Anat.* **22,** 169.
Lewis, W. H. (1922). Endothelium in tissue cultures. *Am. J. Anat.* **30,** 39.
Lewis, W. H. (1923a). Mesenchyme into mesothelium. *J. Exptl. Med.* **38,** 257.
Lewis, W. H. (1923b). The transformation of mesenchyme into mesothelium in tissue cultures. *Anat. Record* **25,** 111.
Lewis, W. H. (1931). Locomotion of lymphocytes. *Johns Hopkins Hosp. Bull.* **49,** 29.
Lewis, W. H. (1940). The role of superficial plasmagel layer in changes of form, locomotion and division of cells in tissue cultures. *Arch. Exptl. Zellforsch. Gewebezücht.* **23,** 2.
Lewis, W. H., and Lewis, M. R. (1924). Behaviour of cells in tissue cultures. *In* "General Cytology" (E. V. Cowdry, ed.), p. 383. Univ. of Chicago Press, Chicago, Illinois.
Likely, G. D., Sanford, K. K., and Earle, W. R. (1952). Further studies on the proliferation *in vitro* of single isolated tissue cells. *J. Natl. Cancer Inst.* **13,** 177.
Mathews, M. B. (1967). Macromolecular evolution of connective tissue. *Biol. Rev. Cambridge Phil. Soc.* **42,** 499.
Maximow, A. A., and Bloom, W. (1930). "Text-book of Histology." Saunders, Philadelphia, Pennsylvania.
Medawar, P. B. (1937). A factor inhibiting the growth of mesenchyme. *Quart. J. Exptl. Physiol.* **27,** 147.
Medawar, P. B., Robinson, G. M., and Robinson, R. (1943). A synthetic differential growth inhibitor. *Nature* **151,** 195.

Morgan, J. F., and Morton, H. J. (1955). Studies on the sulphur metabolism of tissues cultivated *in vitro*. A critical requirement for L-cystine. *J. Biol. Chem.* **215,** 539.

Parker, R. C. (1933a). The races that constitute the group of common fibroblasts. 2. The effect of blood serum. *J. Exptl. Med.* **58,** 97.

Parker, R. C. (1933b). The races that constitute the group of common fibroblasts. 3. Differences determined by origin of explant and age of donor. *J. Exptl. Med.* **58,** 401.

Parker, R. C. (1936). The cultivation of tissues for prolonged periods in single flasks. *J. Exptl. Med.* **64,** 121.

Partridge, S. M., Thomas, J., and Elsden, D. F. (1965). Structure and metabolism of elastin and resilin. 1. The nature of the cross-linkages in elastin. *In* "Structure and Function of Connective and Skeletal Tissue" (S. F. Jackson *et al.*, eds.), p. 88. Butterworth, London and Washington, D.C.

Pasieka, A. E., Morton, H. J., and Morgan, J. F. (1956). The metabolism of animal tissues cultivated *in vitro*. A. Amino acid metabolism of chick embryonic heart fibroblasts cultivated in synthetic medium. *J. nat. Cancer Inst.* **16,** 995.

Pomerat, C. M. and Willmer, E. N. (1939). Carbohydrate metabolism and mitosis. *J. Exptl. Biol.* **16,** 232.

Puck, T. T. and Fisher, H. W. (1956). Genetics of somatic mammalian cells. 1. Demonstration of the existence of mutants with different growth requirements in the human cancer cell strain (HeLa). *J. Exptl. Med.* **104,** 427.

Puck, T. T., Marcus, P. I. and Cieciura, S. J. (1956). Clonal growth of mammalian cells *in vitro*. Growth characteristics of colonies from single HeLa cells with and without a "feeder" layer. *J. Exptl. Med.* **103,** 273.

Puck, T. T., Cieciura, S. J. and Fisher, H. W. (1957). Clonal growth *in vitro* of human cells with fibroblastic morphology. *J. Exptl. Med.* **106,** 145.

Randall, J. T. (1953). "Nature and Structure of Collagen". Butterworth, London and Washington, D.C.

Rinaldini, L. M. J. (1958). The isolation of living cells from animal tissues. *Intern. Rev. Cytol.* **7,** 587.

Royle, J. (1945). Some effects of hexenolactone on tissue cultures. *Growth* **9,**275.

Sanford, K. K., Earle, W. R., and Likely, G. D. (1948). The growth *in vitro* of single isolated tissue cells. *J. Natl. Cancer Inst.* **9,** 229.

Sanford, K. K., Likely, G. D., and Earle, W. R. (1954). The development of variations in transplantability and morphology within a clone of mouse fibroblasts transformed to sarcoma-producing cells. *J. Natl. Cancer Inst.* **15,** 215.

Seed, J. (1965). Deoxyribonucleic acid and ribonucleic acid synthesis in cell cultures. *In* "Cells and Tissues in Culture" (E. N. Willmer, ed.), Vol. 1. p. 317. Academic Press, N.Y.

Shaffer, B. M. (1956). The culture of organs from the embryo chick on cellulose acetate fabric. *Exptl. Cell Res.* **11,** 244.

Simms, H. S., and Stillman, N. P. (1937). Substances affecting adult tissue *in vitro*. 1. The stimulating action of trypsin on fresh adult tissues. *J. Gen. Physiol.* **20,** 603.

Trowell, O. A. (1954). A modified technique for organ culture *in vitro*. *Exptl. Cell Res.* **6,** 246.

Trowell, O. A., and Willmer, E. N. (1939). The effects of some tissue extracts on the growth of periosteal fibroblasts. *J. Exptl. Biol.* **16,** 60.

Vaubel, E. (1933a). The form and function of synovial cells in tissue cultures. (i). Morphology of the cells under varying conditions. *J. Exptl. Med.* **58,** 63.

Vaubel, E. (1933b). The form and function of synovial cells in tissue cultures. (ii). The production of mucin. *J. Exptl. Med.* **58,** 85.

Walker, P. M. B., and Yates, H. B. (1952). Nuclear components of dividing cells. *Proc. Roy. Soc.* (*London*) **B140,** 274.

Waymouth, C. (1955). Simple nutrient solutions for animal cells. *Texas Rept. Biol. Med.* **13,** 522.

White, P. R. (1949). Prolonged survival of excised animal tissues *in vitro* in nutrients of known constitution. *J. Cellular Comp. Physiol.* **34,** 221.

Willmer, E. N. (1933a). An analysis of the growth of chick heart fibroblasts in a hanging drop of fluid medium. *J. Exptl. Biol.* **10,** 323.

Willmer, E. N. (1933b). An analysis of the growth of chick heart fibroblasts in fresh cultures in a plasma coagulum. *J. Exptl. Biol.* **10,** 340.

Willmer, E. N. (1942). Carbohydrate metabolism of chick fibroblasts *in vitro*. *J. Exptl. Biol.* **18,** 237.

Willmer, E. N., and Jacoby, F., (1936). Studies on the growth of tissues *in vitro*. (iv). On the manner in which growth is stimulated by extracts of embryo tissues. *J. Exptl. Biol.* **13,** 237.

Willmer, E. N., and Kendal, L. P. (1932). The utilisation of proteoses by chicken heart fibroblasts growing *in vitro*. *J. Exptl. Biol.* **9,** 149.

Chapter 3

THE GROWTH OF EPITHELIOCYTES

When tissue cultures are made from the skin or from the intestine, in addition to the growth of mechanocytes which emerges freely from the dermis or from the muscle coats and from any connective tissue present, there also appear, sooner or later, cells which behave quite differently. These are the cells of the epithelium, and they emerge from the cut edges and spread out on any available surface in a single layer, as a continuous sheet of more or less flattened cells, i.e. as a closely packed pavement epithelium (Fig. 3.1). In this behaviour the epithelial cells are doing exactly what they may do in the body, as for example, during the first stages in the repair of a wound in the skin. When a portion of skin is removed, the more basal cells of the epidermis all around the cut very soon flatten out and migrate centripetally over the damaged surface, provided that the latter is in a condition suitable for their activity. In doing this the cells undergo few, if any, divisions; they simply tend to spread as a continuous sheet over the wound; at first, this sheet is only one layer thick, and the cells are all closely adherent to each other; though they may change their relative positions the cells remain as a membrane. Only later do cell divisions occur round the margins of the wound and then eventually spread centrally and assist in thickening up the membrane originally formed by cell migration. Cells behaving in this cohesive way have been observed in cultures of skin, intestine, liver, pancreas, kidney, pigment epithelium in the eye, thyroid, and many other tissues, but always from those tissues which originally contained cells of true epithelial type, i.e. cells which formed part of the primary surface of the animal. Thus epithelial cells from many, if not all, true epithelia are found to produce similar membranous sheets of cells in tissue cultures, always provided that a suitable surface is present upon which they may creep. In contrast to the behaviour of mechanocytes, where the cells normally emerge as a network, it is highly charac-

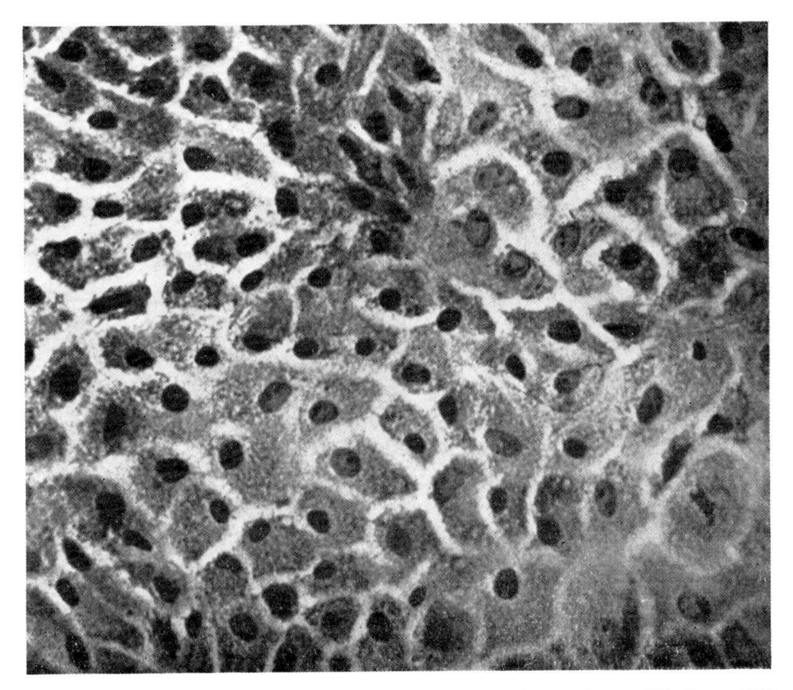

FIG. 3.1. Growth of epithelium from the iris of the chick embryo. (Fischer, 1922).

teristic of these cells of epithelia that they maintain close connections with each other all along their edges of contact and no spaces normally develop between them. There may be a "cement substance" or at least a specialized region of the cell surface between the cells which becomes black when treated with silver nitrate. This, however, is not a diagnostic feature since it also occurs between endothelial cells, which, as discussed in the last chapter, are probably related to mechanocytes. There may even be fibrous or pseudo-fibrous structures running between cell and cell in much the same way as they appear to do in the prickle-cell layer of the human epidermis. In the body, epithelial cells are held together by desmosomes, tight junctions, or by interdigitating folds, causing an interlocking of the cell surfaces. In spite of their close contact, however, the cells remain as discrete entities, and damage to or death of one cell does not involve changes, except of a mechanical nature, in the adjacent cells, as it would do if the cells were actually connected by protoplasmic bridges. Chambers (1924), for example, has shown that so long as the daughter cells after mitosis are still connected by a protoplasmic thread, then puncture of the nucleus of one of the cells causes coagulation of the cytoplasm of both cells, but similar damage to single cells in an epithelium is not transmitted to neighbouring cells.

Although superficially similar, true epithelial growth can usually be distinguished from the epithelioid growth of most vascular endothelia. The cells of the latter are generally larger and appear to have a somewhat different internal organization. However, the nature of the relationship, if any, between the cells of the endothelia of blood vessels, the cells of the so-called mesothelia (e.g. serous membranes of pericardium and peritoneum), and the cells of the epithelia discussed here is obviously a matter of importance, and at present is anything but clear. Further evidence, extending beyond the more purely morphological, is urgently needed as to their specific properties, potentialities and general behaviour, and the chapters that follow may suggest lines of investigation.

As emphasized already, the form of growth described above for the true epithelia is naturally dependent upon the provision of a suitable surface along which the membrane of cells may extend. This is normally, in tissue culture, provided by the glass surface of the coverslip, flask, or culture vessel, or alternatively by the surface of a plasma clot or even an agar gel. The surface of the plasma clot initially forms an excellent basis along which the cells readily extend, but sometimes the cells may liquefy the plasma, and then the whole membranous structure may break down, or the epithelium may roll up into tubular structures and irregular masses. In any case, the stretched membrane is always a rather unstable structure and epithelial cultures are very apt to distintegrate into separate nodules or rolls of tissue. The membrane may quite suddenly break, as if it had been under considerable tension; it then rolls up on itself, and this rather tiresome property from the point of view of tissue culture has made the detailed analysis of the growth in these membranes lag behind that of the growth of cells of other types, for the necessarily long-continued observations on a small group of cells under various conditions is almost impossible. On the other hand, when epithelial cells pile up in this way and do not persist as a flattened sheet, there is obviously more scope for them to form tubular or glandular structures, and this they often tend to do under these conditions.

When epithelial cells are buried in the substance of the plasma clot the growth is much more restricted, and in this situation tongue-like and tubular structures often appear (Fig. 3.2), the latter probably following the cavities produced by liquefaction of the coagulum. With some epithelia there is, under such conditions, a tendency for isolated cells to break away from the membrane and to adopt a more mechanocyte-like type of activity. Here again more observations and analysis are necessary. Are the cells, which thus break away, true mechanocytes or only mechanocyte-like in appearance? Reasons will be advanced on a later page for the belief that the break-up of an epithelium into separate cells, which may resemble mechanocytes or cells of other types, may be an inherent and latent property of epithelia in general.

As already outlined, successful growth of epithelial cells has been obtained from such widely different sources as the pigment epithelium of the iris and retina, lens, skin, thyroid gland, liver, kidney, intestine, pancreas, and uterus. In each case the growth takes the usual form of a continuous membrane of cells, and is naturally favoured when the culture is placed on the surface of the coagulum bathed with sufficient fluid to keep the cells in a

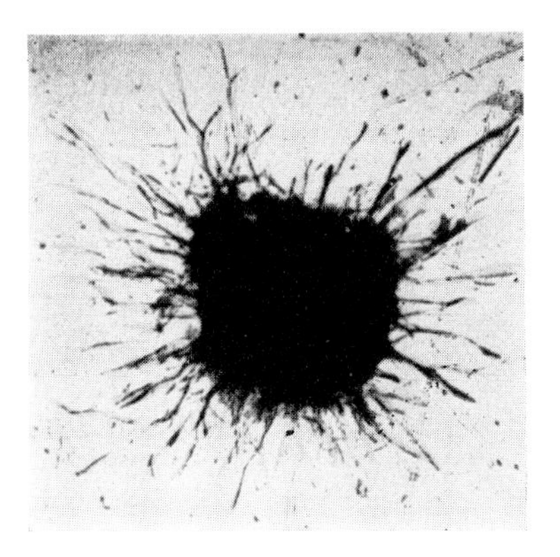

FIG. 3.2. Tubular and cord-like growths of epithelium from the iris when growing in a solid coagulum. (Fischer, 1924.)

moist environment. It is clear from a study of this list that it is of no significance whether the original tissue contained a stratified, glandular, or any other sort of epithelium, whether it belonged to an endocrine gland or an exocrine gland, or whether it was derived from ectoderm (e.g. skin), endoderm (e.g. intestine), or mesoderm (e.g. kidney). The general form and pattern of growth *in vitro* is essentially the same, and fundamentally different from that of the growth of mechanocytes.

Once again, as in the study of mechanocytes, the mechanical qualities of the medium and substratum have their effects in determining cell morphology, and the findings on the behaviour of epithelia in culture would have little significance if they rested only on morphological characteristics. Although the tendency of the cells to stick together so as to form a continuous membrane, a tubular structure, or cords of cells is extremely characteristic of the epitheliocyte type of growth, yet these cells may not always display this behaviour, and it has already been noted that at the edge of the growth zone of epithelia there are often to be found cells breaking way from it and assuming a more independent mode of existence. The classification of such cells on their appearance alone may, as with mechanocytes, be a difficult task, but once again their movement, as seen in cine-recordings, is sometimes diagnostic.

The Products of Epitheliocytes

When cells from various epithelia are kept in culture, there is often a tendency for them to undergo changes which are almost certainly connected with the process of keratinization (Fischer, 1924; Miszurski, 1937), i.e. the formation of the particular type of sclero-protein known as a keratin (Fig. 3.3). This, of course, is the class to which the natural protein-product of the

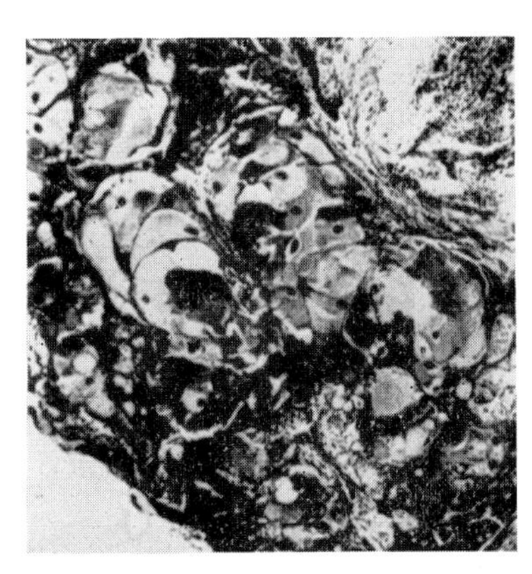

FIG. 3.3. Cells of iris epithelium undergoing changes resembling keratinization. (Fischer, 1924.)

epidermis in man and indeed of most vertebrates belongs. It is a type of protein with a characteristic X-ray diffraction pattern which is significantly different from that of the corresponding protein-product of mechanocytes, namely collagen. The amino-acid constitution of keratins is also quite different, and much of their structural character depends upon the cross-linking of the peptide chains by sulphur bonds. The protein arises within, and remains far more intimately associated with, the epithelial cells than does collagen with the mechanocytes, for the latter is certainly liberated into, if not actually formed in, the surrounding medium. Keratinization generally occurs within the cytoplasm, being preceded by an increased concentration of SH groups in the cells, and it eventually involves the death of the cell. In tissue cultures, cells which are undergoing these changes generally become granular and then vacuolated in a characteristic manner. In the body, keratinization is normally confined to the epidermis and its immediate derivatives, like the hair follicles and the enamel organs of the teeth. It also occurs periodically in the vagina. In conditions of severe deficiency of vitamin A, other epithelia may assume this activity, e.g. the conjunctival epithelium, and the tracheal epithelium; in malignant disease, keratinous or prekeratinous changes may be found in a variety of other epithelial structures. Some birds have keratin in the gizzard and some have hair-like processes which guard the entry to the pylorus. *In vitro*, keratinization has been observed in a variety of other epithelia, e.g. the pigment epithelium behind the iris, and it seems justifiable to conclude that keratin formation is about as typical of epithelia as collagen formation is typical of mechanocytes. It seems to be a latent potentiality of epithelia, at least among the vertebrates, even when it does not normally manifest itself.

Another characteristic of epithelial behaviour, although not one which is often observed in tissue cultures except under such conditions as lead to the differentiation of cells, is the formation of mucoproteins and mucopolysaccharides. This, of course, was also true of mechanocytes, but while mechanocytes produce these substances in the form of a matrix between the cells, many epitheliocytes more conspicuously produce them as droplets within the cytoplasm. Typically, the secretions of the mechanocytes contain chondroitin sulphuric acid, containing galactosamine, while the corresponding sulphated epithelial mucin (mucoitin sulphuric acid), which may of course also be finally liberated from the cells, more often contains glucosamine in place of the galactosamine. It would be interesting to know whether the stereochemical difference between the sugars is in any way related to the location of the product with respect to the cell membrane. In lower animals, chitin, as a glucosamine compound, is also interesting as being an essentially epithelial product, and the chemical properties of the "cuticles", "terminal bars", and "cement substances" between the cells of the epithelial tissues of the vertebrates call for further investigation along these lines. It is perhaps significant that such structures tend to be more prominent between those epithelial cells which are not themselves engaged in "mucin" storage. On the other hand, the hyaluronic acid of the connective tissue "ground substance" is also a glucosamine derivative, so it appears from this that glucosamine may be liberated not only by epitheliocytes but by mechanocytes also, or at least by some cells in the connective tissue. Some epithelial glands, e.g. the rectal gland in elasmobranchs, can produce mucin without its appearing as droplets within the cells. Many epithelia with cilia and microvilli also produce PAS positive material in the neighbourhood of these structures.

Both in mechanocytes and in epithelia the formation or persistence of mucin is affected by the vitamin A content of the medium but the results are different. In normally keratinizing epithelia, e.g. the skin of the chick, an excess of vitamin A in the medium turns many of the cells into mucus-producing cells (Fell and Mellanby, 1953) (Fig. 3.4); in the opposite direction, i.e. in cases of vitamin A deficiency, the mucous cells in the trachea *in vivo* may cease to function as such and begin to keratinize. In these epithelia then the presence of vitamin A in some way favours mucoprotein production and

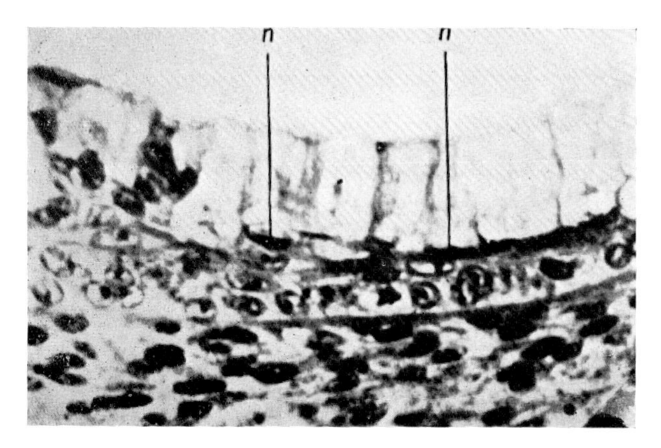

FIG. 3.4. Mucous cells in the skin of the chick produced by high doses of vitamin A in tissue cultures. n, nucleus of mucous cell. (Fell and Mellanby, 1953.)

depresses keratinization. These effects contrast rather strongly with those already mentioned in connexion with mechanocytes, in which an excess of vitamin A, when added to cultures of cartilage and bone, led to the complete disappearance of the mucoprotein matrix, presumably as the result of some new activity on the part of the mechanocytes, since it was found that these must remain present as healthy cells for the process to go on. The activity of an enzyme, forming soluble sulphated mucopolysaccharides instead of chondroitin sulphate has been suggested (Fell *et al.*, 1956) and this enzyme may be liberated from the lysosomes of the cartilage cells and remove the polysaccharide molecules from their attachment with protein (Dingle and Lucy, 1965). Other work also indicates that the changes in the epithelial cells may be connected with the manner in which sulphur is picked up and metabolized by the cells (Wolf *et al.* 1960, 1961). Flesch (1953), also, has suggested that the action of vitamin A is to inactivate the SH groups in the cell. Moreover, autoradiographs obtained after treating tissues in culture with radioactive S show that mucus-secreting cells can pick up their sulphur in the inorganic form whereas keratinizing epithelia do not appear to be able to do so. Presumably they must get their sulphur in the form of cystine or methionine. Perhaps the mechanocytes have to get at least some of their sulphur in the same way, for it has already been observed that cystine is an essential constituent of the medium for the growth of mechanocytes. The ability to use the inorganic sulphur may be connected with the direction of movement of glucosamine or its presence intra- or extracellularly.

It seems likely that one of the actions of vitamin A is to increase the permeability of phospholipid membranes, and this may be its mode of action on the lysosomes.

With reference again to the "cement substance" between epithelial cells, it is interesting to notice that trypsin alone does not cause the separation of the cells of an epithelium as satisfactorily as it does those of the connective tissue when these tissues are similarly treated. On the other hand, when trypsin treatment is combined with Ca^{2+} and Mg^{2+} deficiency, epithelial tissues fall completely apart; this method is widely used for disintegrating tissues for cell culture (Moscona, 1952a, b; Rinaldini, 1958). The direct disintegrating action of Ca^{2+}-free media on the cement substance between epithelial cells and the subsequent rounding up of the cells has been frequently observed in blastulae, endothelia of blood vessels, and in tissue cultures of the skin. Moreover it is a matter of some interest to the present discussion to note that L-ascorbic acid is not necessary for the production of the intercellular substance by epithelial cells (Chambers and Cameron, 1943), though it is necessary if mechanocytes are to produce their intercellular ground substance properly as in teeth, bones, and repairing wounds. However, in the latter case the ascorbic acid is perhaps involved more with collagen fibre formation than it is with the mucoid material, though it is sometimes difficult to separate the early collagen from the carbohydrate or polysaccharide which usually surrounds it. Indeed the polysaccharides are thought by some to have an orientating action on the protein fibres though their action in this direction has been questioned by Gross (1956). It may be recalled that many marine blastulae and embryos in the early cleavage stages may be broken up into their constituent cells by shaking with Ca^{2+}-free sea-water. The cells of such embryos and blastulae are in many ways comparable with those of an epithelium in that they are normally coherent and, at least in blastulae, extend as a single layer around the organism.

When isolated from each other and returned to a medium containing Ca^{2+}, epithelial cells tend to adhere to each other and to aggregate. This was well shown by Holtfreter (1947) for the isolated ectodermal cells of amphibian larvae (Fig. 3.5). It is equally true of the flagellated epithelial cells (choanocytes) of sponges and of human epithelial cells from the skin and also from the liver and kidney. The calcium ion may therefore be regarded as essential for maintaining the adhesive quality of epithelial cells. Nevertheless, a low Ca^{2+} concentration favours the growth of skin epithelia *in vitro*, and so does a high phosphate content (Parshley and Simms, 1950). This may perhaps be so because of the greater motility allowed to the cells by the decrease in rigidity of the intercellular substance; but it should be emphasized that the biological actions of the calcium ion are many and various. Electron-microscope studies of the boundaries between epithelial cells do not suggest that the mechanism of the adhesion between the cells is always identical; nevertheless, as soon as artificially isolated epithelial cells find, in the course of their almost random movement, others of their own type, they remain mutually adherent and seldom leave each other again. Epithelial cells of two different origins, however, do not tend to adhere to each other, not at least for so long or so permanently as those of similar origin. Liver cells recognize other liver cells, and lung cells cohere with lung cells, but liver cells and lung cells are soon divorced. Liver epithelium and lung epithelium, however, are both genuine epithelia whose cells grow as continuous sheets (Weiss, 1959).

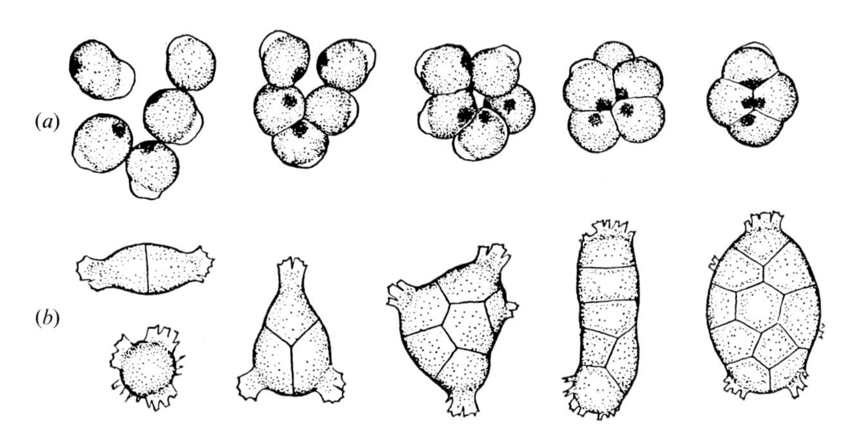

FIG. 3.5. (*a*) Aggregation of ectodermal cells after their isolation; (*b*) ectodermal cells aggregating while creeping on a glass surface. (Holtfreter, 1947.)

Requirements for Growth

Unlike the growth of mechanocytes, the growth of epithelia does not appear to be so specifically dependent on embryo juice or extract, and successful growth of membranes of kidney, liver, and intestinal epithelia can be obtained on the surface of clotted plasma alone. Indeed, concentrations of embryo juice higher than 20% have been shown to inhibit the growth of membranes of hepatic epithelium (Doljanski, 1929). This point, however, as mentioned earlier, needs further investigation because there are practically no measurements of mitotic rates in cultures of epithelia in different media comparable with those carried out on mechanocytes. Many authors certainly use embryo juice in the medium for epithelia, and Doljanski (1930a), has found that pigment formation (i.e. a process dependent upon differentiation and function) is encouraged in iris epithelium in media washed with heparin plasma, while "growth" occurs faster in plasma with 30% extract, thus indicating that in this case the extract does actually stimulate the growth. Puck and his school (1957) have shown that clones of fibroblasts only grow successfully when embryo extract is present in the medium, while "epithelioid" cells grow well in its absence. All epithelial cultures, however, seem to benefit from good conditions of oxygenation and from a high glucose content of the medium, and the speed with which glycogen disappears from the cells in liver cultures, unless sugar is supplied in a concentration greater than 0.5%, suggests an active carbohydrate metabolism (Doljanski, 1930b; Gill, 1938). This is even greater under anaerobic conditions when the glycogen disappears very rapidly.

As with mechanocytes, it seems probable that the nitrogen requirements of epithelial cells may be partially satisfied by proteoses and peptones, for there are reports of epithelial growth being greatly enhanced by proteoses, e.g. such as those in Witte's peptone.

Several strains of cells derived from epithelial tissues, e.g. liver, conjunctiva, and intestine have been grown on mixtures of amino acids with vitamins, glucose, salts and a small addition of serum protein. Under these conditions they all required thirteen amino acids (arginine, cystine, glutamine, histidine, isoleucine, leucine, lysine, methionine, phenylalanine, threonine, tryptophane, tyrosine, and valine) and tended to degenerate when any one of these was missing (Eagle *et al.*, 1957). The same amino acids were found to be necessary for a strain of cells originally established from a culture of mouse fibroblasts (strain L) but these results have to be interpreted with caution since the fibroblast strain is known to have become malignant during its life *in vitro* and the cells are thus no longer to be considered as "normal" fibroblasts. This same criticism can be applied to the epithelial cell strains, and it is again possible that they too have lost some of their original properties and perhaps acquired others in becoming "tissue-culture" cells, though morphologically they remained epithelial and reasonably "normal" in appearance.

On the whole, epithelia are somewhat more tolerant of changed conditions than are mechanocytes; for example, the malt extracts, which inhibit the growth of mechanocytes, leave the growth of epithelial cells unchecked (Medawar, 1937) (see Fig. 2.18). Heated embryo juice, while somewhat harmful to mechanocytes, allowed epithelial cultures from the liver to grow quite freely (Doljanski, 1929).

Differentiation of Epithelia

So far, the growth of epithelial cells has only been considered in its "unorganized form", as sheets or membranes, and it has been pointed out that many types of epithelia can grow in this way, and that they all look very similar, though speeded up cine-films do allow some differences in behaviour to be observed. Once again, however, the cells still retain something of their individual characteristics, for they can be caused to redifferentiate when the proper conditions are provided. The apparent "dedifferentiation" to the simple membrane is therefore, at least for a time, again only a superficial and morphological dedifferentiation. Thyroid (Ebeling, 1924, 1925) (Fig. 3.6), salivary gland (Grobstein, 1955a) (Fig. 3.7), intestine (Törö, 1933) (Fig. 3.8), and kidney epithelia (Grobstein, 1955b) have all been made to resume some at least of their original properties after being reduced to simple membranes in tissue culture, and some of them after being grown in pure culture. Interestingly enough, a common factor in most—perhaps all—cases, in which the differentiation of the epithelium has occurred, has been the presence of mechanocytes. When an epithelium is kept in isolation no differentiation

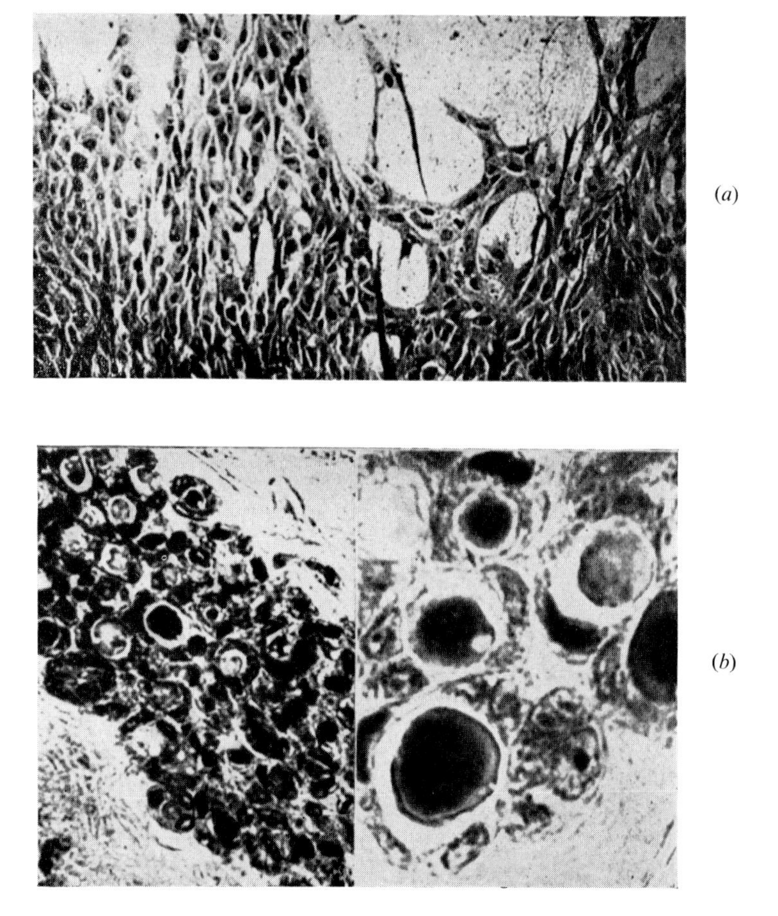

FIG. 3.6. (*a*) Epithelial cells from the thyroid gland in unorganized growth; (*b*) sections of thyroid tissue redifferentiated in vitro. (Ebeling, 1925.)

usually occurs. In the case of iris epithelium, however, the formation of pigment, which can perhaps be regarded as a function of the cells, is dependent only on the growth rate of the cells (Doljanski, 1930a), and there are no necessary and accompanying morphological changes of any importance such as are necessary when most other epithelia start to function. The same

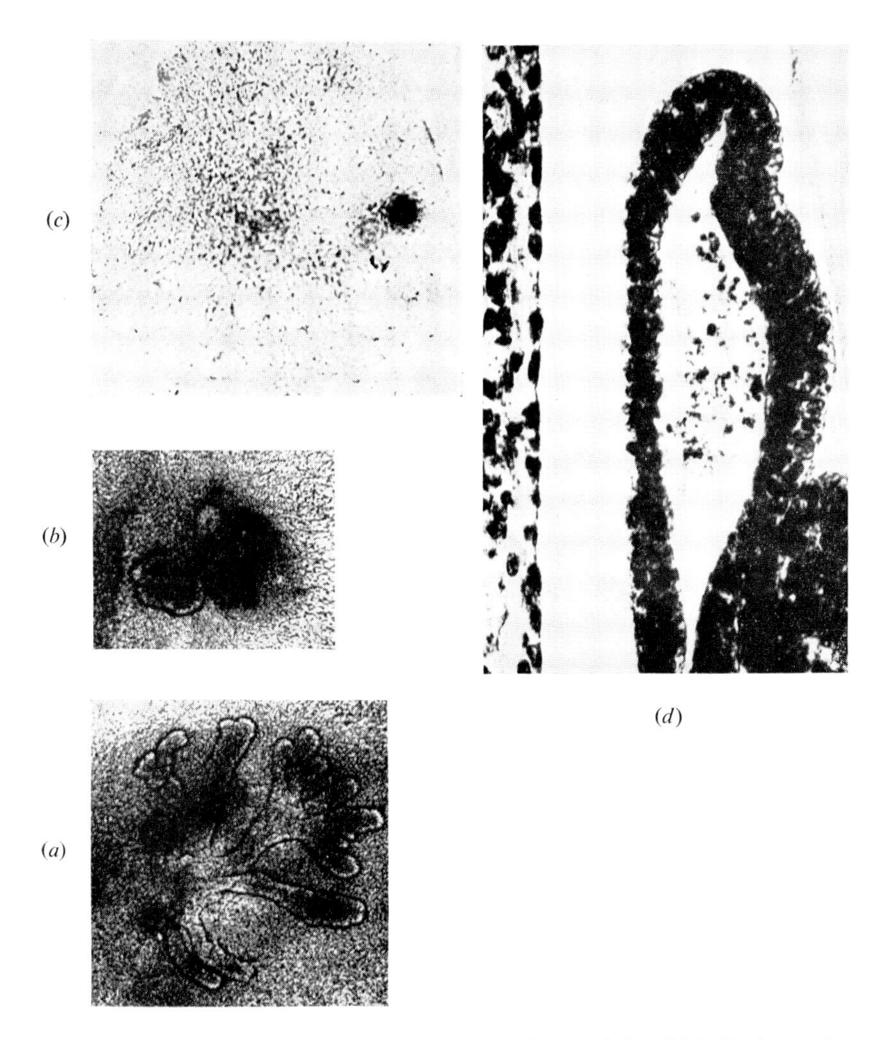

FIG. 3.7. Salivary gland epithelium developing into ducts and alveoli (*a*) after having been cultured as a simple membrane (*c*). The presence of autogenous mesenchyme is necessary for this differentiation. (*b*) Lack of differentiation with heterogenous mesenchyme; (*d*) section of "tubule" on right of "millipore" membrane and inducing mesenchyme on left of membrane. (Grobstein, 1955a.)

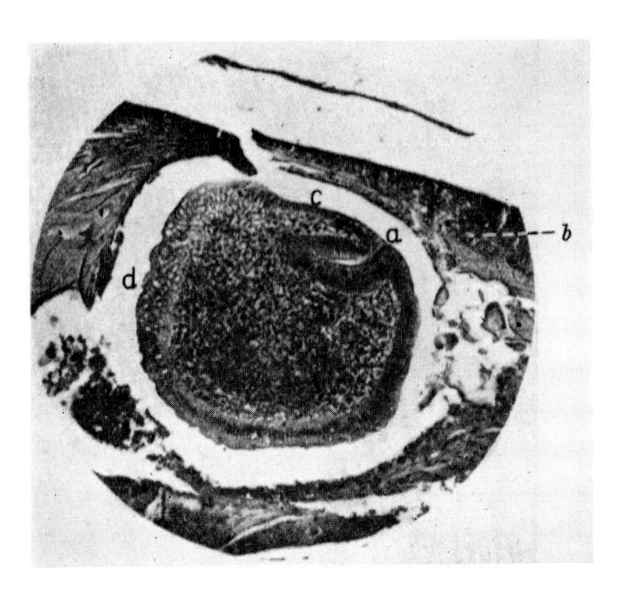

FIG. 3.8. Intestinal epithelium differentiating round a core of fibroblasts in the lens capsule of the eye after each had been grown as a pure culture *in vitro*. (Törö,1933.) (*a*) crypt-like structure of epithelium; (*b*) host tissues; (*c*) columnar epithelium; (*d*) irregular epithelium.

can be said about the appearance of glycogen in pure cultures of liver epithelium when the growth rate is sufficiently reduced (Doljanski, 1930b).

When pure colonies of intestinal epithelium are grown with colonies of mechanocytes from the same source, the epithelium tends to surround the mechanocytes and to differentiate into a columnar epithelium, sometimes with gland-like pockets (Törö, 1933) (see Fig. 3.8). The iris epithelium, after growth in pure culture, has been found, when mixed with a pure culture of fibroblasts, to differentiate in the form of tubular and gland-like structures within the central mass of tissue, thus indicating not only the action of mechanocytes but also the versatility of epitheliocyte form (Fischer, 1930) (Fig. 3.9). Moreover, some of these glands appear to be mucous and others serous, a difference which is somewhat surprising in view of the apparently uniform character of the iris epithelium. Epithelium from the thyroid gland after many passages *in vitro* has also developed vesicles in the central mass of the tissue and produced secretion within the vesicles, particularly when the cultures were planted below the surface of the plasma clot (Ebeling, 1924) (see Fig. 3.6). It is not clear whether fibroblasts were present in these so-called pure cultures, but the illustrations suggest the presence of interacinar tissue. Kidney epithelium has produced tubules *in vitro* under similar conditions, i.e. when connective tissues were added to pure cultures of epithelium obtained by isolation of islands of kidney epithelium in the outgrowth of tissue cultures by protecting them with mercury droplets while the rest of the culture was killed with ultraviolet light. Some reservation is necessary in the interpretation certainly of the last and perhaps of all these experiments, because of the general tendency of epithelia to form "tubes" in tissue cultures; but the peculiar character of the tubes produced is suggestive of true differentiation (Drew, 1923), and Grobstein (1955a, b) has shown the dependence of metanephric rudiments on other tissues for their differentiation, and in this case mechanocytes are included among these other tissues. Moreover, when skin fragments (dermis and epidermis) are grown in a fluid medium in the presence of antibiotics, the epithelium (epidermis) quickly surrounds the dermal part and differentiates towards normal epidermis (Medawar, 1948). There thus seem to be fairly clear indications that epithelia of various sorts are

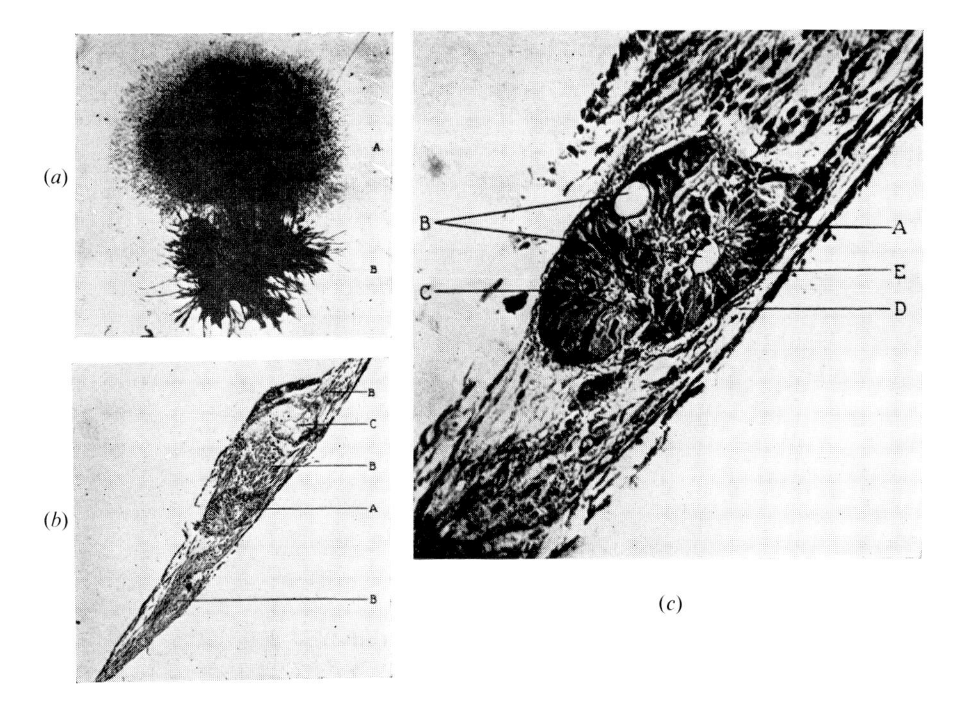

FIG. 3.9. Influence of fibroblasts on differentiation of epithelia. (*a*) Colonies of fibroblasts A and of iris epithelium B shortly after explanation; (*b*) Histological section of combined tissues. A, serous type of glandular structure; B, connective tissue; C, mucous type of glandular structure. (*c*) High power view of epithelial growth in (*b*): A, glandular arrangement of cells; B, lumina; C, lumen filled with secretion; D, connective tissue fibres; E, nucleus of secreting cell. (Ebeling and Fischer, 1922.)

caused or allowed to undergo histological differentiation by the presence of mechanocytes; whether these act mechanically, by orientation of intercellular "ground substance" or the like, through their secretions or their products of metabolism, or because they extract certain substances from the environment of the epithelial cells remains for future research. Submandibular gland epithelium only undergoes differentiation when combined with certain specific types of mesenchyme tissue (Grobstein, 1953a, b), (see Fig. 3.7), and the indications are that mesenchymal cells produce their effects by the intervention of activities or "secretions" which may acquire rather specific qualities in some cases. There may even be two or more stages of action, one to cause a check to membrane formations, others to cause actual differentiation. These effects on epitheliocytes *in vitro* may perhaps be compared with "induction" (or "evocation") and "organization" in embryonic development.

Mechanocytes may not be specific "evocators" in the sense that they cause completely indifferent cells to differentiate into a kidney epithelium or an intestinal epithelium, but they so contribute to the environment of epithelial cells that the latter can then display certain potentialities which may be latent within them or which they have acquired in their previous development (cf. Waddington, 1956). Kidney epithelial cells may grow as a continuous sheet of cells when provided with only a glass surface upon which to creep. In a plasma coagulum, or if allowed by some other means (e.g. by mechanocytes providing substances necessary for the formation of a basement membrane) to assume a three-dimensional pattern of growth, they may be able to develop tubular or other patterns. Mechanocytes may further contribute to the picture by so orientating the base upon which the epithelial cells creep, and in other ways interacting with the epithelial cells, that the latter may subsequently be able to display their full potentialities. In such a system there are obviously various steps and stages, so that there is room for specificities to show themselves. Some mechanocytes, interacting with epitheliocytes, may be able to evoke the full pattern of development, as, for example, their own capsular mesenchyme can do with the epitheliocytes of the salivary gland rudiment; mechanocytes from other sources may only be able to evoke some aspects of the epithelial development but not the whole. Mechanocytes should not therefore be regarded as themselves inducers or evocators, but rather as contributors to that particular environmental situation for the epitheliocytes which allows them subsequently to develop their full potentialities. Evocators are thus not necessarily "substances" but may be "situations", and differentiation should be regarded as being dependent on the establishment of these local situations which may sometimes involve extremely complex and specific patterns. The molecular or structural orientation of the actively inducing agents may well be very important in addition to their physicochemical constitution. Moreover, the situations should not be

regarded as static; almost certainly they are dynamic: gradients of concentration, and the flow of water, ions, and other constituents may be contributing to the behaviour patterns of the cells. Undoubtedly, however, the "proper" behaviour of many epithelial cells can, under these conditions, be allowed to develop by the presence of the proper mechanocytes.

The experiments of McLoughlin (1961) are very illuminating on this point. By means of trypsin digestion she separated the epithelia of skin, stomach, etc. from their underlying mesenchymes, and then recombined them, sometimes interchanging the mesenchymes. The result was that the epithelium tended to develop in the manner characteristic of the mesenchyme, i.e. independently of the origin of the epithelium. It developed as a stratified epithelium on skin mesenchyme or as a glandular epithelium on stomach mesenchyme.

The behaviour of cells on a substrate of collagen is very different from their behaviour on glass (Ehrmann and Gey, 1956; Bornstein 1958; Hillis and Bang, 1962; Hauschka and Konigsberg, 1966), and it is probable that the interaction between epithelial cells and mechanocytes produces a basement membrane appropriate to a particular form of activity of the epithelial cells. Indeed, some of the experiments of McLoughlin seemed to indicate this.

The problem, however, is certainly wider than this treatment perhaps indicates, for epithelial organization may sometimes be brought about by other epithelia of a different type and by their derivatives. For example, the metanephric rudiment has been caused to differentiate in culture by the presence of submandibular gland tissue in its vicinity and also by neural tissue from the dorsal part of the neural tube (Grobstein, 1955b). Here again, probably these "foreign" cells contribute something to the environment of the metanephric rudiment which allows further development. The metanephric rudiment itself is, of course, not a simple epithelium, and the environment of each of its constituent cells is therefore already very complex. The whole question of these intercellular actions will need further discussion. The observations of Weiss, to which reference has already been made, show clearly that epithelial cells of various types have their own specificities, as is seen, for example, when isolated epithelial cells from the epidermis recognize other epidermal cells and form associations and membranes with them, but ultimately always reject liver epithelial cells or kidney cells from such associations.

It is obvious from what has been said that epitheliocytes *in vitro*, although they may sometimes look very much alike in conditions of unorganized growth, still maintain some, at least, of the characteristics of their particular "species" or group, though often in latent form. Indications were given in connection with mechanocytes that this family of cells could be divided into subfamilies, genera and even species of cells. The situation with epithelia

appears to be somewhat similar, though we need to know more of the metabolic properties of the different types of epithelia before the classifications can be made on a very satisfactory basis. Moreover, since epithelia grow as continuous sheets, the presence of cells of different types may often escape notice because the methods which readily identify types of cells in isolation, e.g. the nature of their pseudopodia or their gait, are no longer quite so easy when applied to cells in a sheet. Puck *et al.* (1957) have suggested that the form assumed by cells is partly a "genetic" quality, since in the same medium clones of mechanocytes and epitheliocytes continue to breed true, and that it is partly dependent on environmental factors, since the form of the cells can be varied in different media.

While it is abundantly clear that the various epithelia described, and indeed many others, produce *in vitro* a sheet-like type of outgrowth and that the cells have a natural tendency to adhere to each other, it has already been mentioned that there have been cases described in which the epithelial membrane tends to "fray" at its edges and for cells to leave it and assume a form resembling that of the mechanocytes, for example. Further information is, however, required on the external factors which cause this apparent transformation. We do not yet know how far such cells, which become isolated from epithelial sheets, acquire the physiological properties of mechanocytes, as for instance, collagen formation instead of keratin formation, inhibition by hexenolactone (which, it will be recalled, is probably the active substance in those malt extracts which differentiate between mechanocytes and epitheliocytes), or extracellular production of mucoproteins and other similar characteristics.

During the early stages of embryonic development all the cells of the organism are essentially epithelial cells in that they hold together as membranes. Sooner or later some of these cells acquire other properties and leave the primary layers to become mesenchyme etc. It is therefore possible that similar changes may occur when epithelial cells are cultured as membranes. Some of the cells may undergo this latent form of differentiation and become true mechanocytes. This is only a suggestion which might fit with the morphological change, and as yet there is no evidence to prove or disprove the hypothesis. Incidentally, this tendency for the edge of the membranes to fray is more marked in some epithelia than in others.

If this change towards the mechanocyte form is a genuine differentiation, then it is probably more or less irreversible. Apart from the formation of the rather special endothelial membranes by mechanocytes, these cells certainly do not normally show any tendency to become epithelia with closely packed adherent cells. The change in the other direction must occur in embryology and may be the change which is occurring in the observations just described; it is at least worth bearing in mind as a possibility—though it must be remem-

bered that the true change from epitheliocyte to mechanocyte is not only morphological but also involves radical metabolic reorganization.

Several epithelia (e.g. liver, conjunctiva, and intestine) have now been grown as more or less pure strains in tissue culture, in fairly well-defined media and for long periods of time. Although they were certainly derived from epithelia in the first instance, some of these strains have now lost the power to grow as sheets, and they now multiply as colonies of individual cells (Fig. 3.10). In some cases the cells are spindle shaped, in others they may

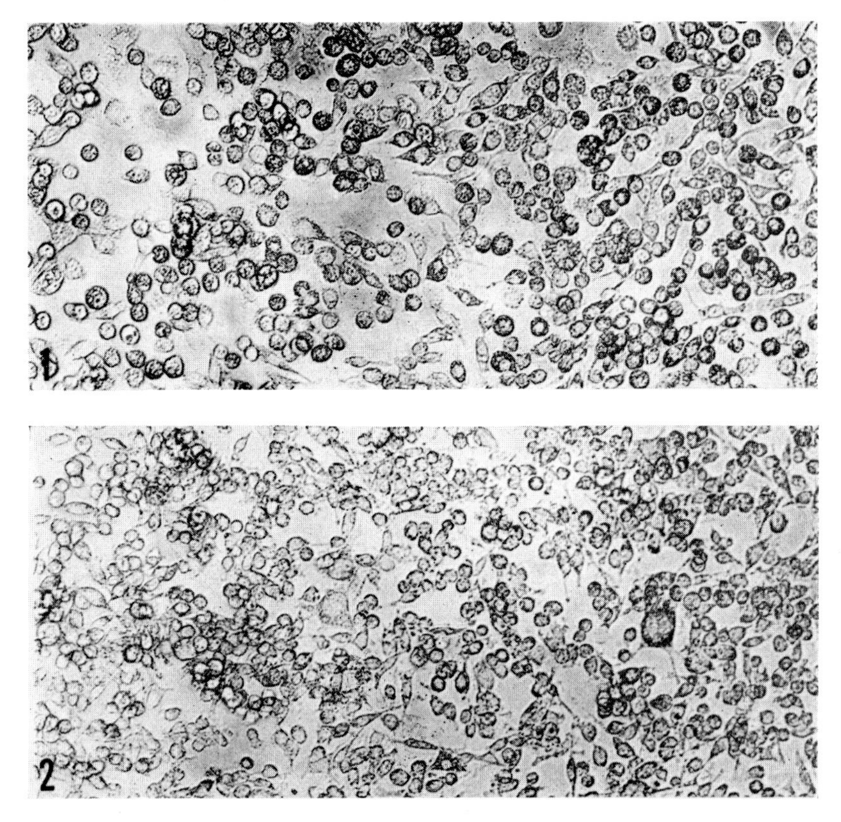

Fig. 3.10. Two strains of mouse liver cells after 2,590 days (1) and 1,239 days (2) *in vitro*. (Evans *et al.*, 1958.)

have more or less membranous pseudopodia, like macrophages. The classification of these "products of tissue culture" certainly raises some interesting and important problems.

One point which has not been systematically followed up in tissue cultures, is the behaviour of the different types of epithelial cells from those glands or

organs where there are originally several types of cells present. For example, it has been stated that in certain cultures of the adult kidney the cells of the convoluted tubules degenerated while those of the collecting tubules grew well (Nordmann, 1930). One would like to know if this was inevitable or if the particular culture medium and conditions of culture determined this selection. Would the tissue have behaved differently in another medium? In the normal intestinal epithelium there are goblet cells and ordinary columnar cells (excluding the Paneth cells and argyrophil cells, which produce other secretions), but it is not known if these two types are detectably present in the membranes which grow from cultures of intestine *in vitro*. In the submandibular salivary glands there are mucous-secreting cells, serous cells, and "demilune" cells; in the skin there are the ordinary squamous cells and there are dendritic cells. There does not, however, seem to be much reliable evidence as to how these different groups of cells behave *in vitro*. It would be most valuable to know whether they all assume the common epithelial type (which as shown above, is somewhat characteristic for each organ) or whether they have different metabolic properties persisting into the tissue-culture state. Putting the problem in another way, it has been shown that an excess of vitamin A converts a squamous epithelium (e.g. the skin of the chick) into one which indulges in mucin production. Is this a form of differentiation open to all epithelial cells given the right evoking agent, e.g. vitamin A? Or is it that certain cells can easily take this path while others cannot do so,

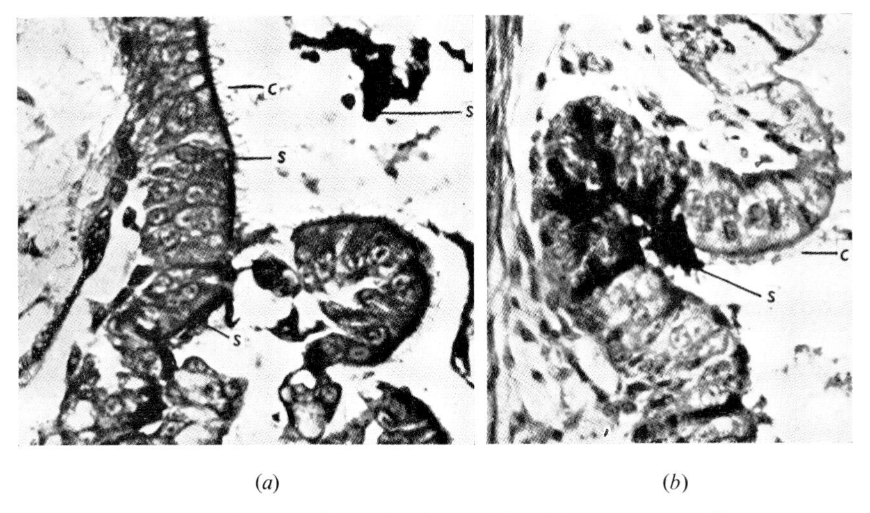

(*a*) (*b*)

FIG. 3.11. (*a*) Epithelial cells of the skin of the chick which have become ciliated under the influence of high doses of vitamin A in tissue culture; (*b*) normal nasal mucosa for comparison. c, Ciliated cells; s, mucous secretion. (Fell and Mellanby, 1953.)

or can only do so with great difficulty. Ciliated cells often appear among the mucous cells in such cultures (Fig. 3.11), and it is not yet clear whether this is a variant type of behaviour or whether there are always two types of cell in the epithelium, i.e. potential ciliated cells and potential mucous cells. In other words, are the ciliated and goblet cells of a normal trachea, for example, fundamentally different types of epithelial cells, or can one change into the other? In tissue cultures, ciliated and goblet cells are generally sloughed off before a stratified epithelium takes their place and *vice versa*; It would therefore seem to be unlikely that a ciliated cell can change into a goblet cell or that the reverse can happen. Both probably develop *de novo* from undifferentiated precursor cells, but it is not known whether the two types are derived from one uniform stock of basal cells or from two separate stocks which are not yet easily separable. The probability that there are two stocks is perhaps suggested by the fact that in those cultures of the chick's skin which were treated with excess of vitamin A many of the cells becameciliated while others in the same culture became mucoid. We do not know, however, to what extent conditions may differ locally in such cultures and cause these differences in the behaviour of cells, so it is possible that the so-called basal cells may be able to differentiate along one of several lines according to conditions, i.e. to produce keratinizing cells on the one hand or to yield mucoid or ciliated cells on the other. The concentration of the vitamin A may be important. Cultures of trachea have been made, in which unorganized growth occurred, but the question of the difference between the differentiated types of cells receives, as yet, no clear answer from the published account of these cultures.

The duality, or in some cases plurality, of cell types in epithelia is such a universal phenomenon that it possibly has some deeper meaning than the mere convenience of the differentiation of the particular organ or tissue showing it. The accompanying Table 3.1 shows the cells which are present in some of the more conspicuous epithelia in the body, and there are certainly very few epithelia in which at least two types of cell cannot be readily demonstrated. It can, of course, always be argued that epithelia seldom have only one task to perform and therefore this histological complexity is solely determined by functional needs. On the other hand, in its evolutionary processes Nature tends to build upon what is already present, and she does so largely by continual processes of modification; it would thus be quite reasonable to suppose that for some reason or other epithelial cells in the vertebrates are not all fundamentally alike, but fall into two or more functional and perhaps structural groups, typified, for example, by serous salivary cells and mucous salivary cells or goblet cells and brush-border cells. It will be seen later that this duality or plurality is something which may arise as a consequence of some important and basic embryological processes which probably have, themselves, an earlier phylogenetic and originally functional origin.

TABLE 3.1

Epithelium	*Cells*
Skin	Keratinizing, dendritic
Sweat glands	"Dark", "light"
Mammary gland	Acinar, duct
Salivary glands	Mucous, serous, duct
Stomach	Surface mucous, neck mucous, peptic, oxyntic, argentaffin, argyrophil
Intestine	Goblet, brush-border Paneth, argentaffin, mucous cells of Brunner's glands
Pancreas	α, β, exocrine, duct
Liver	Duct, hepatic (note also Kupffer cells)
Trachea and bronchi	Ciliated, goblet, basal, non-ciliated
Choroid plexus	Ciliated, polypoid
Kidney	Capsule, neck tubule, proximal tubule (segments 1 and 2), "thin" loop of Henle, "thick" loop of Henle, distal tubule, collecting tubule
Fallopian tube	Ciliated, nonciliated
Vas efferens	Ciliated, nonciliated
Uterus	Ciliated, nonciliated, mucoid
Retina	Rod, cone
Olfactory organ	Hair, supporting
Gustatory organ	Hair, supporting
Auditory organ	Hair, supporting
Anterior pituitary	Basophil, acidophil, chromophobe
Thyroid	Chief, PAS-positive
Parathyroid	Chief, eosinophil

References

Bornstein, M. B. (1958). Reconstituted rat-tail collagen used as substrate for tissue cultures on coverslips. *Lab. Invest.* **7,** 134.

Chambers, R. (1924). The physical structure of protoplasm as determined by micro-dissection and injection. "General Cytology", p. 237. Univ. of Chicago Press, Chicago, Illinois.

Chambers, R., and Cameron, G. (1943). The effect of L-ascorbic acid on epithelial sheets in tissue culture. *Am. J. Physiol.* **139,** 21.

Dingle, J. T., and Lucy, J. A. (1965). Vitamin A, carotenoids and cell function. *Biol. Rev. Cambridge Phil. Soc.* **40,** 422.

Doljanski, L. (1929). Cultures pures de tissu hépatique, *in vitro*. *Compt. Rend. Soc. Biol.* **101,** 754.

Doljanski, L. (1930a). Sur le rapport entre la prolifération et l'activité pigmentogène dans les cultures d'épithélium de l'iris. *Compt. Rend. Soc. Biol.* **105,** 343.

Doljanski, L. (1930b). Le glycogène dans les cultures de foie. *Compt. Rend. Soc. Biol.* **105,** 504.

Drew, A. H. (1923). Growth and differentiation in tissue cultures. *Brit. J. Exptl. Pathol.* **4,** 46.

Eagle, H., Oyama, V. I., and Levy, M. (1957). Amino acid requirements of normal and malignant human cells in tissue culture. *Arch. Biochem. Biophys.* **67,** 432.

Ebeling, A. H. (1924). Cultures pures d'épithélium thyroidien. *Compt. Rend. Soc. Biol.* **90,** 1383.

Ebeling, A. H. (1925). A pure strain of thyroid cells and its characteristics. *J. Exptl. Med.* **41,** 337.

Ebeling, A. H., and Fischer, A. (1922). Mixed cultures of pure strains of fibroblasts and epithelial cells. *J. Exptl. Med.* **36,** 285.

Ehrmann, R. L., and Gey, G. O. (1956). The growth of cells on a transparent gel of reconstituted rat-tail collagen. *J. Natl. Cancer Inst.* **16,** 1375.

Evans, V. J., Hawkins, N. M., Westfall, B. B., and Earle, W. R. (1958). Studies on culture lines derived from mouse liver parenchymatous cells grown in long-term tissue culture. *Cancer Res.* **18,** 261.

Fell, H. B., and Mellanby, E. (1953). Metaplasia produced in cultures of chick ectoderm by high vitamin A. *J. Physiol.* (*London*) **119,** 470.

Fell, H. B., Mellanby, E., and Pelc, S. R. (1956). Influence of excess vitamin A on the sulphate metabolism of bone rudiments grown *in vitro*. *J. Physiol.* (*London*) **134,** 179.

Fischer, A. (1922). A three months old strain of epithelium. *J. Exptl. Med.* **35,** 367.

Fischer, A. (1924). The differentiation and keratinization of epithelium *in vitro*. *J. Exptl. Med.* **39,** 585.

Fischer, A. (1930). "Gewebezüchtung." Müller & Steinicke, Munich.

Flesch, P. (1953). Mode of action of vitamin A. *J. Invest. Dermatol.* **21,** 421.

Gill, P. M. (1938). Effect of adrenalin on embryonic chick glycogen *in vitro* as compared with its effect *in vivo*. *Biochem. J.* **32,** 1792.

Grobstein, C. (1953a). Analysis *in vitro* of the early organization of the rudiment of the mouse submandibular gland. *J. Morphol.* **93,** 19.

Grobstein, C. (1953b). Epithelio-mesenchymal specificity in the morphogenesis of mouse sub-mandibular rudiments *in vitro*. *J. Exptl. Zool.* **124,** 383.

Grobstein, C. (1955a). Tissue interaction in the morphogenesis of mouse embryonic rudiments in vitro. *In* "Aspects of Synthesis and Order in Growth" (D. Rudnick, ed.), Chapt. 10. Princeton Univ. Press, Princeton, New Jersey.

Grobstein, C. (1955b). Inductive interaction in the development of the mouse metanephros. *J. Exptl. Zool.* **130,** 319.

Gross, J. (1956). The behaviour of collagen units as a model in morphogenesis. *J. Biophys. Biochem. Cytol. Suppl,* **2,** 261.

Hauschka, S. D., and Konigsberg, I. R. (1966). The influence of collagen on the development of muscle clones. *Proc. Natl. Acad. Sci. U.S.* **55,** 119.

Hillis, W. D., and Bang, F. B. (1962). The cultivation of human embryonic liver cells. *Exptl. Cell Res.* **26,** 9.

Holtfreter, J. (1947). Observations on the migration, aggregation and phagocytosis of embryonic cells. *J. Morphol.* **80,** 25.

McLoughlin, C. B. (1961). The importance of mesenchymal factors in the differentiation of chick epidermis. II. Modification of epidermal differentiation by contact with different types of mesenchyme. *J. Embryol. Exptl. Morphol.* **9,** 385.

Medawar, P. B. (1937). A factor inhibiting the growth of mesenchyme. *Quart. J. Exptl. Physiol.* **27,** 147.

Medawar, P. B. (1948). Culture of adult mammalian skin epithelium *in vitro. Quart. J. Microscop. Sci.* **89,** 187.

Miszurski, B. (1937). Researches on the keratinization of the epithelium in tissue cultures. *Arch. Exptl. Zellforsch. Gewebezücht.* **20,** 123.

Moscona, A. (1952a). Cell suspensions from organ rudiments of the early chick embryo. *Exptl. Cell Res.* **3,** 535.

Moscona, A. (1925b). The dissociation and aggregation of cells from organ rudiments of the early chick embryo. *J. Anat.* **86,** 287.

Nordmann, M. (1930). The behaviour of adult mammalian kidney in tissue cultures. *Arch. Exptl. Zellforsch. Gewebezücht.* **9,** 54.

Parshley, M. S., and Simms, H. S. (1950). Cultivation of adult skin epithelial cells (chicken and human) *in vitro. Am. J. Anat.* **86,** 163.

Puck, T. T., Cieciura, S. J., and Fisher, H. W. (1957). Clonal growth *in vitro* of human cells with fibroblastic morphology. Comparison of growth and genetic characteristics of single epithelioid and fibroblast-like cells from a variety of human organs. *J. Exptl. Med.* **106,** 145.

Rinaldini, L. M. J. (1958). The isolation of living cells from animal tissues. *Intern. Rev. Cytol.* **7,** 587.

Törö, E. (1933). Organisation und Selbstdifferenzierung der an die Stelle der Linse implantierten Gewebekulturen. 1. Darm. *Arch. Exptl. Zellforsch. Gewebezücht.* **14,** 495.

Waddington, C. H. (1956). "Principles of Embryology." Allen & Unwin, London.

Weiss, P. (1959). Interactions between cells. *Rev. Mod. Phys.* **31,** 449.

Wolf, G., and Varandani, P. T. (1960). Studies on the function of vitamin A in mucopolysaccharide biosynthesis by cell-free particle suspensions. *Biochem. Biophys. Acta* **43,** 501.

Wolf, G., Varandani, P. T., and Johnson, B. C. (1961). Vitamin A and mucopolysaccharide synthesizing enzymes. *Biochem. Biophys. Acta* **46,** 59.

Chapter 4

THE GROWTH OF AMOEBOCYTES

In cultures of many kinds of tissue, in addition to any growth of mechanocytes or of epitheliocytes which may occur, there often appear cells of yet another type. These cells are usually distinguishable in living cultures by the rounded form of their pseudopodia and by their irregular and membranous processes (Fig. 4.1), but they are most easily recognized by their movement when seen in cine-films of the growing cells. They are usually very actively amoeboid, sending out pseudopodia in any direction, so that they move in an apparently random manner, though in a general direction away from the central explant. Their more rapid movement usually causes the formation of a halo of these cells round the margin of any culture which contains them (see Fig. 2.1). Sometimes they remain motionless in a rounded condition for quite long periods. In active movement they are thus clearly distinguishable from the more static and more polarized or orientated mechanocytes on the one hand, and from the membrane-forming epitheliocytes on the other. It is a notable feature of these cells, and one which probably depends in part upon their manner of movement, that they mostly remain isolated from each other and normally show no tendency to stick together. They avoid each other's society as it were, so that in a culture in which these cells are numerous the cells are more or less evenly distributed over the surface upon which they are growing, and, it need hardly be said, that like the other cell types, they require a "wettable" surface before they display their amoeboid activity. Unlike the pseudopodia of mechanocytes, those found in this class of cell are less often pointed and are more usually either rounded or lobose; they may often extend around a large part of the circumference of the cell, or even surround the whole cell in the form of an undulating membrane. On the other hand, under some conditions, the cells may lengthen out along one axis and assume a somewhat mechanocyte-like form; in this state they may easily be mistaken for mechanocytes though the rounded form of the ends of the pseudopodia generally remains and affords a means of distinction, when it can be seen. Very small and fine filamentous pseudopodia are occasionally seen in active movement. It is obvious therefore that cells of this type are extremely variable in their morphology and very dependent on the properties of the medium in which they are placed (Parker, 1938) (Fig. 4.2).

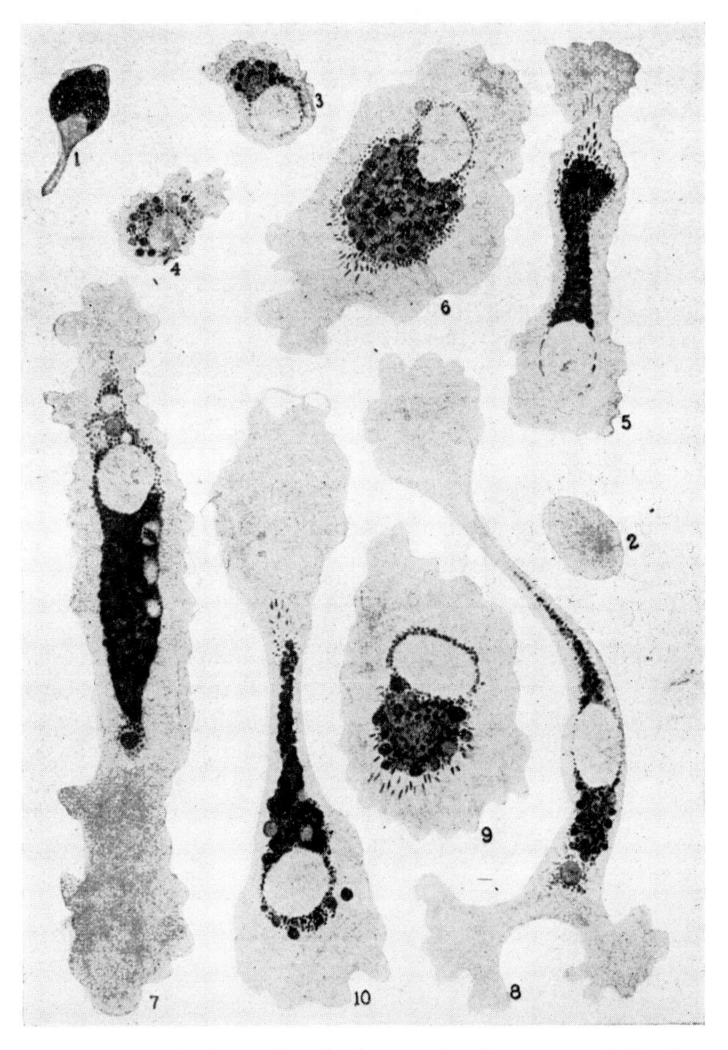

FIG. 4.1. Amoebocytes in tissue culture in plasma and embryo extract. 1, Blood monocyte. 2, erythrocyte, 3–8, blood monocytes in cultures of increasing age up to four days; 9 and 10, Tissue macrophages, after 24 hours in culture. (Carrel and Ebeling, 1926.)

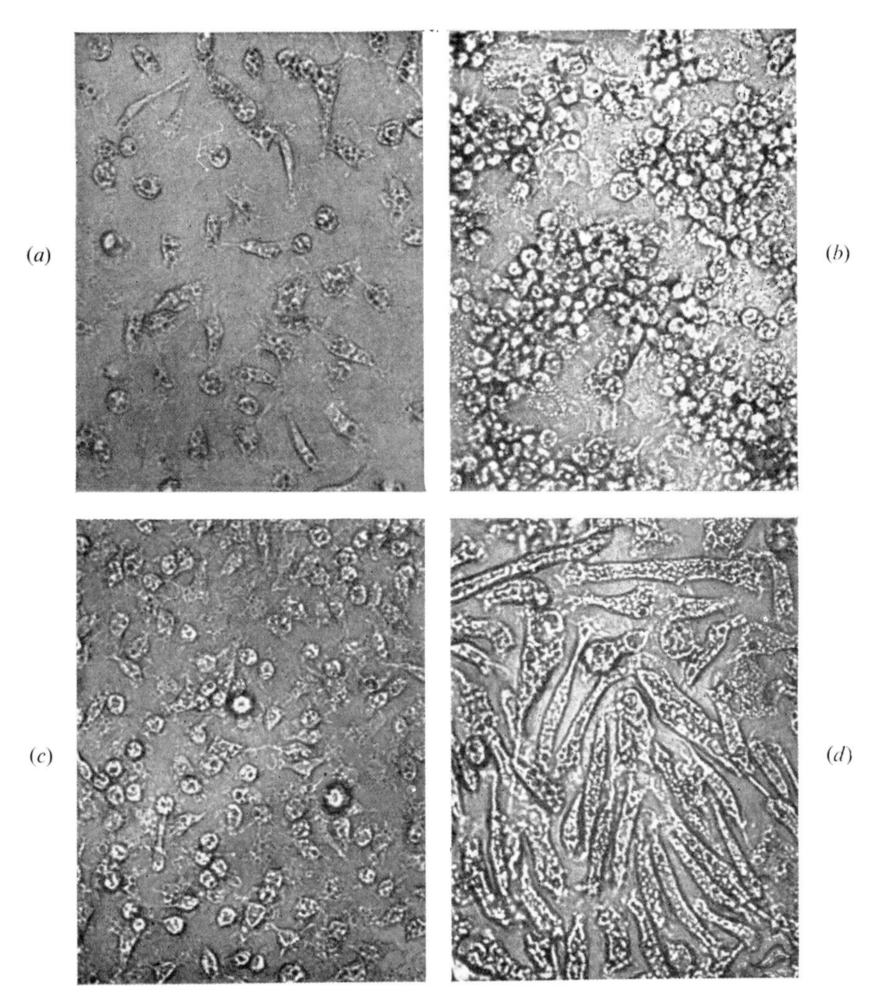

FIG. 4.2. Various forms assumed by chick monocytes when growing under different conditions. (*a*) In 20% serum; (*b*) in 80% serum; (*c*) in 50% serum; (*d*) in 50% serum + 2·5 mg/100 ml thyroxin. (After Carrel, see Parker, 1938.)

Cells of this class are appropriately called amoebocytes and possess other recognizable characters in addition to those already outlined. Amoebocytes are likely to turn up in cultures of almost any tissue; this is not altogether surprising in that many of them are, in all probability, directly derived from the macrophages, which are well known to be almost universally distributed in the tissues of the bodies of the vertebrates. Like the macrophages they are in some conditions highly phagocytic, and they can segregate such colloidal vital-dyes as trypan blue, lithium carmine and erythrolitmin. In tissue culture, they often have a great affinity for the basic dye, neutral red, which they can actively concentrate into their cytoplasm in a very few moments to a surprising degree (Carrel and Ebeling, 1926). This can be used as a ready guide to their presence in mixed cultures, as for example those obtained from the skin.

Numerous amoebocytes emerge from cultures of spleen and lymph nodes. In cultures of lymph from the thoracic duct the few monocytes that are present are clearly distinguishable by their larger size and type of movement from the far more numerous lymphocytes and it is usually only the monocytes which survive after a few days in culture. By then they have become indistinguishable from amoebocytes derived from other sources. Amoebocytes also emerge from tissue cultures of the retina (Pomerat and Littlejohn, 1956) and brain, in which tissue they are normally present as the "microglia".

A very convenient source of these cells for cultures is the "buffy coat" obtained by centrifuging blood. If the blood plasma is carefully removed from the surface of the sedimented blood corpuscles and a drop of tissue extract dropped on to their surface, the buffy coat is then rapidly enclosed in a clot and can be removed as a whole, cut into fragments, and these fragments can then be cultured in plasma, just like the little pieces normally explanted from other tissues, either in flasks or hanging drops. Cells immediately wander out in all directions; granulocytes (i.e. neutrophil, eosinophil, and basophil leucocytes), lymphocytes, and monocytes all emerge actively and many erythrocytes are carried passively into the medium. After a few days, however, the granulocytes, erythrocytes, and in most cases the lymphocytes are necrotic or already dead and disintegrated. The monocytes, however, persist and actively multiply. Most of them adhere to the surface of the glass or to that of the plasma and the former remain adherent to the glass after the plasma has been removed. In this position they may afterwards be kept in a fluid medium consisting of diluted serum alone, and in this, unlike fibroblasts, they continue to multiply so that their growth and general activity can be readily studied (Baker, 1933). Exactly similar cells can be obtained more directly from blood by albumen flotation, i.e. by suspending the cells in an albumen solution of the correct specific gravity and then centrifuging at the appropriate speed. If this is done from fowl blood, the monocytes are at first somewhat mixed with other blood cells and with thrombocytes, but they are easily separated from these since the blood cells and thrombocytes do not stick to the surface of the glass as firmly as do the monocytes (Weiss and Fawcett, 1953) (Fig. 4.3).

These monocytes from the blood behave in exactly the same way as the macrophages described above as emerging from cultures of all sorts of other tissues, for these also survive well on glass in a fluid medium. In fact, under tissue-culture conditions there are no apparent differences between monocytes and macrophages (Jacoby, 1938), for the former readily develop the phagocytic properties of the latter. Furthermore, tissue culture shows up an amusing paradox concerning the classical terminology of the tissue macrophages. These are often—and somewhat arbitrarily—divided into active or free macrophages and fixed macrophages, the latter name referring to those phagocytic cells which appear in tissues in a more or less extended form as though they were part of an established order in the tissue. The former refers to the roundish cells which may be seen loose, or unattached to the tissue

(*a*) (*b*)

FIG. 4.3. (*a*) Culture of monocytes and thrombocytes from chick blood; (*b*) "pure culture" of monocytes now adhering to a glass surface. (Weiss and Fawcett, 1953, courtesy of The Williams & Wilkins Co.)

framework, as though they were moving about or being carried in the tissue fluid or lymph. In fact, the "fixed" macrophage may often be really in a state of active amoeboid movement, phagocytosing particles as it goes, and it may be anything but "fixed"; whereas the round "active" macrophage may simply be indulging in one of the more or less prolonged periods of rest which are typical of these cells, at least in culture conditions where periods of active migration alternate with periods of stagnation, often with the cell in a rounded-up condition. This is but one of the many instances where the opportunity to see the cells alive has largely altered the interpretation to be placed upon a given fixed, and thus falsely static, histological situation. The only other comparable conditions for observing cells alive are those found in such places as the tadpole's tail, in the artificial culture chambers established in rabbits' ears (Sandison, 1928; Clark *et al.*, 1936; Ebert *et al.*, 1940) or in the dorsal skin of mice (Algire, 1943). In such preparations, macrophages can again be clearly seen if trypan blue is used; and, interestingly enough, under these conditions, even the extended macrophages are much more static than they are in cultures.

The fact that these cells not only live but multiply freely in a medium of serum alone, without any period of acclimatization, distinguishes them at once from the mechanocytes which only multiply extremely slowly under such conditions, unless they are specially "trained to it." Fowl macrophages or amoebocytes continue to multiply for several days in serum, and when they have ceased to divide, renewal of the serum is sufficient to start up growth and division once more. The lag period after feeding and before the first cell divisions appear is longer than the lag period when mechanocytes are treated with embryo juice and is generally about 20 hours, or rather less if the serum is frequently renewed (Jacoby, 1941) (Fig. 4.4). There are, however, some

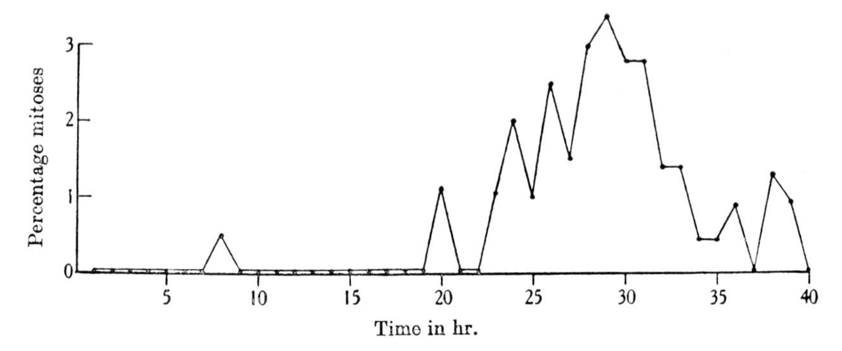

FIG. 4.4. Effect of feeding with 25% serum on the growth of tissue macrophages *in vitro*. The serum was added at zero time and mitoses start again at 20 hours. (Jacoby, 1938.)

races of amoebocytes, e.g. those obtainable from the peritoneum of the guinea-pig, which are somewhat different in morphology and apparently have no power of multiplication *in vitro* (Jacoby, 1957) (Fig. 4.5). Indeed there are indications that the macrophages of mammals may have somewhat different properties from those of birds, and the manner of their multiplication is rather uncertain. Treatment of amoebocytes with embryo juice produces fatty and rather round cells and many of them degenerate; only a very limited growth occurs in its presence. By contrast, mechanocytes may become slightly more fatty than normal in embryo juice, but they certainly thrive in its presence, and under ordinary conditions need it, or some tissue extract like it, for their multiplication.

The property which amoebocytes possess of separating from each other in a given colony has further repercussions, for when several colonies are planted in the same dish, they show, unlike those of mechanocytes, no tendency to fuse, and in fact there often remains a cell-free zone between the colonies (Carrel and Ebeling, 1922) (Fig. 4.6). This is made even more remarkable by the fact that the average velocity of movement of amoebocytes is higher than

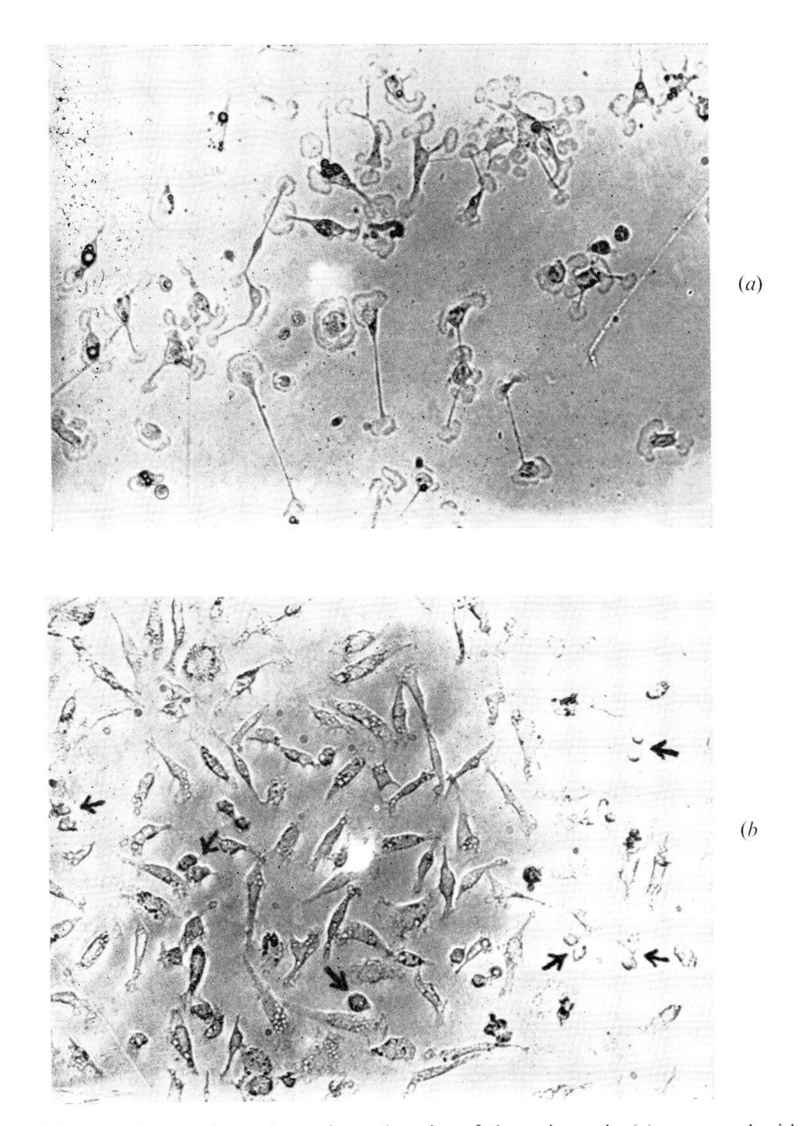

FIG. 4.5. Amoebocytes from the peritoneal cavity of the guinea-pig (*a*) compared with those from the hen (*b*). (Jacoby, 1957.)

that of mechanocytes and they quickly extend all over a vacant surface. Not only that, but colonies of amoebocytes can grow actively within the substance of a plasma medium and invade it to almost any depth. Mechanocyte colonies grow with less and less vigour the deeper they are planted in a tube of coagulated medium, and they are highly dependent on an adequate

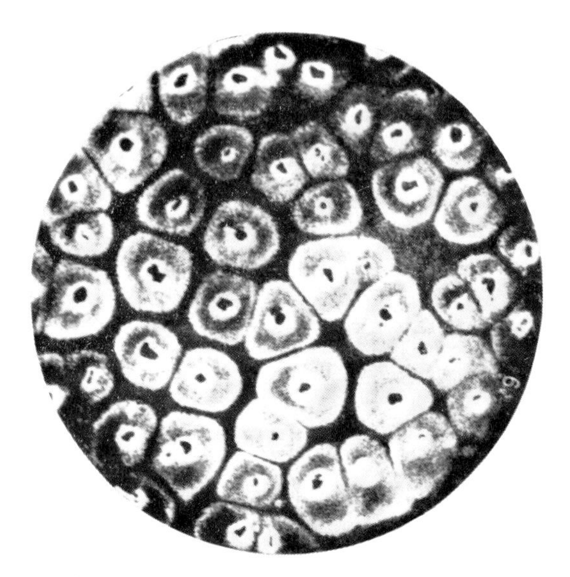

FIG. 4.6. Colonies of monocytes in a Petri dish maintaining zones of separation between the colonies. (Carrel and Ebeling, 1922.)

access to oxygen; amoebocytes, on the other hand, extend widely into the medium more or less independently of their depth in the coagulum (Fischer, 1928) (Fig. 4.7).

In colonies planted in the substance of a clot, it quite often happens that tubular cavities are formed while the plasma is coagulating. These cavities are the seat of an interesting piece of behaviour on the part of the amoebo-

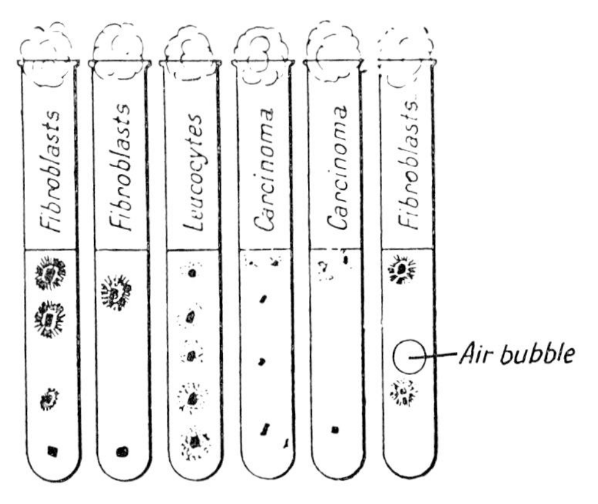

FIG. 4.7. Diagram showing the manner in which different types of cells grow at different depths in a plasma coagulum. (Fischer, 1928.)

cytes, for these cells may flatten on the surface of such cavities and line them with cells of "epithelioid" form (Hueper and Russell, 1932) (Fig. 4.8), rather in the same way as some mechanocytes can behave to form endothelia. Generally, however, the cells are not so closely adherent unless they are very much crowded together. Since amoebocytes are, in general, capable of segregating vital dyes, this behaviour in lining tubules is perhaps suggestive of the manner of formation of those endothelia in the body which are

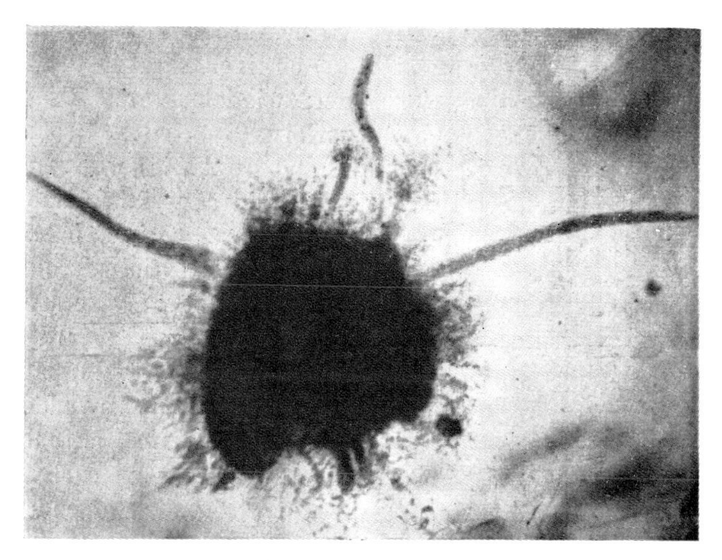

FIG. 4.8. A culture of leucocytes with seven capillary-like formations lined with cells extending into the plasma clot. (Hueper and Russell, 1932.)

exceptional in the rather loose and mesh-like distribution of their cells and in their capacity to segregate lithium carmine and the like, namely the endothelia of the bone-marrow sinusoids, of the liver sinusoids (the Kupffer cells), and of vessels in the adrenal cortex. When amoebocytes adopt the epithelioid form (Fig. 4.9a) they acquire a very conspicuous centrosphere, and this is strongly positive for acid phosphatase (Fig. 4.9b) and also reacts positively to the periodic Schiff reaction, even after digestion with saliva (Weiss and Fawcett, 1953). This, incidentally, is an interesting example of cells increasing the amount of an enzyme during culture *in vitro*, for the monocytes freshly drawn from the blood do not give any reaction for acid phosphatase. Perhaps the enzyme is correlated with the increased phagocytic activity in the tissue culture conditions. It is related to the lysosomes in the cells.

Amoebocytes in the "epithelioid" form may often have several nuclei (Fig. 4.10), and the formation of giant cells with many nuclei by progressive fusion of individual cells occurs readily *in vitro* (Lewis, 1927) particularly in

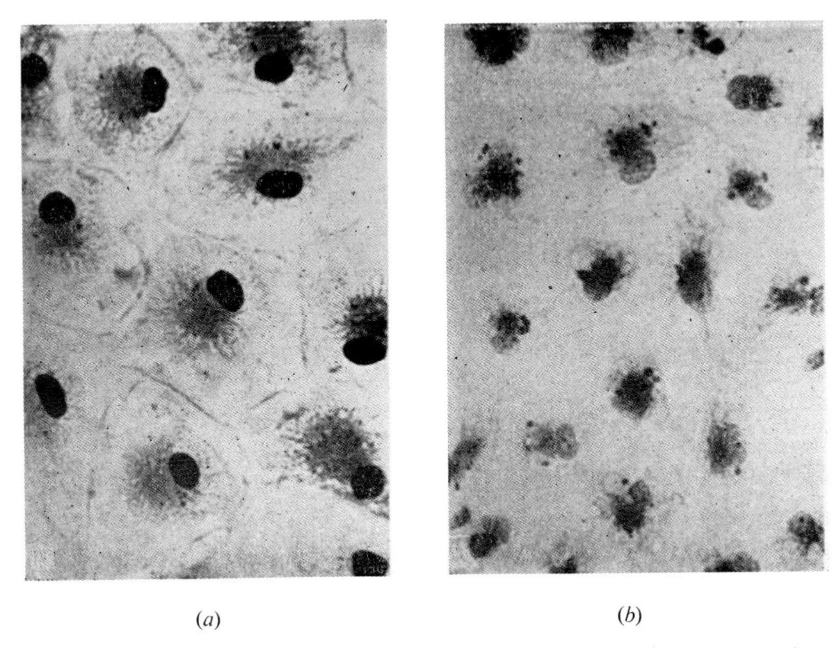

(*a*) (*b*)

FIG. 4.9. (*a*) Amoebocytes in the epithelioid form, each with a conspicuous centrosphere; (*b*) acid phosphatase in the region of the centrosphere in "epithelioid" monocytes. (Weiss and Fawcett, 1953, courtesy of The Williams & Wilkins Co.)

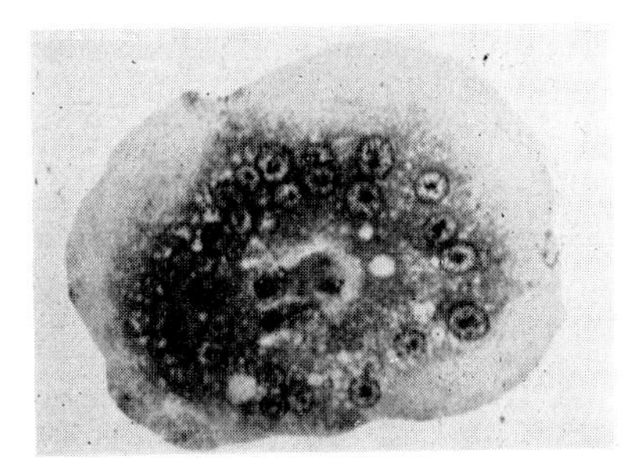

FIG. 4.10. Giant amoebocyte with many nuclei. (Weiss and Fawcett, 1953, courtesy of The Williams & Wilkins Co.)

response to foreign bodies (e.g. tubercle bacilli, lycopodium powder), to low oxygen tension (Barta, 1925), and possibly to low pH (Fig. 4.11). It is curious that these cells, which normally behave as though they repelled each other, should fuse their cytoplasm when they actually make contact in this way, particularly since as epithelioid sheets they seem able to make some sort of contact with their neighbours without fusing with them. When fragments

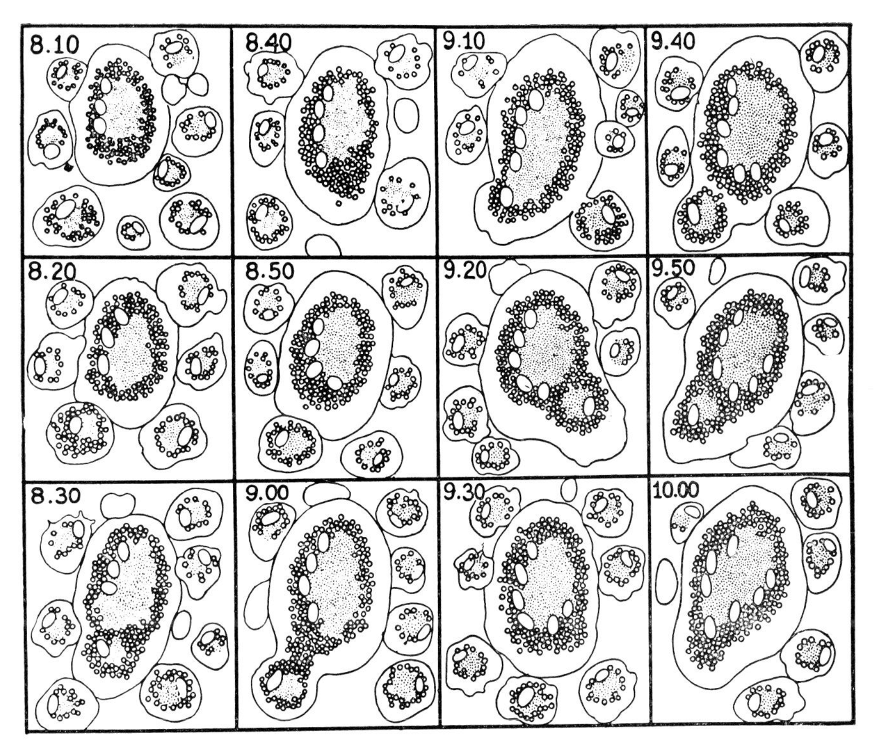

FIG. 4.11. Semidiagrammatic drawings of amoebocytes fusing to form a giant cell. Numbers represent times. (Lewis, 1927.)

of bone are explanted in plasma, there wander out from them into the medium, within the first few hours, not only osteoblast cells, of mechanocyte form, but also cells which rapidly get larger by fusion with other cells of the same type and which may eventually have as many as an hundred nuclei (Hancox, 1946) (Fig. 4.12). These cells, which sometimes emerge in the multinucleate form, presumably correspond to the osteoclasts of the bone. Cinephotography shows that they move with the characteristic gait of amoebocytes rather than with that of mechanocytes, the margins of the cells often showing well-developed undulating membranes (Hancox, 1949).

It is, of course, well known that osteoclasts appear when bone is being broken down *in vivo*, though the proof that they are the causative agents is lacking. Nevertheless, experiments *in vivo* and *in vitro* with vitamin A may be illuminating. It has already been noticed that high concentrations of vitamin A in the medium of bone and cartilage cultures cause the disappearance of the bone matrix without the appearance of numerous osteoclasts (Fell and Mellanby, 1952). On the other hand, restoration of vitamin A to the diet of

A-deficient puppies leads to a sudden thinning of the bones in certain regions of the skeleton where it had previously not been eaten away as it would have been in a normal puppy. In these regions numerous osteoclasts immediately appear (Mellanby, 1947). In the light of the experiments on the metaplasia of epithelia with high doses of vitamin A in the medium (Fell and Mellanby, 1953) and of the suggestion that mucus-secreting cells absorb sulphur as sulphate (Jennings and Florey, 1956; Kent *et al.*, 1956), it seems possible that

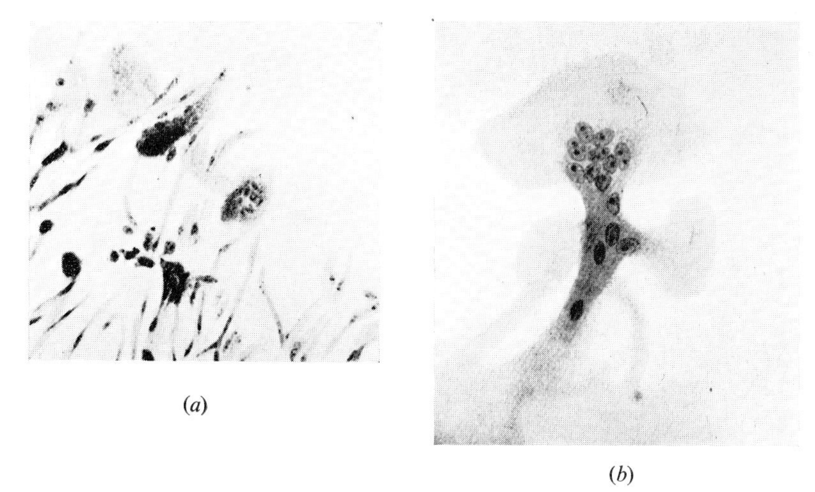

(*a*) (*b*)

FIG. 4.12. Ostoclasts and osteoblasts emerging from fragments of bone in tissue culture (*a*). The former readily form giant cells (*b*). (Hancox, 1946.)

amoebocytes may perhaps do something similar and that the sudden accumulation of osteoclasts when bones are being dissolved away or remodelled in the body as a whole, where there are, of course, numerous circulating monocytes, etc., may be connected with the sulphur-containing breakdown products of the bone or cartilage matrix. In cine-films of tissue cultures of bone grown with pieces of parathyroid tissue the matrix of bone disappears and numerous osteoclasts can be seen in the regions of bone removal (Gaillard, 1955). Striking as this picture is, it does not in fact differentiate between cause and effect, and has to be interpreted in terms of the other experiments in which cartilage resorption, and in some cases bone resorption, occur in response to high doses of vitamin A in tissue cultures of long bones and when few osteoclasts or giant cells appear.

Pure colonies of amoebocytes have not been observed to produce any fibres or "ground-substance" into the medium. However, when cultures are made from the buffy coat of the blood it sometimes happens that some colonies begin after a time to produce mechanocyte-like cells and a tissue with collagen fibres (Maximow, 1928; Hulliger, 1956; Paul, 1958). The capillary-like processes sometimes produced in cultures of blood also occasionally have a fibrous backing (Parker, 1934). This has often been taken as evidence for the transformation of amoebocytes into mechanocytes, but since the same thing does not occur among the amoebocytes cultured from lymph (Hulliger, 1956) the presence of some unrecognized mechanocytic cells in the blood (endothelial cells?) is to be suspected (Paul, 1958), though the possibility of a transformation certainly cannot be excluded.

It is said that colonies of leucocytes planted near colonies of mechanocytes encourage the growth of the latter by the production of food substances or "trephones" (Carrel, 1922), which are advantageous to the fibroblasts, but it is not really clear which cells among the "leucocytes" are responsible for this effect or whether it is more than the effect of the autolysis of the less viable leucocytes, since extracts of tissue which have been allowed to autolyse produce notable stimulation of the growth of mechanocytes. The occurrence of numerous macrophages in a repairing wound after the first invasion of the neutrophil leucocytes suggests a possible stimulating action of these cells on the main tissue-forming cells or fibroblasts, but until the nature and origin of trephones is more firmly established some caution is necessary in the matter. Meanwhile amoebocytes, unlike both epitheliocytes and mechanocytes, have not been shown to form any staple product, comparable with keratin and collagen. The origin of fibrinogen and serum proteins is still problematical: that they arise in the liver is probable, but whether the hepatic cells or the Kupffer cells are responsible still remains uncertain though for the production of the albumens the latter are probably responsible (Best and Taylor, 1966).

Again unlike mechanocytes, amoebocytes are able to start colonies moderately easily from isolated cells, and these can be well maintained in fluid media. Such media may be entirely synthetic; in that case amoebocytes will survive in them for some months, but unless there is at least 10% serum in the medium the cells tend to go slowly downhill, and survival beyond a few months is impossible (Jacoby and Darke, 1948). With some serum present, however, growth is apparently unlimited. Serum is also known to increase the phagocytic capacity of macrophages.

It has been noticed in colonies of mechanocytes and particularly in those from muscle, that cells which have all or at any rate many of the characteristics of macrophages may appear quite suddenly in the outer growth-zone, and this has led to the belief that mechanocytes can become macrophages (Fig. 2.1) (Chèvremont, 1940, 1942). This transformation may or may not be the correct explanation for this phenomenon, and the whole problem will be discussed in some detail later, but it should be emphasized that "pure" colonies of mechanocytes are extremely difficult to obtain with certainty; nests of macrophages, or even individual macrophages, might well remain enclosed in the tissue for long periods and, under conditions favouring the growth of mechanocytes, they would themselves not be much encouraged to proliferate. However, if the embryo juice for some reason became diluted or less potent, then the macrophages might well flourish and migrate out. It has been found that the numbers of amoebocytes appearing in muscle cultures can be increased by the addition of choline and of certain quaternary ammonium compounds (e.g. trimethyl-ethyl-ammonium chloride) to the medium (Thomas, 1936; Chèvremont, 1943; Chèvremont and Chèvremont-Comhaire, 1945). The interpretation of this phenomenon may not be the simple one, namely that mechanocytes turn into amoebocytes. Indeed, the application of certain quaternary ammonium compounds to fibroblasts during their culture, *in vitro*, while cinematograph records of them were being made, was not observed to bring about any morphological changes in the growing cells or any change in their movement except those concerned with toxicity at the higher dose levels (Pomerat and Willmer, unpublished observations).

A physiological feature frequently and easily seen in the amoebocyte, though not exclusive to it, is the ability of the cell to engulf some of the surrounding fluid and to enclose it within the cytoplasm as a vacuole (Lewis, 1931). After a time this vacuole slowly decreases in size and becomes more highly refractile (Gey *et al.*, 1954; Rose, 1955); eventually it disappears. The process, known as pinocytosis, seems to occur especially when there are lamelliform membranes or "ruffles" round the main body of the cytoplasm. It is not known what significance is to be attached to the process, but it is suggestive of a feeding method, if the imbibed water is then excreted. It is interesting to note than pinocytosis also occurs in free-living amoebae, and in *Amoeba proteus* it can be induced by a variety of means including that of raising the NaCl concentration above about $M/16$, so that it may be concerned with the ionic equilibrium of the cell. (Chapman-Andresen, 1958). Electron-microscope pictures of macrophages often show invaginations of the cell surface that give the impression of penetrating deeply into the cell in such a way that their ends may become nipped off as vacuoles (Palade, 1955). These invaginations may well be concerned with pinocytosis.

It is possible that this process is something which is more easily observed in tissue cultures, since the cells are then often excessively flattened and the normal traffic of fluid in and out of the cell may thus be adversely affected. The leading edge of migrating mechanocytes often shows cytoplasmic ruffles and pinocytosis on a limited scale.

In some cells in tissue culture, e.g. Hela cells, (which were actually derived from epithelial cells of a carcinoma of the cervix of the uterus) it has been observed (Rose, 1957) that pinocytotic vacuoles are often "met" in the cell periphery by refractile granules (microkinetospheres) originating near the nucleus and passing outwards to the periphery (Fig. 3.44). These granules, which are perhaps lysosomes, become associated with a vacuole and then accompany it back to the centre of the cell where the vacuole itself may be

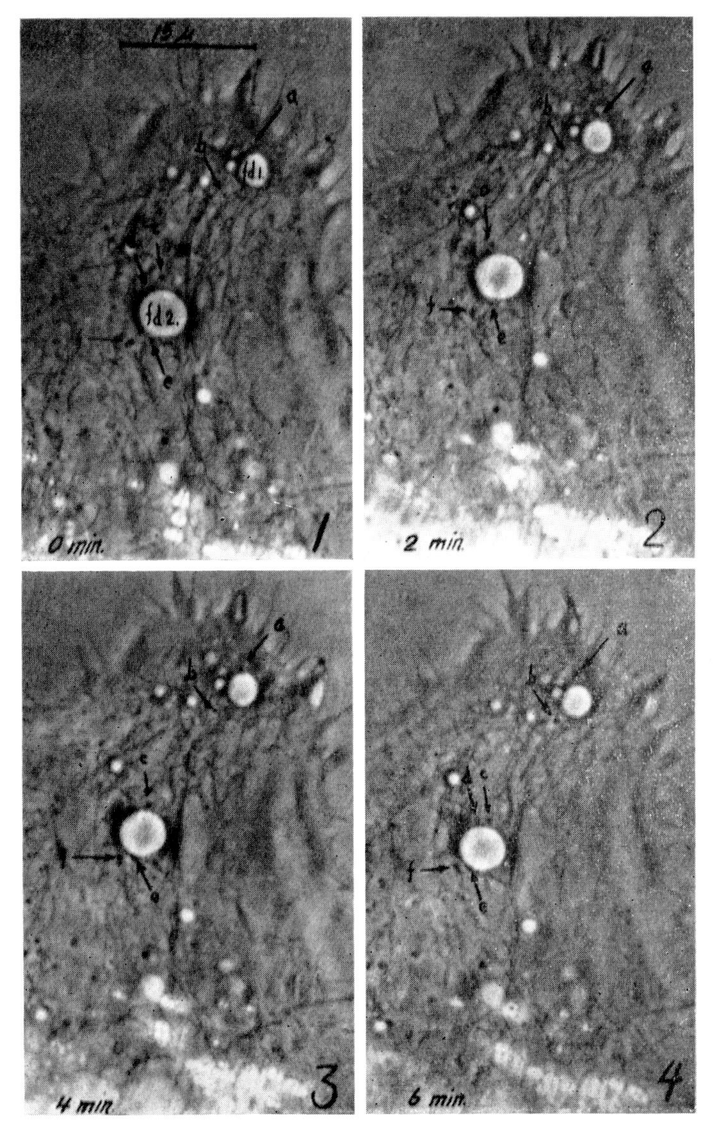

FIG. 4.13. Pinocytosis, with microkinetospheres (*a–f*), in "Hela" cells. Photographs at 2 minute intervals. Phase contrast. fd1, fd2, Pinocytotic vacuoles. (Rose, 1957.)

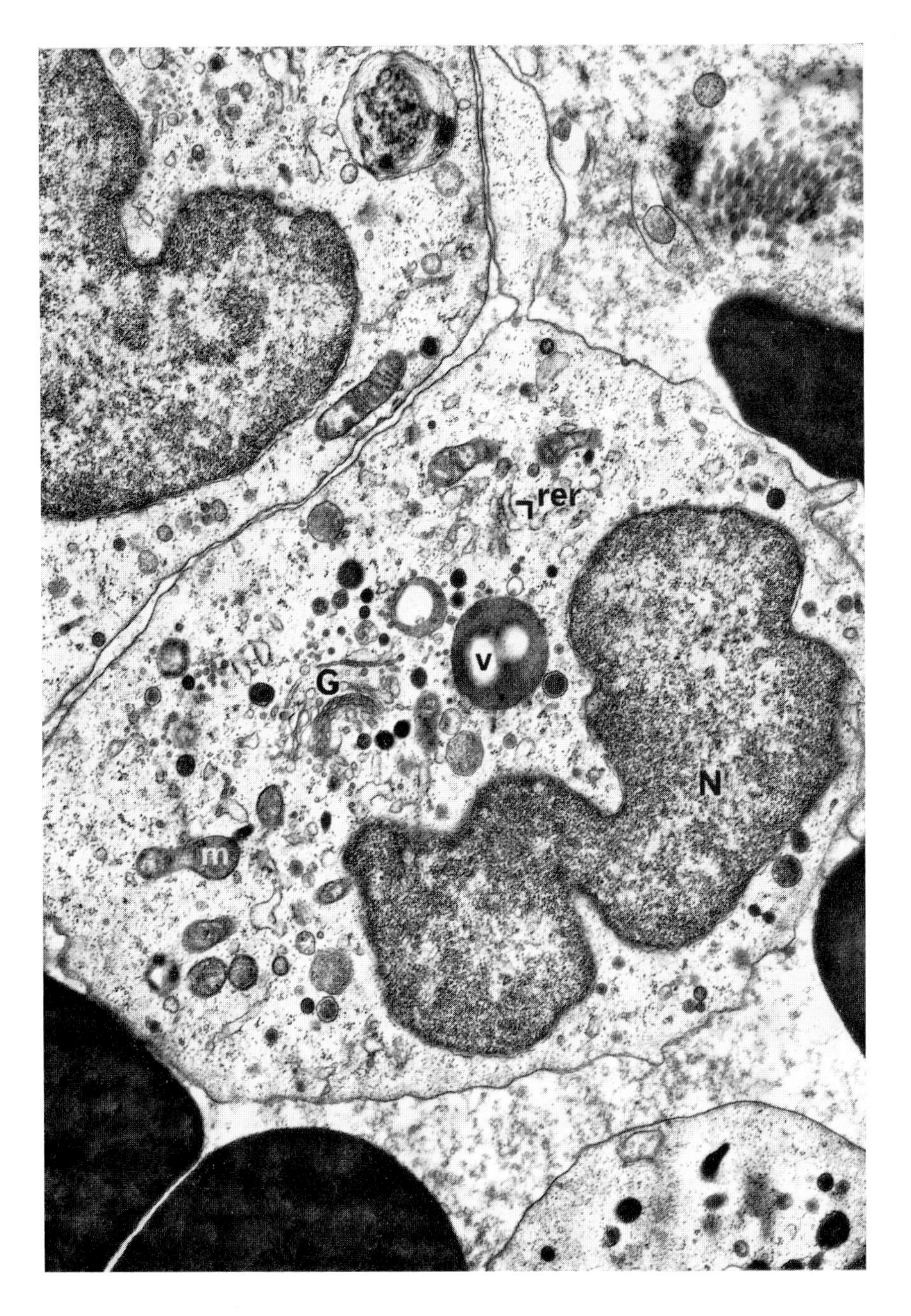

FIG. 4.14. Macrophages in a healing wound (× 14,000). G, Golgi complex; m, mitochondrion; N, nucleus; rer, rough endoplasmic reticulum; v, vacuole. (Photo by R. ROSS.)

greatly reduced in size and disappear as such. While this phenomenon was first observed in the rather abnormal Hela cells, it has since been seen to occur in normal macrophages.

Amoebocytes, therefore, appear by derivation, behaviour, growth-potency and many other characteristics, to constitute a third clear-cut family of cells and in general they can be readily distinguished from both mechanocytes and epitheliocytes. Their electron-microscope structure is shown in Fig. 4.14 and should be compared with that of the fibroblasts shown in Fig. 2.25.

It may be difficult immediately to classify any one particular cell on purely visible characters alone, as definitely belonging to this family or to that, but when all the characteristics of the cell's behaviour, including its relationship with other cells, are noted, then the cell can be classified with a greater degree of certainty. The great majority of cells, as they are seen in the early stages of unorganized growth in cultures from a wide variety of tissues, would then be found to belong to one of the three main categories which have been discussed so far.

References

Algire, G. H. (1943). An adaptation of the transparent chamber technique to the mouse. *J. Natl. Cancer. Inst* **4,** 1.

Baker, L. E. (1933). The cultivation of monocytes in fluid medium. *J. Exptl. Med.* **58,** 575.

Barta, E. (1925). Deficient oxidation as a cause of giant cell formation in tissue cultures of lymph nodes. *Arch. Exptl. Zellforsch. Gewebezücht.* **2,** 6.

Best C. H., and Taylor, N. B. (1966). "Physiological Basis of Medical Practice". Williams & Wilkins, Baltimore, Maryland.

Carrel, A. (1922). Growth-promoting function of leucocytes. *J. Exptl. Med.* **36,** 385.

Carrel, A., and Ebeling, A. H. (1922). Pure cultures of large mononuclear leucocytes. *J. Exptl. Med.* **36,** 365.

Carrel, A., and Ebeling, A. H. (1926). The fundamental properties of the fibroblasts and the macrophages. (ii). The macrophage. *J. Exptl. Med.* **44,** 285.

Chapman-Andresen, C. (1958). Pinocytosis of inorganic salts by *Amoeba proteus* (*Chaos diffluens*). *Compt. Rend. Trav. Lab. Carlsberg.* **31,** 77.

Chèvremont, M. (1940). Le muscle squelettique cultivé in vitro. Transformation d'éléments musculaires en macrophages. *Arch. Biol.* (*Liège*) **51,** 313.

Chèvremont, M. (1942). Recherches sur l'origine, la distribution, les caractères cytologiques et les propriétés biologiques des histiocytes et des macrophages par la methode de la culture des tissus. *Arch. Biol.* (*Liège*) **53,** 281.

Chèvremont, M. (1943). Recherches sur la production expérimentale de la transformation histiocytaire dans les cultures *in vitro*. *Arch. Biol.* (*Liège*) **54,** 377.

Chèvremont, M., and Chèvremont-Comhaire, S. (1945). Recherches sur le déterminisme de la transformation histiocytaire. *Acta Anat.* **1,** 95.

Clark, E. R., Clark, E. L., and Rex, R. O. (1936). Observations on polymorphonuclear leukocytes in the living animal. *Am. J. Anat.* **59,** 123.

CHAPTER 5

NERVE CELLS, NEUROGLIA, AND SCHWANN CELLS

Although the technique of tissue culture was developed in order to investigate the behaviour of nerve cells, and it immediately solved some fundamental problems concerning the neurone, it has not, since those early days, been so extensively used for the investigation of the behaviour of nerve cells as it has been for that of mechanocytes and of the other types of cells already discussed. Perhaps one of the main reasons for this lies in the fact that "unorganized" growth, in the sense of the proliferation by mitosis, of nerve cells, hardly occurs in tissue culture. It is said that nerve cells, while still in the neuroblast stage, grow as an epithelial sheet, but as soon as their differentiation has started, then this no longer happens. On the other hand, in old cultures of brain there do sometimes emerge sheets of epithelial cells, the origin of which has not been definitely established. It is possible that they may represent the undifferentiated ependymal cells (Olivo, 1927).

Nerve cells *in vitro* send out long and branching processes, neurites, each ending in amoeboid pseudopodia. Speeded-up cine-films show very active movement in these outgrowing processes which, in some cases, is extremely similar to the movement of the pseudopodia of amoebocytes even to the extent of the rapid enclosure of droplets of fluid by pinocytosis (Hughes, 1953) (Fig. 5.1). These droplets of fluid are passed quickly up the extended nerve process towards the cell body and if the nerve fibre is cut off from its parent cell, the absorption of fluid and the passage of the vacuoles up the fibre may continue for some time, before the whole disconnected terminal portion degenerates in the expected manner. During this time numerous vacuoles collect at the cut, so that it is quite clear that, at any rate under these conditions of outgrowth *in vitro*, these nerve-cell processes are constantly collecting fluid from their surroundings and passing it up towards the cell body. Since

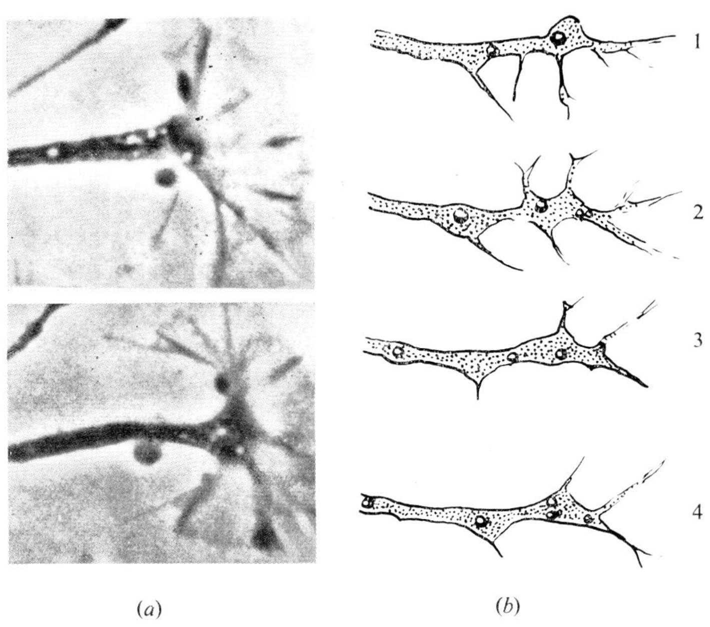

FIG. 5.1. "Pinocytosis" by the processes of outgrowing neurites. (*a*) Membranous neurites with accumulating droplets; (*b*) drawings of a neurite showing the positions of the vacuoles at different times. Interval between 1 and 2, 48 seconds, between 2 and 3, 12 seconds, and between 3 and 4, 12 seconds. (After Hughes, 1953.)

it has also been shown that when nerves are regenerating, after section *in vivo*, they tend to force fluid or cytoplasm outwards down the nerve fibre towards the cut, so that a constriction acts like a dam (Weiss, 1944a) and so that the end of the nerve fibre proximal to the cut may enlarge into a distinct bulge (Young, 1945; Weiss and Hiscoe, 1948), it is difficult to reconcile these two apparently diametrically opposed observations. Either there must be movement simultaneously in both directions along a nerve fibre, and, after all, there is nothing inherently unlikely about this since it is obviously occurring in the less elongated pseudopodia of other types of cells, or else it may be that such factors as the age of the tissue, external conditions, rate of growth, and even the type of nerve cell under observation are the determining factors. There is as yet no information as to whether dendrites and axons behave alike or differently in this way in tissue cultures. In fact, it is not always easy to decide which is which, though there are some nerve cells which in tissue culture show one process, generally posteriorly situated, which is different in kind from the other and anterior processes (Fig. 5.2). The anterior processes, as the cell advances into the medium, are presumably the dendrites in these cells, and it is probably these which show pinocytosis. It would be

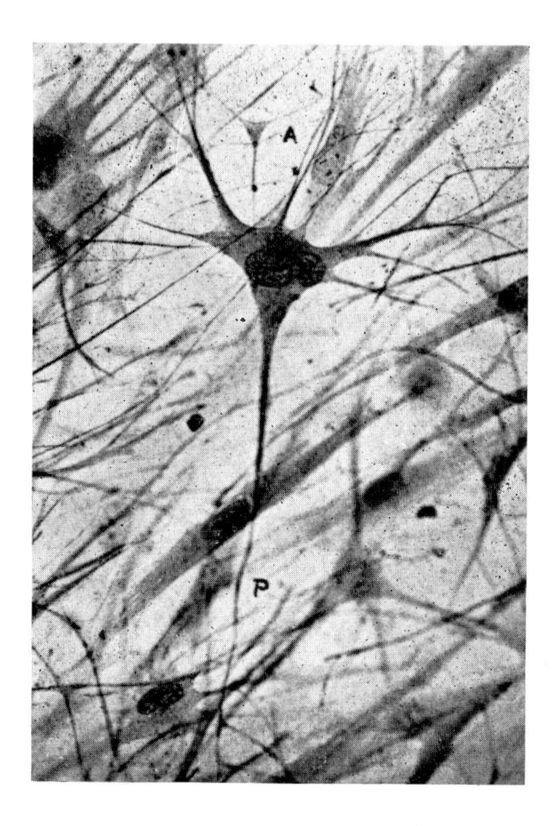

FIG. 5.2. Sympathetic ganglion cell in tissue culture showing distal or advancing processes, A, and a trailing proximal process, P. (Murray and Stout, 1947.)

interesting to know how the axon behaves in this respect and what relationship, if any, exists between these pinocytotic vesicles and synaptic vesicles.

During prolonged culture under conditions which favour survival, both in hanging drops of cerebrospinal fluid (Martinovic, 1932) and in flasks and roller tubes with heparin plasma or serum and with low concentrations of embryo juice (Costero and Pomerat, 1951), nerve tissue survives and differentiates to a remarkable degree (Figs. 5.3, 5.4). The cells may slowly move out into the medium under these conditions, and in some cases the central explant may eventually thin out sufficiently to enable the living cells to be studied in some detail. Cell division does not usually occur in such cultures, although, curiously enough, it has been described in cells, which are apparently nerve cells, emigrating from explants of adult human brain (Geiger, 1958) and sympathetic ganglia (Murray and Stout, 1947) (Figs. 5.5, 5.7). Some of these cells also become multinucleate giant cells (Fig. 5.6).

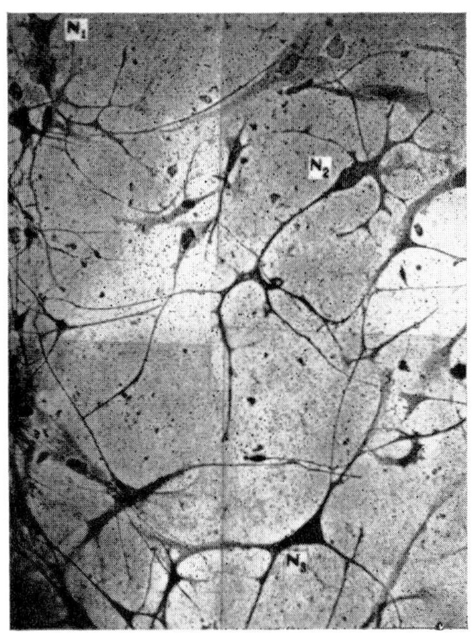

FIG. 5.3. Nerve cells differentiating in the outgrowth from prolonged culture of the cerebellum. Three nerve cells (N_1, N_2, and N_3), can be seen with their processes. Silver preparation. (Costero and Pomerat, 1951.)

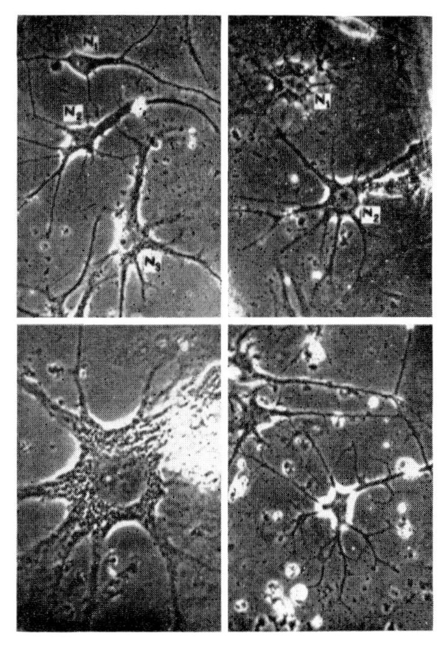

FIG. 5.4. Phase-contrast pictures of living nerve cells in prolonged cultures of cerebellum. (Costero and Pomerat, 1951.)

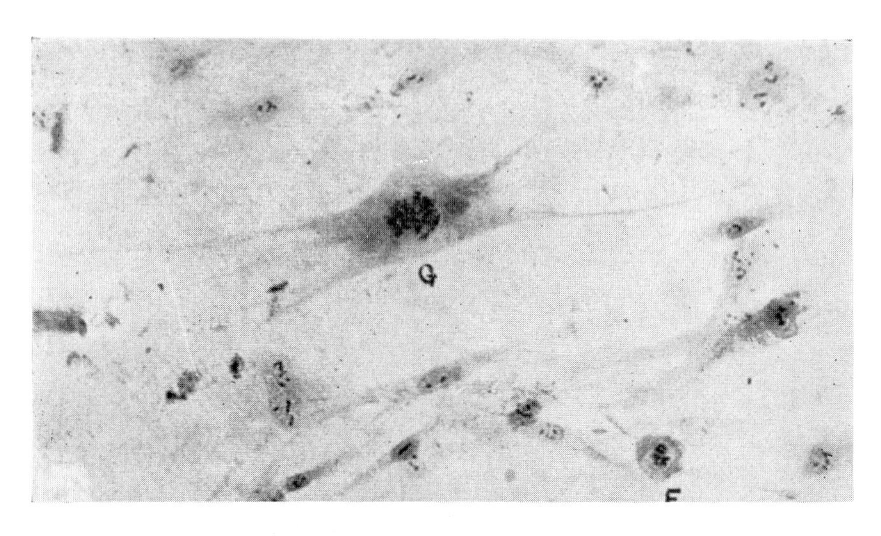

FIG. 5.5. Dividing neurone in culture of adult human sympathetic ganglion. Note the polyploidy. (Murray and Stout, 1947.)

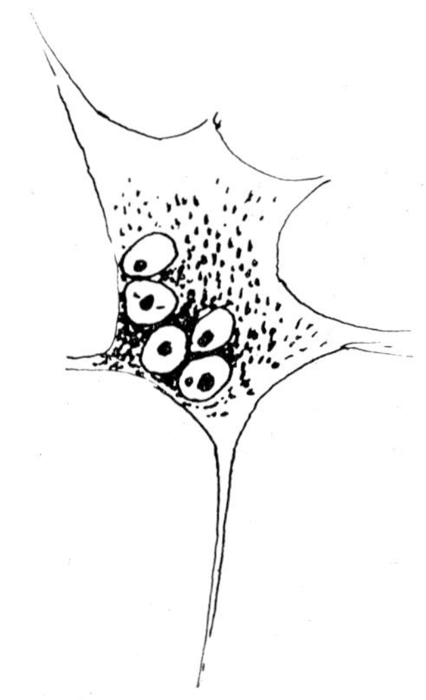

FIG. 5.6. Multinucleate giant cell from neurone growing from human sympathetic ganglion in tissue culture. (After Murray and Stout, 1947.)

This case emphasizes the importance of examining different parts of the nervous system, perhaps rather more carefully than has been done in the past, with respect to the sort of behaviour which can be expected from the different histological types of cells. Sympathetic ganglion cells, derived embryologically from the neural crest, may have somewhat different properties from those cells which are developed from the neural tube proper. Their very migration during embryonic development suggests greater powers of movement than those possessed by many nerve-cell bodies. It certainly seems to be true that the sympathetic ganglion cells also migrate outwards in tissue cultures, and that many of them assume the form of rather large, essentially fan-shaped cells with a definite polarity (see Fig. 5.2). In this way, their general form appears to be very like that of mechanocytes. The cell has a series of sharp and pointed distal (or anterior) processes and it may or may not drag a long proximal (posterior) process behind it. This sort of behaviour has also been observed in Purkinje cells from the cerebellum and in some cells which emerge from cultures of the cerebral cortex (Martinovic, 1932; Hogue, 1947;

FIG. 5.7. Nerve cells from brain cortex in culture. Note neurones in metaphase and telophase. Bodian stain and gold chloride (Geiger, 1958.)

Geiger, 1958). The distal processes, again, are presumably the dendrites, and the proximal process, which is single and somewhat different in appearance, is the axon. Nerve cells *in vitro*, which incidentally have, like those in the body, a strong affinity for methylene blue and also for silver stains when fixed (Fig. 5.7) have, as mentioned above, been shown upon one or two occasions to withdraw their processes to a greater or less extent and then to divide in tissue culture. After division, as with mechanocytes and other cells, their rate of migration is increased. In some of these cells there is a definite increase in the amount of chromosome material as compared with that in the accompanying fibroblasts (see Fig. 5.5).

No definite Nissl's substance can be detected in nerve cells during the first phases of their life in tissue culture, and ultraviolet light shows a rather uniformly high absorption by the whole cytoplasm (Koenig and Feldman, 1953). Later, however, the typical arrangement of Nissl's substance may develop (Deitch and Murray, 1956) (Fig. 5.8). Neurofibrillae are probably structures

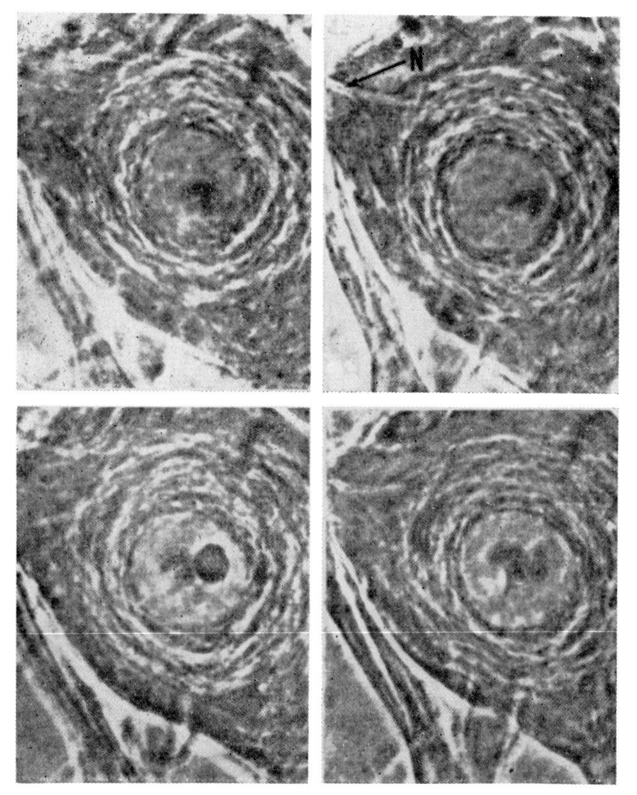

FIG. 5.8. Nissl's substance in spinal ganglion cell in tissue culture. N, Neurofibril. (Deitch and Murray, 1956.)

which are produced by an intensification of the orientation of long molecules or micelles in the cytoplasm during the process of fixation or other treatment, and are thus probably similar, in this respect, to the fibres of the mitotic spindle which can be made to appear and disappear within the living cell. Some orientated material is usually visible in spinal ganglion cells (Weiss and Wang, 1936), and it may be significant that these cells do not readily migrate from the central explant under tissue culture conditions. The mitochondria or similar granules in nerve cells often show a concentric arrangement around the nucleus (Murnaghan, 1941) which could easily help to give rise on fixation to the neurofibrillar pattern of the cytoplasm which characterizes many nerve cell-bodies (Levi and Meyer, 1937). In other systems, e.g. sea-urchin embryos, (Gustafson and Lenicque, 1952) the development of mitochondria often precedes the elongation or stretching of protoplasm as though some phase of mitochondrial activity favoured the production of the fibrous or orientated proteins. A curious feature of many nerve cells, and also of some epithelial cells, is that the nucleus may often be seen, with the help of time-lapse cine-photography, to be rotating within the cytoplasm (Murnaghan, 1941).

As mentioned above, some nerve cells and fibres growing in tissue culture have, as elsewhere, a special affinity for methylene blue, and this allows them to be distinguished fairly readily from the other types of cells which may also grow out from nerve tissues, both *in vivo* and particularly *in vitro*. Neutral red is also copiously picked up by some cells both *in vivo*, e.g. the large ganglion cells in the retina, and *in vitro*, e.g. the spinal ganglion cells (Murnaghan, 1941), and it would be interesting to see if this property could be correlated with other features in the behaviour of nerve cells. Not all nerve cells do it.

After staining with methylene blue, nerve fibres often show, both *in vivo* and particularly *in vitro*, very marked globular swellings at intervals along their length. The significance of these is not yet known; they may be enlarged and made more conspicuous by the treatment with methylene blue, but they are certainly detectable also in the untreated nerve fibres. It is possible that they may be connected with the passage of fluid droplets along the fibre. It may be remarked, however, that similar swellings are a common feature in the elongated threads into which myelin figures can readily be pulled. Such myelin figures are abundantly formed when the lipids extracted from brain tissue by alchohol are exposed to a dilute solution of NaCl (e.g. about $M/5$). Many of the fibres which grow out in culture are undoubtedly naked cell processes—dendrites and axons—but quite often neuroglia cells or Schwann cells may accompany the true nerves, creep along them and sometimes invest them in much the same way as they have been observed to do in the tail of the living tadpole (Speidel, 1933). Myelin formation has been observed in tissue cultures on several occasions though it usually only occurs after prolonged cultivation (Peterson and Murray, 1955). Again the rotation of the Schwann cell around the axon to form the concentrically arranged layers of myelin has been confirmed by direct observation (Murray, 1965).

While the property of conduction is the essential and distinguishing feature of nerve cells and one which is very difficult to investigate by tissue-culture methods, there is no reason to believe that it is the exclusive property of one type of cell only; *a priori*, not all nerve cells should necessarily be regarded as being alike except in the fact that they can transmit excitation or inhibition, i.e. in the fact that they conduct. Many of them conduct "impulses", but it is by no means certain that they all do so. Tissue-culture studies have already emphasized some of the differences between different groups of nerve cells, but there has so far been very little serious attempt at any comprehensive classification according to their general physiological properties. For example, in the retina, only the large ganglion cells pick up methylene blue (Francis, 1953), and in fixed preparations certain silver methods also pick out only the large ganglion cells while leaving the small ones unstained. A similar silver method stains the nerve cells in parasympathetic ganglia but not in sympathetic ganglia (Nonidez, 1939), so there are clearly differences between the cells in these groups, but as yet the nature of these differences is not known. In the amphibia the Mauthner cells degenerate under the action of thyroxin while other cells in the neural tube are stimulated to increased mitotic activity (Weiss and Rosetti, 1951). There are indeed many morphological types of nerve cells but it is possible that they may be found to fall into a more limited number of main groups when investigated more thoroughly from the metabolic and physiological angle, e.g. adrenergic, cholinergic, fast or slow conducting, showing "all or none" or graded activity, etc.

Schwann Cells and Neuroglia

The Schwann cells have been studied *in vitro* by making tissue cultures of portions of peripheral nerve trunks after these have previously been severed from all connection with their nerve-cell bodies, and after sufficient time has been allowed for degeneration of the nerve fibres to begin. This degeneration brings the Schwann cells into a state of migratory activity and under these conditions they wander out from the cut ends and grow in the tissue-culture medium generally in the form of a network. They do not readily emerge from normal nerve fibres (Abercrombie and Johnson, 1942). Cultures have also been successfully made from Schwann-cell tumours and it seems to be fairly clear from these cultures that Schwann cells may behave in one of two ways (Murray *et al.*, 1940) (Fig. 5.9). Perhaps there are, indeed, two types of Schwann cells. Those which are associated with amyelinate fibres are charac-

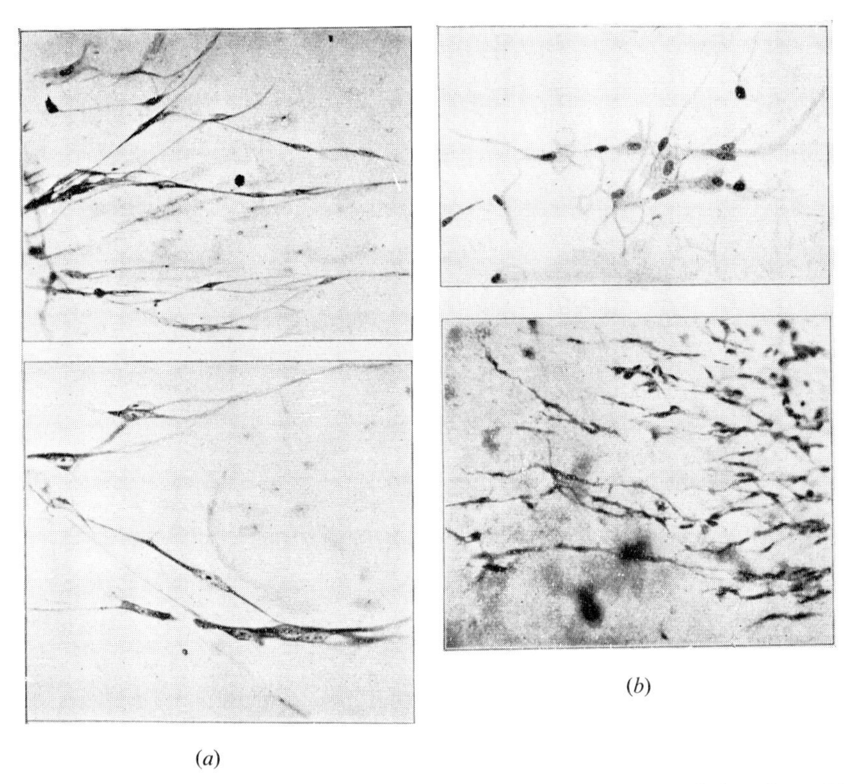

(*a*) (*b*)

FIG. 5.9. Figures illustrating the difference in behaviour of Schwann cells in culture. (*a*) Mechanocyte-like growth; (*b*) amoebocyte-like growth. (Murray *et al.*, 1940, courtesy of The Williams & Wilkins Co.)

terized by a distinct bipolarity and the possession of very long thin processes. The nucleus is generally broader than the average width of the cell, so that its presence and position is marked by a bulge. Branches from these cells tend to sweep gracefully away from the cell body and form contacts with neighbouring cells of similar type or they may creep along nerve fibres (Fig. 5.9 *a*).

The Schwann cells which are associated with myelinate fibres are generally more branched cells and their processes, which are not so slender and fine, tend to be expanded into membranes (Fig. 5.9 *b*). This is particularly noticeable in the cells which migrate out from the spinal ganglia and dorsal roots in a medium of serum and Tyrode (Weiss, 1944b) (Fig. 5.10). These cells closely resemble amoebocytes in appearance (see Fig. 4.5), and the serum medium which was used is one which would favour the growth of such cells. Unlike amoebocytes they form a network with processes from other similar cells, but the whole net appears to be rather irregular. The significance to be

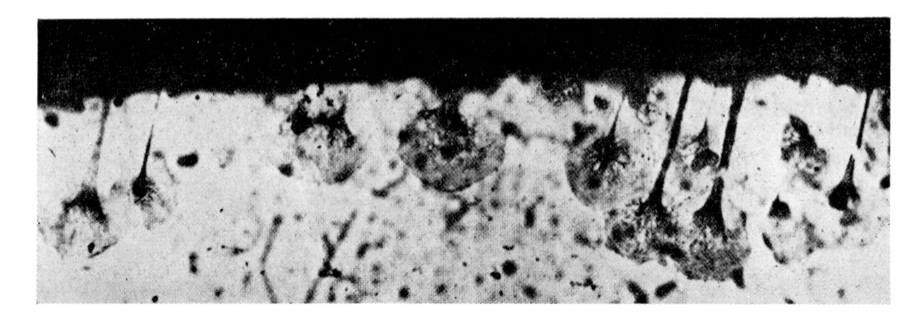

FIG. 5.10. Amoebocyte-like cells emerging from spinal nerve in culture. (Weiss, 1944b.)

attached to these observations is by no means certain. It may simply be that there are two forms of behaviour possible for the Schwann cells, either of which may be set up by external factors. Alternatively, it may be much more fundamental and indicate two separate races of cells both of which are usually classed under the generic term of Schwann cells. It has been known for many years (Langley and Anderson, 1904) that when nerves are cut and then sutured together again, the preganglionic fibres in the autonomic nervous system will not grow down postganglionic paths, nor will postganglionic fibres regenerate down preganglionic or somatic fibres except to supply blood vessels. Most efferent somatic fibres behave like preganglionic fibres, but afferent somatic fibres do not make functional connections with efferent somatic fibres. All these observations indicate differences in character between the Schwann tubes in the various nerves. In the autonomic system the relationship can probably be summarized by the statement that cholinergic fibres do not regenerate down the tubes of adrenergic fibres nor *vice versa*.

In comparing the illustrations of Schwann cells, shown in Fig. 5.9, one is reminded of the patterns of growth of mechanocytes on the one hand and of amoebocytes on the other. A notable difference, however, in the second case is in the fact that the Schwann cells normally form a network, while the amoebocytes tend to remain discrete. Schwann cells are clearly different both from fibroblasts and from macrophages as such, but it seems possible that the two forms of their behaviour may be in some ways similar to the sort of patterns of behaviour seen respectively in mechanocytes and amoebocytes. Weiss and Wang (1945) have, indeed, described a series of changes of form from Schwann cells to fibroblasts and to macrophages in tissue cultures.

Among the other "supporting" cells of the nervous system, mention may first be made of the astrocytes (Fig. 5.11). These appear, in long-continued tissue cultures, to be cells normally characterized by very large undulating membranes; these membranes are of peculiarly unstable character, and they very easily break down into a series of radial processes with more or less

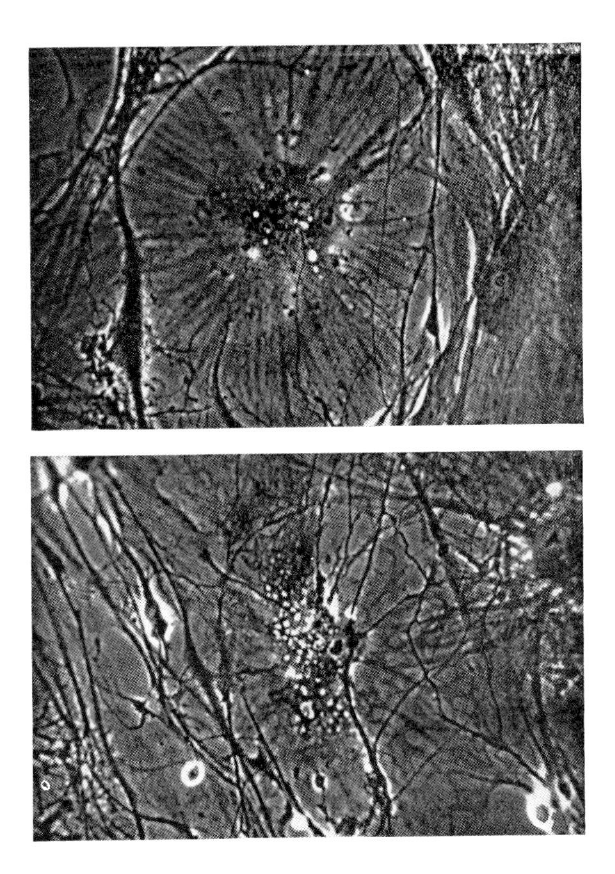

FIG. 5.11. Living astrocytes in tissue culture. Damage leads to dissection of the membranes, as in the lower figure. (Pomerat, 1952.)

amoeboid (Costero and Pomerat, 1951). The changes may perhaps lead to the production of "protoplasmic" and "fibrous" astrocytes as they appear in histologically prepared sections of the central nervous system. A similar state of affairs on a more modest scale is seen in the modification of the lamelliform membranes of amoebocytes. However, before any serious attempts are made in the interpretation of the nature of neural and neuroglial tissue from the appearance of the cells alone, attention should perhaps be called by way of caution to the extraordinary branching cells (illustrated in Fig. 2.11) which appeared in old cultures of fibroblasts (Parker, 1933); morphology without physiology can be very deceptive.

The oligodendroglia are readily recognizable in tissue cultures (Fig. 5.12). They emerge mainly as bipolar cells or sparsely branching cells with small refractile cell bodies and they often collect into groups (Hogue, 1950). They

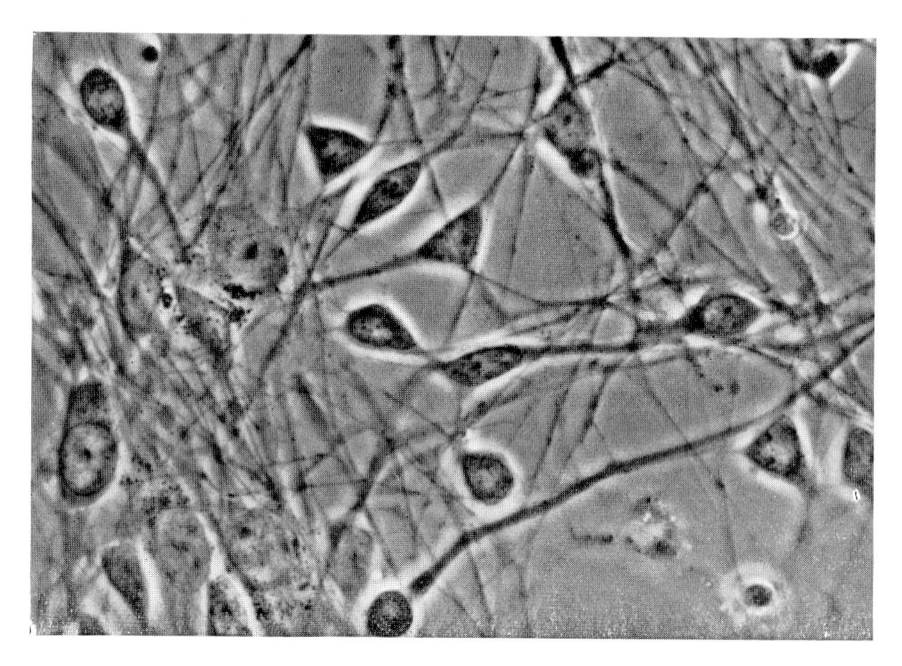

FIG. 5.12. Oligodendroglia cells in tissue culture of cerebellum. (From Murray, 1965, after Pomerat.)

have also been frequently observed to lie with their cell bodies adjacent to those of nerve cells. In fact, it has been suggested on these grounds that there is some kind of symbiosis between the cells in these two classes. Once again the speeded cine-film has brought to light an unexpected form of behaviour, for oligodendroglia cells are slowly and rhythmically contractile, with a period of about 5 minutes (Lumsden and Pomerat, 1951). Recent studies with the electron microscope (Luse, 1956a,b) suggest that oligodendrocytes are partly responsible for the myelination of the axons in the central nervous system; their processes, in this respect, behave in a very similar manner to those of the Schwann cells; that is to say they wrap themselves around or fold themselves repeatedly against the axon and so form a series of more or less concentric layers (Geren, 1954). Astrocytes appear to have a similar function, and this again raises the question of why there are two sorts of cell for this function.

The multipolarity of astrocytes and the bipolarity of oligodendroglia again suggest comparisons with amoebocytes and mechanocytes respectively. So far as the available figures for the relative frequency of astrocytes and oligodendrocytes (Glees, 1955) and those for the extent of acetylcholine synthesis in different parts of the central nervous system (Feldberg and Vogt,

1948) can be compared, there does not appear to be any immediately obvious correlation between them. It is not possible, therefore, to correlate on these grounds either type of neuroglia cell specifically with a particular type of nerve fibre or of chemical transmitter. Perhaps such a correlation should not be expected since nerve cells themselves may contribute to the distribution of these substances (transmitters and their associated enzymes), and probably in an inverse relationship with the neuroglia cells. However this may be, the neuroglia cells and Schwann cells certainly have common points of interest; they each show two forms of behaviour and are each concerned with myelination and there is some sort of "symbiosis" between them and the nerve cells.

The microglia have already been mentioned in the discussion on macrophages, with which they seem to be identical or, at any rate, very closely related.

References

Abercrombie, M., and Johnson, M. L. (1942). The outwandering of cells in tissue cultures of nerves undergoing Wallerian degeneration. *J. Exptl. Biol.* **19,** 266.

Costero, I., and Pomerat, C. M. (1951). Cultivation of neurons from the adult human cerebral and cerebellar cortex. *Am. J. Anat.* **89,** 405.

Deitch, A. D., and Murray, M. R. (1956). The Nissl substance of living and fixed spinal ganglion cells. *J. Biophys. Biochem. Cytol.* **2,** 433.

Feldberg, W., and Vogt, M. (1948). Acetylcholine synthesis in different regions of the central nervous system. *J. Physiol.* (*London*) **107,** 372.

Francis, C. M. (F. M. Chalissery). (1953). A histological study of neurons with special reference to the retina. Ph.D. Thesis, Cambridge Univ., *Cambridge, England.*

Geiger, R. S. (1958). Subcultures of adult mammalian brain cortex *in vitro*. *Exptl. Cell Res.* **14,** 541.

Geren, B. B. (1954). The formation from the Schwann cell surface of myelin in the peripheral nerves of chick embryos. *Exptl. Cell Res.* **7,** 558.

Glees, P. (1955). "Neuroglia, Morphology and Function." Blackwell, Oxford.

Gustafson, T., and Lenicque, P. (1952). Studies on mitochondria in the developing sea-urchin egg. *Exptl. Cell. Res.* **3,** 251.

Hogue, M. J. (1947). Human fetal brain cells in tissue cultures. Their identification and motility. *J. Exptl. Zool.* **106,** 85.

Hogue, M. J. (1950). Brain cells from human fetuses and infants, cultured *in vitro* after death of the individuals. *Anat. Record* **108,** 457.

Hughes, A. (1953). The growth of embryonic neurites. A study of chick neural tissues. *J. Anat.* **87,** 150.

Koenig, H., and Feldman, D. (1953). Structure of living neurons grown *in vitro* as revealed by ultraviolet photomicrography. *Anat. Record* **115,** 336.

Langley, J. N., and Anderson, H. K. (1904). The union of different kinds of nerve fibres. *J. Physiol.* (*London*) **31,** 365.

Levi, G., and Meyer, H. (1937). Die Struktur der lebenden Neuronen. *Anat. Anz.* **83,** 401.

Lumsden, C. E., and Pomerat, C. M. (1951). Normal oligodendrocytes in tissue culture. *Exptl. Cell Res.* **2,** 103.

Luse, S. A. (1956a). Electron-microscopic observations of the central nervous system. *J. Biophys. Biochem. Cytol.* **2,** 531.

Luse, S. A. (1956b). Formation of myelin in the central nervous system of mice and rats, as studied with the electron microscope. *J. Biophys. Biochem. Cytol.* **2,** 777.

Martinovic, P. M. (1932). Survival *in vitro* of explants of the cerebral cortex of the cat cultivated in cerebrospinal fluid of the young animal. *Arch. Exptl. Zellforsch. Gewebezücht.* **12,** 249.

Murnaghan, D. P. (1941). Studies on living spinal ganglion cells. *Anat. Record* **81,** 183.

Murray, M. R. (1965). Nervous tissues *in vitro*. *In* "Cells and Tissues in Culture" (E. N. Willmer, ed.), Vol. 2, pp. 373–455. Academic Press, New York.

Murray, M. R., and Stout, A. P. (1947). Adult human sympathetic ganglion cells cultured *in vitro*. *Am. J. Anat.* **80,** 225.

Murray, M. R., Stout, A. P., and Bradley, C. F. (1940). Schwann cell versus fibroblast as the origin of the specific nerve sheath tumour. Observations upon normal nerve sheaths and neurilemmomas *in vitro*. *Am. J. Pathol.* **16,** 41.

Nonidez, J. F. (1939). Studies on the innervation of the heart: distribution of the cardiac nerves with special reference to the identification of the sympathetic and parasympathetic post-ganglionics. *Am. J. Anat.* **65,** 361.

Olivo, O. M. (1927). Differenziazione e sdifferenziazione del tessuto nervoso embrionale di pollo coltivato per più settimane "*in vitro*". *Arch. Exptl. Zellforsch. Gewebezücht.* **5,** 46.

Parker, R. C. (1933). The races that constitute the group of common fibroblasts. (ii). The effect of blood serum. *J. Exptl. Med.* **58,** 97.

Peterson, E. R., and Murray, M. R. (1955). Myelin sheath formation in cultures of avian spinal ganglia. *Am. J. Anat.* **96,** 319.

Pomerat, C. M. (1952). Dynamic neurogliology. *Texas Rept. Biol. Med.* **10,** 885.

Speidel, C. C. (1933). Studies of living nerves. (ii). Activities of amoeboid growth cones, sheath cells, and myelin segments, as revealed by prolonged observation of individual nerve fibres in frog tadpoles. *Am. J. Anat.* **52,** 1.

Weiss, P. (1944a). Evidence of perpetual proximo-distal growth of nerve fibres. *Biol. Bull.* **87,** 160.

Weiss, P. (1944b). *In vitro* transformation of spindle cells of neural origin into macrophages. *Anat. Record* **88,** 205.

Weiss, P., and Hiscoe, H. B. (1948). Experiments on the mechanism of nerve growth. *J. Exptl. Zool.* **107,** 315.

Weiss, P., and Rosetti, F. (1951). Growth responses of opposite sign among different neuron types exposed to thyroid hormone. *Proc. Natl. Acad. Sci. U.S.* **37,** 540.

Weiss, P., and Wang, H. (1936). Neurofibrils in living ganglion cells of the chick cultivated *in vitro*. *Anat. Record* **67,** 105.

Weiss, P., and Wang, H. (1945). Transformation of adult Schwann cells into macrophages. *Proc. Soc. Exptl. Biol. Med.* **58,** 273.

Young, J. Z. (1945). The history of the shape of a nerve-fibre. *In* "Essays on Growth and Form" (W. E. Le Gros Clark and P. B. Medawar, eds.), Oxford Univ. Press (Clarendon), London and New York.

Chapter 6

THE BASIC CELL TYPES

The behaviour of the different types of mammalian and avian cells and tissues in culture has been described in some detail in order to give a general picture of cell behaviour and to emphasize the outstanding physiological qualities of the cells. All that has been said so far is thus in the nature of an introduction to an approach to cell physiology which has been much neglected in the past, but which now can be tackled with more hope of advantage than ever before. Classical cytology (meaning by this cell physiology rather than cytological genetics) necessarily had to base most of its conclusions and its hypotheses on the knowledge gained by the study of fixed and stained materials or by histochemical methods on either sections or teased preparations of tissues. It was thus essentially cell morphology. So too, are the results obtained by electron microscopy. In interpreting these it must be remembered that each photograph shows, for example, the transmission and absorption of electrons through an extremely thin section of the desiccated framework of a minute fragment of tissue, once containing about 80% water and now impregnated with metallic deposits after fixation, more or less instantaneously at a given phase of the tissue's activity. It must equally be remembered that cine-films of living cells may show the cytoplasm to be in a constant state of flux. Thus the patterns seen in electron micrographs may often bear more resemblance to the patterns seen on photographs of military parades than to the structural patterns of buildings. It has also been observed that particles can move freely and apparently in straight lines through living cytoplasm that the electron microscope shows (when fixed) to possess a highly elaborate organization of endoplasmic reticulum and the like. Finally, the higher the magnification, the more the movement is being "frozen." That is to say, other things being equal, the quicker the fixation, the better will be the definition of the electron micrograph because the organelles have less time to change.

The situation is exactly comparable with that in other photographic fields, where, for example, the shorter the exposure, the clearer will be the pictures of a race-horse jumping a fence. In both cases, the clearer the photograph the more the movement is "frozen", so that the complete event has to be reconstructed by mental processes. In electron microscopy the higher the magnification, the more important do these considerations become. Tissue culture, however, not only accentuates the vital qualities of cells but it allows them to be studied under conditions over which the experimenter has rather more control, though this amount of control is by no means as complete as is desirable. Tissue culture has the advantage that processes can often be studied while they are actually in progress, and not by the interpretation of a series of stages alone. Cells can be seen in action and, what is more important, direct tests can be made upon them in isolation without involving secondary effects which may emanate from the responses of other tissues in the body to the experimental or to other, irrelevant, conditions.

The results of tissue culture work which have received most comment in this account are purposely those which have been obtained by the techniques encouraging "unorganized" growth, since these allow the individual cells to be more easily and directly studied. Under these conditions, it must again be emphasized, the cells are not usually functioning exactly as they do *in situ*, and sometimes they are indeed behaving very differently. There is certainly modification, and probably some simplification of their normal behaviour. All the data gained by tissue-culture methods have, like others, to be sifted and weighed against data from other sources before there is any hope of developing what may be regarded as a balanced view of cell behaviour. In many instances, of course, the results of tissue-culture experiments involving "organized growth" have been able to make extremely valuable contributions and now that somewhat simpler methods are available for these experiments, organized growth is likely to be more studied in the future and to provide more and more relevant information about cell behaviour. Certainly the cells function more normally under these conditions, but at the same time the degree of isolation of the cells is much less and the problems and effects of cellular interaction, on the one hand, and the necessity for finally fixing and sectioning the tissues for examination on the other, both put severe limitations on the method for studies on the physiology of the individual cell. Nevertheless, data obtained by "organ culture" methods must also be woven into the tapestry of knowledge obtained by other methods, and for many purposes such data are likely to be the most reliable.

Before proceeding to discuss the import of the tissue-culture data in a more constructive manner, it will be advisable to review some of the main concepts that have emerged so far.

The salient fact for the purpose of this discussion is that when tissues from the various organs and structures of the human body, and indeed of all the vertebrates and invertebrates which have been investigated, are grown *in vitro*, under conditions which favour unorganized growth, then the emigrating cells simplify their form and their behaviour so that they approach one or another of a very limited number of patterns. If it were only the outward appearance of the cell which tended to conform to a type, then, in the light of the great variability which is noted in tissue cultures under experimental conditions, this tendency to adopt one or other of a few main patterns, though interesting in itself, would not perhaps have much general significance. However, when the characteristic forms of cells in tissue culture are also accompanied by specific biochemical differences, by peculiar methods of locomotion, by distinct patterns of cell contacts, and by somewhat parallel types of behaviour under pathological conditions in the body, it then becomes obvious that the meaning of these "tissue-culture" groups lies deeper and is a matter worthy of investigation since it may throw new light on fundamental cell properties and on essential processes in cell differentiation.

The Nature of Differentiation

The characters and properties of the cell groups have been collected and summarized in Table 6.1; this table deals only with the three main groups: the mechanocytes, epitheliocytes, and amoebocytes. The question of the cells that constitute the neural tissues will be taken up again later. It is fairly clear from the table that there are sufficient distinctive properties to justify this broad classification into three groups, though it is not really possible to say wherein the main differences lie, and there is no one feature which can be made absolutely diagnostic for purposes of classification. One may hazard a guess that the nature of the surface membrane is probably a primary factor in causing some of the changes. This same conclusion was reached by Loeb (1920) as the result of his extensive investigations into the behaviour of the amoebocytes from the blood of *Limulus*, the king crab. These fascinating and very versatile cells have at times many of the characteristics of the amoebocytes of vertebrates but are probably rather more plastic. In support of the idea that the cell surface is important it is known that such a simple but "surface active" substance as sodium oleate in the growth medium of fibroblasts encourages their phagocytic powers and alters the type of movement of the cells, thus rendering them more like macrophages (Fischer, 1930). Similarly, it is tempting to think that the liberation of mucoprotein by mechanocytes into the surrounding medium, as opposed to its storage within the cell by certain epitheliocytes, is perhaps related to stereochemical properties of the cell membranes. Incidentally, electron-microscope photo-

TABLE 6.1

Epitheliocytes	*Mechanocytes*	*Amoebocytes*
Grow as a sheet with intimate cell contact.	Grow as a network exhibiting contact inhibition. Isolated cells survive with difficulty.	Grow as isolated cells, but the cells may fuse.
Desmosomes may be present	Desmosome-like structures may be present (Ross and Greenlee, 1966).	Desmosomes absent.
Pseudopodia, at edge of growth only, mostly lamelliform but sometimes pointed.	Characteristically pointed pseudopodia.	Characteristically lamelliform and rounded pseudopodia.
Endoplastic reticulum variable	Endoplasmic reticulum with flattened cisternae, generally "rough" i.e. with ribosomes.	Endoplasmic reticulum with vesicular cisternae, often smooth, i.e. with few ribosomes.
Cells attached to coverslip locally only—tendency for attachments to break.	Cells in culture attached to coverslip at extremities only. Tough cell membrane. (Chambers and Fell, 1931)	Cells may be attached to surface more completely. Less rigid cell membrane. (Chambers and Fell, 1931)
Colonies fuse.	Colonies fuse.	Colonies remain isolated (Carrel and Ebeling, 1922)
Gliding movement of the sheet, following extension by the peripheral cells.	Gliding movement of cells with polarity.	Amoeboid movement often without marked polarity.
Best growth on a surface.	Limited penetration into a plasma coagulum. (Fischer, 1930)	Penetrate plasma coagulum to all depths. (Fischer, 1930)
Some epithelial cells phagocytic *in vivo*. Not normally so *in vitro*.	Not normally phagocytic. Do not segregate vital dyes very actively.	Usually phagocytic. Active segregation of vital dyes.
Can pinocytose.	Can pinocytose.	Active pinocytosis.
Do not normally require embryo juice. (Liver inhibited by more than 20%.)	Require embryo juice or the like for cell multiplication. Embryo juice increases movement.	Become fatty and quiescent with embryo juice.
Grow in serum.	Cell multiplication not normally possible for long in serum. Survive and differentiate in serum. (Parker, 1938.)	Growth and multiplication of at least some, though not all, amoebocytes in serum. (Jacoby, 1937, 1957)
Produce Carcinomata.	Produce Sarcomata.	Produce Leucaemia. (Round cell sarcomata?)
—	Cause differentiation of epithelial cells (e.g., Grobstein, 1953).	No evidence that they cause epithelial differentiation.

Epitheliocytes	*Mechanocytes*	*Amoebocytes*
Mucosubstances may collect in the cytoplasm. In other cases, they may be extra-cellular.	Mucosubstances mostly produced outside the cell.	Mucosubstances collect in the cytoplasm.
Tend to produce keratin.	Tend to produce collagen.	Perhaps produce albumen or fibrinogen. Acid phosphatase in numerous lysosomes.
Acid or alkaline phosphatase may be present.	Alkaline phosphatase.	
Uninhibited by hexenolactone. (Medawar, 1937; Medawar et al., 1943)	Inhibited by hexenolactone (Medawar, 1937; Medawar *et al.*, 1943.)	—
—	—	Encouraged by choline and quaternary NH_3. (Thomas, 1937)
—	Inhibited by 1/3,000,000 As_2O_3. Killed by 1/800,000 As_2O_3. (Fischer, 1930)	Unaffected by 1/800,000 As_2O_3. (Fischer, 1930)
—	Do not support Rous chicken sarcoma virus. (Carrel, 1926)	Support Rous chicken sarcoma virus. (Carrel, 1926)
Some epithelia support fowl-pest virus. (Hallauer, 1931–1932)	Do not support fowl-pest virus. (Hallauer, 1931–1932)	Do not support fowl-pest virus. (Hallauer, 1931–1932)

graphs of the boundaries between epithelial cells suggest that in some cases the actual cell membranes may be in contact and desmosomes, terminal bars, or tight junctions are formed, while in other cases the membranes interdigitate in a most complex pattern, as in the tubule cells of the kidney. The binding together of cell to cell in many epithelia certainly depends on the action of calcium ions and perhaps of magnesium also. When Ca^{2+} and Mg^{2+} are removed from the medium the binding between the cells becomes weaker and the cells tend to fall apart, a process which can be assisted in some tissues by the presence of trypsin and elastase (Moscona and Moscona, 1952; Rinaldini, 1958). The actions of these substances are in all probability on the cell surface in the first instance.

Sudan Black and Baker's haematein test (combined with the use of suitable solvents) indicate the presence of phospholipins at the cell boundaries in the epidermis in the form of bead-like structures, but their significance is not yet established. Furthermore, the cause underlying the blackening of the boundaries of contact by silver nitrate is not yet established, and in many cells the blackening may be restricted to certain very limited regions of the cell's surface which persist when the contact is broken (Robinow, 1936).

All these observations point to the cell surface as a dominant feature of epitheliocyte behaviour. The contact inhibition described for fibroblasts

strongly favours either an adaptation on the part of the original mechanocyte to the special conditions of culture and of survival in isolation, or, but with less probability, an initial selection of a cell type with properties somewhat different from any of the types discussed here. Until such changes occur, however, the cells within each family (amoebocytes, mechanocytes, or epitheliocytes) apparently breed true, and do not readily change from one form to the other, though the possibility of such interchange is certainly to be considered.

This property of breeding true within the family could be purely cytoplasmic, although in view of the adaptability of the cytoplasm to changed conditions and in view of the apparently random distribution of such cytoplasmic elements as mitochondria at mitosis, a feature which contrasts so strongly with the elaborately equalized nuclear partition, it is perhaps more likely that these races of somatic cells in tissue culture are distinguished by relatively permanent nuclear characteristics. This does not necessarily mean anything so crude as a different number of chromosomes, though the counts of chromosomes in somatic cells are not yet sufficiently accurate or thoroughly known to preclude this possibility absolutely. All that it may mean is, that by some metabolic stimulus received during development, the nuclear pattern may be altered, i.e. the activity of one gene or group of genes more than another may be allowed to predominate in a particular group of cells.

Every histologist is familiar with the differences of nuclear form among the various somatic cells that exist, some of which are illustrated in Fig. 6.3, and it is obvious that not all nuclei in the differentiated tissues are structurally identical, though in some early embryonic stages they may be shown to be physiologically identical or at least not irreversibly modified. The physiological meaning to be attached to these differences in appearance of the nuclei of different groups of cells has never been much investigated, yet it is clearly a very important phenomenon. Unfortunately, the detectable nuclear difference between the simplified epitheliocytes, mechanocytes, and amoebocytes of tissue cultures are much less striking, and in such matters as the number and size of nucleoli there is considerable variation within each group.

There is now evidence that, while in the early stages of amphibian develop-

---

FIG. 6.3. Diagram to emphasize the different forms of nuclei in various somatic cells.

1. Fibroblasts	2. Plain muscle	3. Columnar epithelium
4. Lymphocytes	5. Pancreas cells	6. Striated muscle
7. Liver cells	8. Nerve cell	9. Neuroglia cell
10. Neutrophil leucocytes	11. Monocyte	12. Megacaryocyte
13. Spermatocytes	14. Spermatids	15. Ovum

See facing page→

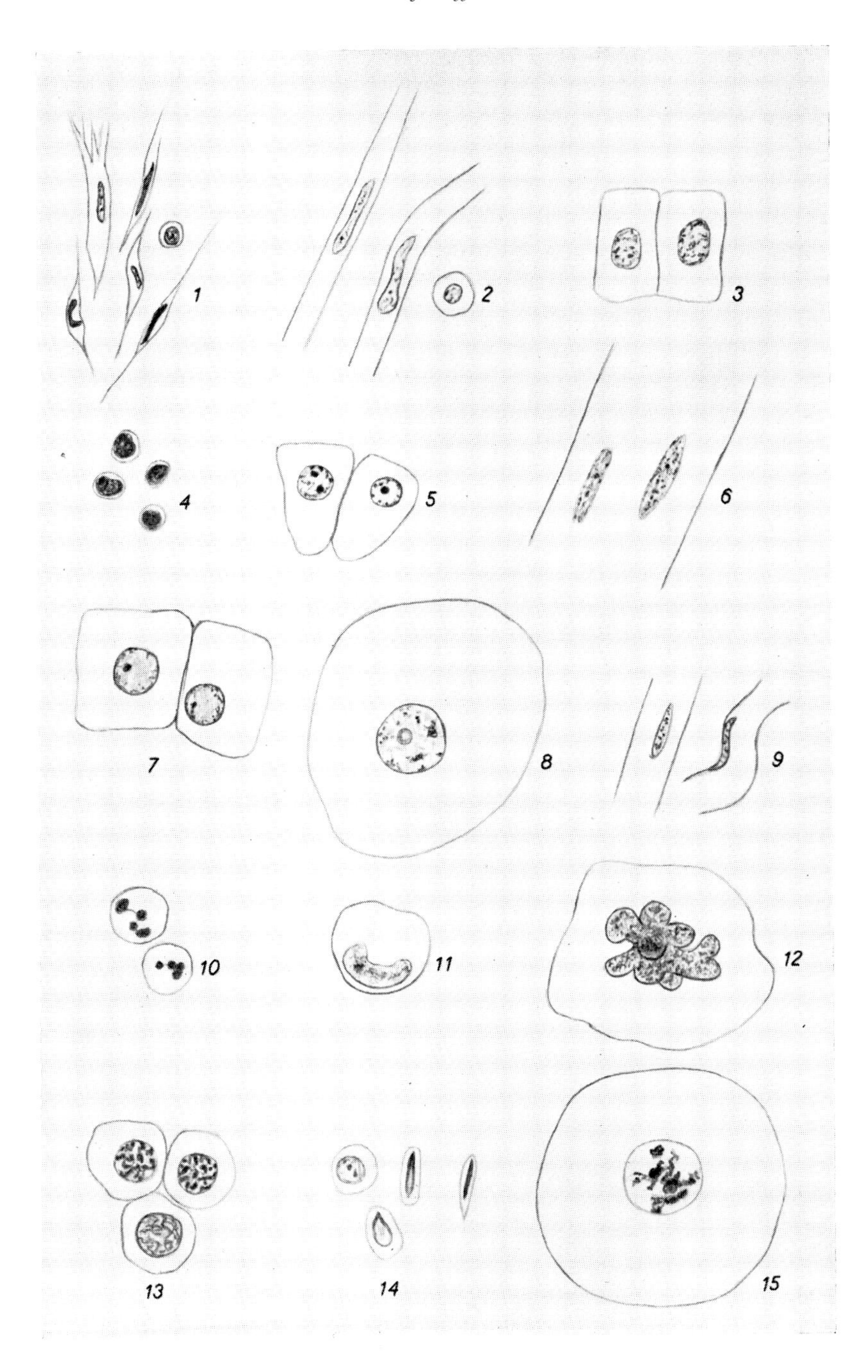
1
2
3
4
5
6
7
8
9
10
11
12
13
14
15

ment all nuclei appear to be functionally identical, nuclei in the endoderm of the late gastrula may be different. This follows from the observations that, while nuclei extracted from cells in the early stages can be injected into and can generally replace the nucleus of the unsegmented ovum and still allow normal development to take place, nuclei extracted from the endoderm cells of the late gastrula have frequently lost this power and an increasing number of abnormal embryos results, when such nuclei are injected into enucleated but activated ova (King and Briggs, 1956). More recent observations by Gurdon (1962, 1966), however, indicate that when better techniques are used the nuclei of endodermal cells can in fact sometimes be used successfully for obtaining whole new embryos when injected into enucleate eggs. This being so, it must be assumed that each nucleus at this stage still carries the full complement of genetic information to provide for the whole organism, though it does not necessarily use it and would not normally do so. Similarly, a mammalian ovum does not itself make either haemoglobin or rhodopsin, but after a series of divisions it eventually produces progeny that can do so. It must, therefore, have all the necessary information in its nuclear record cabinets, and the experiments with enucleate eggs show clearly that, in that case at least, it is the cytoplasm that calls the tunes and the nucleus that has to play them.

It now seems likely that the differentiation of a cell depends on the resultant of three interacting systems. A functioning cell has a particular structural (i.e. nuclear and cytoplasmic) pattern involving the molecular organization of its cell surfaces, mitochondria, endoplasmic reticulum, Golgi complex, nuclear membranes, and the like. The exact form which this pattern takes at any one moment depends both on the molecules composing it and on the environmental conditions. Of the molecules which compose it, representatives of all types, proteins, carbohydrates, lipids, ions, etc. all contribute; but the proteins occupy a special position since they alone, so far as we know, can be directly and specifically determined genetically. The three interacting systems are therefore (1) the molecular organization of cell structures, (2) the genetic constitution, and (3) the environment of the cell. The cell organization is acted upon, from one side, by the active genes supplying the proteins contributing to the system and, from the other side, by the environment which may alter the pattern of organization in a hundred and one ways, by adding or subtracting active molecules, by ionic and osmotic effects, temperature, etc. All these interactions with the molecular organization are of course mutual. The genetic system reacts to changes in its environment by changing its production of RNA and hence of the proteins which can be produced, while the local environment of the cell may also be altered by the metabolic activities of the molecular organization of the cell itself. The recent experiments in which the "dormant" nuclei of bird erythrocytes have been incorporated into HeLa* cell cytoplasm and have started again to produce RNA (Harris, 1967) and the experiments in which intestinal cell nuclei of amphibia have been injected into enucleate ova and have brought about the differentiation of whole embryos (Gurdon, 1966) are clear examples of the "cytoplasm" altering the activity of the genes in its own direction.

If these mutual interactions are followed through still further, it is clear that the environment may determine, albeit indirectly, the behaviour of the genes. Hormone action may sometimes become effective in this way. Hypothetically, a hormone, e.g. a steroid, could alter the character of lipid membranes in the cell and thus alter the molecular organization of cell surfaces, including mitochondria and the like. Such effects on the membranes could well affect the proteins associated with them and alter the behaviour of the whole system. For the cell to regain stability, different proteins might be required to patch the membranes, and in consequence the genome could have one part of its activity closed down and another one opened up so that the necessary different proteins are produced. By such means a condition of stability could be achieved once again, but with the formation of a temporarily different sort of cell. This readjusted cell, by producing different metabolites, could again modify the environment and bring about further changes both in itself and in its neighbours. Thus, in accordance with the above assumptions, cellular activity must inevitably be in a constant state of flux, continually changing and continually adapting.

As a direct consequence of the mutual interaction between these three systems, the regulation of the "internal environment" of an organism is a *sine qua non*, but within that environment, which is constant in a temporal sense, there must inevitably be innumerable local environments which, though held as nearly constant as possible in time differ markedly from each other from place to place. Each of these local environments is probably self-regulating in the manner outlined above.

This simple hypothesis presupposes that at least some of the genes are always accessible and continually producing RNA's in such a way that new proteins can be formed reasonably quickly at any time. This would certainly seem to be an efficient system and would allow cells to react to their surroundings at all times, but it tends to underestimate the determining action of the genome itself, which often seems both to be able and necessary to determine general patterns of behaviour. For example, a liver cell may for long periods carry out its essentially hepatic functions with all the necessary minor fluctuations and modifications of its activities in response to feeding, starvation, diet, and the like. If, however, part of the liver is removed, all this temporarily stops, and each remaining cell turns its attention to the produc-

*HeLa cells are cells originally derived from a human cervical carcinoma which are now extensively grown as cell strains and used for a variety of experimental purposes.

tion of more liver cells like itself. Similarly, myoblasts may reproduce themselves freely as myoblasts, but there comes a time when they unite with each other to form an elementary muscle fibre, wherein they then devote themselves to the production of actin and myosin and, normally, they never divide again. Blood cells undergo innumerable cell divisions as "stem cells" and ultimately one of the products of a division starts to make haemoglobin, after which it may only divide about five more times and the progeny lose their nuclei, circulate for a few months, and then die. The invariant quality and longevity of any particular nerve cell (a cell which throughout life makes immense quantities of ribonucleoproteins and does not normally divide, though it may function for even a hundred years or more) stands in sharp contrast to the relatively short lives, measured in months or even days, of such cells as the erythrocytes, leucocytes, and intestinal epithelial cells. The problem, therefore, is to determine how much of each of these behaviour patterns is determined from within the cells and how much from without.

It is fairly clear that, in the differentiation of some cells, a cell division ultimately becomes unequal, in the sense that one daughter cell starts to differentiate while the other remains in the growing phase. This frequently happens in early embryonic development, in the development of blood cells, and it has actually been seen to happen in the formation of a thymocyte in tissue cultures of the thymus epithelium (Murray, 1947), but there seems to be little information as to what determines this sudden development of inequality. Unequal divisions frequently occur in the differentiation of plant tissues with the formation of different functional types, and these are generally preceded by unequal distribution of cytoplasmic constituents. One suspects that local gradients or environmental differences are at work in these examples, but it is less easy to see how these would operate at particular times in such systems as the seminiferous tubules or the bone marrow (see also p. 208).

If it is agreed that, in general, every cell in an animal has the complete complement of DNA (in insects some chromosomes are eliminated from the somatic cells but this is probably exceptional) characteristic of the species to which it belongs, with, of course, the minor variations that occur from individual to individual, then it is clear that in most cells of the organism a large amount of this DNA is never functional and is presumably inactivated or blocked for a large part of the life of that cell. Since, however, environmental factors are certainly able to alter cell behaviour within limits, the logical conclusion seems to be that the various parts of the genome can be made available for producing appropriate RNA, and hence protein, as and when required. Naturally some genes, or groups of genes, may at any one time be more easily activated than others, and these may determine the behaviour of broad classes of cells. Others may be more heavily repressed, or blocked, and only activated in specific tissues. This blocking and unblocking of protein-synthesizing activities may, of course, occur not only at the DNA level, as has been shown to be the case in bacteriophage, where a specific repressor (λ phage) will bind to the specific DNA that it blocks (Ptashna, 1967); it could equally well occur at one of the RNA steps or even later, though it would be less economical to do it at these later stages. The times at which this blocking and unblocking occur and the mechanisms involved are of fundamental importance, but, as yet, there is very little information about them in the cells of higher animals. In some cases differentiation does not occur unless new cells are produced, i.e. mitosis seems to be necessary, while, in others, the cell activity can be directly modified by external factors, and new proteins can be produced without the intervention of any mitotic division; in fact, the latter may be positively inhibitory.

All these problems are clearly relevant to the observations on the forms of behaviour of cells in tissue culture that are under discussion, as they are to the behaviour of cells in general. Some modification of cell activity is evident in every tissue culture, whether it is growing or not, but there is no doubt that the cells adapt themselves more quickly and completely to the conditions of culture when they are stimulated to divide. In the absence of such a stimulus the cells tend to carry on with what they were doing before they were placed in culture. This suggests that much of the determination of the class of RNA to be formed occurs during the division process, though clearly not all of it. Isolated heart cells provided with a uniform environment first lose their specifically muscular properties and "revert" to mechanocytes; this they do more easily if cell division is activated by embryo-juice, for example. By prolonged activation and adaptation the cells may eventually emerge as a strain of "tissue culture" cells almost indistinguishable, morphologically, nutritionally and enzymatically from other cell strains in tissue culture which may have started as liver epithelium or kidney cells.

In this way the behaviour of cells in tissue culture is comparable with the behaviour of cells during ontogenesis. In both cases the cells adapt to their immediate surroundings. These adaptations often involve the activation or release of particular parts of the genetic code and the process of adaptation is facilitated by cell division. Between each division, the cells modulate to their surroundings, and their RNA metabolism must vary in consequence; it may be these modulations which determine the pattern of RNA release in the next generation of cells. In this way the DNA activity has ultimately to be modified. In conditions favouring rapid and unlimited growth, as, for example, in cell cultures in flasks, the rate of adaptation may be rapid; in conditions not favouring growth and preserving the original histological pattern of the cells the modification and unification of cell behaviour is much less. Thus cells in primary culture which can maintain their immediate surroundings relatively unchanged (as in organ culture) may continue with their normal physiological functions, but if the conditions are seriously changed then the cells adapt, if they can. If they cannot, they either die or divide. If they divide they may be able to produce something out of the genetic store which allows them to continue alive and to make another series of adaptive modulating steps. These in turn may make further adaptive steps possible at the next division and so the process goes on. Thus, after a number of generations in standard tissue-culture conditions various cells which differed widely from each other when in the animal may become more and more alike as they attune themselves to life *in vitro*.

This problem of adaptation in tissue culture has been treated at some length and at the risk of some later repetition because it is of such cardinal importance to the understanding of the sort of mechanisms which have been at work and must still be at work in the gradual modification and development of cellular activities which have occurred during the course of evolution and which have culminated in the necessary information being incorporated into the genome for perpetuation. Some of these problems will be further discussed in other contexts.

To return again to the immediate question of the races of cells as seen in primary cultures of various tissues, we need to know the factors which are at work in determining the behaviour of epitheliocytes, mechanocytes, and amoebocytes. It is notable that cells in culture preserve their species characteristics very tenaciously, as can be shown by immunological tests (Franks *et al.*, 1962). On the other hand, they lose their particular functional characteristics very readily, but having done so they then cling more firmly to the characteristics of the cell "family" to which they belong, i.e. they remain as mechanocytes, amoebocytes, etc., and in this condition they may breed true for a considerable time and produce many generations of similar cells. Moreover, it is not only in tissue cultures that this happens but it also occurs in wound healing, where, once again, the same main types of behaviour are clearly evident. A healing skin-wound shows epitheliocytes, mechanocytes (fibroblasts—see Fig. 2.25), and amoebocytes (macrophages—see Fig. 4.14) as clearly as any tissue culture. It is with the nature and significance of these cell families that we must now be concerned.

Cell Types in Relation to the Germ Layers

Perhaps one of the more unexpected—and certainly one of the most significant—features of the existence of these three main families or races of cells, is the fact that the boundaries between them do not seem to be related to the dividing lines between the germ layers. Epitheliocytes can emerge from ectoderm (skin), mesoderm (e.g. kidney), and endoderm (e.g. intestine), and so far as tissue-culture tests go, there is no general character by which to differentiate between them according to their origin. Epitheliocytes arise in culture from those tissues which are more or less directly connected with the original surface epithelium of the body, including, of course, the alimentary canal and its glands and diverticula, much of the urinogenital system, and certain endocrine glands. Mechanocytes and amoebocytes, on the other hand, derive mostly from those cells which form the inner mass of cells in the body, i.e. those cells which leave the primary surface at a very early stage of development; these cells have been classed together as the "mesohyl" by Hadži (1949); the word means the "middle timber or substance", and it describes the place of origin of mechanocytes and amoebocytes far more nearly than does "mesoderm". Mechanocytes and amoebocytes can be fairly accurately described as emanating from the "mesenchyme" but this is not now a very satisfactory term, as it has been used with many different implications, e.g. it is often applied to the epithelial cells of the kidney rudiments; mesohyl is preferable.

This idea of surface and mesohyl cells as the origin of epithelial cells and of the other cell types, respectively, at first seems very simple; but the immediate question then arises from it as to why there should be two types of behaviour among the cells of the mesohyl and why all the mesohyl cells should not behave identically. In a general way the various epitheliocytes seem to behave similarly in tissue culture, but the differences between mechanocytes and amoebocytes are certainly obvious and deep rooted if not actually irreversible. Perhaps the behaviour of epitheliocytes *in vitro* should be studied in more detail. The plurality of types of epithelial cells *in vivo* (see p. 84) is very evident and could profitably be investigated in relation to possible differences among epithelial cells when living *in vitro*.

Before pursuing this question further, however, it will be preferable to discuss briefly the other groups of cells which do not conform with the three main types as described.

In the embryology of most vertebrates a portion of the surface epithelial cells are early marked off from the rest as neural plate tissue; this tissue sooner or later folds over, and eventually the folds join together longitudinally to produce a tubular form, though no one has yet satisfactorily explained why this happens. But, be that as it may, some cells, of course, fall just outside the neural plate and neural folds and so just miss being incorporated in the tube. Such tissue is called the neural crest. It is from these parts of the embryo, the neural plate and the neural crest, that the majority of nerve cells, neuroglia, and Schwann cells arise, i.e. it is from a region of the surface epithelium which at a comparatively late date in embryological development becomes specially modified. It is perhaps not surprising, therefore, that these nerve cells fall (as shown in Chapter 5) into a somewhat different category from the rest, and for the present, until the origin of the other categories has been more fully studied, it will be convenient to treat them as cells apart. They are clearly derivatives of epitheliocytes that emerge at a later stage in ontogenetic history, probably as the result of the local changes in their environment, some of which are brought about by the proximity of the underlying mesohyl cells.

The neural crest itself is a tissue of great embryological significance, and one which has been the cause of much controversy; it is the fly in the ointment of the classical germ-layer theorists. From it, there arise a number of interesting cell types which on classical lines are very difficult to reconcile with a rigid germ-layer theory. In the first instance the neural crest is essentially an epithelial sheet lying at the edges of the neural folds, and when the folds meet and the superficial epithelium closes over the surface, the neural crest cells are those which fail to become incorporated either in the neural tube itself, which is still fundamentally epithelial, or in the true epithelium of the surface. They are thus really epithelial cells forced out of the surface; in other words, they become mesohyl, though perhaps with special properties consequent on their location and on their late date of entry. It is therefore relevant to the present discussion to note that from these ex-epithelial cells arise not only nerve cells, but cells of the pia mater, Schwann cells, cartilage cells, perhaps even bone cells and odontoblasts (Hörstadius, 1950) (i.e. mechanocytes), and also all the chromatophores, melanophores, and dendritic cells of the body (which are all cells showing many properties characteristic of amoebocytes). As noted earlier, Schwann cells should perhaps be subdivided into two groups according to their behaviour, the spindly mechanocyte-like type being possibly distinguishable from the more diffuse type with irregular pseudopodia. A similar difference between the bipolar nerve cells of the dorsal root ganglia and the more diffuse type of the autonomic ganglia, may perhaps reflect a similar duality. Without delving into these unknown and rather complex regions at this stage however, it is clear that in the "epithelium" of the neural crest there are already the seeds of both mechanocytic behaviour and of amoebocytic behaviour; this epithelium, therefore, seems to contain the same sort of duality, or limited plurality, within itself as has been suggested as being inherent in nearly all the other epithelia of the body, when these are judged by the manner in which their cells may differentiate.

It is therefore of some interest to observe that the neural crest, which is such a very difficult tissue to incorporate neatly into any "classical" germ layer theory, is perhaps the very tissue which shows most clearly the manner in which the epitheliocytes, as soon as they leave the external surface, can acquire new properties and that such cells can then display one of two main lines of development. The final properties of the cells could be considered as variations on one of these two main themes—the mechanocytic theme and the amoebocytic theme—but the main problem obviously lies in the fact that there are again these two major lines of behaviour, and two only. Are epitheliocytes always divisible into two main groups, and if so, what is the nature of the difference between the groups? These seem to be the basic questions arising from this discussion.

Finally it should be emphasized that mechanocytes, amoebocytes, and epitheliocytes are not the restricted property of the higher vertebrates. Similar behaviour has been noted also in the lower vertebrates and in insects and molluscs (Jones, 1966) in so far as successful culture experiments have been performed on their tissues. Furthermore, there are reasons to believe that animals very much lower in the scale would, under similar conditions, produce the same general subdivision of behaviour amongst their main cell types.

Since the origin of vertebrates is shrouded in mystery, it is pertinent to ask at what level of animal organization the subdivision of the families of cells takes place, because, if it can be shown as a fairly general phenomenon in very much simpler organisms, then the details and uncertainties of the origin of vertebrates become less important for the solution of these particular problems of cell behaviour. Unfortunately, tissue culture has not yet been extensively practised on the invertebrates, and it is in any case difficult to know which invertebrates can be profitably regarded as giving direct information relevant to the stream of organisms whose evolution culminated in the vertebrate stem and which would be worth studying from this point of view. From the studies of the amoebocytes of the king crab (*Limulus polyphemus*, Arachnidae) it would appear that these cells are less irrevocably differentiated and are more nearly totipotent, and thus capable both of forming epithelial-like sheets and of assuming mechanocytic characters, than are the amoebocytes of higher forms.

Liebman (1950) too, has studied the leucocytes of *Arbacia punctulata* (Echinoderm). Here again the cells are versatile and seem to be capable of giving rise to cells of other types, e.g. mechanocytes (fibroblasts), petaloid (? epithelioid) cells, amoeboid macrophages, and storage trephocytes. The plasticity of the amoebocyte is probably a rather general characteristic of these cells and reasons for this will be given later in considering the significance to be attached to this phenomenon.

Very interesting information is also available as the result of the histological and cytological study of some of the most primitive of all groups of metazoa, e.g. the Porifera, and there are indications from this study that amoebocytes, mechanocytes, and epitheliocytes are indeed extremely primitive and fundamental as cell types and that they even precede the germ layers. In fact, there are indications that the basic patterns and forms of behaviour characteristic of these types may have originated even before the development of Metazoa. In the following pages it will be shown that there is evidence that the basic differentiation is already present, though perhaps in a more reversible form, in the Porifera, or sponges. This being so, the seeds of mechanocytes and amoebocytes may have their origin even earlier, namely in the Protozoa, and, as will be seen, there are reasons for believing that this too is indeed so.

References

Abercrombie, M., and Heaysman, J. E. M. (1952). Observations on the social behaviour of cells in tissue culture. 1. Speed of movement of chick heart fibroblasts in relation to their mutual contacts. *Exptl. Cell Res.* **5,** 112.

Carrel, A. (1926). Some conditions of the reproduction *in vitro* of the Rous virus. *J. Exptl. Med.* **43,** 647.

Carrel, A., and Ebeling, A. H. (1922). Pure cultures of large mononuclear leucocytes. *J. Exptl. Med.* **36,** 365.

Chambers, R., and Fell, H. B. (1931). Micro-operations on cells in tissue cultures. *Proc. Roy. Soc.* (*London*) **B109,** 380.

Evans, V. J., Bryant, J. C., McQuilkin, W. T., Fioramonti, M. C., Sanford, K. K., Westfall, B. B., and Earle, W. R. (1956). Studies of nutrient media for tissue cells *in vitro*. II. An improved protein-free chemically-defined medium for long-term cultivation of strain L-929 cells. *Cancer Res.* **16,** 87.

Fischer, A. (1930). "Gewebezüchtung." Müller & Steinicke, Munich.

Franks, D., Gurner, B. S., Coombs, R. R. A., and Stevenson, R. (1962). Results of tests for the species of origin of cell lines by means of the mixed agglutination reaction. *Exptl. Cell Res.* **28,** 608.

Grobstein, C. (1953). Epithelio-mesenchymal specificity in the morphogenesis of mouse sub-mandibular rudiments *in vitro*. *J. Exptl. Zool.* **124,** 383.

Gurdon, J. B. (1962). Adult frogs derived from the nuclei of single somatic cells. *Develop. Biol.* **4,** 256.

Gurdon, J. B. (1966). The cytoplasmic control of gene activity. *Endeavour* **25,** 95.

Hadži, J. (1949). Problem mezoderma in celoma v luči turbelarijske teorije knidarijev. *Razprave. Razreda za prirodoslovne in medicinske vede slovenske akademije znanosti in utmetnosti v. Ljubljani.* Knjiga IV.

Hallauer, C. (1931–1932). Uber das Verhalten von Hühnerpestvirus in der Gewebekultur. *Z. Hyg. Infektionskrankh.* **113,** 61.

Harris, H. (1967). The reactivation of the red cell nucleus. *J. Cell Sci.* **2,** 23.

Holtfreter, J. (1947). Changes of structure, and the kinetics of differentiating embryonic cells. *J. Morphol.* **80,** 57.

Hörstadius, S. (1950). "The Neural Crest." Oxford Univ. Press, London and New York.

Jacoby, F. (1937). The rate of cell division of hen monocytes *in vitro*. *J. Physiol.* (*London*) **90,** 23P.

Jacoby, F. (1957). Discussion on the future of tissue culture in relation to physiology. *J. Natl. Cancer Inst.* **19,** 653.

Jones, B. M. (1966). Invertebrate tissue and organ culture in cell research. *In* "Cells and Tissues in Culture" (E. N. Willmer, ed.), Vol. 3, p. 397. Academic Press, New York.

King, T. J., and Briggs, R. (1956). Serial transplantation of embryonic nuclei. *Cold Spring Harbor Symp. Quant. Biol.* **21,** 271.

Liebman, E. (1950). The leucocytes of *Arbacia punctulata. Biol. Bull.* **98,** 46.

Loeb, L. (1920). The movements of the amoebocytes and the experimental production of amoebocyte (cell-fibrin) tissue. *Wash. Univ. Studies Sci. Ser.* **8,** 3.

Medawar, P. B. (1937). A factor inhibiting the growth of mesenchyme. *Quart. J. Exptl. Physiol.* **27,** 147.

Medawar, P. B., Robinson, G. M., and Robinson, R. (1943). A synthetic differential growth inhibitor. *Nature* **151,** 195.

Moscona, A., and Moscona, H. (1952). The dissociation and aggregation of cells from organ rudiments of the early chick embryo. *J. Anat.* **86,** 287.

Murray, R. G. (1947). Pure cultures of rabbit thymus epithelium. *Am. J. Anat.* **81,** 369.

Parker, R. C. (1938). "Methods of Tissue Culture." Hamish Hamilton, London.

Ptashna, M. (1967). Specific binding of the λ phage repressor to λ DNA. *Nature* **214,** 232.

Rinaldini, L. M. J. (1958). The isolation of living cells from animal tissues. *Intern. Rev. Cytol.* **7,** 587.

Robinow, C. (1936). On the structure of epithelial membranes in tissue cultures. *Protoplasma* **27,** 86.

Ross, R., and Greenlee, T. K., Jr. (1966). Electron microscopy. Attachment sites between connective tissue cells. *Science* **153,** 997.

Thomas, J. A. (1937). La transformation des cellules en histiocytes. *Arch. Exptl. Zellforsch. Gewebezücht.* **19,** 300.

Chapter 7

CELL TYPES IN SPONGES AND THEIR ORIGIN

With the ideas and observations of the preceding chapters in mind, it will be instructive to examine the histology and cell behaviour in the Porifera and, in particular, in certain calcareous sponges, and then to study how these animals develop embryologically in order to see how the various types of cells differentiate in these primitive organisms. Before doing this, however, it may be desirable to justify the choice of this material rather than any other. This is especially necessary since it is probable that the sponges are on a sideline or off-shoot from that stream of evolutionary development which finally led to the emergence of the vertebrates.

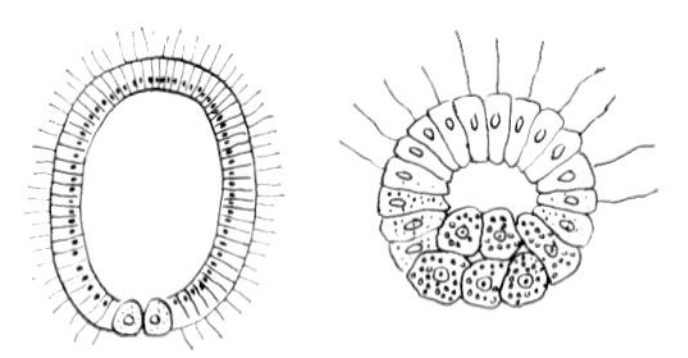

FIG. 7.1. Typical blastulae of calcareous sponges.

Sponges have certainly become highly specialized in their own way and to suit their own particular environment. Nevertheless, their histological and cytological structure, especially in the early embryonic stages, will be seen to have much in common with that of other animal groups. Furthermore, the calcareous sponges develop from relatively simple and uncomplicated blastulae (Fig. 7.1). Though sometimes formed rather differently (Fig. 7.2) perhaps because they develop within the maternal tissues, these blastulae appear to be similar in organization to those of many other organisms whose blastulae also appear to be simple and unspecialized.

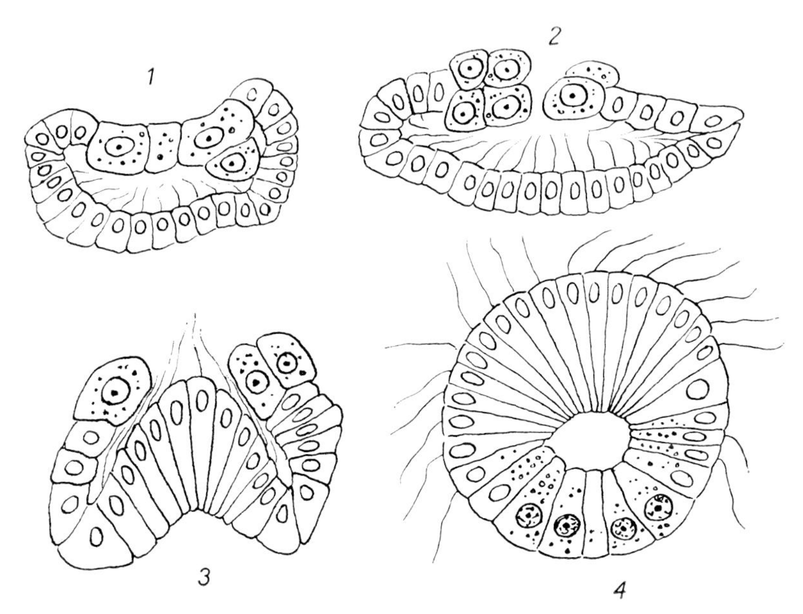

FIG. 7.2. Stages in the development of *Grantia compressa* leading to the formation of the free-swimming amphiblastula (4). Stages 1, 2, and 3 are passed within the parent tissues. (After Duboscq and Tuzet, 1937).

In these calcareous sponges the adult emerges from the embryo as the result of a rather complex metamorphosis of the larva after it has settled down on a suitable substratum. Nevertheless, the embryological origin of most of the cell types present in the adult is fairly well understood.

In its simplest form the adult sponge is essentially a vase-like structure which is lined with a continuous epithelium composed of collar cells or choanocytes (Figs. 7.3, 7.4). Each of these cells is provided with a long flagellum, and the flagellum arises from a basal granule at the bottom of an apparently membranous collar (Fig. 7.5). The latter appears to be composed of a ring of stereocilia (Rasmont *et al.*, 1958). The outer surface of the vase or olynthus is covered by a pavement epithelium of large cells. Every so often in this epithelium there are ring-like cells, each of which surrounds the entrance to a "pore canal" which connects the cavity of the vase with the world outside (Fig. 7.3). The beat of the flagella of the collar cells or choanocytes causes a current to enter through these pores and to emerge through the mouth of the vase, the osculum. Both the pores and the osculum can be modified in diameter, the former by contraction of the pore cells, the latter by contraction of epithelio-muscular cells surrounding it. The tissue composing the substance of the vase and lying between the epithelial layers has many of the properties of a simple connective tissue.

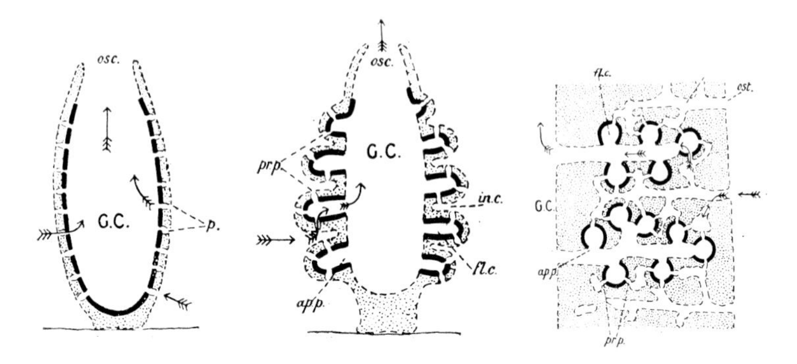

FIG. 7.3. Diagrams of the canal system of sponges. ap.p., Apopyle; fl.c., flagellated chamber; G.C., gastral cavity; in.c., in-current canals; osc., osculum; ost., ostium; p., pores; pr.p., prosopyle. Arrows indicate direction of flow of currents. Thick black represents the gastral layer (choanocytes). Dashed line represents dermal epithelium. (After Minchin, 1900.)

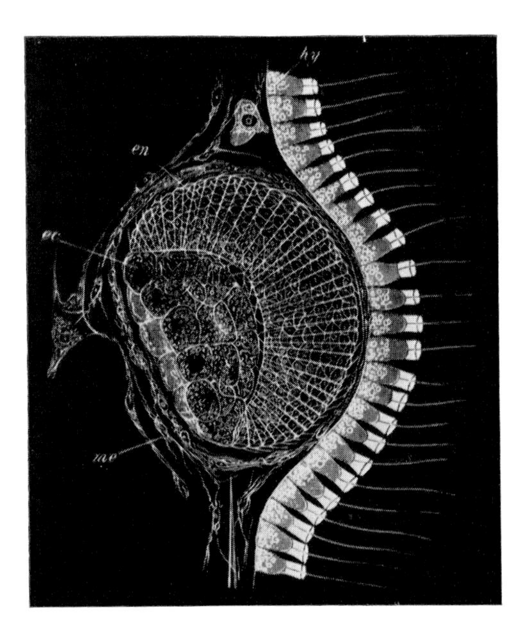

FIG. 7.4. Embryo of *Sycandra raphanus* enclosed within the maternal tissues. Note the two types of cell in the embryo and the layer of maternal choanocytes. ec, amoeboid cells; en, flagellate cells; hy, maternal choanocytes; me, maternal mesoblast. (Balfour, 1885; after Schultze.)

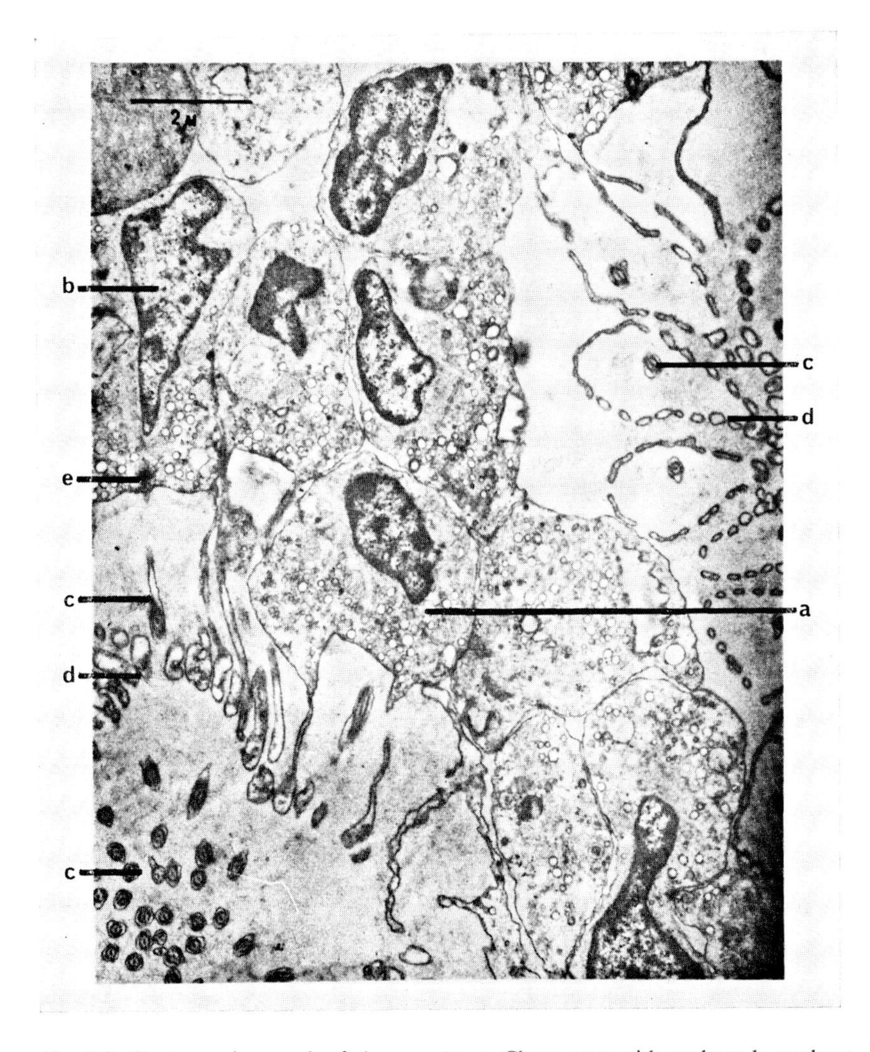

FIG. 7.5. Electron-micrograph of choanocytes. a, Choanocyte with nucleus; b, nucleus; c, flagellum; d, microvilli (stereocilia) forming the collar; e, basal granule. (Borojevic, 1966.)

It contains several sorts of cells and consists of fibres and an amorphous ground-substance (Fauré-Fremiet, 1931) (Fig. 7.6).

Such is the structure of the sponge in its simplest form. More commonly, however, the walls of the olynthus are thicker, and the pore canals may be expanded into chambers which are lined with choanocytes, and the activity of the cells in these flagellated chambers causes the passage of fluid through the olynthus. The structure of the wall of the sponge is clearly shown in

Figs. 7.6 and 7.7. Attention is drawn to the presence of spongin fibres, calcareous spicules, and to the occurrence of several classes of cells in the supporting tissues. In Fig. 7.7a the inner epithelium is seen to be very different from the outer epithelium. In the packing of its cells and in general behaviour, the inner one is similar to typical epithelia as seen in the columnar epithelia of the higher vertebrates and in tissue cultures of most epithelia. The outer one, however, differs from this; its cells are flatter and contain many perinuclear inclusions so that, like the endothelia and mesothelia of higher forms, it more closely resembles the "secondary epithelia". The contractility of the pore cells, too, is reminiscent of the behaviour of the

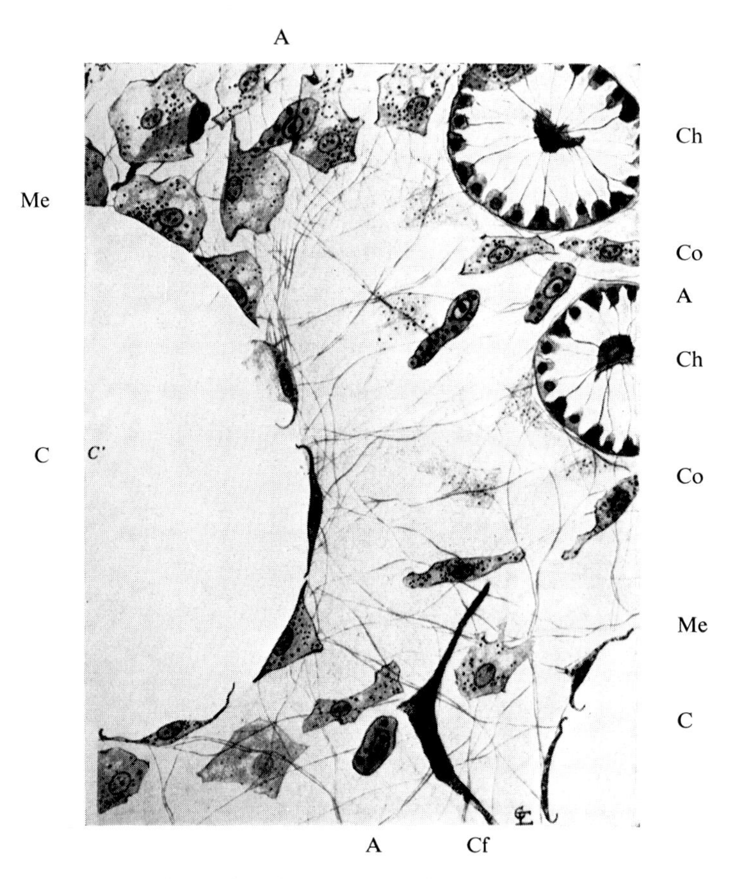

FIG. 7.6. Drawing of a section through the wall of a sponge (*Ficulina ficus*) showing flagellated chambers with choanocytes (Ch). Surface or dermal epithelial cells (Me) line the canals (C). Scleroblasts or collencytes (Co), amoebocytes (A), and fuchsinophil cells (Cf) occur in the substance of the wall. The groundwork of collagen fibres is clearly shown. (Fauré-Fremiet, 1931, Masson & Cie.)

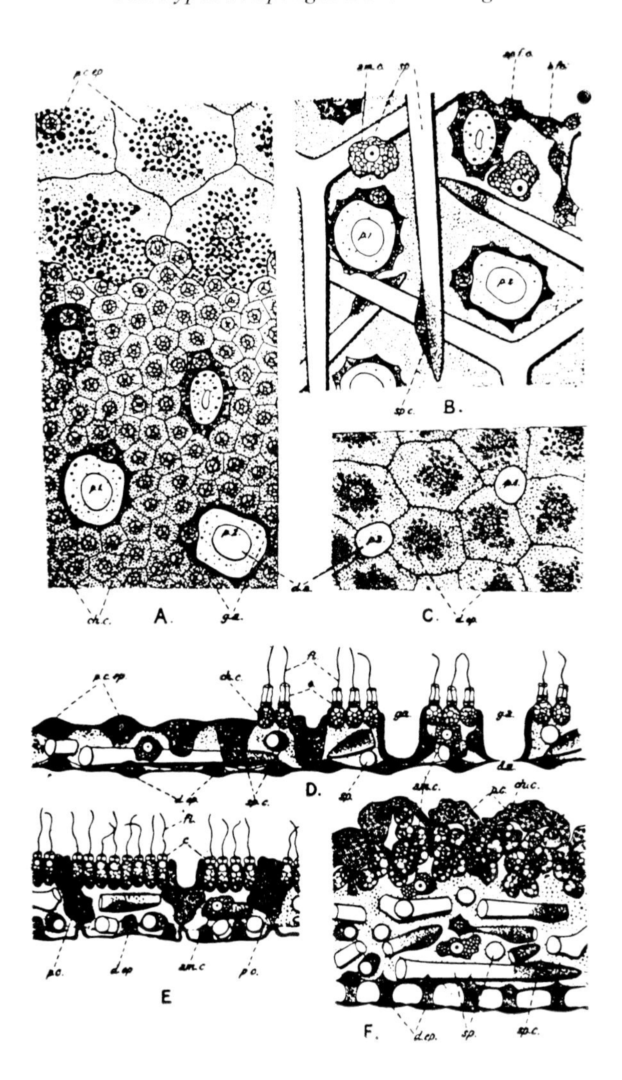

FIG. 7.7. Histology of the body wall of *Clathrina coriacea*, Mont. (*a*) Body wall seen from the inside, near the oscular rim. (*b*) The cells of the parenchyma in the same region. (*c*) The same region from the outside. (*d*) The body wall in longitudinal section, stretched. (*e*) The body wall in longitudinal section, slightly contracted. (*f*) The body wall in longitudinal section, very contracted. am.c., Amoebocytes; ap.f.c., apical formative cell; b.f.c. basal formative cell; c., collars; ch., choanocytes; d.a., dermal aperture of pore; d.ep., dermal epithelium; fl., flagella; g.a., gastral aperture of pore; p^1, p^2., pores; p.c., porocyte; p.c.ep., porocytic epithelium; sp., spicule; sp.c., spicule cell or scleroblast. (Minchin, 1900.)

endothelium of capillary vessels, and the pore cells are probably a modification of the cells of the outer or dermal epithelium. This is of special interest because the external epithelium of the adult sponge is not itself a primary epithelium. It emerges during the rather complicated larval metamorphosis as a reconstituted epithelium from cells which have been part of the "inner mass" of the larva. The inner epithelium, composed of choanocytes, on the other hand, remains a primary epithelium and is probably responsible for the nutrition of the animal (Minchin, 1900). Carmine particles are swept by the flagella into the cytoplasm of the choanocytes.

Between these two epithelia are found cells that belong to two main classes, namely a collection of different sorts of wandering cells or amoebocytes, and the scleroblasts and collencytes, the latter two being concerned with fibre formation and with the production of spicules. The wandering cells or amoebocytes may be specialized in one of several directions, and they often contain granules of one sort or another, e.g. storage or secretory products. The more mechanocyte-like cells, the scleroblasts and collencytes, are often associated with the spicules of calcium salts which strengthen the wall of the olynthus and frequently distort its surface by their continued growth. Associated with these structures may be a special spicule-cell, perhaps somewhat different from the more ubiquitous scleroblasts (collencytes). The latter are apparently responsible for laying down the connective tissue that lies between the epithelia and which has been shown to be composed of the same ingredients as the connective tissues of higher forms, namely collagen fibres (Fig. 7.8) and hexosamine-containing polysaccharides (Gross *et al.*, 1956).

An informative property of calcareous sponges is the ability of their relatively hardy cells to be easily separated. In consequence, the sponge can be forced through a fine sieve of bolting silk, so that isolated cells and small groups of cells can by this means be suspended in a fluid medium (e.g. sea-water for marine sponges; Wilson, 1907; Huxley, 1911). The activity of the various cell types can thus be clearly followed as the cells form an aggregation and begin to reconstitute new individuals. The choanocytes (primary epitheliocytes) generally lose their collars but swim about actively until they meet another of their kind; then they immediately join forces. As a result of this aggregation, groups of flagellated cells form ever-enlarging clusters that have the flagella all directed outwards. This adhesiveness and the subsequent manifestation of polarity are properties that are characteristically epithelial. Sometimes these cells adhere to the substratum; they may then show amoeboid processes at one end and a beating flagellum at the other. Wandering about on the substratum, on which the cells quickly settle, amoeboid cells of various sizes can also be observed, including some very large ones, which have many of the characters of the amoebocytes of the vertebrates. They wander about in apparently random fashion with pseudopodial processes emerging at any point, forming lamelliform membranes and branching processes. Each cell has a large nucleus, generally with a single nucleolus, and conspicuous perinuclear granules. Like amoebocytes, they show an ability to fuse to form multinucleate giant cells (Figs. 7.9, 7.10), and have a tendency to segregate

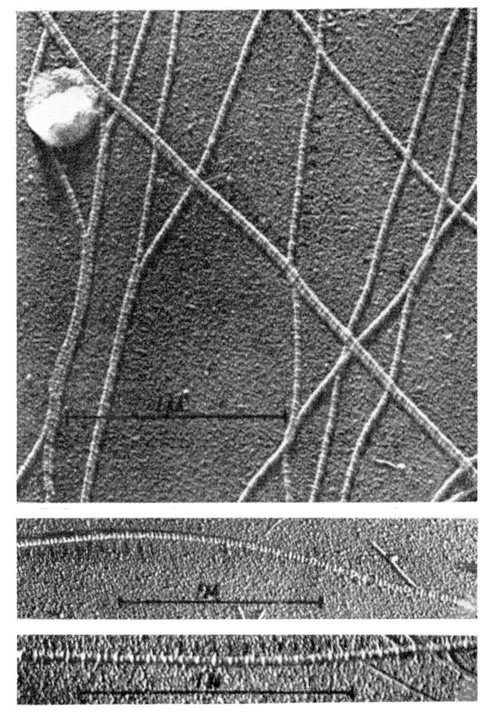

FIG. 7.8. Electron-micrograph of spongin A fibres from a sponge, showing 600–700 Å axial periodicity as in collagen, with 3 or 4 intra-period bands. (Gross *et al.*, 1956.)

trypan blue and similar particulate matter that is offered to them. The germ cells are derived from the amoebocytes, and the maturing ova are very large and actively amoeboid.

In addition to these two main classes of cells seen in the suspensions, there are a number of smaller and less granular cells which the cine-camera shows to be moving about with a gliding motion and to be much more directional in their movement than the amoebocytes. These are the scleroblasts, collencytes and spicule cells (see Figs. 7.6, 7.7); in this respect, they are similar to the mechanocytes of higher forms. They have not been definitely shown to be responsible for the formation of the fibres or ground-substance, but the similarity of their behaviour to that of mechanocytes is very suggestive, and the constitution of the matrix, as already mentioned, is basically similar to that of the connective tissues of higher forms. The collagen fibres, which, the electron-microscope shows, have the same cross-banded structure (see

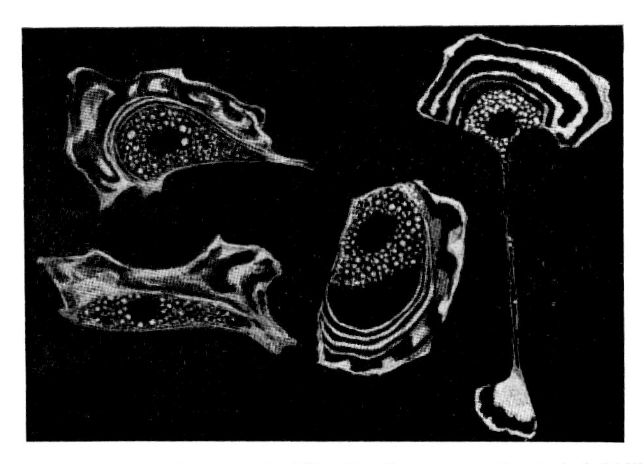

FIG. 7.9. Amoebocytes (archaeocytes) of *Ficulina ficus*, as seen by dark-field illumination. Compare these cells with those shown in Fig. 4.1. These cells segregate trypan blue. (Fauré-Fremiet, 1932, Masson & Cie.)

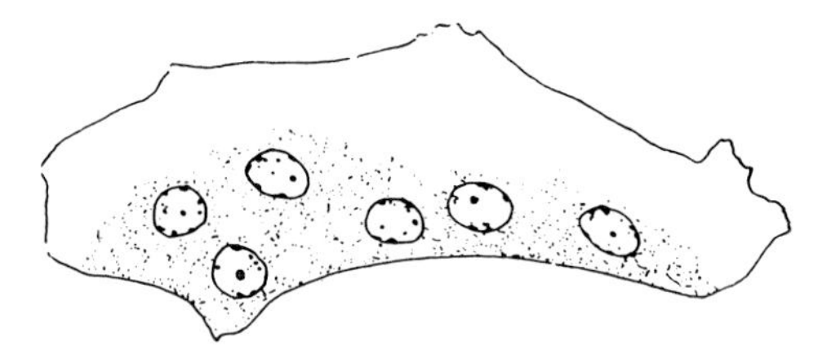

FIG. 7.10. A multinucleate giant cell formed by fusion of amoebocytes from the sponge *Ficulina ficus*. Compare Fig. 4.10. (Fauré-Fremiet, 1932, Masson & Cie.)

Fig. 7.8), are somewhat thinner than those of the connective tissues of the vertebrates, and the exact nature of the polysaccharides has not been determined beyond that they contain the same amino-sugars and carbohydrate constituents as those of the vertebrates. The scleroblasts are less abundant than might be expected in the cell suspensions which have been passed through the silk membrane, but this may simply be due to the fact that the ground-substance and the spicules are mostly held back by the silk and the cells may be held in the matrix. Tissue cultures of sponge tissues show an abundance of mechanocyte-like cells as well as amoebocytes (Fauré-Fremiet, 1932) (Fig. 7.11).

An interesting feature of sponge physiology which has considerable cytological importance has been revealed recently in *Sycon* by Lentz (1966). When subjected to the appropriate histochemical tests, the bipolar cells and the multipolar cells of the "mesenchyme" of this calcareous sponge both showed the presence of various "neurohumours" or of enzymes associated

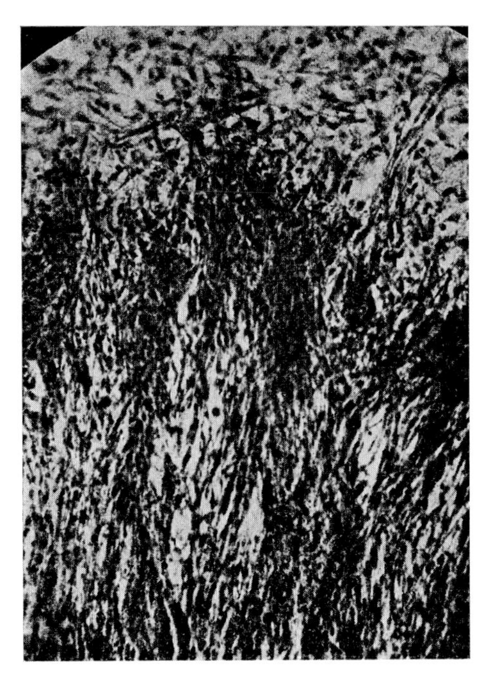

FIG. 7.11. Tissue culture of cells of *Ficulina ficus*. Note the mechanocyte type of growth in the lower half with more amoebocytic cells at the top. Compare with Fig. 2.1. (Fauré-Fremiet, 1932, Masson & Cie.)

with them. No very sharp distinction was found between the two classes of cells, but acetylcholinesterase and noradrenaline were more prominent in certain bipolar cells, whereas monoamine oxidase, adrenaline, and 5-hydroxytryptamine were more conspicuous in the multipolar cells. These observations, though not very easy to interpret, are consistent with the view that sponge cells while not sharply differentiated from each other, probably have activators and inhibitors similar to those in higher animals, though these agents are not yet so specifically related to special groups of cells as they become in higher animals. It is interesting too that, in general, the cells of the inner mass showed more reactivity than either the choanocytes or the dermal epithelial cells. It should also be noted that the distinction between cells only on the basis of bipolarity or multipolarity is not in itself very reliable.

It is clear from this brief account that in the calcareous sponge the same broad classes of cells already exist (epitheliocytes, mechanocytes, amoebocytes, and perhaps the beginning of neural cells) as are found when tissues of higher animals are cultured *in vitro*.

Cell Lineage in Sponges

Sponges, biologically complex and specialized organisms as they are, are structurally very simple as compared with vertebrates. Thus the differentiation of their cells into various types is likely to be less complete, less irreversible, and less overlaid with irrevocable specializations and differentiations than it is in more highly organized creatures. In other words, the differentiation into primary and secondary epithelia and into fibroblasts (mechanocytes) and amoebocytes is almost as far as the process has gone, and even these groups probably retain a measure of plasticity. It is true that there are various subtypes, e.g. thesocytes, fuchsinophil cells, and pore cells, but they can be considered comparatively simple variants (modulations) of the main types. Thus it seems probable that the classes of cells in higher animals, which are shown up so clearly by tissue-culture methods and in wound-healing, are really very archaic in their origin and may even exist throughout the whole kingdom of multicellular animals.

As already pointed out, as an object for experimental study the calcareous sponge has the advantages of possessing a fairly simple embryology, at least in the earlier stages, and of displaying, almost in diagrammatic form, a pattern which is also characteristic of a very large number of invertebrates. Thus in these sponges it is possible to trace the embryological origin of the various groups of cells. Minchin (1900) originally followed this lineage, and his results are summarised in Table 7.1.

The blastulae of such species as *Clathrina*, *Leucoselenia*, and their relatives are composed of closely adherent, flagellated cells occupying more or less of one "hemisphere", while less closely adherent phagocytic and non-flagellated cells occupy less or more of the other "hemisphere" (see Fig. 7.1). The phagocytic cells were termed archaeocytes, partly because they were observed to give rise to the germ cells and therefore were considered to be totipotent and primitive. They also give rise to the wandering amoebocytes and all the subtypes derived from them. The anterior, flagellated cells were observed to give rise to the inner lining of choanocytes, as might be expected. In addition it was observed that cells also migrated inwards from these

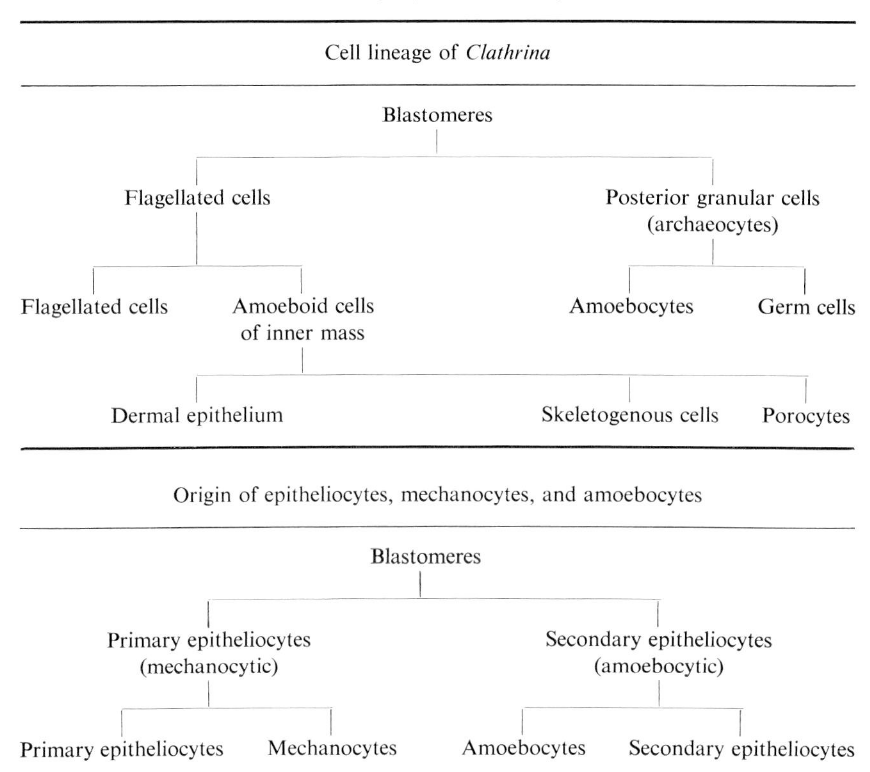

TABLE 7.1

Comparison between the Cell Lineage in Sponges and the Origin of Epitheliocytes, Mechanocytes, and Amoebocytes

cells into the blastula cavity and gave rise to the cells that we have classed as mechanocytes and which in turn gave rise not only to the fibre-forming cells (scleroblasts) but also to the secondary epithelial cells of the outer or dermal layer, including the contractile pore cells.

An interesting feature of the flagellated choanocytes is that, when they become functional as digestive cells, they develop their collars of long microvilli (Fig. 7.5), and then are capable of engulfing particles brought to them by the action of the flagella. Thus on these differentiated cells two parts of the surface display different sorts of activity; this is a phenomenon of cardinal importance and one which must be borne in mind in all discussions of cell behaviour, particularly in differentiated tissues.

A slightly different lineage in a siliceous sponge has recently been suggested by Borojevic (1966), but it leads to the same result in the end. In this sponge the collencytes (mechanocytes) are derived not from the choanocytes but directly from the blastomeres or from archaeocytes which, as in the calcareous sponges, maintain their totipotency and also develop into the germ cells. The two different lineages can be seen in Table 7.1 and Fig. 7.12. They can be reconciled on the basis of different timing of the two differentiative steps,

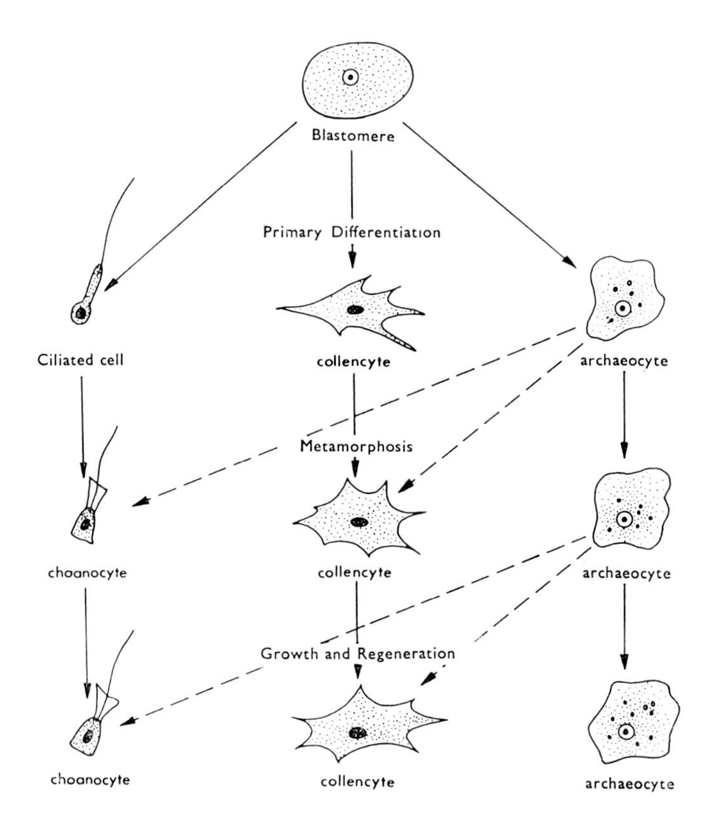

FIG. 7.12. Cell lineage and potentialities in the sponge *Mycale contarenii* (Martens). (Borojevic, 1966.)

the second step, leading to the formation of the mechanocyte, having been advanced in time in the siliceous sponge. The amoebocyte may also be more versatile than Minchin supposed.

One factor, however, which may have to be taken into account in these two lineages is that the formation of calcareous spicules may be dependent on different properties in the formative cells from those required for siliceous spicules. Thus the spicule cell of the siliceous sponge may be somewhat different from the corresponding cell in the calcareous sponge.

If the cell lineage as described for Calcarea is valid and applicable to other invertebrates in general, (and there are no reasons for supposing that this is not so), then it is clear that the fundamental difference between fibroblasts and macrophages, or more generally between mechanocytes and amoebocytes, is already established or becoming established in organisms as simple and primitive as the embryos of calcareous sponges. Moreover, if the situation in the calcareous sponge is indeed more primitive than that in the siliceous sponge, then the essential difference is probably to be sought in the initial difference between the flagellated cells and the archaeocytes.

These cells form two types of essentially epithelial cells in that they adhere to each other, though probably by different methods, to form the blastula, and one of them gives rise to mechanocytes, while the other produces amoebocytes. Attention must therefore now be focussed on the differences between the ciliated and flagellated type of cell, on the one hand, and those cells which display extensive pinocytosis and phagocytosis, on the other. Because cells with all these characteristics exist as separate organisms among the protozoa (e.g. as ciliates, flagellates, and amoebae), we ought to begin our study of differentiation in this phylum if it is to become meaningful and something more than merely descriptive.

In the pages that follow, an attempt is made to investigate the principles underlying the process of differentiation of cell types, not as it occurs in any one embryo, but as it may have occurred throughout the ages, by slow evolutionary steps, with each succeeding generation evolving from the preceding one. It is our purpose to examine the processes by which the various forms of cell and tissue, which are found in the evolutionary series of animals starting from unicellular organisms and leading to and including the vertebrates and man, have gradually come to be what they are and to function in the various ways that they do.

For obvious reasons the accent is always placed on that particular line of evolution which seems most likely to have led to the vertebrates and to man. Only passing mention is made of the points at which the evolutionary process has apparently diverged and at which other groups of animals have begun to evolve separately and along their own special lines. The early steps, or variations of them, may have been common to many phyla, but gradually the physiological processes and specializations in each group became more and more different; thus the patterns of cell behaviour change characteristically in each phylum and diverge more and more from each other.

Tissue culture, by virtue of its property of generalizing cell behaviour, has established the existence of certain basic patterns, perhaps even basic cell types. Although these behaviour patterns have been studied mostly in the vertebrates, we have seen that they were probably established at a very much earlier evolutionary stage, and some of the main traits are in consequence very widespread in the animal kingdom. This observation gives us confidence that the study of the phylogenetic origin of these generalized "tissue-culture" types can throw light on the basic functions and organization of primitive organisms, then it may be that, from these primitive beginnings, the various further specializations can be followed one by one as the evolutionary story unfolds.

References

Balfour, F. M. (1885). "Comparative Embryology", Vol. 1. Macmillan, London.

Borojevic, R. (1966). Étude expérimentale de la differenciation des cellules de l'éponge au cours de son développement. *Develop. Biol.* **14**, 130.

Duboscq, O., and Tuzet, O. (1937). L'ovogénèse, la fécondation et les premiers stades du développement des éponges calcaires. *Arch. Zool. Exptl. Gen.* **79**, 158.

Fauré-Fremiet, M. E. (1931). Étude histologique de *Ficulina ficus* L (Demospongia). *Arch. Anat. Microscop. Morphol. Exptl.* **27**, 421.

Fauré-Fremiet, M. E. (1932). Morphogénèse expérimentale (reconstitution) chez *Ficulina ficus* L. *Arch. Anat. Microscop. Morphol. Exptl.* **28**, 1.

Gross, J., Sokal, Z., and Rougvie, M. (1956). Structural and chemical studies on the connective tissue of marine sponges. *J. Histochem. Cytochem.* **4**, 227.

Huxley, J. S. (1911). Some phenomena of regeneration in *Sycon:* With a note on the structure of its collar cells. *Phil. Trans. Roy. Soc. London* **B202**, 165.

Lentz, T. L. (1966). Histochemical localization of neurohumours in a sponge. *J. Exptl. Zool.* **162**, 171.

Minchin, E. A. (1900). The Porifera. In "A Treatise on Zoology" (E. R. Lankester, ed.), Vol. 2. Black, London.

Rasmont, R., Bouillon, J., Castiaux, P., and Vandermersche, G. (1958). Ultrastructure of the choanocyte collar-cells in fresh-water sponges. *Nature* **181**, 58.

Wilson, H. V. (1907). On some phenomena of coalescence and regeneration in sponges. *J. Exptl. Zool.* **5**, 245.

Chapter 8

THE NAEGLERIOID STAGE

It has been suggested that the two main groups of cells in the sponge embryo, the flagellated and the phagocytic, give rise to the mechanocytes and the amoebocytes, respectively, or at least to cells displaying predominantly mechanocytic or amoebocytic behaviour. If this is true, the essential difference between these two great classes of cells may lie in some features connected with the activity of flagella or cilia, on the one hand, and phagocytic and pinocytic behaviour, on the other. It may be profitable therefore to study those protozoa which during their normal life history can change from being creeping, phagocytic, and amoeboid organisms to become free-swimming flagellates which then may (or may not) feed by methods other than phagocytosis. In addition to the many protozoa which have a sexual phase, in which there are flagellated gametes, there are also some species whose members can reversibly assume the flagellated or the amoeboid and phagocytic form. Several organisms of this type have been investigated in some detail, and the results of these investigations into their physiology are very suggestive.

Histomonas meleagridis (Fig. 8.1) has been investigated by Tyzzer (1934) and Wenrich (1943). It is an amoeboid organism found in the liver of turkeys and is responsible for the disease known as "blackhead". It also exists as a flagellate organism, with a varying number of flagella, in the caecum of pheasants or of the domestic fowl, and it can assume the same form under certain conditions of culture. The detailed cause of this change of form is not known, but the liver of the turkey and the caecum of pheasants clearly provide very different environments to which the organism presumably makes the appropriate adaptations.

Tetramitus rostratus (Fig. 8.2) can exist as a free-living flagellate which, again under some conditions which are as yet ill defined, can lose its flagella

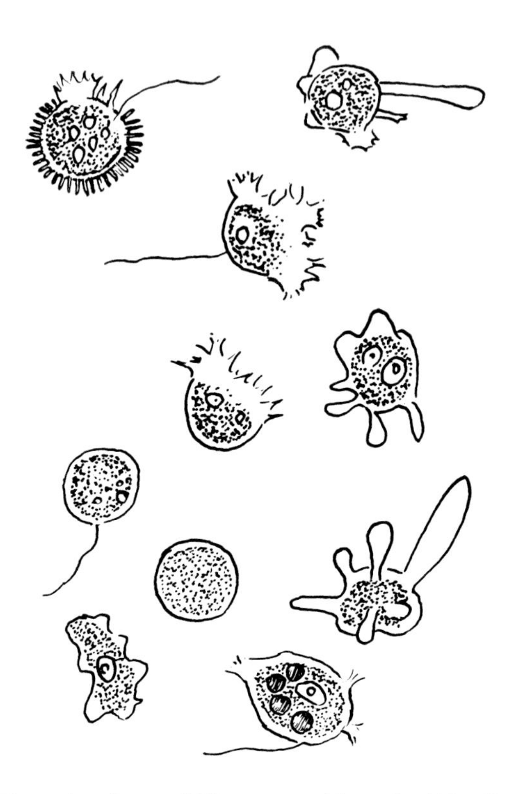

FIG. 8.1. The various forms of *Histomonas meleagridis*. (After Tyzzer, 1934.)

and can temporarily become typically amoeboid (Bunting, 1926; Balamuth and Outka, 1962). As the amoeboid form, but not the flagellate form, can be cultured axenically, presumably there are distinct metabolic differences between the two forms, or differences in the permeability of their surfaces.

A third dimorphic organism, *Naegleria gruberi* (Fig. 8.3), was first described by Schardinger (1899) and has been studied intermittently ever since. It is normally amoeboid and lives in the soil. It is without definite polarity and feeds on bacteria, with which it can be easily cultured on an agar surface. It has not been possible so far to culture *Naegleria* on a synthetic medium. If water is added to a culture of these organisms, or if they are transferred to water, they immediately become polarized, in that lobose pseudopodia continue to form anteriorly while a uroid or tail region develops at the other end. The tail region then contains the contractile vacuole and tends to leave trailing cytoplasmic processes, liberating mucus, behind it as the amoeba creeps about. Bacteria are caught up in these trailing mucoid threads. Depending on the ambient temperature, the cell, after about 30

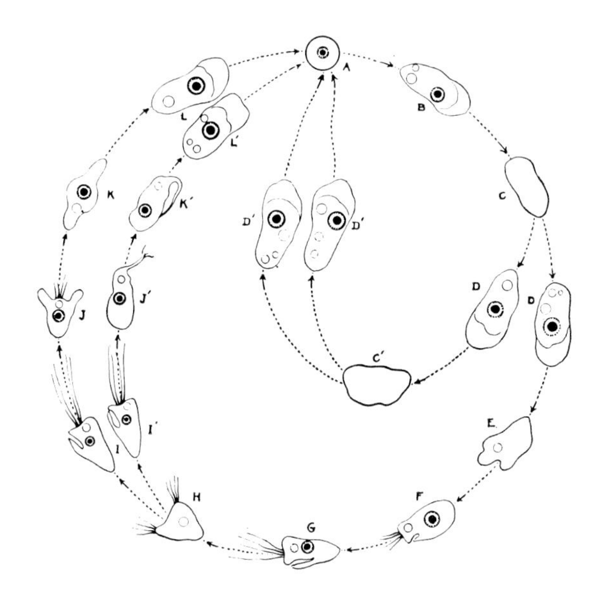

FIG. 8.2. The life cycle of *Tetramitus rostratus*. A, cyst; B, vegetative amoebae; C, C′, division; D, D′, amoebae after division; E, F, phases in transformation to flagellate; G, flagellate; H, flagellate before division; I, I′, flagellate after division; J, K, L, and J′, K′, L′, series of transformation stages from flagellate to amoeba. (From Bunting, 1926.)

minutes, has developed one or more flagella beating from the posterior end, has become round or oval in shape, and has lost, or almost lost, the power to put forward pseudopodia.

Some doubt has been cast on this development of polarity (Pittam, 1963), and it has been suggested that the polarized state occurs only when the cells are transforming from flagellate to amoeba, but film records of cells creeping between coverslips show the sequence described here. When rounded off, the cells appear turgid, as if they were enclosed in a much tougher membrane than those in the amoeboid form. The cells become relatively free from the substratum and start to rotate and spin round their remaining point of attachment to the substratum. During this time the point of origin of the flagellum moves from a position near the contractile vacuole at the posterior end to lie at the opposite, or anterior, end. The beating of the flagella, which are tractella, then accelerates, and the spinning movements increase rapidly so that the organism soon detaches itself from the substratum and swims away, the flagella leading and the contractile vacuole in the stern. If a little sodium chloride is added to the medium at this stage the flagella are rapidly lost, the organism settles down again on the substratum and immediately throws out lobose pseudopodia and returns to its normal feeding habits.

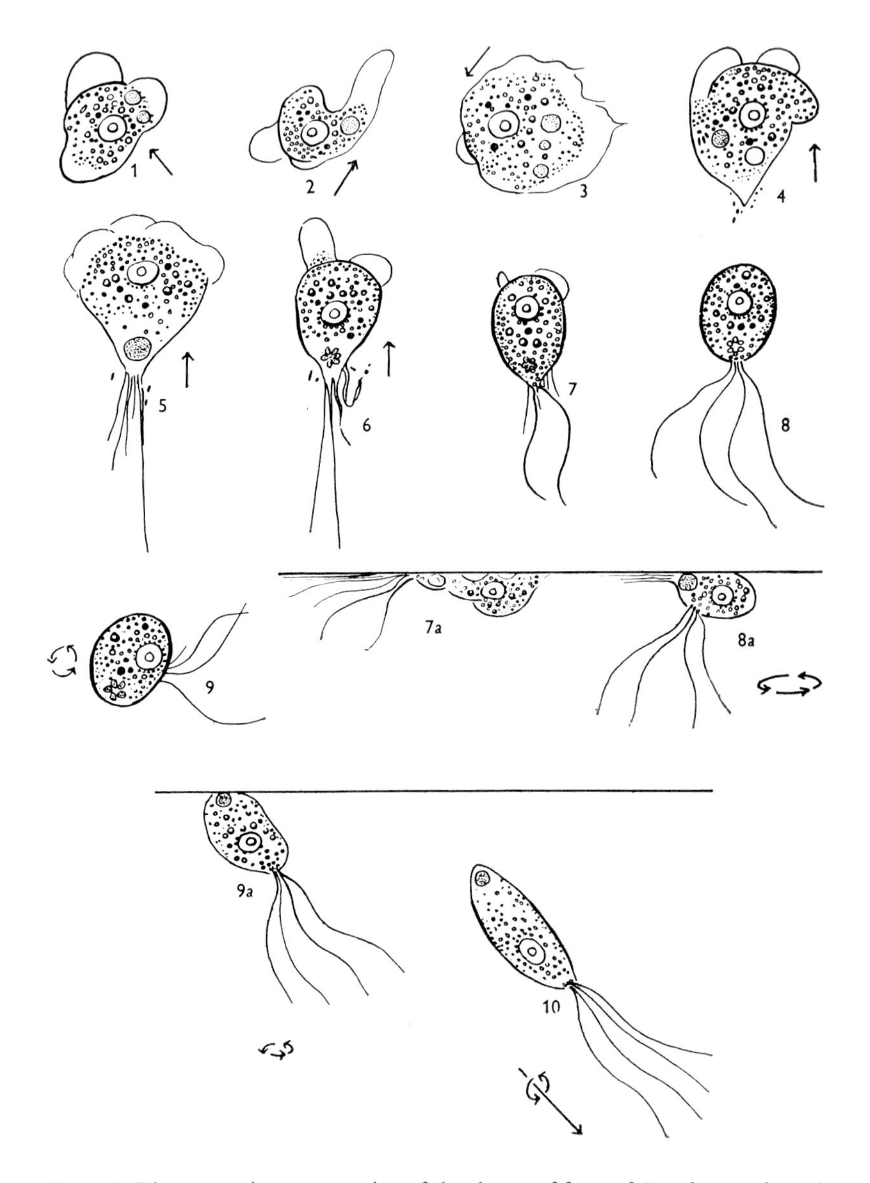

FIG. 8.3. Diagrammatic representation of the change of form of *Naegleria gruberi* when the organism is placed in distilled water. The arrows indicate the direction of motion. 1–3, amoeboid form; 4, polarized form; 5, 6, filiform pseudopodia present at posterior pole; 7–10, development of active flagella; 7a–9a, schematic diagrams of the way events would appear if seen from the side when the amoeba leaves the surface of the coverslip, In stages 4–10 the contractile vacuole, in its different phases, is situated at the posterior pole. (Willmer, 1956.)

A similar environmental change produces similar but less drastic alterations in the morphology of the ciliate, *Espejoia mucicola*. When normally feeding in the presence of mucin this organism has a large and complicated "mouth", but when placed in distilled water it loses all these structures and becomes a very active migratory form with only a trace of the original mouth region (Fauré-Fremiet and Mugard, 1949) (Fig. 8.4). Lewin (1953) has also shown

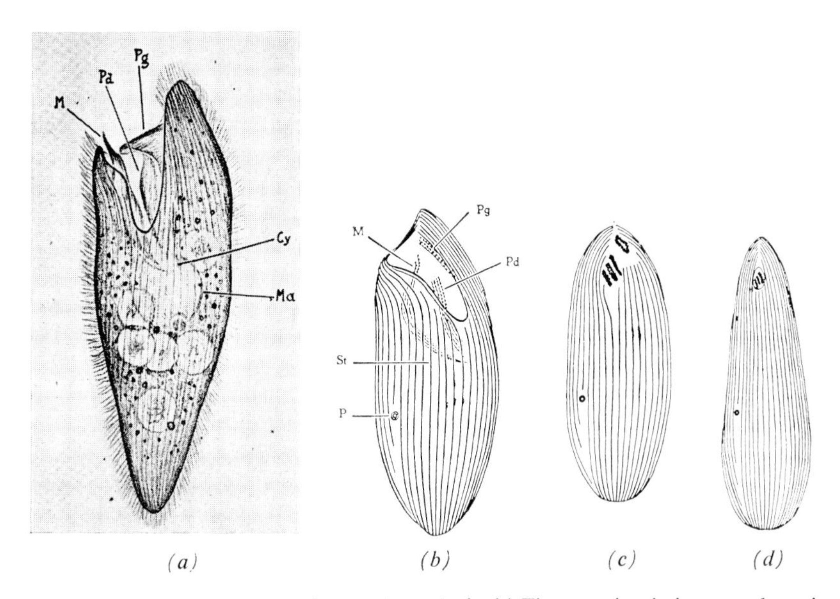

FIG. 8.4. The different forms of *Espejoia mucicola*. (*a*) The organism in its normal mucin-containing environment with fully developed organelles; (*b*)–(*d*) progressive simplifications of form in distilled water as they appear after silver impregnation. Cy, Cytostome; M, quadripartite membrane; Ma, macronucleus; P, pore of contractile vacuole; Pd, dorsal peniculus; Pg, left peniculus; St, ciliated bands. (After Fauré-Fremiet and Mugard, 1949.)

that in the alga, *Chlamydomonas moewusii*, the formation of flagella is induced by water.

The formation of the flagellum and basal body in *Naegleria* seems to take place entirely *de novo* (Fig. 8.5), and the final structure has the 9+2 pattern of internal fibres and the usual type of basal granule which is attached by well developed root-fibres to the nucleus (Fig. 8.6). There is no trace of the flagellum in the amoeboid phase (Schuster, 1963; Dingle and Fulton, 1966).

Tetramitus has a gullet (cytostomal groove) and feeds and reproduces while it is in the flagellate form, but there is no evidence that *Naegleria* can do likewise. In *Naegleria* the flagellate form appears to be adopted as

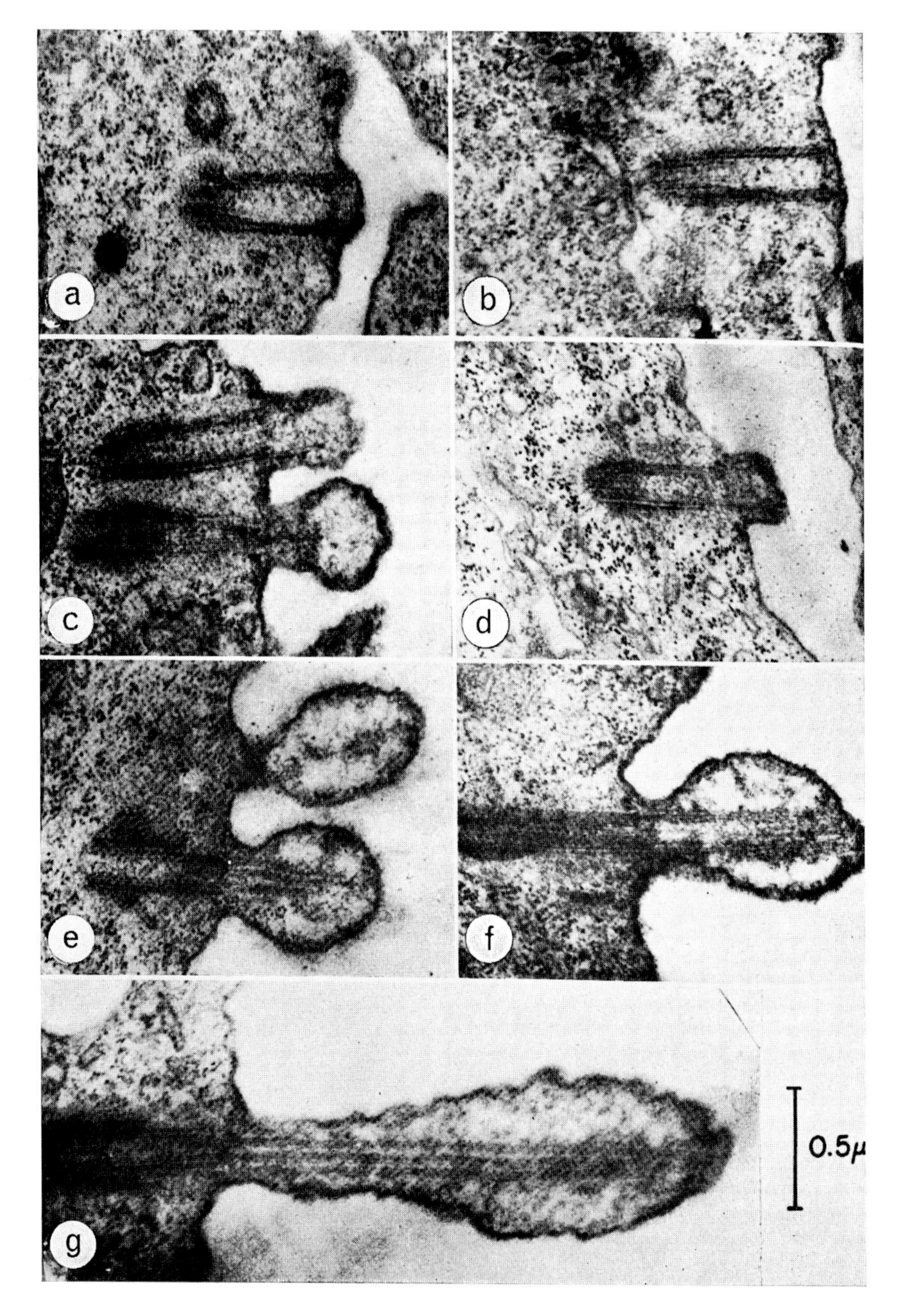

FIG. 8.5. Electron-micrographs of progressive stages (a-g) in the development of the flagellum in *Naegleria*. (Dingle and Fulton, 1966.)

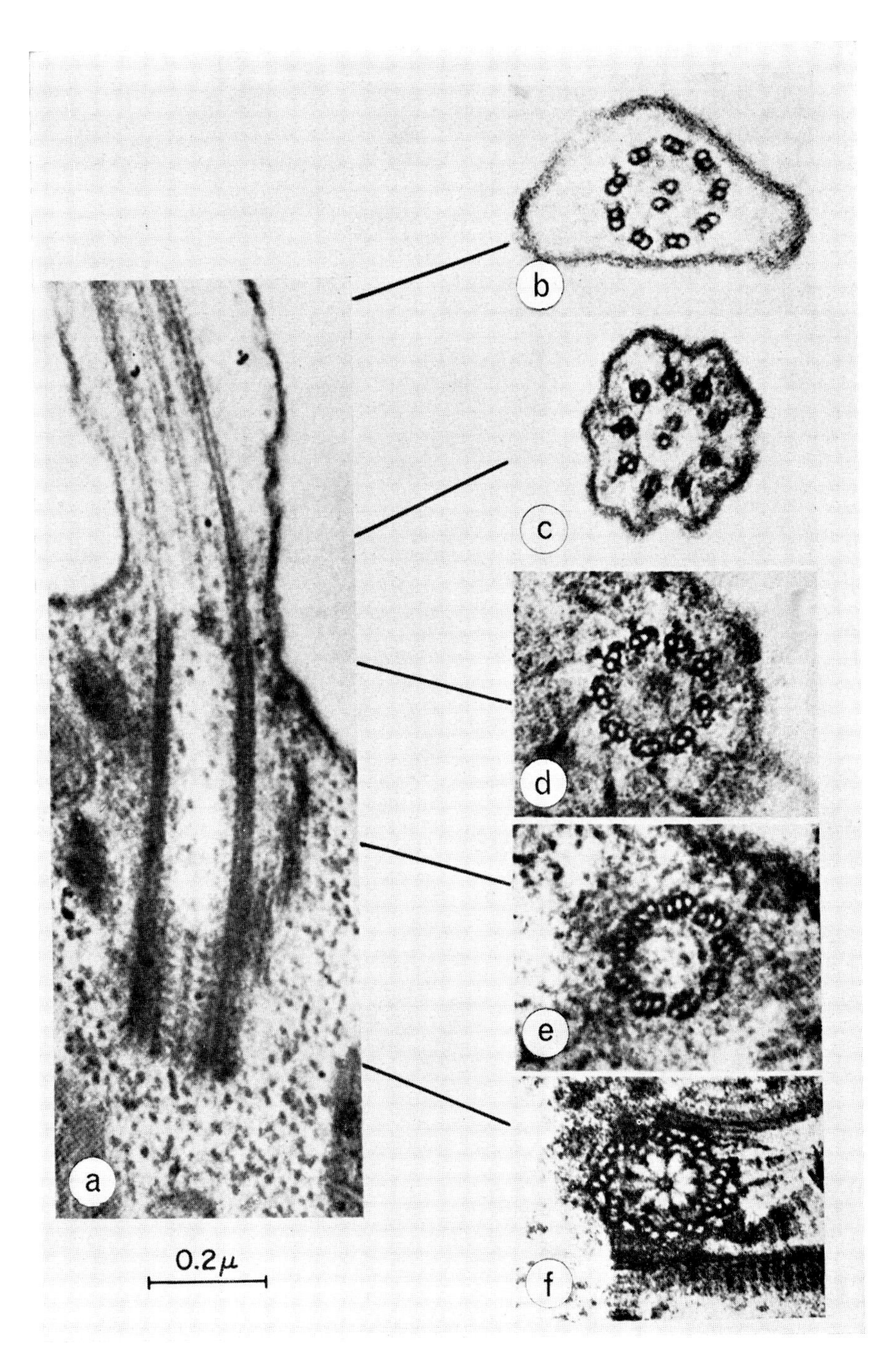

FIG. 8.6. Electron-micrographs of the fully developed flagellum of *Naegleria* showing the characteristic pattern of fibres. a, longitudinal section; b–f, transverse sections at levels indicated. (Dingle and Fulton, 1966.)

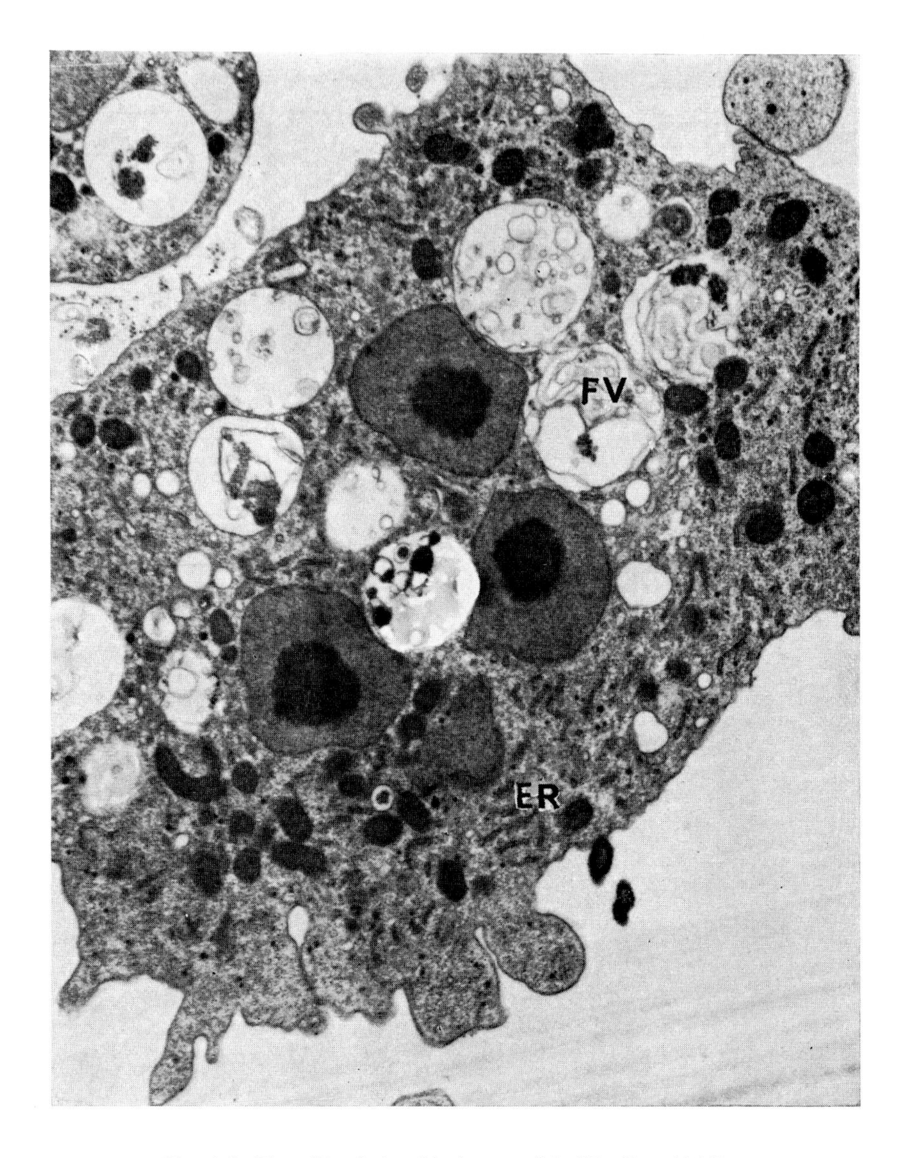

FIG. 8.7. Giant *Naegleria* with three nuclei. (Hartline, 1966.)
ER, endoplasmic reticulum; FV, food vacuole.

a temporary measure and, unlike *Tetramitus*, *Naegleria* does not multiply freely, if it does at all, in the flagellate form. Since *Naegleria* is normally an inhabitant of the soil, it seems likely that the flagellate phase is, in fact, a specialization and an adaptation on the part of the organism which, perhaps, has occurred as a means of extending the range of the creature when moist conditions favour it.

Under adverse conditions, *Naegleria* can also encyst within a tough capsule and become dormant for long periods. But perhaps more relevant to the present discussion of the possible connection between *Naegleria* and sponges, on the one hand, and between sponges and other metazoa, on the other, is the observation (Hartline, 1966) that at relatively high temperatures (30°C) *Naegleria* can form giant cells, like the amoebocytes of higher forms (Fig. 8.7).

Another very interesting aspect of the behaviour of *Naegleria*, and one which is also relevant to the present discussion, is that while in their normal state of activity, either as amoebae or flagellates, the individuals remain entirely independent of each other; under other conditions, e.g. when food is scarce on a relatively dry surface of agar, the amoebae may lie close together like a sheet of cells, and take up a distinctly epithelial or colonial sort of arrangement (Figs. 8.8, 8.9). Under these conditions the organisms do not appear to be rigidly attached to each other; they behave similarly to epithelial

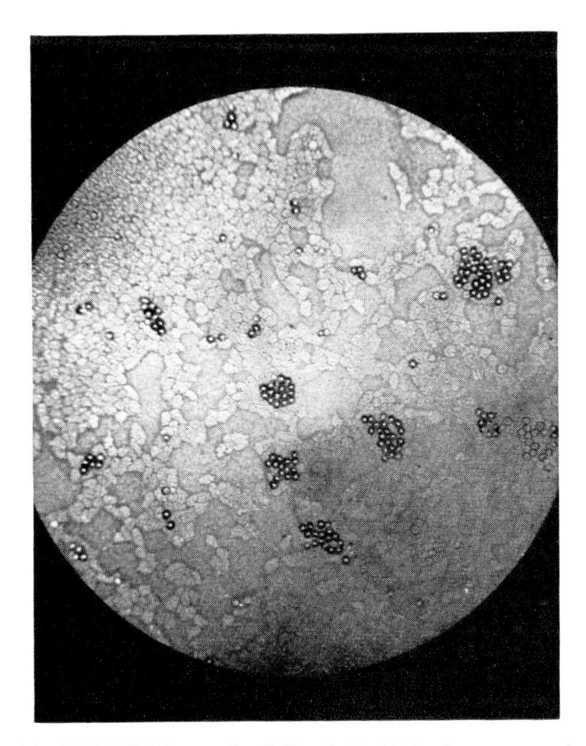

FIG. 8.8. "Epithelial" growth of *Naegleria*. Note also groups of cysts.

FIG. 8.9. "Epithelial" growth of *Naegleria*, under higher magnification. Each amoeba is about 14μ in diameter.

cells in tissue culture. They are in constant slight movement around each other, but they never completely lose contact with each other and the intercellular spaces are uniformly small.

An important feature of the behaviour of both *Naegleria* and *Tetramitus* in transforming from one phase to the other concerns the polarity of the cell. In both cases the flagella act as tractella, and there is some reorganization of the polarity during the change of phase. This problem of polarity is clearly fundamental and must have an important bearing on the problem of orientation of epithelial cells in such a system as the blastula of the sponge (see p. 187), in which the directional activities of the cells must play a large part in determining the stability of the organism. In *Naegleria* itself the flagella form at the posterior pole and then migrate to the anterior end, but the cell still moves with its contractile vacuole at the posterior end, i.e. in its original position. Because in *Tetramitus* the contractile vacuole migrates to the anterior end before the flagella and mouth appear at that end (Bunting, 1926; Hollande, 1942), the reversal appears in one sense to be more radical. Both cells continue to move with their original anterior ends still leading. Perhaps for reasons connected with these reorganizations, *Tetramitus* continues to divide freely in the flagellate phase while *Naegleria* apparently does not. The relationship between the centrioles and the basal bodies of the flagella would perhaps be worth further investigation.

Naegleria is thus a very relevant organism in connexion with the present discussion, since it can exist in either of the two forms which are characteristic of the cells of the sponge blastula; either as a strongly polarized and flagellated cell which is not phagocytic, or as a much less polarized cell which is both amoeboid and phagocytic. Furthermore, in the amoeboid form at least, *Naegleria* can display the property of cohesion which is a necessary property of the cells of the sponge embryo and indeed of all epithelia. The flagellated form of *Naegleria*, however, seems to lack this power of cohesion and thus differs in an important manner from the flagellated cells of the sponge embryos. The nature of the cellular cohesion within the sponges and in *Naegleria* is still a matter for investigation, and the method of cohesion of the amoeboid cells may well be different in kind from that between the flagellated cells. In the epithelia of higher organisms, cell cohesion may depend on specialized structures like desmosomes, tight junctions, etc., or it may simply rely on extensive folding and interlocking of the cell margins, whereby the cells are locked together like a three-dimensional jig-saw puzzle. Desmosomes and terminal bars certainly occur in the blastulae of some invertebrates, e.g. in those of echinoderms (Balinsky, 1959), though how extensively these forms of cohesion are used is not known. One suspects from electron-microscopic pictures of other aggregating amoebae (Band, 1966) that the amoeboid form of *Naegleria* coheres by interdigitations and that special types of desmosome, etc., have not developed in this organism.

It will be remembered that, if the collar cells of the sponge are suspended in water, they lose their collars but maintain their flagella and swim around in much the same way as does the flagellate *Naegleria*, but, when they make contact with each other, unlike *Naegleria*, they stick together, so that cell aggregates are quickly formed. The development of cohesive mechanisms between flagellate cells and of the ability of the cells to recognize their own kind may well have been a decisive evolutionary step. Cohesion certainly occurs in several colony-forming protozoa, e.g. *Gonium*, and perhaps in the choanoflagellates, e.g. *Codonosiga botrytis* (Fig. 8.10), but possibly not by the same mechanism as in blastulae,

It is extremely unlikely that *Naegleria* itself ever had any direct part to play in the evolution of the sponges, but it can nevertheless act as a model for a hypothetical stage of organization which had the features necessary for the formation of a colonial multicellular organism much like the blastula

In the preceding paragraph it was suggested that any one particular protein might only appear, at least in any quantity, in some of the cells of an organism. Haemoglobin, for example, is present in the erythrocytes but is not normally detectable in other cells; similarly, myoglobin appears to be confined to certain classes of muscles. With one or two exceptions, neither the ovum nor the early blastomeres of vertebrates display either of these proteins (at any rate, in significant amounts). However, the ovum and at least some of the blastomeres must certainly be capable of handing on the necessary coded information because some of their offspring produce, not only haemoglobin, but also the right haemoglobin, i.e. the haemoglobin which is characteristic of the species, for there are many haemoglobins which differ in the make-up of their globins. It follows that the fertilized ova carry a great deal of genetic information which they themselves do not directly use, but which must be handed on intact to the blastomeres and to all those derivative cells which under some conditions can be shown to have properties characteristic of that species and which are dependent on the genetic information transmitted in the fertilized ovum. All cells, including the fertilized ova, therefore probably carry a very large amount of suppressed or inactive genetic information. This conclusion applies even to the most differentiated and specialized somatic cells, though it is possible that during differentiation and development the somatic cells may progressively lose some of the information which they do not themselves use. Nevertheless, even in this connexion, it is salutary to remember that, from a single cell extracted from a culture of the callus tissue of a carrot, a whole new carrot plant has been grown (Steward *et al.*, 1958a, b). Not much, therefore, can have been lost from this single somatic cell. Much the same must apply to those cells which are concerned with the regeneration of animals, as, for example, whole planaria from small fragments. It is difficult to see how any genetic information can be lost from the cells from which the regeneration occurs. It must be remembered, too, that whole frogs have resulted from the injection of nuclei isolated from the cells of the tadpole's gut into enucleated frog's eggs (Gurdon, 1962).

The problem of this latent or suppressed genetic information is one of the major problems not only in differentiation but also in the interpretation of evolutionary changes. For example, vertebrates normally possess haemoglobin in their blood, and adult eels are no exception. The leptocephalus larvae from which these eels develop are entirely devoid of haemoglobin, however. Certain antarctic fish, e.g. *Chaenocephalus aceratus*, never have haemoglobin in their blood (Ruud, 1954), and there are cases on record in which adult individuals of *Xenopus laevis* possess no blood pigment, though this condition may be pathological (Ewer, 1959; de Graaf, 1957). What happens to the genetic code for making haemoglobin in such situations?

Is it merely suppressed so that its action remains latent but capable of being called out again should the "suppressor" be removed, as seems likely in the eel, or can it drop out of the system altogether? If the latter, should some future descendent have need of haemoglobin, the only way in which it could get it would be to wait until the necessary base sequence occurred again as a mutation in the genome and in a part of the code which was not suppressed. It is interesting to consider what the consequences might be if, by some form of neoteny, new species or even genera were to arise from leptocephalus larvae. If the production of haemoglobin is merely suppressed in them, then haemoglobin might turn up quite sporadically in any members of the new genera, and would be quite likely to do so. If, on the other hand, the code for haemoglobin was lost, the pigment would be relatively unlikely to reappear.

It is tempting to think that no part of the code would normally be lost unless the results of its presence were directly or indirectly harmful or lethal to the animals possessing that part. Absence of selection pressure could lead, bit by bit, to random modification of the genome, but, since many of these modifications would, more likely than not, lead to other harmful effects, they would tend to be suppressed and would not gain ground or become permanent in the genome. Thus the stability of the genome determining the production of useful, or at least harmless, proteins would be to some extent assured. Moreover, since, from an evolutionary point of view, suppression can be as effective as absence and much less irrevocable, it is not unreasonable to suppose that genetic information, once acquired, is not readily lost, but is normally handed on from one generation to the next in either an active or a suppressed form. In both forms it may be subject to random mutation, though the effects may be different. This hypothesis has far-reaching implications in the evolutionary history of organs and structures and may sometimes account for the apparently capricious occurrence of what seem to be similar characters in animals not normally regarded as close relatives.

To the idea of permanence of the genetic information it may be objected that animals living in caves tend to lose their sight and their pigmentation, and that there are many examples of atrophy of structures related to their disuse. It must be remembered, however, that selection is necessary not only to improve and evolve but also to maintain such complex functional systems. Otherwise, random mutations or recombinations may occur and lead to deterioration. Thus, in the example above, unless the eye was needed as a functional unit, it could easily become inefficient or might cease to develop beyond a certain stage without the animal being any the worse for the change. This would not, however, mean that that animal had lost all the genes that contribute to the making of a functional eye. Some could be lost, probably by change into some other pattern which might or might not be useful; others would simply become ineffective as the result of change or of the continuation of the suppression which must occur in any case during early embryonic stages. One or two false steps in early development could lead to atrophic eyes and thence to almost any degree of reduction; but most of the genetic information necessary to produce a normal eye could still be there, either in a suppressed form or in a form so altered as to be useless in the new situation.

If the absence of a gene confers some direct benefit on the individual in which it has occurred, then the gene may be permanently lost. In terms of the DNA, loss of a gene might mean actual loss of certain sequences or their replacement by others, in most cases probably the latter. The modification (mutation) may be useful, and thus be selected, may be neutral, and may survive, or it may be harmful and therefore may be eliminated. To take two examples of the permanence of genetic information: haemoglobins and rhodopsins turn up more or less sporadically throughout the animal kingdom and often only in some of the species of a closely related group. They do not appear in identical forms in all species; in fact, their protein moieties in different species are probably always slightly different. The production of haemoglobin requires at least the whole machinery for making haem and the DNA code and requisite RNAs for the globin itself. The haem-forming appears to be widespread and perhaps indeed universal in animal cells, since haem is required for the various cytochrome systems. The globin is specific and determines the manner in which the haemoglobin can function. Thus, within certain limits, variation occurs in the genetic code for this part of the system and allows the haemoglobin to be modified and adapted to particular circumstances. A similar situation arises in the case of rhodopsin; the retinene, either I or II, is constant, and the machinery for its production, from vitamin A_1 or A_2 or indirectly from carotene, seems to be built into a large number of cells, but the protein "opsin" is again the variable, determining both the species-specificity and the photosensitivity of the rhodopsin. The suppression of the protein-producing machinery is, in each case, probably the factor that determines whether or not a cell contains the active substance.

Thus, since these substances are so important, it seems unlikely that every time they occur a new mechanism has been specially devised for their construction. It is much more likely that, once the code for the globin of haemoglobin or for the opsin of rhodopsin was developed, from that time onwards it was always handed on with only minor modifications, and that whether or not haemoglobin appears in any particular species depends mainly on the absence or presence of suppressors and on the conditions in the cell which determine their action.

Therefore, in general, if a character, particularly a complex character, appears in two species which are not in the same genus or even in the same family, its appearance in both suggests that a common origin for the genera or families is a more likely cause than the occurrence of a second mutation, or series of mutations, which would lead to the development of the character for the second time. Consanguinity in this respect may, however, run back a very long time, and the term "common origin" is meant to imply that the two genera or families have arisen from ancestors that either possessed the character itself or the means for producing the character. Moreover, a character thus suppressed could potentially appear in any member of any of those families which were derived from the creatures in which the character first appeared. Its failure to appear in all members is then probably due to suppression rather than absence; the factors causing suppression are many and varied, some probably genetic and others environmental. Failure to appear could, of course, in some cases be accounted for by an actual absence or ineffectiveness of the gene, caused, for example, by heterozygosity or selection of recessives.

This concept of the continued possession of unused genetic information, subject of course to change by random mutation, is a very important concept and one that is in need of thorough investigation, because it colours our whole approach to the interpretation of homology, analogy, and interspecific relationships. Its implications will become apparent during the subsequent discussions and will affect our outlook on much of what follows.

Another matter of interest in connexion with this problem is perhaps worthy of mention. In the examples given above, the globin of the haemoglobin is a very specific protein, containing four polypeptide chains. When attached to haem, this protein gives certain unique features to the complex. One wonders how this arose. Did it arise because the globin or a very similar protein was already in the cell doing something else? The alternative, that the globin suddenly appeared in the appropriate form as soon as an organism needed to use haemoglobin for respiration, seems unlikely. This clearly raises the question of what other function or, indeed, functions in the cell are affected by the presence or absence of the globins (or their polypeptide chains) related to those used in haemoglobins. Combination with or proximity to active groups other than haem could perhaps give the protein other functions. The search for features of structure or function that are shared by cells which produce haemoglobin might perhaps indicate a more primitive role for the globin nucleus, though it may of course have gained its importance in the metabolism of extremely primitive organisms, since haemoglobin also occurs in Protozoa (Keilin and Ryley, 1953) and in the root nodules of leguminous plants (Keilin and Wang, 1945). It is interesting that the position of very few amino acids in the protein chains of different haemoglobins is invariable; it is presumably these few that determine the main activity. At the same time the substitution of valine for glutamic acid at position 6 in two of the chains of human haemoglobin is, as already mentioned, sufficient to produce sickle-cell anaemia with its abnormal blood corpuscles and probably important biological consequences in relation to malaria (Lehmann and Huntsman, 1966). One final point of interest in relation to the transcription of the genetic code is that some mammals have one haemoglobin in the embryonic stages and another in the adult. What happens to the DNA to bring this about? Different portions of the code are presumably activated and suppressed in different cells. Mammals, in general, have at least three normally occurring haemoglobins: fetal haemoglobin, adult haemoglobin, and myoglobin. These globins differ in possessing different polypeptide chains and are antigenically separable (Dan and Hagiwara, 1967).

From the Unicellular to the Multicellular

To return now to the main theme, let it be made clear once again that the Protozoa living today have had as much time (and thus probably as many opportunities for mutations to occur) to evolve and specialize as has man himself; thus we have no right to consider any of them primitive in all respects. Even those that seem, on the surface, to be the most simple and primitive have been living and adapting for just as long as most other living creatures.

Protozoa are, however, basically unicellular and thus have not, with one or two exceptions, undergone any of those changes in organization that depend on cellular interaction; thus further changes have clearly been limited. Protozoa also tend either to live in reasonably constant environments or to be active only under a given and restricted set of conditions; they either escape from less favourable conditions or endure them by encystment or the like. This confinement to a particular environment, often by retreat from others, has limited their evolutionary change.

Thus, by examining Protozoa as a whole, we can pick out certain common traits and establish a general level of development and organization. We can also pick out several processes which may have been relevant to the initiation of different paths of evolution. To take the particular example of *Naegleria*: this organism would seem to contain within its make-up the potentialities for evolving as an amoeboid unicellular form, as a ciliated or flagellated unicellular form, as a colonial amoeboid form, as a colonial flagellate or ciliate form, or as a mixed organism composed of an aggregation of flagellated cells and amoeboid cells, not unlike the blastula of the sponge. *Naegleria* itself clearly has not evolved in any of these ways—it is what it is, somewhat more versatile than other amoebae, and it may have been very much the same for many millions of years. It has features in its genetic code, however, which could have been used for different purposes, and may well have been so used, by organisms derived from the same general stock as *Naegleria* or even from a primitive *Naegleria* itself, if *Naegleria* is still the same today as it used to be and has otherwise not significantly evolved over the intervening millions of years. Such static behaviour of organisms does occur; for example, the brachiopod, *Lingula*, has apparently continued to prosper without much change since palaeozoic times. On the other hand, the particular evolutionary path (or paths) which led to the initiation of the metazoa and to the development of a blastula-like organism (a stage through which the majority of invertebrates pass during the course of their embryonic development and one that can probably be regarded as primitive and very ancient; Fig. 9.1) has to be elucidated. For this it is obvious that an organism like *Naegleria*, with its potentialities to behave as a single flagellate cell, as an amoeboid cell, or to adhere to its own kind in such a way as to form potential colonies, is a very likely starter for the evolution of the Metazoa from the Protozoa. It need not, of course, be the actual starter or even the only one, and, in order to exonerate *Naegleria* itself from this role, the hypothetical founder organism of the Metazoa will be referred to as a naeglerioid organism.

The first step from such an organism to a viable Metazoon would involve the cohesion of individuals into a cluster, either as a sheet or as a mass, and, if the latter, then either as a hollow shell like a blastula or as a solid mass like a morula. From its widespread occurrence among the embryos of today (see Fig. 9.1), it seems likely that the hollow blastula was the form preferred, presumably because of the advantages of maintaining stability and maximal contact with the outside world for the acquisition of food, oxygen, etc. Moreover, it is probable that both forms of the naeglerioid cell, with their different ways of responding to the environment, could profitably be used in the formation of such a blastula. The two forms of cell could work together, or in opposition, to assist in stabilizing the spherical form and the contents of the cavity, particularly with respect to their water and ionic constituents. There may still be a few organisms living to-day that have not progressed much beyond this stage. Yet it is very significant, as already pointed out, that the vast majority of metazoan animals alive to-day do pass through an essentially similar blastula stage in their ontogenetic history. It probably happens that, by itself, such a blastuloid organism cannot effectively compete for long with the many other more highly

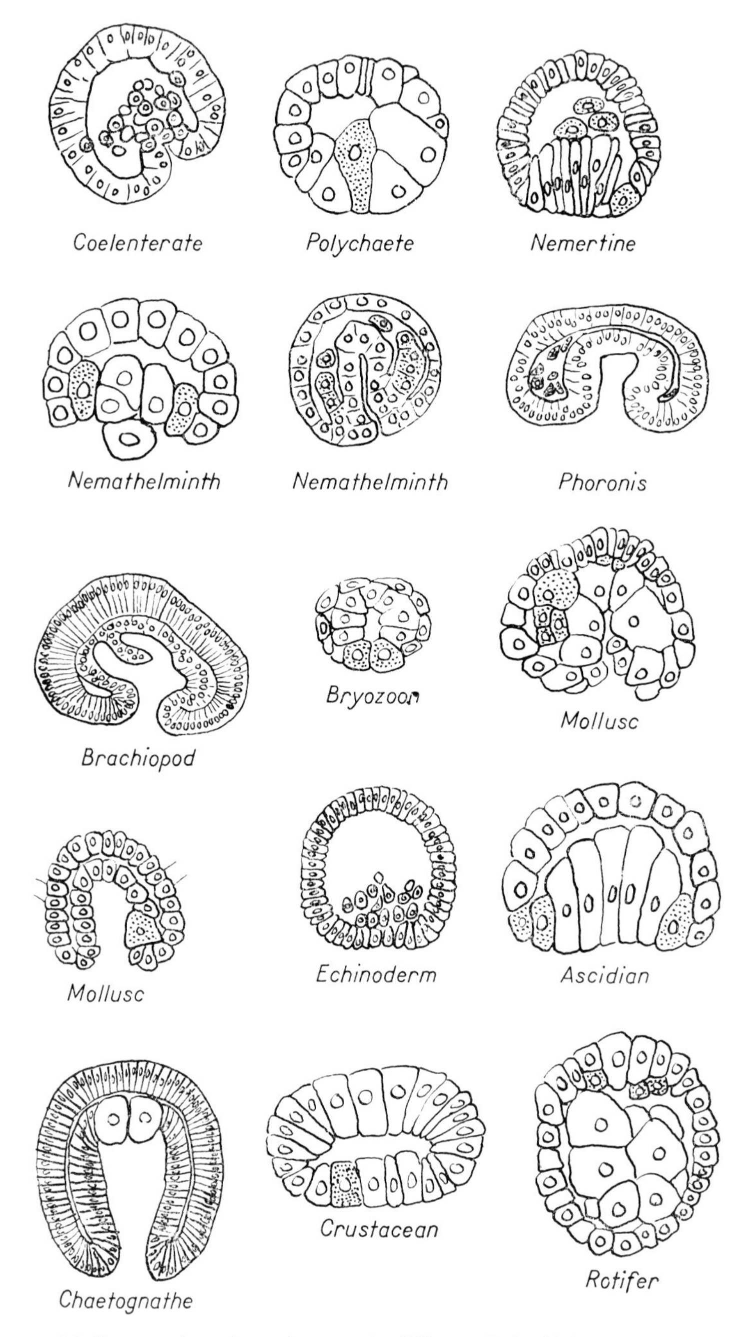

FIG. 9.1. Some early embryonic stages in different phyla. (After Davydoff, 1928.)

organized forms which have readily evolved from the blastuloid form, so that the blastuloid stage itself is now seen only as a transitory phase in the ontogenetic development of the Metazoa. Organisms like *Volvox*, *Gonium*, and a few others have some of the properties of an elementary blastula, but they have other specializations, such as photosynthesis, and are not strictly relevant to the present story.

The blastulae of existing calcareous sponges (see Fig. 7.1) can be taken as illustrative of this particular evolutionary stage, though this is not meant to imply that the evolution of vertebrates actually proceeded through the sponges. Sponges may have gone their own way from the blastuloid organism; other stocks may have made different modifications that have led to other forms of organization.

It is impossible to decide whether, in evolution, the flagellated cell preceded the amoeboid cell, or *vice versa*. It is clear, however, that both forms must have preceded the development of the blastuloid organism of the type under discussion. Moreover, although in ontogeny the epitheliocytes precede the independent mechanocytes and amoebocytes, in phylogeny the independent flagellated and amoeboid cells must clearly have come first. Single cells must at some stage have either united, or subdivided and remained adherent, in order to produce a multicellular form. Cell cohesion in the epitheliocyte manner is thus very important, though the actual mechanisms of cohesion may vary.

It is unlikely that all the existing Protozoa have been developed from Metazoa, though it is certainly possible that some may have been. It is, however, extremely probable that Metazoa developed from some form or forms of Protozoa.

The Properties of a Blastuloid

In the blastuloid organism there must have developed not only a coherence between the cells but also an orientation and a polarity of their activity. For example, all the flagella in the free-living blastulae of the calcareous sponges of to-day point towards the external environment. The formation of a blastula also entails the development of at least two different sorts of membrane on the same cell, the outer in contact with the external environment and the inner in contact with the internal cavity. This step is indeed foreshadowed in both *Tetramitus* and *Naegleria*, where, during the acquisition of the flagellated form, the posterior end of the cell, at which the flagellum develops, is clearly different from the anterior end with its lobose pseudopodia. It is, of course, a fundamental concept of cytology that the nature of a cell surface can vary with the medium in which the cell is placed. If a cell is simultaneously in contact with several environments, it is likely to have at least as many different regions on its surface. Besides having inner and outer surfaces of different kinds, the cells of the blastuloid organism must make contact with and adhere to their neighbours on all other surfaces. In an organism at this stage, therefore, it is probable that some form of cohesive mechanism, e.g. desmosomes, tight junctions, or terminal bars (*zonae adherentes* or *occludentes*) (see Balinsky, 1959; Devis and James, 1964; Dewey and Barr, 1962; Eccles, 1963; Farquhar and Palade, 1963; Kelly, 1966; Locke, 1965; Ross and Greenlee, 1966; Wood, 1959), or interdigitating cell processes must have made their first appearance (Fig. 9.2). It might be assumed that, in the early colonial forms, contact between

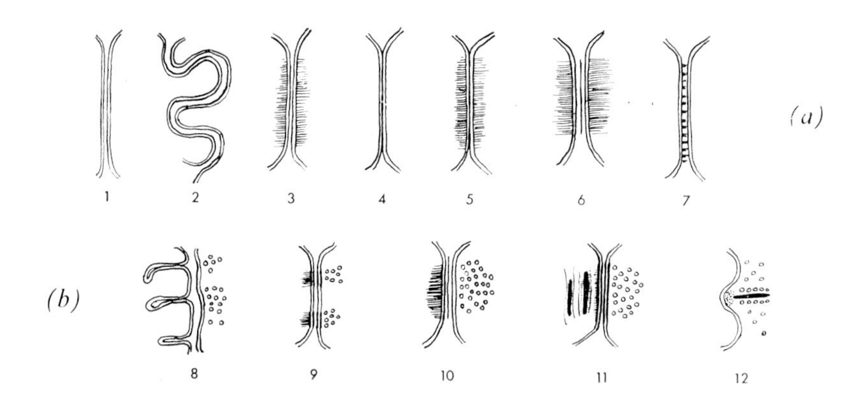

FIG. 9.2. Types of cell contact and special modifications of the cell surface. (*a*) "Tight" junctions. (*b*) Synapses. 1, Normal cell contact, with two unit membranes separated by intercellular space; 2, interdigitating and interlocking membranes; 3, normal cell contact with intracellular fibrils (*zona adherens*); 4, tight junction (*nexus*); 5, tight junction with fibrils (*zona occludens*); 6, desmosome (*macula adherens*); 7, septate desmosome; 8, neuro-myal junction with synaptic vesicles; 9, synapse type 2; 10, synapse type 1; 11, synapse with "spine", pyramidal cell; 12, synapse with "ribbon", retinal cell.

the epithelial cells would be somewhat loose and haphazard (in the chick embryo, desmosomes do not appear very early; Overton, 1962), but there would be a premium on efficient bonds between the cells, not only to keep the cells together but also to keep communication between the constituent cells, if other advantages would be gained thereby (Katsuma Dan, 1960). This primarily mechanical consideration could bring with it far-reaching consequences. For example, it is known that in certain columnar epithelia the passage of ions laterally from one cell to another is often far easier than it is across the inner and outer membranes of the same cell. In addition, sometimes there are special channels of easy communication between definite groups of cells (Loewenstein *et al.*, 1965; Penn, 1966). As is shown later, these special conducting paths can have important implications (see p. 197).

If the contact between the cells of a blastuloid organism were loose, the interchange between the external and internal environments would not be interrupted and the internal would not greatly differ from the external, any more than a coral lagoon, for example, differs from the main ocean with which it remains connected. The embryos of echinoderms may be in this condition, because the observations of Chambers and Pollack (1927) of the pH of the blastocoel fluid of echinoderm embryos and of the ready diffusion of dyes injected into the cavity indicate that in these creatures the epithelial wall offers very little resistance to the passage of relatively small molecules. The formation of tighter junctions and the production of such things as terminal bars (*zonae occludentes*) would necessarily close off the internal environment more or less completely and could be advantageous in preventing the dilution of metabolic pools. Such closure would, however, force the epithelial cells into maintaining the contained fluid in a condition compatible with the life of the cells surrounding it. Thus, in order to maintain this equilibrium, the epithelial cells would be forced to use their powers of moving water, ions, and other constituents across their inner and outer surfaces in an orientated manner. As suggested earlier, it may be that the more successful organisms at this stage would use two classes of cells that were oppositely orientated and so make use of a balance in the directional activity of their respective water and ionic movements. Given sufficient flexibility in the character and permeability of the inner and outer surface membranes, a blastula composed of only one class of cells could theoretically be stable if it balanced the transport across the inner and outer surfaces, but there are reasons (see p. 198) for believing that this hypothesis presents difficulties. In blastulae, as they are found to-day, there is no trace of a basement membrane giving them cohesion or specific permeability, though there are sometimes traces of mucoproteins in the fluid in the blastocoel (Monné and Härde, 1950). It is unlikely, therefore, that the original blastuloid organism had any such membrane; its permeability properties probably depended entirely on the epithelial layer, and, if the epithelial cells were tightly joined, this permeability was likely to have been an active one. Unfortunately, as already indicated, there is little information on how the cells of calcareous sponges do cohere. Septate desmosomes occur in the blastulae of echinoderms (Balinsky, 1959), in the tissues of Cnidaria including those of *Hydra* (Wood, 1959) and desmosomes are present in the blastoderms of chicks (Overton, 1962). Silver-staining material occurs between the cells of epithelial sheets growing from fresh-water sponges; this is perhaps indicative of cohesive mechanisms of some sort.

The successful blastuloid organism can thus be assumed to have displayed in addition to the properties acquired by the naeglerioids:

1. Two classes of epitheliocytes with means for maintaining mutual contact and the integrity of the organism.
2. An organized polarity of the cells in each of the two groups with cell membranes differing at the inner and outer surfaces.
3. Some control over the composition of the "internal environment".

For reasons which follow, two further properties must be added:

4. Some capacity for each group of cells to breed true as essentially flagellated or amoeboid; the difference must be mainly intrinsic to the cells and presumably, though not necessarily, nuclear in origin.
5. An organized "gradient" system of activity from one pole to the other (see pp. 192, 196).

Some of these developments are clearly fore-shadowed in the behaviour of *Naegleria*, because the flagellated form is a highly polarized cell in which the two ends differ very markedly. Moreover, during the change from the amoeboid to the flagellated form the cell passes through a phase in which it can be seen to acquire this polarity, because flagella may beat at one end while amoeboid processes are still active at the other. This behaviour raises the possibility that in the development of a successful blastuloid organism it is not necessary to have cells which are either wholly of the flagellated type or wholly amoeboid. Provided that one surface of each cell, let us say the outer, shows directional properties, while the other acts passively, the proper balance of activities could be maintained.

The blastula stages of the calcareous sponges are of special interest in this connexion. In *Sycon raphanus* and *Grantia compressa*, as described by Duboscq and Tuzet (1937), the ovum is fertilized within the parent tissues and then develops *in situ* surrounded by a capsule of connective tissue. Within this capsule, cleavage takes place and the embryo becomes a small blastula (stomatoblastula). In this stage the flagella point inwards and the amoeboid cells lie towards the choanocytes of the parent. One or two of the maternal choanocytes then lose their flagella, sink inwards, and proliferate. These cells then flatten and surround the embryo in a new capsule (placenta). When the embryo is completely surrounded, it ruptures at the pole occupied by amoeboid cells and then literally turns itself inside out in the manner shown in Fig. 7.2, so that the flagella now point outwards and the amoeboid cells join up again at the opposite pole. During this phase the flagellate cells produce what are probably stereocilia or very long microvilli directed towards the cavity of the new blastula, or amphiblastula as it is called (Fig. 9.3). As soon as this reversal is accomplished, the capsule and the layer of maternal choanocytes rupture and the blastula emerges into the sea-water (see Figs. 7.2, 7.4). The whole operation suggests that the enclosing placental cells alter the immediate environment of the embryo and it responds by reversing the polarity of its two groups of cells, not by altering the polarity

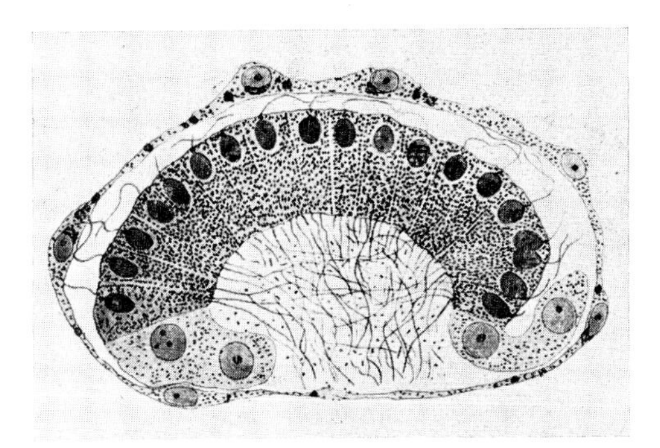

FIG. 9.3. Development of sponge embryo (*Sycon raphanus*) showing method of inversion. Note the microvilli on the bases of the flagellate cells (see also Fig. 7.2). (Duboscq and Tuzet, 1937.)

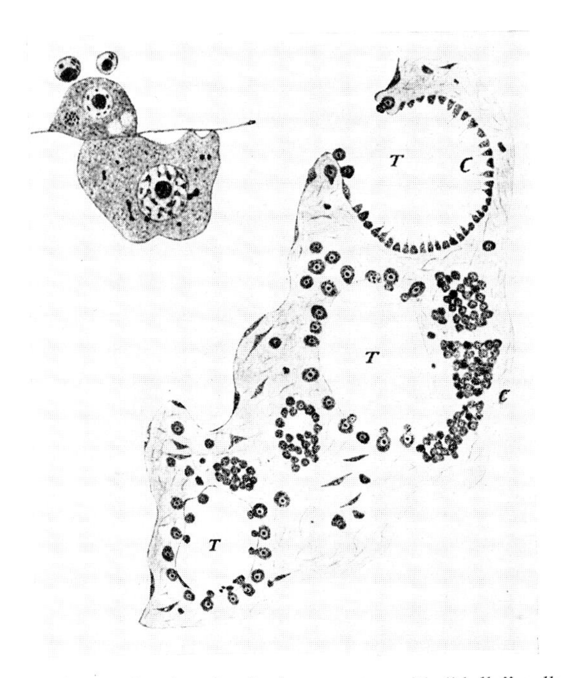

FIG. 9.4. Sponge tissues showing developing oocytes with "dolly" cells and satellites. (*Bottom, right*) Three atria (T) lined with choanocytes (C) or with modified choanocytes (dolly cells and satellites) and gonocytes attaching themselves to the dolly cells. (*Top, left*) Gonocyte with dolly cell and two satellites. (Duboscq and Tuzet, 1937.)

of each cell but by changing the environments towards which they point. A study of the maturation of the ovum and of the relationships between the maternal tissues and the unfertilized ovum, the fertilized ovum, and the developing embryo in terms of ionic balance and symbiotic relationships would be of extreme interest. There seem to be some very special cellular interactions, as, for example, that between the developing ovum and its "dolly" cells as well as that between the blastula and its capsule (Fig. 9.4).

Finally, it should be emphasized that the cells of the blastuloid could be arranged in the form of a gradient, proceeding from cells with outwardly directed flagella exposed at one pole (e.g. the "animal" pole) to cells showing the more amoebocytic characters at the other pole (e.g. the "vegetal" pole). The transition from one type of cell to the other need not be sudden in its spatial arrangement any more than it is in time. Existing blastulae often show cells of intermediate type (see Fig. 10.4).

Gradients

Blastulae in living organisms are not formed by the association of adult cells, but by proliferation of cells from a single ovum. The proliferation occurs through subdivision of the ovum itself, and it is probable that the gradient which is finally established in the blastula is related to the original axial gradient in the ovum, itself dependent at least partly on the properties of the surface membrane of the cell. In such a situation, therefore, it is difficult to know which is cart and which is horse. The gradient in the blastula may result from the gradient in the ovum, but the latter may have been determined by selection in order to ensure the development of the former. The occurrence and significance of the primary axial gradient is thus still an outstanding problem, but at least a consideration of the "stability" of the blastula and its contained fluid does indicate one possible explanation for its almost universal occurrence.

The recent work of Jaffe (1966) in establishing the existence of an electrical gradient in the developing ova of *Fucus* is interesting in this connexion. It could indicate an opposite polarity of the membranes at the "animal" and "vegetal" poles, and these differences could well persist after cleavage. Differences of potential have been observed at the two ends of active amoebae (Allen, 1968).

It will have occurred to the reader that, if the axial gradient in the ovum determines the distribution of the flagellate cells and of the amoebocyte cells in the blastula, and if this gradient is associated with changes in the surface layer of the ovum, it must follow that the types of nuclear behaviour that produce the flagellate cells on the one hand or the amoebocyte cells on the other, and allow them to reproduce their kind must be determined by the cell surface. This situation is thus comparable to that described by King and Briggs (1956) and especially by Gurdon (1962) in which the nuclei taken from gastrulae and even later larvae can have their activity determined by the cytoplasm of the ova into which they are inoculated.

In considering the nature of the gradient of epithelial cells it may be of interest to point out at this stage that, in higher animals, epithelial cells can be arranged in a series according to the character of their external surface (Fig. 9.5). At one end of the scale are the ciliated and flagellated cells, They can perhaps be classed together since, in the flagellated cell, the single flagellum makes up in length of its flagellum for the greater number of shorter cilia in the ciliated cells. Among the best developed flagellated cells forming an epithelium may be mentioned those which line the cephalic grooves in nemertine worms (see pp. 288, 366). These cells have exceptionally long flagella with well developed "root-fibres", but the part which these and similar organelles play is still problematical.

The revealing analysis of ciliary and flagellar movement by Sleigh (1968) raises many interesting problems in this context. It has shown that, although the 9+2 pattern of internal fibres (see Fig. 1.1) is common to most cilia and flagella and, apart from the occasional absence of the central pair, remarkably constant in design, the actual performance of the cilia and flagella can be very different. Their performance can differ particularly in relation to the temporal and spatial relationships of the bending process and consequently of the form, frequency, and character of the effective and recovery strokes. The beat of some cilia approaches that of flagella in that several waves of bending may be present simultaneously; in other cases the effective and recovery strokes are clearly differentiated from each other and confined to a single plane. At present there is insufficient evidence to indicate the evolutionary relationships between the various types of beat, i.e. to determine which is primitive and which derivative. In view of the localized bending of filiform pseudopodia, which, because they lack the internal structure of true cilia and flagella, might be regarded as primitive, the simple ciliary beat with its "stiff" effective stroke might be considered primitive also. On the other hand, the sculling movement of a single flagellum is probably more effective in moving a unicellular organism through the water than is the rowing action of a single cilium. The latter obviously becomes the more effective when serried ranks perform in metachronal rhythm. At this stage and in the present context, it is probably better to class ciliated and flagellated cells together. Ciliated cells, and indeed some flagellated cells, may also have more or less numerous and well developed microvilli or stereocilia among the motile units. In the epididymis of mammals and in the lens cells of the pineal eye of lizards,

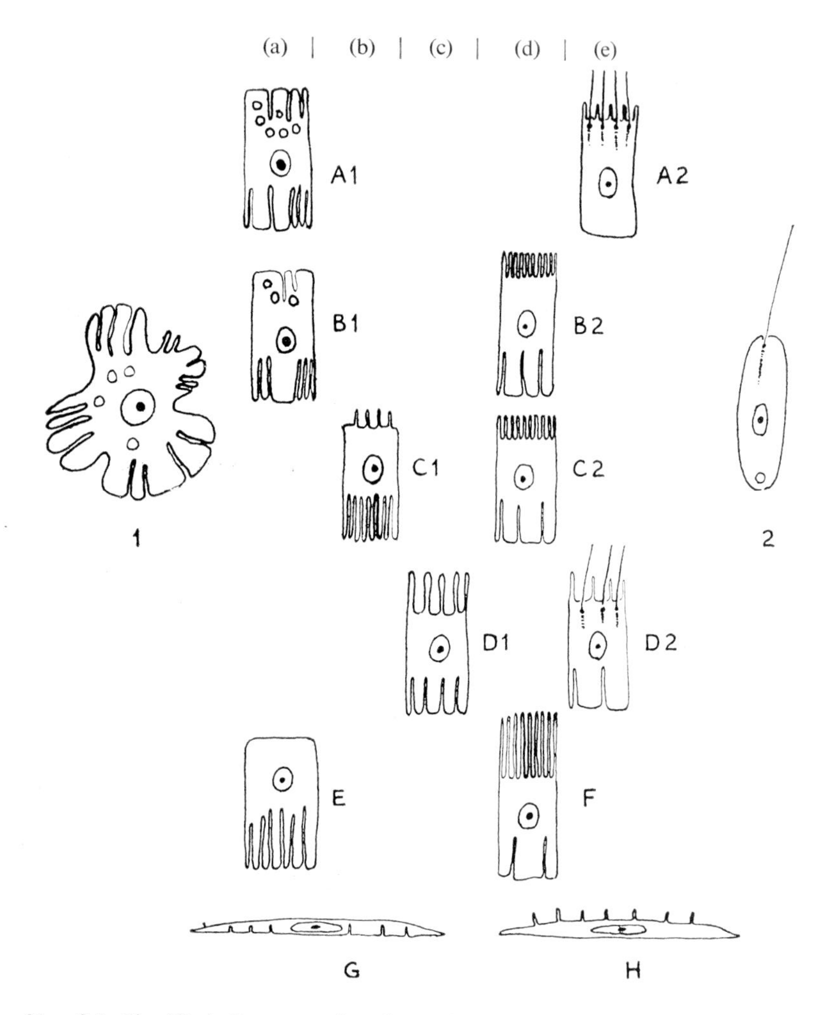

FIG. 9.5. Simplified diagrams of main gradations (*a–e*) of cell surfaces in various epithelia in relation to the amoebocytic (emballic*) cell on the left (1) and the flagellate (ecballic*) cell on the right (2). The interdigitations with neighbouring cells have been omitted. In the epithelial cells the upper surface is the free or distal surface, and the lower surface is the basal or proximal surface adjacent to the tissue fluids. In E the infolded membranes lie next to the aqueous humour from which they are separated by a basement membrane. A, Trachea (a cell with microvilli only has also been described); A1, goblet cell; A2, ciliated cell. B, Intestine; B1, goblet cell; B2, brush-border cell. C, Kidney; C1, distal tubule cell; C2, proximal tubule cell. D, Choroid plexus; D1, "polypoid" cell; D2, ciliated cell. E, Cell of ciliary process (adjacent to vitreous body). F, Cell from epididymis with stereocilia. G, Endothelial cell. H, Cell from peritoneal surface.

*"Ecballic" and "emballic" are used throughout instead of the more cumbersome terms hydrecballic and hydremballic defined on p. 175.

such stereocilia occur alone, whereas in the sensory areas of the labyrinth and lateral line they accompany motile cilia (kinocilia) (see Fig. 15.5). If the microvilli become very numerous, they may constitute a "brush border" as in the proximal tubules of the kidney or the chief cells of the mammalian intestine (Fig. 9.6). Cells with microvilli are generally capable of pinocytosis, which leads to the cells with lobose pseudopodia, flattened surfaces, and finally to indented surfaces as seen in the distal tubules of the

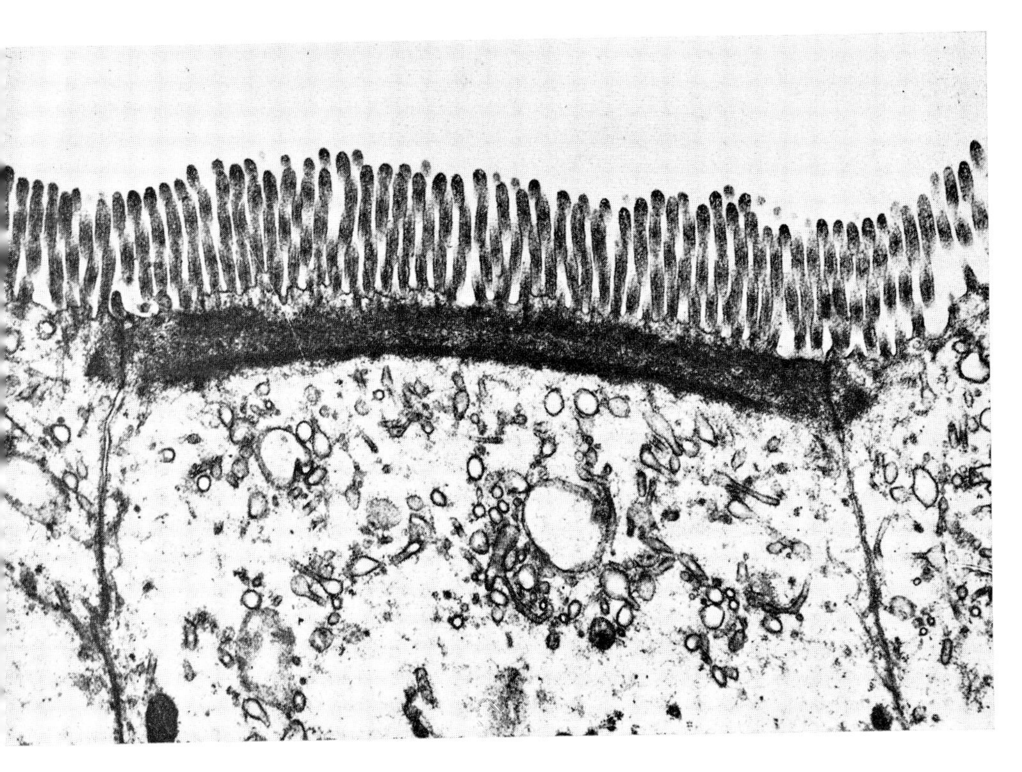

FIG. 9.6. Brush border of intestinal cells showing microvilli. The section passes near the margin of the central cell and shows the structure of the terminal bar in longitudinal section. In the cells at the sides the terminal bars are seen in transverse section. (Photograph by Hockaday.)

kidney, in the oxyntic cells of the stomach, or in the "chloride" cells of the gills of certain fish. Cells at this end of the scale often have a propensity to store carbohydrates or mucopolysaccharides within their cytoplasm. Ciliated cells should perhaps be regarded as cells retaining the essential characters of the original flagellated cells (cf. *Naegleria*), and the pinocytosing and mucoid cells as more closely related to the original amoeboid cells. There are all gradations between them. This classification is clearly arbitrary and takes no account of the basal surface, but in metazoa, in which epithelia separate external or luminal cavities from the tissue fluid and blood, the external or luminal surface generally can be considered dominant. Nevertheless, a similar ordering of epithelial cells could be based on the properties of the internal surfaces also, and in obtaining a complete picture of epithelial function both surfaces must indeed be taken into account.

In hypothetical physiological terms (based on the lessons to be learnt from the behaviour of *Naegleria*) the original gradient through the blastula, e.g. the animal-vegetal gradient, might mean that cells strongly organized for the ejection of water and for the conservation of ions would concentrate at one end, while the cells at the other end would do exactly the opposite. The intermediate cells could well act less strongly in one direction or the other, so that somewhere, though not necessarily exactly equatorially, there would be cells which were more or less neutral in this respect. By some such means an axial gradient would inevitably develop; indeed, when we examine the blastulae of organisms living at present, the rigid separation into two absolutely sharp groups of cells, as seen in some calcareous sponges, is by no means common. Far more often, in fact almost universally, we find blastulae whose cells grade in morphology more or less sharply from animal pole to vegetal pole, and blastulae with such primary axial gradients are of more frequent occurrence than the sharply subdivided blastulae like those illustrated in Fig. 7.1.

In *Naegleria*, which is an exceptional organism, the form of the cell is readily reversible and largely determined by the environment. This reversibility may have been possible initially for the cells of early blastuloid organisms also, but it is fairly clear that, in the successful forms, there must be some rigid control of activity to allow the organisms to survive in a varying environment without necessarily changing their whole way of life as does *Naegleria*. A gradient system embodying some degree of fixation of the type of activity in each cell probably gets the best of both worlds. At the extremes, i.e. at the poles, the flagellated cells or amoeboid cells are probably firmly fixed in behaviour, would be expected to breed true as such, and are unlikely to be changed radically by the range of environmental factors which are compatible with life. Equatorially, however, there could be a group of cells more or less in equilibrium with their environment, but not so determined towards activity in one direction that they might not be able to make adjustments in the other. Their directional behaviour, never very strong, could possibly be reversed according to conditions and thus act as a fine adjustment to the system.

If this problem is considered at the genetic level, it presumably means that a *Naegleria* has all the DNA mechanism necessary to produce either amoeboid activity or flagellate activity. It can turn on one form of activity or the other. If different proteins are required in the two forms, then there must be suppressors of some sort to minimize the production of the unwanted proteins. When naeglerioid organisms combined for colony formation, the activity of one group of suppressors became greatest at one pole, that of the other at the other pole. This establishment of a gradient system suggests that the action of suppressors is not entirely "all or none", but that it can be graded and so act with more or less efficiency; presumably, large or small quantities of protein can be made at will. There are several places at which this gradation could occur, e.g. at the level of the DNA, at the messenger RNA, the soluble RNA, or on the ribosomes.

Since the behaviour of *Naegleria* is determined by the environment, it could be suggested that changes in the cell surface are perhaps the effective agents in calling for different proteins and so redirecting, by some means, the activity of the nucleus to meet that demand. Thus, after cleavage in the early multicellular organisms, the nuclear material (or ribosomes, etc.) would be subjected to different suppressors at the animal and vegetal poles on account of the different nature of the cell surfaces at the two poles, which may itself have been at least partly determined in the ovum by the gradient within the ovum or over its surface.

Elementary blastuloid organisms such as those under discussion have not developed any discrete nervous system as such. However, it is unlikely that the cells would remain completely unco-ordinated if the organisms were to remain viable. In those existing blastulae that are ciliated, the beat of the cilia is co-ordinated into a metachronal rhythm. In sponges and in many ciliated membranes this has been shown (Parker, 1919) to depend on "neuroid transmission" from cell to cell and not directly on the mechanical action of the cilia on each other. The structural basis for this neuroid transmission has not been established with certainty. Within the cell the various "root-fibres" and "spurs" suggest themselves. If a special structure is necessary between cells, then desmosomes (and perhaps other specialized connections) are obvious candidates for this function in view of their similarity to some of the structural characteristics of synaptic connexions (see Fig. 9.2), both chemical and electrical, between the nerve cells of higher organisms. As indicated on p. 187, the development of such features may therefore have been "part and parcel" of the formation of the blastuloid organism, and the study of such structures may eventually help towards the understanding of the more sophisticated synapses of the human central nervous system. The free ionic connection between neighbouring epithelial cells discussed on pp. 188–9 is probably very important in this transmission of information.

The Stability of the Blastuloid and the Internal Environment

The possible advantages of the juxtaposition of two cell types in the formation of a primitive organism were considered and emphasized many years ago by Metschnikoff (1886). He pointed out that an organism able to move from place to place, either by the aid of flagella or by amoeboid movement, must have advantages over both a sessile organism and an organism possessing only one type of movement. Different forms of nutrition by two types of cell (e.g. saprophytic and phagocytic) could also be advantageous and so could different means of obtaining energy (e.g. respiration and glycolysis). Glycolysis has been considered the more primitive and probably the original means of obtaining energy, but a combination of both methods might be very helpful. There is indeed some suggestion that the amoeboid form of activity relies more on glycolysis than on respiration while the flagellate or ciliate probably does the reverse. If the flagellated cells can act as sensory mechanisms and the phagocytic cells can store food to be used in times of shortage, these abilities are again advantageous. In addition, if the amoeboid and flagellate forms also differ in their response to ionic environments (using this expression in its widest sense), still other advantages would follow from their combined activity. Indeed, such combined activity may be a necessity for the efficient control of the composition and volume of the internal fluid. Since this internal fluid forms the "internal environment" for all the cells of the organism, its control is a matter of the utmost importance.

It is probably not impossible for a blastula whose cells are all of one type to regulate the composition of the contained fluid. Protozoa can, after all, regulate their ionic content within certain limits. Nevertheless, there are numerous examples in physiology where regulation and constancy are achieved by balancing one activity against another, and it seems likely that the blastula may be another example to be added to this list.

The experiments of Chambers and Kempton (1933) on tissue cultures of the chick mesonephros are an object lesson in the importance of ionic and water regulation in closed systems. When portions of the cortex of the kidney are cut up and explanted in a plasma medium, the tubules that were cut during the dissection tend to seal their ends and become closed systems. Thereafter, the units that were derived from the original proximal tubules become increasingly distended with fluid, while those derived from the distal tubules lose all their contained fluid and collapse into cords of cells (Fig. 9.7). In view of what has been said about the grading of epithelia, it is pertinent to remark that the proximal tubules have brush borders of microvilli, whereas the cells of the distal tubules have only a few microvilli and a very indented basal cytoplasm [see Fig. 9.5 (C1 and C2)].

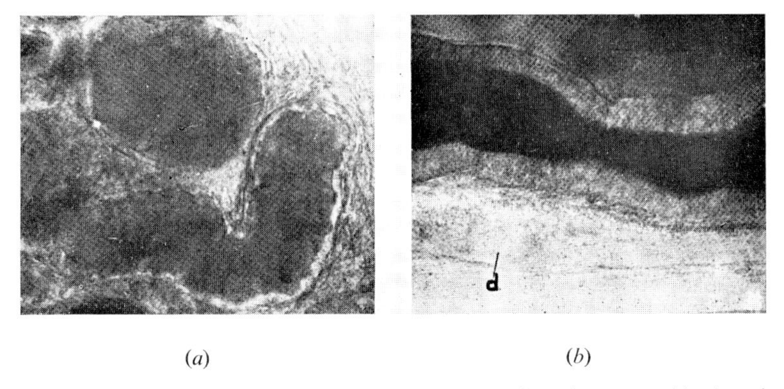

(*a*) (*b*)

FIG. 9.7. Kidney tubules from the chick mesonephros cultured *in vitro* with phenol red. (*a*) Proximal tubules distended with fluid containing phenol red; (*b*) proximal tubule distended (the lumen is filled with red fluid) and distal tubules, d, collapsed and empty. (Chambers and Kempton, 1933.)

Similar observations have been recorded by Beadle and Booth (1938) on isolated fragments of ectoderm and endoderm of the coelenterate, *Cordylophora lacustris*. Fragments of ectoderm were seen to form hollow vesicles, while fragments composed of the neutral-red-accumulating cells of the endoderm formed solid masses which subsequently disintegrated (Fig. 9.8). When some endoderm was included with the ectoderm, a normal two-layered hydranth was formed. These workers also noted that the interstitial cells were formed from the ectoderm, a point of interest in connexion with the origin of cnidoblasts from interstitial cells (see p. 215).

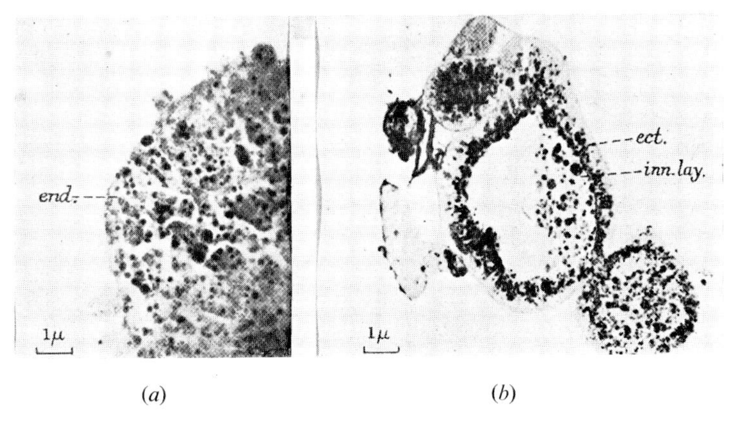

(*a*) (*b*)

FIG. 9.8. Reorganization of ectoderm and endoderm masses isolated from the hydroid, *Cordylophora lacustris*. (*a*) The endoderm forms a solid mass; (*b*) The ectoderm forms a hollow mass. (Beadle and Booth, 1938.)

Of still more relevance are the observations of Hörstadius (1937) on the fate of cells isolated from the early cleavage stages of the nemertine worm, *Cerebratulus*. Cells isolated from the animal pole developed into hollow ciliated blastulae, while those isolated from the vegetal pole either did not survive or produced clusters of adhering cells like morulae (Fig. 9.9). Zeleny (1904) had previously observed that at the 8-cell stage the isolated dorsal cells, if allowed to develop, formed a large blastocoel, while the four ventral cells formed a very small cavity.

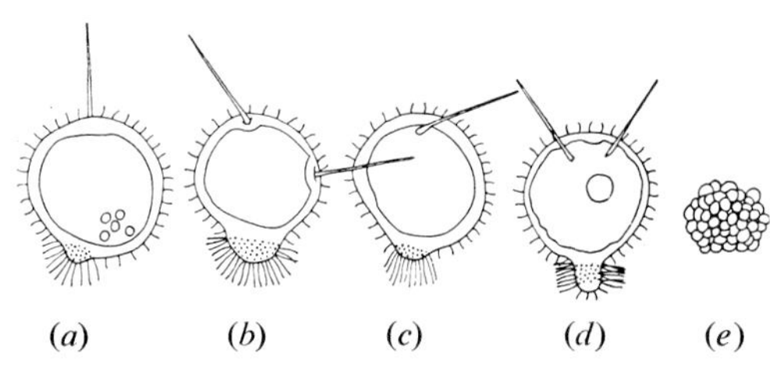

FIG. 9.9. Potentialities of animal and vegetal pole cells of the nemertine, *Cerebratulus lacteus*. (*a*)–(*d*) Blastula-like structure developed from the isolated four most-animal cells of the 16-cell stage; (*e*) Morula-like structure from the four most-vegetal cells of the 16-cell stage. (Hörstadius, 1937.)

As another example of the idea of balance in the embryo, there are the observations of Paterson (1957) on the potentialities for development of portions of amphibian embryos revealed by sectioning the embryo so as to divide the animal pole and the vegetal pole from a central (equatorial) region. The only embryos which can repair and continue development from these segments are those from the equatorial part. Presumably the cells in the two "polar" segments are too biassed in one direction or the other to be able to reorganize sufficiently.

The unbalanced passage of fluid across epithelia is also very well shown by the observations of New (1956) on the chick blastoderm. If the blastoderm is explanted in tissue culture, with the ectoderm downwards on vitelline membrane spread over albumen, fluid collects above the tissue. If it is planted with the endoderm downwards, the ectoderm curls downwards and grows round the endoderm, eventually enclosing it in the form of a two-layered hollow sphere. The sphere then accumulates fluid and enlarges (Fig. 9.10). Similar observations have been made by Tuft (1961) on the blastula of *Xenopus* and by Jurand and Tuft (1961) on the developing chick. In the blastulae of *Xenopus* there is good evidence for fluid movement inwards through the cells of the animal pole into the blastocoel, and for movement outwards through the cells of the vegetal pole. The latter movements, through the cells of the vegetal pole, can be independently abolished by β-mercaptoethanol (Fig. 9.11).

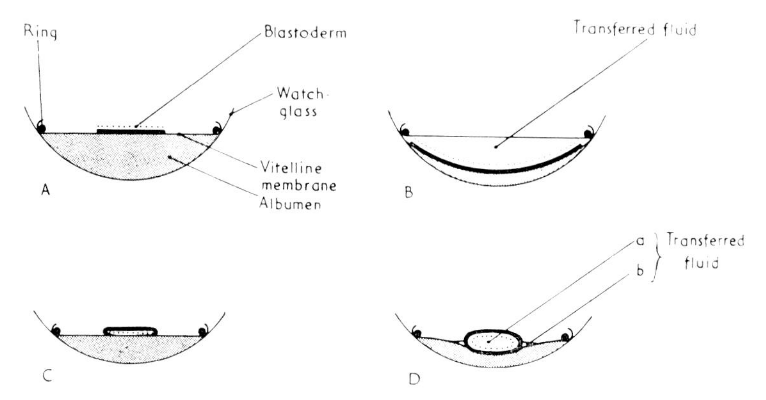

FIG. 9.10. The formation of the sub-blastodermic fluid in hen's eggs. (*A*), Blastoderm explanted ectoderm downwards: (*A*) condition at time of explantation; (*B*) condition after a total incubation time of 48 hours. (*C*), (*D*) Blastoderm explanted endoderm downwards: (*C*) the edges curl under; (*D*) closed vesicle formed; the transferred fluid is mostly within the vesicle (a), though a little remains outside (b). Ectoderm: continuous thick lines; endoderm: dotted lines. (New, 1956.)

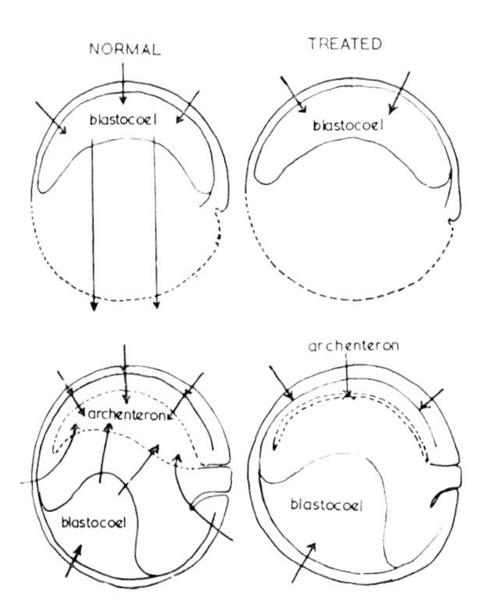

FIG. 9.11. The effect of the net water fluxes in normal embryos of *Xenopus* and embryos treated with β-mercaptoethanol. (Tuft, 1961.)

All these findings dramatically demonstrate the problems which arise when sheets of cells enclose confined spaces. Though osmotic, ionic, and hydraulic problems may not be of much importance for completely marine blastulae, the products of cellular activity diffusing into the cavity of such organisms may raise problems of the same sort, particularly if the cells are bound together in such a way that a closed system is established. Problems of buoyancy also have to be solved, and this can sometimes be done by providing the appropriate ionic concentrations of the internal fluid. It has already been emphasized that the formation of a closed cavity like the blastula cavity establishes to a greater or lesser extent a new environment for the inner surface of the cells surrounding it. It thus forms the first internal environment and, as such, demands regulation and constancy as a condition for "la vie libre" of the organism.

It is relevant here to note that the blastula of *Echinus*, while it is still surrounded by its fertilization membrane, is relatively shut off from its external environment and appears to consist of only one type of cell; thus it is an exception to the usual structure of blastulae. When, however, the ciliated embryo becomes free-swimming and the enclosing membrane ruptures, there is evidence that the permeability of the system to glucose, for example, increases. At that stage the "basal" cells are seen to become morphologically different from the rest, and they are the cells that lead to the flattening of the "base" and that initiate the ensuing processes of invagination and gastrula formation.

The observations that have been made on the animalization and vegetalization of embryos are not without interest in this connexion. For many years, lithium salts, for example, have been known to be powerful vegetalizing agents (see Needham, 1942), i.e. they favour the production and activity of cells at the vegetal pole of embryos and tend to suppress it at the animal pole. Magnesium ions behave somewhat similarly, and lack of sulphate ions favours animalization and the development of flagellated and ciliated cells at the expense of the vegetal cells (Lindahl and Stordahl, 1937; Immers and Runnström, 1965). A comparison of these effects, mostly found on echinoderm blastulae, with the effects of ions of *Naegleria* (see Figs 8.11–8.13 and Willmer, 1956) shows great similarity between the effects of these ions on the two systems. Lithium, magnesium, and sulphate all favour the amoeboid form in their own ways and tend to suppress the flagellate form.

When all these points are considered, the lesson to be learnt is that the simplest colonial organisms were probably blastuloid, and, as such, they embarked on an irrevocable course of further differentiation. Moreover, it is probable that the majority of the cellular modifications that have made the blastuloid organism viable became permanently stamped into both their own make-up and that of their offspring. From the development of the blastuloid organism onwards, one modification after another has been introduced. It is the aim of the succeeding chapters to note how and when in this process the various special features of vertebrate cytology have made their appearance and how they, in their turn, may have been modified, with regard to both form and function, with succeeding evolutionary steps.

With the establishment of the blastuloid form, two important events took place. First, two groups of cells with mutually opposing activities combined in a sort of symbiosis by which the whole organism achieved a stability towards the environment impossible for either group of cells alone. By combining in space what *Naegleria* had achieved in time, a new constancy of activity could be achieved with less subjection to environmental change. Second, the internal environment was initiated, and many of the ensuing steps in evolution will be shown to result directly or indirectly from the development of mechanisms concerned in maintaining its constancy. The possibilities for evolution from the simple blastuloid form are obviously many and varied, and only certain of them have relevance to the ancestry of the vertebrates. Thus the subsequent chapters are concerned mainly with these possibilities, though references are made in passing to some of the other modifications which may have led in such directions as have blossomed into the production of other phyla.

TABLE 9.1

Features of Significance to the Evolution of Vertebrates that appear in or before the Blastuloid Stage

1. A hollow blastuloid structure with spatial differentation and polarity of its cells
2. An epithelum composed of epitheliocytes (flagellated, intermediate, and phagocytic) incorporating the advantages of symbiosis between these various classes of cells
3. Stable contacts or adhesions between cells, with the possibility of developing lateral communication and neuroid transmission
4. A stabilized internal environment
5. Axial gradients, at least in the form of an animal-vegetal gradient

References

Allen, R. D. (1968). Differences of a fundamental nature among several types of amoeboid movement. *Soc. Exptl. Biol. Symp.* **22**, 151.

Balinsky, B. I. (1959). An electron microscopic investigation of the mechanisms of adhesion of the cells in the sea-urchin blastula and gastrula. *Exptl. Cell Res.* **16**, 429.

Beadle, L. C., and Booth, F. A. (1938). The reorganization of tissue masses of *Cordylophora lacustris* and the effect of oral cone grafts, with supplementary observations on *Obelia gelatinosa*. *J. Exptl. Biol.* **15**, 303.

Chambers, R., and Kempton, R. T. (1933). Indications of function of the chick mesonephros in tissue culture with phenol red. *J. Cellular Comp. Physiol.* **3**, 131.

Chambers, R., and Pollack, H. (1927). The pH of the blastocoele of echinoderm embryos. *Biol. Bull.* **53**, 233.

Dan, M., and Hagiwara, A. (1967). Detection of two types of haemoglobin (HGA and HGF) in single erythrocytes by fluorescent antibody technique. *Exptl. Cell Res.* **46**, 596.

Davydoff, C. (1928). "Traité d'Embryologie comparée des Invertebrés". Masson, Paris.

de Graaf, A. R. (1957). A note on the oxygen requirements of *Xenopus laevis*. *J. Exptl. Biol.* **34**, 173.
Devis, R., and James, D. W. (1964). Close association between adult guinea-pig fibroblasts in tissue culture, studied with the electron microscope. *J. Anat.* **98**, 63.
Dewey, M. M., and Barr, L. (1962). Intercellular connection between smooth muscle cells: the nexus. *Science* **137**, 670.
Duboscq, O., and Tuzet, O. (1937). L'ovogenèse, la fécondation et les premiers stades du développement des Éponges calcaires. *Arch. Zool. Exptl. Gen.* **79**, 158.
Eccles, J. C. (1963). "The Physiology of Synapses". Springer, Berlin.
Ewer, D. W. (1959). A toad, *Xenopus laevis*, without haemoglobin. *Nature* **183**, 271.
Farquhar, M. G., and Palade, G. E. (1963). Junctional complexes in various epithelia. *J. Cell Biol.* **17**, 375.
Gurdon, J. B. (1962). Adult frogs derived from the nuclei of single somatic cells. *Develop. Biol.* **4**, 256.
Hörstadius, S. (1937). Experiments on determination in the early development of *Cerebratulus*. *Biol. Bull.* **73**, 317.
Immers, J., and Runnström, J. (1965). Further studies of the effects of deprivation of sulphate on the early development of the sea urchin, *Paracentrotus lividus*. *J. Embryol. Exptl. Morphol.* **14**, 289.
Jaffe, L. (1966). Electrical currents through the developing *Fucus* egg. *Proc. Natl. Acad. Sci. U.S.* **56**, 1102.
Jurand, A., and Tuft, P. (1961). Distribution of water in the chick blastoderm. *Nature* **191**, 1073.
Katsuma Dan (1960). Cytoembryology of echinoderms and amphibia. *Intern. Rev. Cytol.* **9**, 321.
Keilin, D., and Ryley, J. F. (1953). Haemoglobin in Protozoa. *Nature* **172**, 451.
Keilin, D., and Wang, Y. L. (1945). Haemoglobin in root nodules of leguminous plants. *Nature* **155**, 227.
Kelly, D. E. (1966). Fine structure of desmosomes, hemidesmosomes, and an adepidermal globular layer in developing newt epidermis. *J. Cell Biol.* **28**, 51.
King, T. J., and Briggs, R. (1956). Serial transplantation of embryonic nuclei. *Cold Spring Harbor Symp. Quant. Biol.* **21**, 271.
Krahl, M. E. (1961). "The Action of Insulin on Cells". Academic Press, New York.
Lehmann, H., and Huntsman, R. G. (1966). "Man's Haemoglobins". North-Holland Pub., Amsterdam.
Lindahl, P. E., and Stordahl, A. (1937). Zur Kenntnis des vegetativen Stoffwechsels im Seeigelei. *Arch. Entwicklungsmech. Organ. Wilhelm Roux* **136**, 44.
Locke, M. (1965). The structure of septate desmosomes. *J. Cell Biol.* **25**, 166.
Loewenstein, W. R., Socolar, S. J., Higashino, S., Kanno, Y., and Davidson, N. (1965). Intercellular communication: Renal, urinary bladder, sensory and salivary gland cells. *Science* **149**, 295.
Metschnikoff, E. (1886). "Embryologische Studien an Medusen". Hoelder, Vienna.
Monné, L., and Härde, S. (1950). On the formation of the blastocoele and similar embryonic cavities. *Arkiv Zool.* **1**, 463.
Needham, J. (1942). "Biochemistry and Morphogenesis". Cambridge Univ. Press, London and New York.
New, D. A. T. (1956). The formation of sub-blastodermic fluid in hen's eggs. *J. Embryol. Exptl. Morphol.* **4**, 226.
Overton, J. (1962). Desmosome development in normal and reassociating cells in the early chick blastoderm. *Develop. Biol.* **4**, 532.

Parker, G. H. (1919). "The Elementary Nervous System". Lippincott, Philadelphia, Pennsylvania.

Paterson, M. C. (1957). Animal-vegetal balance in amphibian development. *J. Exptl. Zool.* **134**, 183.

Penn, R. D. (1966). Ionic communication between liver cells. *J. Cell Biol.* **29**, 171.

Ross, R., and Greenlee, T. K., Jr. (1966). Electron microscopy: Attachment sites between connective tissue cells. *Science* **153**, 997.

Ruud, J. T. (1954). Vertebrates without erythrocytes and blood pigment. *Nature* **173**, 848.

Sleigh, M. A. (1968). Patterns of ciliary beating. *Soc. Exptl. Biol. Symp.* **22**, 131.

Steward, F. C., Mapes, M. O., and Mears, K. (1958a). Growth and organised development of cultured cells. II. Organisation in cultures grown from freely suspended cells. *Am. J. Botany* **45**, 705.

Steward, F. C., Mapes, M O., and Smith, J. (1958b). Growth and organised development of cultured cells. I. Growth and division of freely suspended cells. *Am. J. Botany* **45**, 693.

Tuft, P. (1961). A morphogenetic effect of beta-mercaptoethanol. Distribution of water in the embryo. *Nature* **191**, 1072.

Willmer, E. N. (1956). Factors which influence the acquisition of flagella by the amoeba, *Naegleria gruberi*. *J. Exptl. Biol.* **33**, 583.

Wood, R. L. (1959). Intercellular attachment in the epithelium of *Hydra* as revealed by electron-microscopy. *J. Biophys. Biochem. Cytol.* **6**, 343.

Zeleny, C. (1904). Experiments on the localization of developmental factors in the nemertine egg. *J. Exptl. Zool.* **1**, 293.

Chapter 10

THE PLANULOID AND ACOELOID STAGES

The Planuloid Stage

From the blastuloid stage, several possible courses are open for further development. In one course the organism remains as a single shell of cells, but by a process of invagination takes on the form of a double cup, or simple gastraea (Fig. 10.1a), the new cavity so formed may remain hollow and become concerned with digestive processes. The original blastocoel is more or less occluded and the "internal environment" is reduced to a minimum.

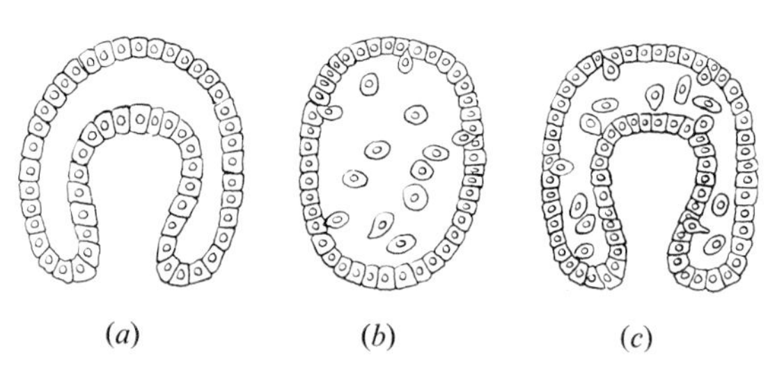

Fig. 10.1. Possible developments from the blastuloid. (*a*) Invagination (gastraea formation); (*b*) Dehiscence; (*c*) Invagination and dehiscence (gastrula formation).

Such diploblastic organisms enjoy the benefits that accrue from having all the cells near the surface and in direct contact with the outside world; indeed, they are to be found with little modification among the Cnidaria of to-day. From the frequency with which they occur as a transitory stage in the embryological development of invertebrates, it is likely that this course is biologically advantageous and so may have been followed more than once. Without further modification it inevitably limits the differentiation and functions of the cells to those that are possible in epithelial cells, i.e. cells which still have contact with the external medium.

Another course of development open to the blastuloid is the colonization of the interior of the hollow shell with cells that leave the surface and sink inwards, as in calcareous sponges (Fig. 10.1b). These cells enter a new environment which, in theory at least, can be controlled by the enclosing epithelial cells. Problems of nutrition, respiration, and excretion are inevitably raised for the inner cells by this course; these problems may be solved to some extent if the invasion of the cavity is also accompanied by a flattening of the organism to a "planula". Such flattening may also facilitate contact with the substratum as it does in sponges, and this may then facilitate locomotion over such a substratum.

A third course is the adoption of both of these procedures and the formation of a gastrula which also has cells between the two layers; this obviously occurs in Echinoderms and many other groups (Fig. 10.1c). As already indicated, examples of all these courses are found in the living embryos of different phyla, though it appears that some variant of the third course has usually been adopted by the majority of animals. This is not surprising, because the third course seems to get the best of both worlds, and once again the gastrula with its contained cells may flatten into a planula.

Since three distinct processes are involved in these manoeuvres, namely the dehiscence of cells individually, the invagination of groups of cells collectively, and the flattening of the organism, the number of possible variations that could be produced by altering the timing and the extent of the three processes could be extremely large. Thus there is little to be gained by attempting to construct a strictly historical hypothesis to account for the actual form of the embryos of the various phyla, for it would have to be largely based on guesswork. It must be remembered that in some cases the embryos may have specialized at some later stage for their own immediate advantage, and such specializations may have no relevance to the main trend towards vertebrate evolution. Thus it is more important to distinguish between and appreciate the effects of the three separate processes than to arrange them in chronological order or to consider them mutually exclusive.

The actual method of dehiscence is important; two distinct processes have been described (Fig. 10.2). In the echinoderm, for example, and indeed in the calcareous sponges, cells have been described as withdrawing their contacts from their neighbours and simply migrating out of the epithelial layer. This must, of course, involve changes not only in the migrating cells themselves, but also in the epithelial cells that are left behind and which are now compelled to reorganize themselves in such a way as to restore the original cellular cohesion effected by desmosomes or some other form

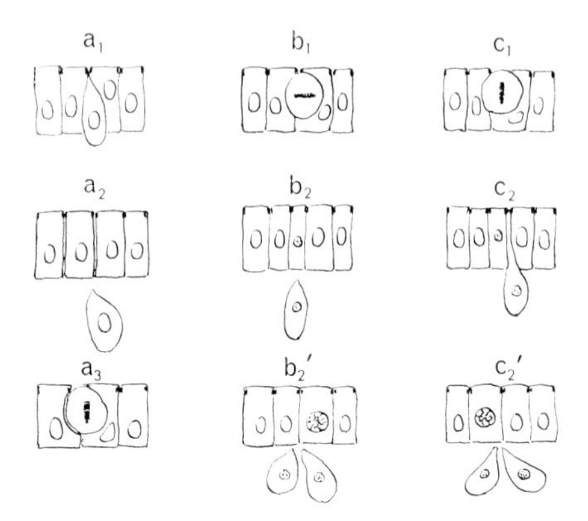

FIG. 10.2. Methods of dehiscence of cells from an epithelium. a_1, a_2, a_3, Direct dehiscence, which may or may not be immediately accompanied by cell division; b_1, b_2, b'_2, dehiscence of one (b) or both (b′) daughter cells after division, with "horizontal" metaphase plate; c_1, c_2, c'_2, dehiscence of one (c) or both (c′) daughter cells after division with "vertical" metaphase plate.

of adhesive mechanism. In other cases, cells leaving an epithelial surface do so only after a division and then usually after one in which the metaphase plate is orientated parallel to the epithelial surface instead of at right angles to it. Something akin to the latter form of dehiscence but with the plate orientated at right angles to the surface is seen in the formation of neuroblasts in the vertebrates (Fig. 10.3). However, in that situation some of the cells that apparently migrate are actually held by their terminal bars and are drawn back into the epithelium before the next division (Sauer, 1935; Watterson, 1965). No matter how the actual dehiscence is effected, it must involve changes in the character of the cell surface of the separating cell. This surface change is most important in determining subsequent behaviour.

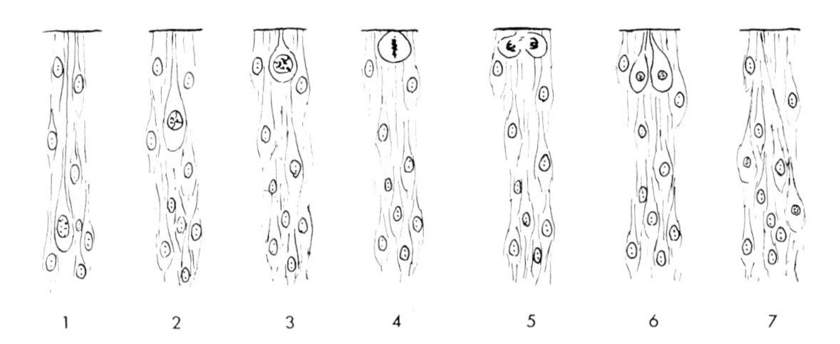

FIG. 10.3. Successive stages in the division of a neuroblast. Late prophase, metaphase, and anaphase occur when the cell has been drawn up to the luminal surface of the neural tube. (After Sauer, 1935.)

Sometimes the change of character of the cell is determined from within, and sometimes it is imposed on the cell by external conditions. The following examples show something of the range of events. In sponge blastulae, cells have been described as leaving the epithelium spontaneously. Presumably, in this case, the change occurs within the cell in question. In the stratified epithelium of the oesophagus of the rat, cells enter the stratum spinosum after division. Sometimes both daughters enter and differentiate; sometimes only one daughter enters and differentiates and the other remains behind and perhaps divides again; sometimes both daughters remain behind as basal cells. In this tissue, therefore, cell division ceases and differentiation begins when the cells enter the stratum spinosum, and it would thus seem that the position of the cell, i.e. its local environment, determines its course of action (Marques-Pereira and Leblond, 1965). In the testis of the monkey, spermatogonium type A_1 may divide and produce two daughter cells of type A_1 or two of A_2. When an A_2 cell divides, it always produces two type B_1 cells. In this tissue, differentiation seems to be initiated at division (Clermont and Leblond, 1959).

In the ontogeny of vertebrates the dehiscence of cells, from either the outer (ectodermal) or inner (endodermal) layer of cells or from both, and also direct gastrulation, by some form of invagination, may all play a part. Moreover, one of the effects of flattening a spherical organism that already has an antero-posterior (animal-vegetal) axis is the accentuation or production of a dorso-ventral axis and bilateral symmetry. Since both of these properties are possessed by the embryos of vertebrates, it becomes directly relevant to the present discussion to consider a hypothetical "planuloid" state as a possible step in vertebrate phylogeny (Fig. 10.4).

By planuloid stage (Fig. 10.5) is meant a flattened organism with a superficial epithelium and a contained inner mass of cells or "mesohyl",

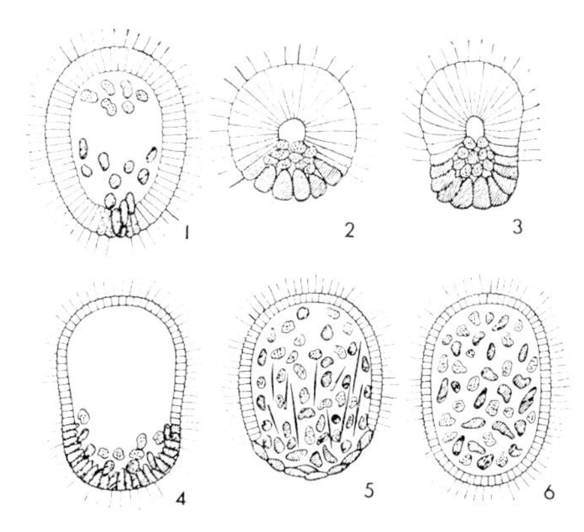

FIG. 10.4. Cells entering the cavity of sponge blastulae (Minchin, 1900.)

to use the term coined by Hadži (1949). Such a planula-like organism could be produced by dehiscence of cells from the surface, followed by flattening, or the two processes could occur simultaneously, or indeed in the opposite order. Actual invagination may or may not be involved in the formation of a simple planuloid, and it should perhaps be emphasized again that the planulae living to-day should not be regarded as necessarily related in any

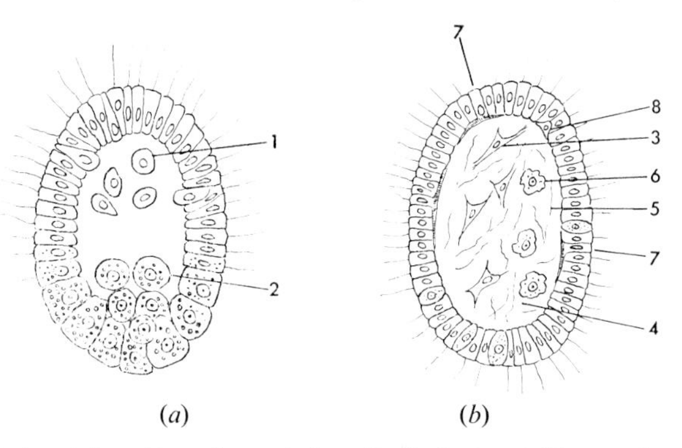

FIG. 10.5. Horizontal sections through hypothetical planuloid organisms. (*a*) Simple planuloid. The cavity is filling with dehisced cells derived from two classes of epitheliocytes; (*b*) more advanced planuloid with some specialization of the interior and the formation of epithelio-muscular cells. 1, premechanocyte; 2, preamoebocyte; 3, mechanocyte, 4, mesohyl; 5, collagen fibre; 6, amoebocyte; 7, epithelio-muscular cell; 8, Nerve cell.

way to the hypothetical planuloid organism under discussion. Existing planulae (see Fig. 10.13) are found mostly as transitory stages in the development of more elaborate creatures, and they have solved their physiological problems sufficiently to survive to the present day. The hypothetical planuloid was, however, superseded by the more successful organisms to which it gave rise.

In the conversion of the blastuloid organism outlined in the preceding chapter into the planuloid, the entering cells could be derived from the ecballic flagellate blastomeres and become potential mechanocytes, from the emballic phagocytic blastomeres and become amoebocytes, or from both groups together, either simultaneously or in some definite time sequence. This process of dehiscence is clearly seen in the calcareous sponges (see Fig. 10.4) in which cells of both types can break contact with the surface layer and leave it to enter the inner mass or mesohyl. Once inside, the flagellate cells undergo further differentiation to become scleroblasts and also the dermal epithelium and porocytes, while the phagocytic cells maintain their amoebocytic qualities and some, at least, remain totipotent and give rise to the germ cells. The term mesohyl, or middle substance, is really much more expressive and meaningful than the earlier term mesoderm, since the latter, if only by analogy with ectoderm and endoderm, implies a definite layer of cells such as would be formed, for example, by invagination; this layer may never be there. Moreover, tissue culture has shown (see p. 134) that the rigid concept of ectoderm, mesoderm, and endoderm is not, physiologically speaking, very useful. The really fundamental step is taken when an epithelial cell changes its properties and loses its ability to adhere to its neighbours in true epithelial fashion, and thereby acquires either the characters and semi-independent existence of the mechanocyte or the complete independence of the amoebocyte. Independence is perhaps not the best term to use in this connexion because it obscures the important fact that the mesohyl cells, since they are enclosed in a shell of superficial epitheliocytes, must always be to some extent dependent on the latter for their nutriment and, indeed, for the whole of their immediate environment.

It is important to accentuate the fact that the cells that have left the superficial epithelium and have acquired a quasi-independent existence need not be regarded as reverting to the state they were in before they could cohere into an epithelium. It is much more likely that they have achieved a secondary independence, which may be limited to the particular environment reigning inside the organism of which they form a part and to which they will sooner or later contribute. This change involving dehiscence may in the first instance have been reversible, but sooner or later it may have become so standardized or deep-seated that it has led to a more or less irreversible form of differentiation and a characteristic pattern of development. It is, however, relevant to notice in this connexion that cell replacement in the epidermis of some living planarians is by recruitment from the cells of the underlying mesohyl (Skaer, 1965). This may be an example of cells from the mesohyl being still in a reversible or uncommitted state, but, before it is possible to be certain of this, it will be necessary to know more about the physiological properties and embryological history of the cells in question. Planarians are efficient enough to have survived until the present day, and the manoeuvres by which to-day's planarian has survived may now be different from those which were adequate for the earlier organisms that have since been replaced by "higher" forms.

We have referred (p. 208) to cells which appear to leave an epithelium yet actually remain connected with the surface layer by their terminal bars, and may at times be able to return to the epithelium temporarily. The completeness of any separation of cells from an epithelium must clearly be firmly established before true dehiscence and its consequences can be assumed.

To return to the main theme, the essential historical sequence of events up to this point may have been somewhat as follows. A cell leading an independent existence became first a cell capable of uniting with its neighbours to form a blastuloid organism. Some of these cells then changed and again acquired a limited freedom, this time to live in the internal environment. The last cells, the cells of the mesohyl, must be expected to have been much changed during the course of this evolutionary history and to have acquired properties very different from those of the initial free-living cell. It must be emphasized that this is merely a suggested historical sequence of evolutionary events; it should not be regarded as a direct ontological one such as might be applicable to any organism now living. Ontological events are frequently coloured by the requirements of the local environment in which the animal has to survive, and no living organism to-day is likely to correspond exactly to the organisms extant when life had progressed only to the stage when the planuloid organism still represented the top of the evolutionary tree.

In this account so far, the vital steps in forming the mesohyl of the planuloid have been considered to be the changes in the epithelial cells that caused them to lose contact with the neighbouring epithelial cells. It has also been assumed that these changes, occurring in the two main classes of cells, are identical with those that have produced the mechanocyte and the amoebocyte. Certainly, in tissue cultures of certain epithelia (e.g. skin) this sort of behaviour (i.e. the splitting off of individual cells) has been observed to produce mechanocyte-like cells. It is, however, possible that the process is in each case divisible into two, namely loss of contact with neighbouring epitheliocytes and subsequent differentiation. This subdivision, with its possibility of altering the time sequence of the two events, would make some of the differentiations that occur in higher animals more intelligible. For example, in the formation of the sex cords in the ovaries of vertebrates (see p. 469) epithelial cells still seem to maintain their epithelial potentialities in spite of having entered the mesenchyme, i.e. in spite of having lost contact with the surface.

As already mentioned, it is also theoretically possible that a planula-like organism could arise as the result of gastrulation by invagination followed by flattening and by occlusion of the blastopore. This, however, would produce initially an organism of very different type, because the epithelial cells of the inner layer presumably would not have undergone the essential change that would convert them into mechanocyte or amoebocyte. They could still remain strictly epithelial, adhering to each other, or they could lose contact with their neighbouring epitheliocytes and yet not differentiate further. If, on the other hand, invagination were combined with some degree of dehiscence, the result might be almost indistinguishable from that produced by direct dehiscence alone (see Fig. 10.13). Thus, again, the timing of the various processes inevitably produces different passing phases. The closure of the blastopore and the consequent separation of the invaginated cells from the external environment could well be a potent stimulus to further differentiation.

Some sort of planuloid organism, formed essentially by dehiscence and perhaps partly by invagination, can thus be considered to be the next stage along the line of vertebrate descent and could have acquired the new and relevant characters listed in Table 10.1.

TABLE 10.1

New Characters in the Planuloid

1. Cells of the mesohyl, divisible into two main classes as mechanocytes and amcebocytes corresponding to the two main classes of primary epitheliocytes from which they are derived.
2. The possibility of the development of the internal environment, e.g. by the formation of an intercellular matrix
3. The development of a transport system, e.g. to carry food substances, etc., by the use of amoeboid wandering cells in the mesohyl
4. Accentuation, by flattening, of the development of a dorso-ventral axis and of directional movement of the whole animal, e.g. by cilia, leading to an antero-posterior axis and when both occur together, to bilateral symmetry.

On the Nature of Mechanocytes

It is probable that mechanocytes in general are derived (by analogy with the origin of similar cells in existing calcareous sponges) from the dehiscing of the flagellated, ecballic blastomeres. This is borne out by the observation (Scherft and Daems, 1967) that such undoubted mechanocytes as chondroblasts and osteoblasts of vertebrates often possess cilia, both *in vivo* and *in vitro*. It is also likely that in the process of becoming mechanocytes the ecballic cells have acquired or developed the potentialities which they display in culture, i.e. to form reticulin (Levi, 1931) and collagen fibres (McKinney, 1930; Green and Goldberg, 1965; Jackson, 1966) and lay down an extracellular mucoprotein matrix (Grossfeld *et al.*, 1957; Fitton Jackson and Randall, 1956; Morris and Godman, 1960). Certainly the existing adult calcareous sponges which develop from simple blastulae and planulae show this capacity for forming collagen fibres and ground-substance. Therefore, it must be assumed either that the capacity has evolved at least twice, quite independently, or that it is latent in the planuloid stage whether or not collagen, etc., actually appear in any existing planula.

At this stage, therefore, it is pertinent to examine more closely the claim that the flagellated or ciliated epithelial cell is specially related to the mechanocytes. Since collagen and hexosamines are such important products of mechanocyte activity, it is relevant to ask if there are any traces of similar activity in epithelial cells or even in earlier forms. On p. 48 it was shown that collagen is composed very largely of amino acids that are not required as nutrients by the cells forming them. Thus these amino acids or combinations of them could be regarded as cell products, perhaps even the products of some cellular activities initially having nothing specific to do with fibre production, e.g. the preservation of ionic balance or some form of cell movement. It is thus possible that these amino acids may be characteristic metabolites of the flagellate or ciliate form of cell activity and might thus make their appearance as cell products in other forms. If these metabolites were formed by an epithelial or free-living cell, they could be excreted directly and lost, unless they were required for some specific purpose within the cell. For example, it is a matter of common observation that ciliated cells frequently have a layer of PAS-positive material trapped between the cilia which perhaps represents a similar excretion of the polysaccharide by-products. On the other hand, if the metabolites continue to be formed and passed out of the cells after the cells have left the superficial epithelium and entered the mesohyl, their further excretion may present some difficulties. Under these circumstances some better use may have been found for the originally waste material.

That these are not idle thoughts is indicated by the following observations, the first examples of which are drawn from the plant world. When growing fast as suspension cultures under the influence of added coconut milk, cells of carrot, potato, bean, sycamore, and crown-gall, and therefore presumably of many other types also, show an increase in water content and changes in salt absorption. At the same time they produce a protein in their cell walls that is very rich in hydroxyproline (Steward *et al.*, 1958a, b; Lamport and Northcote, 1960); and this hydroxyproline is probably formed from proline made by the cells (Pollard and Steward, 1959). Furthermore, when different amino acids are added to such cell cultures, it is found that hydroxyproline, in particular, is strongly inhibitory to growth (Steward *et al.*, 1958a). Fries (1951) has also shown that hydroxyproline inhibits the growth of roots and particularly of shoots at concentrations far below the inhibitory concentrations of other amino acids. It would thus appear to be important that the hydroxyproline should not accumulate as such in the neighbourhood of the growing cell producing it.

It has already been noted on p. 48 that, when mammalian and avian fibroblasts are grown *in vitro*, they do not normally require, and may indeed be inhibited by, the presence in the surrounding medium of glutamic acid, proline, hydroxyproline (Biggers *et al.*, 1957; Morgan and Morton, 1957), and glucosamine (Ely *et al.*, 1953). Sarcoma cells, which are generally mechanocytic, are also inhibited by glucosamine (Quastel and Cantero, 1953, Rubin *et al.*, 1954). There is, however, one notable exception: a cell-line from the Chinese hamster has been produced that specifically requires proline in the medium (Ham, 1963). The exact nature of this cell, which was derived from the ovary, is of course important and is, unfortunately, dubious.

These observations are consistent with the view that hydroxyproline is a cell product which, in soluble form, does the cell no good, and, therefore, that there are advantages in incorporating it into a relatively insoluble and inert material in the cell wall in the case of plants, or into structurally useful extracellular collagen fibres in the case of animals. Alternatively, it may simply be that its incorporation into the proteins makes the proteins suitable for wall material on the one hand, or for strengthening and supporting fibres on the other.

In the Cnidaria the cnidoblasts are characteristic and conspicuous cells that develop from the interstitial cells and move about below the superficial epithelium (Fig. 10.6). These cells develop a vacuole that is converted into

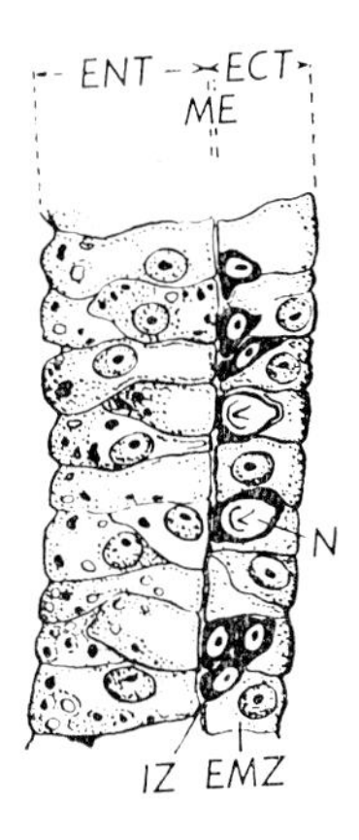

FIG. 10.6. Interstitial cells and cnidoblasts in the wall of a Hydrozoon. ECT, ectoderm; EMZ, ectoderm cell; ENT, endoderm; IZ, interstitial cell; ME, mesoglea; N, nematocyst. (Tardent, 1963.)

a capsule generally containing an invaginated and twisted thread of intricate structure. The whole constitutes the nematocyst (Fig. 10.7) whose thread, armed with elaborately arranged barbs, can be forcibly everted by a rise of pressure in the capsule. This is a most remarkable structure in many ways, but, from the present point of view, its particular interest lies in the chemistry of its capsule, thread, and fluid. The capsule and thread contain a collagen-like protein and the fluid contains mucoprotein (Johnson and Lenhoff, 1958). About 22% of the protein is hydroxyproline, and it also has a high glycine and high alanine content. Blanquet and Lenhoff (1966) regard the capsule as being composed of a di-sulphide-linked collagenous protein. In the sulphide bonding it has some resemblance to the collagen of the cuticle of the nematode, *Ascaris* (Mathews, 1967). This is interesting because the substance is not quite typical collagen and because epithelial cells have some part to play in its formation. The situation in the cnidoblast,

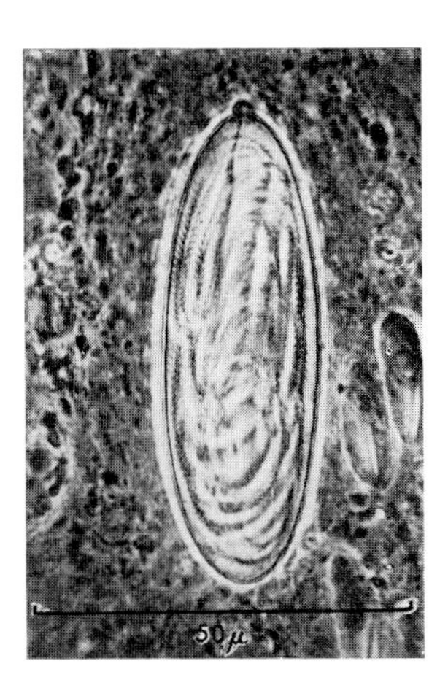

FIG. 10.7. Nematocysts from *Corynactis viridis*. The large structure is a holotrichous isorhiza. The two small structures are microbasic mastigophores. The capsule and the thread are collagenous; the fluid contains polysaccharides. (Photo by R. J. Skaer.)

which is not strictly a mesohyl cell, but more a modified epithelial cell, appears to be similar. The fluid within the nematocyst capsule is rich in glutamic and aspartic acids and has a chondroitin-like substance composed of galactosamine and glucuronic acid (Phillips, 1956; Picken and Skaer, 1966). The whole structure therefore has the same main constituents as the matrix of connective tissues, the product of mechanocytes.

Cnidoblasts often bear cilia in addition to nematocysts and are probably derivatives or modifications of the ciliated type of cell. Is it therefore possible that, when the ciliated cell sinks into the mesohyl or leaves the outer surface of the organism, it is faced with the problem of doing something with its normal metabolic products, namely its hydroxyproline, glycine, etc., on the one hand and its hexosamines and uronic acids on the other? In the Cnidaria the problem has been partly solved by using them for the formation of offensive and defensive weapons, the nematocysts, whereas in higher animals

the same metabolites have been utilised more for the formation of a structural matrix for support. With the exception of the gelatinous umbrella of the medusae, the relatively sparse connective tissues (mesoglea) of the Cnidaria and the abundance of their cnidoblasts, in contrast to the absence of cnidoblasts and the abundance of fibrous connective tissue in higher animals, are certainly suggestive of some relationship of this sort. *Hydra* and other Cnidaria, in addition to their cnidoblasts, also possess some collagen and mucopolysaccharides in their mesoglea (Shostak *et al.*, 1965). Thus the production of cnidoblasts and the formation of mesoglea or other supporting matrix appear to be alternative methods of utilizing the metabolites of dehisced cells. In some groups, one method prevails, in some, the other.

It is also interesting to note that the acoels have sagittocysts and the rhabdocoels have rhabdites (Fig. 10.8). Though differently composed and

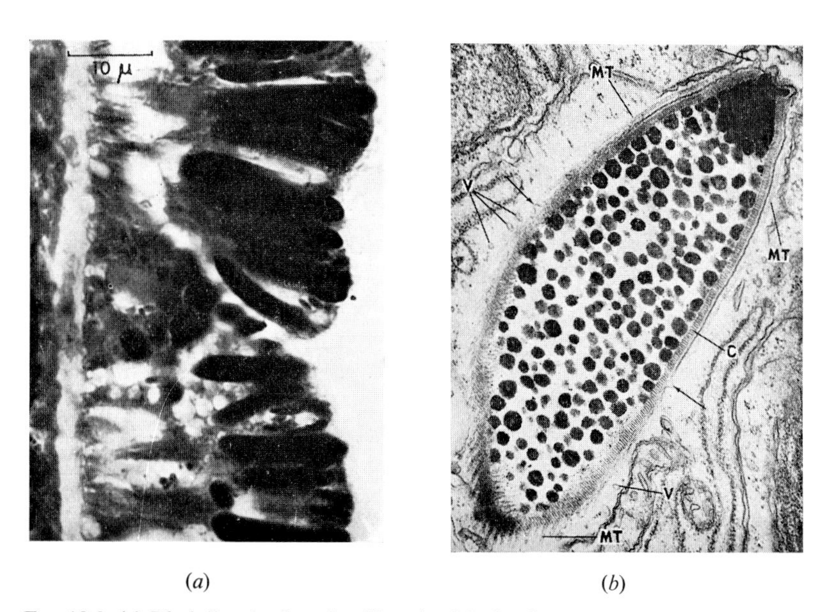

(*a*) (*b*)

FIG. 10.8, (*a*) Rhabdites in dorsal epidermis of *Polycelis tenuis*. (Photo by J. R. Skaer.) (*b*) Osmic-gallate preparation of a rhabdite in longitudinal section. C, Capsule; MT, microtubules; V, vesicles. Arrows indicate vesicles fusing with the membrane. (Lentz, 1967.)

organized from both nematocysts and collagen fibres, they may represent other methods of solving a very similar problem, namely that of using the metabolites produced by what would otherwise be ciliated cells or their derivatives and turning them to some biological advantage. In neither acoels nor rhabdocoels is connective tissue abundant, though the latter

animals certainly possess some. Unfortunately, little is known about the composition of rhabdites except that, since they absorb ultraviolet light rather strongly at $\lambda = 265$ mμ and are very basic, neither collagen nor mucopolysaccharide appears to be very likely as major components. They have a high arginine content (Skaer, 1961).

Another observation of interest is that, in many ciliate and flagellate protozoa, trichocysts are found (Fig. 10.9). They are small protein-containing

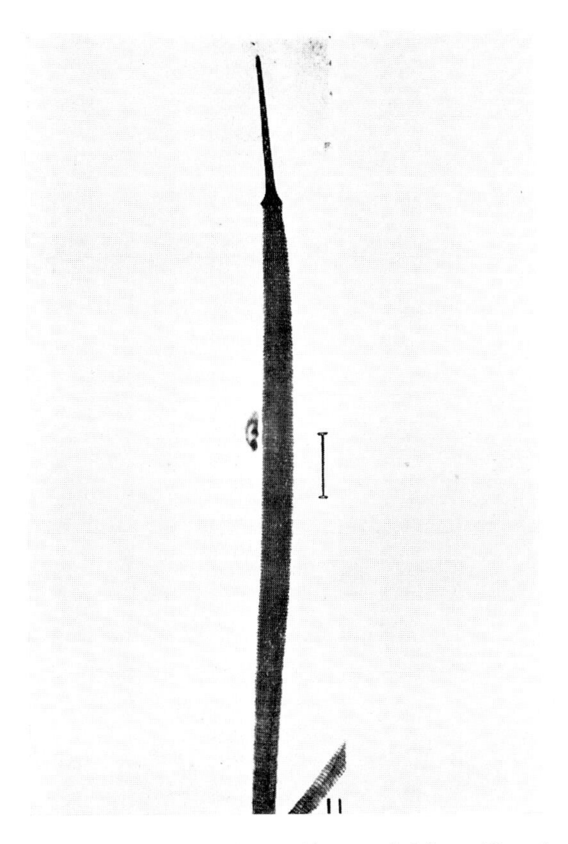

FIG. 10.9. Trichocyst of *Paramecium*, isolated in extended form. Note the cross-banding. (Jakus, 1945.)

bodies lying just within the surface of the cell whose tips can be ejected into the medium, often by a lengthening and narrowing of the inner part that may lead to the production of an elongated and cross-banded fibrous structure in some ways resembling collagen. The whole mechanism is rather like a miniature nematocyst. It would thus be interesting to know more about

the composition of these structures in view of their connection with cilia. The central body stains with basic dyes and could be mucoprotein, though there does not seem to be any direct evidence that it is.

Finally, it may be pointed out that ciliated and flagellated cells often possess well-developed "root-fibres" penetrating into the cytoplasm from the basal bodies of the cilia or flagella (Fig. 10.10). The electron-microscope reveals that such root-fibres and related structures are again often strongly cross-banded in a manner reminiscent of collagen. Though the spacing of the bands is not identical with that of most naturally occurring collagens,

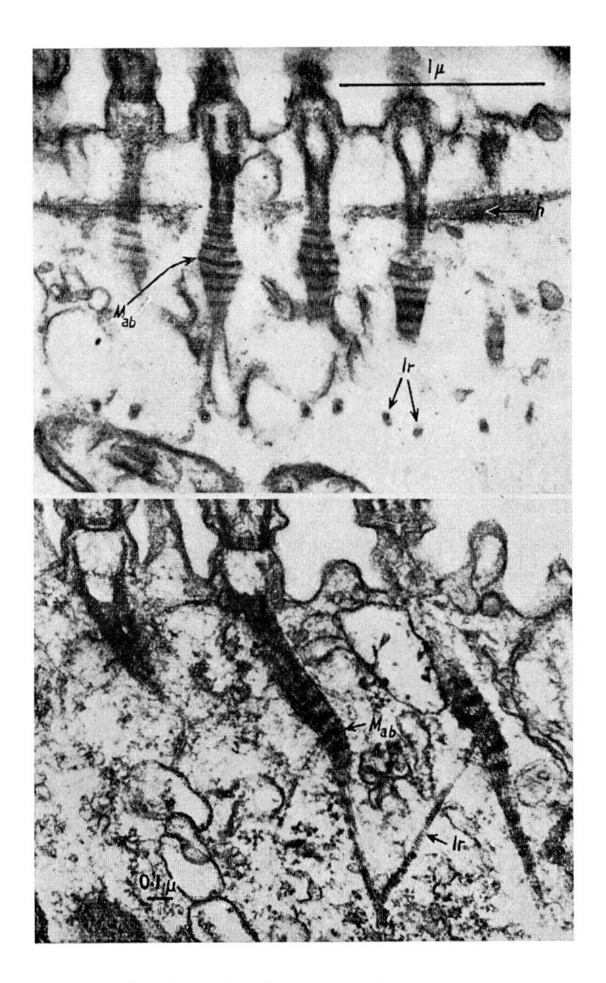

FIG. 10.10. Root-fibres (M_{ab}) in ciliated cells, showing the striated pattern. Note also the lateral branching (lr) and the interconnecting fibres (h). (Dorey, 1965.)

it is not impossible that these intracellular fibres are related to collagen and built from similar ingredients (Fawcett and Porter, 1954; Sedar and Porter, 1955; Rouiller *et al.*, 1956; Fawcett, 1958). As yet, no analyses appear to have been made of their amino acid composition.

The observations on *Naegleria* (see p. 168) emphasized the close relationship between cell form and ionic balance. The hollow blastula (see p. 203) was envisaged as making use of both ecballic and emballic cells to act symbiotically to maintain the volume and composition of the internal fluid of the blastocoel. Derivatives of these two classes of cells, the mechanocytes and the amoebocytes, thus probably still maintain remnants of these original ionic-balancing properties. Their differences in metabolism and activities may thus still be determined by their ionic relationships with their immediate local environments. In other words, the equilibrium of a connective tissue may well depend on the balance of activities of its contained mechanocytes and macrophages or amoebocytes. Mechanocytes, then, should perhaps be considered primarily as ecballic cells, and their production of proline, hydroxyproline, glycine, glutamic acid, hexosamines, and uronic acids as secondary to this form of activity.

Fibroblasts probably make polypeptides as precursors of tropocollagen (Jackson and Smith, 1957; Smith and Jackson, 1957) and it seems likely that the transport of them through the cell membrane is in some way linked to the change of proline to hydroxyproline (Ross, 1968). Hulliger *et al.* (1957), on the other hand, have stated that free hydroxyproline increases in the medium in which chick-heart fibroblasts are growing.

The formation of polypeptides containing a high proportion of proline raises an interesting question concerning the origin of the rather widely distributed defensive and offensive agents related to bradykinin. This biologically very active substance is a nonapeptide with the sequence, arginine-proline-proline-glycine-phenylalanine-serine-proline-phenylalanine-arginine (see Collier, 1968). It thus seems to have affinity with collagen on account of its high proline content, and with rhabdites on account of its high arginine content. Is the formation of this polypeptide perhaps another useful method of disposing of the same metabolites? The idea receives some support in that, just as elastin is a collagen-like protein containing important lysine groupings, so also is there a lysyl-bradykinin.

Observations on the metabolism of undoubted fibroblasts are scarce, but there are some observations on various cells in tissue culture and on "natural" tissues that may be relevant to the relationship between metabolism and ionic equilibria. For example, ascites tumour cells (derived from cervix uteri and growing *in vitro*) lose potassium and gain sodium when they are treated with glycine; when treated with glutamate, they gain both potassium and sodium (cf. *Naegleria*, p. 170–1). Furthermore, when the cells are treated

with potassium, glycine uptake (which is normally very rapid in these cells) is inhibited. All of these observations indicate that ionic balance and these amino acids are in some way interconnected (Christensen and Riggs, 1952). Glutamate uptake and potassium uptake have also been shown to be closely connected in nerve cells and retinal cells (Krebs and Eggleston, 1949; Terner *et al.*, 1950), and, as will be shown later (p. 410), many retinal cells, as derivatives of ependymal cells, are closely related to flagellated cells.

The observation that the fluid in the blastocoel of sea-urchin embryos contains sulphated mucopolysaccharides (Monné and Härde, 1950; Immers, 1961) suggests that the epithelial cells, which in these early stages are ciliated, liberate these substances directly into their surroundings without first storing them in their cytoplasm and subsequently liberating them as a "secretion".

The question that needs to be answered therefore is: What advantages are gained by ecballic cells when they liberate hexosamines and uronic acids, together with polypeptides containing substantial proportions of glycine, proline, hydroxyproline, and glutamic acid? Similarly, it may be asked why glutamine is so beneficial to fibroblastic growth in cultures as it is often found to be (Eagle and Levintow, 1965; Biggers, 1965; Waymouth, 1965). Is it possible that some or all of these substances are related to the ionic input and output of fibroblasts? Glutamate is certainly concerned with potassium intake in some nerve tissues (Terner *et al.*, 1950). Also, there are some remarkable changes in glutamine synthetase activity in the developing retina at about the time that the visual processes (modified cilia) of the receptor cells are developing (Piddington and Moscona, 1965). Glutamine can provide the NH_3 for the formation of amino sugars (Lowther and Rogers, 1953; Leloir and Cardini, 1953). Since glutamic acid is a product of this reaction, it would not be impossible for this substance both to alter the ionic distribution and to contribute to the production of proline and hence of hydroxyproline.

Not only could the formation and ejection of amino sugars from cells have been turned to advantage by combining them with uronic acids to produce matrices like those of the connective tissues, but also their production must have been an early step in the production of chitin, a characteristic product of many epithelia, though, curiously enough, not among the vertebrates. All these substances, i.e. chitin (acetyl glucosamine polymers), hyaluronic acid (acetyl glucosamine and glucuronic acid polymers), and related chondroitins (see Fig. 10.11), should thus perhaps be considered useful by-products resulting from some essential metabolic activity of ecballic cells. The final stages in the formation of these polysaccharide chains depend on uridine triphosphate, a nucleotide specially associated with cell membranes. It combines with the carbohydrate moieties to form uridine-diphosphate-acetyl glucosamine, uridine-diphosphate-hexuronic acid,

etc. When the carbohydrates are so combined, an enzyme, also associated with particulate matter (probably cytoplasmic membranes) effects the linkage of the carbohydrate fragments necessary for the formation of the polymers, hyaluronic acid, chondroitins A, B, and C, cellulose, chitin, etc. The enzyme is strongly activated by Mg^{2+} ions (Dorfman, 1965), and it is pertinent to notice that polysaccharides are apparently formed in the smooth endoplasmic reticulum.

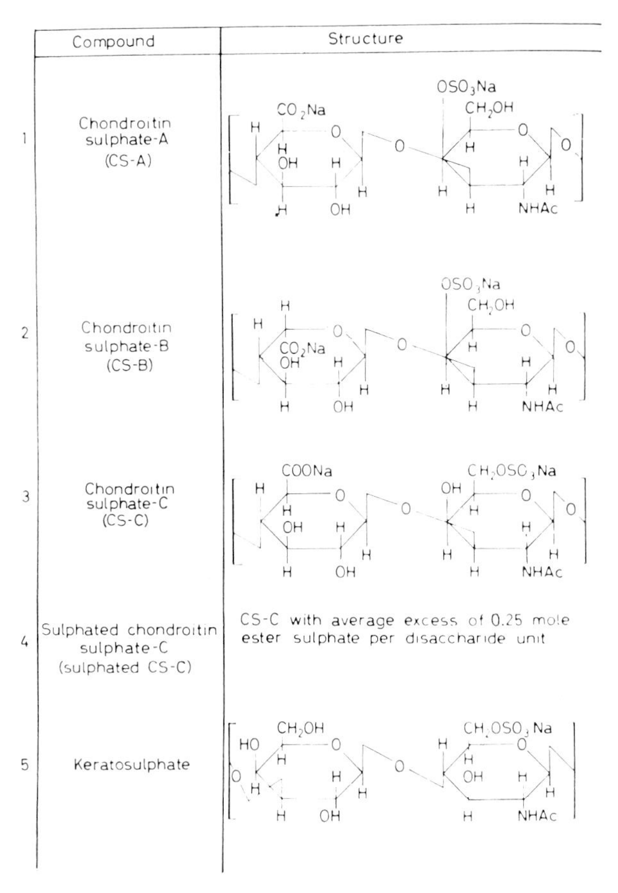

FIG. 10.11. Repeating disaccharide units of acid mucopolysaccharides. (Mathews, 1965.)

It is interesting that these processes are associated with the cell surface to such an extent, though there are of course obvious structural advantages to be obtained from producing these polymers. Nevertheless, it is also worth considering how the processes started. The utilization of some

metabolite (e.g. the hexosamine), which might otherwise be lost, seems a possible solution.

The water-binding capacity of hyaluronic acid is remarkable, and the manner in which it is affected by ions, e.g. HCO_3^-, could be very important in relation to the maintenance of ionic equilibria. Moreover, the specific affinity towards Na^+ rather than K^+ ions (10:1) which can be displayed by hyaluronic acid is another observation of particular relevance to ion transport and balance (Kulonen, 1952).

The absence of sulphate ions from the surrounding medium is one of the conditions that favour the animalization of embryos such as those of the sea-urchin (Fig. 10.12) (Needham, 1942; Immers and Runnström, 1965),

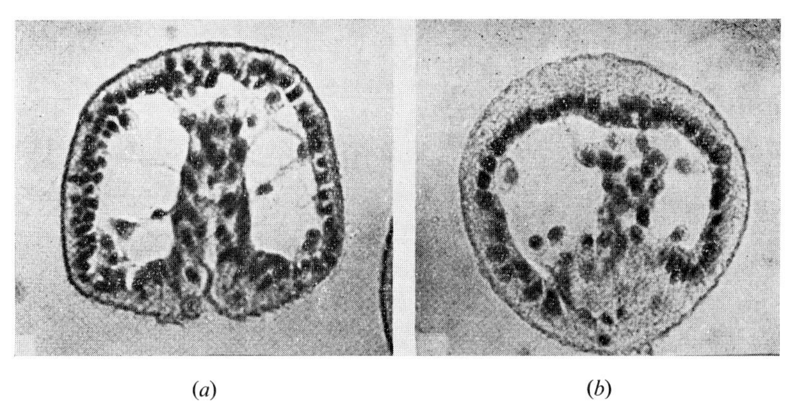

(*a*) (*b*)

FIG. 10.12. The effect of absence of sulphate in "animalizing" the embryo of the sea-urchin. (*a*) Embryo in normal sea-water; (*b*) embryo in sea-water $-SO_4^{2-}$. (Immers and Runnström, 1965.)

and, on the other hand, the chondroitins of connective tissues tend to be heavily sulphated. It is thus possible that this sulphation could have arisen as a means of removing the sulphate ion from the vicinity of the mechanocyte, whose activities might otherwise have been impeded. The relationship between uronic acids, sulphate, and cellular activity, however, comes up again in another context and, there also, it is shown to concern ionic movement (p. 498).

In contrast to these very characteristic features of mechanocytes and their metabolism, amoebocytes, derived from cells that need to conserve water or eject ions (i.e. they are emballic cells), presumably also have their metabolism geared to these activities. However, apart from the frequent intracellular storage of polysaccharides and sulphated polysaccharides and often a filmy coating of mucopolysaccharide attached to the cell surface,

they do not seem to have consistently produced by-products like those of the mechanocytes. Their active pinocytosis and phagocytosis and their possession of numerous lysosomes and phagosomes suggest that traffic through the cell membrane of these cells may be organized on a basis different from that of the mechanocytes.

While it is fairly evident that mechanocytes and amoebocytes are fundamentally different, the exact nature of these differences is still far from being understood. The reason lies partly in the general failure on the part of biochemically minded cytologists to appreciate the existence of this difference, and partly in the fact that most tissues contain representatives of both families of cells. Thus, although some of the observations recorded above are exlusively relevant to ecballic cells (mechanocytes), i.e. they were obtained from studies of these cells in relatively pure cultures, the majority of metabolic studies have used whole tissues, e.g. brain, bone, etc. In consequence the measurements that have been made are likely to be some resultant of the activities of both mechanocytes and amoebocytes. Intensive studies of the metabolism of pure cultures of mechanocytes and of pure cultures of amoebocytes are greatly needed, together with studies of how their respective metabolisms alter, if they do alter, when the ionic balance of the system is changed. Probably, if more information were available concerning the metabolism of pure cultures of macrophages, the nature of the symbiosis that appears to exist between fibroblasts and macrophages would emerge more clearly. Such information could be of the utmost practical value in helping to solve the problems of the rheumatic and arthritic diseases in which the normal balance of fibre and matrix formation is more or less disturbed.

Planuloid Organization

To return once more to the original theme: it is unlikely that the form of organization found in the adult calcareous sponge of the present day has much relevance to that which eventually led to the organization characteristic of the vertebrates. On the other hand, the form of organization exemplified by the blastula and even the early planula of such sponges does seem to have great relevance.

Planuloid organisms display certain definite features which are of great importance. They have a flattened structure and have internal cells that are no longer epithelial and are enclosed within a continuous epithelial covering. In contrast, it should be emphasized that the gastraea, produced by direct invagination of a blastula (see Fig. 10.1), is still, from the present point of view, a blastuloid organism.

A planuloid organism could acquire an internal cavity either by invagination (Fig. 10.13), before or after the formation of the mesohyl, or by hollowing out a space among the dehisced cells and lining it with a secondary epithelium. A cavity formed in the first way might be considered to be an intestine lined with endoderm, whether or not it remained open to the exterior, though to regard it as such would seem to suggest that it should play some special part in digestion and absorption. There are no *a priori* reasons for supposing that either ecballic or emballic blastomeres should invaginate preferentially in producing such an invagination, except that the amoebocytic (emballic) cells have greater powers of phagocytosis and the

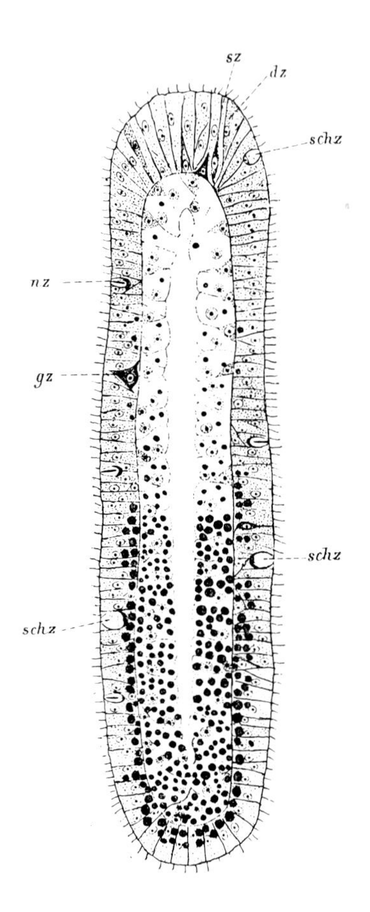

FIG. 10.13. Longitudinal section of the planula of *Gonothyraea loveni*. dz, Gland cell; gz, "ganglion" cell; nz, nematocyst; schz, mucous cell; sz, sensory cell. Note the gradient of cell types in the "endoderm" and the mixture of cells in the "ectoderm". (Wulfert, 1902.)

mechanocytic (ecballic) cells may have other means of absorbing nutrient materials. Their cohesive powers are also different. A cavity formed by hollowing out the mesohyl might be called a coelom, that formed by invagination, an enteron. At this point, however, we are in danger of getting into the deep waters of controversy and prejudice concerning the meanings and definitions of such terms as coelom, archenteron, blastocoel, schizocoel, endoderm, mesoderm, and mesenchyme. Such discussions have bedevilled embryology for far too long and are, in general, not really very helpful because they take little account of the nature, physiological properties, or past history of the actual cells that constitute or surround the structures or cavities in question, or of the nature of the contents of the cavities. Certain aspects of some of these structures must inevitably be considered later, in their respective contexts, but at this stage there is no need to cloud the main issue by discussing them here.

Sexuality exists in a variety of forms of the Bacteria; it is more developed in the Protozoa. At some time in the very early stages of evolution towards the vertebrates the basic sexual process with elaborate meiotic divisions must have become stabilized and firmly established, though the details of its actual timing and mechanism seem always to have remained peculiarly labile and variable from group to group. For this reason the sexual apparatus of animals frequently constitutes one of the main criteria for establishing specific and generic differences. In general, however, a sessile, reserve-storing, passive, but essentially amoeboid form characterizes the ovum, while active motility, generally brought about by flagellar movement, characterizes the sperm. In the early stages, both in phylogeny and in ontogeny, germ cells which are initially amoebocytic seem, like *Naegleria*, to have the potentiality of developing along either of these paths. In several cases the immediate environment of the cell has been shown to determine the direction of the differentiation. The behaviour of *Naegleria* in different media may thus be relevant to any discussion of the differentiation of germ cells (see p. 475), and so also may the idea of symbiosis between the two forms of blastomere in the formation of stable blastulae. Planuloid organisms and, indeed, blastuloid organisms must have had some system of sexual reproduction. Male and female gametes can be assumed to have been of normal occurrence at this stage of organization. The details of their differentiation, maturation, and release, however, can only be surmised and may well have been many and varied.

In spite of what has just been said about the terminology of the germ layers and the like, the significance to be attached to the terms ectoderm and endoderm does need some immediate clarification. In the simplest situations, as in the formation of the gastraea of the Scyphozoa, the terms are applied, respectively, to the superficial epithelial layer and to the

epithelial layer which invaginates. This really tells us nothing of the nature of the cells in the two layers, however. In the sponge blastula depicted in Fig. 10.4,2, the "endoderm" is synonymous with cells of the vegetal pole and is entirely composed of amoeboid emballic cells, and the ectoderm (animal pole) consists of flagellate ecballic cells. This is a simple and probably atypical case, but it illustrates the confusion of thought which inevitably occurs if cells of the vegetal pole and cells of the endoderm are considered to be necessarily homologous. There is more often a gradient of transition of ecballic cells to emballic cells, instead of a sudden difference, and cell movements often mingle the types of cells. The centre of invagination need not be the centre of the emballic cells. In the gastrula of amphibia, for example, the centre of invagination is in the region of the grey crescent, so that ecballic and emballic cells could invaginate more or less equally. Electron-microscope studies of this region certainly demonstrate invaginating cells with differing internal organization (Baker, 1965).

Moreover, in epithelia in general, there is always considerable shuffling of position of the individual cells, so that the original gradient of ecballic cells progressing to emballic cells may become very obscured by the mingling of the cells of the two types. This mingling can be seen in the frog embryo, for example, where parts of the ectoderm of the neurula show a complete mosaic of ciliated and non-ciliated cells (Fig. 10.14) (Assheton, 1896).

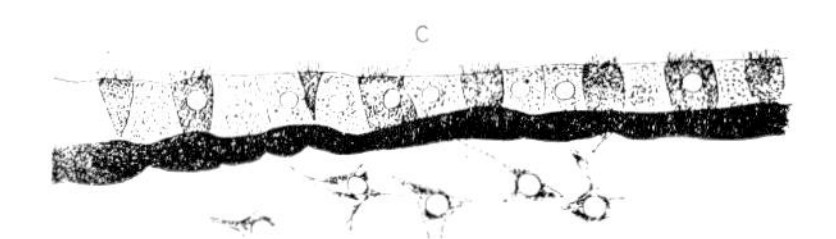

FIG. 10.14. Epidermis of frog neurula, showing a mosaic of ciliated (C) and non-cilitated cells with mesenchyme cells below the basement membrane. (Assheton, 1896.)

In the simple gastrula the ectoderm and endoderm both separate the outside world from the mesohyl or the blastocoel cavity. Though there may be some difference in their relative activities because the ectoderm cells mainly, though not necessarily, cover a convex surface and the endoderm cells habitually cover a concave surface, and because in some cases the endoderm may, as in the Cnidaria, be composed largely of emballic blastomeres while the ectoderm consists predominantly of cells of ecballic type, there is no *a priori* reason for believing that endoderm cells always function in any fundamentally different way from ectoderm cells. Both layers of cells may be formed from mixed populations of cells; certainly in all the observations that have been made on cells from the two layers in cultures of vertebrate tissues, there is none to suggest that ectoderm and endoderm cells differ in any essential, physiological way. Any differences that are

observable are almost certainly differences in their own right, are caused by adaptation to local environments, and are not differences caused by basic ectodermal or endodermal properties. Both epithelia are layers of cells that, like all epithelia, separate two different environments, e.g. in these simple embryonic stages, the outside world and the internal fluid. For this purpose, as in the original blastuloid organism, two types of cells in each layer are probably better than one. It would therefore be advantageous if both ectoderm and endoderm were mixed epithelia including both ecballic and emballic cells. This, in general, is what they appear to be.

All these considerations therefore reduce the value, in physiological terms, of the distinction between ectoderm and endoderm, a distinction which had already become blurred by anomalies in development and by the behaviour of the cells from the two stocks in tissue cultures (see p. 67). Nevertheless, it is probably true that, in the more primitive forms like the sponge embryos, the amoeboid cells, being phagocytic and less coherent, were invaginated first for digestive, nutritional, and protective purposes. An important factor could be the nature of the attachments between the epithelial cells. Tight junctions and special attachment zones are perhaps better developed in the ciliated (ecballic) epithelia than in those composed of emballic cells; thus, the outer coat might be stronger if composed of the ecballic cells. As already noted, there is little information as yet on the nature of the junctions between the cells of various blastulae. The blastomeres of the sea-urchin embryo show both interdigitations of the cell boundaries and septate desmosomes (Balinsky, 1959). Septate desmosomes are also found in the epithelium of such "primitive" forms as *Hydra* (Wood, 1959). By analogy with ciliated epithelia elsewhere, some sort of desmosomal connexion is probable between the originally flagellated cells. It is interesting that mechanocytes also show desmosome-like structures when they make stable contacts (Devis and James, 1964; Ross and Greenlee, 1966), while amoebocytic cells seem to have either no contacts or interdigitating surfaces. In mixed epithelia the amoebocytic cells can take part in specialized junctions with ciliated cells.

In the hypothetical planuloid organism, envisaged here as derived from some organism resembling the sponge embryo, the invaginated portion might properly be thought to be concerned mainly with the acquisition and digestion of food. However, since most of the feeding of such organisms could have involved not only some sort of phagocytosis with intracellular digestion but also saprophytic activities, this need not have been so. Ciliary or flagellar currents may of course have assisted, as in the sponges, in directing food particles to the digesting or absorbing cells. Such currents may also assist in oxygenation and in the movement of the organism.

Muscle, as such, is unlikely to have developed at this stage. The planulae

of the present day (e.g. of *Gonothyraea*; see Fig. 10.13) which lack muscle have considerable powers of movement on a flat surface by virtue of the cilia on their under surface. Presumably the ordinary changes of shape shown by almost all cells in the course of their everyday activities would also lead to some changes of form of the organisms as a whole. The beat of the cilia is sooner or later co-ordinated into metachronal rhythms; this presumably implies the existence of neuroid transmission between the ciliated cells (Parker, 1919) as well as integration within each cell, perhaps by root-fibres and spurs (see Fig. 10.10) (Dorey, 1965). It is probable that, in most of those embryonic stages of existing organisms that could be considered comparable to the hypothetical planuloid, neither true muscles, nor nerve cells, nor any of their associated cells are present. For this reason it is unlikely that nerve and muscle had differentiated at the stage of evolution corresponding to the state of complexity characteristic of the planuloid form, unless perhaps the formation of myoepithelial cells preceded the complete dehiscence of cells from the epithelium. Similarly, the histochemical observations of Lentz (1966) (see p. 149) perhaps indicate that the beginnings or neurohumoral mechanisms might well be present among the mesohyl cells at this sort of stage of organization. On the other hand, because many planulae do acquire muscle and nerve cells later, the prerequisites for these structures may in fact be present in the planuloid stage. The origins of nerve and muscle systems form an important step forward in evolution. They are more appropriately considered in connexion with later stages than the planuloid.

The Acoeloid Stage

From the planuloid stage a logical step forward would be to an organism somewhat resembling the Acoela of the present day, though with certain importance differences. This organism might be described as an "acoeloid"; it resembles a more sophisticated planuloid (Fig. 10.15).

The epidermis has begun to differentiate overtly into a mixture of cells of different types, e.g. mucus-secreting "goblet" cells are mingled with ciliated cells. Other epidermal cells also send long processes which run between the bases of the main cells in the epithelial sheet or immediately below them. These processes are contractile, and they constitute the beginnings of a muscular system. Other cells have processes which are specialized for conduction and are thus the rudiments of a subepithelial nervous system. Moreover, during the early embryonic stages of living acoels, certain superficial cells sink into the inner mass, especially at the anterior end, and then develop into the brain and main nervous system.

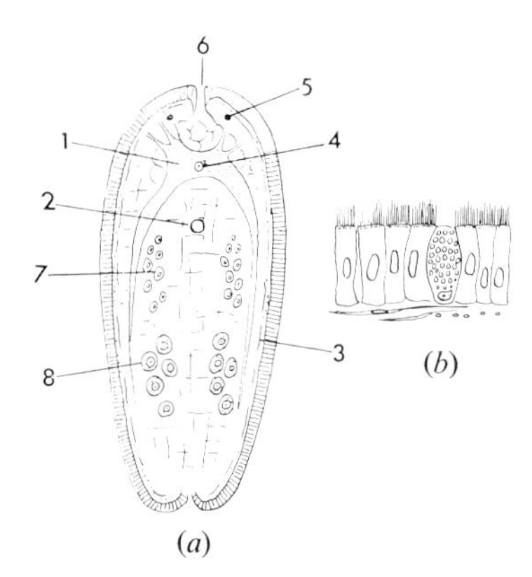

FIG. 10.15. Diagram of acoeloid organization. (*a*) Dorsal view; (*b*) epidermis. 1, Nervous system; 2, pharynx; 3, muscle; 4, statocyst; 5, eye; 6, frontal organ; 7, male germ cells; 8, female germ cells.

If this origin of the nerve cells that form the main nervous system occurred in the acoeloid, it is interesting because it suggests that the strictly epithelial character of the cells is lost at an early stage and that the cells forming the nervous system may share properties with either mechanocytes or amoebocytes, depending on the character of the epithelial cells before they left the surface. This contrasts somewhat with the more direct development of the cells of the central nervous system of vertebrates from the neuro-epithelium, and, as will be seen later, is a matter of some importance. In the vertebrates, only the cells of the ganglia migrate away from the neural ectoderm and wander in the mesohyl until they reach their final destination. In the acoeloid, the sensory cells in the epithelium presumably sent their processes to join the main nervous system, though their cell bodies may have remained strictly epithelial.

The mechanocytes in the acoeloid organisms probably had the capacity for producing a coherent ground-substance together with collagen fibres, at least to separate the epithelium from the mesohyl, thus forming a basement membrane. It is, however, interesting to note that, in some acoels of the present day, collagen fibres are in fact absent or nearly so (Dorey, 1965). The whole organism in living acoels is essentially cellular, and nowhere, except in the wall of the otocyst, does the ground-substance amount to anything in the way of a skeletal or supporting structure. The relative absence of collagen raises an interesting point because the acoeloid stage is being judged to be a somewhat higher state of organization than the larva of the sponge (the blastuloid stage). Yet the sponge larva, as already noted (see p. 146), goes on to develop collagen fibres and ground-substance in the adult sponge, whereas in existing acoels these inert materials are

produced in only one place. Either collagen fibres have been evolved twice in relation to mechanocytes or, more likely, both stocks, i.e. Porifera and Acoela, have diverged from some blastuloid creatures in which the genetic mechanism for collagen-fibre formation was latently present, probably in connexion with the ecballic epitheliocytes. In the living acoels the necessary conditions for the appearance of collagen fibre are not generally fulfilled, or its production is suppressed for some other reason which remains obscure for the present. Collagen fibres and ground-substance appear freely in the rhabdocoels, animals which are, as we shall see, just that much more differentiated than the acoels; they also occur in the Cnidaria, where again they may be formed by the ecballic epitheliocytes. Alternatively, the living acoels could be regarded as aberrant and possibly derived, as others have suggested, not in the manner favoured here, i.e. by aggregation from protozoa, but by the splitting up of some multinucleate ciliate protozoon into separate cells leading to the formation of tissues and organs from what were initially intracellular organelles (see Hanson, 1958). The similarity between living acoels and organisms described here as acoeloid would then be largely fortuitous. For this reason the term acoeloid has been preferred, and it should not be interpreted to mean that existing acoels are closely related to the hypothetical organisms under discussion.

The simple way to interpret the acoeloid pattern seems to be to assume that cells which leave the epidermis tend to differentiate in the first instance either into mechanocytes or into amoebocytes, depending on their origin (and these differentiations have certainly occurred in acoels). Then it may be assumed also that the mechanocytes have the potentialities to produce the connective tissue elements, though they do not happen to do so, in any existing acoels, and probably both types may undergo further differentiation into nerve cells and muscle cells.

With regard to the production of connective-tissue matrices it may be relevant to point out that it is not yet known what factors determine whether or not mechanocytes *in vitro* produce collagen fibres or mucoproteins. Under some conditions of culture, both are produced in large quantities and, under other conditions, neither appear. Sometimes the one is produced without the other, at other times the reverse. In view of the activities of *Naegleria*, the ionic relationships of these cultures might be worth investigation, particularly the Na/K ratio and the Mg^{2+} availability (the latter is necessary for polysaccharide synthesis; see p. 222).

To return to the hypothetical acoeloid organism relevant to the main line of vertebrate evolution: it would probably possess ciliated and goblet cells in its epidermis, mechanocytes and amoebocytes in a mesohyl containing a ground-substance of collagen and mucoprotein, the beginnings of a nervous system, and the beginnings of a muscular system. With regard

to the function of nerve and muscle, it is difficult to know by inspection when a pseudopodial process should be called a nerve process, and when a muscle process (see p. 32). All such processes are probably capable of some degree of conduction and also of contraction, though whether this contraction is similar to muscular contraction is not known. It is notable, both in the tube feet of echinoderms and in the so-called ventral nerve roots of *Amphioxus*, that each "nerve" fibre is in fact an extension of the muscle, i.e. a long process of the muscle cell which may even be striated (Cobb, 1967); Flood, 1966). Thus it is quite possible that the mechanocytes and amoebocytes could both contribute to the formation of nerves and muscles; if they both did so, each would probably impart something of its own character to the emergent nerve or muscle.

One of the main difficulties in investigating the structure of the living acoels is the extraordinarily complex pattern of the parenchymal cells which interdigitate with each other in a most intricate manner (Dorey, 1965; Pedersen, 1964). It could be that this is the manner in which these creatures have evolved a system of mutual coherence in the absence of the structural scaffolding of collagen and mucoprotein. Continued cell mobility within this mass may thus solve other problems of oxygenation, nutrition, and excretion which would face cells in a more permanently organized system resting on a structural framework. The living acoels are thus different from the hypothetical acoeloid.

The acoeloid would have some motility other than that caused by cilia, but it would probably not have much in the way of an organized, epithelium-lined intestine. However, in living forms there is a pharynx leading from the outside into a space in the parenchyma or mesohyl where digestion goes on intracellularly and absorption is direct, phagocytosis playing a large part. This arrangement in the living acoels could be taken as indicative of the manner in which an elementary gut might subsequently develop.

In addition to nerve cells (conducting), there are also in living acoels, and, indeed, in some living planulae, certain cells in the epithelium which are thought to be sense cells of one sort or another, e.g. both single sensory cells and also flagellated cells grouped together, as in the frontal organ, which may mediate some form of chemical sense.

The acoeloid stage, like the planuloid, has definite bilateral symmetry and well developed antero-posterior and dorso-ventral axes. These properties make an acoeloid organism relevant to the present discussion and representative of a stage definitely more developed than the planuloid stage. Creatures at about this level of organization presumably went on to evolve in a variety of ways which, though fascinating in themselves, are probably irrelevant to the present discussion which is primarily seeking a line of evolution that could reasonably be supposed to culminate in the vertebrate

stock. From this point of view, Porifera and Cnidaria fall by the wayside at about the planuloid or acoeloid stages and follow their own specializations. So far as certain traits may be common to Porifera, Cnidaria, and Vertebrata, the likelihood is that these traits were actually or potentially present in the ancestors of all three groups. This is not to deny the possibility of convergent evolution or of the second, *de novo*, occurrence of certain features. It seems probable, however, that convergence is far more likely to occur by adaptation of similar raw materials or of existing biochemical processes than it is to occur by the development of entirely new and different materials. In mundane human affairs, given a basic supply of bricks, timber, tiles, etc., there is bound to be a certain similarity in the sort of building that can be produced, though they may be constructed by different races in different parts of the world. By and large, Nature, too, must use what is at hand and modify it; the organization of living matter is so complex that the introduction of some basically new material or method is fraught with even more difficulties than are the introductions of new processes into human affairs. Things that do not immediately work in Nature tend to be eliminated, and existing materials and processes tend to be adapted to new functions as and when it is profitable to be so adapted. Carotenoid pigments, for example, are used by Nature to provide light-sensitivity to man, fish, birds, insects, crustacea, molluscs, and probably to cnidaria. This does not mean that these creatures are all closely related to each other, nor that their eyes are directly comparable. It means simply that the application of carotenoids to mechanisms increasing photosensitivity was developed at a stage long before all these forms of life diverged, and these mechanisms have since been modified for use in a variety of different situations. This explanation of the sporadic occurrence of visual pigments is made the more probable by the observations that many plants also have their photosensitivity determined by carotenoids, as do several flagellate protozoa. The particular way in which the carotenoid is used and the particular organization of the

TABLE 10.2

New Characters in the Acoeloid

1. A mixed epidermis (though this is only a matter of degree) containing ciliated cells, goblet cells, and modifications thereof
2. Basement membrane between epidermis and mesohyl
3. Epithelio-muscular and muscular cells
4. Neuro-epithelial cells, separate nerve cells, and a separate nervous system
5. A pharynx and elementary gut, probably largely dependent on phagocytosis and intra-cellular digestion

photoreceptors could, however, be much more informative concerning the possible relationships between their possessors. Indeed they are shown to be on a later page (see p. 379), though they are so basic to the requirements of the visual process as not to have very much practical significance.

Thus, by the time that the acoeloid stage has been reached, several more vertebrate characters had probably appeared. They are listed in Table 10.2.

References

Assheton, R. (1896). Notes on the ciliation of the amphibian embryo. *Quart. J. Microscop. Sci.* **38**, 465.

Baker, P. C. (1965). Fine structure and morphogenetic movements in gastrula of the tree-frog. *Hyla regilla. J. Cell Biol.* **24**, 95.

Balinsky, B. I. (1959). An electronmicroscopic investigation of the mechanisms of adhesion of the cells in the sea-urchin blastula and gastrula. *Exptl. Cell Res.* **16**, 429.

Biggers, J. D. (1965). Cartilage and bone. *In* "Cells and Tissues in Culture" (E. N. Willmer, ed.), Vol. 2, p. 197. Academic Press, New York.

Biggers, J. D., Webb, M., Parker, R. C., and Healy, G. M. (1957). Cultivation of embryonic chick bones on chemically defined media. *Nature* **180**, 825.

Blanquet, R., and Lenhoff, H. M. (1966). A disulphide-linked collagenous protein of nematocyst capsules. *Science* **154**, 153.

Christensen, H. N., and Riggs, T. R. (1952). Concentrative uptake of amino acids by the Ehrlich mouse ascites carcinoma cell. *J. Biol. Chem.* **194**, 57.

Clermont, Y., and Leblond, C. P. (1959). Differentiation and renewal of spermatogonia in the monkey, *Macacus rhesus. Am. J. Anat.* **104**, 237.

Cobb, J. L. S. (1967). The innervation of the ampulla of the tube foot in the starfish *Astropecten irregularis. Proc. Roy. Soc. (London)* **B168**, 91.

Collier, H. D. J. (1968). Bradykinin and its allies. *Endeavour* **27**, 14.

Devis, R., and James, D. W. (1964). Close association between adult guinea-pig fibroblasts in tissue culture, studied with the electron microscope. *J. Anat.* **98**, 63.

Dorey, A. E. (1965). The organization and replacement of the epidermis in acoelous turbellarians. *Quart. J. Microscop. Sci.* **106**, 147.

DORFMAN, A. (1965). The biosynthesis of acid mucopolysaccharides. *In* "Structure and Function of Connective and Skeletal Tissue" (S. F. Jackson *et al.*, eds.) p. 297. Butterworth, London.

Eagle, H., and Levintow, L. (1965). Amino acid and protein metabolism. *In* "Cells and Tissues in Culture" (E. N. Willmer, ed.), Vol. 1, p. 277. Academic Press, New York.

Ely, J. O., Tull, F. A., and Schanen, J. M. (1953). The effect of glucosamine on culture of embryo chicken heart cells. *J. Franklin Inst.* **255**, 561.

Fawcett, D. W. (1958). *In* "Frontiers in Cytology" (S. L. Palay, ed.), pp. 19–41. Yale Univ. Press, New Haven, Connecticut.

Fawcett, D. W., and Porter, K. R. (1954). Fine structure of ciliated epithelia. *J. Morphol.* **94**, 221.

Fitton Jackson, S., and Randall, J. T. (1956). Fibrogenesis and the formation of matrix in developing bone. *In* "Bone Structure and Metabolism." (C. E. W. Wolstenholme and C. M. O'Connor, eds.) *Ciba Found. Symp.*

Flood, P. R. (1966). A peculiar mode of muscular innervation in *Amphioxus*. Light and electron microscopic studies of the so-called ventral roots. *J. Comp. Neurol.* **126**, 181.

Fries, N. (1951). The influence of amino-acids on growth and lateral root formation in cotyledonous pea seedlings. *Experientia* **7**, 378.

Green, H., and Goldberg, B. (1965). Collagen synthesis by cultured cells. *In* "Structure and Function of Connective and Skeletal Tissues" (S. F. Jackson *et al.*, eds.), p. 288. Butterworth, London.

Grossfeld, H., Meyer, K., Godman, G., and Linker, A. (1957). Mucopolysaccharides produced in tissue culture. *J. Biophys. Biochem. Cytol.* **3**, 391.

Hadži, J. (1949). Problem mezoderma in celoma v luči turbelarijske teorije knidarijev. *Razprave. Razreda za prirodoslovne in medicinske vede slovenske akademije znanosti in utmetnosti v. Ljubljani.* Knjiga IV.

Ham, R. G. (1963). An improved nutrient solution for diploid Chinese hamster and human cell lines. *Exptl. Cell Res.* **29**, 515.

Hanson, E. D. (1958). On the origin of the Eumetazoa. *Syst. Zool.* **7**, 16.

Hulliger, L., James, D. W. and Allgower, M. (1957). Hydroxyproline production in chick fibroblast cultures. *Experientia* **13**, 454.

Immers, J. (1961). Comparative study of the localization of incorporated ^{14}C-labeled amino acids and $^{35}SO_4$ in the sea-urchin ovary, egg and embryo. *Exptl. Cell Res.* **24**, 356.

Immers, J., and Runnström, J. (1965). Further studies of the effects of deprivation of sulphate on the early development of the sea-urchin *Paracentrotus lividus*. *J. Embryol. Exptl. Morphol.* **14**, 289.

Jackson, S. F. (1966). The molecular organization of cells and tissues in culture. *In* "Cells and Tissues in Culture" (E. N. Willmer, ed.), Vol. 3, p.1. Academic Press, New York.

Jackson, S. F. and Smith, R. H. (1957). Studies on the biosynthesis of collagen. 1. The growth of fowl osteoblasts and the formation of collagen in tissue culture. *J. Biophys. Biochem. Cytol.* **3**, 897.

Jakus, M. A. (1945). The structure and properties of the trichocysts of *Paramecium*. *J. Exptl. Zool.* **100**, 457.

Johnson, F. B., and Lenhoff, H. M. (1958). Histochemical study of purified *Hydra* nematocysts. *J. Histochem. Cytochem.* **6**, 394.

Krebs, H. A., and Eggleston, L. V. (1949). An effect of L-glutamate on the loss of K-ions by brain slices suspended in a saline medium. *Biochem. J.* **44**, vii.

Kulonen, E. (1952). On the relation of hyaluronic acid to the water and electrolyte metabolism. *Acta. Physiol. Scand.* **27**, 82.

Lamport, D. T. A., and Northcote, D. H. (1960). Hydroxyproline in primary cell walls of higher plants. *Nature* **188**, 665.

Leloir, L. F., and Cardini, C. E. (1953). The biosynthesis of glucosamine. *Biochim. Biophys. Acta* **12**, 15.

Lentz, T. L. (1966). Histochemical localization of neurohumours in a sponge. *J. Exptl. Zool.* **162**, 171.

Lentz, T. L. (1967). Rhabdite formation in planaria. The role of microtubules. *J. Ultrastruct. Res.* **17**, 114

Levi, G. M. (1931). Ricerche sulla istogenesi delle fibre collagene e reticolari nelle colture *in vitro*. *Arch. Exptl. Zellforsch Gewebezücht.* **11**, 189.

Lowther, D. A., and Rogers, H. J. (1953). The relation of glutamine to the synthesis of hyaluronate or hyaluronate-like substances by Streptococci. *Biochem. J.* **53**, xxxix.

McKinney, R. L. (1930). Studies on fibres in tissue culture. The development of reticulum into collagenous fibres in cultures of adult rabbit lymph nodes. *Arch. Exptl. Zellforsch. Gewebezücht.* **9**, 14.

Marques-Pereira, J. P., and Leblond, C. P. (1965). Mitosis and differentiation in the stratified epithelium of the rat oesophagus. *Am. J. Anat.* **117**, 73.

Mathews, M. B. (1965). Molecular evolution of connective tissue. A comparative study of acid mucopolysaccharide-protein complexes. *In* "Structure and Function of Connective and Skeletal Tissue" (S. F. Jackson *et al.*, eds.), p. 181. Butterworth, London.

Mathews, M. B. (1967). Macromolecular evolution of connective tissue. *Biol. Rev. Cambridge Phil. Soc.* **42**, 499.

Minchin, E. A. (1900). The Porifera. *In* "Treatise on Zoology" (E. R. Lankester, ed.), Vol. 2, pp. 1–178. Black, London.

Monné, L., and Härde, S. (1950). On the formation of the blastocoele and similar embryonic cavities. *Arkiv Zool.* **1**, 463.

Morgan, J. F., and Morton, H. J. (1957). The nutrition of animal tissues cultivated *in vitro*. IV. Amino acid requirements of chick embryonic heart fibroblasts. *J. Biophys. Biochem. Cytol.* **3**, 141.

Morris, C. C., and Godman, G. C. (1960). Production of acid mucopolysaccharides by fibroblasts in cell cultures. *Nature* **188**, 407.

Needham, J. (1942). "Biochemistry and Morphogenesis," p. 486. Cambridge Univ. Press, London and New York.

Parker, G. H. (1919). "The Elementary Nervous System". Lippincott, Philadelphia, Pennsylvania.

Pedersen, K. J. (1964). The cellular organization of *Convoluta convoluta*, an acoel turbellarian: a cytological, histochemical and fine structural study. *Z. Zellforsch. Mikroskop. Anat.* **64**, 655.

Phillips, J. H. (1956). Isolation of active nematocysts of *Metridium senile* and their chemical composition. *Nature* **178**, 932.

Picken, L. E. R., and Skaer, R. J. (1966). A review of researches on nematocysts. *In* "The Cnidaria and Their Evolution" (W. J. Rees, ed.), p. 19. Academic Press, New York.

Piddington, R., and Moscona, A. A. (1965). Correspondence between glutamine synthetase activity and differentiation in the embryonic retina *in situ* and in culture. *J. Cell Biol.* **27**, 247.

Pollard, J. K., and Steward, F. C. (1959). The use of C^{14}-proline by growing cells: Its conversion to protein and to hydroxyproline. *J. Exptl. Bot.* **10**, 17.

Quastel, J. H., and Cantero, A., (1953). Inhibition of tumour growth by D-glucosamine. *Nature* **171**, 252

Ross, R. (1968). The fibroblast and wound repair. *Biol. Rev. Cambridge Phil. Soc.* **43**, 51.

Ross, R., and Greenlee, T. K. Jr. (1966). Electron microscopy: Attachment sites between connective tissue cells. *Science* **153**, 997.

Rouiller, C., Fauré-Fremiet, E., and Gauchery, M. (1956). Origine ciliaire des fibrilles scléroprotéiques pédonculaires chez les ciliés peritriches. Étude au microscope électronique. *Exptl. Cell Res.* **11**, 527.

Rubin, A., Springer, G. F., and Hogue, M. J. (1954). The effect of D-glucosamine hydrochloride and related compounds on tissue cultures of the solid form of mouse sarcoma 37. *Cancer Res.* **14**, 456.

Sauer, F. C. (1935). Mitosis in the neural tube. *J. Comp. Neurol.* **62**, 377.

Scherft, J. P., and Daems, W. T. (1967). Single cilia in chondrocytes. *J. Ultrastruct. Res.* **19**, 546.

Sedar, A. W., and Porter, K. R. (1955). The fine structure of cortical components of *Paramecium multimicronucleatum*. *J. Biophys. Biochem. Cytol.* **1**, 583.

Shostak, S., Patel, N. G., and Burnett, A. L. (1965). The role of mesoglea in mass cell movement in *Hydra*. *Develop. Biol.* **12**, 434.

Skaer, R. J. (1961). Some aspects of the cytology of *Polycelis nigra*. *Quart. J. Microscop. Sci.* **102**, 295.

Skaer, R. J. (1965). The origin and continuous replacement of epidermal cells in the planarian *Polycelis tenuis* (Iijima). *J. Embryol. Exptl. Morphol.* **13**, 129.

Smith, R. H., and Jackson, S. F. (1957). Studies on the biosynthesis of collagen. 2. The conversion of ^{14}C-L-proline to ^{14}C-hydroxyproline by fowl osteoblasts in tissue culture. *J. Biophys. Biochem. Cytol.* **3**, 897.

Steward, F. C., Pollard, J. K., Patchett, A. A., and Witkop, B. (1958a). The effects of selected nitrogen compounds on the growth of plant tissue cultures. *Biochim. Biophys. Acta* **28**, 308.

Steward, F. C., Thompson, J. F., and Pollard, J. K. (1958b). Contrasts in the nitrogenous composition of rapidly growing and non-growing plant tissue. *J. Exptl. Botany* **9**, 1.

Tardent, R. (1963). Regeneration in the Hydrozoa. *Biol. Rev. Cambridge Phil. Soc.* **38**, 293.

Terner, C., Eggleston, L. V., and Krebs, H. A. (1950). The role of glutamic acid in the transport of potassium in brain and retina. *Biochem. J.* **47**, 139.

Watterson, R. L. (1965). Structure and mitotic behaviour of the early neural tube. *In* "Organogenesis" (R. L. De Haan and H. Ursprung, eds.), p. 129. Holt, Rinehart & Winston, New York.

Waymouth, C. (1965). Construction and use of synthetic media. *In* "Cells and Tissues in Culture" (E. N. Willmer, ed.), Vol. 1, p. 99. Academic Press, New York.

Wood, R. L. (1959). Intercellular attachment in the epithelium of *Hydra* as revealed by electronmicroscopy. *J. Biophys. Biochem. Cytol.* **6**, 343.

Wulfert, J. (1902). Die Embryonalentwicklung von *Gonothyraea loveni* Allm. *Z. Wiss. Zool.* **71**, 296.

Chapter 11

THE RHABDOCOELOID STAGE

The next important stage along the line to the vertebrates can be conveniently taken as the rhabdocoeloid stage (Fig. 11.1), which has many features of the present-day rhabdocoels and allied turbellarians. Again, as with the acoeloid stage, it is important not to consider any one living rhabdocoel (or other turbellarian) as a potential ancestor. The characters

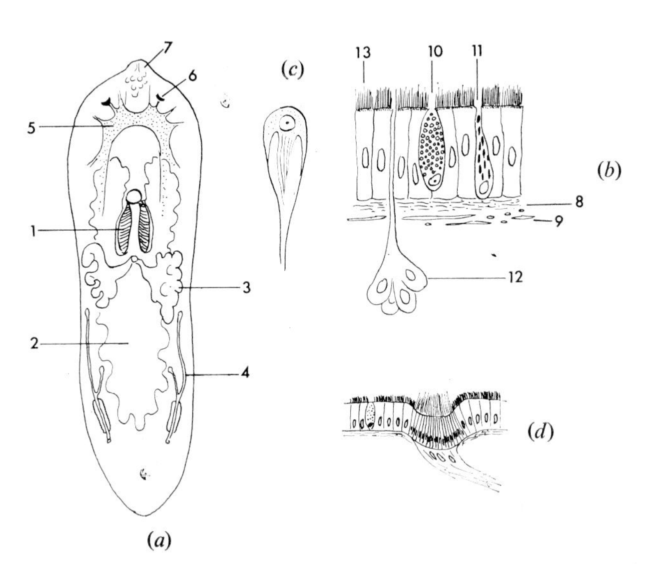

Fig. 11.1. (*a*) Diagram of the main features of the rhabdocoeloid stage. 1, Pharynx; 2, intestine; 3, gonad; 4, nephridia; 5, brain and nerve cords; 6, eye-spot; 7, frontal organ. (*b*) Section of epidermis. 8, Basement membrane; 9, muscle fibres; 10, mucous cell; 11, rhabdite-containing cell; 12, long-necked mucous cells; 13, ciliated cell. (*c*) Flame cell at head of nephridial tube. (*d*) Sensory area in epidermis.

to consider are all those that are relevant to the development of vertebrates and which may be found in the rhabdocoels as a whole and for which it may therefore be presumed that a genetic basis has already developed at this stage. Whether or not any one particular character finds expression in any one particular species must depend on a number of factors. The existence of even one species possessing the character, however, means that a genetic basis for it exists in organisms at about this level of organization. Moreover, it is worth emphasizing again that the genetic basis for even the simplest morphological character is probably extremely complex and involves the consequences of the orderly production of several proteins. It is not the sort of thing that can be produced at the wave of a wand, though the event is not, in other respects, unlike the proverbial production of a rabbit out of a hat. In both cases, the truth lies in the fact that rabbit and character can be produced only when they are already preformed, or, in the case of the character, almost preformed.

When a "vertebrate" character appears in a more primitive organism, it is important to examine the species in which it occurs in order to determine whether or not any similarity with the vertebrates is fortuitous, is caused by convergent evolution of unrelated mechanisms, or is based on a genuine similarity and derivation from a common stock. Furthermore, it is important to establish whether the species bearing the particular character, or any of its near relatives, has other features of similar significance.

Epidermis

Among the characters of interest in connexion with the evolution of the pattern of tissues and organs of vertebrates and which make their appearance among living rhabdocoels or their near relatives, the first is perhaps the epidermis.

The epidermis is basically of the "respiratory" type; it consists of several classes of cells, resting on a definite basement membrane of collagen fibres embedded in a mucoprotein matrix (Fig. 11.1b). The cells in the surface layer belong to at least three classes: ciliated cells, mucous cells, and cells carrying inclusions called rhabdites. These rhabdites, which are probably a specialization for defence (their origin was briefly discussed on p. 217), are apparently not represented as such in the vertebrates themselves. Their occurrence here is significant from an evolutionary point of view, since the appearance of structures which are possibly related to rhabdites is a conspicuous feature in the epithelia of other possible intermediate forms (see p. 263). A further study of their function and properties could be very illuminating.

Some of the glandular epithelial cells have their main cell bodies situated below the basement membrane but opening to the surface by long processes. This, again, may be a defensive feature which persists into higher forms—defensive both in the sense that the cell bodies, being below the basement membrane, are not so readily subject to abrasive insults and in the sense that the cells provide a protective mucous coating.

This mixed type of epithelium, or "respiratory" epithelium, is an important development. In some form or other, it persists throughout all those creatures which seem to be on the line of vertebrate descent and into the vertebrates themselves as far as man himself, where it is of course present in the nasal and respiratory passages. It occurs especially on those surfaces which separate a fluid of rather variable salt composition from the more constant internal environment. Thus it is probably always actively concerned with preserving the integrity of the latter. The ideas developed earlier (p. 185) suggest that the ciliated cells are activated by dilute external environmental conditions, while the mucous or goblet cells cope effectively with more concentrated conditions. When certain nemertine worms, e.g. *Lineus* (see p. 267), are treated with dilute sea-water, the ciliated cells of the pharynx appear healthy and active, while the supporting mucoid cells shrink and lose their affinity for stains. On the other hand, in sea-water to which extra NaCl has been added, the ciliated cells are quickly damaged and in high concentrations may be sloughed off, while the mucoid cells thrive and become loaded with secretory materials (Fig. 11.2). The ciliated cells of echinoderm embryos also lose their cilia in double-strength sea-water (Auclair and Siegel, 1966) and, even in the vertebrates, the beat of the ciliated cells of the tracheal epithelium of the frog is inhibited by hypertonic NaCl solution (Richardson, 1937). All these observations are in keeping with the reactions of *Naegleria* to changed salt concentrations (see p. 168).

Just as in the Acoela, the recruitment of new cells for the epidermis from the cells of the parenchyma (i.e. from the mesohyl) is a normal process in some living rhabdocoels (Skaer, 1965). It may again indicate that there is still some degree of reversibility in the differentation of cells, a concept which is made all the more plausible by the remarkable capacity for regeneration which is possessed by these organisms. The renewal of the rhabdite-containing cells from the mesohyl (Skaer, 1961; Lentz, 1967) is particularly interesting in relation to the significance of the rhabdites themselves (see

FIG. 11.2. Pharyngeal epithelium of the nemertine, *Lineus*. (*a*) After the worm has been for 1 hour in sea-water diluted with an equal quantity of distilled water; (*b*) after the worm had been for 1 hour in sea-water to which 2 gm NaCl/100 ml had been added. In (*a*) the ciliated epithelium is in good shape and the supporting cells are reduced. In (*b*) the ciliated cells are "tatty", but the underlying cells are engorged with secretions.

See facing page→

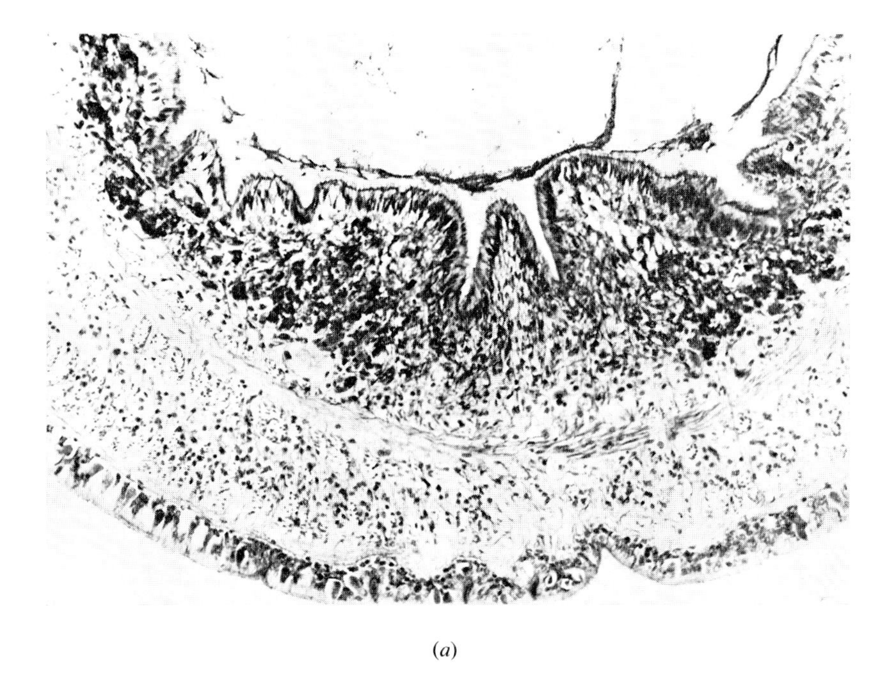

(*a*)

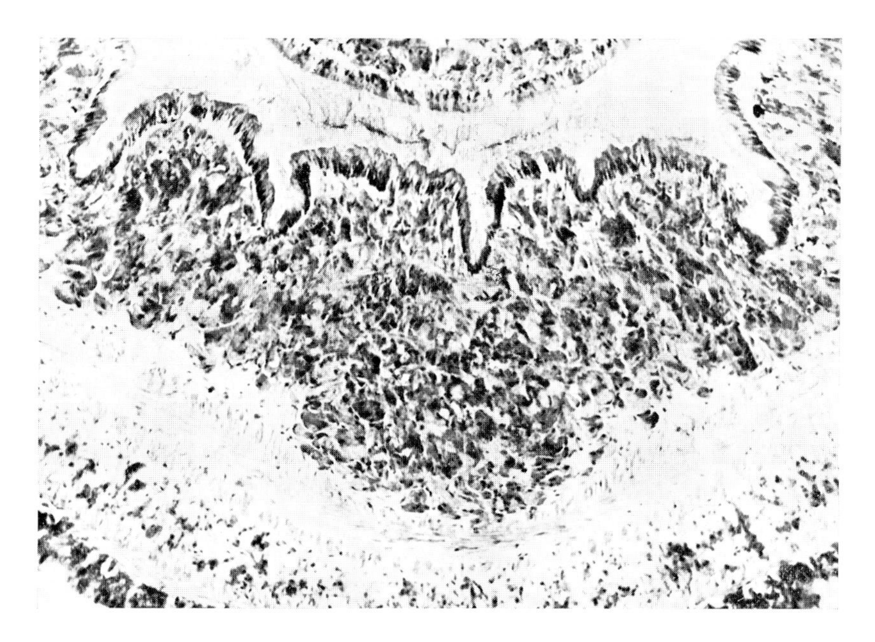

(*b*)

p. 217). It is possible that renewal of the epidermis may also occur by division of the cells already in it, just as in other epithelia, but there is no evidence that it does.

This type of epidermis, composed of ciliated and mucous, or similar, cells and related to a definite basement membrane, essentially remains the dominant "skin" until animals begin to emerge from an aquatic to a terrestrial or aerial environment. Among the vertebrates, for example, this type of skin occurs in the cyclostomes and in many groups of fish. Its function, in addition to those of locomotion and cleanliness which it effects by means of its cilia, is, as already mentioned, to maintain a degree of controlled permeability to water, ions, and dissolved substances. Like the rhabdites when they occur, the mucous secretion may also be used defensively or offensively against other creatures, and the maintenance of currents of sticky mucous-containing fluid over the surface clearly assists in keeping invading bacteria and other organisms at bay and sometimes in directing food particles towards the mouth. It may also minimize the effects of ionic changes.

There is one other interesting feature in relation to epidermal cells. During development, when epidermal cells are being recruited from the parenchyma, cells appear in the epidermis that contain vacuoles in their cytoplasm into which short cilia project (Skaer, 1965). The significance of these structures is quite obscure, and is it not known how widely distributed such structures are in other animals and tissues. Perhaps it is not mere coincidence that similar vesicles with cilia have been described in the respiratory epithelium of mammals (Mihálik, 1935). However, not too much weight should be attached to this observation because cilia are being found with increasing frequency in situations where their presence had been previously unknown, but in which it might have been foreseen on the basis of the hypothesis concerning the origins of cell types outlined on these pages.

Neural Tissues

Sensory cells, which are fore-shadowed in the acoeloid, and which sometimes occur as tightly packed clusters of ciliated cells, also develop among the epithelial cells (Fig. 11.1d), particularly along the lateral margins of the worms (Fig. 11.3). They may send processes below the basement membrane through two layers of muscles, which have now developed and lie below the basement membrane, to join a submuscular nerve plexus which tends to concentrate into longitudinally running nerve cords. These nerve cords may have transverse ring-like connections, and they lead to a definite and somewhat more deeply situated brain at the anterior end. The pattern of architecture in this brain and neural tissue is a fairly clear derivative of the corresponding structures in the acoeloid. It is basically similar to that which persists into animals that for other reasons must represent stages more nearly related to the vertebrate stock, stages which are considered in more detail later. The main nervous system, as in the acoeloid, develops more or less *in situ* from a group of cells that have already lost contact with the epidermis and lie in the inner mass. Although there seems to be some

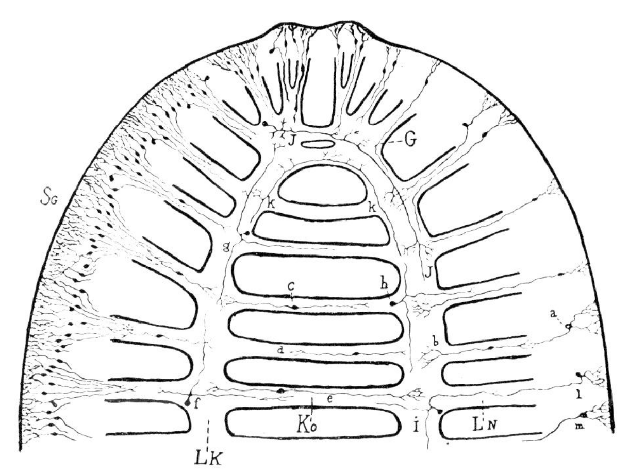

FIG. 11.3. Diagram illustrating the types of nerve cells present in the Turbellarian *Bdelloura candida*. SG, sensory cells; G, brain; LK, lateral nerve cords; KO, commissure; LN, lateral nerve branch. Small letters indicate different sorts of nerve cells as demonstrated by Golgi methods. (After Hanström, 1926.)

plasticity in their origin, the nerve ganglia and cords in the rhabdocoeloid stage are essentially solid structures. There is no sign of a tubular invagination of the ectoderm, such as is found in the formation of the central nervous system of the vertebrates. The cells are either derivatives of the mesohyl or are cells which have left the epidermis at a somewhat later stage than the cells of the original mesohyl. In the latter case they have undergone changes similar to those leading to the formation of mesohyl cells but have entered an environment already altered by the presence of the earlier invaginating mesohyl cells and thus have acquired different properties.

In the rhabdocoeloid stage it is probable that the nerve cords grew backwards from the "brain" (Hanström, 1928) and the nervous system became enclosed in a connective tissue capsule, e.g. as in the polyclads of to-day (Turner, 1946). In existing Turbellaria there are several sorts of nerve cell—unipolar, bipolar, and multipolar, and cells which could be called neuroglial cells have also been described in the polyclads. Thus the nervous system is now reaching some degree of complexity.

At this point it may be appropriate to consider some of the factors involved in the formation and evolution of a nervous system as the result of cells leaving an epithelium and entering a mesohyl. Before doing this, however, it is worth noting that the epithelia of *Hydra* and other Cnidaria show spontaneous spike potentials with a typical overshoot beyond zero potential. That these may be conducted from cell to cell (Josephson and Macklin, 1967; Mackie, 1965) indicates that the raw materials for the development of a nervous system may already be latent in such epithelia.

The polarity of the entering cell and the directions of its axes must be considered first. Epithelial cells, which essentially separate two phases, are of necessity different at their two exposed ends (e.g. ciliated surface and basal end). If such a cell sinks below the surface to become an internuncial neurone, is this polarity maintained, lost, or reversed? Perhaps the possibilities can most easily be visualized by reference to the diagrams in Fig. 11.4. Figure 11.4a indicates some of the possibilities when an uniform population of epithelial cells gives rise by cell division to a second order of cells in a neurone chain. In Fig. 11.4b it is supposed that the first internuncial neurones are established from an epithelium composed initially of two main opposing cell types without altering their initial polarities, but merely making possible connections. The symbols indicate the properties of the cell surfaces with respect to such things as transmitter substances.

In the formation and growth of neural tissue from the neural ectoderm of the vertebrates, the majority of the dividing cells are orientated as in cells c in Fig. 10.2, though a few are also found orientated with the metaphase plate parallel to the epithelial surface (Sauer, 1935). Undoubtedly, most of the multiplication of cells and the consequent thickening of the epithelium is accounted for by the process illustrated in Fig. 10.3. In fact, however, nothing is known about the orientation of the cells which finally lose contact with the terminal bars of the ependymal cells and take up their positions as definitive neurones in relation to their original orientation in the neural ectoderm. It is possible that when, a cell is actually going to become a neurone, as distinct from multiplying as a neuroblast, it has been produced from a cell division in which the metaphase plate is parallel to the surface (cells b, Fig. 10.2). The problem is considered again in relation to the development of the retina (see p. 410).

In the sort of situations depicted in these diagrams, the problem of the nature of the transmitter substances naturally arises. It is known, for example, that many ciliated epithelial cells are affected by acetylcholine and have cholinesterase near their cilia (Seaman and Houlihan, 1951; Kordik *et al.*, 1952). This is an observation of great significance in relation to the origin of nerve cells. That acetylcholine is formed by ciliated cells is also indicated by the fact that it can be formed by the nerve-free tissue of the gill-plates of the mussel (*Mytilus edulis*). Moreover, when applied in low concentrations, both acetylcholine and eserine increase the ciliary activity of this tissue, but depress it in larger concentrations as though the formation of acetylcholine was a necessary part of the activity of the cells (Bülbring

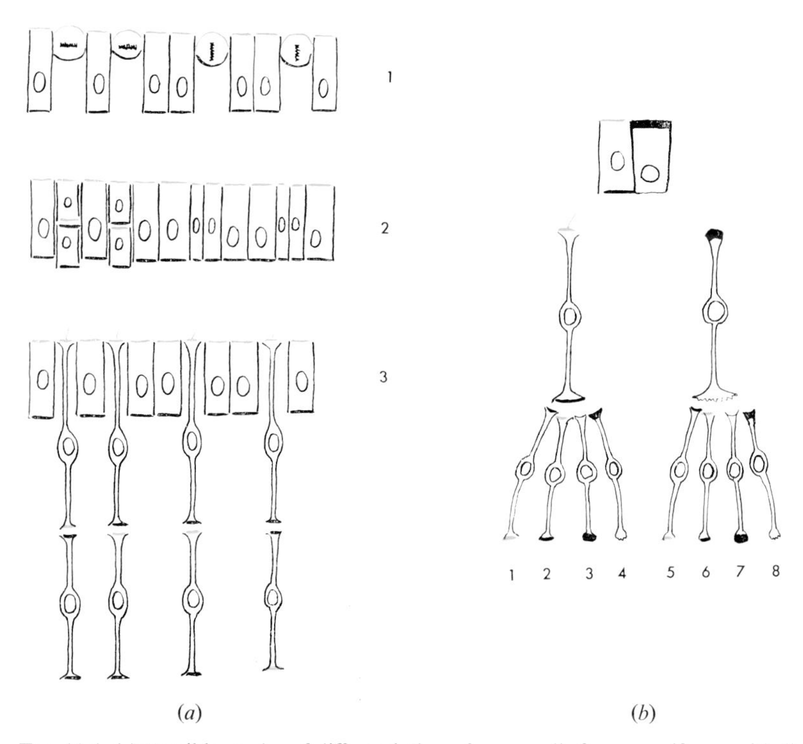

FIG. 11.4. (*a*) Possible modes of differentiation of nerve cells from a uniform epithelium, and methods of connexion with each other. 1, Cell divisions. Note the orientation of the metaphase plates and the nature of the cell surfaces reflecting the orientation of the cells. 2, Immediately after the cell divisions. 3, The differentiation of the nerve cells to form the first "synapse". Note the possible orientation of the cells. (*b*) A few of the possible modes of differentation of nerve cells (1–8) from an epithelium composed of cells of two balancing types, e.g emballic and ecballic cells, and the ways in which they could connect. The colours and thickness of lines indicate different properties of the cell surfaces, emphasizing cell polarity.

et al., 1953). Similar observations have been made (Seaman and Houlihan, 1951) on the tracheal epithelia of the frog and rabbit (Kordik *et al.*, 1952; Corssen and Allen, 1959). Unfortunately, in these experiments on frog and rabbit the ciliated cells are mingled with mucous cells to some extent, so that the actual location of the formation of acetylcholine is left in some doubt. The human placenta produces large quantities of acetylcholine, which has been localized to the fine processes on the cells of the chorionic villi (Chang and Wong, 1933; Wen *et al.*, 1936). Cholinesterase is increased in sea-urchin embryos when they become free-swimming by means of cilia (Augustinsson and Gustafson, 1949). Perhaps it is significant that lithium salts lower the cholinesterase activity and at the same time vegetalize the embryo, i.e. favour the non-ciliated cells of the vegetal pole. The choline-esterase activity also rises in amphibian embryos when the neural folds are formed (Boell and Shen, 1944). Not only do these folds show large numbers of flagellated cells, but also the rest of the surface of the embryo at that stage develops a mosaic of ciliated and non-ciliated cells (Assheton, 1896; Twitty, 1928).

From all these observations it seems probable that acetylcholine production and cholinesterase activity preceded neural activity as such, and that the presence of these substances in the nervous system may indicate derivation of the cells from the ecballic, flagellate, or ciliate type of epitheliocyte, or from the mechanocyte derived from it. There is even evidence for an earlier origin. Cholinesterase has been reported in flagellate trypanosomes (Bülbring *et al.*, 1949) and in the ciliates *Tetrahymena* (Seaman and Houlihan, 1951) and *Paramecium* (Bayer and Wense, 1936), though its presence in the ciliates has also been doubted or denied (Mitropolitanskaya, 1941; Tibbs, 1960).

Cholinesterase is present in certain fish sperm, and the suggestion has been made that it is concerned with wave propagation on the flagellum or cilium rather than with the actual beat (Tibbs, 1960).

Relevant to this topic are some observations which connect the presence of cholinesterase (particularly butyrylcholinesterase) with sodium transport, e.g. in the ascending limb of the loop of Henle in the vertebrate kidney (Fourman, 1965), in the gills of the crab, *Eriocheir sinensis* (Koch, 1954), and in frog's skin (Koblick *et al.*, 1962), though in none of these examples are ciliated cells specifically involved.

If it were supposed, for the sake of argument, that acetylcholine was normally liberated at the ciliated end of the cell, the transmitter substance at the second "synapse" in cases 1 and 5 of Fig. 11.4b would presumably be acetylcholine. By arguing along these lines and assuming that an active product exudes from each end of the cell as a potential transmitter, it becomes clear that there might well be at least four classes of transmitter. Furthermore, the effects of these transmitter substances might either be excitatory or inhibitory on the next cell in the chain. Indeed, any surface could well liberate more than one metabolite, and each of them could have a different effect on the cell surface with which it came in contact, e.g. surface A could liberate a substance activating surface C but inhibiting surface B or *vice versa*. There is no point in pursuing this argument further at this stage, however, except to indicate that, given a minimum of two groups of polarized cells already integrated into a viable epithelium, a complex neural system could well be built up without involving any change other than the morphological and positional changes necessary to form chains of neurones from the original epithelial cells. It may also be emphasized that, since the initial activities of the two groups of cells are probably co-ordinated in some way by mutual feed-back mechanisms, the seeds for the production of stimulating and inhibitory mechanisms are likely to be already present in the epithelial cells. An investigation of the exact mechanisms involved in producing the nervous system, both phylogenetically and ontogenetically, in such creatures as the acoels and the rhabdocoels could well throw light on neural interaction, organization, and transmission.

Cells derived from epithelial cells that were themselves strongly polarized, like the ciliated cells, would probably be more likely to produce uni- or bipolar cells, and the multipolar type would be more likely to result from the less polarized and more randomly amoeboid type of cell. However, the number of processes that cells have is only partially an intrinsic property and is often dictated by their environment. Similarly, if cells of the mesohyl were to give rise to conducting neurones, some would be expected to have properties more akin to those of the mechanocyte, others would resemble the amoebocytes. Among these properties required for conduction the nature of the cell surface would probably be more fundamental than the degree of polarity. Since the surface properties of these two classes of cells are so very different, they could contribute in different ways to neural activity.

The observations of Lentz (1966) on the cells of the sponge (see p. 149) are relevant here too, for he observed that acetylcholinesterase was associated more with the bipolar cells, and that monoamine oxidase, adrenaline and, 5-hydroxytryptamine were more prominent in the multipolar cells. Though these results cannot immediately and without question be transferred to mechanocytes and amoebocytes, they are suggestive.

In any case, a consideration of the phylogenetic origins of conducting tissues, which must have occurred in different places and at different times, points strongly to a degree of variation in cell behaviour among neurones of different groups. This variety is seldom fully appreciated by electrophysiologists.

To return to the emergence of a pattern of tissues in the rhabdocoeloid organisms: the separation of definite muscle groups and of sensory, internuncial, and motor nerve cells into an integrated neural system controlling the organisms is a major step forward in the organization of tissues. In addition to the several types of nerve cells distinguishable in various parts of the system (see Fig. 11.3) (Hanström, 1926), there are also cells which

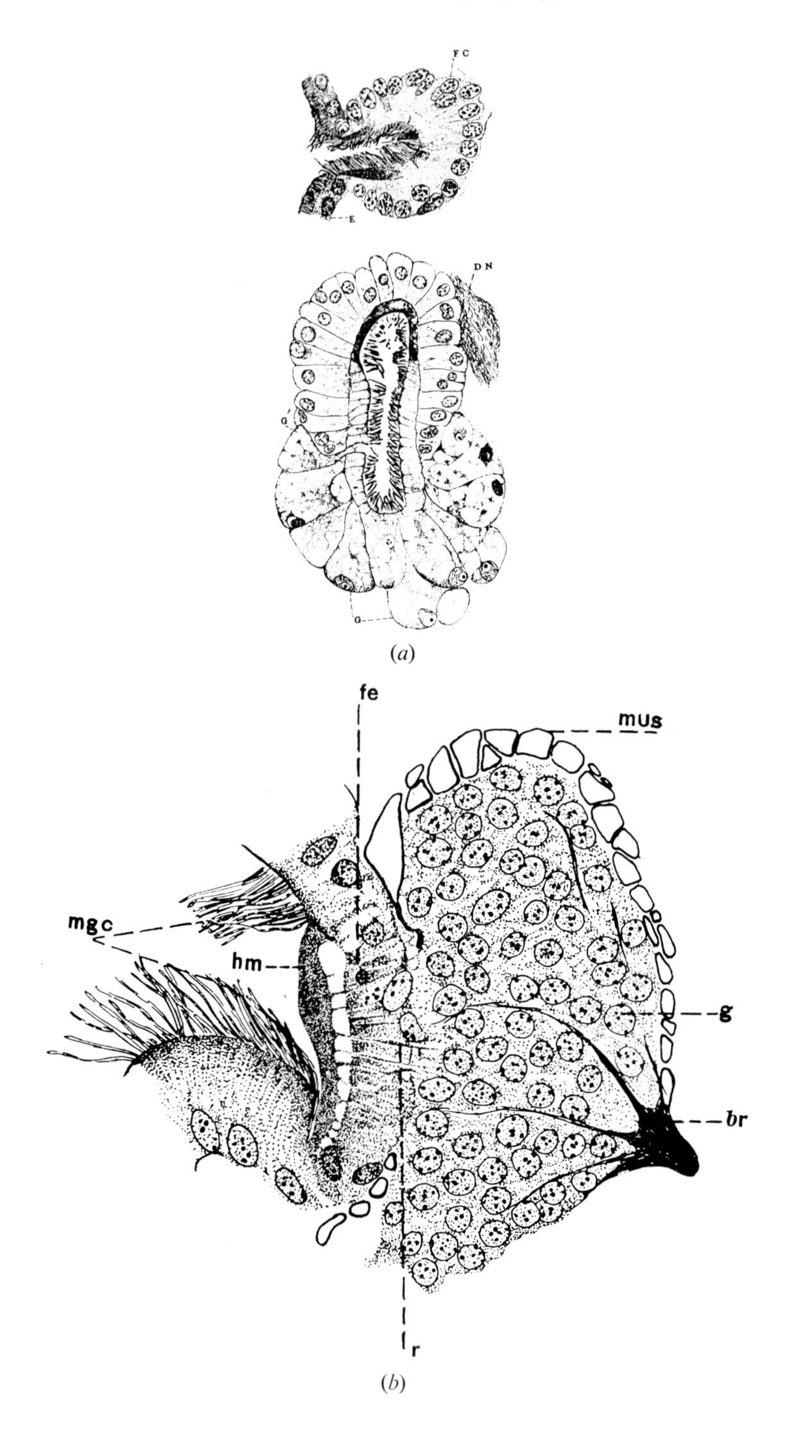

(*a*)

(*b*)

have neurosecretory properties (Lender and Klein, 1961). At least, they contain varying amounts of secretion giving positive reactions to such staining tests as the paraldehyde-fuchsin test. The amount of secretion in the cells varies with the conditions and is affected by light and darkness and by the tonicity of the medium, properties which will be seen (p. 386) to have particular significance (Ude, 1964).

Of great interest, too, are certain patches of highly flagellate cells in the epidermis sometimes forming ciliated pits and grooves (see Fig. 11.1d). In these, there is evidence for the existence of rheoceptors and chemoceptors, and as such they may well fore-shadow the development of similar sensory systems in higher organisms, e.g. the lateral-line organs, vestibular organs, and taste buds. In some species, cephalic pits are developed, one on either side of the cephalic ganglia (Fig. 11.5a) They are lined with cells bearing long flagella, and there is evidence (Kepner and Taliaferro, 1912) that they assist the animal in its responses to changes in the salinity of the medium. Associated with the flagellate cells in *Microstomum* are a few cells attached to the margins which contain some sort of secretory material in discrete masses in their cytoplasm. These secretory cells probably fore-shadow similar cells which are developed on a more lavish scale in some higher forms (see p. 370). In *Stenostoma* aganglionic mass is associated with the pit (Kepner and Cash 1915) and probably has an origin different from that of the main neural tissues (Fig. 11.5b). These cephalic pits will be shown to be important structures capable of considerable further evolutionary development and particularly relevant to the ancestry of the vertebrates. They are discussed again (see p. 370).

Muscles

As mentioned earlier (p. 229), the strictly epitheliomuscular cells are now less important. Discrete muscle cells now run as separate fibres or in sheets, approximately longitudinally and circumferentially in well organized fascicles, with intervening connective tissue containing definite collagen fibres. There are groups of muscle fibres running dorso-ventrally. This system of muscles and connective tissue fibres constitutes the so-called fluid-skeleton system.

FIG. 11.5. (*a*) The ciliated pits of *Microstomum caudatum*. (*Above*) Epidermis (E) and ciliated cells of the pit (FC); (*below*) a section deeper in the pit showing ciliated cells and nerve fibres (DN) as well as large gland cells (G, G'). (Kepner and Taliaferro, 1912). (b) Cephalic pit of *Stenostoma sp.* br, nerve to brain; fe, fundus epithelium; g, ganglion cells; hm, homogeneous mass; mgc, marginal epithelium with cilia; mus, muscle; r, sensory rods. (Kepner and Cash, 1915).

←*See previous page*

Since the volume of the animal remains constant (collagen fibres being relatively inextensible and the body being filled with fluid), the relaxation of the one main set of muscles must depend on the contraction of the other. The intervening fibres of the connective tissue tend to be arranged in alternating layers criss-crossing in spirals that intersect at angles which vary with the extension or contraction of the worm (Clark 1964) but which are approximately at right angles in the resting condition. The position is more complicated than it appears to be, because the worm may be flattened dorso-ventrally to different extents when it lengthens or shortens. The two main sets of muscle fibres, which could be broadly described as the circular and longitudinal, in spite of the generally flattened form of the organism, must be essentially antagonistic. If they depend on chemical transmitters for their activation, it is unlikely that they depend on the same transmitter, unless that transmitter can be adequately prevented from diffusing from its site of action. Cholinesterase is certainly present in the platyhelminthes, but its exact location has not yet been specified (Bacq, 1947). Almost all the muscle is "plain" or unstriated, and each fibre is a single mononucleate cell, but all the fibres are integrated in such a way that co-ordinated movements are made in response to stimuli. Some of the responses are remarkably quick. Relatively simple as these rhabdocoeloid creatures are, the muscular system is in fact a very highly organized system. The rapid manner in which we have passed from a simple blastuloid creature to the highly organized rhabdocoeloid does not do full justice to the manner in which, during this phylogenetic progression contractile cells must have become integrated into efficient mechanical systems and activated by appropriate neural systems so that the observed complex and co-ordinated movements can occur. Again, the properties of these early muscles and nerves deserve much more detailed study than they have so far had. For example, how do muscle fibres become attached to inert collagen fibres in the correct situations, and how do the primitive nerve fibres find the correct muscle fibres to activate? All such problems have been solved by animals no more complex than the rhabdocoeloid. If we understood how they had been solved, we might be much more knowledgeable about the neuromuscular mechanisms of higher animals, many of which must have evolved from these or similar early beginnings. It may be a significant observation that a substratum of collagen fibres is a peculiarly effective means of inducing the differentiation of muscle fibres *in vitro* (Hauschka and Konigsberg 1966).

Alimentary System

In the rhabdocoeloid stage, it is assumed that a definite alimentary canal has been acquired as a branching sac-like invagination comprised of two main classes of cells. The mouth is on the ventral surface, presumably in adaptation to the manner of feeding, but there is no anus and much of the digestion of the food is intracellular. There is generally a more or less elaborate muscular pharynx which can be eversible. In addition to the gut, there has developed in one particular group, the Kalyptorhynchia, an eversible proboscis, which in the resting condition is essentially an invagination of the anterior end of the body (Fig. 11.6). When everted it is covered

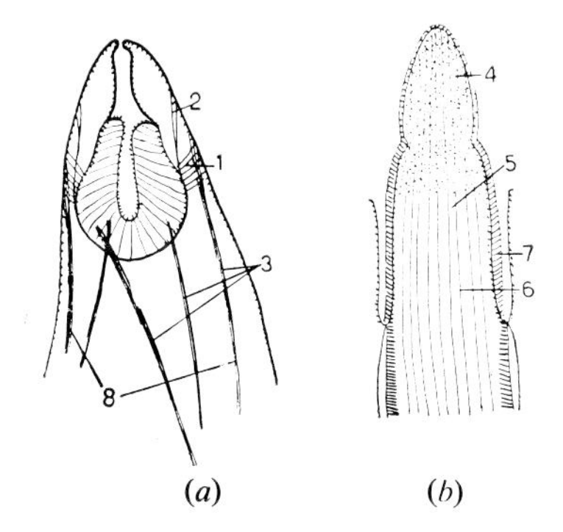

FIG. 11.6. The eversible proboscis of a kalyptorhynchid (*Gyratrix hermaphroditus*). (*a*) Inverted; (*b*) everted. 1, Fixator muscles; 2, protractors; 3, retractors; 4, end cone; 5, muscle cone; 6, longitudinal muscles; 7, transverse muscles; 8, integumental retractors. (From "The Invertebrates", by Hyman, 1951, McGraw-Hill, with permission of McGraw-Hill Book Company.)

with papillae, is sticky, and is used in the capture of prey (Fig. 11.7). This organ probably fore-shadows the much more elaborate structure which is such a characteristic feature of the nemertine worms (see p. 260) and an organ of some importance in relation to the arguments that follow. In existing Kalyptorhynchia, it is independent of the alimentary canal, but the biological problem of the development of this organ, both phylogenetic and ontogenetic, is an intriguing one. Its solution could be helpful in elucidating the why and wherefore of the nemertine proboscis and its derivatives.

Urinogenital Systems

Apart from the gut, the body of the rhabdocoeloid is still an essentially solid structure, consisting of an epithelial covering filled with a connective tissue parenchyma containing several types of cell in a mucoprotein ground-substance sometimes supported by collagen fibres. No definite coelomic or vascular cavities appear. The germ cells, however, are generally housed in separate pockets or gonocoels. How these pockets arise and the nature of

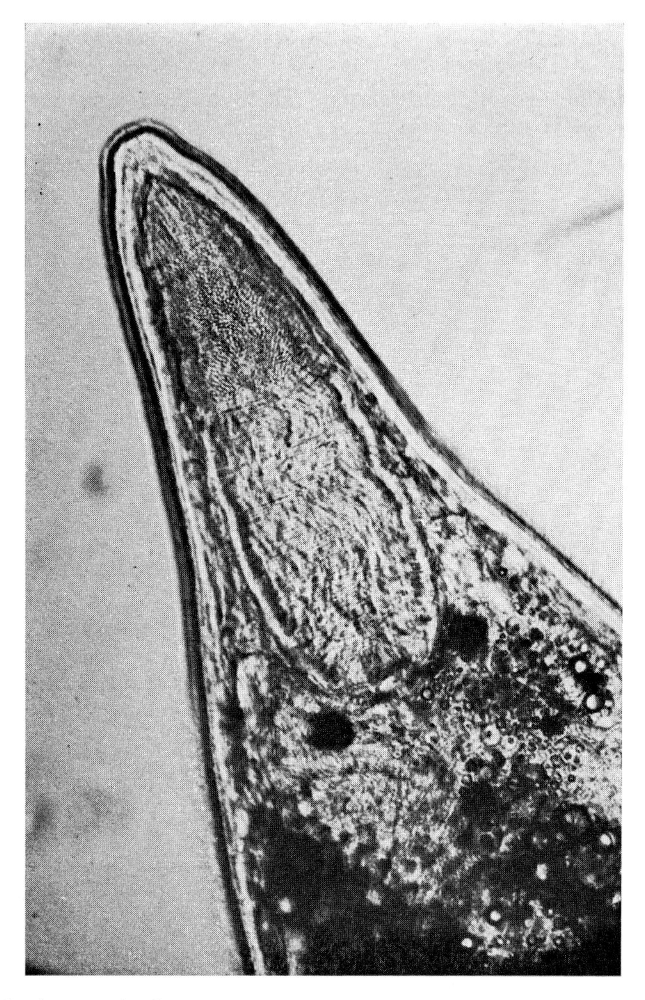

FIG. 11.7. Photograph of proboscis of *Gyratrix* showing organ everted but not extruded. Note papillae on the proboscis cone. (Photograph by Skaer.)

the cells surrounding them are, unfortunately, obscure. The male gonads are nearly always situated separately from the female gonads, i.e. more anteriorly or laterally. A study of the differences in the immediate local environments of the developing ova and sperm in these creatures could make a very important contribution to our understanding of the mechanisms leading to the differentiation of the gametes into motile sperm or sessile ova. It is of interest to note (Fig. 11.8) that in some existing rhabdocoels the developing ova are entirely surrounded by close-fitting follicle cells or nurse cells which differ in different groups. The sperm, on the other hand, develop in fluid-filled sacs (Hyman, 1951). This sort of difference is also seen in the gonads of many higher forms and is presumably of some physiological importance (see pp. 455, 474). Details of the other parts of the genital apparatus in living species are sometimes very complex and are largely irrelevant to the present theme because so often they are clearly adapted to the particular circumstances and way of life of the species concerned. As

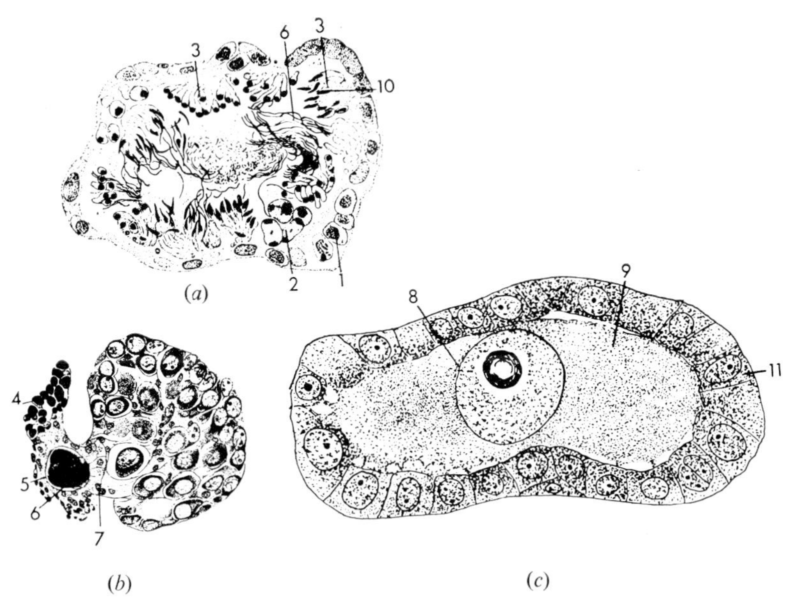

Fig. 11.8. Gonads of fresh-water planarians. (*a*) Testis, with spermatogenesis. Note the cells developing within the cavity of the organ. (*b*) Ovary with developing ova surrounded by "nurse" cells. (*c*) An ovarian follicle in *Prorhynchus*. 1, Spermatogonia; 2, spermatocytes; 3, spermatids; 4, yolk glands; 5, seminal receptacle; 6, mass of sperm; 7, ovarian membrane; 8, nucleus of ovum; 9, cytoplasm of ovum; 10, sperm nuclei; 11, follicular epithelium. (From "The Invertebrates", by Hyman, 1951, McGraw-Hill, with permission of McGraw-Hill Book Company.)

mentioned before, it is well known that the separation and definition of species often depends on differences in the genital apparatus; thus, only major trends are important in relation to the direction of evolution. Such differences in genitalia give rise to a form of isolation similar to that imposed by geographical barriers, thereby favouring the development of separate races. When adaptation to other factors in the environment leads to modifications of form, these modifications may then be made more permanent by parallel changes in the genitalia or by separation of the adapting race by geographical isolation or the like, both of which prevent dilution with the parent stock.

The solid parenchyma of the rhabdocoeloid, together with the increase in size of the organism as a whole, seems to have necessitated the development of an excretory or regulatory system in addition to the main body surface, i.e. the skin and gut, which may still carry out these functions. It is difficult to know what the principal determining factor in the development of this new system can have been. The main nitrogenous excretory product of most lower organisms is ammonia, which, like CO_2, has little difficulty in permeating and escaping through cells and would hardly demand the development of a special excretory system unless the organisms grew very large. The production of other toxic and non-volatile metabolites may have been the operative factor, or an increase in the capacity to regulate the ionic concentrations of the parenchyma may have been required. The first special excretory or regulatory system, whatever its primary function, is the nephridial system; in its primitive form it consists of tubular invaginations of the superficial epithelium (Fig. 11.1a). As might be expected, at least two classes of cells are usually present in it. The tubules are generally headed by flagellated "flame cells" or solenocytes (Fig. 11.1c), which convey water from the parenchyma through narrow tubules of squamous or flagellate cells to a more "secretory" section of the tubule and thence into a simple tubule leading to the exterior. In some ways these primitive nephridia can be considered extensions of the original external surface into the interior, and the arrangement of the cells, along the length of the tube, is reminiscent of that between the flagellate pole and the amoeboid pole of the original blastuloid organism; i.e. there are flagellate cells at one end and amoeboid and phagocytic cells at the other. Moreover, as in the blastuloid, the gradient of cell form, and therefore presumably of function, may be sharp and steep or it may be more gradual. If this interpretation of nephridia is the correct one, it is probable that, in the rhabdocoeloid organism, fluid and ionic regulation are carried out not only by the entire surface epithelium, including that of the skin and gut, but also by the nephridia. Some authors (e.g. Beadle, 1934) have denied a regulatory function to the nephridia in these organisms on the grounds that the cells of the gut may become vacuolated when the animals are subjected to a water load, i.e. by immersion in fluid of lower ionic content than normal. However, the superficial layers of the organism probably remain active as ion and fluid regulators until animals much higher in the evolutionary scale make themselves relatively impervious by the development of a chitinous cuticle, by forming a superficial layer of keratinized cells, or by some other means. Even then the skin may not entirely lose its regulatory powers (e.g. sweat glands in mammals), and the intestine probably never does.

Pigmentation

An interesting feature in animals at about this stage of organization is the frequent occurrence of pigment spots on the lateral margins of the body, particularly in the anterior region. The pigment is often melanin, and quite frequently the pigment-containing cells are arranged in a cup-like manner with sensory cells in connexion with them. These sensory cells are sometimes "inverted" and have some sort of striated border composed of parallel lamellae (Press, 1959) or microvilli (Röhlich and Török, 1961) (Fig. 11.9). Their cytoplasm shows changes in response to light (Kepner and Foshee, 1917; Taliaferro, 1920). In other species the sensory cells point towards the light and interdigitate with the pigment cells. Such systems have naturally been called eyes, but only in one or two cases is anything known about their actual function. It must be remembered that there is considerable difference between a cell that responds to light and the highly specialized, image-forming, recording, and processing mechanism that most of us understand by the term "eye". The cells that respond to light could, for example, be true light receptors dependent on a photosensitive pigment, such as rhodopsin, as they appear to be in *Dendrocoelum lacteum* (Marriott, 1958), or they could merely detect thermal changes resulting from the absorption of a wide band of radiant energy by the melanin. Little is known about the extent to which such sense organs are integrated into the nervous system: whether in such manner that effective motor responses are made in response to the combined action of light and other stimuli on the animal, or in such manner that the animal can respond in some way to length of day and the like. The relationship between these primitive "eyes" and the visual mechanisms of higher organisms is discussed more fully later, because the origin of the eyes of vertebrates is one of the main enigmas in the story of evolution. Carotenoid and semicarotenoid pigments are present in some, if not all, flatworms, but, since vitamin A (retinol) and retinene (retinal) probably have other functions in cells besides those connected directly with photoreception, this does not necessarily mean that even primitive visual mechanisms are present. Such pigments are indeed of widespread, if not universal, occurrence in all phyla from the protozoa upwards, but their original functions are still problematical.

A melanin-like pigment is present subcutaneously in some planarians. It has recently been noticed that treatment of the worms with an anti-fungal agent (Fungichromin) or with oleic acid causes the cells containing this pigment to migrate to the gut and be excreted (Johnson *et al.*, 1962). These cells may represent the beginnings of the melanophore system of higher forms, though the distribution of the pigment within the planarian cells may be more complex than this simple account suggests. The power of

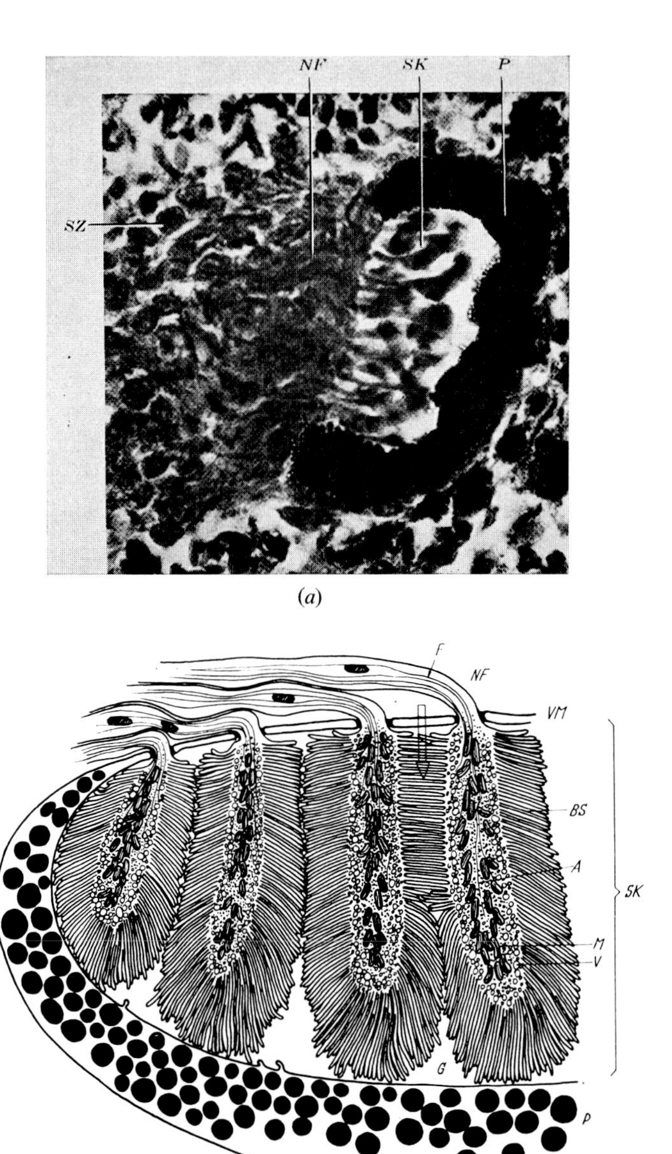

FIG. 11.9. (*a*) Eye-cup of planarian. Sagittal section showing (P) pigment layer, (SK) sensory processes, (NF) nerve fibres, (SZ) sensory cell nuclei. (*b*) Diagram from electron-microscope section of eye-cup, showing sensory processes. P, Pigment; A, cell process; M, mitochondria; V, vesicles; BS, microvilli; SK, sensory processes; NF, nerve fibres; F, fibre; G, cavity of the cup; VM, limiting membrane, Arrow indicates direction of light. (Röhlich and Török, 1961.)

independent movement possessed by these pigmented cells is interesting in this general context because of the extraordinary mobility of the melanophores and chromatophores in higher animals. Significant also is the fact that the melanin of the eyes of the planarians is not affected by these agents because, in the higher animals, the melanin-containing cells of the eye likewise act quite independently of the melanophores and respond to different stimuli.

Once more, it may be interesting to take stock, and to add to the earlier lists of the main vertebrate characters those that have appeared or been fore-shadowed at the level of rhabdocoeloid organization (see Table 11.1).

TABLE 11.1

Further Features Relevant to Vertebrate Evolution which appear in or before the Rhabdocoeloid Stage

1. More organized surface epithelium ("respiratory" epithelium) with ciliated cells, goblet cells
2. Basement membrane with collagen fibres
3. Plain muscle in circular and longitudinal arrangement; muscle antagonists
4. Alimentary cavity with a lining composed of at least two types of cell
5. Pigmented eye-spots with photoreceptors; melanophores
6. Peripherally (laterally) situated sensory cells; ciliated sense-patches
7. Ciliated, cephalic pits
8. Nervous system (in parenchyma) with anterior ganglia and lateral nerve cords, the whole being enclosed in a connective tissue capsule; "neurosecretory cells"; neuroglia
9. Eversible pharynx
10. Germ cells in definite gonocoels (♂ and ♀ types differ)
11. Nephridia—flame-cells and "secretory" cells
12. Beginning of a "serial" arrangement—e.g. in gonads and gut diverticula
13. Proboscis (in *Kalyptorhynchia*)
14. Chemical transmitters, etc.

Features Present but Subsequently Lost

1. Rhabdites
2. Cell replacement in epidermis from sub-epithelial cells

References

Assheton, R. (1896). Notes on the ciliation of the ectoderm of the amphibian embryo. *Quart. J. Microscop. Sci.* **38**, 465.

Auclair, W., and Siegel, B. W. (1966) Cilia regeneration in the sea-urchin embryo: Evidence for a pool of ciliary proteins. *Science* **154**, 913.

Augustinsson, K. B., and Gustafson, T. (1949). Cholinesterase in developing sea-urchin eggs. *J. Cellular Comp. Physiol.* **34**, 311.

Bacq, Z. M. (1947). L'acétylcholine et l'adrenaline chez les Invertebrés. *Biol. Rev. Cambridge Phil. Soc.* **22**, 73.
Bayer, G., and Wense, T. (1936). Über den Nachweis von Hormonen im einzelligen Tieren. 1. Cholin und Acetylcholin im *Paramecium. Arch. Ges. Physiol. Pflügers* **237**, 417.
Beadle, L. C. (1934). Osmotic regulation in *Gunda ulvae. J. Exptl. Biol.* **11**, 382.
Boell, E. J., and Shen, S.C. (1944). Functional differentiation in embryonic development. 1. Cholinesterase activity of induced neural structures in *Amblystoma punctatum. J. Exptl. Zool.* **97**, 21.
Bülbring, E., Lourie, E. M., and Pardoe, A. V. (1949). The presence of acetyl choline in *Trypanosoma rhodesiense* and its absence from *Plasmodium gallinaceum. Brit. J. Pharmacol.* **4**, 290.
Bülbring, E., Burn, J. H., and Shelley, H. J. (1953). Acetylcholine and ciliary movement in the gill plates of *Mytilus edulis. Proc. Roy. Soc.* (*London*) **B141**, 445.
Chang, H. C., and Wong, A. (1933). Studies on tissue acetylcholine. (1) Origin, significance and fate of acetylcholine in human placenta. *Chinese J. Physiol.* **7**, 151.
Clark, R. B. (1964). "Dynamics in Metazoan Evolution". Oxford Univ. Press (Clarendon) London and New York.
Corssen, G., and Allen, C. R. (1959). Acetyl choline: Its significance in controlling ciliary activity of human respiratory epithelium *in vitro. J. Appl. Physiol.* **14**, 901.
Fourman, J. (1965). Cholinesterase in the mammalian kidney. *Nature* **209**, 812.
Hanström, B. (1926). Über den feineren Bau des Nervensystems der Tricladen Turbellarien auf Grund von Untersuchungen an *Bdelloura candida. Acta Zool.* (*Stockholm*) **7**, 101.
Hanström, B. (1928). Some points on the phylogeny of nerve cells and of the central nervous system of invertebrates. *J. Comp. Neurol.* **46**, 475.
Hauschka, S. D., and Konigsberg, I. R. (1966). The influence of collagen on the development of muscle clones. *Proc. Natl. Acad. Sci. U.S.* **55**, 119.
Hyman, L. H. (1951). "The Invertebrates. Vol. 2: Platyhelminthes and Rhynchocoela. The Acoelomate Bilateria". McGraw-Hill, New York.
Johnson, W. H., Miller, C. A., and Brumbaugh, J. E. (1962). Induced loss of pigment in planarians. *Physiol. Zool.* **35**, 18.
Josephson, R. K., and Macklin, M. (1967). Transepithelial potentials in *Hydra. Science* **156**, 1629.
Kepner, W. A., and Cash, J. R. (1915). Ciliated pits of *Stenostoma. J. Morphol.* **26**, 235.
Kepner, W. A., and Foshee, A. M. (1917). Effects of light and darkness on the eye of *Prorhynchus applanatus* (Kennel). *J. Exptl. Zool.* **23**, 519.
Kepner, W. A., and Taliaferro, W. H. (1912). Sensory epithelium of pharynx and ciliated pits of *Microstoma caudatum. Biol. Bull.* **23**, 42.
Koblick, D. C., Goldman, H. H., and Pace, N. (1962). Cholinesterase and active sodium transport in frog's skin. *Am. J. Physiol.* **203**, 901.
Koch, H. J. (1954). Cholinesterase and active transport of sodium chloride through the isolated gills of the crab. (*Eriocheir sinensis* M. Edw.). *In* "Recent Developments in Cell Physiology" (J. A. Kitching, ed.). Butterworth, London.
Kordik, P., Bülbring, E., and Burn, J. H. (1952). Ciliary movement and acetyl choline. *Brit. J. Pharmacol.* **7**, 67.
Lender, T., and Klein, N. (1961). Mise en évidence de cellules sécrétrices dans le cerveau de la Planaire *Polycelis nigra*. Variation de leur nombre au cours de la régéneration posterieure. *Compt. Rend.* **253**, 331.
Lentz, T. L. (1966). Histochemical localization of neurohumors in a sponge. *J. Exptl. Zool.* **162**, 171.
Lentz, T. L. (1967). Rhabdite formation in planaria. The role of microtubules. *J. Ultrastruct. Res.* **17**, 114.

Mackie, G. O. (1965). Conduction in the nerve-free epithelia of siphonophores. *Am. Zoologist* **5**, 439.

Marriott, C. H. (1958). The absolute light-sensitivity and spectral threshold curve of the aquatic flatworm *Dendrocoelum lacteum*. *J. Physiol.* (*London*) **143**, 369.

Mihálik, P. V. (1935). Flimmerblasen im Epithel der Luftwege. *Z. Zellforsch. Mikroskop. Anat.* **23**, 510.

Mitropolitanskaya, R. L. (1941). On the presence of acetyl choline and cholinesterase in Protozoa, Spongia and Coelenterata. *Compt. Rend. Acad. Sci. URSS* **31**, 717.

Press, N. (1959). Electron microscope study of the distal portion of a planarian retinular cell. *Biol. Bull.* **117**, 511.

Richardson, A. P. (1937). Toxic potentialities of continued administration of chlorate for blood and tissues. *J. Pharmacol, Exptl. Therap.* **59**, 101.

Röhlich, P., and Török, L. J. (1961). Elektronenmikroskopische Untersuchungen des Auges von Planarien. *Z. Zellforsch. Mikroskop. Anat.* **54**, 362.

Sauer, F. C. (1935). Mitosis in the neural tube. *J. Comp. Neurol.* **62**, 377.

Seaman, G. R., and Houlihan, R. K. (1951). Enzyme systems in *Tetrahymena geleii.* (iii) Acetyl cholinesterase activity. Its relation to motility of the organism and to co-ordinated ciliary action in general. *J. Cellular Comp. Physiol.* **37**, 309.

Skaer, R. J. (1961). Some aspects of the cytology of *Polycelis nigra*. *Quart. J. Microscop. Sci.* **102**, 295.

Skaer, R. J. (1965). The origin and continuous replacement of epidermal cells in the planarian *Polycelis tenuis* (Iijima). *J. Embryol. Exptl. Morphol.* **13**, 129.

Taliaferro, W. H. (1920). Reactions to light in *Planaria maculata*, with special reference to the function and structure of the eyes. *J. Exptl. Zool.* **31**, 59.

Tibbs, J. (1960). Acetyl cholinesterase in flagellated systems. *Biochim. Biophys. Acta* **41**, 115.

Turner, R. S. (1946). Observations on the central nervous system of *Leptoplana acticola*. *J. Comp. Neurol.* **85**, 53.

Twitty, V. C. (1928). Experimental studies on the ciliary action of amphibian embryos. *J. Exptl. Zool.* **50**, 319.

Ude, J. (1964). Untersuchungen zur Neurosekretion bei *Dendrocoelum lacteum* Oerst. (Platyhelminthes, Turbellaria). *Z. Wiss. Zool. Abt. A.* **170**, 224.

Wen, I. C., Chang, H. C., and Wong, A. (1936). Studies on tissue acetyl choline. (IV) Cytological considerations of the chorionic villous epithelium of the human placenta. *Chinese J. Physiol.* **10**, 559.

CHAPTER 12

THE NEMERTEOID STAGE

A logical development from the basic structure of the rhabdocoeloid stage leads to the somewhat more elaborate and differentiated nemerteoid stage, which can be represented by an hypothetical organism embodying features exhibited by at least some members of the phylum, e.g. Rhynchocoela or Nemertea (Fig. 12.1) (Nemertina). The nemertines or nemerteans still possess the majority of the characters of the rhabdocoeloid stage, but they also developed many new properties and features. The monograph by Bürger (1895) gives some indication of their widely varying properties and potentialities.

One of their most conspicuous and, at first sight, distracting features is the development of an eversible proboscis which is sometimes an elaborate and specialized structure and which, at rest, is housed in a fluid-filled cavity, the rhynchocoel. There are, however, as mentioned on p. 251 (see Figs. 11.6, 11.7), among the rhabdocoels one or two representatives, e.g. *Gyratrix hermaphroditus*, in which an eversible proboscis is already present on a more modest scale; the link between the two groups could well have come through such creatures. The eversible proboscis is a different and separate structure from an eversible pharynx, which may be part of the alimentary canal, but there may be a greater similarity between them than is generally

---

FIG. 12.1. Characteristic features of a nemerteoid, illustrated by transverse sections at various levels (*a*)–(*f*). 1, Frontal organ; 2, eye-cup, with nerve; 3, ciliated groove; 4a, blood vessel associated with rhynchodaeum; 4b, lateral blood vessel; 4c, lateral vessel with commissural vessel; 5, rhychodaeum; 6, dorsal ganglion; 7, cephalic organ; 8, ventral ganglion; 9, lateral nerve cord; 10, proboscis; 11, nephridium with nephridiopore; 12, rhynchocoel; 13, pharynx; 14, gonocoel; 15, armature; 16, intestinal diverticulum; 17, gonopore; 18, retractor muscle in rhynchocoel; 19, dorsal nerve commissure; 20, Anus.

See facing page→

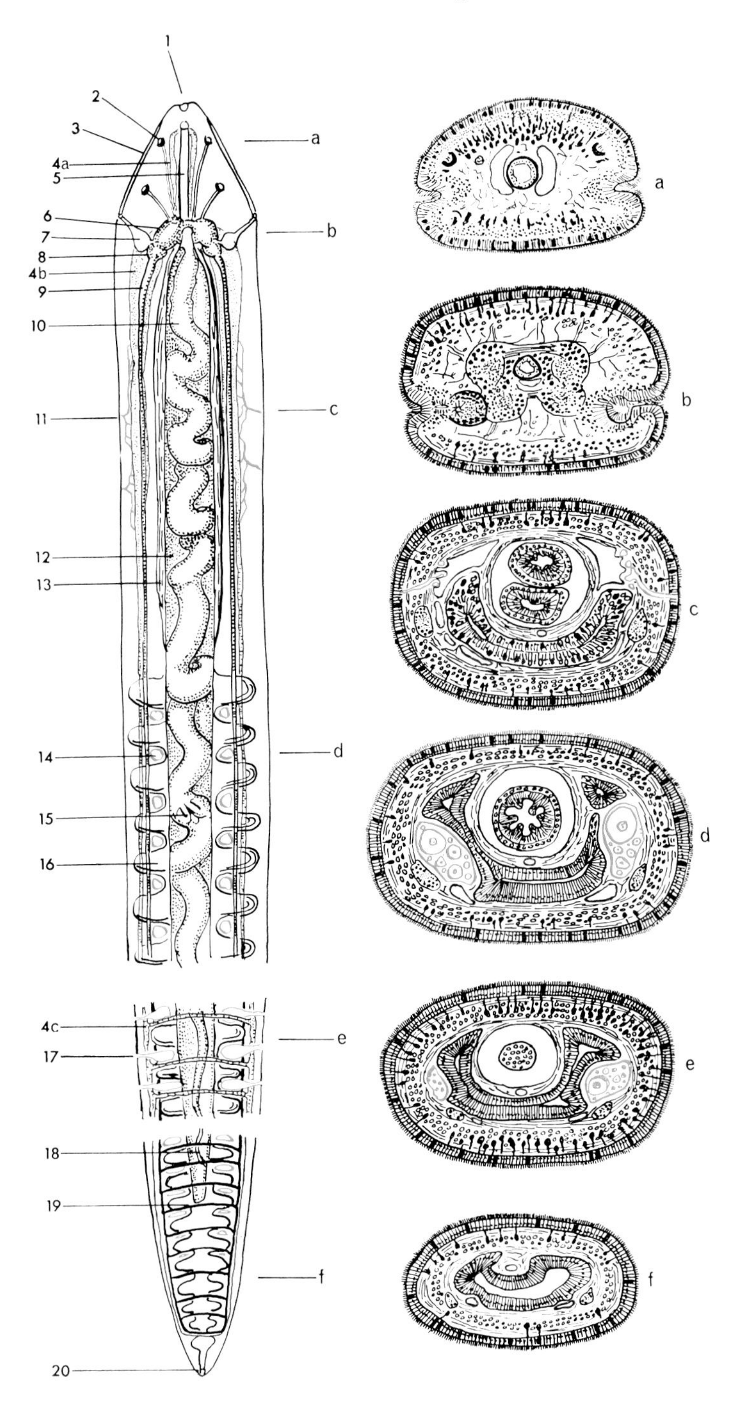
1
2
3
4a
5
6
7
8
4b
9
10
11
12
13
14
15
16
4c
17
18
19
20
a
b
c
d
e
f

supposed. Curiously enough, the fully developed nemertine proboscis has cytological features which are closely paralleled in the gut. In each case the structure is an elongated invagination lined near the orifice with a complex epithelium containing several sorts of cells, and, further in, the epithelium changes abruptly to a much more simple and uniform type. The interesting idea has been expressed that a mutation, or a developmental variation, leading to a duplication of the gut during embryological development, could have brought about the development of a proboscis. For reasons to become apparent later, there may be more than a grain of truth in this idea.

The superficial epithelium, or skin, covering the nemertine maintains much the same characteristics as that in the rhabdocoels, except that the

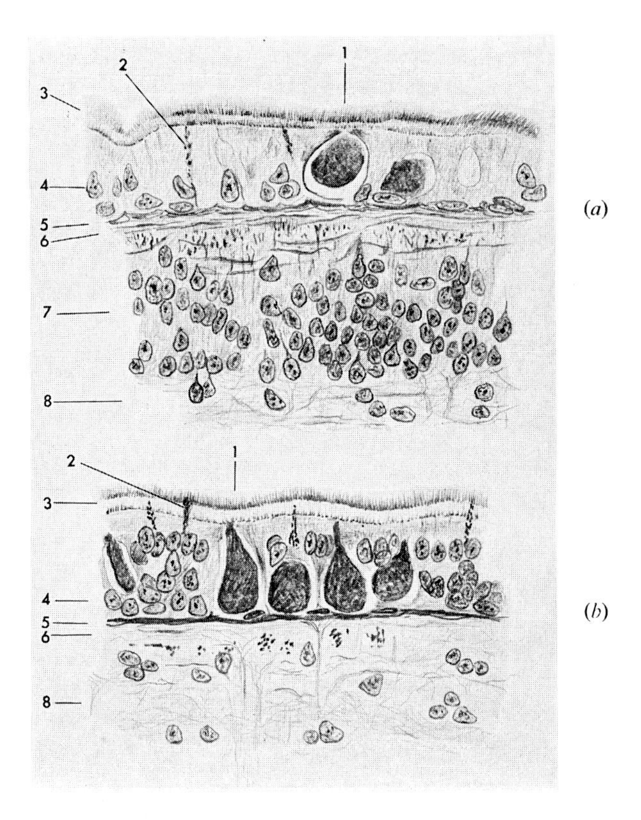

FIG. 12.2. Epidermis of a nemerteoid. (*a*) From the anterior end; (*b*) From the dorsal surface. 1, "Serous" gland cell; 2, neck of mucous cell; 3, cilia and microvilli; 4, nuclei of ciliated cells; 5, basement membrane; 6, muscle fibre layer; 7, nuclei of "sensory" cells; 8, connective tissue fibres.

rhabdites as such are usually a much less conspicuous feature, though their place may be taken by a different secretion product. At least three classes of cells are present: ciliated, mucous, and what have been described as serous gland cells (Fig. 12.2). The last-mentioned cells produce a protein secretion which can be ejected from them in a manner similar to the ejection of mucus from goblet cells. The secretion which is sometimes in granular form within the cells stains red with azan and also with Masson's trichrome stain, but its composition is not known. The cellular organization of the epithelium is variable and is difficult to determine with certainty, since the worm can extend or contract to a prodigious extent; some of the cellular peculiarities and arrangements are probably related to such movements. For example, under certain conditions there seem to be spaces in the epithelium between the bases of the cells into which fluid can penetrate, as there may be in the intestine of vertebrates during absorption of fluid. Many of the epithelial cells contain pigments which are probably carotenoids. In many species the skin secretes a thick mucus which can restrain and paralyse the animal's prey. In two species the worms have been shown to contain a substance (amphiporin) which has nicotine-like properties and a similar substance, "nemertine" (Bacq, 1936, 1937), but the cells from which these substances originate have not been identified. Furthermore, as in the rhabdocoeloids, there are often cells that penetrate below the basement membrane and whose cell bodies may cluster to form small gland-like structures. The contents of these cells vary in character and amount. They are sometimes alcian-blue positive. The presence of these glands again suggests an elaboration of a feature which was already present in the rhabdocoels.

The ecological demands on the skin in different groups of the living nemertines presumably determine some of the detailed characters. It is noteworthy that existing nemertines occupy habitats ranging from the wholly pelagic to the terrestrial and even to the arboreal, provided that there is sufficient moisture in the immediate environment to prevent too great desiccation. This wide variation in environments has probably been conducive to specializations and adaptations which may have led to several divergent paths of evolution.

The movement and general behaviour of most of the nemertines are basically similar to those of the rhabdocoels, namely a combination of creeping by ciliary action and lengthening and shortening by muscular movement. Some species are free-swimming and progress rather in the manner of a swimming snake, except that the undulations are in the vertical plane. Such swimmers have flattened bodies whose margins are sometimes extended by fin-like extensions. The muscles act on a fluid skeleton (Clark, 1964), and their relaxation and lengthening are probably brought about by the tensions exerted on the contained fluid by the contraction of the opposing muscles, e.g. the contraction of the circular muscle is compensated by the extension of the longitudinal muscles. As in the rhabdocoeloid, the body fluids are contained in what is essentially a non-distensible collagenous cylinder whose fibres criss-cross approximately at right angles (Cowey, 1952), i.e. a vessel whose volume cannot be altered beyond a certain limit, though its shape may change considerably. The muscle coats are arranged similarly to those in the rhabdocoels, i.e. they mainly run circularly and longitudinally (Cowey, 1952), but there is much variation and elaboration in their detailed arrangement from species to species, presumably much of it of an adaptive nature and suited to the different modes of life, e.g. free-swimming or floating, creeping in water, burrowing. Many species can both creep and swim and also burrow in the sand. As in the rhabdocoeloid, all the muscle is "plain", though some of it is remarkably rapid in its contraction and relaxation. The fact that, for the reasons given above, circular and longitudinal muscles are essentially antagonistic in these creatures suggests that either neural control predominates or, alternatively, if hormones or chemical transmitters are used, that the circular and longitudinal groups probably respond to different transmitters or hormones, or that they respond differently to the same hormone. Some circular muscles, in *Lineus* for example, give characteristic responses to applied acetylcholine and are inhibited by atropine. Longitudinal muscles respond rhythmically

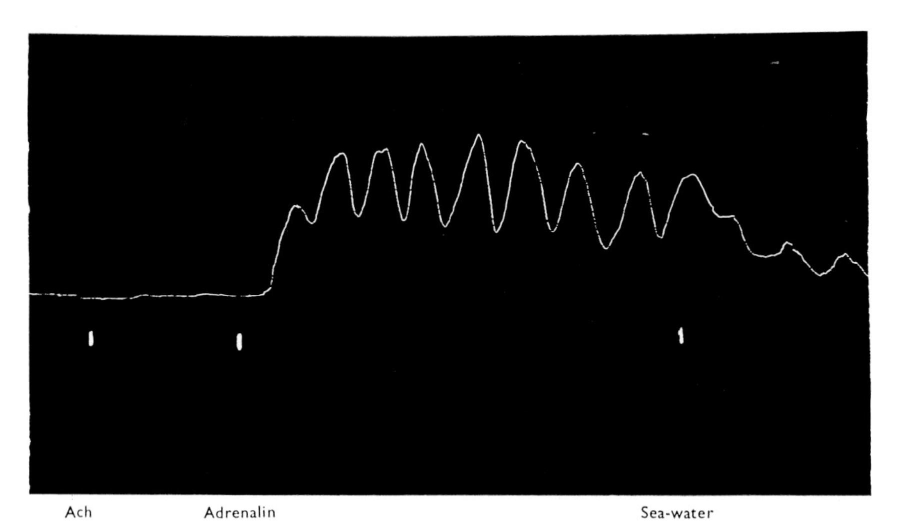

FIG. 12.3. Tracing showing the rhythmic type of contraction initiated by adrenalin (10 γ/ml) in the longitudinal muscle of *Lineus*. Note the absence of any response to acetylcholine (100 γ/ml).

to adrenalin (Fig. 12.3), and some of them respond to oxytocin with a sharp contraction. These observations do not necessarily mean that these substances are the natural excitants, but they indicate that different muscles probably have different excitants. The longitudinal muscle of the proboscis contracts to acetylcholine (Fig. 12.4), whereas the retractor of the proboscis responds vigorously to oxytocin (Mitchell and Willmer, 1967) (Fig. 12.5) as do some of the longitudinal muscles of the body wall. The presence of cholinesterase has been recorded in nemertines (Bacq, 1947), and recently Ling (1969) has established its presence in the ciliated epithelium as well as in several nerve tracts, including those going to the proboscis. In considering these various. groups of muscles, the strictures discussed in the preceding chapter (p. 247) with reference to different classes of nerve cells, apply. Because cells contract, they are not necessarily the same in other respects; a study of the derivation of muscle types from particular cells of epithelia or mesohyl could prove to be very instructive.

The alimentary canal, though generally similar to that in the rhabdocoeloid stage, has several new features. In the first place, it is an actual canal and

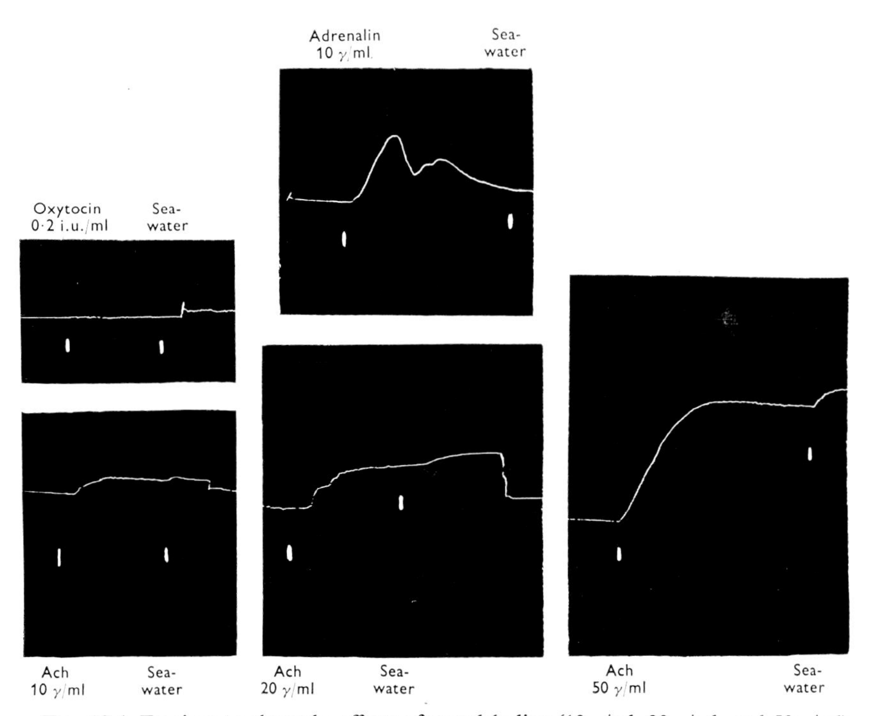

FIG. 12.4. Tracings to show the effects of acetylcholine (10 γ/ml, 20 γ/ml, and 50 γ/ml), adrenalin (10 γ/ml), and oxytocin (0.2 unit/ml pitocin) on the longitudinal muscle of the proboscis.

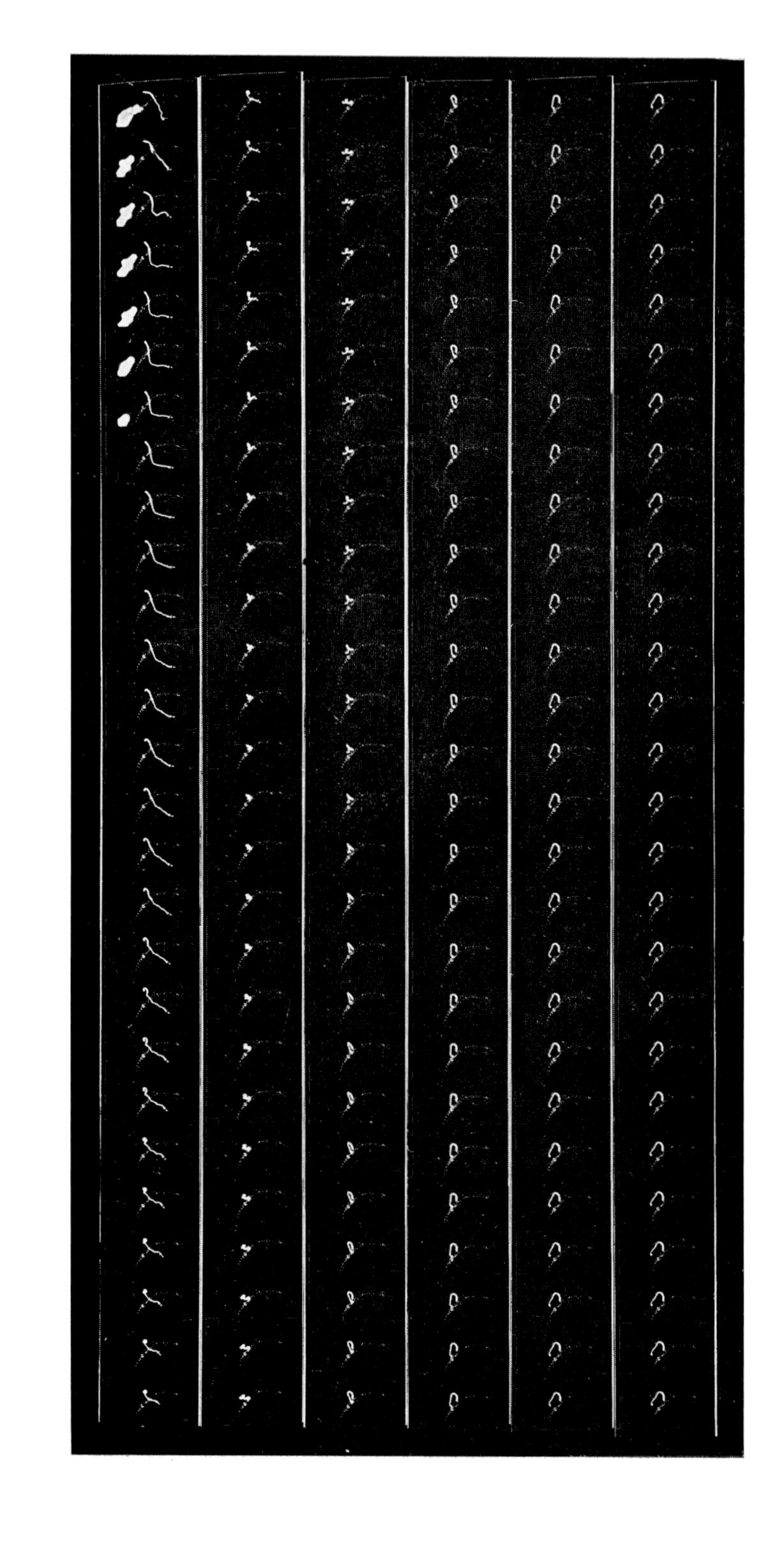

is open at both ends, i.e. a definite anus has appeared. The mouth has moved forward, nearer the anterior end of the body: it opens into a ciliated pharynx, or fore-gut, which may have a complex pouched structure and a complicated and variable epithelium containing several visibly different classes of cell. The epithelium has many points of similarity to that covering the external surface and appears to contain modifications of the same types of cells (Fig. 12.6). It is basically a ciliated epithelium with mucous and

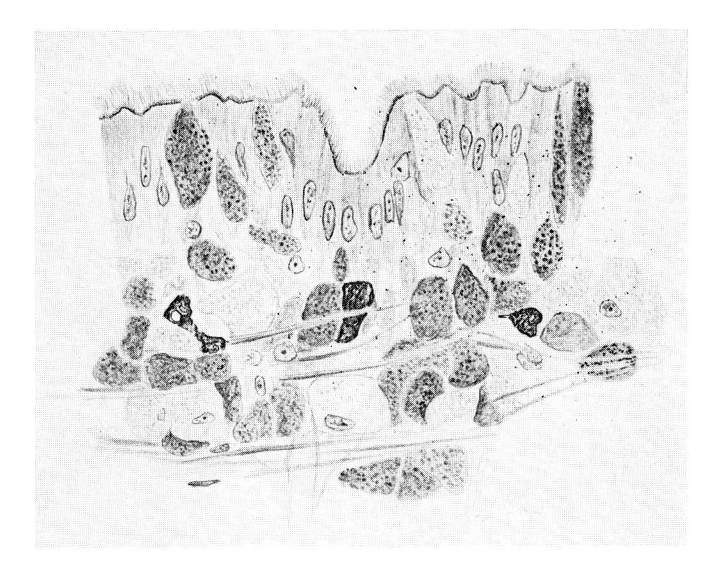

FIG. 12.6. Epithelium of the pharynx of *Lineus* stained with Masson's stain. Note the ciliated epithelium with several sorts of "mucoid" and "serous" cells. Dark shading indicates red stain, light shading indicates green stain. Note also the muscle fibres running between the cells.

eosinophil cells, and it is backed by varying proportions of basophil and eosinophil cells of which there are sometimes several different classes. The cells in this region do not always appear to be in a strictly epithelial arrangement. The large spheroidal and eosinophil cells, for instance, seem not only to occur in the epithelium but also to be loosely attached to the base of other epithelial cells. They contain large nuclei with a conspicuous nucleolus, and they are shown on p. 351 to be of some interest in relation to

FIG. 12.5. Cinematograph record (16 frames/sec) of the contraction of the retractor muscle of the proboscis in response to a drop of solution of oxytocin (0.2 unit of pitocin) added to the 2 ml of sea-water in which the muscle was immersed. Read from top to bottom and from left to right. The white marks in the first row are made by the delivery pipette.
←*See facing page*

further evolutionary developments of the pharyngeal wall, since they may contain carbonic anhydrase (Jennings, 1962). Other cells, like those noticed in the skin, maintain contact with the free surface and pharyngeal lumen by means of long processes penetrating between the ciliated cells. The elaboration of the pharyngeal epithelium bears witness to a diversity of functions, and the epithelium will be shown to have many intrinsic or latent potentialities.

At its posterior end the pharynx opens abruptly into an intestine which usually, though not always, has elaborate and, often, regularly placed lateral diverticula. In some species there are also one or more blind diverticula running forward from the intestine just posterior to its origin from the pharynx and thus overlapping the pharyngeal or forward part of the gut. The alimentary canal, apart from the possession of an anus, is anatomically a less complicated and more regularly organized structure than that seen in the rhabdocoeloid, though its cytology may be more complex.

In general, not much muscle is associated with the alimentary canal; presumably the general body movements provide for the distribution and movement of the gut contents, and it may even be a function of the

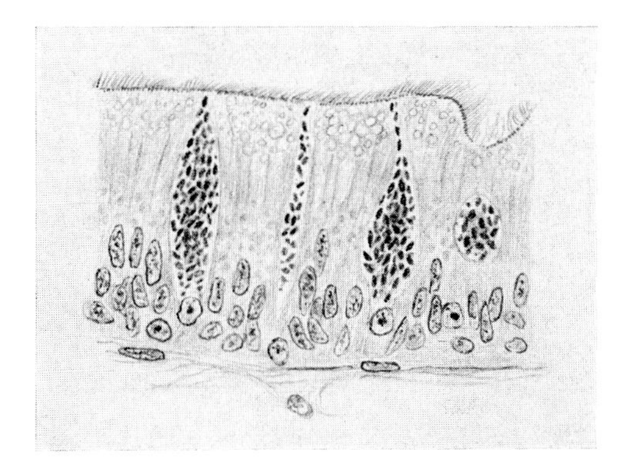

FIG. 12.7. Epithelium of the intestine showing the chief cells (here ciliated) and cathepsin-containing gland cells.

diverticula to restrict this movement to some extent. Whereas the epithelium of the pharynx or fore-gut has ciliated cells, mucous cells, and cells of several other types, the epithelium of the intestine or hind-gut appears to be much simpler. In fact, once again there are usually three main classes of cells (Fig. 12.7). The majority of cells are simple columnar cells whose free surface may have cilia, stereocilia, or microvilli, and this surface may vary with the state of nutrition and digestion. These main cells often show conspicuous vacuolation of the luminal cytoplasm, and, when digestion is in progress, they may be filled with food vacuoles. Under some conditions there may be, among the main cells, more or less mucous cells which give a positive reaction with alcian blue, and, particularly in the region of the opening of the pharynx, there are cells with conspicuous nucleoli and granules in their cytoplasm which stain intensely red with azan and with Masson's trichrome stain. These cells are rich in cathepsin (Jennings, 1962).

The nephridial system, though basically similar to that in the rhabdocoeloid, is tremendously variable and often much more elaborate in its organization (Fig. 12.8). It is even entirely absent in the bathypelagic forms (Coe, 1930), perhaps indicating that its main function is ionic regulation, since the latter must be less important to the bathypelagic forms than to the littoral or estuarine forms. In most cases, flame-cells or solenocytes

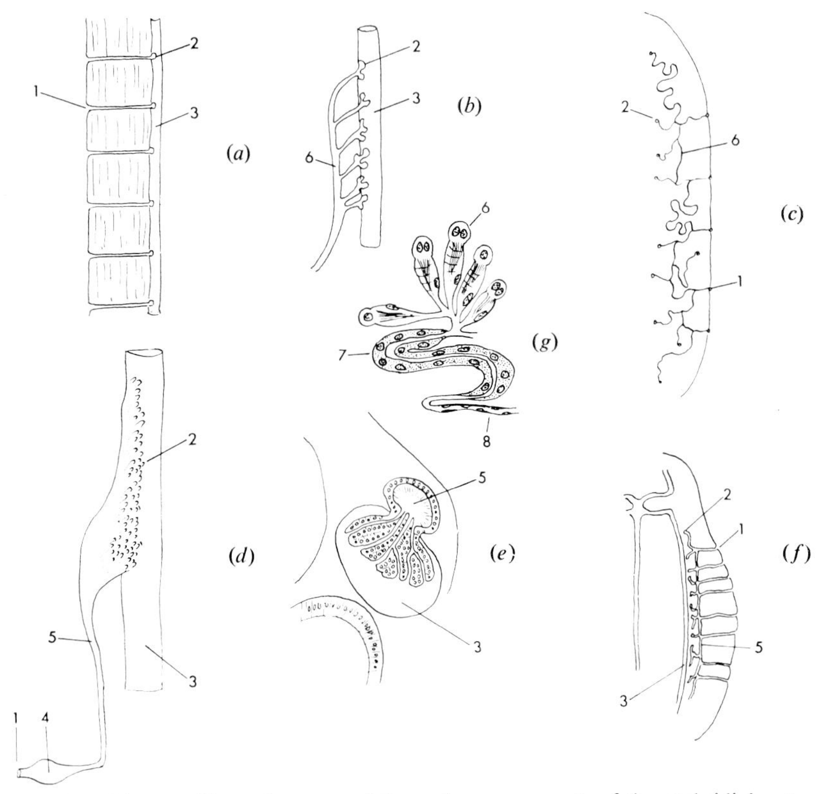

FIG. 12.8. Diagram illustrating some of the main arrangements of the nephridial system in nemertines. (*a*) *Cephalothrix*; (*b*) *Tubulanus*; (*c*) *Prostoma*; (*d*) *Procarinina*; (*e*) *Procarinina* (section through nephridium and blood vessel); (*f*) *Amphiporus*; (*g*) *Geonemertes*. 1, Nephridiopore; 2, nephridium; 3, blood vessel; 4, ampulla; 5, nephridial duct; 6, flame cells (solenocytes); 7, "secretory" epithelium; 8, conducting tube.

lead into a tube subdivided into various sections, each section being lined with a different class of cell (Fig. 12.9). These tubes may join and reach the surface by a common excretory pore or pores, or they may open separately. There is great variation among the living nemertines, and some of the variations, to be discussed later, have significance in relation to the organization of excretory systems in the vertebrates. One such feature is the development of a capsular type of nephridium, so far found only in *Cephalothrix major* and then only in the females. In this nephridium the

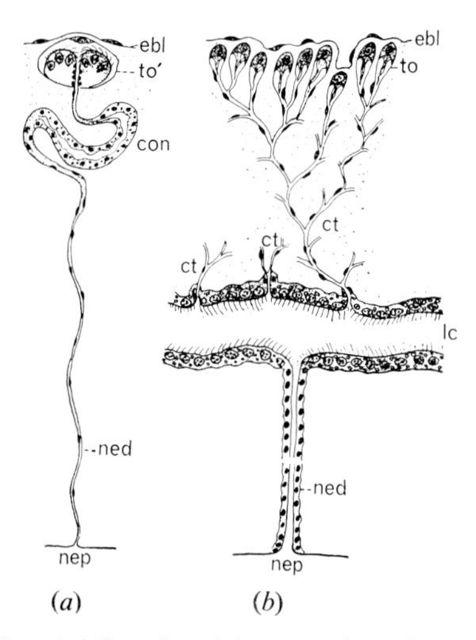

FIG. 12.9. Two types of nephridium found in nemertines. (*a*) As found in *Cephalothrix* (females); (*b*) as found in other nemertines. ebl, Epithelium of blood vessel; to, nephridial capsule; con, convoluted tube; ct, collecting tube; lc, lateral communicating vessel; ned, excretory duct; nep, excretory pore. (From Coe, 1930.)

tube begins in a dilated and thin-walled capsule, the floor of which is lined with flagellate cells which presumably drive the fluid through an orifice in the floor into the next section of the tube, whence it proceeds through a secretory tubule to a collecting tubule (Coe, 1930).

Sometimes the nephridial system is restricted to one or two pairs of large nephridia. Sometimes it is almost serially arranged as a number of repetitive units.

A great advance in the transition from the rhabdocoeloid to the nemerteoid is the acquisition of a closed vascular system (Fig. 12.10). There is little in the way of a propulsive heart, and presumably the contained fluid, like the gut contents, is circulated by the contractions and extensions of the worm as a whole. This could well be adequate since, apart from the rhynchocoel,

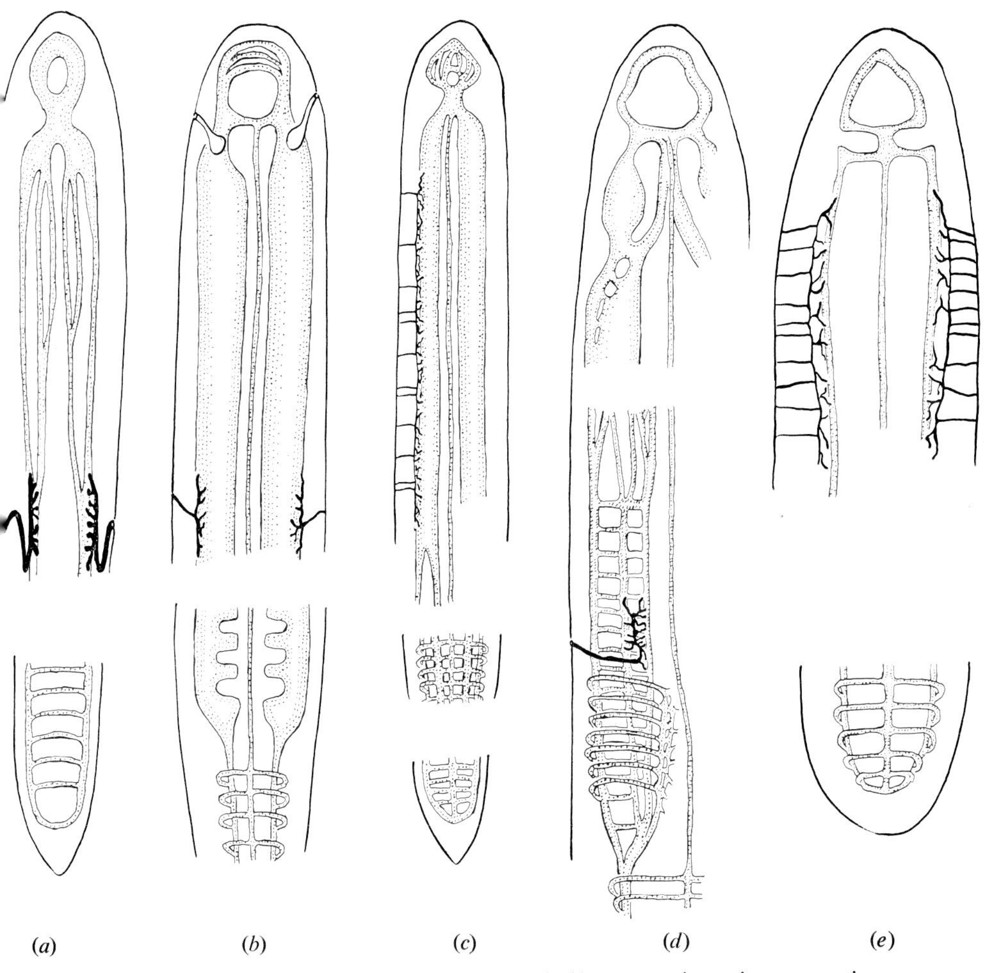

FIG. 12.10. The vascular (red) and nephridial (black) systems in various nemertines. (*a*) *Carinoma*; (*b*) *Hubrechtia*; (*c*) *Valencinia*; (*d*) *Cerebratulus*; (*e*) *Amphiporus*. (After Bürger, 1895.)

which, unless the proboscis is being extended or withdrawn, probably remains constant in volume, there is no separate coelom or coelomic fluid (see, however, p. 432). In some existing species the endothelium of the blood-vessel walls is contractile, and muscles have been described in the walls of the larger vessels (Böhmig, 1897; Riepen, 1933). Haemoglobin is sometimes present in nucleated corpuscles, and several classes of "white cells" have been described as present in the blood; they include cells with eosinophil granules, with basophil granules, and with no visible granules.

Like the comparable cells in the vertebrates, the cells with granules give positive reactions to the test for peroxidase systems (Ohuye, 1942). Thus the types of cells in the blood of nemertines are strikingly similar to those in the blood of vertebrates, though their relative numbers are vastly different (Fig. 12.11).

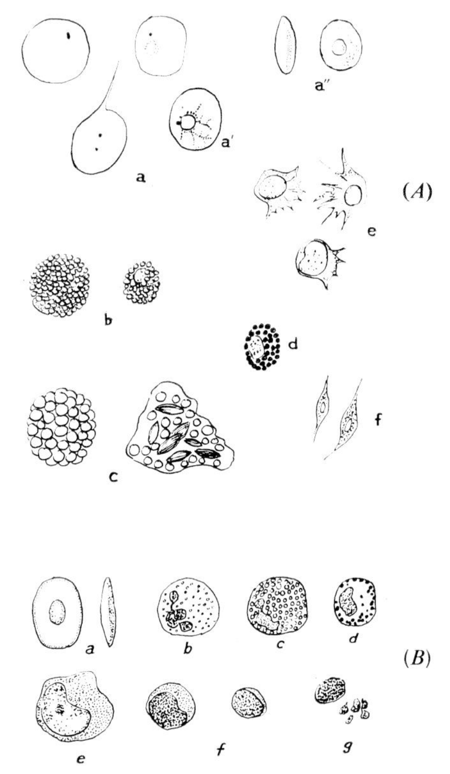

FIG. 12.11. Blood cells in nemertines (*A*) and in vertebrates (*B*). a, a″, Haemoglobin-containing corpuscles; a′, haemoglobin-containing cell with basophil reticulum; b, eosinophil leucocytes with small granules; c, eosinophil leucocytes with large granules; d, basophil leucocyte; e, "lymphocytes"; f, spindle cells; *a*, red corpuscles of frog; *b*, neutrophil leucocyte; *c*, eosinophil leucocyte; *d*, basophil leucocyte; *e*, monocyte; *f* lymphocytes; *g*, platelets, from frog and man. (After Ohuye, 1942 and Bürger, 1895.)

The actual pattern of the vascular system is widely variable among the living nemertines, and the osmotic relationships of the worms, at least in such littoral forms as *Lineus*, can change the amount of fluid in the system extensively. An interesting feature of these worms is that the nephridia are often closely related to the blood vessels, and sometimes direct connections may be formed between the vascular and nephridial cavities (Fig. 12.8e)

(Nawitzki, 1931; Hubrecht, 1885; Oudemans, 1885). These connections probably reflect the co-ordinating action exerted by the blood in the maintenance of a constant internal environment and the function of the nephridia as regulators of that environment.

Experiments in which worms are placed in solutions of differing salt concentrations, however, make it quite clear that many other types of cells, besides those of the nephridia, react to such changes. Osmoregulation is clearly not centralized in any one organ, e.g. kidney, in worms at this level of organization. Let it suffice to say, at this point, that the littoral worms like *Lineus* have considerable powers of adaptation to changes in salinity, and that there are characteristic cytological responses in several groups of cells to both high (up to 5%) NaCl and low (down to 0·6%) NaCl. Both the skin and the pharyngeal epithelium react characteristically, the ciliated cells, in general, standing up well to dilution of the medium but being readily damaged by high concentrations (see p. 240).

The sense-organs in the nemerteoid are mostly straightforward elaborations of those already represented in the rhabdocoeloid stage. Pigmented "eyespots", anteriorly situated, are well developed in some species; their detailed structure is discussed later (see p. 379) in connection with the eyes of vertebrates and other animals. In other species they may be more numerous but individually less well developed, and they may be situated not only anteriorly but also on the lateral margins of the body and, occasionally, throughout almost the whole of its length. Ciliated grooves, with rheoreceptors or chemoreceptors or both, are often elaborately developed, especially on the lateral margins of the head region. In particular, a ciliated pit, like that described in the rhabdocoels, on the lateral margin of the head at about the level of the posterior ganglia often develops into a very complex cephalic organ and becomes closely integrated with the neural tissue of the brain (Fig. 12.12). Its structure and functions are still problematical, but they may have great importance in relation to further evolutionary trends, which are discussed on p. 384). It will have been noticed that the word "head" has been used in this description, and with some justification, since the mouth, frontal organ (see Fig. 15.1), chemo-receptors, ciliated grooves, ciliated pits, cephalic organ, brain, "eyes", and the opening of the proboscis are now all concentrated anteriorly (see Fig. 12.1).

The brain in nemertines (Figs. 12.12, 12.13) appears to be a direct descendant of that of the rhabdocoeloid and takes the form of paired dorsal and ventral ganglia, connected by commissures and thus forming a ring-like structure round the point of attachment of the eversible proboscis. The nerve cells are situated round the periphery of the brain, and the centre is occupied by neuropil and fibres. Neuroglia cells surround the central neuropil, and there are some neurosecretory cells among the large nerve cells near the insertion of the proboscis.

Two large lateral nerves, with accompanying nerve cells, run down the length of the body; in addition, there may be other longitudinal trunks including, in some of the pelagic and swimming species, a dorsal nerve trunk. The latter trunk is peculiar in that it may be ganglionated (Coe, 1927) (Fig. 12.13). In some species the lateral cords remain within the epidermis, but in most groups they have sunk into the mesohyl. The cords seem to consist of fibres enclosed in a collagenous capsule or sheath with

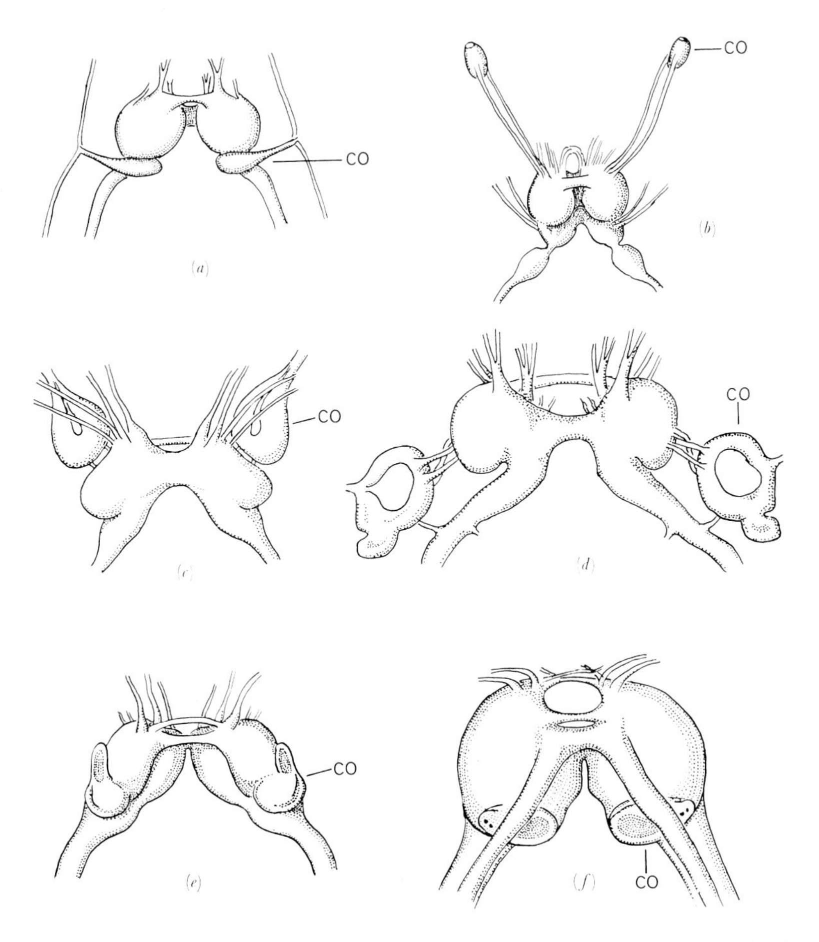

FIG. 12.12. Diagrams of the cephalic ganglia of nemertines and the relationship of the cephalic organ (CO) to them. (*a*) *Valencinia*; (*b*) *Eunemertes*; (*c*) *Tetrastemma*; (*d*) *Drepanophorus*; (*e*) *Micrura*; (*f*) *Eupolia*. (After Bürger, 1895.)

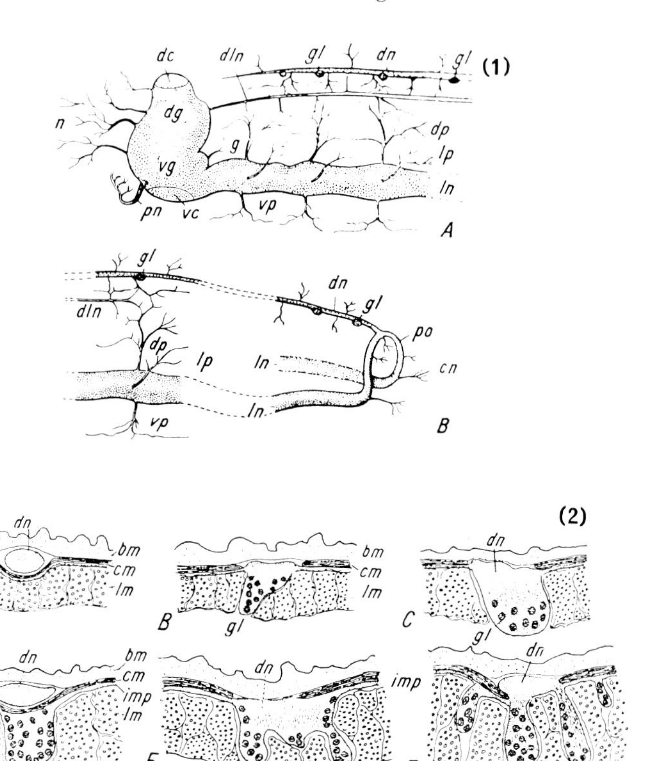

FIG. 12.13. Nervous system in pelagic nemertine (*Neuronemertes*). (1) Main nerves. (*A*) Anterior end; (*B*) posterior end. (2) Arrangement of cells into ganglia in the dorsal nerve cord of pelagic nemertines. (*A*), Section between two ganglia; (*B*), (*C*), (*D*), undivided ganglia; (*E*), (*F*), lobed ganglia. bm, basement membrane; cm, circular muscle; cn, caudal nerve; dc, dorsal commissure; dg, dorsal ganglion; dln, dorsal-lateral nerve; dn, dorsal nerve; dp, dorsal peripheral nerve; g, gastric nerve; gl, ganglion, ganglia; imp, intermuscular nerve; lm, longitudinal muscle; ln, lateral nerve; lp, lateral peripheral nerve; n, cephalic nerves; pn, proboscidial nerve; po, posterior commissure; vc, ventral commissure; vg, ventral ganglion; vp, ventral peripheral nerve. (Coe, 1927.)

neuroglial nuclei just inside, and unipolar nerve cell bodies mostly outside. Structurally, they are essentially prolongations of the neural tissues of the brain. The intimate relationship between the neuroglia cells and the rest of the tissue does not seem to have been very thoroughly investigated, but the processes of the neuroglial cells probably surround the nerve fibres much as they do in amyelinate fibres elsewhere.

The proboscis itself, as already indicated, is formed by an extensive invagination of the anterior end of the animal and is thus, basically, lined with a modification of the superficial epithelium and enclosed in muscle coats that are directly related to the muscle coats of the body wall as a whole (Fig. 12.14). It is very well supplied with nerves, though their nature and function are obscure. These nerves run longitudinally in a series of bundles, and the bundles are connected by ring-like cross-connections. The proboscis, when everted, is an exploratory organ, very motile, and used in the capture of prey, e.g. small worms, etc., around which it can

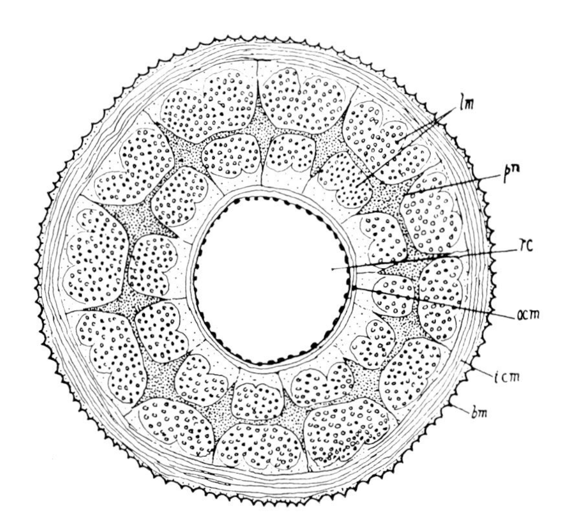

FIG. 12.14. Transverse section of the everted proboscis of a pelagic nemertine, showing the muscle groups and nerve supply. bm, Basement membrane; icm, inner circular muscle; lm, longitudinal muscle; ocm, outer circular muscle; pn, proboscis nerve; rc, rhynchocoel. (Coe, 1927.)

rapidly coil and which in some cases it can surround with mucus and poison, When everted, its surface is covered with papillae like those seen in *Gyratrix*. (see Fig. 11.7), and the cells may contain rhabdite-like bodies.

During the development of the proboscis, a split develops in the muscle coats surrounding the primary invaginations so that the epithelial tube, backed by one group of muscles, is separated from a second group by a fluid-filled space, the rhynchocoel, which is lined with a flat mesothelium. This fluid-filled space is thus a schizocoel and comparable with some "coelomic" cavities in higher animals. Probably by a combination of increased pressure applied to this fluid by the external muscles and the progressive contraction of the muscles of the proboscis itself, the invaginated epithelial sac can be turned inside out. It would thus be forcibly everted through the anterior proboscis-opening (rhynchodaeum) (Fig. 12.15), so that it extends for a long distance in front of the worm itself. In some cases at least, it can be withdrawn again by the contraction of a remarkable muscle attached to the tip of the invagination at one end and to some point near the posterior end of the rhynchocoel at the other. This is done rather in the way that a glove finger might be invaginated by the attachment of an elastic thread between the inside of the finger tip and the inside of the wrist of the glove. The extraordinary change in the length of this muscle will be appreciated.

The proboscis, as a functional unit, is apparently restricted to the Kalyptorhynchids and nemerteoid creatures and, at first sight, may seem to be an irrelevant specialization, placing the nemertines out on a limb of the evolutionary tree. As will be shown later, however, probably this view is erroneous and the proboscis is really a highly significant structure that

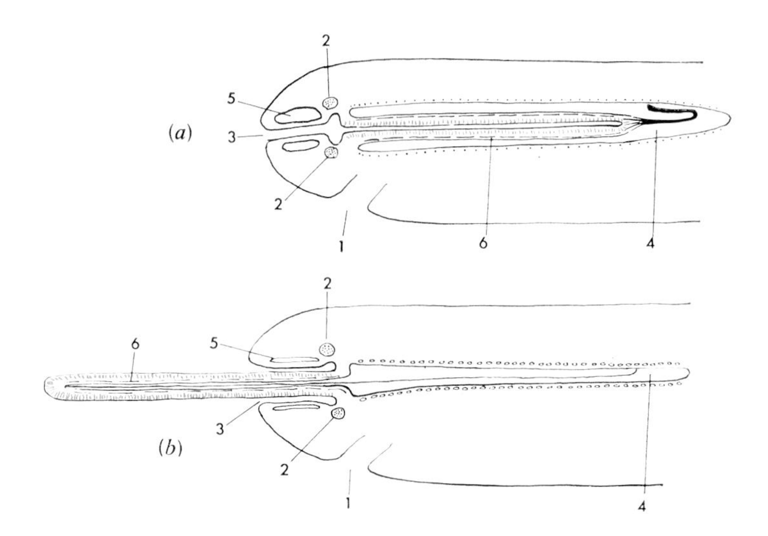

FIG. 12.15. The proboscis (*a*) at rest and (*b*) everted. The papillated epithelium is shown in green, and the retractor muscle in red. 1, Mouth; 2, brain commissure; 3, rhynchodaeum; 4, rhynchocoel; 5, cephalic blood space; 6, proboscis muscle.

is by no means irrelevant to the further evolution of the group. As already mentioned, the proboscis, like the alimentary canal, is sometimes divided into two distinct regions. The divisions between the two may become the site of development of a so-called armature that consists of tooth and bristle-like structures that are sometimes supplied with poison glands. The armature has been used extensively for taxonomic purposes. Anterior to this division, the lining epithelium is generally complex, having several classes of cells, one of which has already been noticed for its cytoplasmic inclusions which have much in common with rhabdites. The cells in the anterior half are often arranged into closely fitting bunches that give the proboscis its round and beaded appearance when it is ejected. On a different scale, one is reminded of the barbs on the cnidocil of the nematocyst. Posterior to the division, the epithelium is much simpler and it terminates in a low columnar type.

The sex glands in nemerteoids are generally arranged serially as separate gonocoels, often alternating with the diverticula of the gut. If the animal is hermaphrodite, the testes are usually anterior to the ovaries, but there is great variation in the different species. The gonocoels can open to the exterior by separate gonoducts, but in some cases the germ cells escape by other routes.

One has the impression that the organization of the tissues of the nemertines is in general very labile, and some features, such as the extrusion of the germ cells, are organized in a somewhat haphazard manner. A worm may be so distended with sperm that even the nerve cord may be surrounded by ripening sperm apparently lying in the interstices of the connective tissue and no longer enclosed in a gonocoel sac. Other tissues also show this lack of compartmentalization; muscle fibres, for example, may traverse the brain, connecting the cephalic organs of the two sides. An impression of general plasticity is created by such arrangements. Although collagen fibres occur freely in the nemertines, their distribution is not so highly organized as in the vertebrates, for example. It may be for this reason that the cellular arrangement of these animals is apparently much more complex and ill-defined than in higher forms. Cohesion of the cells is maintained largely by interdigitations of the cytoplasmic membranes, so that the whole body appears to be more confused in its cellular organization and the various cell groups are much less clearly separable from one another than they are, for example, in the most primitive of the vertebrates.

On the one hand, it is not difficult to imagine that the nemertines have arisen by direct evolution from the rhabdocoeloids: on the other hand, as the following pages show, the nemerteoids have great potentialities for development in several directions and exhibit many features that could fore-shadow those which are characteristic of the vertebrates. Other lines of evolution could also have originated at this level. Moreover, modifications of the rhabdocoeloid stock itself other than that leading specifically to the nemerteoids have undoubtedly occurred besides. These modifications may have led, among other variations, to the development of the trematodes, the cestodes, and probably the nematodes. However, the extant members of such groups do not show as many features of interest in the direction of vertebrate organization as do the nemerteoids, and from the present point of view they must be regarded as aberrant specializations.

The next problem therefore is to consider the directions in which evolution is likely to have proceeded from the nemerteoid and to study in particular the evidence for and against the view that a nemerteoid stage lay on the line of evolutionary development of the vertebrates. The probable positions of the nemerteoids and rhabdocoeloids with respect to the origin of such groups as the molluscs, annelids, arthropods, echinoderms, and other groups are also briefly indicated.

Finally, we may summarize in Table 12.1 the features relevant to vertebrate organization which are developed at or before the nemerteoid stage.

TABLE 12.1

Characters Developed in or before the Nemerteoid Stage Relevant to the Evolution of Vertebrates

1. Complex alimentary canal with mouth and anus; a pharynx with complex epithelium; intestine with diverticula and caeca
2. Serial organization involving gonads, gut diverticula, nerves, and blood vessels
3. Closed vascular system; haemoglobin in corpuscles; white corpuscles of several classes, some of which are peroxidase positive
4. Execretory tubes (nephridia) associated with blood vessels; "capsular" nephridia
5. Nerve cells related to dorsal epithelium (in addition to the nervous system as developed in rhabdocoeloids)
6. Proboscis and rhynchocoel
7. Ciliated grooves and cephalic organ (fore-shadowed in the rhabdocoeloids)

References

Bacq, Z. M. (1936). Les poisons des Nemerteans. *Bull. Classe Sci. Acad. Roy. Belg.* **22**, 1072.

Bacq, Z. M. (1937). L'amphiporine et la "némertine", poisons des Vers némertiens. *Arch. Intern. Physiol.* **44**, 190.

Bacq, Z. M. (1947). L'acétylcholine et l'adrénaline chez les *Invertébrés. Biol. Rev. Cambridge Phil. Soc.* **22**, 73.

Böhmig, L. (1897). Excretory organs and blood vascular system of *Tetrastemma. Ann. Mag. Nat. Hist. Ser.* 6 **20**, 324.

Bürger, O. (1895). Die Nemertinen des Golfes von Neapel. *Fauna and Flora des Golfes von Neapel. Monograph* **22**, 1.

Clark, R. B. (1964). "Dynamics in Metazoan Evolution." Oxford Univ. Press (Clarendon), London and New York.

Coe, W. R. (1927). The nervous system of pelagic nemerteans. *Biol. Bull.* **53**, 123.

Coe, W. R. (1930). Unusual types of nephridia in nemerteans. *Biol. Bull.* **58**, 203.

Cowey, J. B. (1952). The structure and function of the basement membrane muscle system of *Amphiporus lactifloreus* (Nemertea). *Quart. J. Microscop. Sci.* **93**, 1.

Hubrecht, A. A. W. (1885). Der excretorische Apparat der Nemertinen. *Zool. Anz.* **8**, 51.

Jennings, J. B. (1962). A histochemical study of digestion and digestive enzymes in the rhynchocoelan *Lineus ruber* (O. F. Müller). *Biol. Bull.* **122**, 63.

Ling, E. A. (1969). The structure and function of the cephalic organ of a nemertine, *Lineus ruber*. *Tissue and Cell*. (In press).

Mitchell, J. F., and Willmer, E. N. (1967). Unpublished observations.

Nawitzki, W. (1931). *Procarinina remanei*. Eine neue Paläonemertine der Kieler Förde. *Zool. Jahrb. Abt. Anat. Ontog. Tiere* **54**, 159.

Ohuye, T. (1942). On the blood corpuscles and the hemopoiesis of a nemertean, *Lineus fuscoviridis*, and of a Sipunculus, *Dendrostoma minor*. *Sci. Rept. Tohoku Imp. Univ. Fourth. Ser.* **17**, 187.

Oudemans, A. C. (1885). The circulatory and nephridial apparatus of the Nemertea. *Quart. J. Microscop. Sci.* **25**, Suppl. 1.

Riepen, O. (1933). Anatomie und Histologie von *Malacobdella grossa*. *Z. Wiss. Zool.* **143**, 323, 425.

CHAPTER 13

FROM NEMERTEOIDS TO VERTEBRATES
I. NEUROMUSCULAR SYSTEM

In the preceding chapters, a reasonably credible sequence of developmental stages has led from a protozooid to a nemerteoid organization. During this progress new features have been acquired that have allowed new ways of life to be followed and fresh environmental niches to be colonized. Each new feature has entailed consequential modifications, and these modifications have in their turn opened up new possibilities and still further developments.

Divergent Paths from the Nemerteoid

As off-shoots of this evolutionary sequence and as diverging branches on their own, such groups as the Porifera, Cnidaria, Trematoda, Cestoda, and Nematoda seem to find reasonable places. In this discussion of the particular trend of evolution towards the vertebrates, such groups need not be considered further.

In the past it has been fashionable to relate the partially segmental arrangement of the vertebrate body to the strict metameric arrangement of the annelids, if in no other way than to suggest that they both had a common origin. It is therefore relevant to note that the serial or repetitive arrangement of organs which first appears in some rhabdocoeloids and is a frequent feature of the nemerteoids could well blossom out both into the more rigid segmentation characteristic of the annelids and the arthropods and into the less rigid type that is seen in the vertebrates. There is nothing in either annelids or arthropods that corresponds in any way to the proboscis of the nemerteoids, and these groups have both specialized in the production of chitin, which is not a particular feature of the nemerteoids. For these reasons and several others it seems more likely that these groups originated from organisms at the upper level of rhabdocoeloid organization rather than from true nemerteoids. Interestingly enough, chitin has been proved to be present in platyhelminthes only in the cysts of some cestodes and in the Kalyptorhynchia, which, because of the presence of a proboscis, may be considered to be related to the nemerteoids and are thus probably among the more advanced rhabdocoeloid types (see p. 251).

Annelids and arthropods have many cytological and other features akin to those of vertebrates, though their segmentation and their nervous systems are, in fact, very differently organized from those of vertebrates. Nevertheless, in their nervous system the myelination of the nerve fibres can be very similar to that seen in the vertebrates. Myelination depends primarily on the capacity of certain mechanocytic or amoebocytic cells to flatten out on nerve fibres and then to migrate around them. The recent observations (Ernyei and Young, 1966) in which Schwann cells and satellite cells from the dorsal root and sympathetic ganglia of mice have been shown to migrate (like mechanocytes) around appropriate glass fibres are revealing in this connection. However, they do not make it clear why, in the body, nerve fibres behave like the glass and provoke this behaviour on the part of the Schwann cells while collagen and elastin fibres, though excellent substrates for mechanocytes, do not. This particular property seems to have developed in annelids and arthropods and in vertebrates, but not in the nemerteoids, though the nerve fibres in the latter are probably enclosed in neuroglial cytoplasm. As another example of the similarities between annelids and vertebrates, some of the behaviour of nephridia and coelomoducts as seen in annelids has a parallel in the urinogenital system of the primitive vertebrates (see p. 436). All these similarities, however, are not inconsistent with the view that the annelids and arthropods emerged from some creatures resembling well developed rhabdocoeloids or primitive nemerteoids lacking the development of a proboscis, and in which, for mechanical reasons connected with movement (Clark, 1964), the serial organization crystallized into strict segmentation.

The molluscs might be considered rather irrelevant to any discussion of the origin of vertebrates, but, for reasons that will become apparent, this may not be entirely true. They are more perplexing in their origin, but they too, could well have developed from rhabdocoeloid or early nemerteoid organisms. Because of certain factors to be discussed later (see p. 381) their origin from creatures akin to the more highly developed rhabdocoeloids seems the more probable, though there are difficulties in this hypothesis which also are to be discussed. Again, the use of chitin by molluscs and the absence of anything which can be obviously related to the nemertine proboscis suggest something prenemerteoid as the more likely stock. For somewhat similar reasons the echinoderms would seem to branch off at about this point: their early embryogeny, though differing in the mode of cleavage, has some similarity to that of the later rhabdocoeloid and also, as Metschnikoff (1869) noted, to that of some of the nemerteoid types (Fig. 13.1). The claims which have been made that the vertebrates originate

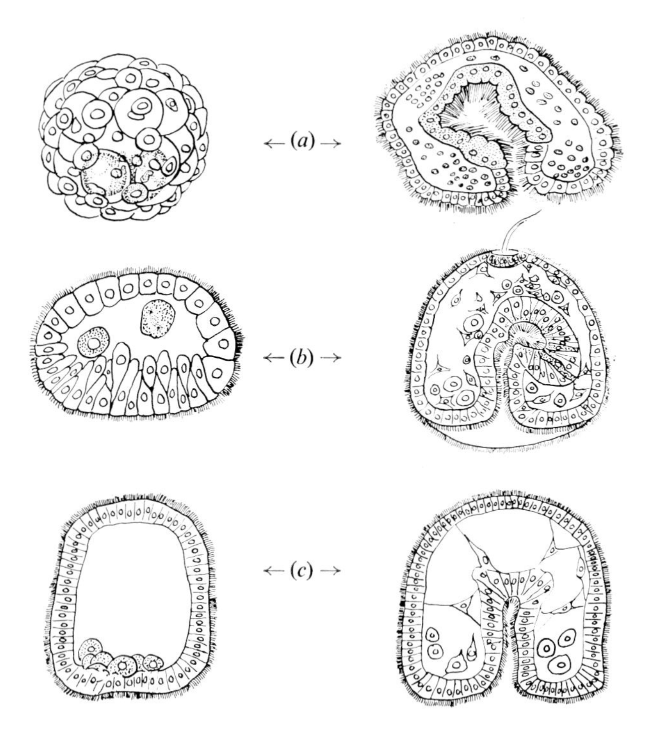

FIG. 13.1. Gastrulation and development of mesohyl in (*a*) turbellarian, (*b*) nemertean, and (*c*) echinoderm.

from echinoderms were partly based on the notion that the distribution of creatine and arginine phosphates had considerable taxonomic significance, a notion that is now believed to be erroneous. Apparent similarities in the embryogeny of the two groups (Barraclough Fell, 1948) might apply equally well and probably more logically between the embryogeny of some existing nemertines and that of the chordates. As Barraclough Fell points out, the larval forms of invertebrates are sometimes specializations within the group to satisfy particular requirements in early life, and they do not necessarily throw any light on ancestral forms, though in most cases they may do so. Each case has to be examined on its own merits.

It should be remembered that the living nemertines belong to a man-made taxonomic group characterized by the presence of a proboscis and classified largely on the details of the proboscis. In the palmy days when worms were the most highly organized creatures in the world and large predators were few or nonexistent, there may well have been a host of other creatures that were similarly organized but which lacked a proboscis. These creatures may then have evolved in various directions, in some of which they have left little trace. One direction of their evolution could have been to the nemertines of today, i.e. to animals dependent on a proboscis; another towards the annelids, i.e. animals specializing in serial organization for mobility; another to the arthropods, i.e. specializing first in segmental organization and then, having acquired locomotory limbs, developing them in conjunction with an exoskeleton; another to molluscs; and so on. Presumably these and other evolutionary paths were followed only because they produced animals more successful in the battle for life than their siblings of parental type: the less efficient stem types would have had to find some sheltered or protected form of existence or they would have become extinct. Unfortunately for us, those intermediate forms and those less efficient forms are now extinct, and they alone could give the real history of events. The living animals whose tissues we are able to study are either the end-products of a successful evolutionary line or rather specialized relics which have succeeded in surviving in some sequestered nook. From the present point of view it may be that the proboscis provided the nemerteoid with just that extra opportunity of feeding on the sea floor, while remaining mostly buried in the sand, which gave it the advantage over its super-rhabdocoeloid parents, and which has since allowed it to continue to occupy an environment in which it can still compete successfully with its rivals.

The question which needs to be answered in the present search for the origin of the vertebrates is: To what animals are we to turn in order to find the beginnings of such features as a neural tube, myelinated nerve fibres, axial skeleton with notochord and myotomes that give co-ordinated side-to-side swimming movements, a heart and gills, a liver, kidneys, eyes with lens and inverted retina, pituitary body, pineal body, thyroid, post-anal tail, and various other features characteristic of the vertebrates in general? All these features are to be found already well developed even in the cyclostomes, those queer creatures which are themselves usually considered to be specialized relics, but which, nevertheless, are probably the most primitive living members of the vertebrate stock.

Most of the groups already mentioned, i.e. the echinoderms, annelids, arthropods, and molluscs, are, for various reasons, probably off the particular line of development that led to the vertebrates and can be forced on to it only by somewhat unlikely reasoning.

Protochordates

The rather heterogeneous group of creatures classed as protochordates contain features indicating that they have some relationship with the vertebrates, though none can be regarded as clearly "ancestral". Their origin and early history are still to be unravelled.

Among them, the Enteropneusta, as exemplified by *Saccoglossus* and *Balanoglossus* (Fig. 13.2), are indeed relevant to the present line of thought (see Barrington, 1965). They are burrowing creatures with a habit of life and general muscular movements organized in a manner similar to those found in some nemertines. Indeed, the posterior part of the body (perhaps the part least likely to change, since it is always buried in the mud) is organized in such a way that it could seemingly have evolved very easily from that of the nemerteoid. The diverticulated gut, the cellular arrangement of the epidermis, and the muscular movements are strikingly similar. Like that of the nemertines, the way of life of the enteropneusts is to lie with most of the body protected in the sand or mud and to protrude only the proboscis and the collar. Like the nemertines, too, they show their resentment of any interference by discarding portions of the posterior parts of the body by autotomy. As always, the problem here is to decide whether adaptation to the particular mode of life accounts entirely for the similarity, or whether the similarity has been produced only because the environment worked on basically similar organizations in both cases.

The proboscis of the enteropneusts, unlike that of the nemertines, is not a retractable structure; nevertheless, it probably gives a clue to the phylogenetic position of the group. The cytology of the epidermis of the proboscis is very similar to that of the nemertine proboscis (Fig. 13.2h); the internal structure of the organ with its muscle and proboscis-coelom is closely paralleled by, and could have developed from, the muscular system and the remnants of the rhynchocoel of the nemerteoid organ. For these reasons the proboscis of the enteropneusts could well be regarded as a permanently everted nemerteoid proboscis. The "glomerular body" at the base of the proboscis is in some ways comparable with the villus body (see p. 348) in the rhynchocoel of nemertines. The so-called collar of the enteropneust could be the metamorphosed head-region of the nemerteoid, where muscles normally run in many directions through a semi-fluid, gelatinous matrix. Modifications of the rhynchocoel, or more probably of the haemocoel cavities surrounding the rhynchodaeum, could also account for the "coelom" of the collar. The ring-like arrangement of the cephalic ganglia could have become the prebranchial nerve ring, and the ganglia could have been drawn out anteriorly with the proboscis nerves to form the so-called neurocord. This structure is interesting in that it is formed by an epithelial invagination

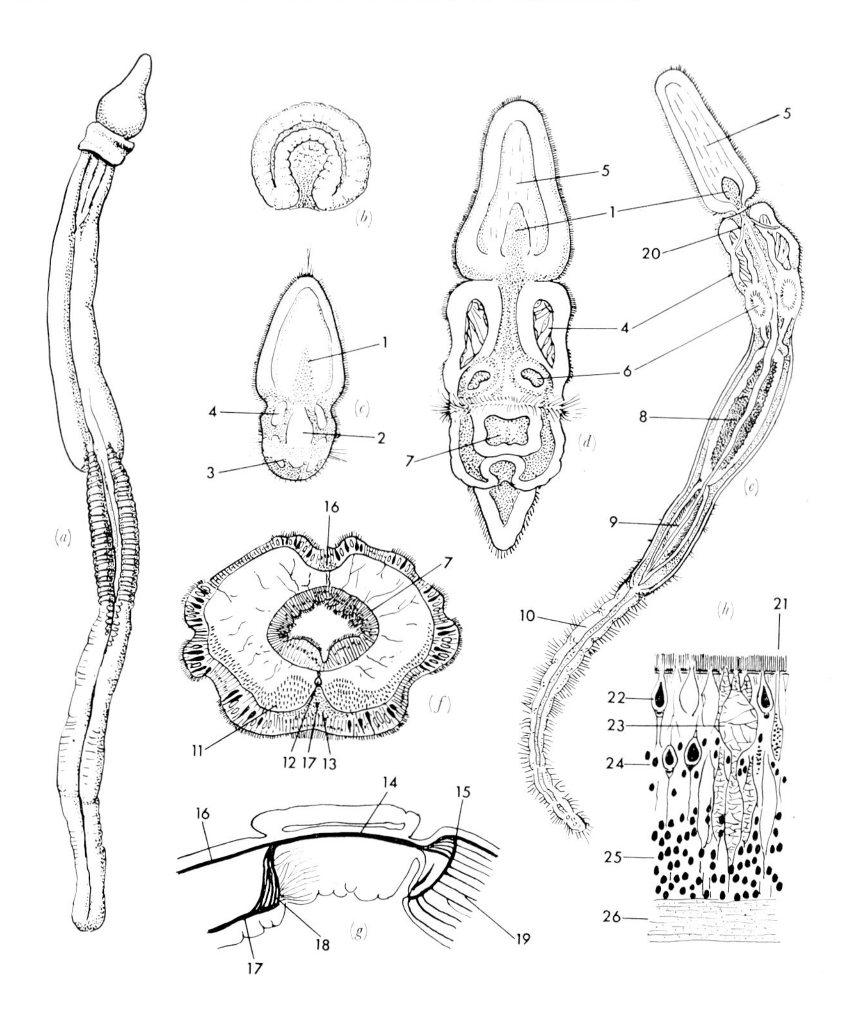

FIG. 13.2. Some features of the Enteropneusta. (*a*) *Glossobalanus minutus*, with proboscis, collar, and trunk; (*b*)–(*e*) stages in the development of *Saccoglossus horsti*; (*f*) transverse section of the body of *Saccoglossus* showing intestine slung in a mesentery and epithelium akin to that of nemerteoids; (*g*) collar region and main nervous tissue in *Saccoglossus*; (*h*) epithelium of proboscis of *Saccoglossus* (cf. epidermis of nemertines). 1, Proboscis complex; 2, archenteron; 3, trunk coelom; 4, collar coelom; 5, proboscis coelom; 6, gill slit; 7, gut, 8, hepatic gut; 9, intestine; 10, segmented region; 11, muscle; 12, mesentery with blood vessel; 13, sensory epithelium; 14, neurocord in collar region; 15, fan of nerves to proboscis; 16, dorsal nerve cord; 17, ventral nerve cord; 18, prebranchial nerve ring; 19, proboscis nerves from anterior nerve ring; 20, proboscis skeleton; 21, mulberry cell; 22, cell with pear-shaped contents; 23, mucous cell; 24, nuclei of ciliated cells; 25, nuclei of mucous cells, etc.; 26, layer of nerve fibres. (After Barrington, 1965; Burdon-Jones, 1952; and Knight-Jones, 1952.)

which becomes rolled up, longitudinally, thus sometimes assuming a partly tubular structure. The neurocord has large (giant) nerve cells in it, reminiscent of those in the ganglia of some of the living nemertines (Fig. 13.3).

The method of feeding of the enteropneusts has changed from that of the nemerteoids in a manner which might well arise from such a creature as *Cerebratulus* is today, i.e. a nemerteoid which fills and empties its pharynx rhythmically for respiratory purposes (Wilson, 1900). If such a pharynx had pouches, as those of many nemertines do (see Fig. 13.4 and p. 349), e.g. *Nemertopsis* and *Prosorhochmus*, and if the pouches of the pharynx were to penetrate the body wall to the outside in the manner of gill-slits, the possibility of filter-feeding would immediately arise. If this occurred,

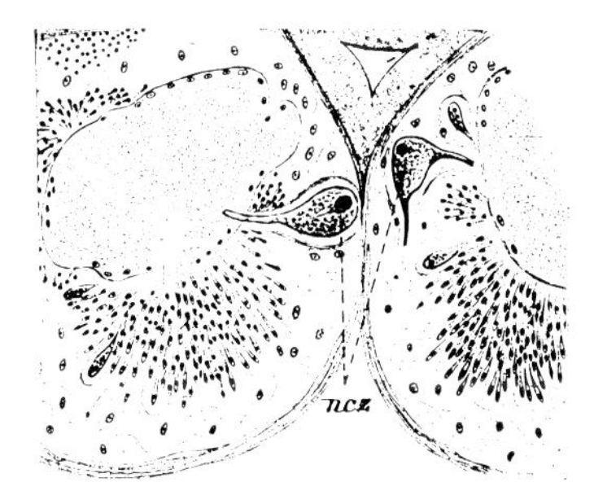

FIG. 13.3. Section through ventral ganglia of nemertine, showing the giant nerve cells ncz). (Bürger, 1895.)

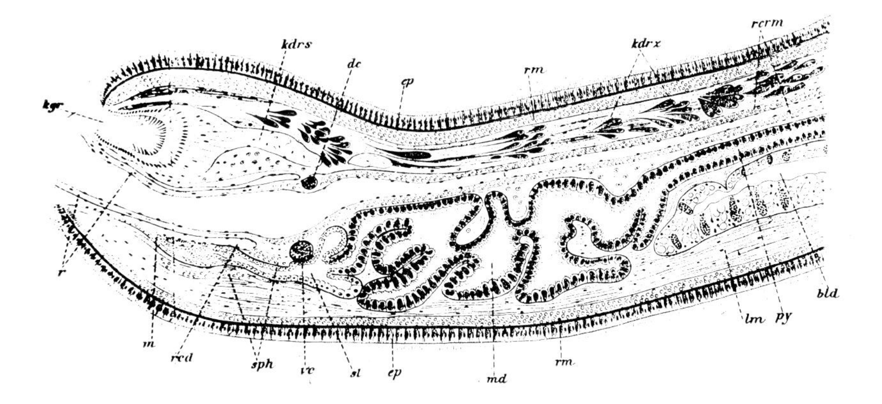

FIG. 13.4. Longitudinal section through the anterior end of *Prosorhochmus* showing the pouching and segmentation of the fore-gut. bld, caecum; dc, dorsal commissure; ep, epithelium; kdrs, cephalic glands; kdrx, gland cells; kgr, frontal organ; lm, longitudinal muscle; m, mouth; md, stomach; py, pylorus; r, proboscis; rcd, rhynchodaeum; rcrm, rhynchocoel muscle; rm, circular muscle; sl, pharynx; sph, sphincter; vc, ventral commissure. (Bürger, 1895.)

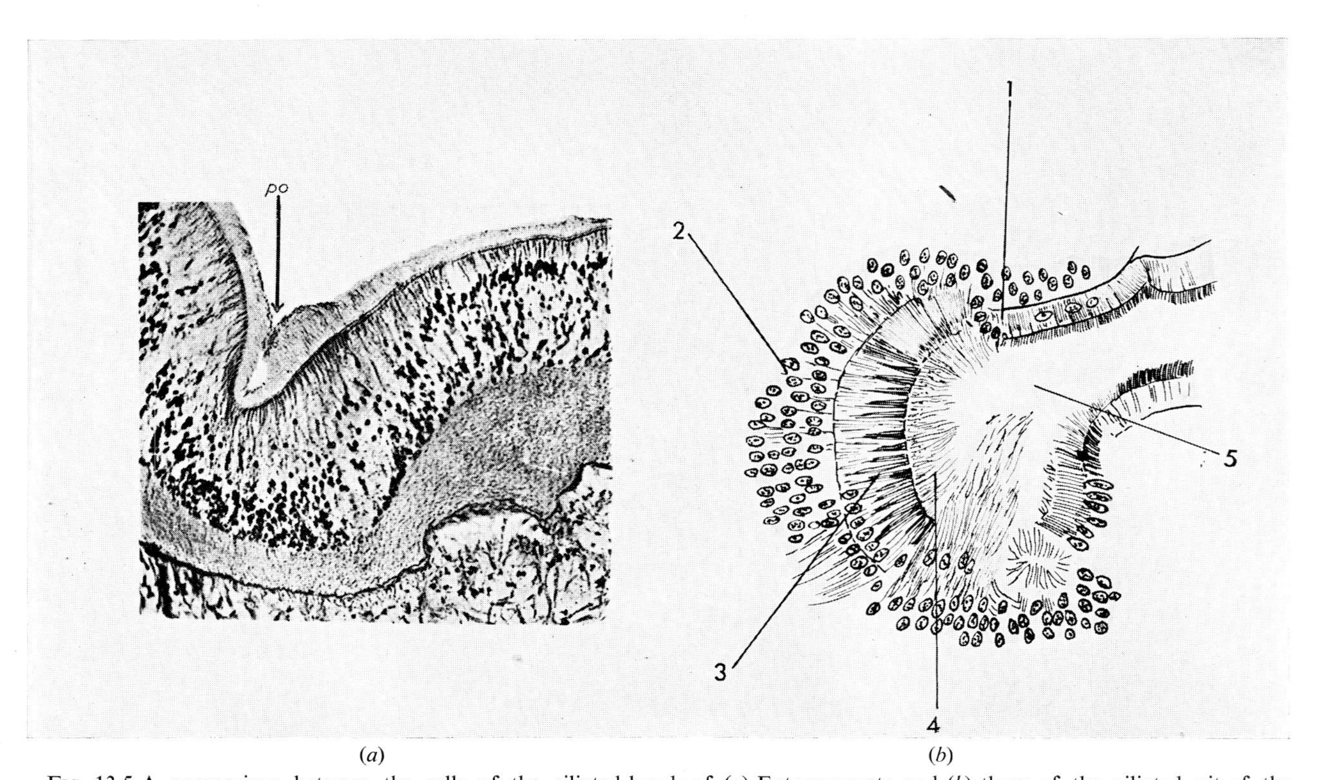

FIG. 13.5 A comparison between the cells of the ciliated bands of (*a*) Enteropneusta and (*b*) those of the ciliated pit of the cephalic organ of nemerteans. po, Pre-oral ciliary organ. 1, Junction between epidermis and sensory epithelium; 2, nuclei of sensory cells; 3, root-fibres; 4, flagella; 5, lumen of pit. (Enteropneust data after Knight-Jones, 1952.)

the proboscis as such could become redundant in terms of its original purpose. Yet, to an animal living in a tube, it could still remain useful in the everted position for immobilizing, by its mucoid and paralyzing secretions, the innumerable small creatures in its vicinity and thus rendering their propulsion by ciliary currents towards the mouth a more certain event. The ciliated grooves of the head region of the nemerteoid could well contribute to these propulsive streams. It is notable that the base of the proboscis in enteropneusts is characterized by bands of cells with extremely long flagella and well developed "root-fibres" similar to those found in the cephalic grooves and cephalic pits of the nemertines (Fig. 13.5) (Knight-Jones, 1952; Brambell and Cole, 1939).

If this were the correct explanation of the general body form of the enteropneusts, then, as already suggested, the "coelom" of the proboscis would be the equivalent of the rhynchocoel cavity of the nemerteoid. The "coelomic" cavities in the collar and trunk together with the other "blood spaces" in this region could well arise from the large "blood spaces" that normally lie along the rhynchodaeum and just posterior to the brain in some nemertines. The "glomerulus", it has already been suggested, has certain similarities to the "villus" structure in the floor of the rhynchocoel (see p. 285). The stomochord with the curious proboscis skeleton (Fig. 13.6) could be related to the pocket of the pharynx, which in some nemerteoids takes over the function of the rhynchodaeum and allows the proboscis to be everted through the mouth. It would thus have features in common with the notochord of the higher forms (see p. 299) but would not be strictly homologous with it. Such diverticula from the pharynx are also conspicuous features in those nemertines in which the proboscis has its own separate rhynchodaeum. These pockets may also have other potentialities, as will be seen on p. 349.

A dorsal nerve cord, as found in the enteropneusts, is found also in some nemertines, but it is curious that the main lateral nerves are not present in the enteropneusts, in view of the great similarity in the bodily movements of the two groups of animals. The lateral cords may, of course, simply have fused ventrally, as they have probably done in the annelids and the arthropods. The arrangements of the gonads and the process of spawning are again very similar to those found in the nemerteoids. The intestine is very like that of the nemerteoid and is ill provided with muscle fibres (Fig. 13.2b). Unlike the intestine of the living nemertines, it is slung in a mesentery in a much more open coelomic cavity than ever occurs in the nemertines. The development of this cavity may have resulted from the absence of the main rhynchocoel cavity, which would no longer be necessary to house the retracted proboscis and would not have the normal stimulus for its development. The fluid in the rhynchocoel cavity in the nemertines undoubtedly

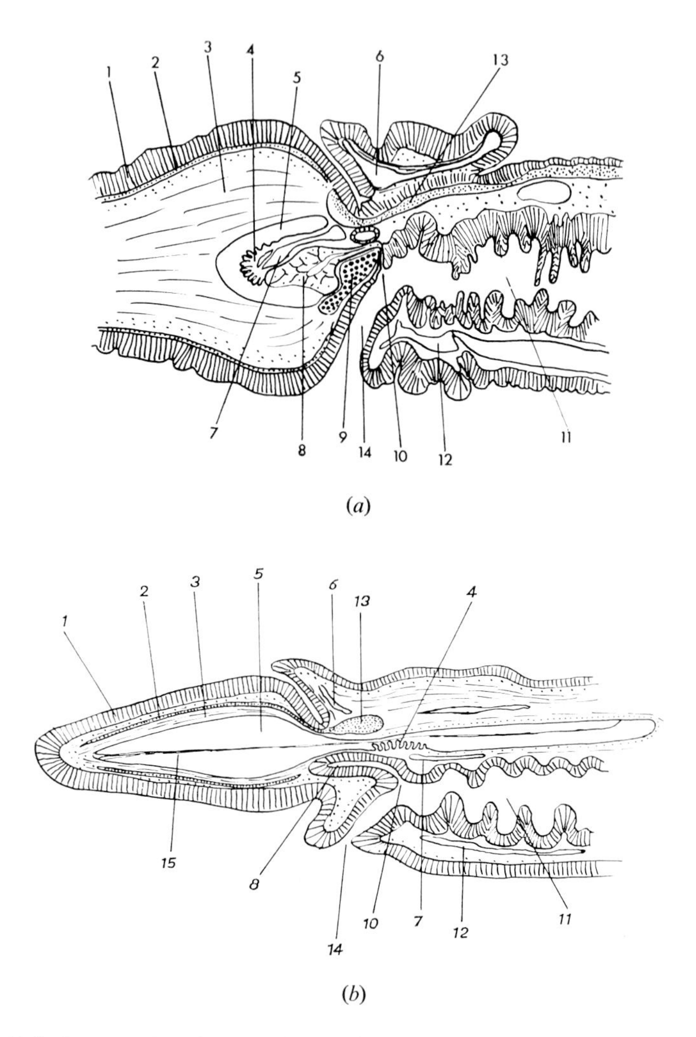

FIG. 13.6. A comparison between the relationships of the proboscis, stomochord, etc., of enteropneusts and possibly corresponding structures in nemertines. (*a*) Sagittal section through peduncle region of enteropneust; 1, Proboscis epithelium; 2, Proboscis nerve; 3, proboscis muscle; 4, glomerulus; 5, proboscis coelom; 6, collar coelom; 7, "heart"; 8, "notochord"; 9, nuchal skeleton cavity; 10, cavity of "notochord"; 11, pharynx; 12, collar coelom; 13, nerve cord; 14, mouth; (*b*) nemerteoid with proboscis ejected. *1*, Proboscis epithelium; *2*, proboscis nerve; *3*, proboscis muscle; *4*, villus organ; *5*, rhynchocoel; *6*, perirhynchodaeal vessel; *7*, dorsal vessel; *8*, stomodaeal invagination (see Fig. 13.10a); *10*, opening of invagination; *11*, pharynx; *12*, blood vessel; *13*, ganglion; *14*, mouth; *15*, retractor muscle. (Enteropneust data after Bullock, 1946; Brambell and Cole, 1939.)

assists in the hydraulic movements of the body and in allowing the fantastic changes of shape that occur in these animals. Its absence might therefore be severely limiting to muscular activity and movement. In its place, by the production of a more fluid ground-substance, the general parenchyma may have become a much more fluid structure in general, and may have eventually acquired the essential features of a coelomic cavity, providing a reservoir of fluid enclosed within a membrane of mesenchymal cells (see also p. 432). It may indeed be this event that, in these creatures, has favoured the fusion

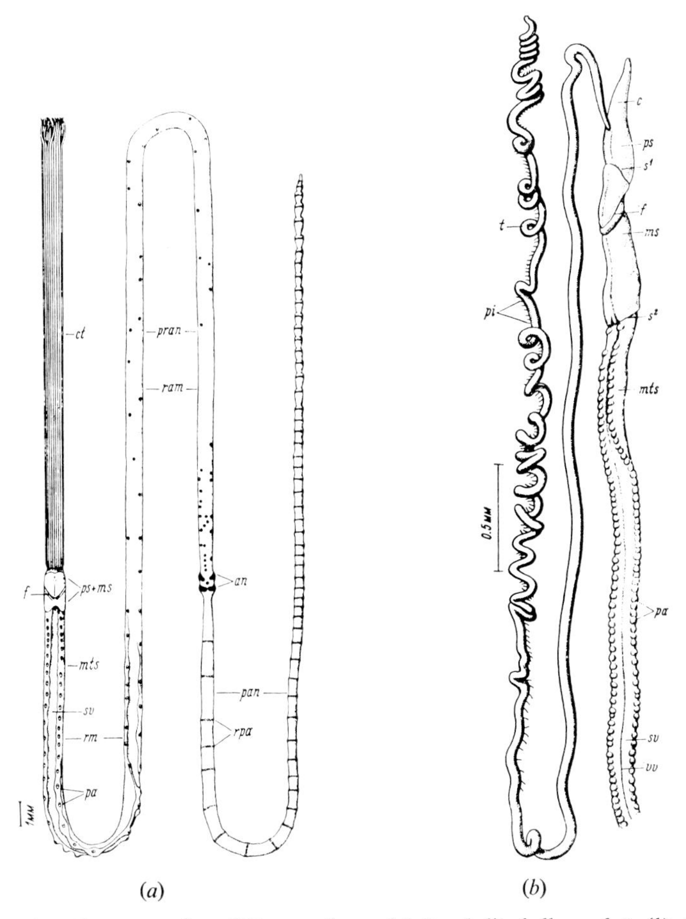

FIG. 13.7. Two examples of Pogonophora. (*a*) *Lambellisabella zachsi*; (*b*) *Siboglinum caulleryi*. an, Girdles; c, cephalic lobe; ct, tentacles; f, frenulum; ms, mesosoma; mts, metasoma; pa, papillae; pan, postanal region; pi, pinnules; pran, preannular region; ps, protosoma; ram, non-metameric region; rm, metameric region; rpa, transverse rows of papillae; s^1, groove between proto- and mesosoma; s^2, groove between meso- and metasoma; sv, ventral sulcus; t, tentacles; vv, ventral blood vessel. (Ivanov, 1963.)

of the lateral nerve cords into the mid-ventral position, though the relationship between the nerve cords and the epidermis suggests a more primitive origin for the positions of the nerve cords. The embryological development and the nature, origin, and activity of the cells surrounding the cavity are matters of great importance in determining the homologies of coelomic cavities, and so far, they have been insufficiently studied.

Although there is evidence for light-sensitivity in the enteropneusts, there is no organ which can immediately be called an eye.

FIG. 13.8. Proboscis of *Gorgonorhynchus*. (Dakin and Fordham, 1936.)

The enteropneusts, if these arguments are in the main correct, should therefore be regarded as a direct modification of the nemerteoid, characterized by the penetration of gill-slits, the acquisition of filter-feeding, and the conversion of the use of the proboscis to a mechanism for permanently increasing the catchment area for small organisms, instead of for occasionally facilitating the capture of larger prey.

Perhaps the enigmatic Pogonophora (Fig. 13.7) found a similar niche, the crown of tentacles being derived by reduplication and elaboration of the proboscis for a similar purpose, i.e. increasing the catchment area for small organisms. The subdivision of the proboscis in the nemertine, *Gorgonorhynchus* (Fig. 13.8), presumably has much the same effect. The resemblance between the *Siboglinum* with its single tentacle (Manton, 1958; Ivanov, 1963) and a nemertine with its proboscis extruded is also very striking (Fig. 13.9). In the Pogonophora it appears that the activity of the

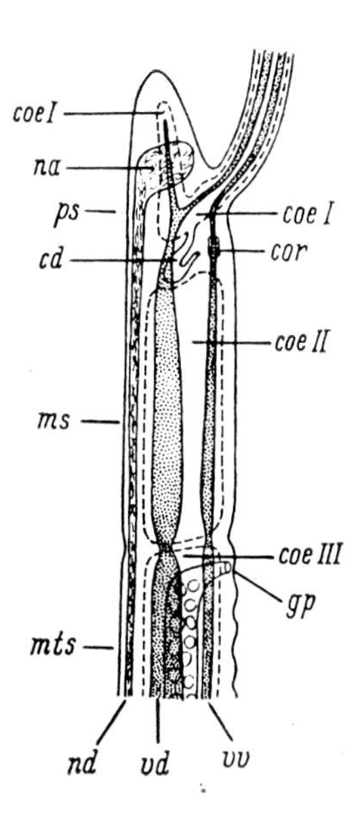

FIG. 13.9. Diagram of the anterior end of a Pogonophoran (cf. Fig. 13.6). cd, nephridium; coe I, II, III, coeloms; cor, heart; gp, genital pore; ms, mesosoma; mts, metasoma; na, cervical nerve ring; nd, nerve trunk; ps, protosoma; vd, dorsal blood vessel; vv, ventral vessel. (Ivanov, 1963.)

proboscis has been extended beyond that of mere capture of prey to that of its digestion. Such a change is in some ways not unexpected in view of the suggestion (p. 262) that the embryological invagination which gives rise to the proboscis may initially have occurred as a teratological duplication of the embryonic rudiment of the gut which was subsequently turned to advantage. With the enormous increase in area of the proboscis in the Pogonophora and the digestion of the small food particles on the spot, the alimentary canal as such has become unnecessary. The body is now more thoroughly devoted to reproduction and requires only a vascular system to convey the digested food materials from the proboscis to the rest

of the body and the gonads. Provided that the food supply is adequate (and in the calm waters of the deep, in which all dead creatures sink to the bottom, it probably is), the arrangement presumably works very well.

These two groups have been discussed at some length, not because they are directly on the line to vertebrates, but because they seem to show the effects that changes of feeding habits and respiratory methods could have on the ways in which the basic structure of the nemerteoid could be modified and have the essential cellular characteristics remain recognizably the same. In both cases the proboscis has been assumed to have changed its function but to have remained recognizable as a proboscis with only relatively minor modifications of its cellular activities.

Other Chordates

In the case of the enteropneusts it was assumed that filter-feeding evolved first from some creature with both an inflatable pharynx, like *Cerebratulus*, and a pouched pharynx, like *Prosorhochmus*, and that the animal otherwise maintained its general habits as a relatively stationary creature based on a burrow in the sand. Protection in this case no doubt compensated for limited food supply, and this protection may have allowed the creature its survival to the present day. The permanently everted proboscis, with its ciliary currents constantly trapping small food particles, ensured a constant, if limited, food supply.

Another way for a filter-feeding nemerteoid to increase its food intake above that possible by ciliary action alone, or by ciliary action assisted by swallowing movements, would be to increase the current through the mouth. This could be done by swimming movements propelling the whole animal through the water or allowing it to hold a fixed position in a current of water. If this "feeding by swimming" could be achieved, as, of course, it has been

FIG. 13.10. The fate of the proboscis rudiment. (a) The embryo nemertine with proboscis rudiment and stomodaeal pouch; (b), (b_1) relation of proboscis to buccal cavity in different nemertines; (b_2) the histology of the rhynchodaeum, proboscis, rhynchocoel, cerebral commissures, and buccal evagination; (c) anterior end of *Amphioxus* with notochord and Kölliker's pit; (d) section through anterior end of a lamprey, showing hypophysial sac, notochord, stomodaeal invagination; (d_1) development of parts of the pituitary gland in lampreys; (e) section through anterior end of myxinoid, showing hypophysial tube and notochord; (e_1) pituitary body of *Myxine*; (f) corresponding section through higher vertebrate showing notochord and Rathke's pouch; (f_1) the pituitary body. 1, Frontal (olfactory?) organ; 2, rhynchodaeum; 3, proboscis; 3′, notochord; 4, rhynchodaeum; 5, stomodaeal invagination (Rathke's pouch?); 5′, Hatschek's pit; 5_1, pars distalis; 5_2, pars tuberalis; 5_3, pars intermedia; 5″, "anterior lobe"; 6, gut; 7, anus; 8, pars nervosa.

See facing page→

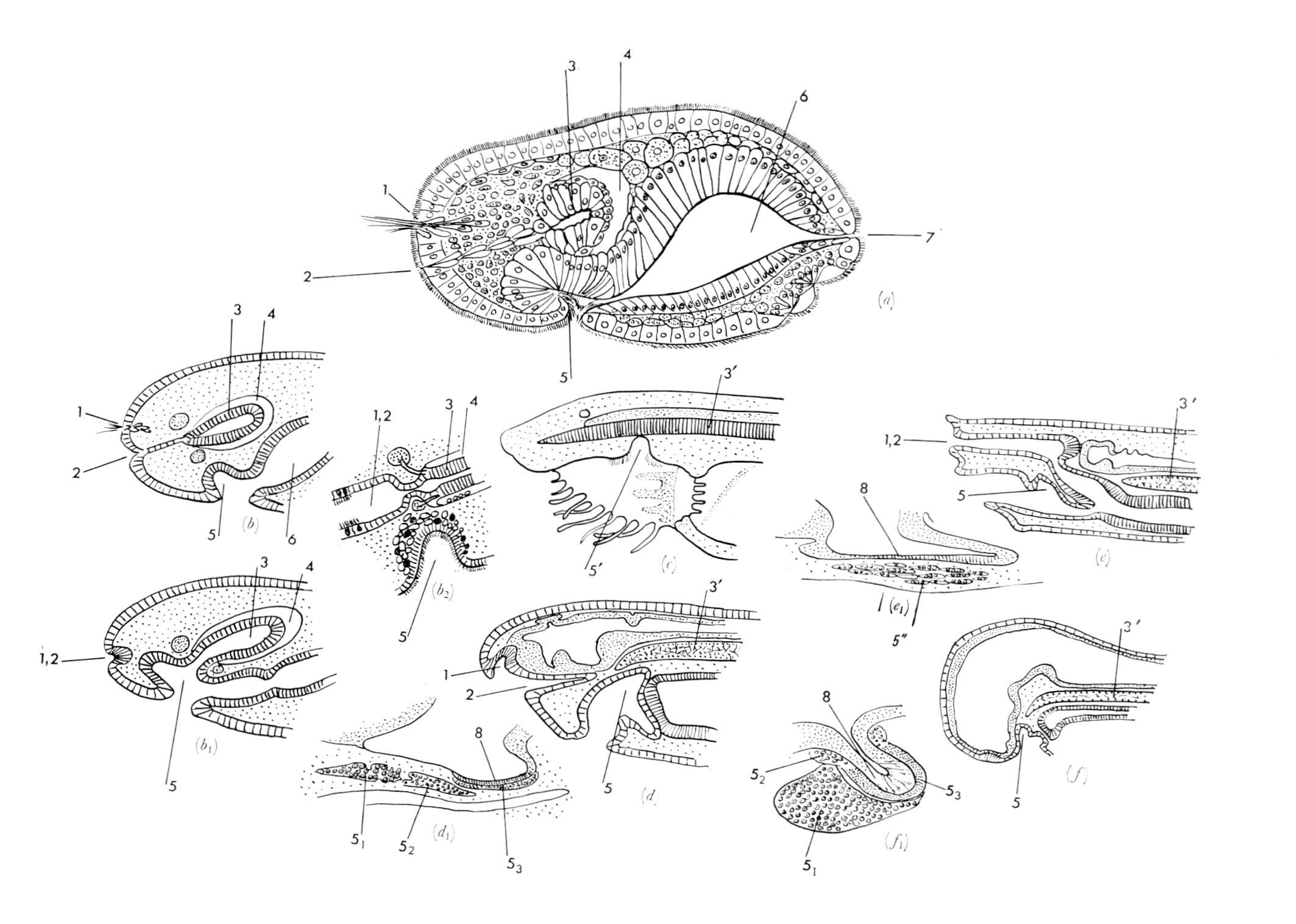

(a)
1
2
3
4
5
6
7
(b)
(b₁)
(b₂)
1,2
(c)
3′
5′
(d)
(d₁)
5_1
5_2
5_3
8
(e)
(e₁)
5″
(f)
(f₁)

in other phyla, e.g. the Cetacea, the proboscis would not only become redundant but might also be positively harmful on account of the navigational effects of its sudden ejection, its lateral movements, or its retraction. What, therefore, would happen if the proboscis were to atrophy? It has a sizeable anlage, in the form of a column of cells within the embryo (Fig. 13.10), running dorsally, above the gut cavity. These cells are active producers of muco-substances almost as soon as they differentiate; so much that the column of cells could easily act as a relatively rigid rod, either at an early state on account of intracellular turgidity, or later by the secretion of the muco-substances into the limited and confined space of the proboscis cavity. In either case a semi-rigid rod could assist in maintaining the stream-lining of the body and helping it to orientate in a current. Moreover, the turgidity of the proboscis could positively interfere with the normal plasticity of the creeping worm. This sequence of events would seem likely to offer another potential mode of evolution for the nemerteoid stock, namely that a nemerteoid should become capable of directional swimming and thus make better use of a filter-feeding mechanism, initially made possible by the perforation of the body wall by gill-slits in the pharyngeal region, itself brought about in response to increased respiratory requirements. It may, of course, be argued that there actually exist in the open oceans of to-day free-swimming nemertines and that neither do they have gill-slits nor do they swim in this way; they still have a proboscis and do not feed by filtration. Their bodies are flattened dorso-ventrally, and they develop lateral and caudal "fins" (Fig. 13.11). These creatures are pelagic, and their swimming is in fact little more than an adaptation to keeping afloat by "treading water". Interestingly enough, in view of what follows, they have, as stated, developed fin-like structures on the sides and the tail of the body, and some of them have developed nerve cords on the dorsal surface, just below the epithelium. These nerve cords may show ganglionation, i.e. the nerve cells are collected into groups (see Fig. 12.13). The fact that some nemertines have specialized in this way does not in the least argue against other nemerteoids specializing in other ways and becoming able to swim by a different method. Hubrecht (1883, 1887) put forward the view that chordates were possibly derived from nemertines, and that the notochord of the chordates had its origin in the proboscis sheath of the nemertines. This extremely sensible suggestion, though recently supported by Jensen (1960, 1963), has not met with much approval from zoologists for a variety of reasons; however, the reasons do not seem to be as cogent now as they once were. Among the objections raised may be mentioned the presence in the nemertine of lateral nerve trunks and the absence of a tubular nervous system; the absence of definite metameric segmentation, gill-slits, coelomic cavity, cartilage and bone rudiments; and the presence of spiral cleavage of the

fertilized ovum. It will therefore be interesting to reconsider some of the ideas put forward by Hubrecht and amplified by Jensen, and to examine afresh the changes in nemerteoid organization that would be required for the evolution of chordate characters from the raw materials present among the nemerteoids. The possibility that a bottom-living creature, if it acquired filter-feeding habits, would have a great deal to gain from the acquisition of directional swimming movements must be kept in mind. Many nemertines, in addition to the bathypelagic forms already mentioned, do in fact swim,

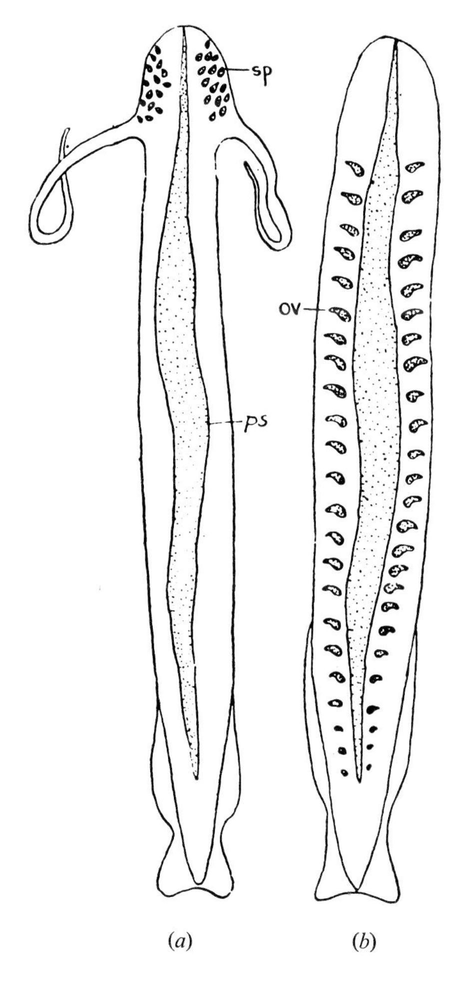

FIG. 13.11. Pelagic nemertine (*Nectonemertes mirabilis*). (*a*) Male with cephalic spermaries, sp; (*b*) female with paired ovaries, ov. ps, Proboscis sheath. (Coe, 1945.)

some by undulatory movements and others by ciliary action, but the latter is possible only with creatures below a certain size. Efficient muscular swimming could be very advantageous.

For the advanced rhabdocoeloids or nemerteoids to increase their capacity for free-swimming in a directional manner, several possibilities suggest themselves. The body could become semi-rigid by becoming strictly segmented, and the segments could become filled with fluid in such a way that only minimal interchange could occur between the segments. Swimming could then be accomplished much as it is in the polychaete worms, where it is brought about by a combination of sinusoidal muscle-produced waves and the beat of parapodia; both of these movements necessitate co-ordinated metachronal rhythmic events in the segments. Thus it is reasonable to suppose that the annelids could have arisen by intensification of the sort of serialization that is present in the advanced rhabdocoeloid or early nemerteoid. Since no trace of the proboscis, nor of certain other special nemerteoid features, is present among the annelids (unless perhaps the typhlosole and the proboscis are related structures!), it is reasonable to suppose that they branched off from the stock under consideration at an earlier stage than the typically nemerteoid stage (see p. 282). They achieved the necessary semi-rigidity of the body by dividing it more or less completely into segments in the manner just indicated. The sort of movement seen in the polychaetes could well have arisen by specialization of the creeping and squirming habits of rhabdocoeloid creatures in contact with the sand into snake-like movements which, when sufficiently perfected, could lead to the animal "taking off" and swimming freely. Some living nemertines have limited powers of progression in very much this way.

A second possible method for producing swimming creatures would be by perfecting the undulatory movements of creatures like the existing pelagic nemertines. This would in fact lead to swimming movements not unlike those of the annelids, except that the oscillations would occur in the vertical plane rather than in the horizontal one. For creatures living on the detritus of the sea floor, it would probably not be such an efficient method, though *Cerebratulus* can actually "take off" and swim freely in this way. It is interesting, however, to observe that rigidity may be given to the body of some pelagic nemertines by a great development of gelatinous mesenchyme. This could act in somewhat the same way as the segmentation of annelids or the notochord of the early chordates. However, it should be noted that the "floating" rather than the swimming pelagic nemertines of to-day have this development, which may therefore be concerned more with buoyancy than with rigidity. The pressure of the rhynchocoel fluid could also be used similarly; this is what Hubrecht and Jensen appear to have had in mind.

The Notochord and the Somites

Another possible method for a creature like a nemerteoid to become free-swimming would be for the movement to start much as it does in the embryos of vertebrates, namely by alternating muscular contractions on either side of a semi-rigid rod running along the length of the body, i.e. a notochord.

The notochord of *Amphioxus*, in which the cells contain parallel fibrous plates (Eakin and Westfall, 1962), and those of tunicates are essentially cellular structures (Fig. 13.12b). Their derivation from a column or cord of mucopolysaccharide-producing cells is fairly easy to envisage as a sort of arrested development or neoteny of the proboscis rudiment of the nemerteoid. Not all notochords are built on this plan, however, but their structure is equally explicable (see p. 296). For example, in elasmobranchs the notochord is composed of a fibrous sheath lined with an epithelium. The epithelial cells have their nuclei at the base, but cytoplasm or cytoplasmic products fill the rest of the cavity. This is exactly the sort of pattern that would be expected to be produced if neoteny occurred at a slightly later embryonic stage than it did in the case of the ancestral *Amphioxus*, i.e. the neoteny occurred when the tubular form of the proboscis was already beginning to appear (Fig. 13.12d). Under some conditions a transverse section of the proboscis of a nemertine can closely resemble in its cellular pattern the transverse section of such a notochord (Fig. 13.12e). These ideas suggest that the semi-rigidity may have been obtained by modification of the cellular structure of the proboscis, rather than by the pressure of the rhynchocoel fluid. Nature may have tried both methods, one surviving in the pelagic nemertines, the other in the chordates.

By controlling the muscles producing the side-to-side bending and straightening of the body, this simple movement can of course develop into a very efficient means of swimming, as it has done, for example, in the eel (Gray, 1933a). It does not even require the development of fins, for it has been shown (Gray, 1933b) to be effective in the whiting in the absence of tail fins. Creatures like *Amphioxus* and the larvae of the tunicates, which are considered to be related to ancestral vertebrates, also have this sort of motion. The main point at issue is, therefore, how the bending motion could have been initiated most successfully. Here we can only guess, but there is a possible clue in the embryology of *Amphioxus*.

In *Amphioxus*, and in the tunicates, the muscles develop from mesodermal pockets (Fig. 13.13) which grow out serially from the wall of the archenteron and then are nipped off. This method of developing muscle is quite unlike that found in rhabdocoeloids and nemerteoids, where the muscle develops directly from mesenchymal tissues separately dehisced from the ectoderm.

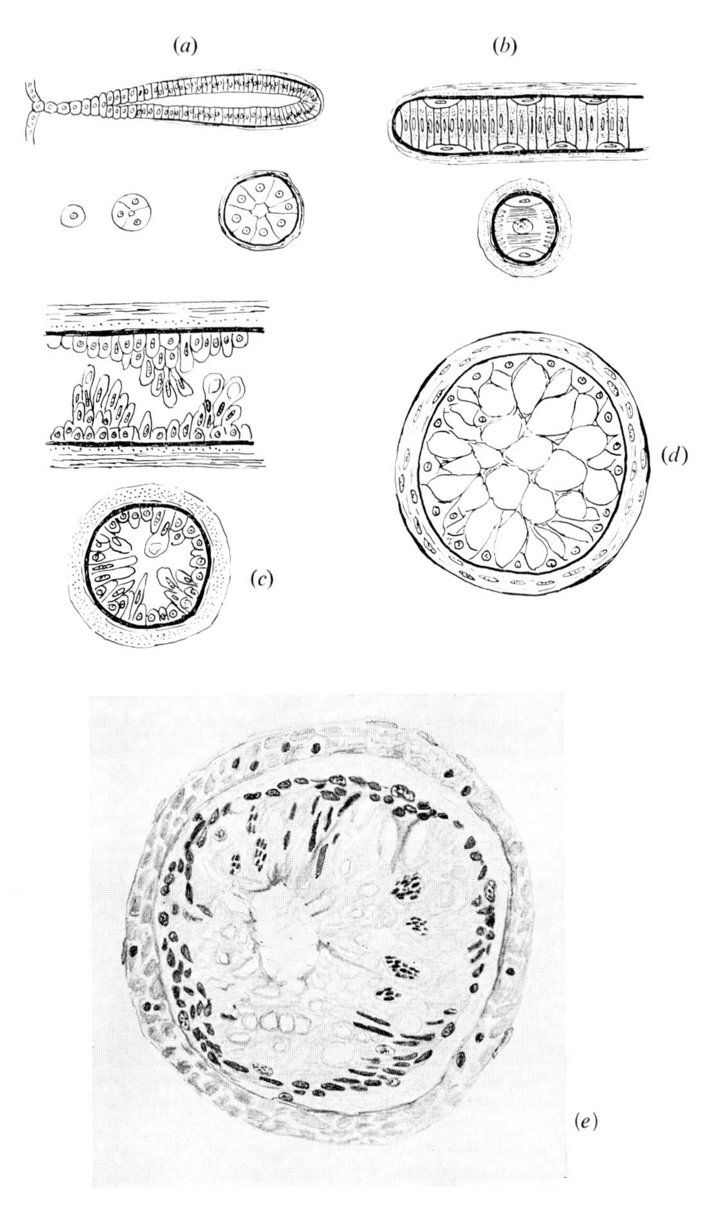

FIG. 13.12. Proboscis rudiment and the development of a notochord. (*a*) Longitudinal and transverse sections of early proboscis rudiment; (*b*) specialization of the early rudiment into a cellular notochord as in *Amphioxus*; (*c*) longitudinal and transverse sections of later proboscis rudiment; (*d*) specialization of the later rudiment into the "tubular" notochord of elasmobranchs, etc.; (*e*) transverse section of a regenerating proboscis of a nemertine; compare with (*d*).

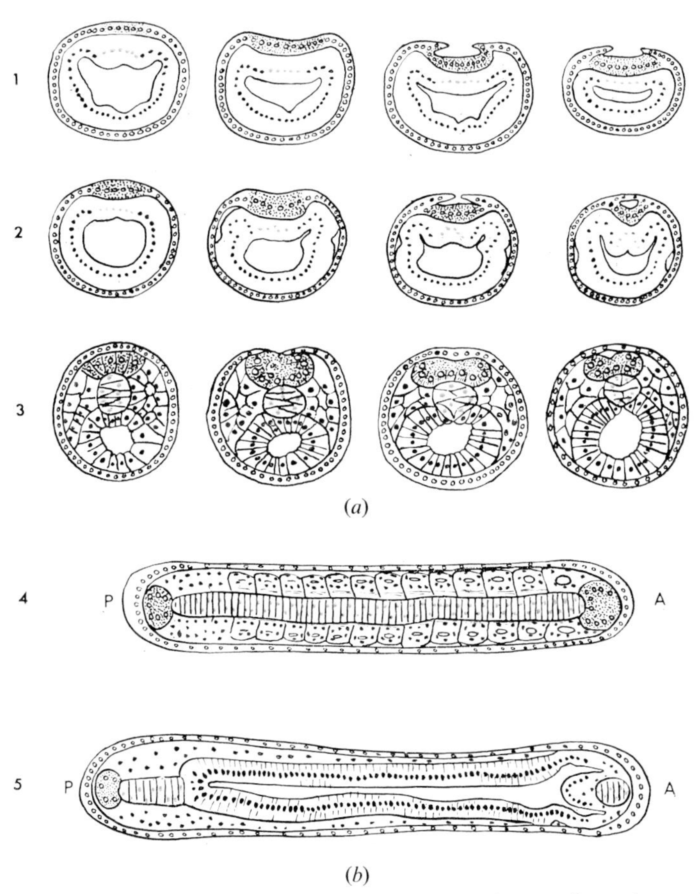

FIG. 13.13. The development of somites in *Amphioxus* from archenteric pouches. (*a*) Transverse sections at intervals down the body; left to right = head to tail. Row 1, 15-hour embryo; row 2, 19-hour embryo; row 3, $24\frac{1}{2}$-hour embryo. (*b*) Longitudinal (coronal) sections of larva at 38 hours. Row 4, at the level of the notochord with muscle fibres developing on mesial side of somites; row 5, at level of gut, with right and left diverticula in "head" region. Nerve cord, dotted; notochord, nuclei green; mesoderm, nuclei red; epidermis, nuclei open; endoderm, nuclei black; A, anterior; P, posterior. (After Conklin, 1933.)

Thus the development of somites is a particularly chordate feature and is a new development. Elsewhere muscle generally appears to develop by dehiscence from the ectoderm or "on the spot" from modification of mesohyl cells. Nevertheless, there is an extraordinary close resemblance between the embryonic forms of *Amphioxus* and the Pogonophora. In the latter the pockets of the archenteron form in the same way, but the gut then fails to develop. What significance can be attached to the sudden appearance of somites in the chordates and to their embryological origin from diverticula of the archenteron? The gut of nemertines perhaps affords another clue, for it is thrown into a series of folds by the development of regular dorsolateral diverticula which in later life interdigitate with the gonads. The reason for these diverticula is obscure. They could be simply an inheritance from the rhabdocoeloid stage where, in the absence of a developed vascular system, the diverticula presumably assist in providing the tissues directly with their nutrients. The intestine of some nemertines is similar to that of some Turbellaria. The original diverticula could have persisted in nemertines to act like the villi of the intestine of vertebrates, primarily to increase the absorptive area of the gut. They could have been there to act as baffle plates to minimize surges of the contents of the gut in relation to the very extensive bodily movements, which involve peristalsis, antiperistalsis, lengthening, and shortening in a system which remains of approximately constant volume (Cowey, 1952). In the last connexion the potentialities of the diverticula may become apparent. If, in an effort to turn the body to right or left, the longitudinal muscles of one side of the body were to contract and there were a semi-rigid rod down the axis of the body, the body would flex to that side and, in so doing, squeeze out the fluid contents of the diverticula. This fluid would then fill the cavities on the other side as they were simultaneously being passively stretched. The response of the latter cells to stretching would probably then be to attempt to restore the *status quo*, and so to contract. In such a system, therefore, unilateral flexion could well set up a state of oscillation which would involve the alternating contraction and swelling of the diverticula on the two sides of the body and which, once initiated, would tend to be self-perpetuating and to propel the animal through the water. Though the movement is initiated by the longitudinal muscle, much of its effective force could subsequently come from the contraction of the cells of the gut diverticula themselves in response to stretch, and these might thus begin to develop into true muscles and eventually acquire striations. It would not be the only example of muscles developing directly from epithelial cells, and the nemertine intestinal cells are certainly very plastic.

It may eventually turn out to be quite irrelevant, and to have some entirely different explanation, but a recent report (Hofmann, 1967) described some extremely early fossil remains lying on mud with ripple marks, whose pattern and position suggest worm-like creatures with bilateral symmetry. They have an axial and possibly rod-like structure in which there are paired diverticula, pouches, or serial myotome-like masses along about two thirds of the sides of the body. These could be the remains of creatures in almost exactly the hypothetical stage under discussion, or they could be merely "red herrings".

Perhaps it is much more relevant here to note how the muscular movements actually develop in embryonic elasmobranchs, in which they have been investigated extensively by Paton (1911), by Wintrebert (1920), and more recently by Harris and Whiting (1954). In these animals the first movements are myogenic and are transmitted directly from myotome to myotome. The nervous system initially plays no part. Because the rhythms of the myotomes on the two sides are independent of each other in this case, flexing movements are maximal when the rhythms of the two sides are most out of phase, and a "shrugging" movement occurs when the two sides are contracting simultaneously. These observations are of interest as indications of the originally myogenic origin of the movements whose control is not taken over by the nervous system until later. This is a recapitulation of the order of events postulated in the present hypothesis for the origin of the somites and, as will be seen later (see p. 325), for the development of the central nervous system. On the other hand, the independence of the two sides does not tally with the idea of one side causing the response from the other. However, in the elasmobranchs the myotomes are separated from the main gut cavity, and there is no movement of fluid from one side of the animal to the other. It was this movement of fluid that was suggested as the initial stimulus for contraction of the opposite site. In elasmobranchs, in the absence of this direct stimulus, the rhythmic mechanism has become built into the developing musculature. It must also be pointed out that this myogenic origin of the first movements is apparently peculiar to the elasmobranchs; lampreys and teleosts do not show it. Nevertheless, it should be remembered also that "a great deal of water must have flowed under the mills" between the time when the nemerteoids began their hypothetical filter-feeding and the time when the elasmobranchs developed their myotomal movements. Acceleration of the development of the neural connexions could well have brought advantages to those animals in which it occurred. It may be that for some not very obvious reason the elasmobranchs have developed this peculiarity, or it could be that they, alone of extant forms, have retained the original and primitive order of events. Although it is a well recognized observation that the cells of the myotomes in the vertebrates initially have all the characters of epithelial cells, rather than of mesohyl cells, it may be objected that the change of cells originally destined to become intestinal epithelial cells into muscle cells would be a very unlikely phenomenon. But would it be so unlikely? The most primitive muscle cells in the Porifera and the Cnidaria and even the muscles of the human iris develop by direct conversion of epithelial cells with almost as many specialized features. The strong mutual attachment of intestinal cells by their terminal bars could well be significant in initiating the molecular orientation necessary for muscle development (See Fig. 9.6).

Then, again, it is a curious fact that somites, with the dubious exception of the anlagen for the eye muscles, develop only posteriorly in most vertebrates, i.e. behind a fairly extensive head region. This could well correspond with intestinal diverticula as they develop in many nemertines. The muscles of the head region and branchial arches of vertebrates could develop by specializations of the muscles already in these regions. The fact that the eye muscles have an innervation somewhat similar to that of the main myotomal muscles may simply mean that the co-ordination of the eye with the other muscular movements was part of the same development that led to the production of myotomes and of a more elaborate sensory-motor system (see p. 325).

The anterior region of the body of vertebrates has been forced from time to time into the straight jacket of a theory of strict metameric segmentation, but it really does not fit very well. Thus the clear prolongation of metamerism to the anterior region of *Amphioxus* may well have some other explanation. That its notochord penetrates to the anterior end is explicable in terms of its having developed from a proboscis that opened anteriorly, while the notochords of vertebrates may have developed from probosces opening into the buccal cavity (see p. 295). The anterior development of its myotomes is, however, less easy to understand, except perhaps for the fact that the first somites of *Amphioxus* send forward hollow prolongations to form the muscles of the anterior region (see Fig. 13.13b). The timing of the formation of the diverticula of the archenteron or intestine may be the important factor here.

It will, of course, be realized that the longitudinal muscles which have already been implicated in starting the process of somite differentiation could themselves have initiated oscillatory bending movements and become modified into myotomal groups themselves. This, indeed, might be considered the most likely interpretation of events, but it has several drawbacks as an hypothesis. First, it leaves the distribution and archenteric origin of the somites (as in *Amphioxus*) unexplained. Second, myotomes in vertebrates are initially epithelial; nemerteoid longitudinal muscle is probably mesenchymal. Third, nemerteoid longitudinal muscle does not appear to be cholinergic (see Fig. 12.3), whereas at least some somitic muscles are. Fourth, the longitudinal muscles of nemerteoids occur throughout the whole length of the body and, if they were the original muscles from which the somites are derived, it must be asked why the latter are present only in the more posterior regions (*Amphioxus* excepted). Fifth, the longitudinal muscles of nemertines are in continuous bundles subdivided by radial septa, so the formation of transverse myocommata would need explanation.

It is fairly clear that swimming movements of the sort described did start in some animal at about the nemerteoid level of organization, and somites appeared, but there is not yet sufficient evidence to determine by what means the somites appeared or why their distribution is as it is. Only further research will allow us to approximate more closely to the truth. It may be that an analysis of some of the factors involved in converting an intestinal diverticulum into muscle can throw light on these problems. In the diagrams in Fig. 13.14, the diverticula (e.g. like those in *Amphioxus*) are represented in transverse and in longitudinal (coronal) section at different horizontal levels as seen from above. The cells on the outer margins, if they were to become muscle fibres, could initially do so in much the same way as epitheliocytes form muscle fibres elsewhere, particularly in the invertebrates.

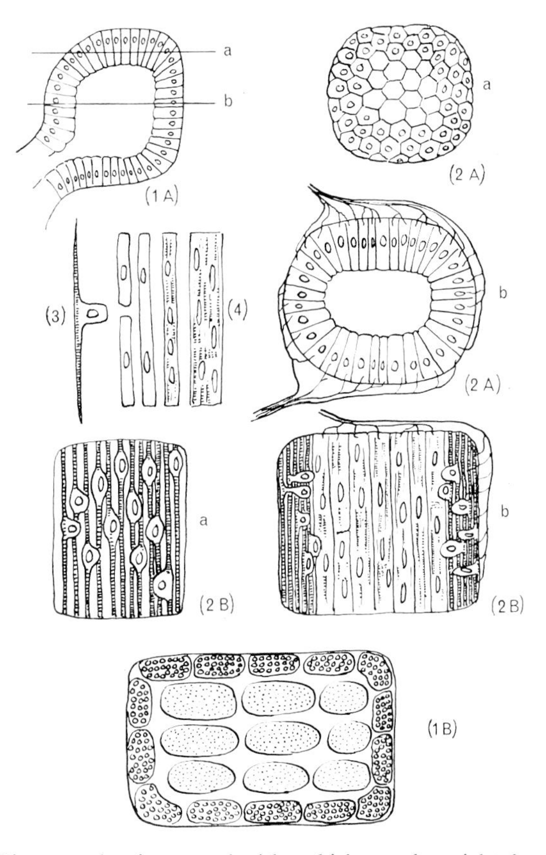

FIG. 13.14. Diagrams showing a method by which muscles might develop and become innervated from a diverticulum of the gut, or from a somitic vesicle as in *Amphioxus*. (1A) Transverse section of diverticulum, showing the position of two coronal sections, a and b; (1B) transverse section after the muscle fibres have developed as shown in the coronal sections; (2A) longitudinal coronal sections at levels a and b; (2B) the same sections after muscles have developed; (3) fibre formation by lateral cells (cf. 2A, b); (4) fibre formation by antero-posterior cells (cf. 2A, b).

The cells could remain attached at their terminal bars and extend basal processes antero-posteriorly in which the myofibrillae would develop [Fig. 13.14(3)]. Subsequently the fibres could fuse in some sort of syncytium or coenocyte. On the other hand, the cells on the anterior and posterior margins could become effective only if their free or luminal ends were to become attached and to fuse either by stretching across the lumen or as a consequence of the withdrawal of fluid from the lumen (Fig. 13.14(4)). The cells most effective in making such fusions would be those belonging to the amoebocytic group, while those with more strongly developed terminal bars, etc., would probably be mechanocytic. Since both classes of cells are probably represented in the diverticula as potential mucous and "chief" cells, respectively, there could well be some selection of appropriate cells for the formation of the peripheral and the central fibres, respectively. Even if this is not so, the peripheral cells would most easily produce their fibres by lateral extension of processes followed by their fusion, while the central fibres would most readily arise by terminal extension and fusion.

If the myotomes of *Amphioxus* and the cyclostomes can be regarded as roughly cubical, as indicated in Figs. 13.13(4) and 13.14, then a consideration of the geometry would indicate that the dorsal, ventral, and lateral margins would be likely to produce fibres by lateral extension, and the cells in the centre would produce them by terminal fusion, or perhaps by a more general

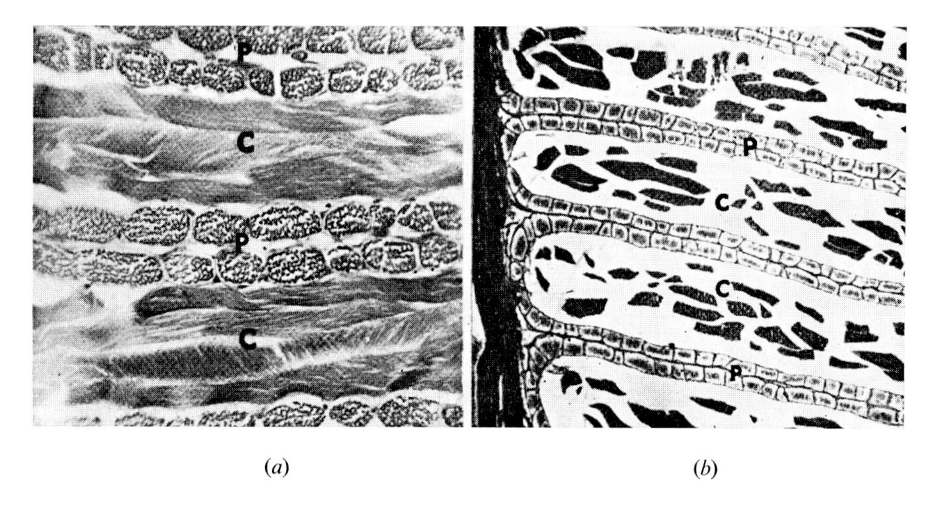

(*a*) (*b*)

FIG. 13.15. Transverse sections of the myotomes of the lamprey. (*a*) Transverse-section of part of myotome (silver impregnation); (*b*) transverse-section of part of myotome (Wilder's "reticulum" method) (central fibres shrunken). P, Parietal fibres; C, central fibres. Compare with Fig. 13.14(1B). (Peters and Mackay, 1961.)

giant cell formation. In transverse section, such a myotome would be expected to have a pattern like that in Fig. 13.14(1b) and, in longitudinal coronal section, one like that in Fig. 13.14(2b). The actual pattern in lampreys is shown in Fig. 13.15. In *Amphioxus*, most of the peculiar, flat laminar muscle fibres would seem to correspond to the central fibres. They have numerous myofibrillae and little glycogen, but other fibres are also present. Perhaps the latter correspond to the peripheral fibres which have fewer myofibrillae and more glycogen (Flood, 1965, 1967), though this difference does not seem to have been observed by Peachey (1961) in his electron-microscopic study of the myotomal muscle of this creature. In lampreys, however, the pattern is very pronounced (Peters and Mackay, 1961), and its vertical repetition could readily be explained if the original diverticulum became subdivided into a series of pockets. Further observations by Peters and Mackay on the innervation of these muscles are also of considerable interest. If it were supposed that nerve fibres supplied the epithelial cells of the original diverticulum as they might those of the intestine, these fibres would presumably have approached the cells from the basal ends near the basement membrane [see Figs. 13.14(2A), 13.14(2B)]. This would mean that the cells along the dorsal, ventral, and lateral margins could continue to be so supplied when they differentiated into muscle fibres, i.e. multiple innervation would be likely to remain. The cells on the anterior and posterior margins, on the other hand, could be innervated only at their ends unless the nerves subsequently grew inwards along the fibres. The expected pattern of innervation would thus be as shown in Fig. 13.14(2B); the actual pattern as depicted by Peters and Mackay in lampreys is shown in Fig. 13.16, and that by Bone (1966) in elasmobranchs in Fig. 13.17. Not only are there different locations for the nerve endings, but also in both cases the lateral nerve endings are "en grappe" while the terminal ones are "en plaque". This is the sort of difference one might expect if mechanocytes form the lateral fibres and amoebocytes the central fibres, since the surface properties of the two types of cell are very different. It is interesting that cholinesterase is involved in both types of innervation. It is not yet clear which cholinesters are important, though there is little butyrylcholinesterase on the lateral fibres. It will be remembered that the longitudinal muscle in *Lineus* responds actively to adrenalin but not to acetylcholine.

In *Amphioxus* the "innervation" of the myotomal muscles is different from that in lampreys and elasmobranchs, but equally informative (Fig. 13.18). Nerve fibres do not approach these muscles, but extensions of the actual muscle fibres approach the spinal cord in two groups, according to their origin, i.e. there is a dorsal bundle of thin fibres from the lateral, glycogen-rich, muscle fibres and a ventral bundle of thick fibres from the main muscle lamellae (Flood, 1966, 1967). When they reach the margin of

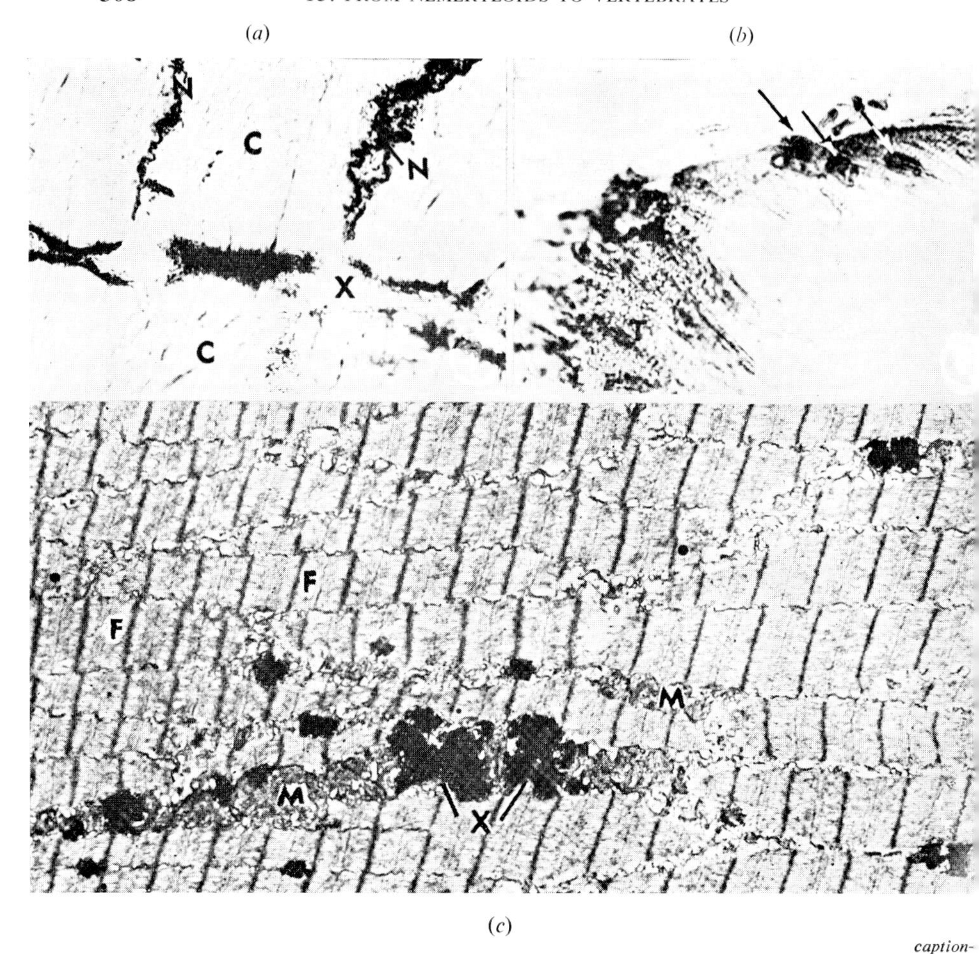

caption-

the spinal cord, they make contact with boutons of two different sorts. The thin fibres meet large boutons with small synaptic vesicles, and the thick fibres meet small boutons with large synaptic vesicles (Fig. 13.18). It was suggested earlier (see p. 299) that in *Amphioxus* the notochord started to develop as such from the proboscis-rudiment before the latter had become tubular. The contractile movements could thus perhaps have started rather early in these animals, and the consequent close proximity of the myotomes to the developing nerve cord (see p. 329) meant that direct contact was easily established between the muscles and their activating cells. Subsequently, as the animals grew, it was the muscle process that became stretched. In the lamprey, where the notochord is of the hollow type, movement may have

(*d*)

FIG. 13.16. Structure and innervation of myotomal muscles of the lamprey. (*a*) Parasagittal section; acetylthiocholine method. C, Central muscle plate; X, myosepta; N, nerve fibres between layers of parietal fibres. (*b*) Central muscle plate; acetylthiocholine method: myotendinous reaction, but end-plates shown as small areas of intense staining (arrows). (*c*) Central muscle plate. F, Myofibrils; M, mitochondra; X, osmiophilic bodies. (*d*) Parietal muscle plate. F, Myofibrils; M, mitochondria; N, nerve; Nu, nucleus; P_1, P_2, two separate parietal fibres; X, osmiophilic bodies. Arrows indicate neuromuscular junctions. (From Peters and Mackay, 1961.

started later, when the intestinal pouches were already innervated or in process of becoming so, so that the pattern of innervation followed the lines explained above.

In more advanced fishes and in higher animals generally, of course, there has been ample opportunity for the relatively simple patterns seen in the cyclostomes and elasmobranchs to become considerably modified and complicated by morphogenetic movements and changes in developmental patterns. Nevertheless, the embryological development of the myotomal muscle in the dipnoan, *Lepidosiren*, as studied many years ago by Kerr (1919), is very illuminating. His illustrations are shown in Figs. 13.19 and 13.20. In this creature the cells of the outer wall of the myotome produce fibres by quite a different process from that adopted by the cells of the inner wall. It may be asked why there are no fibres of the outer (lateral) type on the inner side of the myotome also. No definite answer can be given to this question; it may be noted, however, that in many species the inner wall of the myotome does not contribute to muscle formation, but its cells disperse to become the connective tissues of the axial skeleton. This could mean that the only parts of the myotome that are functional in muscle formation in these creatures are the outer wall and the anterior and posterior walls.

As pointed out above, the cells most likely to fuse, and so to form the main muscle plates (in *Amphioxus*, the cyclostomes, and elasmobranchs), would be

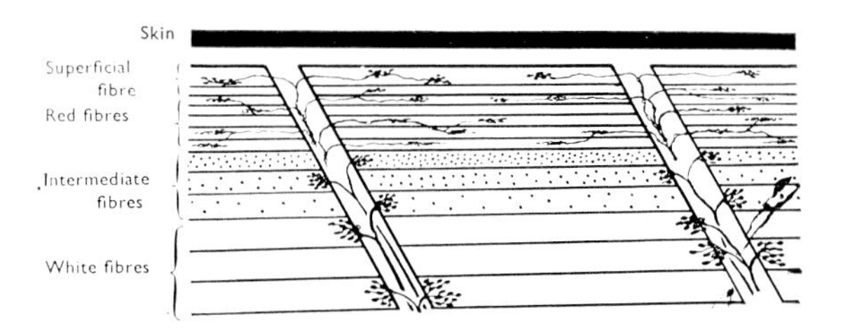

FIG. 13.17. Innervation of different muscle types in the myotomes of an elasmobranch. (Bone, 1966.)

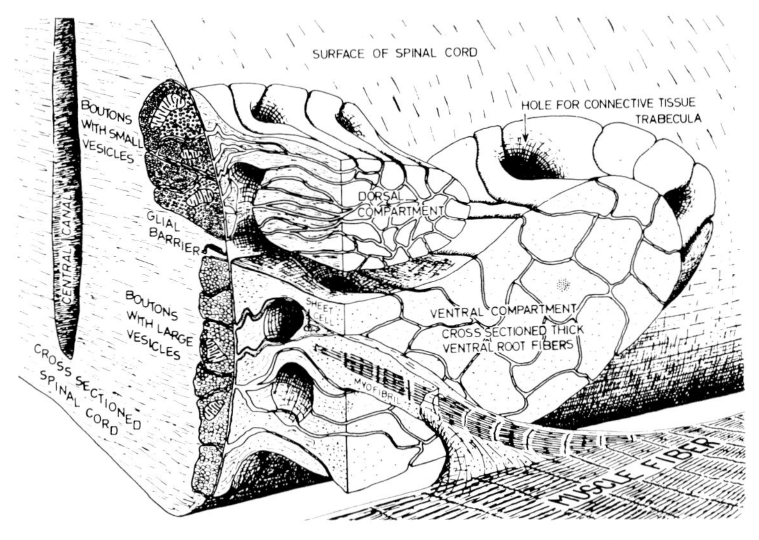

FIG. 13.18. Innervation of myotomes of *Amphioxus*. Structure of ventral root of spinal cord. (Flood, 1966.)

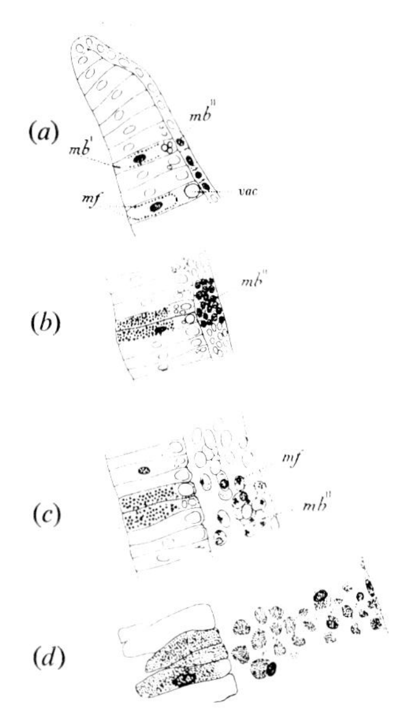

FIG. 13.19. Differentiation of the myotome of *Lepidosiren* as seen in transverse section. (*a*)–(*d*), progressively older stages. mb′, myoblasts of inner wall; mb″, myoblasts of outer wall; mf, myofibrils; vac, vacuoles. (Kerr, 1919.) Compare these vacuoles with those shown in Fig. 12.7.

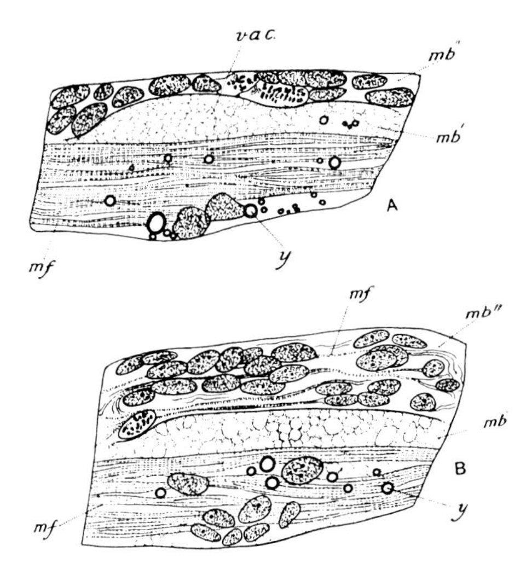

FIG. 13.20. Differentiation of the myotome of *Lepidosiren* as seen in horizontal section. A is at an earlier stage than B. mb′, myoblasts of inner layer; mb″, myoblasts of outer layer; mf, myofibrils; vac, vacuoles; y, yolk. (Kerr, 1919.)

cells with amoebocytic properties, while mechanocytic cells would be more likely to behave like the cells of the lateral wall. Since, in elasmobranchs, Bone (1966) has clearly shown that the lateral fibres are red (myoglobin-containing), "slow", glycogen- and fat-containing fibres that do not propagate action potentials and are furnished with "en grappe" nerve-endings, while the central fibres are "fast" (twitch), white fibres poor in glycogen and fat which propagate action potentials and possess "en plaque" endings, it seems probable that all these properties are initially related to the sort of surface and cytoplasmic differences that exist between mechanocytic and amoebocytic cells. A similar difference in surface properties is also suggested by the difference in membrane organization under "en grappe" and "en plaque" endings noted by Hess (1965) in a snake, and by the observations of Flood (1966) which showed that in *Amphioxus* the two types of muscle fibres are activated by synapses with different sizes of synaptic vesicles. Similarly, the differences in methods of cell fusion in the more amoebocytic and more mechanocytic myoblasts may account in part for the different organization of the T system of tubules in fast and slow fibres (Hess, 1965; Page, 1965) (Fig. 13.21). The selection of cells and the manner of their fusion to form multinucleate muscle fibres are of cardinal importance in relation to the problem of "fast" and "slow" fibres, and they would seem to deserve further investigation. The properties of these fibres are such as could well be related to the extent to which the original myoblasts displayed ecballic or emballic qualities.

There is in some vertebrates a clear distinction between the muscles that develop from the somites and those that develop from the lateral plate (see Konigsberg, 1965), and perhaps this distinction should be accentuated. It may be that the contribution of the lateral plate roughly parallels the musculature already present in nemerteoids and would thus be homologous with the muscular systems of annelids, arthropods, etc., while the strictly myotomal muscle would be a new development in the immediately pre-vertebrate organisms resulting from the changes in the diverticula of the archenteron or gut already outlined. In the chick, however, the trunk muscles from the lateral plate and the muscles developing from the somites do not appear to be distinguishable (Straus and Rawles, 1953).

It is also interesting that, in *Amphioxus*, besides the muscle which develops from the myotomes, non-myotomal muscles develop from the somatopleure and splanchnopleure, and that these muscles, which are not divided into segments, are innervated from dorsal nerve roots, not from the ventral roots which supply the myotomes (Young, 1950). It is tempting to think that these muscles could perhaps correspond to those already existing in nemerteoids, because the ventral roots are essentially structures formed in connexion with the myotomes (Flood, 1966) and thus are possibly new.

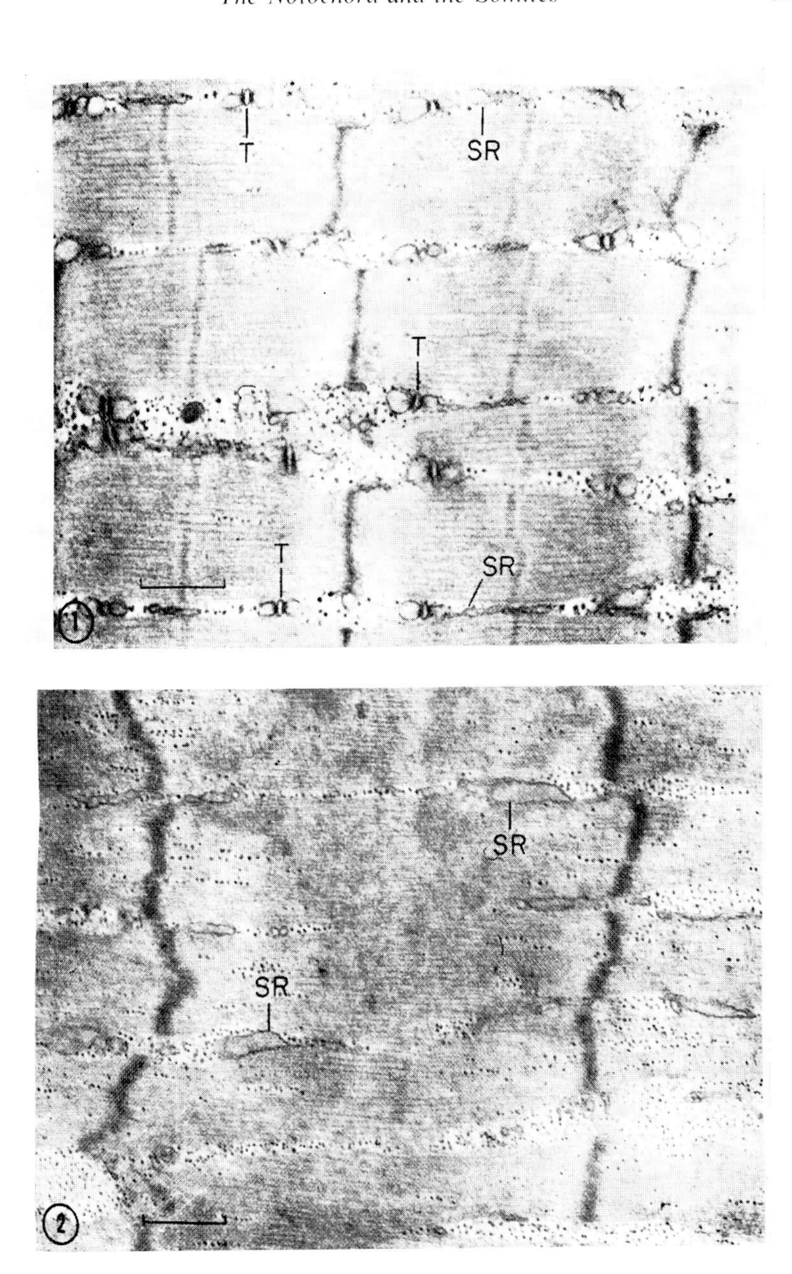

FIG. 13.21. Longitudinal sections of slow and fast fibres in the garter snake. Note the relative absence of the T system of the endoplasmic tubules in the slow fibre. (1) Fast fibre; (2) slow fibre. SR, Sarcoplasmic reticulum. (Hess, 1965.)

Another difficulty of a somewhat different kind that argues against the derivation of striated muscle from the gut diverticula in the manner outlined above is that, in insects, the form, method of development, and microscopic structure of multinucleate muscle fibres can sometimes be almost identical with those seen in the vertebrates, though there is little to suggest their derivation from diverticula of the gut. It is true that the arrangement of actin filaments is somewhat different, but the multinucleate coenocytic arrangement can be very similar. Two obvious and possible explanations for this similarity suggest themselves. First, the actomyosin system is inherent in both the mechanocytic and amoebocytic types of cell; when these cells are called on to act similarly and efficiently as muscles, the pattern of fibrillae which each type develops becomes identical or nearly so. Thus, if both sets of muscles arose from the mechanocytic and amoebocytic cells of the mesenchyme, there would be no difficulty in accounting for their similarity in animals as far apart as bees and bats. The necessity for efficient contraction and relaxation could lead, in the end, to similar structural specializations. The supposition that there is a likeness between insect muscles and vertebrate muscles, however, asks more of coincidence if the muscles of vertebrates are not derived from mesenchymal cells but are derived, as suggested, from the epithelial cells of the somites which arose first as endodermal pouches. Second, it is, in fact, possible that those muscles in the insects which most resemble the muscles in vertebrates may also derive from diverticula of the archenteron. Unfortunately, the embryology of insects is not clear on this point, but it may be noted that some muscles in insects are derived from mesoblastic pouches, sometimes called somites, that may well have an origin similar to the pouches and somites in the vertebrates. In this case the similarity might result from the development of the muscles from essentially the same group of cells in both phyla.

It is true that muscles may be developed to serve different ends. Probably for this reason, even when muscles are derived directly from mesenchyme cells, they do not always specialize in exactly the same way. In molluscs, annelids, and vertebrates the contractile fibres of the muscles are organized differently and for somewhat different functions, and their fine structures, though clearly based on the same elements, are correspondingly differently organized (Hanson and Lowy, 1960). This emphasizes the need for caution in trying to interpret derivation either from structure or from mode of action. It also accentuates the curiosity that some insect muscles, in spite of certain differences in fibrillar pattern, should be so similar in structure to those of the vertebrates.

It is clear from this discussion that the solution to the whole problem of the origin of myotomes and indeed of all striated muscles with their special motor end-plates is still very obscure. When it is approached from the vertebrate side, the questions to be asked are: Where do the myotomes come from, and is their origin directly from the archenteron, as in *Amphioxus*, significant? What is the relationship of the myocoel to the splanchnocoel and hence of the muscles derived from each? What is the origin of the different classes of motor end-plates? What is the origin of "ventral roots" and hence of the somatic motor fibres?

When the matter is approached from the nemerteoid side, the questions are: Is the muscle in the nemerteoid that develops from mesenchyme subsequently organized embryologically in a new manner, namely into myotomes? Was the muscle of the vertebrate produced by the development of new contractile tissue by conversion of the epithelial cells of the gut diverticula, while the existing muscle was modified to provide the smooth muscle and other subsidiary muscle groups in the vertebrates?

The Heterogeneity of Muscle

It cannot be too strongly emphasized that a great disservice has been done to muscle physiology by the concept of muscle as a particular cell or tissue type. Contractility, dependent on some development of the interaction between an actin-type protein and a myosin-type protein, is a very fundamental and early development of cellular activity and is probably potentially, if not actually, present in most classes of cells. For greater efficiency or greater speed of response, for sustained contraction, and so on, there may be several ways in which the actomyosin system can be organized. The same improvements having been made in two or more groups of cells does not mean that these cell types are otherwise identical. For example, in the vertebrates, there are striated fibres that are known to develop from epithelial cells, e.g. iris muscles. Heart muscle in tissue culture gives rise to mechanocyte-like cells, whereas many skeletal muscles give rise to cells with amoebocyte properties (see p. 320). Some muscles develop directly from ectomesenchyme, others from somites. Within the same muscle, some fibres show, under the electron-microscope, a pronounced "fibrillen" structure, and others show a "felden" structure (Fig. 13.22). Actin and myosin fibrillae of similar organization occur in both types, but the grouping of the fibrillae and the character of the endoplasmic reticular system and other properties are different (see Fig. 13.21). Some muscle fibres have enormous numbers of mitochondria, strong succinic and lactic dehydrogenase activity, and high cytochrome oxidase activity; other fibres in the same muscle have only feeble activities in these directions (Fig. 13.23). Similarly, in any given muscle there may be fibres that store glycogen, and others that do not. The same applies to the storage of fat and of myoglobin and the development

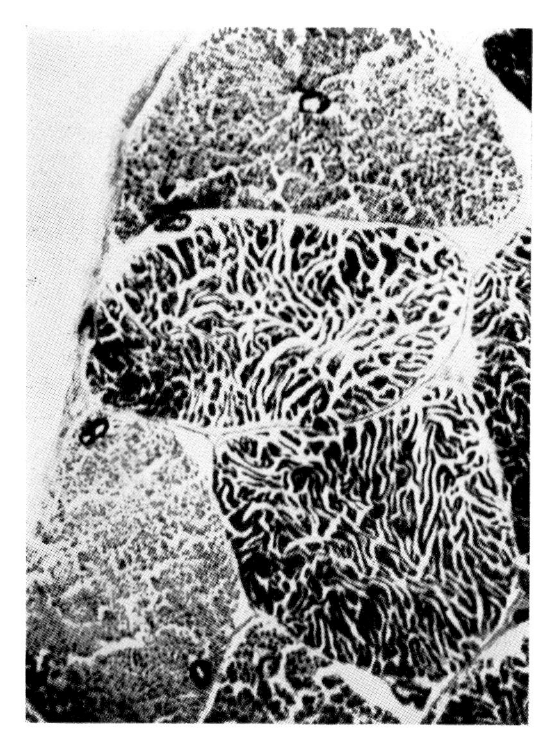

FIG. 13.22. Transverse section of muscle fibre showing "Fibrillenstruktur" and "Feldenstruktur". (Krüger, 1952.)

of several other enzymes (George and Berger, 1966) (Fig. 13.24). These changes could be functional and the histological appearance could be dependent on previous activity, but the evidence does not point in that direction. It points to the existence of at least two and possibly three (Flood, 1967; Stein and Padykula, 1962) or more classes of fibres with different properties, e.g. fast and slow fibres, fibres that can be activated by a single motor end-plate and fibres that require multiple innervation (Kuffler and Vaughan Williams, 1953), fibres with centrally placed nuclei and fibres with nuclei at the periphery, and fibres with intermediate characters.

The muscle fibres of the diaphragm of mammals are illuminating. The cow has a slow-acting diaphragm in which the fibres are large, clear, devoid of lipid granules, and possess few mitochondria. The shrew has a rapidly acting diaphragm composed of red fibres, rich in lipid droplets, and possessing many mitochondria and their associated enzymes. Mammals of intermediate size have diaphragms composed, not of intermediate fibres, but of mixtures of the two distinct types in varying proportions (Gauthier and Padykula,

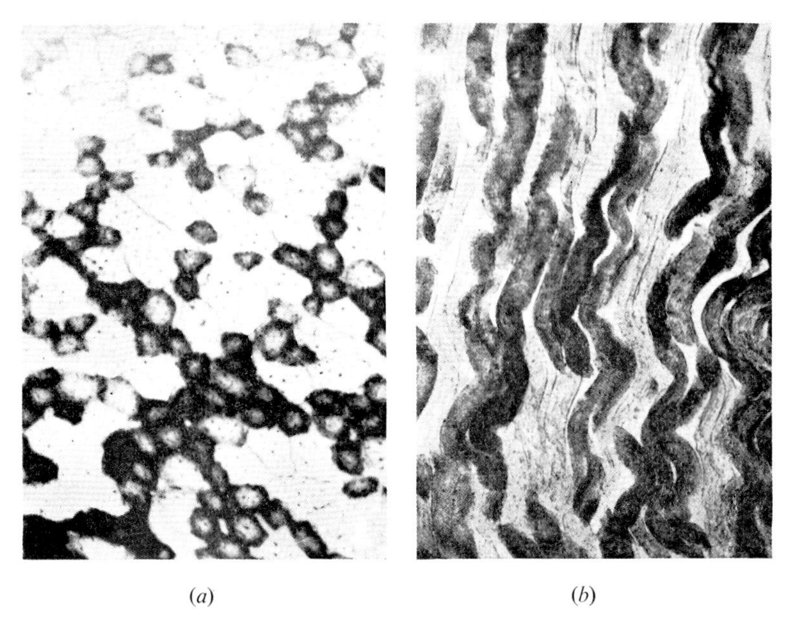

(*a*) (*b*)

FIG. 13.23. Muscle fibres stained to show the presence of succinic dehydrogenase. Note that some fibres are heavily stained while others are almost negative. (a) Transverse section; (b) longitudinal section. (Wachstein and Meisel, 1955.)

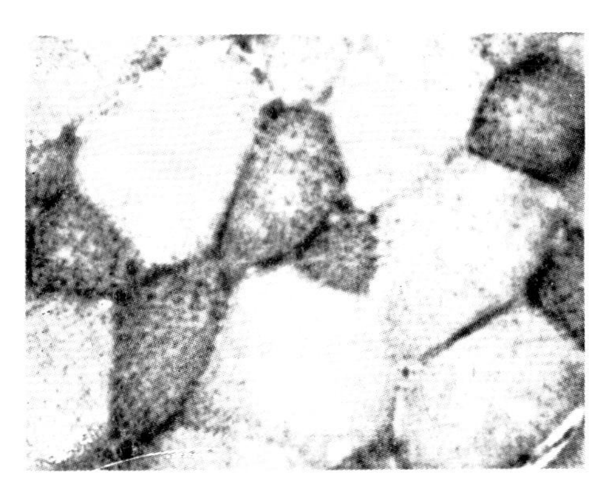

FIG. 13.24. Localization of lipoid (sudan-staining) material in certain muscle fibres of the rat. (Nachmias and Padykula, 1958.)

1966). But, to add to the complication, Smith and Lännergren (1968) suggest that in the iliofibularis muscle of the toad, *Xenopus laevis*, there are both fast and slow fibres belonging to each of the two histological groups and that the enzyme content is concerned more with fatigability than with speed (see also Lännergren and Smith, 1966).

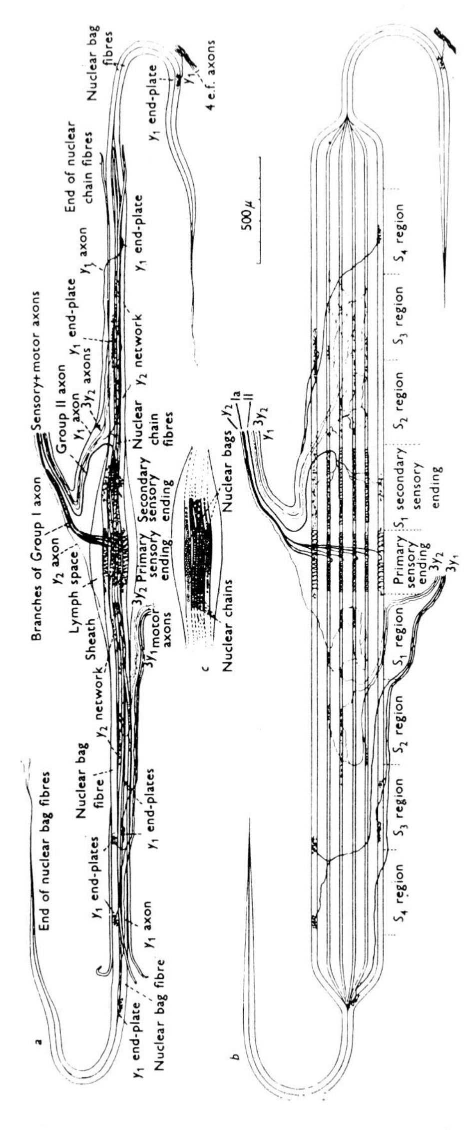

FIG. 13.25. The structure of a mammalian muscle spindle. (*a*) As seen in gold chloride preparations; (*b*) diagrammatic representation of fibres and nerve endings; (*c*) nuclear region as seen in de-afferented preparations. (Boyd, 1961.)

These differences are clearly shown in the muscle fibres within the muscle spindles of cats, where there are fibres with "nuclear bags" and single motor innervation with well-developed "en plaque" end-plates on the one hand, and fibres with nuclear chains and multiple "en grappe" endings on the other (Fig. 13.25) (Boyd, 1961). At least some of these differences could have been determined in the first instance by the character of the original cells in which the contractile properties developed. For example, it is interesting to note that the heart-muscle fibres of vertebrates are joined into a syncytium* from which the constituent cells can be easily separated

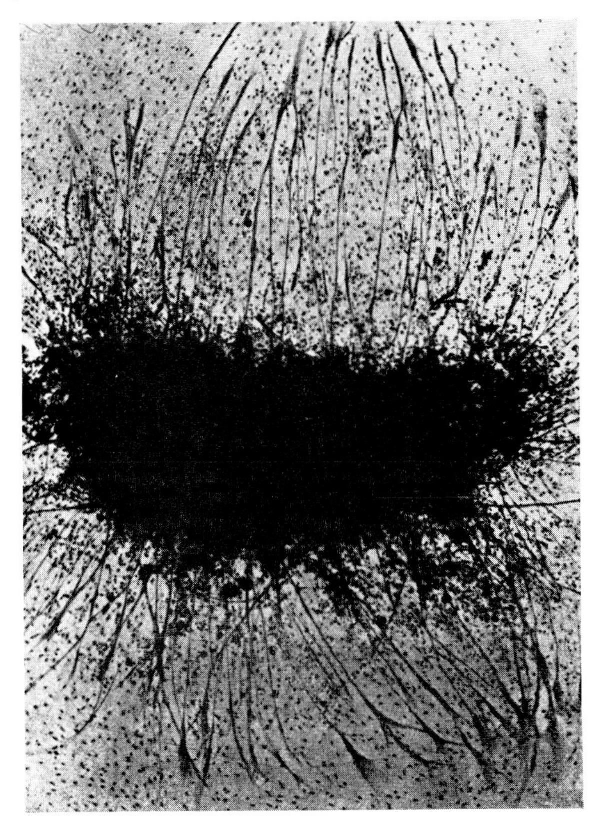

FIG. 13.26. Culture of skeletal muscle showing multinucleate "muscle buds" developing from some fibres and a copious outgrowth of mechanocytes occurring at the same time. (Lewis and Lewis, 1917.)

*A "syncytium" is used here to define a combination of cells into one physiological unit, e.g. in this case, the heart. There is not necessarily any fusion of cytoplasm between the cells of a syncytium. A "syncytium" differs from a "coenocyte" in that the latter is the result of the fusion of cells; striated muscle fibres in vertebrates are coenocytes.

at the inophragmata or intercalated discs, structures that are probably to be considered specialized desmosomes. The resulting cells are essentially mechanocytes which in culture remain as separate cells (see Fig. 2.4), though they may readily link together by means of a kind of desmosomal unit.

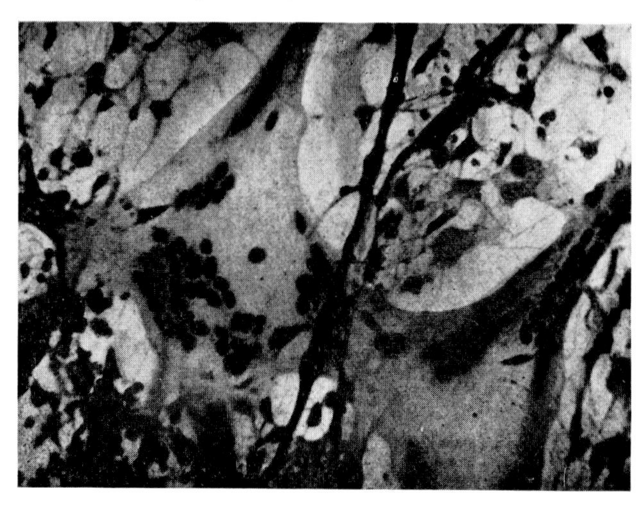

FIG. 13.27. Multinucleate giant-cell mass developing from a muscle fibre. (Lewis and Lewis, 1917.)

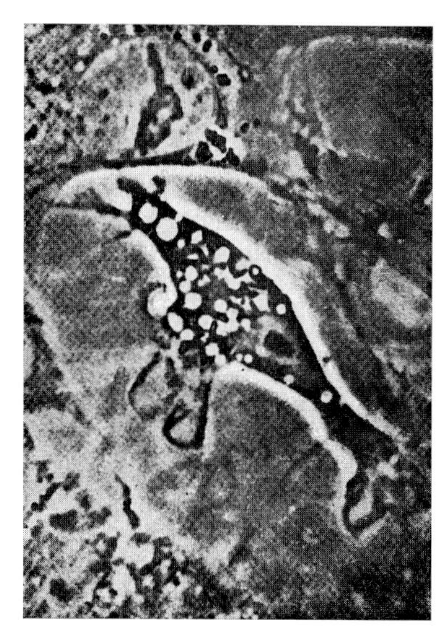

FIG. 13.28. Macrophage developed from a culture of skeletal muscle. (Chèvremont, 1953.)

On the other hand, many skeletal muscles in culture produce large multinucleate masses (Figs. 13.26, 13.27) and may divide into individual cells that display the properties of amoebocytes (Fig. 13.28). We have seen (p. 124) that the surface properties of amoebocytes and mechanocytes are very different; in consequence, muscle fibres developed from them should be expected to have very different properties of excitation, conduction, storage of metabolites, glycolysis, respiration, growth, and so on. We have seen, too, that the methods by which the cells unite to form the fibres may also be very different. These various properties are seldom considered in the classification and physiological investigation of muscle; all too often some unspecified muscle is treated as a uniform and single physiological unit. At the extreme, two muscle fibres might have little more in common than that they both contract. Probably, however, muscle fibres could be grouped more effectively than they now are if the nature and properties of the cells from which they are derived were more clearly understood and if more attention were paid to the character of the nerve cells which innervate them. Some muscle fibres may have come from cells with strong mechanocytic tendencies; others from those not heavily biased as either mechanocyte or amoebocyte; others from strongly amoebocytic cells; still others may derive from cells which still maintain epitheliocytic characters. For example, in tissue cultures of cell suspensions from chick thigh muscles, the "myoblasts" are distinctly bipolar cells clearly distinguishable (Fig. 13.29) from fibroblasts (Fig. 2.3). When they grow into colonies and differentiate into muscle fibres, their behaviour is quite different from that of fibroblasts under similar conditions (Fig. 13.30) (Konigsberg, 1963). On the other hand, cardiac muscle readily gives rise to colonies having the morphology of fibroblasts (see Fig. 2.6c and Fig. 13.31). These properties must almost certainly be reflected in the properties of the functional fibres, particularly those of excitation, conduction, respiration, and metabolism. Just as it is possible to change the strength of the amoebocytic or mechanocytic properties of cells by external means, so it may be that external influences can modify the properties of a muscle fibre, as, for example, in the conversion of slow fibres into fast fibres by changing the innervation, as recorded by Eccles *et al.* (1962). In this connexion, however, a word of warning is not out of place. Most muscles, as seen from the accompanying illustrations, are, in fact, mixtures in different proportions of fast and slow fibres or at least of fibres with different histological properties (see George and Berger, 1966), and the mechanical response of the muscle as a whole is presumably a resultant. There is, so far, little direct evidence that a single slow fibre has been converted into a fast fibre by changing its nerve supply, though the presence or absence of its nerve supply, as might be expected, greatly affects its function and sensitivity, and a cell that is constantly stimulated by small doses of

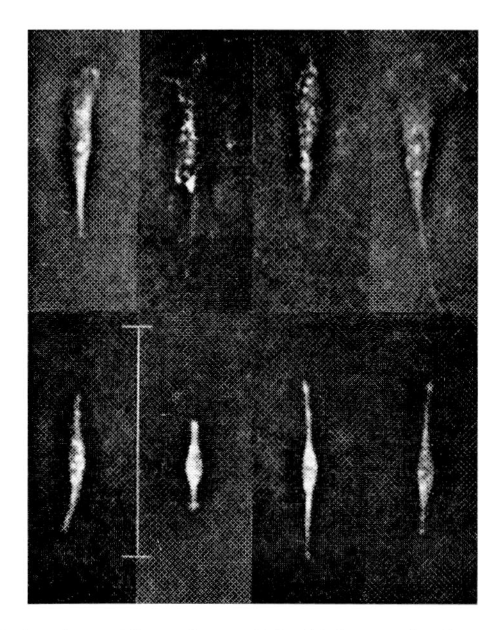

FIG. 13.29. Isolated myoblasts from chick thigh muscle. (Konigsberg, 1963.)

transmitter from its motor end-plates is not likely to remain unchanged after such treatment. Much more understanding might be gained in this connexion by more correlation between the nature of exciting and inhibiting agents and the other properties of individual muscle fibres, e.g. behavioural, structural, trophic, electrical, and biochemical.

The curious case of the insect muscles should perhaps be reconsidered in the light of the discussion above. Meanwhile, one interesting suggestion has been made with reference to cross-striations. Cross-striated muscles occur in widely different situations in the animal kingdom, and clearly they do not constitute a homogeneous group. Nevertheless, the striations tend to be better developed when muscles are attached to rigid supports, i.e. an endo- or exoskeleton, and it is possible that such attachment assists in the regular orientation of the molecules, though it clearly cannot be entirely responsible.

It is easy to understand that, if striated muscles develop more or less directly from epithelial cells (as in myotomes), on the one hand, or from mechanocytes or amoebocytes (as in most of the muscles of the head), on the other, the union between the myoblasts in each case might be differently organized, and the relationship between the cell surface and the contractile elements correspondingly different. If we add to this concept the many degrees of difference between the extremes of mechanocytic and amoebocytic

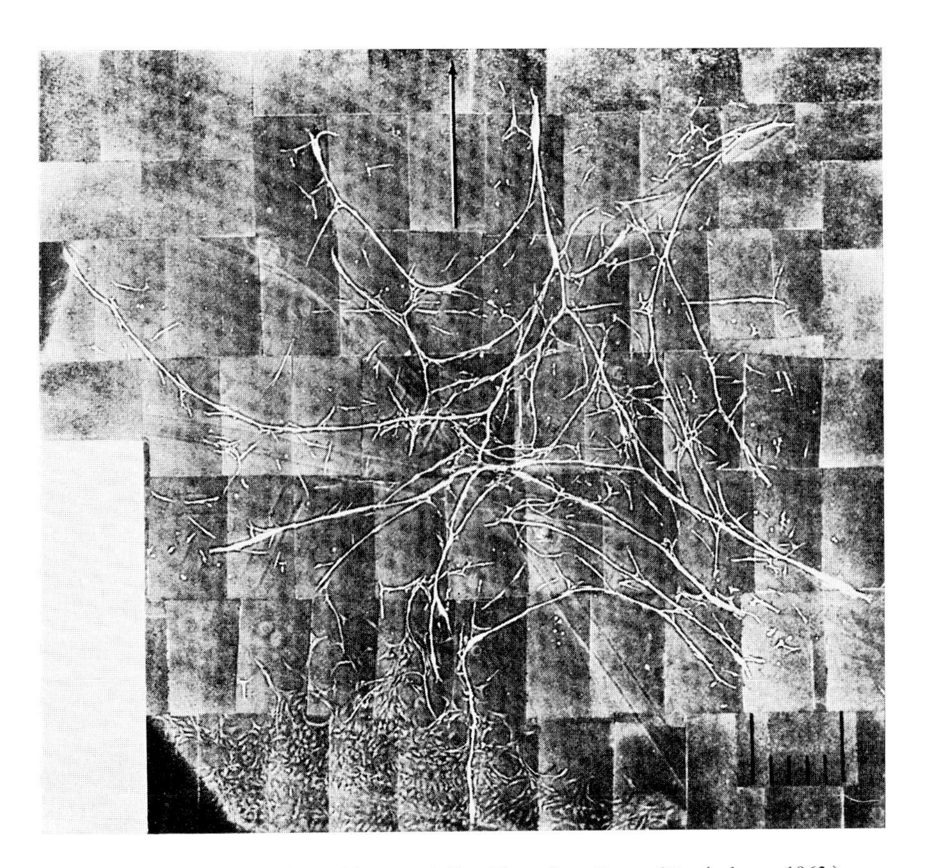

FIG. 13.30. Clones of myoblasts and fibroblasts in culture. (Konigsberg, 1963.)

behaviour, it becomes clear that the term "muscle" is likely to embrace a wide range of tissues whose histological structure may be but a poor guide to classification, and the myograph an even poorer one. When plain muscle develops in the mesenchyme, it probably comes mostly from mechanocytes. Certainly, the muscle coats of blood vessels and the muscles of the alimentary canal both grow as mechanocytes in tissue culture. This does not, however, rule out the possibility that in other situations plain muscle has had a different origin, nor does it mean that all plain muscles must have the same properties, for there is great variety even amongst mechanocytes.

In annelids and molluscs, cells have specialized the actomyosin system in ways quite unlike those in vertebrates, e.g. in the "spiral" muscles found commonly in the molluscs (Hanson and Lowy, 1960). The peripherally striated fibre of the nematodes, with the large separate nuclear compartment, as in *Ascaris*, is probably a modification similar to that which has occurred

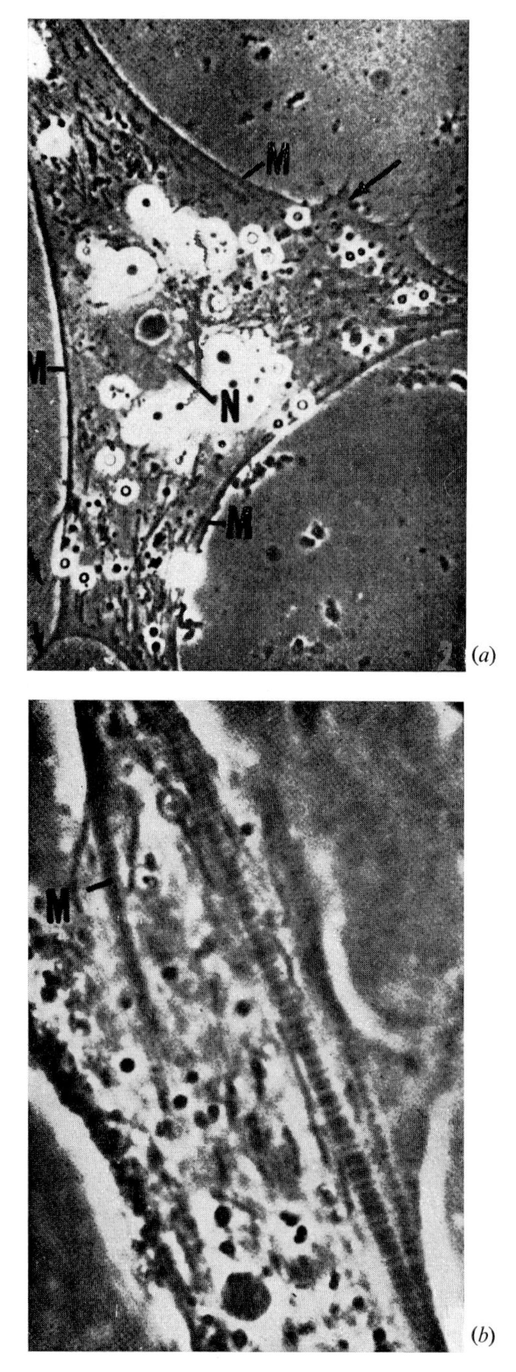

FIG. 13.31. Single cardiac myoblasts with developing myofibrils (M). N = nucleus. (Rumery *et al.*, 1961.)

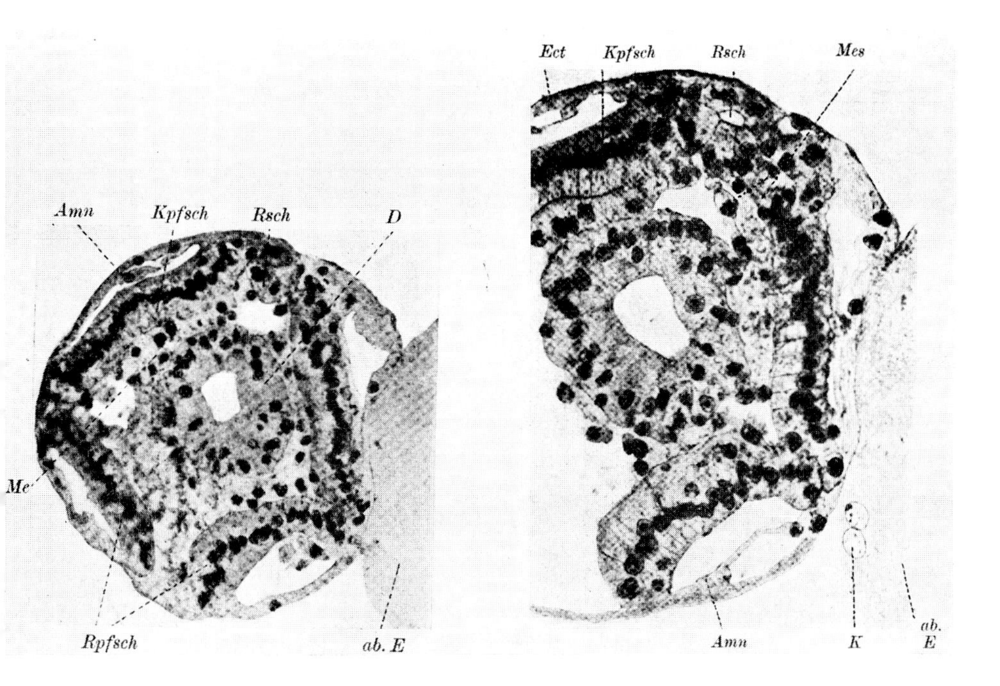

FIG. 13.34. Sections through developing embryos of *Lineus*. ab.E, Dead egg; Amn, amnion; D, gut; Ect, ectoderm; K, nuclei; Kpfsch, head placode with amnion; Me, Mes, mesenchyme; Rpfsch, trunk placode; Rsch, proboscis rudiment. (Schmidt, 1934.)

frequently contains a longitudinally running dorsal nerve, which in the floating pelagic nemertines is actually developed into a ganglionated cord (Coe, 1927), perhaps indicating a motor function related to the swimming movements (see Figs. 12.12, 12.13). It will be remembered that in the neural tube of vertebrates the alar plate is mainly sensory and the basal plate, motor.

In the nerve cord of *Amphioxus* the large *boutons terminaux* are located against the sheath of the cord, and the muscle fibres of the myotomes send processes to make contact with them (Flood, 1966) (see p. 308). This is exactly the sort of system that one might expect to develop if the existing dorsal epidermis, with its contained neural processes, were to be folded into a neural tube so that the basement membrane became the sheath of the tube. The close proximity of the myotomes to the neural canal in the embryo of *Amphioxus* would favour such an union.

An interesting feature of nemerteoid cytology which may be relevant to the origin of neural tissue is that, in the most anterior region of the worm, the epidermis has the peculiar quality that the cell nuclei all lie below what may be described as an external limiting membrane (see Fig. 12.2). This applies also to the cells of the cephalic pit. Posteriorly, the epithelial nuclei begin to appear external to this membrane, which ultimately thickens to become a typical basement membrane, and the epidermis assumes the more usual characteristics. Thus, this anterior epidermis of the nemertine has a distinct similarity to the neuro-epithelium of higher forms.

If these ideas concerning the possible origin of a tubular nervous system are in the main correct, the system so developed would successfully integrate much peripherally situated neural tissue. It would not necessarily interfere, however, with the main nervous system that already exists in the nemerteoids and which looks after the feeding and reproductive activities of the worms. The cephalic ganglia would presumably persist and become integrated into the new system, probably occupying a position in the ventral region of the neural tube in the head region. In other words, they would lie in the correct position to form the "nuclei" in the base of the brain and the hypothalamic region, where they would retain control of all those activities which were already going on in the worm. These activities would, in general, correspond to those carried out by the autonomic system of the vertebrates.

It is interesting to note that many of these cells in the ganglia and associated structures of nemertines (Fig. 13.35) and their relatives are "neurosecretory" (Scharrer, 1941; Gabe, 1954; Lechenault, 1962, 1963), and that the hypothalamus is the part of the vertebrate nervous system where neurosecretion is most conspicuous (see also p. 386).

It should also be noted that the cells from which the new nervous system would develop, according to the preceding hypothesis, would otherwise have formed the epidermis and, in addition to any nerve cells already present, would have consisted of ciliated cells, goblet cells, etc. The enclosed environment of the neural tube may have modified their behaviour and favoured some forms of activity over others. However, the retention of cilia or flagella by the essentially epithelial ependyma (without basement membrane) and even by some nerve cells in the main nervous system of vertebrates is interesting. This, of course, is not meant to imply that only the ciliated cells became nerve cells. The epithelium as a whole would be converted into neural tissue, so that, if the problem is considered in its simplest form, it would appear possible that nerve cells could be developed from potentially ciliated (mechanocytic, ecballic) cells or from potentially goblet (amoebocytic, emballic) cells. The two types of axonal endings (*boutons*) with large and small vesicles seen in the cord of *Amphioxus* (see p. 308) perhaps indicate this sort of difference. Moreover, in the nervous system, as it develops, the neuroglial supporting cells are equally important, and theoretically they too could arise from either ciliated or goblet cell types. In the absence of any elimination by cell degeneration, during development

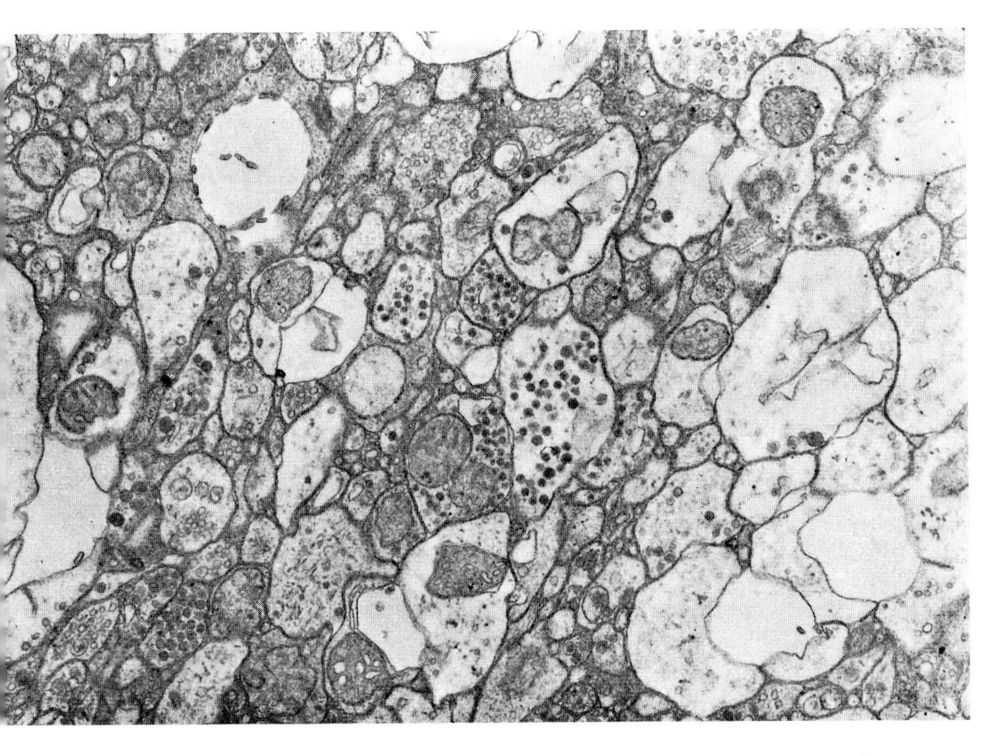

FIG. 13.35. Neurosecretory granules in cells in the dorsal ganglia of the nemertine, *Lineus*. (Photo by E. A. Ling.)

there are thus four main possibilities for the organization of the neural tissue without considering any questions of cell polarity. If the nerve cells are designated M or A (mechanocytic or amoebocytic) and the neuroglia or supporting cells as m or a, then the possible combinations are Mm or Ma and Am or Aa. On the principle of preserving a balance of activities, it might seem that Ma and Am would be the most successful combinations. There is no real evidence for or against this, although attention may be called to the different behaviour of oligodendrocytes and astrocytes *in vitro* as well as to the apparently variable behaviour of Schwann cells (see p. 114). Some of these combinations may simply not be viable, and it must be emphasized that, in spite of many suggestions, we do not yet know the essential features of the relationship between a nerve fibre and its supporting cells (see Kuffler, 1967). Thus, within the nervous system, as with muscle, the cell population can be very heterogeneous, and one nerve cell may differ greatly from another. In the nervous system, however, conduction of impulses appears to be largely a phenomenon of the cell surface; this may restrict the number of classes of cells that can develop the sort of conductivity required by the nervous system. It should also be pointed out that there are many "nerve" cells, e.g. the bipolar cells and probably some of the ganglion cells in the retina, from which no electrical nerve impulses as such have been recorded, but which are presumably necessary for the normal function. As suggested in Chapter 11, an investigation of the electrical properties of mechanocytes and amoebocytes might be a suitable starting point for a reassessment of the organization of the nervous system. It is well recognized nowadays that the nervous system performs many functions other than simply conducting impulses.

That nerve fibres always have neuroglia or similar cells in close relationship with them has already been emphasized. This is perhaps most readily understandable on the ground that the stability and nutrition of such extremely long and tenuous fibres must present great problems from several points of view, and it is the function of the neuroglia or similar cells to minister to these demands. Thus the duality of the nervous system may simply be another example of the way in which Nature has organized stability in physiological systems by the balanced action of opposing forces. Moreover, as already suggested, the neuroglia-nerve cell arrangement may be comparable with the flagellate-amoeboid symbiosis of the elementary blastuloid, but in a more sophisticated form. If this is so, the same question may be asked about muscle, and particularly about striated muscle: Do these large cytoplasmic masses exist without any adjacent compensating mechanisms? In answer to this, there have recently been described (Mauro, 1961) "satellite" cells lying in close proximity to certain muscle fibres in the frog, in very much the same way that satellite cells associate with ganglion cells. Further study of these cells could be very illuminating, though it has been suggested that they are simply myoblasts that have for some reason failed to fuse with the main sarcoplasm.

From the embryological point of view, in the evolving vertebrate the cells of the existing nervous system in the nemerteoid might be expected to differentiate first, i.e. before those which were to develop in connexion with the more recently acquired somites. These cells would still be concerned mainly with the innervation of the functional gut, with the existing plain-muscle systems of the body, e.g. blood vessels, on the motor side, and with the general surfaces of the body on the sensory side. It is thus significant that, even in the nervous systems of higher vertebrates, the sensory cells emerge early from the neuroectoderm to form the dorsal root ganglia. The cells of the sympathetic and parasympathetic systems do so too to form their ganglia on the motor side. It is also interesting that in *Amphioxus*, as mentioned in the discussion of muscle, the ventral roots contain fibres to the myotomes only, whereas the motor fibres to the lateral-plate musculature (which could be the equivalent of the existing musculature in the nemerteoid) emerge from the dorsal roots. The general form of the dorsal root ganglion cells is certainly of the unipolar type most commonly found in the invertebrates (though embryology and the bipolar form of these cells in elasmobranchs suggests that this unipolarity is a secondary phenomenon), and the autonomic system is arranged along lines that could well have evolved from the neural system already present in nemerteoids. Indeed, the serial distribution of transverse neural connexions between the intestinal diverticula may have initially determined the segmental distribution of the sympathetic chain with its contained ganglion cells.

Myelinated fibres make their appearance in connection with the newly developed system, and it may be that the availability of epithelial cells in the dorsal neural plate has been turned to advantage in this direction. The amyelinate character of the final motor fibres of the autonomic system would also be in line with the concept of the autonomic system corresponding essentially to the motor pathways already present in the nemerteoids. In existing nemertines, neuroglia cells and their processes enclose the processes of the nerve cells, but myelin, as such, has not been described.

In the course of forming the neural tube, the epidermal cells would presumably maintain their normal position and polarity. The ciliated surface would line the tube; the neural connexions, and fibrous prolongations of the sensory cells would continue to lie basally, and thus would be brought to the outside of the neural tube. This is, of course, exactly what happens in the development of the neural tube in vertebrates; in lampreys, the outer part of the neural tube is essentially neuropil. In the living nemertines it is probable that the nerve cells are primarily epithelial, and in several places they tend to migrate inwards from the surface and to form layers just below the epithelium. Nerve fibres also run parallel to the surface on the inner margins of the epithelium. Therefore, if the dorsal surface of the animal were folded in the manner suggested into a neural and longitudinally running tube (i.e. in a manner similar to that at present seen in amnion formation in nemertine embryology), the formation of the marginal and mantle layers, as described for the neuro-embryology of vertebrates, would be a direct consequence of the development of increasing numbers of nerve cells and their processes. This would simply be an accentuation of the events which actually occur to a limited extent in living nemertines, and would lead to patterns akin to those seen in the neural tube of the lamprey and, indeed, of higher vertebrates also. New cells would be recruited by division of the undifferentiated epithelial cells, and the neuroblasts would migrate from the epithelial surface to form secondary layers. The problem of the orientation of such cells has already been discussed (p. 245).

The concentration of neural organization in the cephalic region (cephalization), already begun in the more primitive worms, has clearly progressed much further in nemerteoids and must be expected to do so still more as they evolved towards the vertebrates, since directional movement necessitates efficient sensory mechanisms at the anterior end.

Finally, in considering the folding of the upper surface of the nemerteoid and the development of the neural tube, it will not have escaped the notice of the reader that the lateral margins of many nemertines have conspicuous melanin pigmentation, whether or not this is aggregated into definite spots or eye-spots. Such melanin-containing cells would, in the course of the folding, be either just enclosed or just not enclosed in the neural folds. This is perhaps the reason why, in the vertebrates, the majority of the melanin-containing cells, if not all of them outside the retina, have their origin in the neural crest, since this is, of course, the tissue in precisely that position in the vertebrate embryo.

References

Åkesson, B. (1961). The development of *Golfingia elongata* Keferstein (Sipunculidea) with some remarks on the development of neurosecretory cells in Sipunculids. *Arkiv Zool.* **13**, 511.

Balfour, F. M. (1885). "Comparative Embryology", Vol. 1. Macmillan, London.

Barraclough Fell, H. (1948). Echinoderm embryology and the origin of chordates. *Biol. Rev. Cambridge Phil. Soc.* **23**, 81.

Barrington, E. J. W. (1965). "The Biology of Hemichordata and Protochordata". Oliver & Boyd, Edinburgh and London.

Bone, Q. (1966). On the function of the two types of myotomal muscle fibres in elasmobranch fish. *J. Marine Biol. Assoc. U.K.* **46**, 321.

Boyd, I. A. (1961). The motor innervation of mammalian muscle spindles. *J. Physiol. (London)* **159**, 7P.

Brambell, F. W. R., and Cole, H. A. (1939). The preoral ciliary organ of the Enteropneusta: Its occurrence, structure, and possible phylogenetic significance. *Proc. Zool. Soc. London* **109B**, 181.

Bullock, T. H. (1940). The functional organization of the nervous system of Enteropneusta. *Biol. Bull.* **79**, 91.

Bullock, T. H., (1946). The anatomical organization of the nervous system in Enteropneusta. *Quart. J. Microscop. Sci.* **86,** 55.

Burdon-Jones, C. (1952). Development and biology of the larva of *Saccoglossus horsti* (Enteropneusta). *Phil. Trans. Roy. Soc. London* **B236**, 553.

Bürger, O. (1895). Die Nemertinen des Golfes von Neapel. *Fauna und Flora des Golfes von Neapel Monograph* **22**, 1.

Chèvremont, M. (1953). Les potentialités des cellules histiocytaires (S.R.E.). *XV Congr. Soc. Intern. Chir. Lisbonne* p. 277.

Clark, R. B. (1964). "Dynamics in Metazoan Evolution". Oxford Univ. Press (Clarendon), London and New York.

Coe, W. R. (1899). Development of the pilidium of certain naemertens. *Trans. Conn. Acad. Arts Sci.* **10,** 235.

Coe, W. R. (1927). The nervous system of pelagic nemerteans. *Biol. Bull.* **53**, 123.
Coe, W. R. (1945). Plankton of the Bermuda Oceanographic Expeditions. XI. Bathypelagic nemerteans of the Bermuda area and other parts of the north and south Atlantic oceans, with evidence as to their means of dispersal. *Zoologica* **30**, 145.
Conklin, E. G. (1933). The embryology of *Amphioxus*. *J. Morphol.* **54**, 69.
Cowey, J. B. (1952). The structure and function of the basement membrane muscle system of *Amphiporus lactifloreus* (Nemertea). *Quart. J. Microscop. Sci.* **93**, 1.
Dakin, W. J., and Fordham, M. G. C. (1936). The anatomy and systematic position of *Gorgonorhynchus repens* (gen.n.sp.n.): A new genus of Nemertines characterised by a multibranched proboscis. *Proc. Zool. Soc. London* **1**, 461.
Eakin, R. M., and Westfall, J. A. (1962). Fine structure of the notochord of *Amphioxus*. *J. Cell Biol.* **12**, 646.
Eccles, R. M., Eccles, J. C., and Kozak, W. (1962). Further investigations on the influence of motorneurones on the speed of muscle contraction. *J. Physiol. (London)* **163**, 324.
Ernyei, S., and Young, J. Z. (1966). Pulsatile and myelin-forming activities of Schwann cells *in vitro*. *J. Physiol. (London)* **183**, 469.
Flood, P. R. (1965). Skeletal muscle fibre types in *Amphioxus lanceolatus* and *Myxine glutinosa*. *J. Ultrastruct. Res.* **12**, 238.
Flood, P. R. (1966). A peculiar mode of muscular innervation in *Amphioxus*. Light and electron microscopic studies of the so-called ventral roots. *J. Comp. Neurol.* **126**, 181.
Flood, P. R. (1967). Structure of the segmental trunk muscle in *Amphioxus*. *Z. Zellforsch. Mikroskop. Anat.* **84**, 389.
Gabe, M. (1954). La neurosécrétion chez les invertébrés. *Ann. Biol.* **30**, 6.
Gauthier, G. F., and Padykula, H. A. (1966). Cytological studies of fibre types in skeletal muscle. A comparative study of the mammalian diaphragm. *J. Cell Biol.* **28**, 333.
George, J. C., and Berger, A. J. (1966). "Avian Myology". Academic Press, New York.
Gray, J. (1933a). Studies in animal locomotion. i. The movement of fish with special reference to the eel. *J. Exptl. Biol.* **10**, 88.
Gray, J. (1933b). Studies in animal locomotion. iii. The propulsive mechanism of the whiting (*Gadus merlangus*). *J. Exptl. Biol.* **10**, 391.
Hanson, J., and Lowy, J. (1960). Structure and function of the contractile apparatus in the muscles of invertebrate animals. *In* "Muscle" (G. Bourne, ed.). vol 1, p. 265. Academic Press, New York.
Harris, J. E., and Whiting, H. P. (1954). Structure and function in the locomotory system of the dogfish embryo. The myogenic stage of movement. *J. Exptl. Biol.* **31**, 501.
Hess, A. (1965). The sarcoplasmic reticulum, the T system and motor terminals of slow and twitch muscle fibres in the garter snake. *J. Cell Biol.* **26**, 467.
Hofmann, H. J. (1967). Precambrian fossils (?) near Eliot Lake, Ontario. *Science* **156**, 500.
Hubrecht, A. A. W. (1883). On the ancestral form of the chordata. *Quart. J. Microscop. Sci.* **23**, 349.
Hubrecht, A. A. W. (1886). Contributions to the embryology of Nemertea. *Quart. J. Microscop. Sci.* **26**, 417.
Hubrecht, A. A. W. (1887). The relation of the Nemertea to the Vertebrata. *Quart. J. Microscop. Sci.* **27**, 605.
Ivanov, A. V. (1963). "Pogonophora" (D. B. Carlisle, transl. and ed.). Academic Press, New York.
Iwata, F. (1958). On the development of the nemertean *Micrura Akkeshiensis*. *Embryologia (Nagoya)* **4**, 103.
Jensen, D. D. (1960). Hoplonemertines, Myxinoids and deuterostome origins. *Nature* **188**, 649.

Jensen, D. D. (1963). Hoplonemertines, Myxinoids and vertebrate origins. *In* "The Lower Metazoa" (E. C. Dougherty *et al.*, eds.), p. 113. Univ. of California Press, Berkeley, California.

Johannsen, O. A., and Butt, F. H. (1941). "Embryology of Insects and Myriapods". McGraw-Hill, New York.

Kerr, J. G. (1919). "Text-book of Embryology. Vol. II: Vertebrata, with the Exception of Mammalia". Macmillan, London.

Knight-Jones, E. W. (1952). On the nervous system of *Saccoglossus cambrensis*. Enteropneusta). *Phil. Trans. Roy. Soc. London* **B236**, 316.

Konigsberg, I. R. (1963). Clonal analysis of myogenesis. *Science* **140**, 1273.

Konigsberg, I. R. (1965). Aspects of cytodifferentiation of skeletal muscle. *In* "Organogenesis" (R. L. de Haan and H. Ursprung, eds.). Holt, Rinehart and Winston, New York.

Krüger, P. (1952). "Tetanus und Tonus der quergestreiften Skelettmuskeln der Wirbeltiere und des Menschen". Akademische Verlagsgesellschaft, Geest & Portig, K-G, Leipzig.

Kuffler, S. W. (1967). Neuroglial cells: Physiological properties and a potassium mediated effect of neuronal activity on the glial membrane potential. *Proc. Roy. Soc.* (*London*) **B168**, 1.

Kuffler, S. W., and Vaughan Williams, E. M. (1953). Small nerve junction potentials. The distribution of small motor nerves to frog skeletal muscle, and the membrane characteristics of the fibres they innervate. *J. Physiol.* (*London*) **121**, 289.

Lännergren, J., and Smith, R. S. (1966). Types of muscle fibres in toad skeletal muscle. *Acta Physiol. Scand.* **68**, 263.

Lechenault, H. (1962). Sur l'existence de cellules neurosécrétrices dans les ganglions cérébroides des Lineidae (Hétéronémertes). *Compt. Rend.* **255**, 194.

Lechenault, H. (1963). Sur l'existence de cellules neurosécrétrices chez les Hoplonémertes. Caracteristiques histochimiques de la neurosécrétion chez les Némertes. *Compt. Rend.* **256**, 3201.

Lewis, M. R., and Lewis, W. H. (1917). Behaviour of cross-striated muscle in tissue cultures. *Am. J. Anat.* **22**, 169.

Manton, S. M. (1958). Embryology of Pogonophora and classification of animals. *Nature* **181**, 748.

Mauro, A. (1961). Satellite cells of skeletal muscle fibres. *J. Biophys. Biochem. Cytol.* **9**, 493.

Metschnikoff, E. (1869). Studien über die Entwicklung der Echinodermen und Nemertinen. *Mem. Acad. Imp. Sci. St. Petersbourg* **7**, 14.

Nachmias, B. T., and Padykula, H. A. (1958). A histochemical study of normal and denervated red and white muscles of the rat. *J. Biophys. Biochem. Cytol.* **4**, 47.

Page, S. G. (1965). A comparison of the fine structure of frog slow and twitch muscle fibres. *J. Cell Biol.* **26**, 477.

Paton, S. (1911). The reactions of the vertebrate embryo and associated changes in the nervous system. *J. Comp. Neurol.* **21**, 345.

Peachey, L. (1961). Structure of the longitudinal body muscles of *Amphioxus*. *J. Biophys. Biochem. Cytol.* **10**, Suppl., 159.

Peters, A., and Mackay, B. (1961). The structure and innervation of the myotomes of the lamprey. *J. Anat.* **95**, 575.

Riepen, O. (1933). Anatomie und Histologie von *Malacobdella grossa*. *Z. Wiss. Zool.* **143**, 323.

Rumery, R. E., Blandau, R. J., and Hagey, P. W. (1961). Observations on living myocardial cells from cultured 48-hour chick hearts. *Anat. Record* **141**, 253.

Salensky, W. (1896). Bau und Metamorphose des Pilidium. *Z. Wiss. Zool.* **43**, 481.

Scharrer, B. (1941). Neurosecretion. III. The cerebral organ of the Nemerteans. *J. Comp. Neurol.* **74**, 109.

Schmidt, G. A. (1934). Ein zweiter Entwicklungstypus von *Lineus gesseriensis-ruber* (Müll), Nemertini. *Zool. Jahrb. Abt. Anat. Ontog. Tiere* **58**, 607.

Smith, J. E. (1934). The early development of the Nemertean *Cephalothrix rufifrons*. *Quart. J. Microscop. Sci.* **77**, 335.

Smith, R. S., and Lännergren, J. (1968). Types of motor units in the skeletal muscle of *Xenopus laevis*. *Nature* **217**, 281.

Stein, J. M., and Padykula, H. A. (1962). Histochemical classification of individual skeletal muscle fibres of the rat. *Am. J. Anat.* **110**, 103.

Straus, W. L., and Rawles, M. E. (1953). An experimental study of the origin of the trunk musculature and ribs in the chick. *Am. J. Anat.* **92**, 471.

Wachstein, M., and Meisel, E. (1955). The distribution of histochemically demonstrable succinic dehydrogenase and of mitochondria in tongue and skeletal muscles. *J. Biophys. Biochem. Cytol.* **1**, 483.

Wilson, C. B. (1900). The habits and early life of *Cerebratulus lacteus* (Verill). *Quart. J. Microscop. Sci.* **43**, 97.

Wintrebert, P. (1920). La contraction rythmée aneurale des myotomes chez les embryons de Sélaciens. *Arch. Zool. Exptl. Gen.* **60**, 221.

Young, J. Z. (1950). "The Life of Vertebrates". Oxford Univ. Press (Clarendon), London and New York.

Chapter 14

FROM NEMERTEOIDS TO VERTEBRATES
2. THE PITUITARY AND THE PHARYNGEAL COMPLEX

Pituitary

Anterior Lobe

In considering the possible effects of the obsolescence of the eversible proboscis, some other events of great significance to the animal may be revealed. In many nemertines of to-day the proboscis no longer opens at the anterior end of the body but is everted through the mouth by way of a rhynchodaeum which has united with an outgrowth from the dorsal surface of the mouth. Thus the proboscis, as stated earlier (p. 277), is in three parts: a rhynchodaeum opening into the mouth cavity, an anterior proboscis, and a posterior proboscis which lies behind the armature of such forms as have developed this structure. The last two parts are surrounded by the rhynchocoel and can be everted through the rhynchodaeum and mouth. Figure 14.1 shows, in sagittal section, the manner in which the rhynchodaeum has changed its position in different species. Figure 13.10 shows the relationship of what are perhaps corresponding structures in the anterior region of early vertebrates as they would appear in the same plane of section. Attention is called to the position of Rathke's pouch and the anterior limit of the notochord. It will be agreed that the anatomical relationships in the two diagrams would be strikingly similar if the proboscis proper were to become the notochord, as suggested in Chapter 13, and the rhynchodaeum were to become modified into Rathke's pouch. It will also be remembered that the latter is converted during embryonic development into the anterior lobe of the pituitary body, a structure notable for the wide variety of cell types present in it.

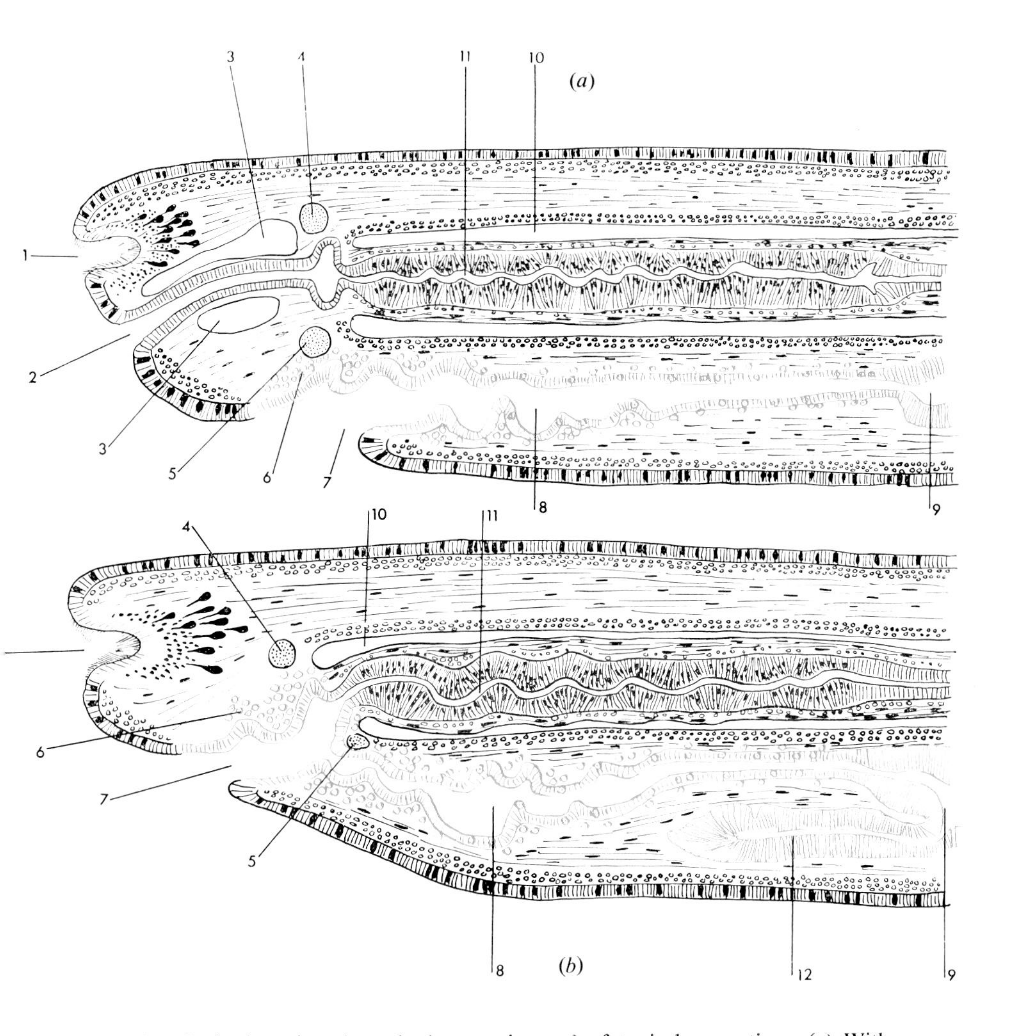

FIG. 14.1. Sagittal section through the anterior end of typical nemertines. (*a*) With separate rhynchodaeum; (*b*) with proboscis opening into the buccal cavity. 1, Frontal organ; 2, rhynchodaeum; 3, blood lacuna; 4, dorsal commissure; 5, ventral commissure; 6, complex buccal epithelium; 7, mouth; 8, pharynx; 9, intestine; 10, rhynchocoel; 11, proboscis cavity; 12, intestinal caecum.

Figure 13.10*b* showed sagittal sections through the anterior end of a nemertine. The rhynchodaeum and the frontal organ occupy the same relative positions as the hypophysial organ and the nasal sac do in the embryonic cyclostomes. It is not difficult to imagine that variations on this theme could account for the events which occur in the development of this part of the head of cyclostomes and indeed of other vertebrates, both ontogenetically and phylogenetically.

A slightly different interpretation is, in fact, suggested by the embryology of the chick (Adelmann, 1926) in which there is a well developed strand of tissue which connects the anterior end of the notochord to Rathke's pouch (Fig. 14.2). This tissue called the prechordal plate, has had a variety of

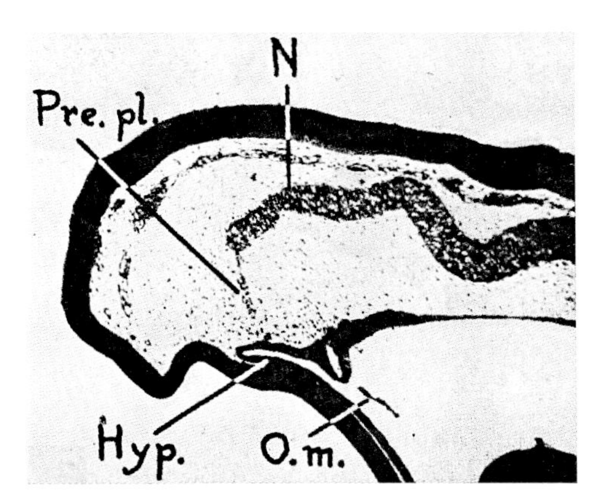

FIG. 14.2. Sagittal section through the anterior end of a developing bird. Hyp., Hypophysial evagination (Rathke's pouch); N, notochord; O.m., oral membrane; Pre., pl., prechordal plate. (Adelmann 1926.)

interpretations placed on its origin, but it seems likely that it may correspond to the vestige of the rhynchodaeum itself, and that Rathke's pouch is actually the evagination of the buccal cavity which originally joined the rhynchodaeum and thus provided the exit for the proboscis. This idea is satisfactory in two ways. First, the rhynchodaeum in many nemertines is surrounded by "blood spaces" and loose mesenchyme; the prechordal plate is thought to be connected in some close way with the rather anomalous "head cavities" and with the development of cephalic mesenchyme. The epithelium surrounding the anterior blood spaces in *Lineus* is peculiar in being neither strictly a pavement endothelium nor a truly columnar structure, and the head cavities in vertebrates share this peculiarity. Second, the rhynchodaeum itself, as it occurs in many nemertines, tends to have a relatively simple columnar epithelium which, structurally, might perhaps form a suitable precursor for the pars intermedia of the pituitary, but would be more difficult to reconcile with the complex cellular structure of the partes distalis and tuberalis. It is particularly relevant to observe, however, that the dorsal surface of the mouth in nemertines, from which Rathke's pouch would derive, can have a very complex epithelium with at least three visibly different classes of cells, together with a superficial layer of ciliated cells. In fact, when stained by azan, Masson's stain, or the performic acid-alcian blue-periodic acid-Schiff stains, sections of the dorsal wall of the mouth cavity of *Lineus* and of the anterior pituitary of a mammal are very similar in many respects (Fig. 14.3), if due allowance is made for changes in the buccal epithelium likely to occur when it ceased to be functional as such and became closed off from the buccal cavity in the manner of Rathke's pouch. The cytological pattern of the two structures is consistent with the idea that the one could transform into the other. As in all other tissues, there is a greater precision in the cellular pattern in the vertebrates, i.e. the cell boundaries are more clearly defined, but, apart from that, the resemblance is striking. Moreover the recent discovery (Barnes, 1961) in electron-microscopic studies of the anterior pituitary that many of the cells are ciliated is particularly interesting in view of the nature of the epithelium in the nemertine, and there are some remarkable similarities in the granules of the various groups of cells in the nemertine and vertebrate tissues. Ziegler (1963) also has noted that many of the cells of the pars intermedia carry cilia. The physiology of the buccal cavity and thus of the potential rhynchodaeum will be considered again (p. 505), and it will be shown to offer some clues to how Rathke's pouch came to give rise to a structure which eventually developed, with adjacent structures, into "the leader of the endocrine orchestra".

Posterior Lobe

In pursuance of this line of thought concerning the anterior body, the origin of the posterior lobe also needs investigation. Unlike the anterior lobe, which is glandular and epithelial, the posterior lobe is essentially a neural structure; in fact, it bears some resemblance to a neuroma.

It will be remembered that the proboscis of the nemertine is well supplied with nerves and can be everted, retracted, and coiled with considerable speed. So far, little is known of the mechanism of ejection; it probably depends on the rise of pressure caused by constriction of the rhynchocoel, partly by its own muscle and partly by the combined contraction of both the circular and longitudinal muscles of the body wall. Retraction, however, is brought about, at least in part, by the shortening of the longitudinal fibres of the proboscis itself. Invagination or inversion is brought about by the contraction of the extremely elongated retractor muscle which stretches in the everted state from the inside of the tip of the proboscis to a point of attachment near the posterior end of the rhynchocoel (see Fig. 12.14). The main nerve to the proboscis enters it at the "hinge" about which the structure everts and inverts, the hinge being situated very close to the brain. When the proboscis is everted, however, the retractor muscle is stretched freely

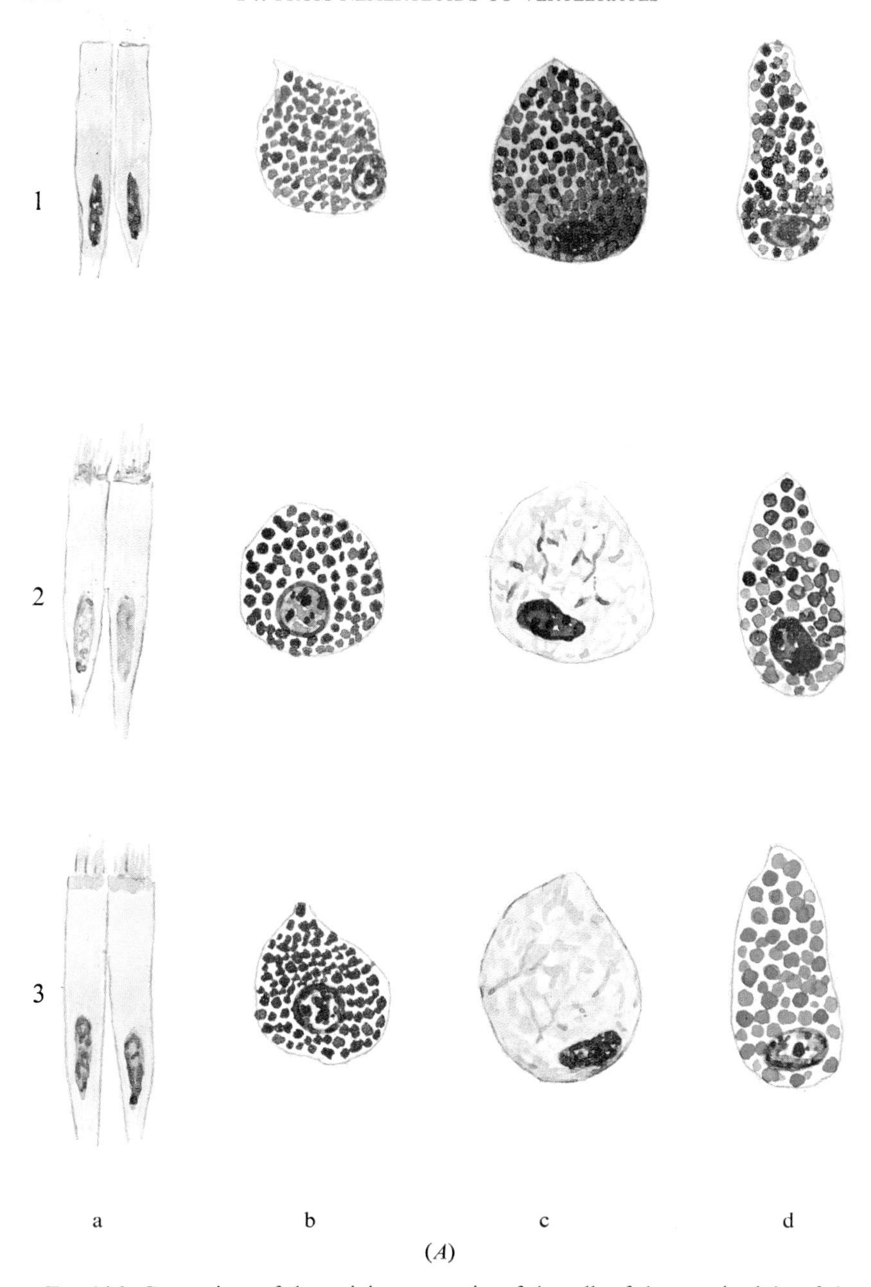

FIG. 14.3. Comparison of the staining properties of the cells of the anterior lobe of the hypophysis of the rat (*B*) and of the dorsal epithelium of the buccal cavity of *Lineus ruber* (*A*). 1, Stained by performic acid-alcian blue, PAS, orange G method; 2, stained with Azan; 3, stained with Masson's stain and haemalum. a–d, four types of cell.

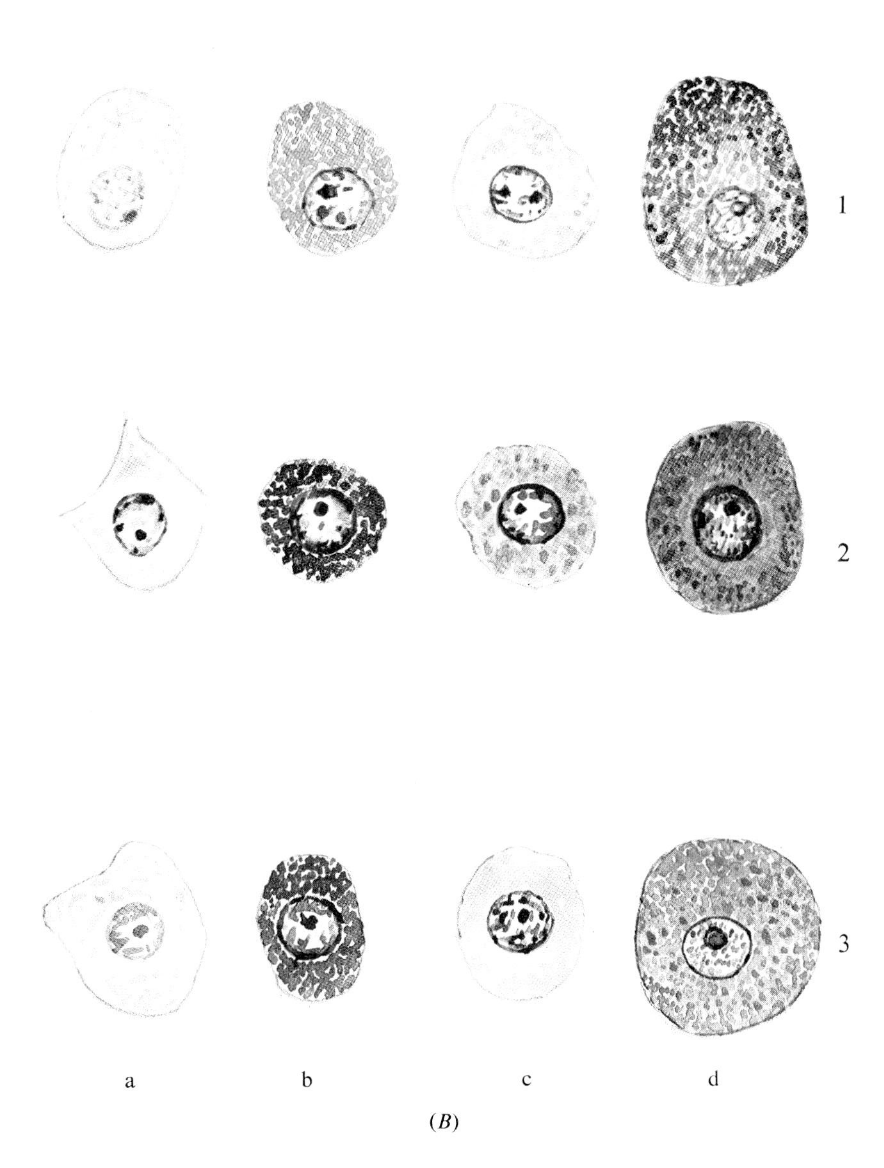

(*B*)

within the lumen of the rhynchocoel and could only be reached, neurally, either at the tip of the proboscis or, posteriorly, at the point of attachment. In neither situation is there any obvious nerve, though this is not to say that such nerves do not exist. If, however, the stimulus to the retractor muscle could be applied near the hinge, perhaps by liberating a transmitter directly from nerve-endings in the vicinity into the fluid of the rhynchocoel, an efficient humoral mechanism for stimulating the muscle could be achieved. Thus the observation mentioned on p. 265 that the isolated retractor muscle, which remains more or less unresponsive to acetylcholine, adrenalin, and 5-hydroxytryptamine, responds with a vigorous contraction to the application of the posterior pituitary hormone, oxytocin, at a dose of about 0.2 unit per ml, may be very significant (see Fig. 12.5). When suspended in sea-water the muscle is not in the best physiological state, and its tenuous fibres tend to disintegrate even without stimulation. This may explain why the effect can be obtained only a limited number of times and with decreasing vigour from any one muscle. Better results might be obtained if the rhynchocoel fluid could be substituted for sea-water. It is specially interesting that acetylcholine (50 λ per ml) did not cause any effect on the retractor muscle, though it can produce sustained contracture of the proboscis muscles themselves (see Fig. 12.4) and also of some, at least, of the circular muscles (though not the longitudinal muscles) of the body wall (Mitchell and Willmer, 1966).* In view of the fairly widespread activity of oxytocin as a muscle stimulant, its action on the retractor muscle may be no more than a curious coincidence. This action may equally indicate that the posterior pituitary owes its origin to the nerve which supplied the proboscis of the nemerteoid and its liberation of oxytocin to the mechanism by which the retractor of the proboscis was stimulated. If this is so, one may ask why it should have persisted as a secretory mechanism after the proboscis became obsolete and the retractor muscle without function. One can only suggest that, since oxytocin does cause other muscles in the nemerteoid (e.g. the longitudinal muscles) to contract, and these muscles have subsequently become involved in other functions, its production may have been a valuable asset. The development of suitable vascular connexions to this potent source of oxytocin could then turn the supply to advantage as a source of hormone. It may be that oxytocin is a fairly widespread neural transmitter and cell activator in the nemerteoids. Incidentally, if oxytocin turns out to be the natural transmitter for the general longitudinal muscles of the nemerteoid body, it would be difficult to reconcile this with these muscles developing into the somitic muscles (see p. 304), since the latter are largely cholinergic, but it could be relevant to its action on milk ejection, uterine contraction, intestinal motility, and vascular tone in higher animals.

*Catechol amines have recently been demonstrated in nerve fibres to the proboscis (Reutter, K. 1969. *Z. Zellforsch. mikr. anat.* **94**, 391.)

There is another relevant point about the ejection and retraction of the proboscis. As stressed earlier, the nemertine worm is essentially a closed system of constant volume. What happens, therefore, when the proboscis is everted? The volume of the posterior rhynchocoel must shrink as its contained fluid fills the emerging part, i.e. the cavity of the proboscis. Similarly, when the proboscis is drawn in, the rhynchocoel must swell, at least temporarily, in order to accommodate the fluid which filled the everted proboscis. If there were coincident changes of permeability in the walls of the rhynchocoel or proboscis or both, the volume changes could gradually be accommodated without permanent swelling or shrinking of the rhynchocoel cavity, and so life could, perhaps, be made more comfortable for the worm. In the vertebrates the posterior pituitary produces variants on two polypeptides, or a larger polypeptide with two actions. These are exemplified by oxytocin, the plain muscle stimulant, and vasopressin, which increases the permeability of membranes, especially to water. Why it produces these two polypeptides is quite obscure. It would seem, however, to be possible that the retraction of the proboscis of the nemerteoid could depend on the production of oxytocin; that the retraction could be accompanied by the simultaneous release of a substance that altered the permeability of the rhynchocoel and perhaps of the skin and nephridia also; that this substance could well be vasopressin or a related polypeptide. The proboscis apparatus of the nemertine would, therefore, seem to be very well worth an investigation from this point of view, since it may provide the solution to the puzzle concerning the origin and significance of the oxytocin-vasopressin system. It may also offer a clue to why the two substances are so closely linked both chemically and physiologically. From what is known of the variability of oxytocins and vasopressins in the vertebrates it is quite likely that similar substances in nemerteoids may be chemically distinct from any actually occurring in the vertebrates, though having essentially similar physiological actions.

There are certain histological features in the neighbourhood of the hinge of the proboscis that are relevant to this discussion. The nerves to the proboscis (in *Lineus*, at least) leave the dorsal lobes or commissure of the brain in two main branches. In the proboscis itself, these nerves are divided into a series of longitudinally running trunks joined by ring connexions. Very little appears to be known about these nerves, not even the extent of their sensory and motor components. Because they are very well developed in the bathypelagic nemertines, which are creatures singularly lacking in other sense organs (Coe, 1927), they may be largely sensory. Nevertheless it is reasonable to enquire whether these nerves merely atrophied when the proboscis became obsolete.

Just posterior to the exit of the proboscis nerves from the ganglia, two groups of large nerve cells make their appearance in the cephalic ganglia, just lateral to the proboscis channel. Some of these cells contain neuro-

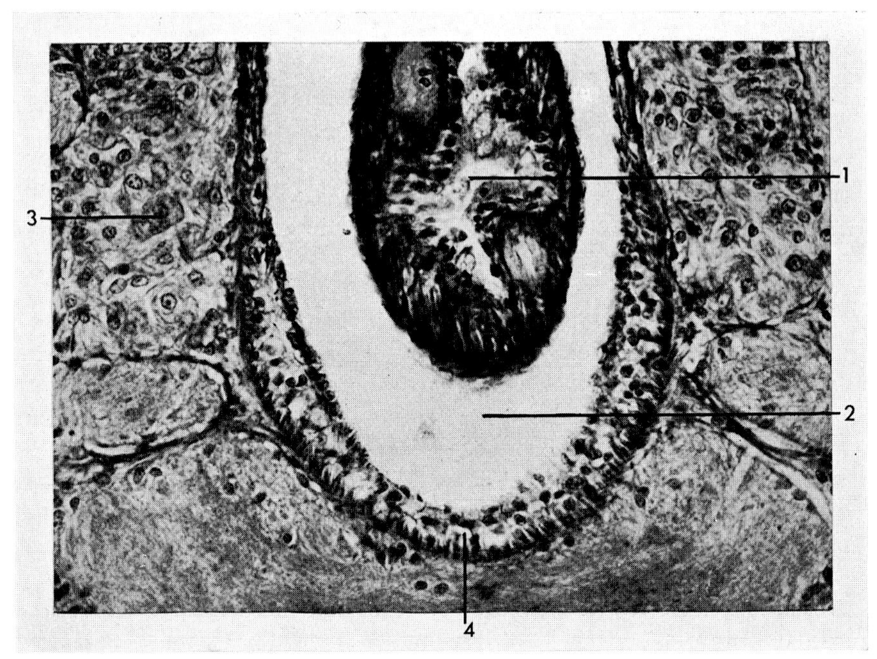

FIG.4.14

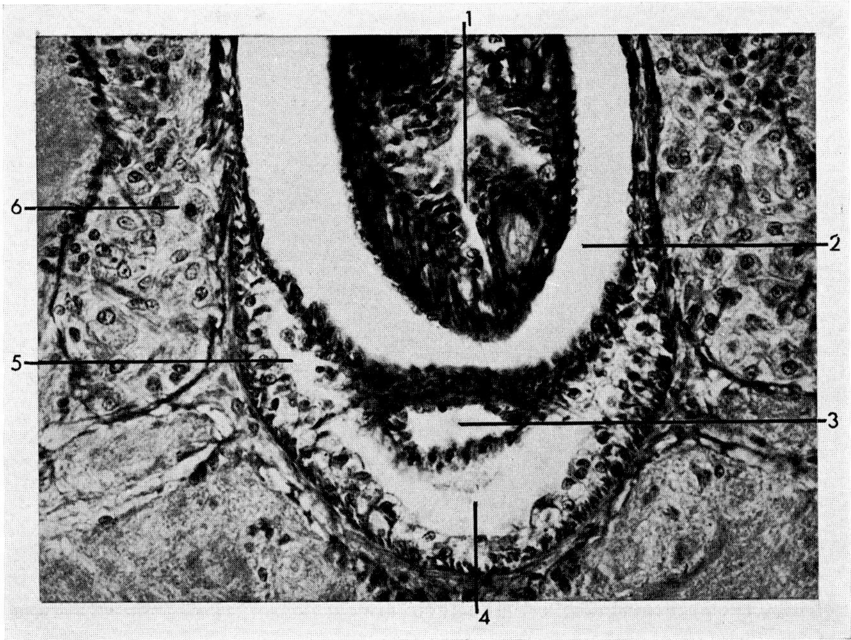

FIG. 14.5

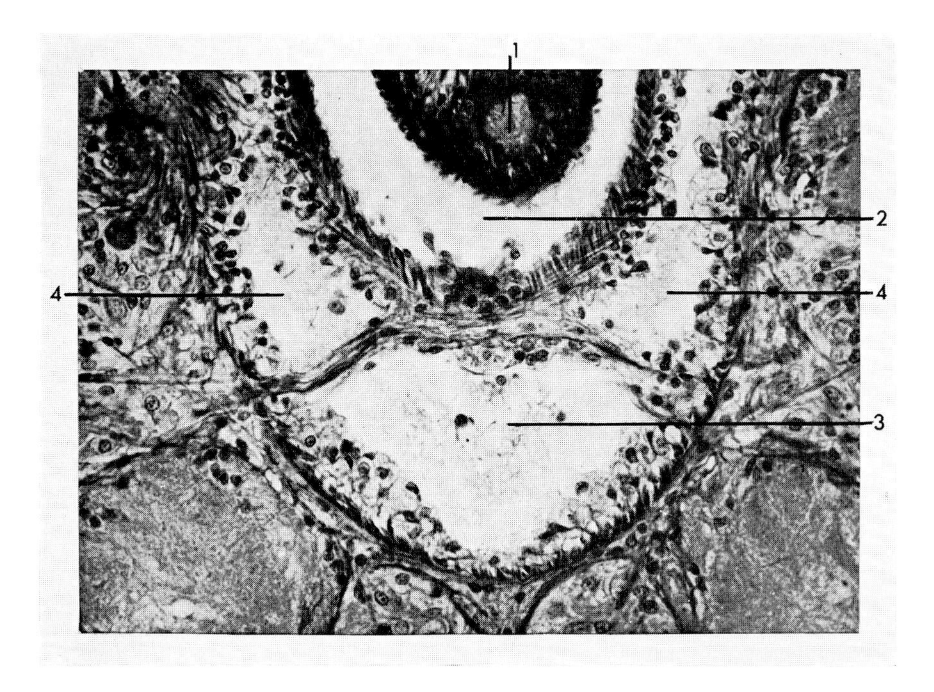

FIG. 14.6. Transverse section of dorsal vessel just anterior to the cephalic organs. 1, Proboscis; 2, rhynchocoel, with beginning of villus on its ventral floor; 3, dorsal blood vessel; 4, lateral vessels.

secretory material which can be stained with the Gomori technique for neurosecretory substances (Ling, 1969) and with paraldehyde fuchsin (see Figs. 13.35 and 14.4). A blood vessel develops ventral to the proboscis at this point, and seems to have some sort of tubular connexion with the rhynchocoel itself, though the exact nature of this connexion is obscure. It appears to be a diverticulum of the rhynchocoel into the blood vessel (Fig. 14.5). The walls of this blood vessel are peculiar in that they have an almost columnar epithelium, and under some conditions there is evidence for the presence of

←*Figs. 14.4 and 14.5 on facing page*

FIG. 14.4. Transverse section through the proboscis (1) and rhynchocoel (2) in the region of the cephalic ganglia. Large nerve cells (3), some of which appear to be "neurosecretory", flank the sides of the rhynchocoel, the floor of which is becoming vascularized (4).

FIG. 14.5. Transverse section of the rhynchocoel and related blood vessels at the level of the dorsal ganglion in *Lineus*. 1, Proboscis; 2, rhynchocoel with flattened epithelium; 3, diverticulum; 4, dorsal blood vessels with "columnar" epithelium. 5, beginning of connection to lateral blood vessel; 6, large ganglion cells, some of which are neurosecretory. (See also Fig. 16.4).

secretory globules both in the cells and in their vicinity; this substance could perhaps be related to that in the neurosecretory cells. Slightly posterior to this, the ventral wall of the rhynchocoel is formed into a longitudinal tubular structure covered, inside and out, with a columnar type of epithelium (Fig. 14.6.) This tube is probably a diverticulum or branch of the dorsal vessel, though its exact relationships have not been determined with certainty, and it runs backward for some distance along the inside of the rhynchocoel and finally ends in a sort of villus (Fig. 14.7). This histological picture suggests fluid transfer, somewhat on the lines of that occurring in the choroid plexus of

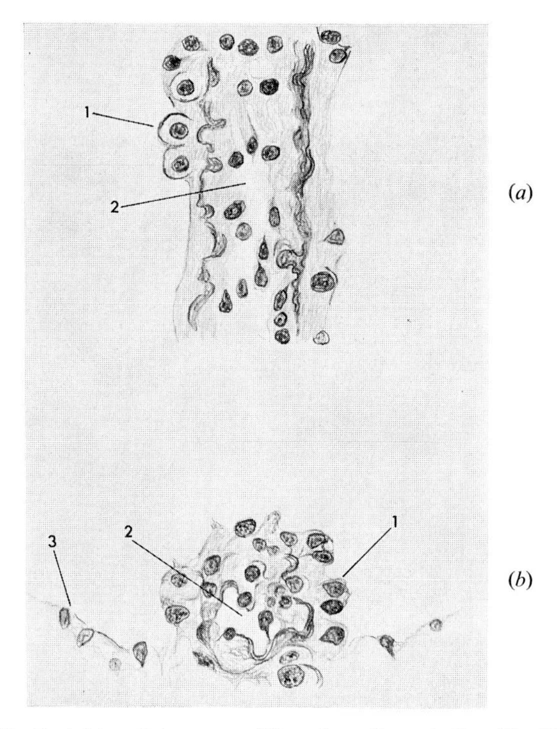

FIG. 14.7. Longitudinal (*a*) and transverse (*b*) sections through the villus-like organ on the floor of the rhynchocoel, just posterior to the cephalic organ. 1, Epithelium facing rhynchocoel cavity; 2, lumen of blood vessel with thickened endothelium; 3, lining of rhynchocoel cavity.

vertebrates; histologically this would appear to be the most likely area in which the rhynchocoel fluid is produced and regulated. Moreover, the whole complex—neurosecretory cells, blood vessel, "choroidal cells", and villus—is very suggestive as a basis from which the whole posterior pituitary function may have been built up. Though obviously differing in many ways, it is structurally not entirely dissimilar to the posterior lobe complex, as seen in *Myxine*, for example, (see Fig. 14.8) where the homology of the so-called adenohypophysis is by no means clear (Brodal and Fänge, 1963). The vascularity of this region in *Lineus* is striking, though the vessels are in the form of large lacunae and not of course in any "portal" arrangement. The

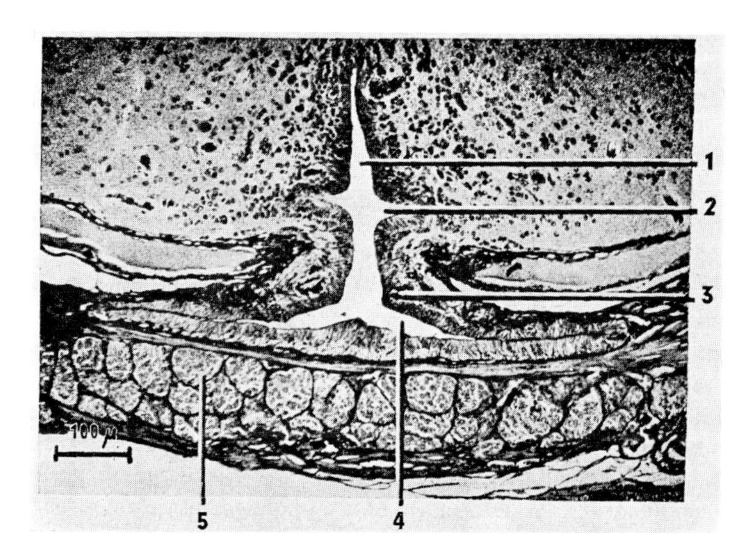

FIG. 14.8. Transverse section through the area of connexion between hypothalamus and neurohypophysis in *Myxine glutinosa*. 1, Hypothalamic ventricle; 2, lateral extensions of ventricular cavity; 3, infundibular region with vascularized wall, and nerve fibres from preoptic-neurohypophysial tract; 4, neurohypophysial recess; 5, adenohypophysis, separated from neurohypophysis by connective tissue. (Adam, 1963.)

large secretory nerve cells, the cephalic organs which also contain "neurosecretory" cells (see Fig. 15.9), and the tubular process of the rhynchocoel are all in such close proximity to large blood vessels, of which some have a secretory type of epithelial lining, that the situation appears to be ideal for the circulation of chemical mediators.

Pharyngeal Derivatives

All the changes which have been discussed so far with reference to the further evolution of the nemerteoids are those which could have been set in train by the assumption of a filter-feeding mode of life made possible by the perforation of the pharyngeal pouches to the outside. It is relevant, therefore,

points of interest when the digestive and absorptive processes are compared with those that go on in the lower chordates as described, for example, by Barrington (1942). The facts that the eosinophil cells of the pharynx swell or shrink with different concentrations of sodium chloride and that they contain the enzyme carbonic anhydrase (Jennings, 1962) both suggest possible relationships with comparable cells in the gills of fishes and the stomach of

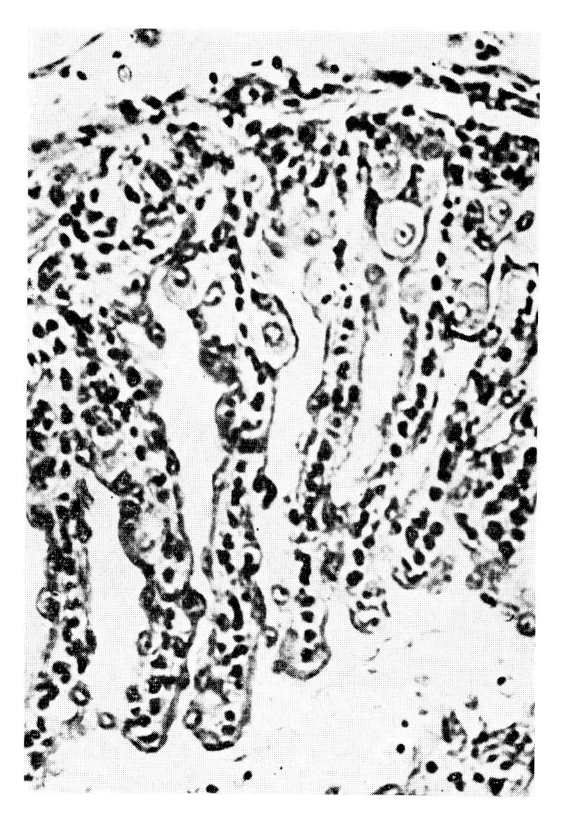

FIG. 14.10. Chloride-secreting cells on the gills of the eel. (Keys and Willmer, 1932.)

higher vertebrates. It should be noted that the chloride-secreting cells on the gills of fishes (Fig. 14.10) are probably of much wider occurrence in different species than was at first supposed (Keys and Willmer, 1932). They are probably present in the gill epithelia of most fish, and they are detectable, for example, in the gills of the ammocoete larvae of cyclostomes. They can swell and become very conspicuous when fish are loaded with salt (Liu, 1942; Copeland, 1948; Burns and Copeland, 1950; Vickers, 1961; Munshi, 1964;

and others), just as the eosinophil cells of *Lineus* become conspicuous under the same conditions.

From the behaviour of the pharynx of *Lineus*, it might be suggested that the ciliated cells are concerned with regulating the salt balance in dilute solutions, while the eosinophil and other cells become involved when the solutions become more concentrated. The latter cells would presumably be concerned with forcibly ejecting salt, while the former would be conservers of salt. Whether the mechanism depends primarily on Cl^-, Na^+, K^+, H^+, HCO_3^-, or other ions is still to be elucidated. Again, however, it is likely that a salt-balancing mechanism based on two oppositely orientated types of cells is present in this tissue, as has been suggested for others.

Thyroid Gland

Another characteristically vertebrate tissue which must have its origin in the pharyngeal region of the ancestors of the chordates is the equivalent of the endostyle and ultimately of the thyroid gland. The thyroid rudiment develops from the epithelium of the pharyngeal floor; in urochordates and cephalochordates this region is already notable for its uptake of iodine (see Barrington, 1965). In tunicates, *Amphioxus*, and the larval lamprey, the iodine is bound by a special region in the longitudinally running pharyngeal groove known as the endostyle, and chromatographic studies indicate that both tri-iodothyronine and thyroxin are formed. In the ammocoete larva of *Petromyzon* the binding of iodine takes place in special groups (Barrington and Franchi, 1956) of ciliated cells that are in close association with groups of gland cells of a rather peculiar type (Fig. 14.11). Although there is no evidence that these gland cells have any part to play in the metabolism of iodine, which seems to be the concern of the ciliated cells, treatment of the animal with thiouracil depletes them of their secretion (Barrington and Sage, 1963). In this animal, the final product, which may be either thyroxin or tri-iodothyronine, may be dependent on the temperature, since in some not very well controlled experiments the former was formed more at 2°C than at 15°C (Leloup and Berg, 1954). Gorbman (1959) has suggested that this iodine-binding function has developed out of some more general and widespread property of iodine-binding which is present in some mucous cells and is perhaps basically similar to that seen in the formation of the iodinated proteins of some sponges and Cnidaria.

Gorbman suggested too that the chordates could have specialized a region of the pharynx for the purpose, but that initially, in some hypothetical ancestral invertebrate, the whole pharyngeal epithelium might have been involved, and that the iodinated mucus was secreted back into the cavity of the alimentary canal. Examination of the pharynx of *Lineus* reveals several interesting features in this respect. First, the anterior region shows rather pronounced longtitudinal folds of the ventral mucosa, and the cellular pattern bears a remarkable resemblance to that in the endostyle of ammocoetes, though much less neatly organized. Second, experiments with radioactive I_2 show that the whole pharnygeal region, and particularly its anterior portion (Fig 14.12), picks up iodine far more rapidly than the rest of the worm, though some mucous cells on the skin and occasionally the proboscis may also give a positive response.

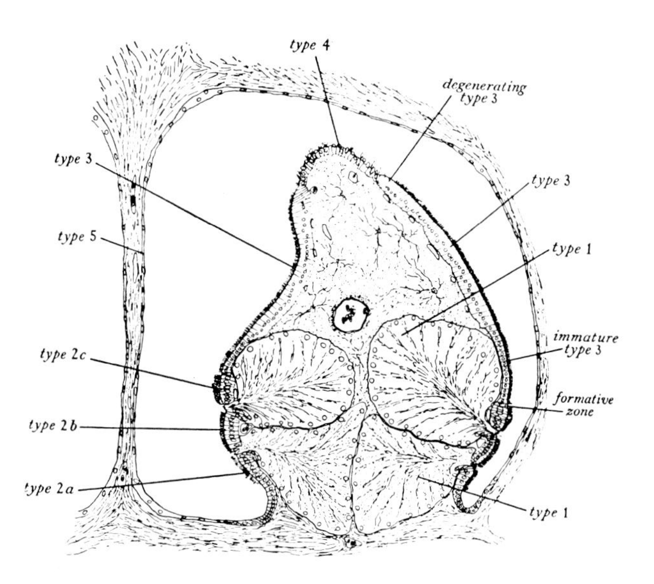

Fig. 14.11. Transverse section of the endostyle of the ammocoete showing the various cell types. Iodine is accumulated by types 2 and 3. (Barrington and Franchi, 1956.)

It may be noted in passing that the notochord of the marine lamprey accumulates iodine (Leloup, 1952), but it is not known whether this is a primary or a secondary phenomenon. In view of the earlier discussion of the origin of the notochord (see Chapter 13 and p. 262), this peculiarity of the notochord is not without interest, though its significance is not yet clear.

Iodine-containing mucus may be found in the lumen of the pharynx and intestine, and after some days the subcutaneous tissues of the worm also contain radioactive material. Third, the iodine, though accumulated and capable of being held in the animal for a considerable time, is only specifically bound to a very limited extent. It is nearly all present as iodide, and only very small traces of thyronine compounds are detectable (Balfour and Willmer, 1967). Fourth, the amount of radioactive iodine picked up is greater,

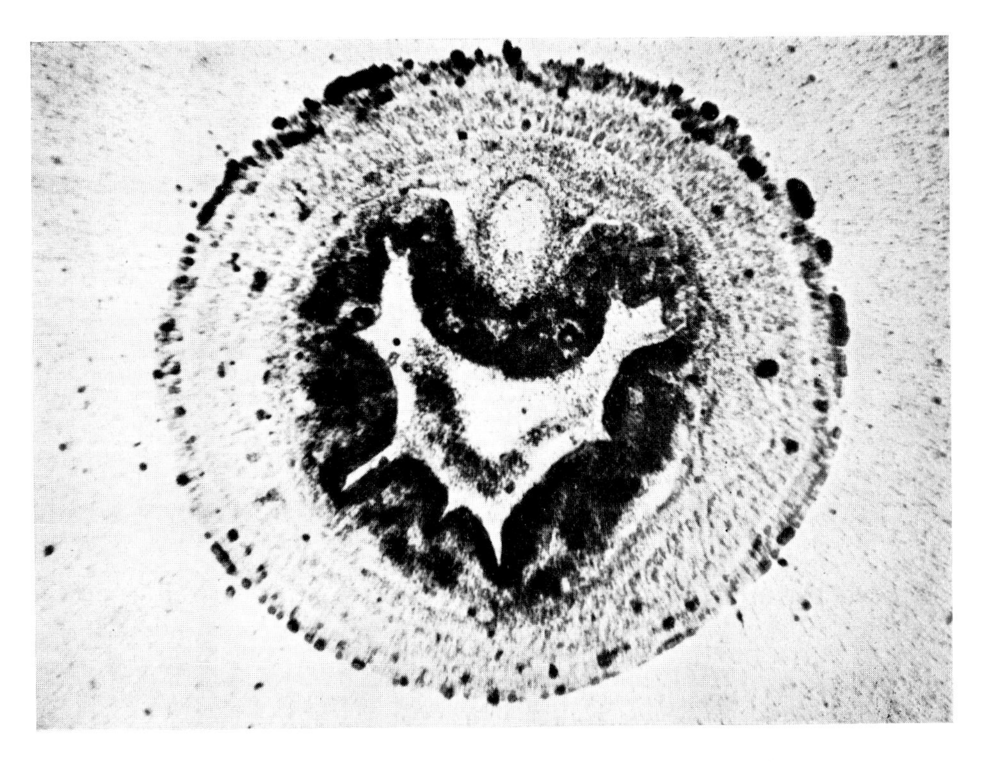

FIG. 14.12. Autoradiogram of the pharyngeal region of *Lineus ruber* after exposure to ^{125}I in sea-water for 48 hours. Note the radioactivity in the pharyngeal epithelium and related cells. Note also the longitudinal grooves in the pharyngeal floor. (Balfour and Willmer, 1967.)

and the amount of thyronine formed larger, when the sea-water is more concentrated by additional NaCl than when it is diluted with distilled water, the amount of available iodine being the same in both cases. This suggests that the epithelium collects iodine in relation to the general salt balance of the medium, and that iodine and chlorine are not necessarily alternatives to be picked up according to their availability. In the endostyle of the ammocoete, on the other hand, the iodine-binding cells (the ciliated, type 3, cells) are also highly radioactive after the creature has been immersed in a medium containing radioactive chloride (Morris, 1967). As comparable experiments have not yet been done on *Lineus*, it is not yet clear whether there are some cells which are general halogen-binders and others which are iodine-specific. Further studies may reveal more certainly the part played by the different cell groups. It could be, for example, that the ciliated cells are concerned with halogen uptake while the eosinophil cells are more concerned with the ejection of these ions, and that neither of these processes necessitates the retention of iodine within the cells. In the ammocoete the iodine-binding is done, in the first instance at least, by the ciliated cells (Gorbman and Creaser, 1942). Treatment of the ammocoete with thiouracil causes cellular changes in the endostyle (Barrington and Sage, 1963) which are very like those noticed in the pharynx of *Lineus* when the latter is subjected to diluted sea-water: the gland cells become depleted of their secretion. In view of the observation that sticklebacks prefer sea-water to fresh water after they have been treated with thiourea (Baggerman, 1959), it would be interesting to know if the gland cells of the ammocoete can be restored by treatment with increased salinity.

Some comparable observations have been made in an entirely different field which may yet throw light on problems connected with thyroid development and function. Fresh-water algae and ordinary land plants only pick up small quantities of iodine, but it has been known for many years that marine algae accumulate large quantities of iodine, and it has more recently been shown (Fowden, 1959) that halophytic plants do the same. Furthermore, in ordinary plants the iodine remains as iodine or iodotyrosine, but in halophytes appreciable quantities of thyronines are formed. Thus the influence of the NaCl concentration (or of the sea-water) on these plants seems to be directly comparable with its influence on the pharynx of *Lineus*. It would be interesting to know which cells in the plants are responsible for these actions.

These observations, coupled with the observation that the enteropneust, *Saccoglossus horsti*, picks up iodine and stores some of it as mono-iodotyrosine (Barrington and Thorpe, 1963) could be interpreted to mean that a primitive capacity for iodine-binding has been localized in the pharyngeal epithelium in nemerteoids, and has been turned to advantage in the enteropneusts. As suggested above, the enteropneusts may have arisen from nemerteoids by making use of the permanently ejected proboscis for capturing small creatures. The capacity for iodine-binding has also been turned to advantage in the urochordates, where Barrington and Thorpe (1965) have found di-iodotyrosine and small amounts of thyroxin, and in the cephalochordates and vertebrates all of which, it has been suggested, arose by converting the introverted proboscis into a notochord. The thyroid gland, with its production of thyroxin and tri-iodothyronine, would thus emerge as a specialization of a cellular function initiated in much simpler form in some very remote ancestors. Again, experiments on living nemertines to determine how, why, when, and where iodine is picked up in relation to salt concentration, temperature, light, and other conditions would seem likely to lead to a better understanding of what the thyroid gland is really doing and how it came to acquire its present functions. The connexion with salt balance is in itself interesting, since thyroxin is known to alter the rate of Na^+ transport in toad membranes (Green and Matty, 1963), and there are often great changes in the activity of the gland in fishes which migrate to more or less saline conditions for spawning (Fontaine, 1954). When treated with thyroxin, the stickleback (*Gasterosteus aculeatus*) prefers fresh water to salt water; when treated with thiourea, it goes for the salt water (Baggerman, 1959). Fresh-water fishes in general, unlike sea-water nemertines, are reported to bind iodine more avidly than do salt-water fishes (Hoar, 1959), though there are exceptions. Some of the experimental observations of the amount of I_2 picked up may be misleading, as they have not always been related to the total amount of iodine available or to the stores of it which may be held in the fish. Hickman (1958) has shown that with unlimited I_2 the flounder picks up more in sea-water, exactly as does *Lineus*. The effects of temperature on the iodine-binding and on its further metabolism in various species of nemertines would be particularly interesting. In fish, the data are confusing. Finally, the discovery in vertebrates of cells in the thyroid (Pearse, 1966) which are probably concerned with the liberation of thyrocalcitonin into the blood, thus causing an increase in blood calcium, again emphasizes the desirability of studying the nemertine pharynx under different conditions of calcium balance. The relationship between calcium-containing water and the incidence of human goitre is of course well known.

The specialization of the pharyngeal region, first towards iodine-binding and subsequently towards "thyroid" activity together with its variations, connected with osmoregulation, salt balance, spawning, and migration as seen in fishes, raises one more point. It has been suggested (p. 338) that the dorsal surface of the mouth region in nemerteoids is the tissue providing the probable origin of Rathke's pouch and, thus, of the anterior lobe of the pituitary. Histologically, there are close similarities between the epithelia of the dorsal and ventral surfaces of the anterior pharyngeal regions of the nemerteoids, though they differ in detail. Both regions also pick up iodine. It is not unreasonable to suppose, therefore, that activity in one part of the pharynx could well lead to some corresponding modification of function in another. This could well be the point at which the thyroid-pituitary relationship was first established. The ventral region is thus envisaged as specializing in iodine-collecting, through iodine-binding to thyroxin production, while the dorsal region specialized in other directions but retained sensitivity to some product of this special iodine metabolism and responded by moderating its production of something which stimulates thyroid activity, i.e. of thyroid-stimulating hormone. Perhaps this could have come about somewhat as follows. The pharyngeal epithelium is composed of several types of cells and acts as a primary barrier between the outside world and the tissue fluids of the body. On the assumptions made earlier, therefore, it can be presumed that these cells co-operate with each other in this general task, and that there are probably some cells whose activities are polarized in one direction and others that are polarized in the opposite way. In the formation of a thyroid, certain of these cells have been sequestered from the others, thereby to some extent producing an imbalance, which is then kept in control by the activity of another part of the same epithelium that responds to the environment (mostly the internal, presumably) created by the activity of the first group. Thus the development of a thyroid from the pharyngeal epithelium may have led to a specialization of another region of the same epithelium, i.e. that of Rathke's pouch to act as a compensator, somewhat on the principle of "setting a thief to catch a thief". Rathke's pouch may have been selected on account of its close juxtaposition both to the important neural tissues whose function it becomes specially concerned in stabilizing as it develops into the anterior lobe of the pituitary and also, in its early stages, to the blood lacunae that bathe the cerebral ganglia.

This concept can only be considered suggestive because, in the course of the evolution of the complex thyroid function from the activity of a simple iodine-binding ciliated epithelium, many changes have obviously occurred. To mention a few: First, in place of an elaborate ciliated epithelium lining the whole pharynx, small closed vesicles are formed from a more limited number of different cell types. The cells lining these vesicles mainly have microvilli, though some cells do have cilia. Second, it is probable that in nemerteoids the iodine is picked up directly from the external medium. In the thyroid it is picked up from the iodine circulating in the blood. This may involve a change in the orientation of the cells. Third, in the thyroid the iodine becomes bound as tri-iodothyronine (T_3) or thyroxin (T_4), and these substances can then be incorporated into thyroglobulin and stored extracellularly in the vesicles. In this connexion it is interesting to note that a "matrix" sometimes appears between the epithelial cells of the nemertine pharynx. Fourth, this store of thyroglobulin can be drawn on, and thyroxin or T_3 liberated from the protein and moved back into the blood. The thyrotrophic hormone seems to favour the uptake of iodine from the blood, and also the projection of thyroxin back into the blood, certainly from the stored "colloid" and perhaps directly as well. In view of the complexity of these events, it is premature to suggest in detail how the thyrotrophic hormone came to assume its functions, but a better understanding of the events and activities in the nemerteoid pharyngeal and rhynchodaeal epithelia could well prove to be illuminating. Attention should, of course, be paid to the activities of the basophil, PAS-positive cells in the nemerteoid when the pharyngeal epithelium is being subjected to conditions that modify its activity and its iodine uptake since the thyrotrophic cells of the pituitary have these qualities. These cells increase in staining intensity in animals taken from concentrated sea-water.

Salivary Glands and Other Derivatives

The idea that the pharyngeal epithelium plays a dominant part in the regulation of the internal environment of the nemerteoid, in addition to its actions in a digestive capacity, is an important one. As soon as the gill arches are formed, they must obviously become key structures because they are constantly exposed to currents of external medium and thus not only do they act as traps for suspended food particles but also they are ideally situated for gaseous exchange and ionic regulation, particularly when they become well vascularized. As the creatures have evolved, traces of this initial regulatory activity would presumably remain and thus provide logical evolutionary explanations for several physiological phenomena that occur in higher animals and which seem, at first sight, to be the result of Nature's caprice.

For example, in considering the origin of salivary glands, the tubular invaginations of the buccal epithelium in some nemertines should be borne in mind, though there are also other possible candidates for the position of precursor of these organs (see p. 524). Thus there are long-necked gland cells which encircle the mouth of some nemertines, e.g. *Cerebratulus*, and obviously there are also other possibilities in the arrangement and specialization of existing cells. In any case, salivary glands become organs of any significance only in those animals which have left the aqueous environment, and they must then be closely integrated into the system for the conservation of water and the regulation of salts. Their evolution has occurred mostly within the vertebrates themselves.

The parathyroids and the thymus are also derivatives of gill-arch structures. So also are the carotid and aortic bodies. These are all structures that are concerned with homoiostasis or defence. They could be direct derivatives of the pharyngeal epithelium; but again there are other possibilities (see p. 541).

The fore-gut of the nemerteoid, which we have so far simply called the pharynx, characteristically opens rather suddenly into the intestine. However, in some existing nemertines (Hoplonemertines), the pharynx is clearly divisible into a buccal region, an oesophageal region, and a stomach region, a subdivision which obviously has much in common with that of the fore-gut of many of the vertebrates. Such variations and specializations are to be expected as the animals evolve and change their environments and feeding habits. However, they are interesting also because the pharyngeal epithelium is initially complex, containing what appear to be fairly direct modifications of the three main classes of cells that constitute the epidermis and the other surface epithelia of their ancestors, and so the epithelium as a whole is capable of being readily modified in several directions.

In the two preceding chapters it has been suggested that the nemerteoid stock is of such a nature that, by more or less direct transformations, it could give rise to several different groups of animals, in particular, the Pogonophora, the Enteropneusta, the Urochordata and the Cephalochordata, in addition to the Vertebrata. Molluscs and arthropods also have some features which could be reconciled with the possible origin of these groups from rhabdocoeloid or nemerteoid stocks. The nemerteoid stage was clearly a stage of evolution which offered great scope for specialization and development in several different directions. The acquisition of a filter-feeding mechanism and the consequences thereof have been shown to provide opportunities for selection to act in such a way as to lead to the sort of cellular modifications necessary to convert nemerteoids into organisms having at least some of the properties and behaviour characteristic of those various groups that have from time to time been associated with the origin of the vertebrates. Three other major problems of vertebrate phylogeny have not yet been discussed from this point of view, however. The vascular and the urinogenital systems of the vertebrates need to be related to this possible course of evolution from the nemerteoid stock, and an origin has to be found for the vertebrate eye, a structure which appears in an almost complete form in the lampreys without having any very obvious antecedents. These topics form the main subjects for the next three chapters.

References

Adam, H. (1963). The pituitary gland. *In* "The Biology of Myxine" (A. Brodal and R. Fänge, eds.) p. 459. Universitetsforlaget, Oslo.

Adelmann, H. B. (1926). The development of the premandibular head cavities and the relation of the anterior end of the notochord in the chick and robin. *J. Morphol.* **42**, 371.

Baggerman, B. (1959). The role of external factors and hormones in migration of sticklebacks and juvenile salmon. *In* "Comparative Endocrinology" (A. Gorbman, ed.), p. 24. Wiley, New York.

Balfour, W. E., and Willmer, E. N. (1967). Iodine accumulation in the nemertine *Lineus ruber*. *J. Exptl. Biol.* **46**, 551.

Barnes, B. G. (1961). Ciliated secretory cells in the pars distalis of the mouse hypophysis. *J. Ultrastruct. Res.* **5**, 453.

Barrington, E. J. W. (1942). Gastric digestion in the lower vertebrates. *Biol. Rev. Cambridge Phil. Soc.* **17**, 1.

Barrington, E. J. W. (1965). "The Biology of Hemichordata and Protochordata". Oliver & Boyd, Edinburgh and London.

Barrington, E. J. W., and Franchi, L. L. (1956). Some cytological characteristics of thyroidal function in the endostyle of the ammocoete larva. *Quart. J. Microscop. Sci.* **97**, 393.

Barrington, E. J. W., and Sage, M. (1963). On the responses of the glandular tracts and associated regions of the endostyle of the larval lamprey to goitrogens and thyroxin. *Gen. Comp. Endocrinol.* **3**, 153.

Barrington, E. J. W., and Thorpe, A. (1963). Comparative observations on iodine binding by *Saccoglossus horsti* (Brambell and Goodhart), and by the tunic of *Ciona intestinalis* (L). *Gen. Comp. Endocrinol.* **3**, 166.

so-called ampulla sys
organs, there are sev
muscle spindles and t
in the vertebrates de
auditory apparatus an
of what pre-existing
materials which coul
organs there are inde
"ripe for developmen

Frontal Organ

Among the first of
essentially a ciliated
In their neighbourhoc
whose cell bodies ma
towards the brain. N
clearly defined, but o
of some living nemer
is that there is no tru
The cell bodies, with t
limiting membrane".
of vertebrates, in wh
the olfactory bulb, th
about the function
occurrence both in tu
importance and hav
Kölliker's pit at the a
organ, and, so far as
vertebrates could dev
cilia) and glandular
in the olfactory orga
with the sensory epit
little appears to be k
Ciliated cells and su
almost universally p
least in dogs, peculi
surface (Fig. 15.2) (C
The position of th
incorporation into t
hypophysial sac of t

Barrington, E. J. W., and Thorpe, A. (1965). The identification of monoiodotyrosine, diiodotyrosine and thyroxin in extracts of the endostyle of the ascidian *Ciona intestinalis* (L). *Proc. Roy. Soc.* (*London*) **B163**, 136.

Brodal, A., and Fänge, R., eds. (1963). "The Biology of Myxine". Universitetsforlaget, Oslo.

Burns, J., and Copeland, D. E. (1950). Chloride excretion in the head region of *Fundulus heteroclitus*. *Biol. Bull.* **99**, 381.

Coe, W. R. (1927). The nervous system of pelagic nemerteans. *Biol. Bull.* **53**, 123.

Copeland, D. E. (1948). The cytological basis of chloride transfer in the gills of *Fundulus heteroclitus*. *J. Morphol.* **82**, 201.

Fontaine, M. (1954). Du déterminisme physiologique des migrations. *Biol. Rev. Cambridge Phil. Soc.* **29**, 390.

Fowden, L. (1959). Radioactive iodine incorporation into organic compounds of various angiosperms. *Physiol. Plantarum* **12**, 657.

Gorbman, A. (1959). Problems in the comparative morphology and physiology of the vertebrate thyroid gland. *In* "Comparative Endocrinology" (A. Gorbman, ed.), p. 266. Wiley, New York.

Gorbman, A., and Creaser, C. W. (1942). Accumulation of radioactive iodine by the endostyle of larval lampreys and the problem of the homology of the thyroid. *J. Exptl. Zool.* **89**, 391.

Green, K., and Matty, A. J. (1963). Action of thyroxine on active transport in isolated membranes of *Bufo bufo*. *Gen. Comp. Endocrinol.* **3**, 244.

Hickman, C. P. (1958). Quoted in Hoar (1959).

Hoar, W. S. (1959). Endocrine factors in the ecological adaptation of fishes. *In* "Comparative Endocrinology" (A. Gorbman, ed.), p. 1. Wiley, New York.

Jennings, J. B. (1960). Observations on the nutrition of the rhynchocoelan *Lineus ruber* (O. F. Müller). *Biol. Bull.* **119**, 189.

Jennings, J. B. (1962). A histochemical study of digestion and digestive enzymes in the rhynchocoelan *Lineus ruber* (O. F. Müller). *Biol. Bull.* **122**, 63.

Keys, A. B., and Willmer, E. N. (1932). "Chloride secreting cells" in the gills of fishes with special reference to the common eel. *J. Physiol.* (*London*) **76**, 368.

Leloup, J. (1952). Fixation selective du radioiode dans le chorde dorsale de la lamproie marine. *Compt. Rend.* **234**, 1315.

Leloup, J., and Berg, O. (1954). Sur le présence d'acides aminés iodés (monoiodotyrosine, di-iodotyrosine et thyroxine) dans l'endostyle de l'Ammocoete. *Compt. Rend.* **238**, 1069.

Ling, E. A. (1969). The structure and function of the cephalic organ of a nermertine, *Lineus ruber*. *Tissue and cell.* (In press.)

Liu, C. K. (1942). Osmotic regulation and "chloride secreting cells" in the paradise fish, *Macropodus opercularis*. *Sinensia* **13**, 15.

Mitchell, J. F., and Willmer, E. N. (1966). Unpublished observations.

Morris, R. (1967). Personal communication.

Munshi, J. S. D. (1964). Chloride cells in the gills of fresh-water teleosts. *Quart. J. Microscop. Sci.* **105**, 79.

Pearse, A. G. E. (1966). The cytochemistry of the thyroid C cells and their relationship to calcitonin. *Proc. Roy. Soc.* (*London*) **B164**, 478.

Vickers, T. (1961). A study of the so-called "chloride-secretory" cells of the gills of teleosts. *Quart. J. Microscop. Sci.* **102**, 507.

Ziegler, B. (1963). Licht- und elektronenmikroskopische Untersuchungen an Pars intermedia und Neurohypophyse der Ratte. *Z. Zellforsch. Microkscop. Anat.* **59**, 486.

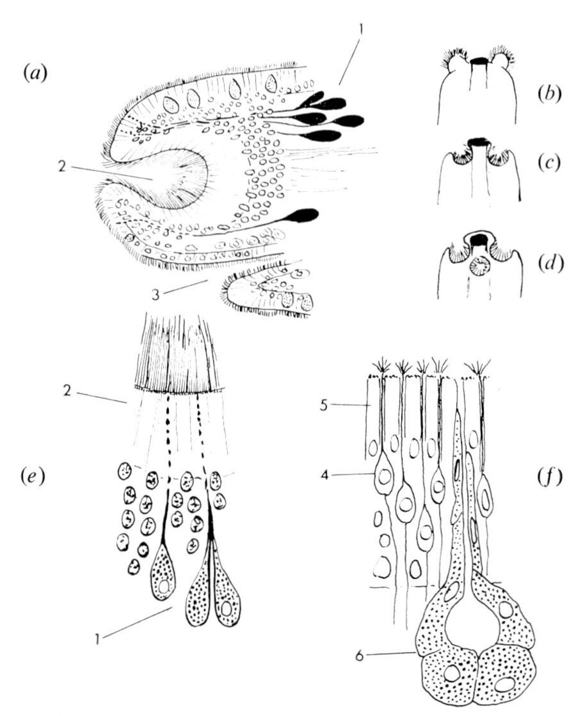

FIG. 15.1. Nemertine frontal organ and vertebrate olfactory organ compared. (*a*) Sagittal section through a nemertine frontal organ. 1, frontal gland cells; 2, flagellate "sensory" cells in pit; 3, rhynchodaeum. (*b*) Anterior end of nemertine with two frontal organs everted. (*c*) Anterior end of nemertine with two frontal organs inverted. (*d*) Anterior end of nemertine with three frontal organs. (*e*) Cell arrangement in frontal organ. (*f*) Cell arrangement in olfactory organ. 4, sensory cell; 5, supporting cell; 6, gland cells.

myxinoids. It will be remembered (see Fig. 14.1) that in some nemertines the proboscis is everted through a rhynchodaeum which opens in the neighbourhood of the frontal organ; in others the proboscis opens into the mouth cavity. In *Petromyzontia* the hypophysial sac may thus be the remnant of the nemerteoid rhynchodaeum. In the *Myxinoidea* the rhynchodaeum, incorporating the nasal sac, may have persisted and joined a secondary evagination of the roof of the mouth which may or may not correspond to Rathke's pouch. In the nemerteoids, there are several variations on the theme of combined or separate openings of the proboscideal and stomodaeal cavities and, in the vertebrates, similar variations on the theme of Rathke's pouch and hypophysial sac also occur. Therefore it is probably not profitable to discuss exactly how the nemerteoid structures may have been modified to give rise to those in the vertebrates until we are in a better position to identify the cell types concerned and determine their

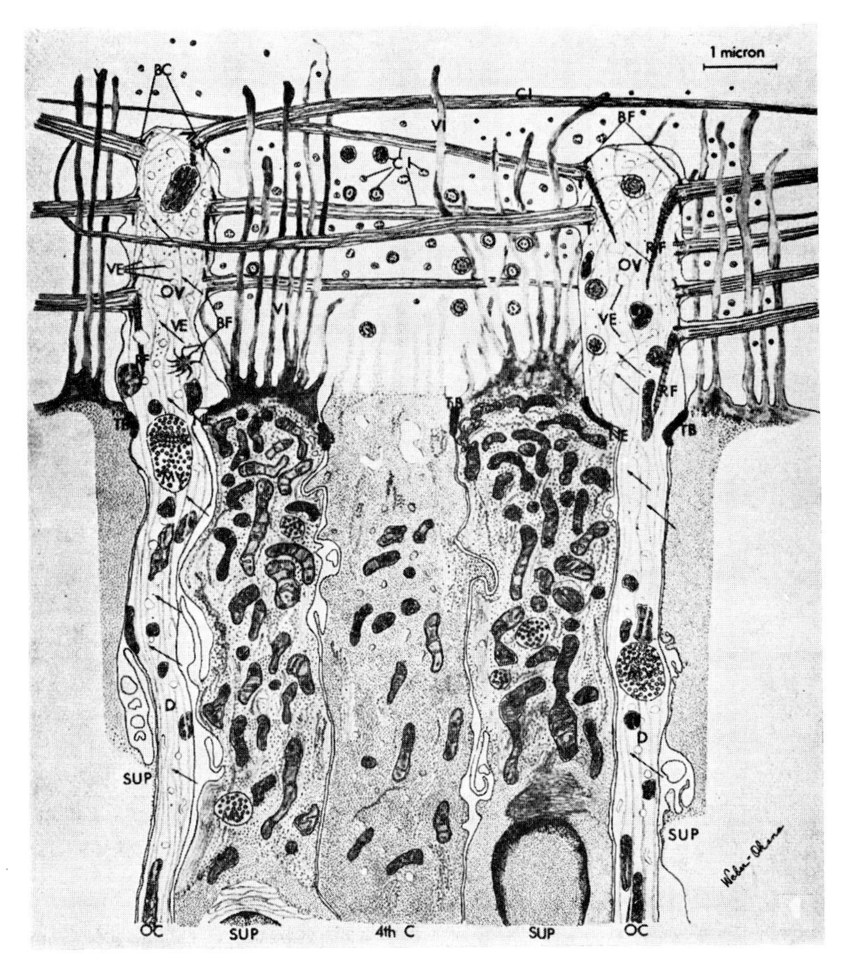

FIG. 15.2. Canine olfactory epithelium. BC, Basal granule; BF, basal fibres; CI, cilia; D, dendrite; OC, olfactory cell; OV, olfactory vesicle; RF, root fibres; SUP, supporting cell; TB, terminal bars; VE, vesicles; VI, microvilli; 4th C, fourth type of cell. (Okano *et al.*, 1967.)

functions. Meanwhile, the general variability of these structures both in the nemerteoids and in the lower vertebrates is an indication of the probable ease with which the required modifications could be brought about.

Ciliated Grooves

We now turn to the ciliated grooves at the sides of the anterior region of nemerteoids (Fig. 15.3). They probably serve some sensory function, though

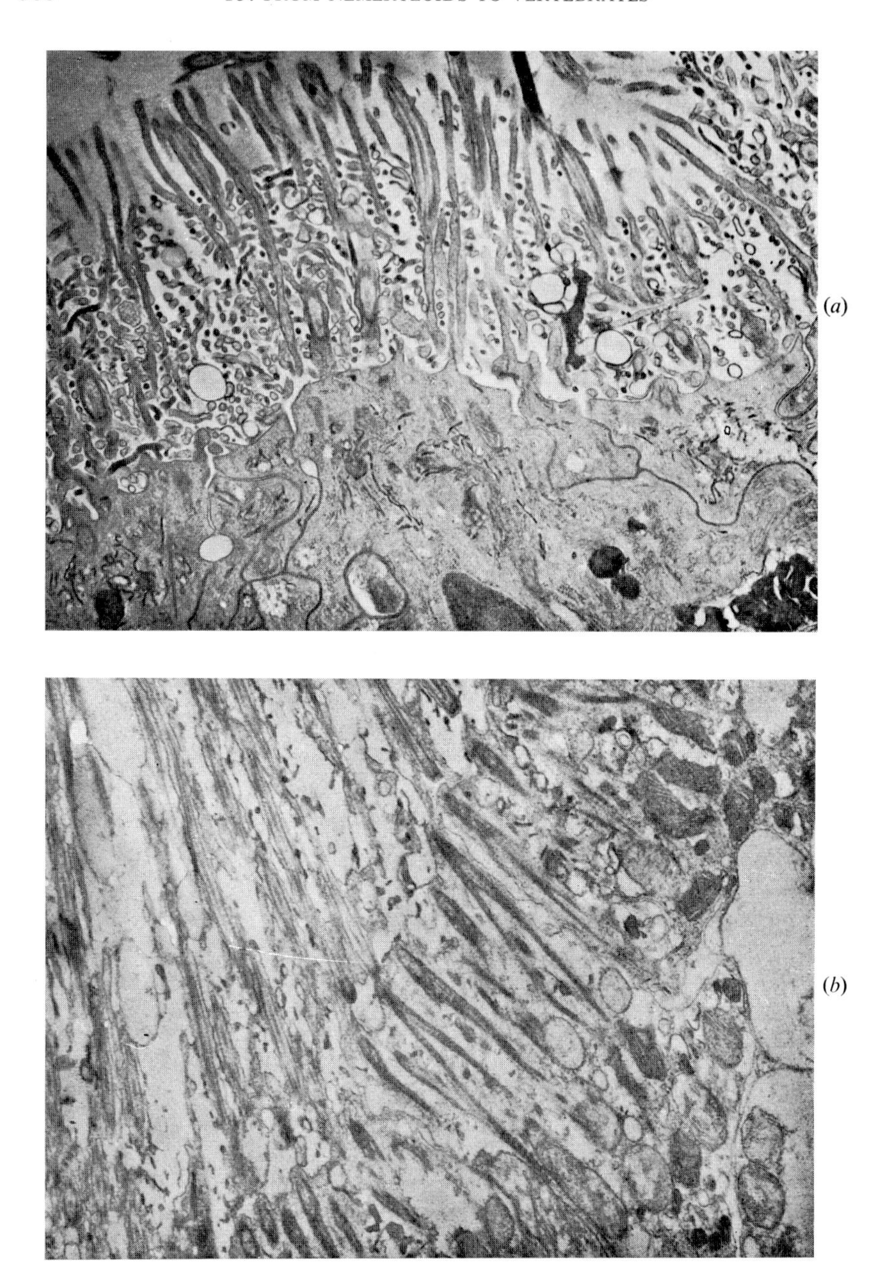

FIG. 15.3. (*a*) Electron-micrograph of epithelium at the entrance to the cephalic pit. Note cilia, stereocilia, and microvilli. (*b*) Electron-micrograph of epithelium in cephalic pit. Note long cilia or flagella, and well developed and cross-banded root fibres. (Photographs by Skaer.)

exactly what that function is remains obscure. The cells often have long and well developed cilia or flagella, long microvilli or stereocilia, and the main cilia sometimes have very conspicuous root fibres. Mucous and glandular cells are rare or absent. Cells similar to those of the cephalic slits are also found in the ciliated grooves of the enteropneusts (Brambell and Cole, 1939; Knight-Jones, 1952) (see Fig. 13.5). The grooves and slits can open up or close over according to circumstances (Wilson, 1900), and currents in their vicinity produce reactions on the part of the worm. In *Cuneonemertes* there are structures which are histologically similar to the lateral line and associated organs of cyclostomes and fish (Herrick, 1901; Young, 1950) or the taste buds of vertebrates (Fig. 15.4) (Coe, 1927). Whether these

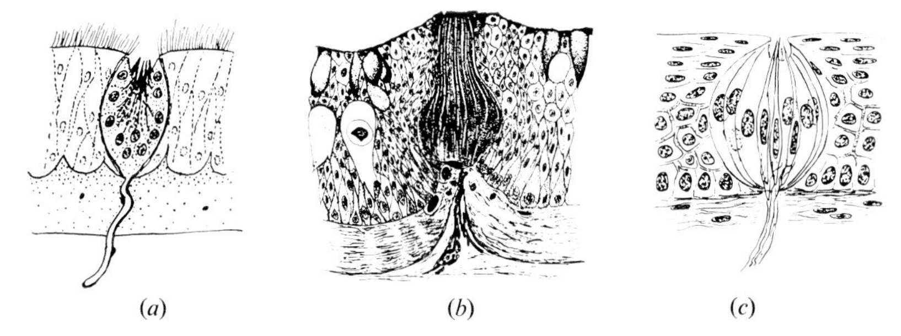

FIG. 15.4. (*a*) Sense organ in *Cuneonemertes* (Coe, 1927). (*b*) Terminal bud in head of siluroid fish (Herrick, 1901). (*c*) Taste-bud in mammal.

organs detect currents or salinity differences and the like, i.e. taste, is at present unknown. Nevertheless, one cannot help being impressed by the possible correspondence between the cephalic slits and grooves, etc., of nemerteoids and the similarly placed lateral-line system of the lower vertebrates. It would be interesting to know how the cilia are arranged on the cells in these organs. Gustatory and rheotactic cells in vertebrates differ in this respect, the latter generally having a single kinocilium on one side of a group of stereocilia (Fig. 15.5) (Löwenstein and Wersäll, 1959; Flock and Wersäll, 1962). In the groups usually classed as among the most primitive fishes, i.e. the Cephalaspida and Pteraspida, though probably not the Anaspida, lateral lines seem to have been well developed; therefore, these structures are probably a specialization of something already existing in the prevertebrate ancestor. Nevertheless, it is worth pointing out that these apparently primitive forms of fish may not actually be so primitive as they are usually thought to be. If creatures like the nemerteoids represented the most highly organized creatures before the vertebrates emerged, or before the molluscs or the arthropods had developed very far, then it is difficult

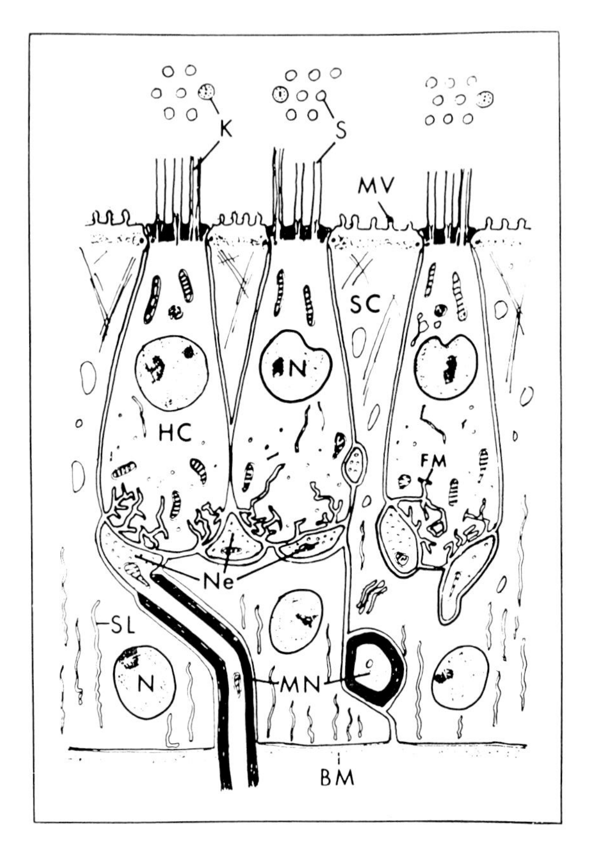

FIG. 15.5. Structure of lateral-line organ (schematic). BM, Basement membrane; FM, folding membrane system; HC, hair cell; K, kinocilia; MN, myelinated nerve fibre; MV, microvilli; N, nucleus; Ne, nerve-ending; S, stereocilia; SC, supporting cell; SL, supporting lamella. (Flock and Wersäll, 1962.)

to understand why such a heavy armour of bony plates should become sufficiently necessary or beneficial to have survival value at such an early stage. From what predators, one wonders, was such heavy protection required? It seems likely that, although these creatures embody many primitive features, they may really represent remnants of rather highly specialized groups of animals, quite a long way along the road of development of the vertebrates and in need of protection from their more aggressive and successful contemporaries. The relative ease with which their bony plates were preserved in the fossil record may give a false impression of their true position.

In the hypothetical intermediate forms between nemerteoids and the early vertebrates there would seem to be plenty of opportunity for the development and specialization of olfactory, gustatory, rheotactic, electrotactic, or galvanotropic systems from the rudiments and materials present in the nemerteoids. It is certainly striking that the corresponding sense organs in the vertebrates structurally resemble those in the nemerteoids much more closely than they do those of the arthropods for example, although the exo-skeleton of the latter manifestly imposes special requirements with regard to their sense organs.

In considering the ciliated grooves as possible precursors of the lateral-line system, it is relevant to point out that there are, in one or two species of nemertines, e.g. *Ototyphlonemertes* and *Procarinina* (Fig. 15.6), statocyst

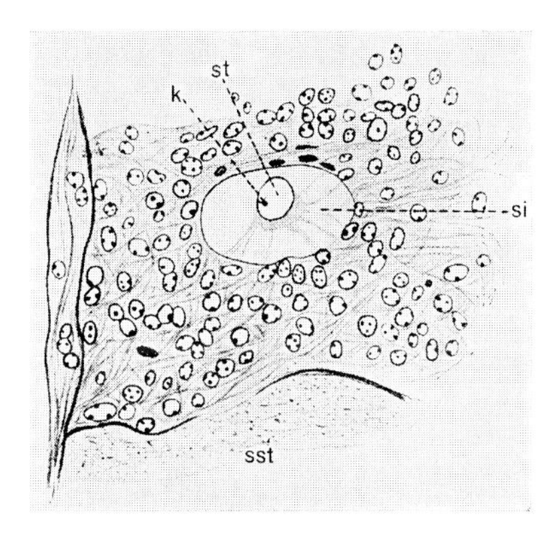

Fig. 15.6. The "statocyst organ" near the lateral nerve of *Procarinina*. si, Statocyst organ; st, crystalline body; k, nucleus; sst, lateral nerve trunk. (Nawitzki, 1931.)

organs (Bürger, 1895). However, apart from these bodies of rather doubtful significance, there are no other structures which indicate a nemerteoid origin for the auditory or vestibular system of the vertebrates with the possible exception of the cephalic organ (see p. 370). In the creeping nemerteoids this absence of a balancing organ is perhaps not surprising, though some creeping platyhelminthes, e.g. *Monocelis*, certainly have them. Tactile impressions from contact with the ground, photoreceptors, rheotactic receptors, and so on can presumably keep the animal reasonably informed about its position in space. This would not be so if the creatures were to become free-swimming but still dependent on localized food supplies, e.g. on the sea floor or river bed. Whether the statocysts of *Ototyphlonemertes* and other nemertines can be regarded as forerunners of the otocyst of higher vertebrates is very suggestive, but dubious. Statocysts and otocysts in various forms occur sporadically in many invertebrates, e.g. in acoels and particularly in arthropods, but their homology with each other or with vertebrate structures is improbable. Cyclostomes have, by contrast, well developed otocysts, with maculae and ampullae, and they originate as invaginations of the surface. They have been considered to be derived from the lateral-line system, and thus may have originated in the ciliated grooves of nemerteoids. These grooves could receive the necessary stimulus for further development if the animals possessing them adopted a free-swimming mode of life. Since the semicircular canals make use of rheodetectors, it might be reasonable to suppose that these developed first superficially in connexion with the acquisition of swimming, and that some of them were subsequently enclosed within the labyrinth.

The Cephalic Organ

It has been suggested (Jensen, 1960, 1963) that the cephalic organ with its invaginating tubular system, which is sometimes very complex and coiled, could be the tissue from which the otocyst and its derivates, the cochlea and vestibular organ of vertebrates, are derived. Furthermore, it is an interesting observation (Ling, 1969) that in *Lineus* there are three right-angled bends in the tubular part of the cephalic organ; thus, an orientating function of some kind is suggested. Nevertheless, an alternative hypothesis for the possible fate of that organ will now be put forward, which leaves the vestibular organ and ear to emerge from some further developments of the ciliated grooves, first into a more extensive lateral-line system and subsequently into a labyrinth.

The cephalic organ of living nemertines, which is, of course, a paired structure, may be anything from a simple ciliated pit, as it is in the rhabdocoels, to the elaborate organ that it is in *Lineus*, *Cerebratulus*, *Drepanophorus*, and some others (Fig. 15.7; Plate 1, Fig. 1). Significantly, it is not present at all in the pelagic nemertines, and it is most elaborately developed in the littoral forms. The simple pit-like organ, which has special ganglion cells at its base, and which is found in the rhabdocoels (Fig. 11.5), responds to changes in the salt concentration in its neighbourhood (see p. 249), and this seems likely to have been its primary function. However, it is difficult to understand why a pit or invaginated tube lined with very long cilia and microvilli should be required for this purpose, when taste-buds of much simpler character perform that function in other animals. Thus it is tempting to think that some other functions must be subserved by the more elaborate organs of the littoral nemertines.

In *Lineus* the ciliated groove first opens up into a wide ampulla (Plate 2, Fig. 2) lined with exceptionally long flagella with highly developed and cross-banded root fibres. Between the flagella there are numerous very

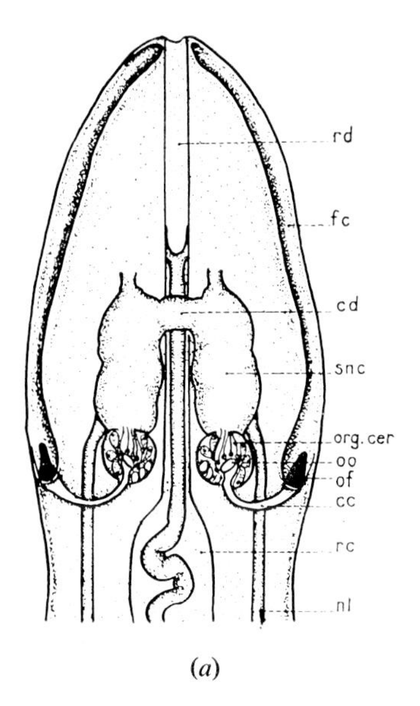

(*a*)

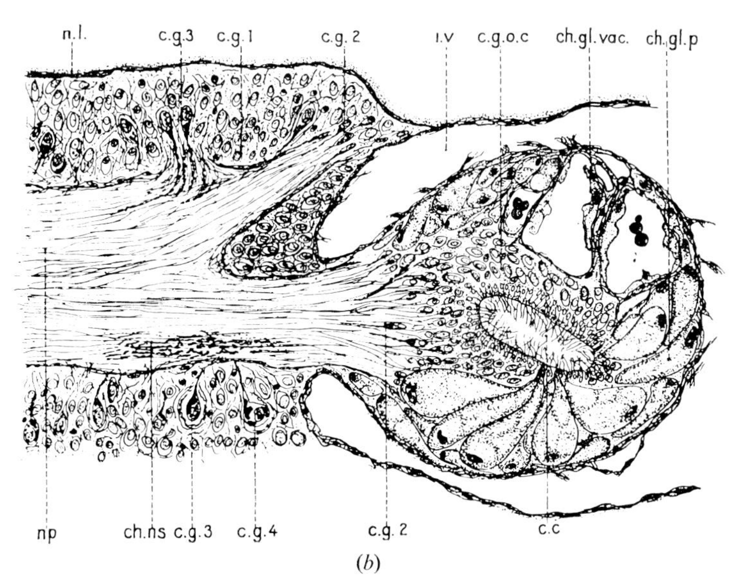

(*b*)

FIG. 15.7. (*a*) Anterior end of *Lineus* showing position and general form of the cephalic organ. cc, Cephalic canal; cd, dorsal commissure; fc, ciliated groove; nl, lateral nerve; of, opening of canal; oo, canal in cephalic organ; org. cer, cephalic organ; rc, rhynchocoel; rd, rhynchodaeum; snc, cephalic ganglion. (*b*) Longitudinal section of cephalic organ. c.c, Cephalic canal; c.g., 1–4, types of ganglion cells; c.g.o.c., ganglion cells in cephalic organs; ch.gl.vac., vesicular tissue; ch.gl.p., secretory cells; i.v, blood vessel; np., neuropil; nl., cephalic ganglion; ch.ns., neuro-secretory fibres. (Lechenault, 1963.)

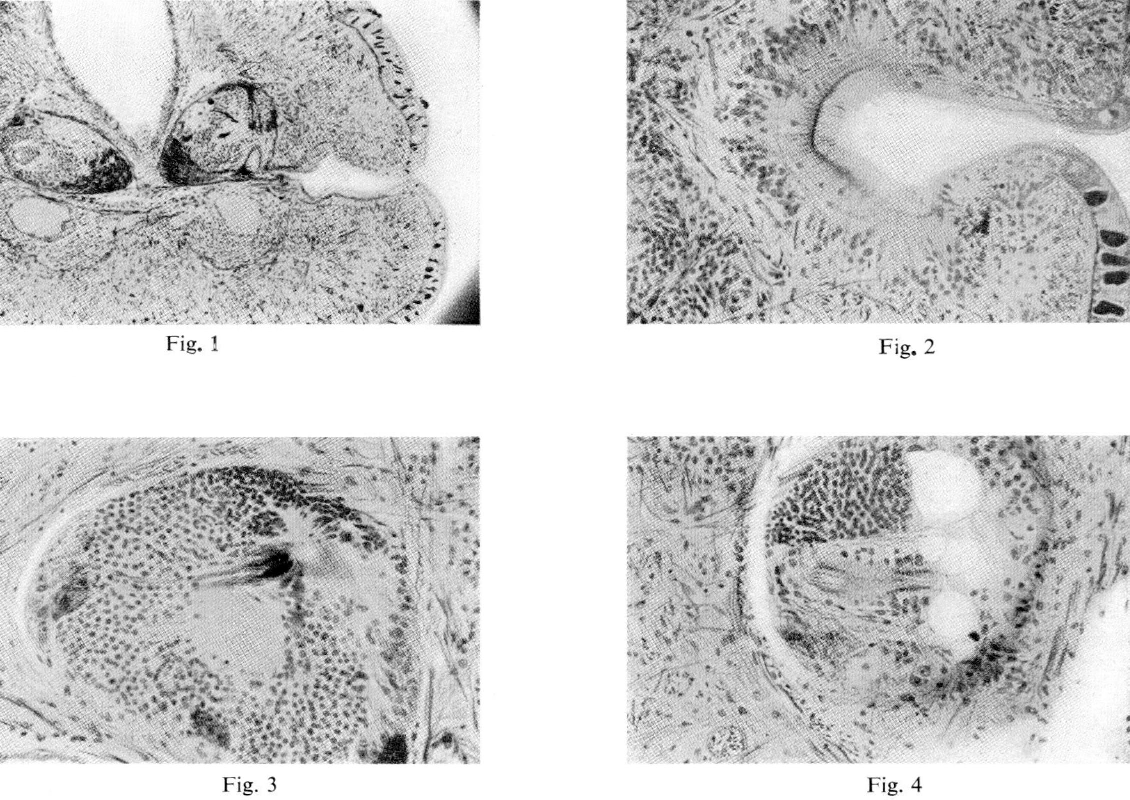

PLATE 1. Fig. 1. Transverse section through the cephalic organs of *Lineus ruber*. Note the inlet tube and ampulla, and the tube bending at right angles in the organ itself. The secretory cells are stained dark. Fig. 2. The ampulla showing the sensory epithelium and root fibres. Fig. 3. The cephalic organ showing secretion (black) entering the tube. The light area in the centre of the organ is the nerve connecting the organ to the brain. Fig. 4. Inner termination of the tube showing the vesicular area.

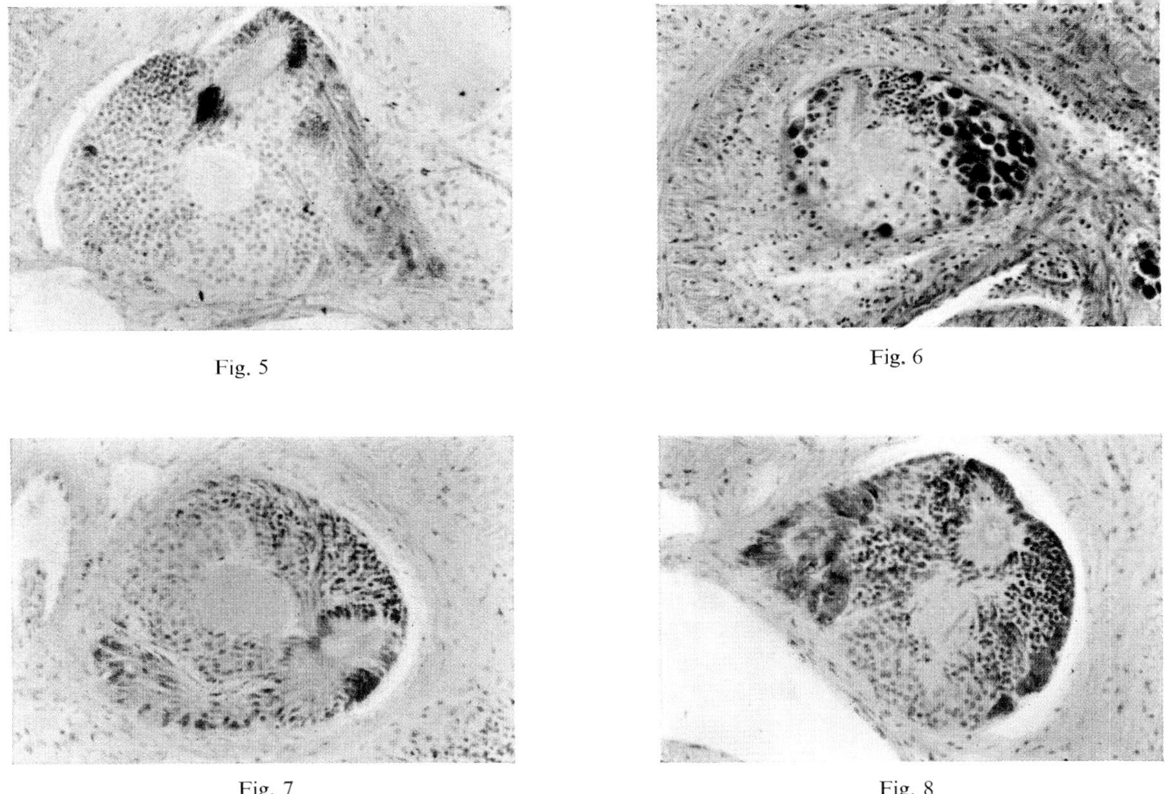

Fig. 5 Fig. 6

Fig. 7 Fig. 8

PLATE 2. Fig. 5. Cephalic organ from an animal kept in 50% sea-water for 24 hours. Secretory cells empty and secretion entering tube. Fig. 6. Cephalic organ from an animal kept in sea-water plus an extra 3% NaCl for 24 hours. Secretory cells damaged, but filled with secretion. Fig. 7. Cephalic organ from an animal kept in s a-water for 2 days in daylight. Secretory cells empty. Fig. 8. Cephalic organ from an animal kept in sea-water for 2 days in darkness. Secretory cells filled with secretion.

long microvilli, in the neighbourhood of which a positive PAS reaction is normally obtained. This appears to be a sensory surface, and the epithelium which, as in the anterior region of the body, lacks a typical basement membrane is backed by rows of nuclei some of which probably belong to nerve cells. The sensory cells have their flagella and much of their cytoplasm on the outside of an "external limiting membrane" somewhat in the manner of the receptors of the vertebrate eye, although in this case it is not yet clear how the "membrane" is formed or what, if anything, corresponds to the Müller's cells.

In fixed preparations the centre of the floor of this ampulla is raised into a conical hump, from the middle of which a narrow ciliated tube penetrates more deeply into the animal, bends thrice at right angles, and finally terminates in a diverticulum surrounded by a group of highly vacuolated cells of ill-defined relationship (Plate 1, Fig. 4). The cilia in this tube are arranged partly radially and partly (i.e. along one side) longitudinally. The two parts are separated by a palisade of very long and large flagella in which the cell membrane covering the flagellum is inflated away from the central column of fibres (Fig. 15.8). At the first two bends in this tube, two groups of elongated pear-shaped secretory cells can pour out their secretion (Plate 1, Fig. 3) into the lumen (Fig. 15.9). When secretion does not occur in this way, the cells become engorged with their own products. Curious globular cell inclusions, probably derived from the Golgi apparatus, may also be present. The secretion, or constituents of it, stain readily with the same staining reactions (PAS, paraldehyde fuchsin, Masson's trichrome, and, less satisfactorily, chrome-alum-haematoxylin) as those of neurosecretary substances elsewhere, e.g. as in the preoptic nucleus of the hypothalamus in vertebrates. The chrome-haematoxylin stains the secreted material more readily than the cell contents, and in the tube itself the staining reactions of the contents among the longitudinal cilia may sometimes be different (e.g. they are alcian-blue positive) from those in the main body of the tube among the radial flagella. Under some conditions, secretory material collects in the vesicles and spaces among the cells at the end of the tube; this material stains like the material among the longitudinal cilia and is alcian-blue positive.

Around this flagellated or ciliated tube are several groups of nuclei corresponding to different classes of cells. The cells which separate the group bearing longitudinally orientated flagella from those with radial flagella are structurally quite distinct and have very swollen flagella. The first group of nuclei around the tube belong to the cells forming the tube, but outside it there is a second group of cells which stain readily with silver stains. For example, Bodian, Cajal, and Nonidez methods all produce impregnated cells. These cells send processes both toward

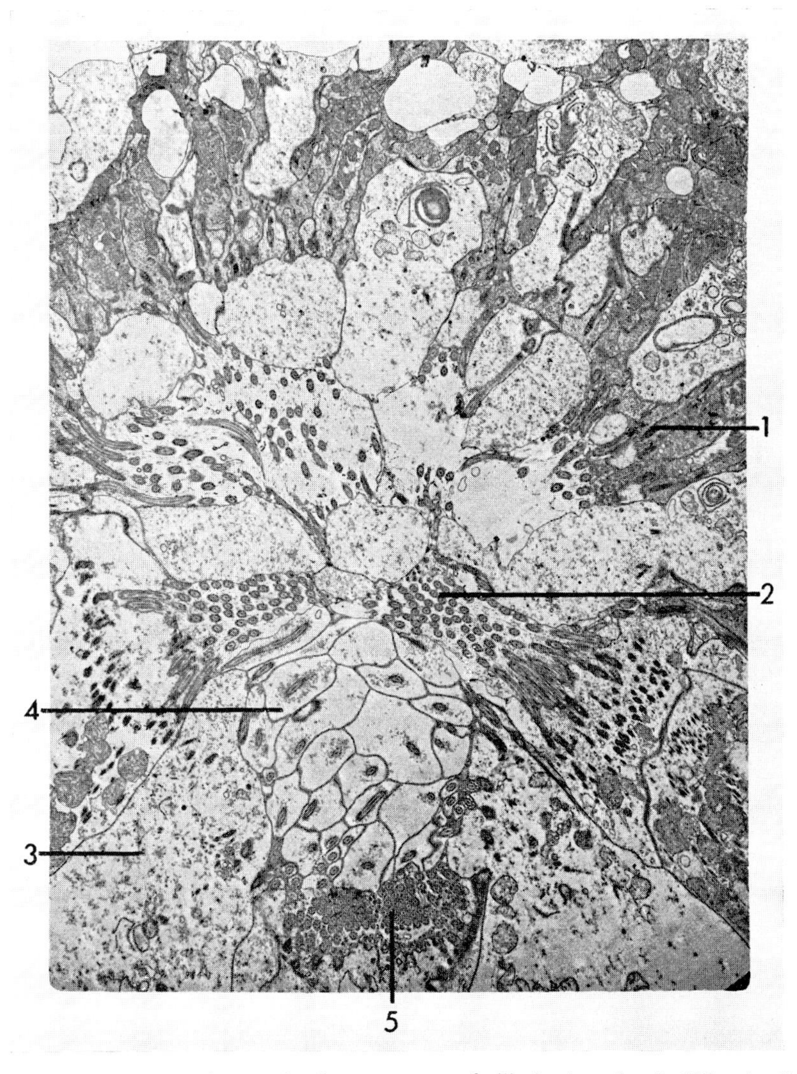

FIG. 15.8. Electron-micrograph of arrangement of cilia in the tube. 1, Ciliated cell of main tube; 2, cilia in main tube; 3, cell with "swollen" cilia; 4, "swollen" cilia; 5, cilia in subsidiary tube. (Photo by Ling.)

the flagellate cells, where they appear to end between the cell bodies in a process bearing a simple cilium, and to join the nerve fibres which run towards the main cerebral ganglia. Three sorts of nuclei are clearly distinguishable among these "bipolar" cells (Ling, 1969) (Fig. 15.10). Another group of nuclei belongs to "ganglion" cells that are similar to

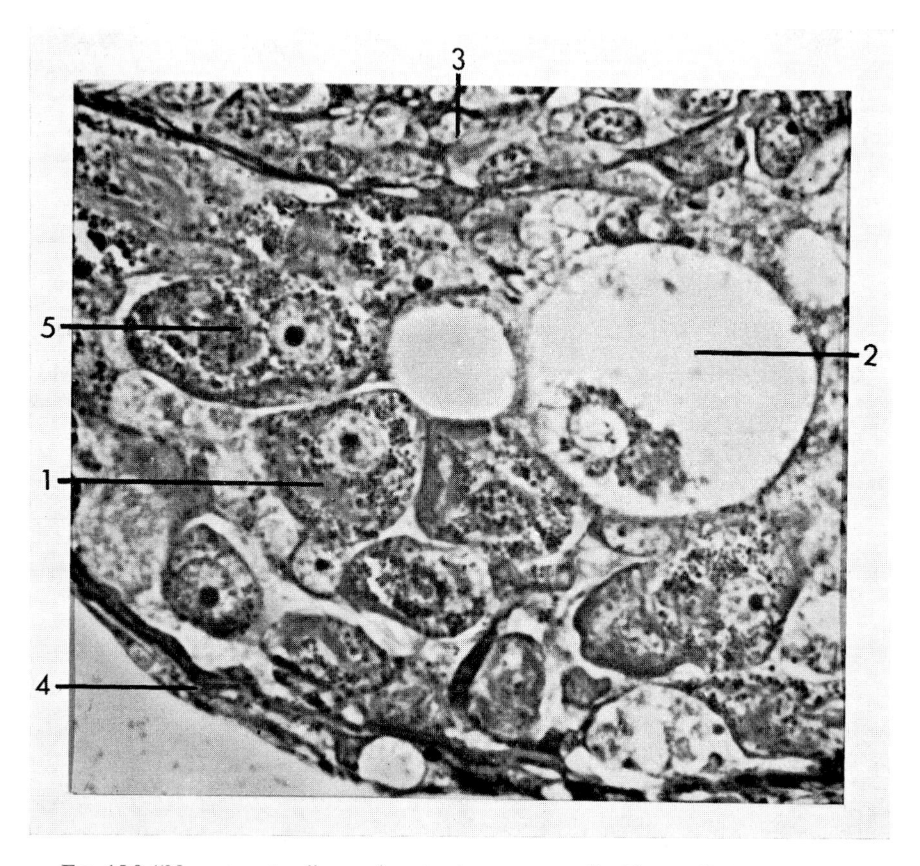

FIG. 15.9. "Neurosecretory" granules, etc., in secretory cells. These cells have nuclei with large nucleoli. 1, Neurosecretory cell; 2, vesicular cell; 3, bipolar cells; 4, capsule of organ; 5, secretory mass (Golgi apparatus?). (Cf. Figs. 15.14 and 15.16.) (Photo by Ling.)

those of the cerebral ganglia themselves. The connexions of these cells are at present obscure, but they certainly send fibres which pass into the main nerve going to the brain. These three main groups of cells (flagellated cells, "bipolar" cells, and ganglion cells) can be reasonably certainly established as separate entities, and the main nerve is surrounded by neuroglia cells whose nuclei are also clearly distinguishable. Better methods may yet reveal other groups of cells and allow a further understanding of the connexions

---

FIG. 15.10. (*a*) Electron-micrographs of nuclei of "bipolar" cells in the cephalic organ (photo by Ling); (*b*) electron-micrograph of nuclei of "bipolar" cells in the primate retina (Villegas, 1960). Compare a, b, and c in both parts of the figure.

See facing page→

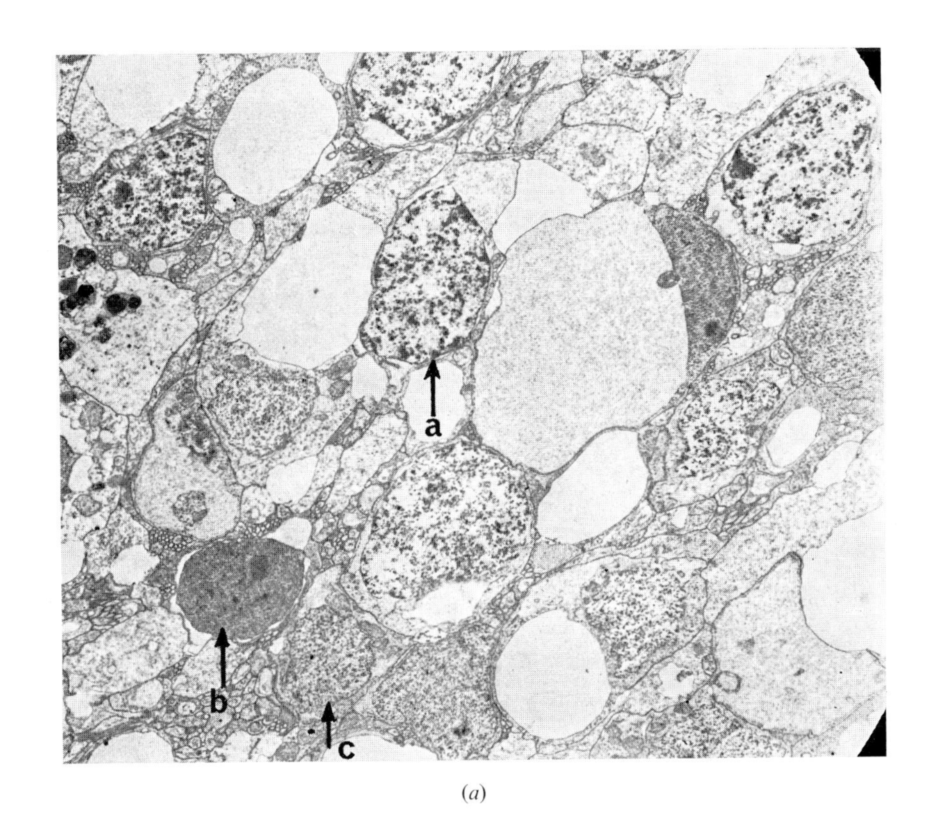

(*a*)

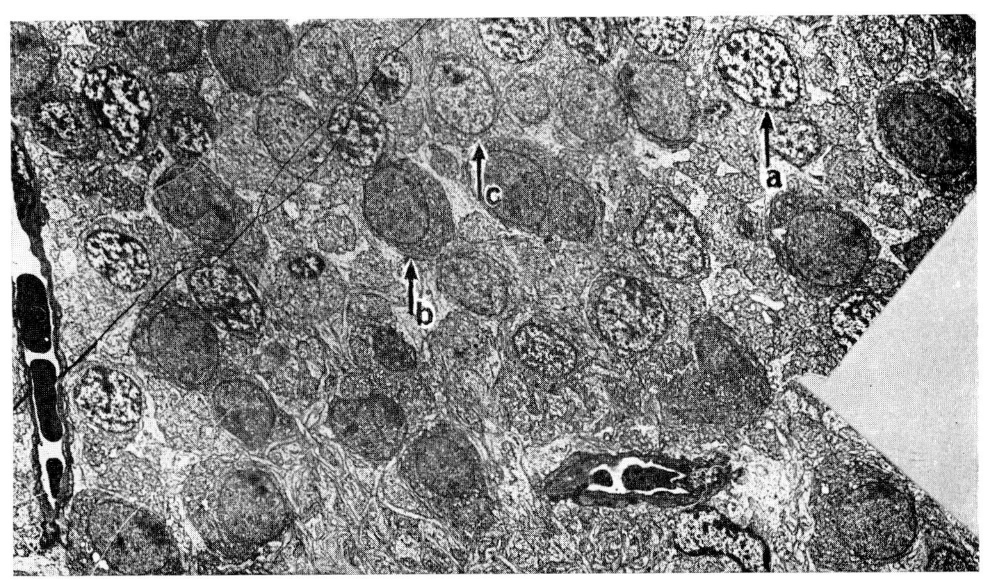

(*b*)

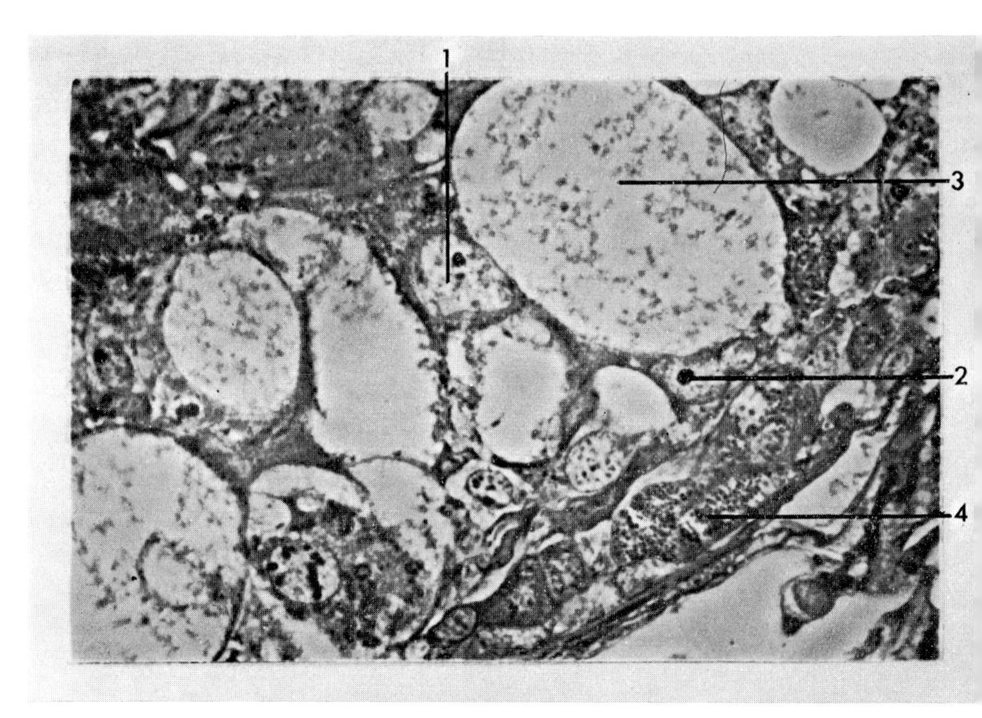

FIG. 15.11. Cells in the glandular and vesicular areas of the cephalic organ. 1, Nucleus of vesicular cell; 2, nucleolus in nucleus of vesicular cell; 3, intracellular vesicle; 4, secretory cell. (Photo by Ling.)

between those already detected. For example, in the vesicular area at the inner end of the tube there are sometimes two classes of nuclei. One nucleus is very large and has a prominent single nucleolus, the other is more like that of a fibroblast (Fig. 15.11). This whole area, however, has a structure which is anything but clear, and it is not known exactly how its cells are related to those of the tube itself, though the arrangement appears to be similar to that of an acinus at the head of a secretory duct.

It may reasonably be asked what is the significance of this elaborate structure and what is its relevance to the present discussion. In answer to these questions two things can be said immediately about the behaviour of the organ, but whether or not they are relevant to the analysis of its normal function is not yet known. First, when the worm is placed in sea-water diluted to 50% with distilled water, within an hour there is a great outpouring of secretion from the glandular cells into the lumen of the tube, and the staining of the secretory cells becomes much less intense. Conversely, when the worm is placed in concentrated sea-water, i.e. sea-water to which 50%

extra NaCl has been added, the reverse happens: the secretory cells become engorged with secretion and little appears in the tube (Plate 2, Figs. 5 and 6). The organ is clearly able to respond to changes in salinity, but whether it responds directly or indirectly, i.e. to the fluid in the ampulla and canal, or to neural influences, has yet to be determined.

Second, the secretory cells change their appearance in a similar manner when the animal is subjected to light or darkness. In daylight and on a white ground the secretory cells rapidly discharge their contents; in darkness, on the other hand, they become engorged (Plate 2, Figs. 7 and 8). Again, it is not known whether this action is direct or indirect or by what means the cells are stimulated to secrete. Prolonged exposure to light appears to be deleterious to the animal.

Eye-spots, Eyes, and Associated Tissues

Many nemertines, including *Lineus*, have pigmented eye-spots, similar to those of the turbellarians (see Fig. 11.9), distributed over the anterior end of the body, and they are often directed towards the lateral margins of the worm. These structures probably have something to do with the reception of radiation, but Gontcharoff (1953, 1956) has shown that *Lineus* still responds to light, in very much the same way as does a normal worm, after the eye-spots have been removed and before they have regenerated. In the "blinded" worm the best reactions to light were obtained when the light fell on the region of the cerebral ganglia. Ling (1969) has confirmed these findings, though the reaction may be somewhat slower in the "blinded" worms. It is therefore possible that the cephalic organ itself is directly concerned with these reactions to light, since it is closely associated with the ganglia. Moreover, there is good evidence for directional sensitivity even in the worms whose eye-spots have been removed, and it could be suggested that the orientation of the flagellated cells in the ampulla and tube might be responsible for this (see p. 374). The three bends in the tube may also be important in this way.

If the cephalic organ really responds to light, what is the meaning of the reactions to changed salt concentrations? There could be a connexion.

The vast majority of visual mechanisms in the animal kingdom depend on the pigment rhodopsin, its congener porphyropsin, or a similar compound of a semicarotenoid and a protein (opsin). The semicarotenoid is usually either vitamin A_1 (retinol$_1$) or vitamin A_2 or, more accurately, the corresponding aldehyde (retinal). For light-sensitivity the molecule has to be in the 11-*cis* form; light converts it into the all-*trans* form. Light may also liberate retinol.

Visual mechanisms also appear to depend on a multilayered lipo-protein complex formed by some repetitive form of the cell-surface, or of an organelle. In the vertebrates, flagella become modified to produce this multilayered system (Sjöstrand, 1949; Tokuyasu and Yamada, 1959; de Robertis, 1960). In planaria (see Fig. 11.9) (Röhlich and Török, 1961), in many arthropods, and in the cephalopod molluscs (Wolken, 1958a), the surface modifications of the sensory cells originate from microvilli or similar structures.

In view of the recent findings on the action of vitamin A (retinol) on the permeability of natural membranes, e.g. of erythrocytes and lysosomes (Dingle and Lucy 1965), and of the finding of retinal in the egg membranes of fish (Plack *et al.*, 1959) and of retinol in bovine olfactory membranes (Bernard *et al.*, 1961), it could be suggested that the flagellated cells of the ampulla and tube of the cephalic organ might also contain retinal-like substances. If this were so, it is conceivable that similar changes of cell permeability or ionic balance could be produced in them either directly by changing the ionic distribution or indirectly by the action of light or darkness on their flagella. In other words, light might act on the retinal-retinol-protein system to produce effects on the cells similar to those caused by a lack of sodium. Further investigation along these lines might be profitable.

In *Chlamydomonas reinhardi* and *C. moewusii*, gamete formation can be induced by adding distilled water to the vegetative phase living on agar slopes. Gamete formation involves the production of flagella, and the process requires, or is hastened by, the action of light. When ammonium nitrate is added to the distilled water, the organisms, which photosynthesize, tend to remain in the vegetative phase, but it is not known whether the ammonium is acting as a source of nitrogen or is acting ionically as it does on *Naegleria* to prevent flagellum formation. In any case, however, light, flagella, and ionic concentrations appear to be involved in gamete formation (Lewin, 1953; Sager and Granick, 1954). Thus the connexion here between flagella, light sensitivity, and salt balance may not be so remarkable as it appears at first sight.

Furthermore, since in the more primitive rhabdocoels the animals react directly, by movement, to changed salt concentration in the region of the cephalic pits, it is probable that these organs are not only directly sensitive themselves to this stimulus, but also that the nervous system is already organized to effect an appropriate response. Thus the change-over from an organ responding to NaCl, or to salinity changes generally, to an organ detecting changes in light-intensity, could in fact be effected rather easily if this change had some biological advantage. It is clear that the cephalic organ in nemertines is structurally very much integrated into the neural tissues, though, physiologically, little is known about the connexions.

Two particular sense organs have been especially puzzling to biologists for many years. They are, first, the eyes of vertebrates which appear to arise, more or less fully developed and without any obvous precursor, in the most primitive members of the phylum and, second, the eyes of some of the cephalopod molluscs which have many close similarities to those of the vertebrates and which again arise unheralded by any obvious earlier forms

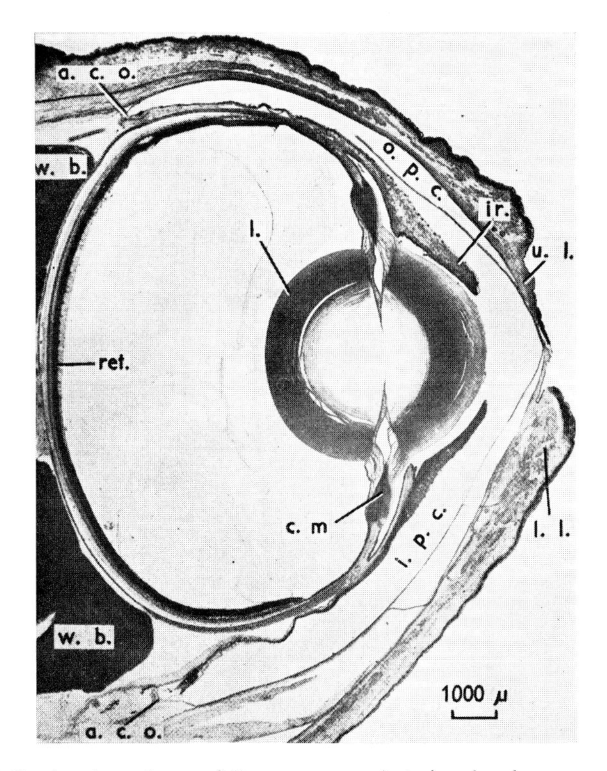

FIG. 15.12. Section through eye of *Octopus*. a.c.o., Anterior chamber organ; c.m, ciliary muscle; i.p.c., inner pseudo-corneal space; ir., iris; l., lens; l.l., lower lid; o.p.c., outer pseudocorneal space; ret., retina; u.l., upper lid; w.b., white body. (Boycott and Young, 1956.)

(Fig. 15.12). It is, of course, true that animals with poor eyes do not stand a chance against animals better endowed with vision; therefore, the evolution of eyes is likely to have been exceptionally rapid on account of the rigorous selection. It is also true that there are only a limited number of ways of constructing an eye that can form an actual or neural image of the outside world. It may not therefore be surprising that, occasionally, as in the cephalopods and man, in the interest of perfection Nature has achieved the same ends by different routes, and that convergent evolution can adequately account for the similarity between these two classes of eyes. This is no more unacceptable in this case than it is in explaining the spectacular examples of mimicry among insects wherein shapes, patterns, and behaviour have been approximated to an extraordinary degree by modifications of structure and patterns that were initially quite different.

Between the eyes of cephalopods and those of vertebrates there are, of course, great differences of detail. For example, the cephalopod has a mucoid lens, a direct retina with retinula cells pointing towards the light and possessing rhabdoms composed of microvilli. It has a separate retinal ganglion. The vertebrate eye, on the other hand, has a cellular lens, an inverted retina and receptors formed from modified flagella, and the retinal ganglion is built into the retina itself. These differences alone would seem to confirm the idea of convergence and suggest that there is no need to probe the matter further. However, there are other considerations, and several unanswered questions. First, from what precursors did the eyes of cephalopods and of vertebrates, respectively, start? Second, how did the lens, in each case, come to be developed from tissues quite separate from the retina? Third, in the amphibia, where these things have been investigated, how is it that, if the lens is removed, some (e.g. *Triturus*, though not *Amblystoma*) can regenerate a new lens from the anterior dorsal part of the iris (Stone, 1967), a tissue which, on the face of it, is ontogenetically unrelated to that from which the initial lens is derived (see Twitty, 1956)? It is unlikely that amphibia lose their lenses so frequently that a special mechanism has been developed for their replacement.

If the nemerteoids are considered to be possible precursors for the vertebrates, then the obvious precursors for the eyes are the eye-spots, some of which have direct retinae with sensory cells interdigitating with pigmented cells, while others have retinula cells which are inverted and enter the eye cup from in front. The latter would appear to offer serious possibilities of being the progenitors of the vertebrate eye, particularly if their rudiments became incorporated as vesicles into the foldings of the neural ectoderm. In some of these eye-spots the external surface has a lens-like epithelium formed of translucent columnar cells (Fig. 15.13). Thus the ingredients for developing eyes, with either direct or indirect retinae, appear to be present in these structures, and the matter could well rest there.

However, these eyes and eye-spots are not provided with much in the way of retinal ganglia, and, as we have seen, they can in some cases be removed without interfering grossly with directional light-sensitivity. Again, this argument, which at first appears to be conclusively against a visual function, may not be important because light-sensitivity can be very widespread in the tissues of lower animals; in *Amphioxus* and in the lampreys it may even extend to the skin of the tail region (Young, 1935, 1950).

If the receptor cells in the eye-spots with the inverted retinae are anything like their precursors in the eye-spots of planarians (Press, 1959; Röhlich and Török, 1961), they are really built on very different lines from the rods and cones of vertebrates. It must be remembered, however, that there are many variations among the eye-spots of nemertines, and some of them are more eye-like than others. Only a few have been investigated in any detail.

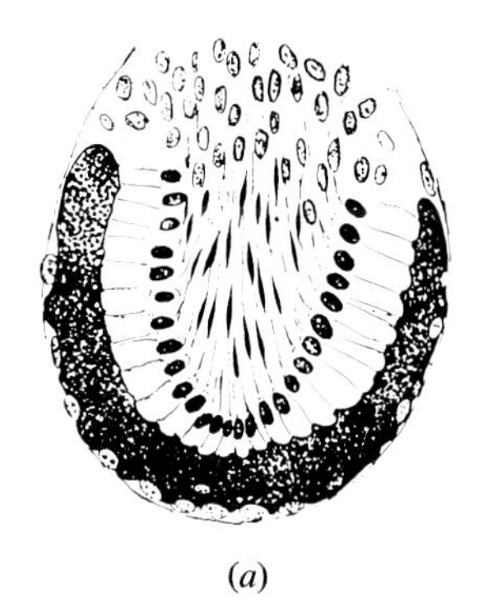

(*a*)

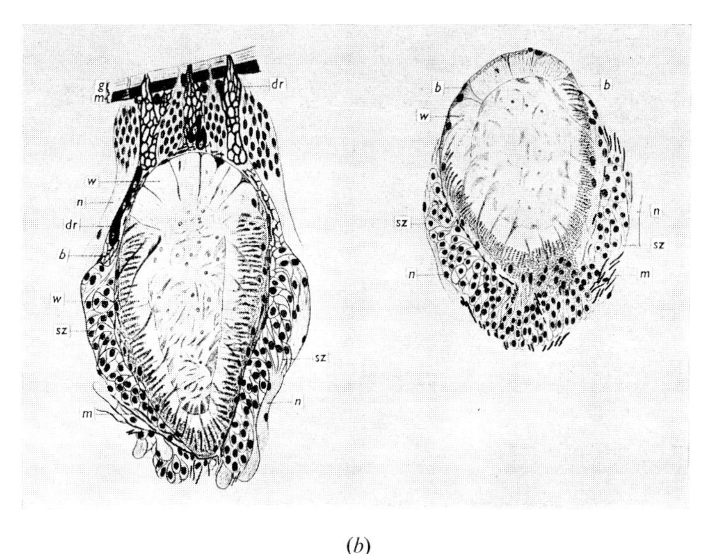

(*b*)

FIG. 15.13. Eyes of nemertines. (*a*) Inverted type in *Drepanophorus*. Note the pigment cup lined with sensory cells leading to nerve cells. (*b*) *Geonemertes*. b, Pigment cells; dr, gland cells; g, epidermis; m, muscle; n, nerve; sz, sense cells; w, clear "lens" cells anteriorly and sensory processes posteriorly. (Schröder, 1918.)

These points indicate that, in looking for the precursors of the vertebrate eye, it may be desirable to examine possible starters other than the apparently obvious ones. The cephalic organ, which is known to change the activity of its glandular constituents in response to light, may be profitably examined in more detail and further discussion of the future of the eye-spots deferred (see p. 391).

First, the existence of a large ampulla lined with closely packed cells already connected to the nervous system suggests a directionally sensitive structure. That the lining cells have both well developed flagella and stereocilia (or very large microvilli) is, of course, pertinent, since the rods and cones of vertebrates are developed from flagella and are sometimes (e.g. in amphibia) surrounded by long microvilli. Second, the tube leading from the ampulla and bending three times at right angles could also be directionally sensitive. Moreover, its cells are immediately connected to at least two groups of nerve cells in the organ before the issuing nerve enters the cerebral ganglia. In other words, there already is a potential retinal ganglion with three distinct groups of sensory and integrating neurones. One wonders if it is mere coincidence that the three types of nuclei present in the second group bear such a strong resemblance to the three types in the layer of bipolar cells in the monkey's retina (see Fig. 15.10). Third, it will be recalled that, ontogenetically, the vertebrate eye develops from a vesicle which at an early stage evaginates from the side of the neural tube. The cephalic organ of nemertines develops from a separate placode (see Fig. 13.33) which invaginates from the surface. During phylogeny, this vesicle or the inner part of it could well have become fused with the invaginating neural tube, if that were formed in the manner outlined on p. 328, and eventually could have become transferred to it. This transference could account for the hitherto unexplained origin and position of the optic vesicle in the lateral wall of the neural tube of the vertebrates and perhaps explain, as satisfactorily as invoking the eye-spots with already inverted retinae, how the retina came to be inverted in this phylum. Such a transference and attachment to another placode does in fact occur during the ontogenesis of some species of nemertines (Salensky, 1896). Fourth, since the cephalic organ is essentially a tubular structure derived by invagination of the surface, the inner part, where flagellate cells preponderate, could develop into the retina as suggested. The outer part, which would be the last to invaginate, could remain in its original position and give rise to the lens either from the reorganization of its cells, perhaps in an attempt to produce another cephalic organ placode (or ampulla), by production of mucoid secretion in the cavity, or by a combination of both. This concept of retina and lens both coming initially from the same invaginated tubular structure would offer a ready explanation for the anomalous regeneration of the amphibian lens from antero-dorsal retinal cells. Furthermore, it would not be difficult to imagine how an eye of the vertebrate type might arise by modification of the cephalic organ in nemerteoid creatures as soon as they became actively swimming and required more directional control of their movements. Such an eye could be produced by direct conversion of an already existing sensory mechanism, responding to salinity changes and with its own built-in neural connexions, into one responding to light in the manner outlined (p. 378). For animals invading estuaries, it is easy to imagine that light and low salinity might both require the same response from the animal, e.g. both might indicate excessive "exposure" and require immediate retreat.

One obvious point is left unexplained by this hypothesis, namely the presence of the pigmented epithelium with its contained melanin in the vertebrate eye. This difficulty, though disturbing, is not insuperable, since melanin is abundantly present in the antero-lateral regions of nemerteoids and is incorporated into the epithelia of eye-spots; hence its incorporation into the epithelium of the cephalic organ would not seem to be an impossible step. Bürger (1895) describes pigment as present in the cephalic organ of one existing species (*Drepanophorus*). The exact nature of this pigment, however, is not specified, though a melanin is suggested.

Eyes of Cephalopod Molluscs

To pursue these ideas in a somewhat different direction: if it were supposed that the cephalopods also arose from creatures related to the nemerteoid stock, then the cephalic organ might equally well be the precursor of the cephalopod eye, but by more or less direct development of the ampulla into the retina (with its direct sensory elements) and with the lens developing from mucoid secretions from the cells in the mouth of the tube. The cephalopod eye develops embryologically by direct invagination of the surface, and such development would be consistent with this idea. In the squids the photoreceptors, unlike those of vertebrates, are modified microvilli (Wolken, 1958b; Zonana, 1961), and cilia do not seem to be present. At first sight, this argues against their development from the cells of the ampulla because the latter have very well developed cilia or flagella. Nevertheless, some of the cells of the ampulla also have very well developed microvilli, while the deeper and more tubular part of the organ from which the vertebrate eye is more likely to have originated has mostly cilia or flagella and fewer microvilli. The different anatomical arrangement of the ganglia in the squids and of the retina in the vertebrates would also be consistent with this view if the nerve cells in the cephalic organ proper became the integrating cells in each case. The absence of any trace of cilia in the cephalopod

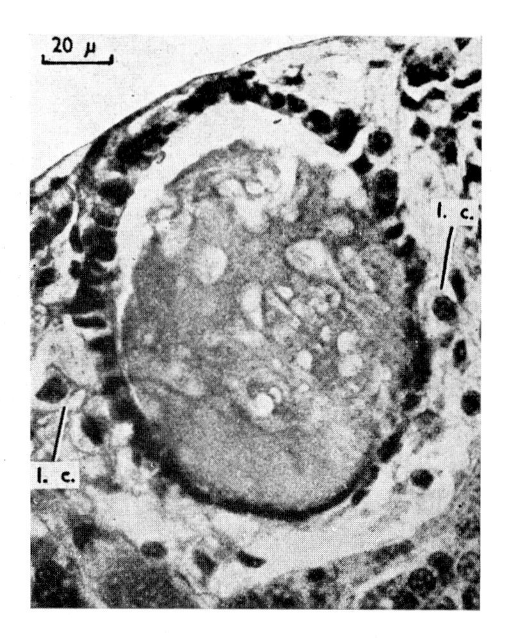

FIG. 15.15. Parolfactory (intercellular) vesicle of *Sepia*. Compare with the vesicular tissue of *Lineus*. (Figs. 15.7b, 15.9 and 15.11). l.c., Large cell with prominent nucleolus. (Boycott and Young, 1956.)

which is mediated via the hypothalamic region. It is perhaps an apt comment that *Lineus lacteus* spawns in captivity only when its head is chopped off (Gontcharoff and Lechenault, 1958)!

On the one hand, the well developed cephalic organs of the nemertines are integrated with the cephalic ganglia by means of a large nerve (this would essentially become the optic nerve). On the other hand, these organs are almost completely surrounded with blood lacunae, which is all that separates them from the roof of the buccal cavity whose epithelium is very complex and has already been cited (p. 343) as a possible precursor for the anterior pituitary tissue. During salt-deprivation the cells in the walls of this blood vessel sometimes become filled with stainable globules, while the secretory cells of the cephalic organ empty (Plate 2, Fig. 5). Whether there is any direct connexion between these two events is not yet known, however. Furthermore, it has already been mentioned (p. 348) that the blood vessels round the cephalic organ are in close relationship with the cavity of the rhynchocoel and with the peculiar secretory vessel that lies in its floor. These connexions were thought to have some relation to the development of posterior pituitary function. While the blood vessels surrounding the cephalic organ are under discussion, it is pertinent to note

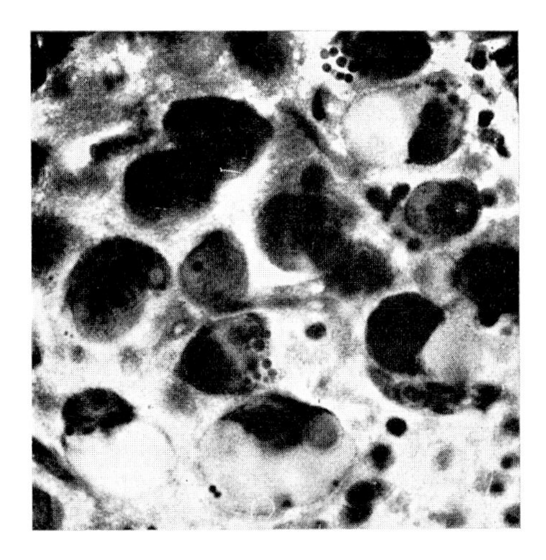

FIG. 15.16. Neurosecretory cells of *Fundulus heteroclitus* (cf. Figs. 15.9, 15.14). Note secretion bodies, granules, and nuclei with prominent nucleoli. (Scharrer, 1941.)

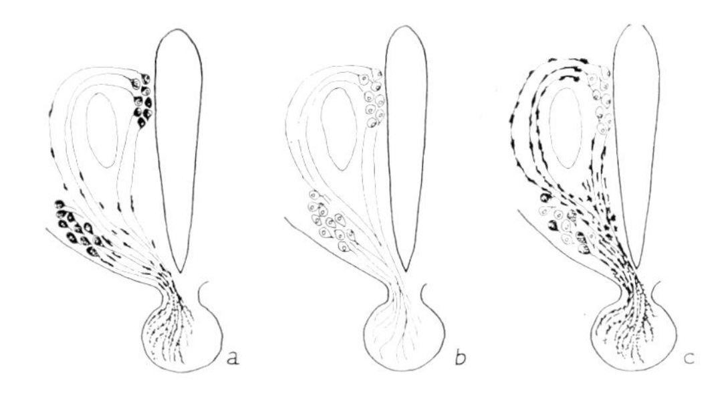

FIG. 15.17. Summary of the effects of changes in the water intake on the neurosecretory material in preoptic nuclei of the dog. (*a*) Normal intake; (*b*) water deprivation; (*c*) excess water. Note that excess water leads to the passage of secretion into the cell processes. Deprivation, on the other hand, leads paradoxically to absence of neurosecretory material. (Andersson and Jewell, 1957.)

that the white body behind the eye of the octopus has been thought to be connected with the production of blood cells (Noel and Jullien, 1933).

Thus the whole complex of the cephalic organ and its neighbouring tissues is remarkably suggestive of the sort of complex which must have been present in those organisms which gave rise to the primitive vertebrates of which the existing representatives (with the exception of the hag-fishes) all have well formed eyes, an hypothalamus, and a pituitary complex containing both anterior and posterior portions. Indeed, one may well put the question in another way: if not in the nemerteoid, then where else is it possible to find such a complex system "ripe for development" along the lines required by the primitive vertebrates?

Eyes and Pineal Eyes

At this point it is relevant to mention that the controversial Silurian fossil, *Jamoytius*, which shows what appears to be a notochord, scales apparently reflecting an underlying myotomal segmentation except in the head region, a dorso-ventrally flattened head with large lateral vesicles that could be interpreted as eyes, and no signs of any limbs, fins, etc. (Fig. 15.18) (White, 1946; Ritchie, 1960). The fossil Anaspid fishes were probably not much more advanced than these creatures, at least in some respects

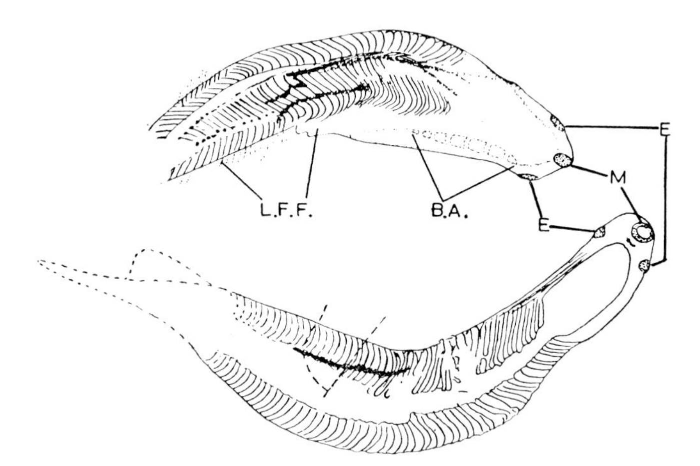

FIG. 15.18. Two specimens of *Jamoytius kerwoodi* as interpreted by Ritchie. E, Eye; L.F.F., lateral fin fold; M, mouth; B.A., branchial apertures. (Ritchie, 1960.)

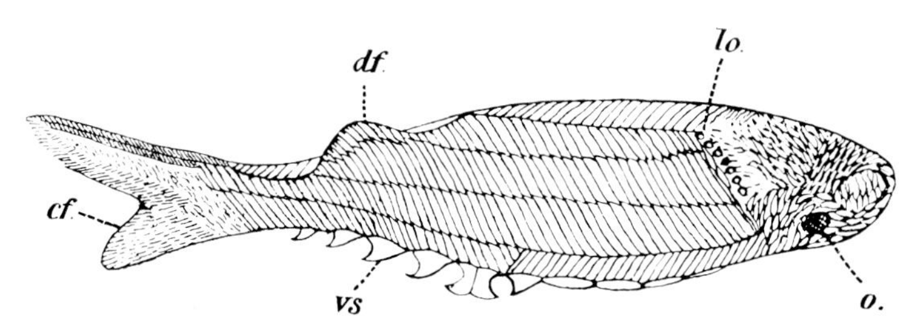

FIG. 15.19. *Birkenia elegans* (Anaspida). cf, caudal fin; df. dorsal fin; lo. lateral openings; o, orbit; vs. ventral scales. (Goodrich, 1909.)

(Fig. 15.19). Although this is only fragmentary information, it does fit in with the general scheme of how a nemerteoid could acquire a swimming habit in the manner suggested (p. 299) and then develop eyes to guide it. It does not, of course, say whence these eyes developed. Nor does the scheme above, which bases the eye on the cephalic organ, throw much light on the fate of the eye-spots of the nemerteoids, which, although they are not the only light-sensitive mechanism, are, by analogy with similar organs in the turbellaria, almost certainly sensitive to light. In planarians, the eye-spots have been shown (Merker, 1934) to be sensitive to ultraviolet light also. This is, of course, not surprising as the vertebrate eye, being dependent on the spectral sensitivity of rhodopsin, becomes sensitive to ultraviolet light when the lens is removed. Nevertheless, it may indicate an important function of these spots, namely that they should provide a warning signal when their owners are not protected by an adequate depth of water. In the nemertines these organs, in so far as they have central neural connexions at all, are connected to the brain by nerves to the dorsal ganglia or commissure, while the cephalic organ nerves tend to enter ventrally and posteriorly.

In vertebrates there are, in addition to the eyes whose nerves penetrate the brain ventrally, what appear to have once been paired structures connected to the brain on its dorsal surface. In some species, one of them has a definitely eye-like structure with a direct retina. These are the pineal eyes, pineal bodies, and epiphysis. They vary in the detail of their structure in the different groups of animals, and in higher forms they are represented by the pineal gland or epiphysis. The eye of the terrestrial nemertine, *Geonemertes palaensis* (Fig. 15.13b), which is special among its fellows, seems to have a structure (Schröder, 1918) which is directly comparable with that of the pineal eye of the lamprey, *Geotria australis*, as illustrated in Fig. 15.20 (Dendy, 1907). It is clearly a vesicular structure with a lens-like body of transparent cells and a direct retina, the vesicle being filled with a transparent secretion.

The last word has not yet been said on the function of the pineal organ, and many, various, and even conflicting statements appear in the literature on the topic. In the lampreys the retina of the pineal eye is a direct one with sensory cells interspersed between supporting cells; certainly it would not be difficult to envisage its derivation from such eye-spots as those seen in *Geonemertes*. When the pineal organ turns up again in the lizards, it is a more elaborate structure (Fig. 15.21) (Eakin and Westfall, 1959), and the sensory cells have an electron-microscopic structure similar to that of the rods (Steyn, 1959; Eakin and Westfall, 1960), or perhaps more exactly the cones (Eakin *et al.*, 1961), of the mammalian retina with the outer processes apparently formed from flagella in exactly the same manner as

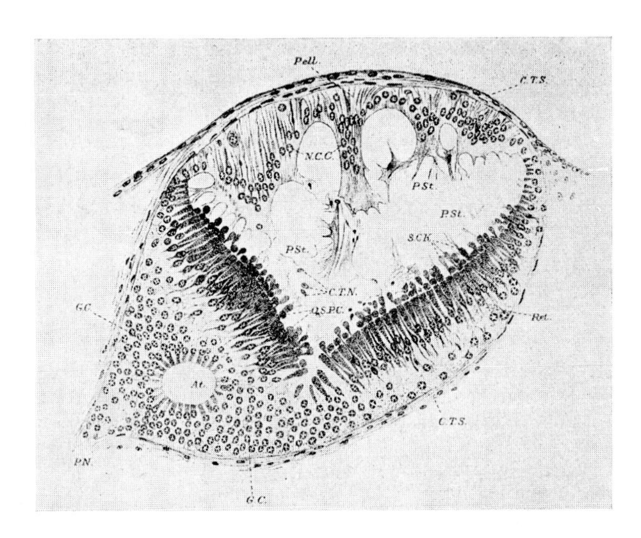

FIG. 15.20. Section through the pineal eye of the lamprey, *Geotria australis*. At., Atrium; C.T.N., nuclei of connective tissue cells; C.T.S., connective tissue sheath; G.C., ganglion cells; P.N., pineal nerve; P.St., protoplasmic strands; N.C.C., nuclei of columnar cells of pullecida; O.S.P.C., outer segments of pigment cells; Pell., pellucida; Ret., retina; s.c.k., sensory cell process. (c.f. Fig. 15.13.) (Dendy, 1907.)

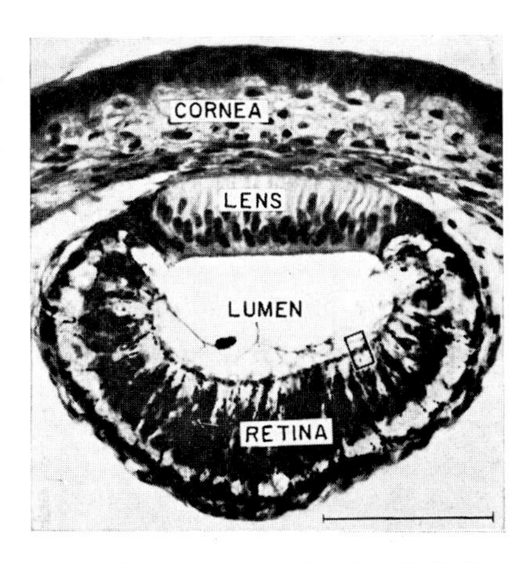

FIG. 15.21. Section through the pineal eye of a lizard. (Eakin and Westfall, 1959.)

FIG. 15.22. (*a*) The receptor and supporting cells in the pineal eye (lizard) (Eakin and Westfall, 1960); (*b*) receptors in the eye of the guinea-pig (Sjöstrand, 1961); (*c*) receptors in the eye of the mud-puppy (Brown *et al*., 1963.)

See facing page→

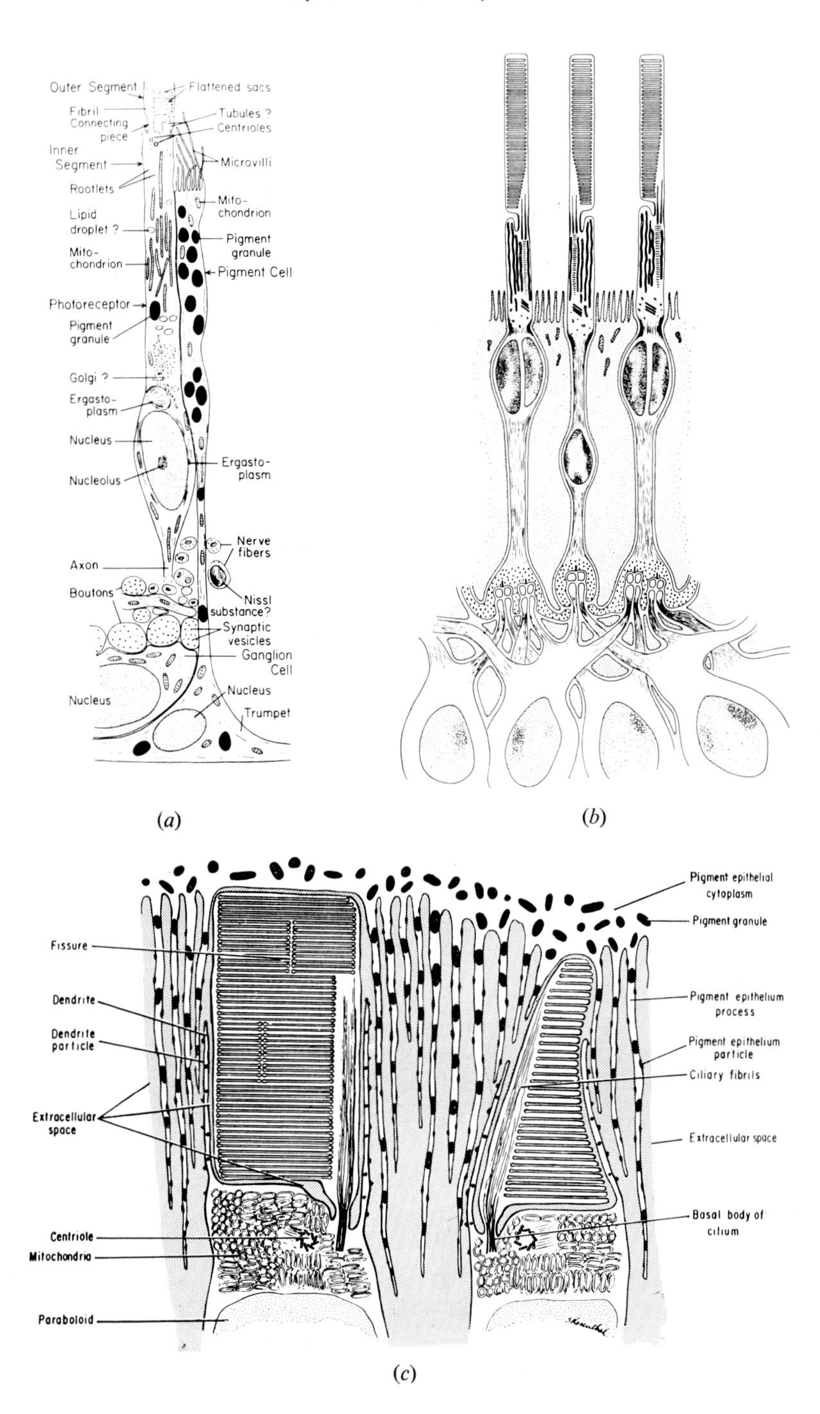
Outer Segment
Flattened sacs
Fibril
Connecting piece
Tubules ?
Centrioles
Inner Segment
Microvilli
Rootlets
Mito-chondrion
Lipid droplet ?
Pigment granule
Mito-chondrion
Pigment Cell
Photoreceptor
Pigment granule
Golgi ?
Ergasto-plasm
Nucleus
Ergasto-plasm
Nucleolus
Nerve fibers
Axon
Boutons
Nissl substance?
Synaptic vesicles
Ganglion Cell
Nucleus
Nucleus
Trumpet
(a)
(b)
Pigment epithelial cytoplasm
Pigment granule
Fissure
Dendrite
Dendrite particle
Pigment epithelium process
Pigment epithelium particle
Ciliary fibrils
Extracellular space
Extracellular space
Basal body of cilium
Centriole
Mitochondria
Paraboloid
(c)

those of the lateral eyes of the vertebrates (Fig. 15.22). Unlike the latter, however, they are interspersed with pigmented cells which have microvilli and may also have cilia. The similarity of the receptor cells to the rods and cones of the lateral eye is, of course, a point of great interest and confusion. It may mean nothing more than that a light-sensitive structure has been made by modifying a ciliated or flagellated cell in the most efficient way, or it may indicate a greater affinity between the direct retina of the pineal eye and the inverted retina of the lateral eye than has been suggested. If the latter is the case, much of what has been discussed above would have to be revalued. Another point of similarity between pineal receptors and rods and cones comes from the observations of Eakin (1964) that vitamin A is concerned with their structure. Animals made deficient in vitamin A suffer the same changes in their pineal cells as occur in rods and cones under the same conditions. On the other hand, it is well known that ciliated epithelia in general depend on an adequate supply of vitamin A for their full function, a fact which is also basic to the whole concept of the cephalic organ becoming converted into an eye.

If we may argue by analogy with similar structures in planaria, the cells in the eye-spots of those nemerteoids which have inverted "retinae" probably have a special modification of microvilli as their photosensitive (?) outer segment. On the other hand, the sensory cells of the eye of *Geonemertes* appear to have flagella or, at least, processes entering the cavity of the bulb. The cells of the anterior or outer wall of the eye of *Geonemertes* form a translucent lens-like body which compares in structure with the better organized lens in the pineal eye of the lamprey and, even more favorably, with that in the Anura and the lizards (Figs. 15.13, 15.21). In the last group these lens cells are well provided with microvilli.

The nerve connecting the pineal eye to the brain arises from a single layer of ganglion cells and consists of a number of fibres enclosed in neuroglial cell processes (Eakin and Westfall, 1960) in a manner characteristic of invertebrate nerves, though in some species there may be some myelinated fibres also (Oksche and Harnack, 1963). Curiously enough, the epiphysis of mammals, which is probably a derivative of the pineal system, has an efferent autonomic nerve supply from the superior cervical ganglion (Kelly, 1962; Kappers, 1965).

In considering the functions of these various organs, it is necessary to remember that the term "eye" is applied rather in the dark. As Gontcharoff (1956), Young (1935), and others have shown, animals, like plants, are not necessarily dependent on eyes in the ordinary sense for at least some of their reactions to light. Indeed, the structure of the "eye" in *Geonemertes* is so different from that of the "eye" in *Lineus* that the two organs may easily have quite different functions. In the ammocoete larvae of *Lampetra planeri*

(Young, 1935), there is a daily rhythm of contraction (in darkness) and expansion (in light) of the melanophores of the body (Fig. 15.23). This rhythm is abolished by removal of the pineal. In larval amphibia also, darkness induces pallor and light brings about expansion of the melanophores. In these animals, too, pinealectomy abolishes the effect, though other factors besides the pineal and epiphysis are involved (Bagnara, 1963).

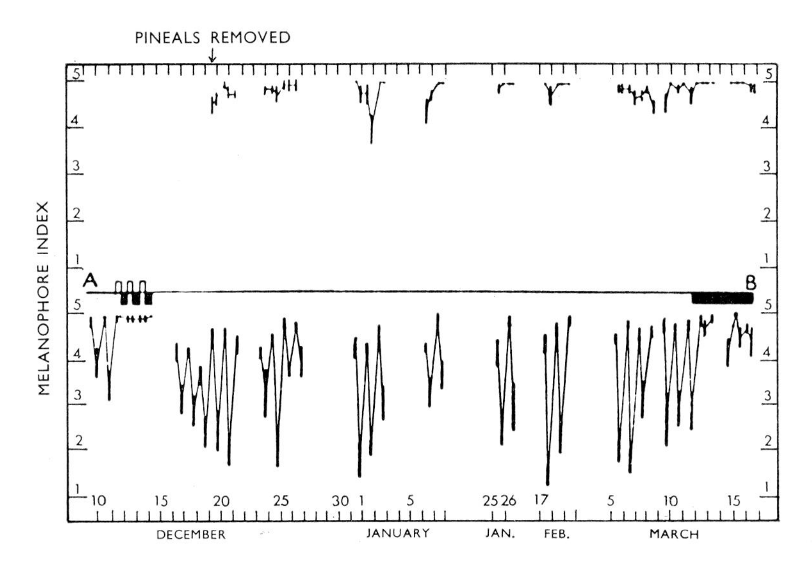

FIG. 15.23. Melanophore movement in lampreys. Expansion (upwards) and contraction (downwards) of melanophores of larval lampreys. Animals kept out-of-doors as shown along the line AB, where rectangles above the line show illumination with electric light and below the line (black), total darkness. Normal animals show a regular daily rhythm, becoming pale at night. Reversal of normal day and night stops the rhythm (10th to 14th day). Removal of the pineal from half the animals on Dec. 19 stopped the response (upper chart). The other half showed normal diurnal rhythm till March 12, when they were placed in continuous darkness. (Young, 1950.)

In the adult lamprey it may be necessary to remove the lateral eyes also before the rhythm is abolished. These results perhaps suggest that the pineal eyes were more primitively organized for this function, but that, with the greater development of the lateral eyes, the function has been gradually taken over by them and the pineal has gone into a secondary position or into relative disuse. In frogs the pineal eye (in which there are cone-like receptors; Van de Kamer, 1965) is sensitive to light of wave-lengths between about 300 mμ and 750 mμ, and it has been observed that violet and ultraviolet light seem to inhibit activity set up by longer wave-lengths

(Dodt and Heerd, 1962; Dodt, 1964). Three photosensitive systems connected with the pineal eye are suggested with λ max. at 350, 500–520, and 560 mμ, respectively. The animals darken in blue light and become lighter after exposure to red light. The pineal body of mammals, though it is probably not sensitive to light in the ordinary sense, is said to store serotonin and to contain high concentrations of melatonin (a hormone that contracts melanophores) and other indole compounds (McIsaac *et al.*, 1965) when the animal has been kept in the dark. The enzyme, hydroxyindole-O-methyl transferase, which is present in pineal tissue, is inhibited by light, and consequently less melatonin is formed during exposure to light (Wurtman *et al.*, 1963b). Extracts of pineal (McCord and Allen, 1917) and melatonin both cause the melanophores of amphibia to contract, and the latter delays the onset of breeding in the rat (Wurtman *et al.*, 1963a). It has also been claimed that the pineal gland contains substances which activate the zona glomerulosa of the adrenal gland (Farrell, 1959).

Sensitivity to light is, of course, a widespread phenomenon and, as already indicated, may be used for different purposes. It does not necessitate any very special sense organs. Absorption of light is theoretically all that is required, provided that the effects produced can be in some way utilized. In the lamprey the whole body is reactive and, indeed, the motor reactions to light are more quickly produced by illumination of the tail than of the eye (Young, 1935)! The facts that the rods and cones of vertebrate eyes are derived from flagellate cells, and that the ependyma of the central nervous system may be photosensitive, suggest that something in the structure of flagella lends itself particularly to this sensitivity. However, as we have seen, microvilli are similarly used in cephalopods and arthropods. Whether or not a response is made in a whole multicellular animal depends not only on the actual light-sensitivity but also on the existence of suitable conducting paths and effector organs. Moreover, eyes may be developed in order to give different sorts of information, e.g. an image of the outside world, a measure of the length of day, of the turbidity of water, of the movement of external objects, or the detection of the presence of such harmful radiations as certain wave-bands of ultraviolet light.

With these thoughts in mind, it is worth examining the state of affairs in nemertines more closely. Melanin pigmentation of the skin and surface layers is commonly present, and the worms can and do alter their colour very noticeably. A cup-shaped structure of melanin-containing cells, however, does not necessarily constitute an eye in the visual sense, and perhaps not even in the sense of an organ specifically sensitive to light, if by light we mean those radiations which are visible to man. Given suitably sensitive cells, it could detect thermal effects caused by the absorption of all the radiations that are absorbed by melanin. A worm in a shallow pool on the

shore could, by possessing such an organ, be advantageously informed of the presence of strong radiation of any kind, particularly thermal. Ultraviolet light might, in general, be less important to submerged creatures, since so much of it would be absorbed by the overlying water. On the other hand, a worm coming into shallow water or meeting terrestrial conditions, like *Geonemertes*, might gain great advantage by being able to detect and avoid ultraviolet light. Planarians do so (Merker, 1934).

The pineal eye, or some part of the epiphyseal complex, also appears to be able to fulfil similar functions. Light stops the production of melatonin, the hormone which contracts the melanophores (Wurtman *et al.*, 1963b), and ultraviolet light in particular inhibits the neural response to light of other wave-lengths (Dodt, 1964). Thus it seems to be quite possible that this organ acts primarily to protect the organism from ultraviolet light. When the light gets too strong, it activates a mechanism for pulling the blinds over the rest of the body in the form of expanding melanophores. In this process it may of course be assisted by the pituitary, which, we know, liberates a malanophore-expanding principle.

It would seem to be quite possible that the eye-spots, which, we are suggesting, have become converted into the pineal, may initially have been concerned with the detection of the total amount of light, largely for protective purposes. Whatever organs gave rise to the lateral eyes (possibly the cephalic organs) used their machinery to give information about direction and pattern of light; in other words, they developed vision in the true sense. In the lower vertebrates and in some lizards (e.g. *Sphenodon*), both systems may still be actively functional. More generally, as the lateral eyes became better developed, they also began to provide the same information as the pineal eye was providing. The latter therefore may have dropped out of use as a primary sense organ, but may have continued to be used as a director of rhythms dependent on light by becoming activated by effects initiated in the lateral eyes (see Charlton, 1966) and transmitted to it either hormonally or neurally, perhaps even via the nerve which in mammals runs from the superior cervical ganglion to the epiphysis (Wurtman *et al.*, 1964; Kappers, 1965).

The difference between the eye-spots of most nemertines and the differently formed eyes of *Geonemertes* and the pineal eye of vertebrates may thus be a change not only in the range of wave-lengths to be monitored but also in the use to be made of the information.

Incidentally, the production of melatonin, or of similar transmitters, in the dark and its cessation in the light could provide the makings of an excellent system for measuring the length of day, initiating diurnal rhythms, and thus of controlling other vital processes affected by night and day and by seasonal changes. Eakin and Westfall (1960) have reported that PAS-positive material accumulates in the parietal eye in the dark-adapted lizard; Wurtman *et al.* (1963b) note that the pineal gland stores serotonin; and both of these observations are consistent with the action of the organ as a clock. For example, if melatonin were made only in darkness, the necessary enzyme being inhibited by light, then a store of melatonin could accumulate during the night and would start to run down during the day. Quay (1963, 1966) has, in fact, shown the existence of diurnal rhythms of serotonin and melatonin in the pineal of rats and the pigeon, though in both of these creatures the direct response of the pineal to light is masked by other factors. In the winter, large stores could be built up and only run down slowly and for a short time each day. As the days lengthened and the light intensity increased, the stores would be less replenished at night and would run down more quickly and completely during the day. In such a way, seasonal breeding activities could be regulated, while, on a day-to-day basis, the animal could use the level of the stored product as a rough measure of the time since dawn, just as on a regular feeding regimen the human stomach is a fairly good guide to the passage of time. It is relevant that the pineal body of some birds is a vesiculated structure, perhaps indicating storage (the pigeon's pineal apparently contains far more 5-hydroxytryptamine than that of the rat; Quay, 1966), and that many birds appear to have a built-in clock somewhere in their economy which assists them in their navigation. It would be interesting to know whether birds with good powers of navigation are those which have well developed vesicles in the pineal body. The cormorant, the pigeon, and the duck certainly have well developed follicles; the fowl has a solid structure with many small vesicles; the pineal of the sparrow is hollow and sac-like (Quay, 1965; Oksche, 1965).

To return again to the cephalic organ, this is an extremely well developed structure in organisms like *Lineus* and *Cerebratulus* and other littoral nemertines and thus probably in those nemerteoids from which it has been suggested that the vertebrate stock arose. It is pertinent, therefore, to ask what has happened to this organ in the various groups which can legitimately be regarded as possible descendants from the nemerteoid stock. The eye of *Myxine*, for example, perhaps merits more consideration than it has so far received in this respect. Apart from a rather generalized light sense (λ max. 500–520 mμ; Steven, 1955) connected with the skin in the anterior and cloacal regions (Newth and Ross, 1955) and apparently used only to encourage burrowing in the mud, *Myxine* is not directionally sensitive to light, and its so-called eye is covered by skin and muscle. Its eye is usually regarded as a degenerate structure consistent with life in the dark depths. The illustrations that have been produced of the histology of the eye of *Myxine* (Dücker, 1924) are curiously similar to certain sections of the cephalic organ of nemertines (Fig. 15.24). If this superficial similarity were found to have a deeper foundation, it might indicate that the eye of *Myxine* is a primitive rather than a degenerate structure. *Myxine*, on the other hand, has well developed neurosecretory cells in its central nervous system. This might indicate that the cephalic organ rudiment, or placode, had incorporated its neurosecretory part into the nervous system, but that the tubular and eye-like part had been kept separate and had not developed much further.

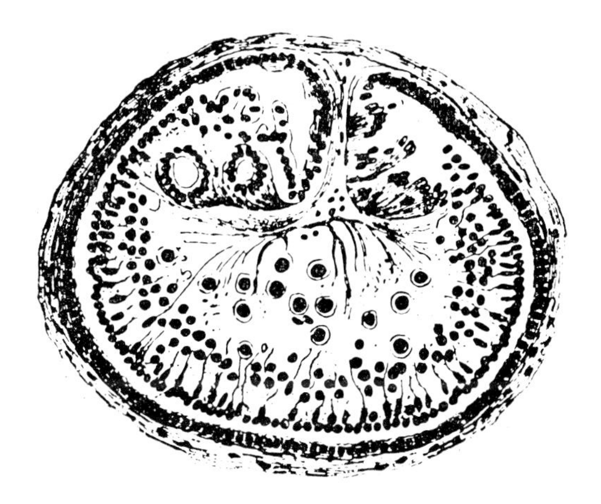

FIG. 15.24. Eye of *Myxine*. (Dücker, 1924.)

If, as suggested earlier, the nemerteoid or late rhabdocoeloid stock has branched in several directions, the cephalic organ could be the precursor not only of the vertebrate eye, the cephalopod eye, and its associated optic gland, etc., but also of the curiously large eyes of the polychaete worms, *Alciopa* and *Vanadis* (Fig. 15.25) (Hesse, 1899) among the annelids and even, by evolving along rather different lines, of the lateral eyes of the arthropods, while the ocelli of the latter more likely had their origin in the eye-spots, especially of the type found in *Geonemertes*. The last suggestion receives some support from the observation of Wald and Krainin (1963) that the median eyes of *Limulus* are sensitive to ultraviolet in the way the pineal eye of the frog is, though again this sensitivity may depend simply on the absence of filtration of the light. If the arthropods were shown to be derived from late rhabdocoeloid ancestors, their neurosecretory cells, e.g. the corpus allatum and associated tissues, would presumably also have the same origin as the neurosecretory cells in other phyla. Neurosecretion is, however, a vague term; it is worth remembering that in the nemertines, in addition to the secretory cells in the cephalic organ, there are also the cells in the main ganglia which have neurosecretory material in their cytoplasm (see

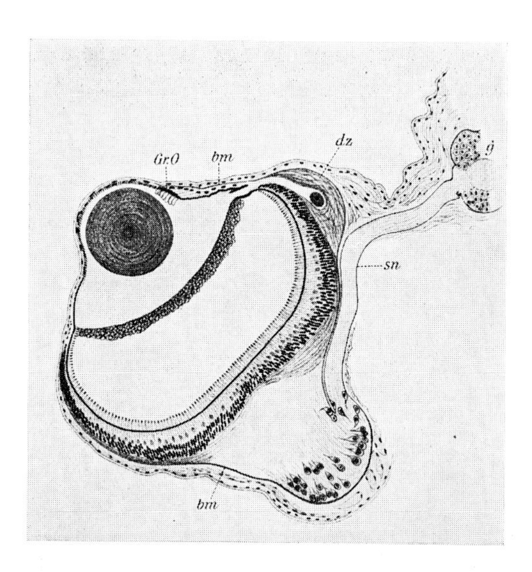

FIG. 15.25 Eye of *Vanadis formosa* (cf. Fig. 15.12). bm, Basement membrane; dz, gland cell; g, brain; Gr.O, Graeff's organ; sn, sensory nerve. (Hesse, 1899.)

p. 274). One suspects that the latter cells are homologous with similar cells in the ganglia of the annelids, for example.

Rods, Cones, and Related Structures

In the development of the vertebrate eye, a feature of considerable interest and importance is the dichotomy between the rods and the cones, and the nature of this difference merits further examination. Whether the retina is considered simply as derived from the neural ectoderm of the optic vesicle, or whether the earlier origin of the optic vesicle from the placode of the cephalic organ is taken into account, the origin is epithelial. As we have seen, the majority of epithelia function as active boundaries between two phases, and, as such, are in general composed of at least two sorts of cells.

Phylogenetically, both the epithelium, which, it has been suggested, gave rise to the neural ectoderm of vertebrates, and the epithelium of the cephalic organ are complex. In the latter, although most of the cells are flagellated and have microvilli, there are, in addition to the clearly different secretory cells, at least two distinct groups of lining cells in the deeper parts of the tube. The cells of one group have the usual elongated flagella, whereas those of the other group stain differently and, in all preparations so far made, have flagella with dilated membranes (see Fig. 15.8) There are also the processes of the bipolar cells whose single flagella project into the lumen. (Ling. 1968).

Ontogenetically, in frogs for example, the ectoderm of the gastrula is a mosaic of ciliated and non-ciliated cells. In the neurula, the neural plate is composed of columnar and non-ciliated cells together with pigmented cells,

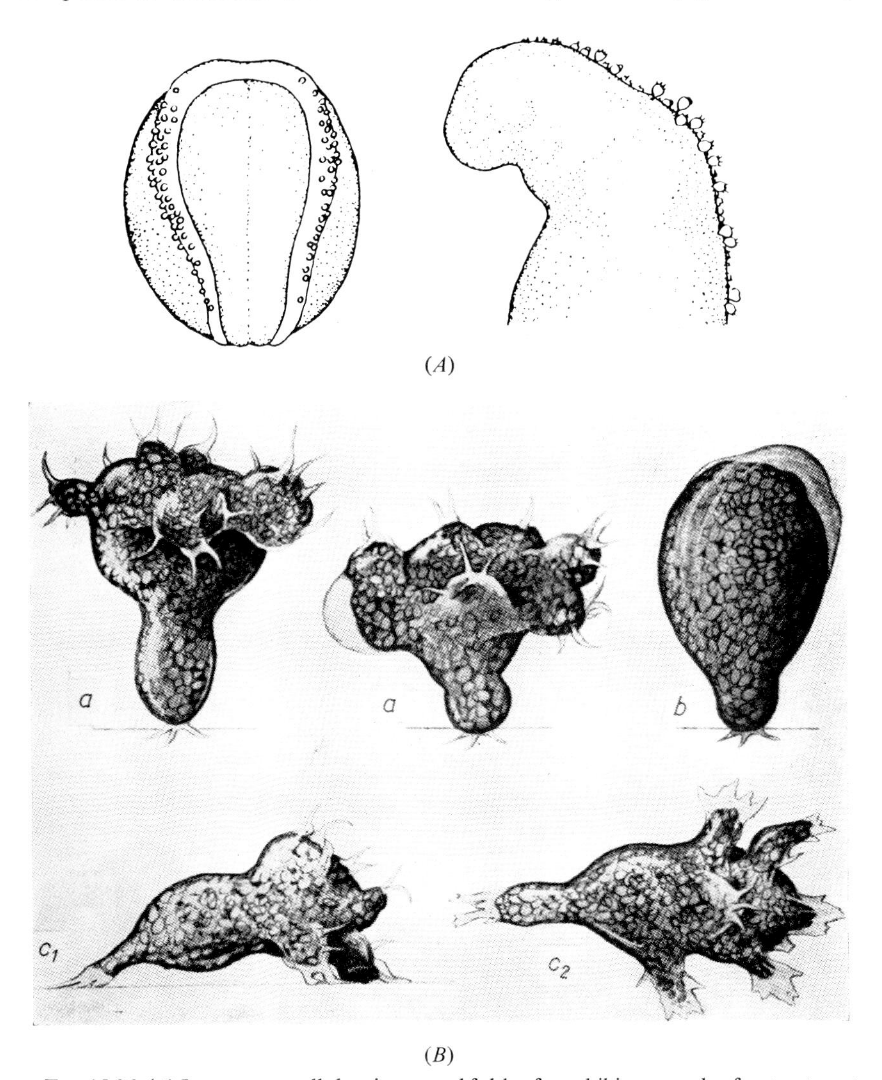

FIG. 15.26. (*A*) Large nerve cells leaving neural folds of amphibian neurula after treatment with hypertonic solution. (*B*) Isolated neuroblasts. (a) Neuroblast standing on its posterior pole and changing its form; (b) neuroblast with rotating hyaline bulge. c_1, migrating neuroblast seen from its side; c_2, the same cell seen from above. (Holtfreter, 1947.)

while the neural folds contain both ciliated cells (Fig. 10.14) (Assheton, 1896) and large nerve cells which can be specifically extracted by treatment with hypertonic sea-water (Fig. 15.26) (Holtfreter, 1947). Thus, whatever the phylogenetic origin, it is likely that the layer of cells which embryologically forms the eye is initially a typical epithelium with at least two sorts of cells. It is thus worth examining the behaviour of this epithelium during the embryonic development of the vertebrate eye, starting with the optic vesicle. The cavity of the optic vesicle is continuous with the cavity of the neural tube. This cavity, however it was formed, was initially part of the outside world which became enclosed during the folding of the tube. From that time onwards it became necessary to control its composition, and thereby provide the necessary stable environment for the developing neural tissues. Thus a major function of the lining epithelium must have been to stabilize the composition of the cerebrospinal fluid. Much of this is undoubtedly done by means of special areas like the choroid plexus, where villi covered with two types of cells effect the necessary exchanges. These cells either possess bulbous microvilli or have narrow microvilli and a few cilia (Millen and Rogers, 1956; Maxwell and Pease, 1956) (Fig. 15.27). Elsewhere the ependyma has flagellated cells, and it is probable that the whole inner surface of the tube is concerned with what may be broadly

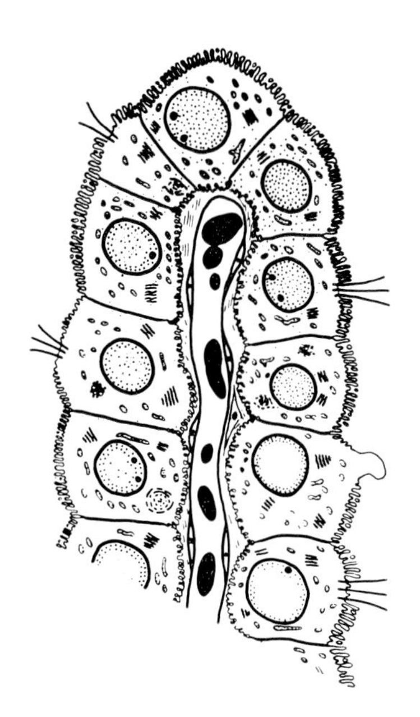

FIG. 15.27. Epithelium of the choroid plexus. Note the ciliated and the nonciliated cells. (Millen and Rogers, 1956.)

termed ionic regulation. When the choroid plexus is studied in tissue culture, it tends to form vesicles with the cilia pointing outwards, while the ependyma, under similar conditions, forms vesicles with the cilia pointing inwards (Cameron, 1953). This is a clear hint of the polarity of the two types of cells. There is no reason to believe that the cells of the optic vesicle are unconcerned with this general ionic balance. If, moreover, we are correct in linking the optic vesicle with the original placode of the cephalic organ, and if the functions of the epithelium of the cephalic organ are considered to be related to salt balance, then the lining cells of the tube would clearly have the major role in maintaining the equilibria.

The diagram in Fig. 15.28 shows the varieties of cell types which arise from the epithelium of the optic vesicle of a vertebrate as it first invaginates and then differentiates. The outer wall of the cup differentiates into the pigment epithelium whose cells may have processes that interdigitate with the rods and cones. In some cases, particularly in association with cones,

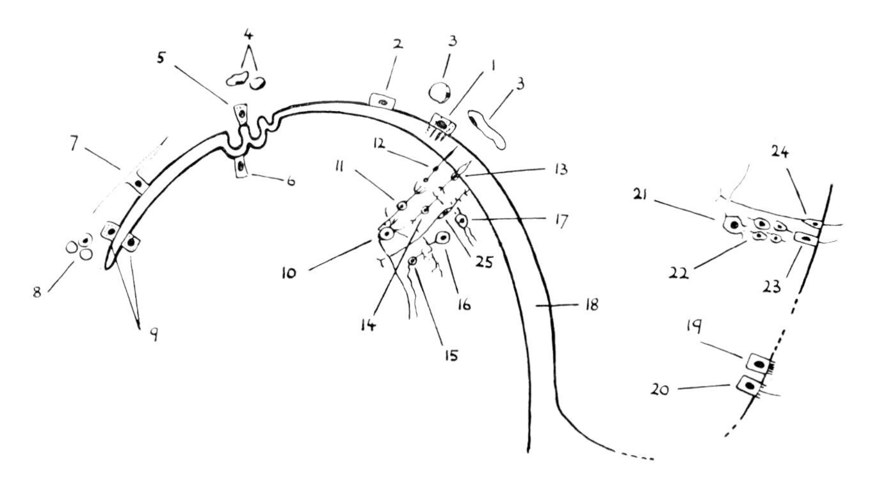

FIG. 15.28. The derivatives from the epithelium of the optic vesicle. 1, Pigment cell with processes and pigment, particularly associated with cones; 2, "pigment cell" without pigment or processes, particularly associated with rods; 3, blood vessels of choroid (derived from mesenchyme); 4, blood vessels of ciliary processes (derived from mesenchyme); 5, pigmented cell on ciliary processes; 6, non-pigmented cell on ciliary processes; 7, epitheliomuscular cell of dilator pupillae muscle; 8, epitheliomuscular cell of sphincter pupillae muscle; 9, double layer of pigmented cells on inner surface of iris; 10, large ganglion cell; 11, rod-bipolar cell; 12, rod; 13, cone; 14, cone bipolar cell; 15, small ganglion cell; 16, amacrine cell; 17, horizontal cell; 18, cavity of optic vesicle, continuous with that of neural tube; 19, cell of choroid plexus with microvilli; 20, cell of choroid plexus with cilium and microvilli; 21, large nerve cell in central nervous system; 22, nerve cells and neuroglia; 23, ependymal cell; 24, ependymal cell with process crossing the wall of the neural tube; 25, Müller's fibre (cf. 24). (Willmer, 1965.)

these processes may be long and may be filled with oval lipofuscin or melanin granules. Alternatively, and particularly in association with rods, the cells may be much simpler in outline and devoid of pigment. Whether

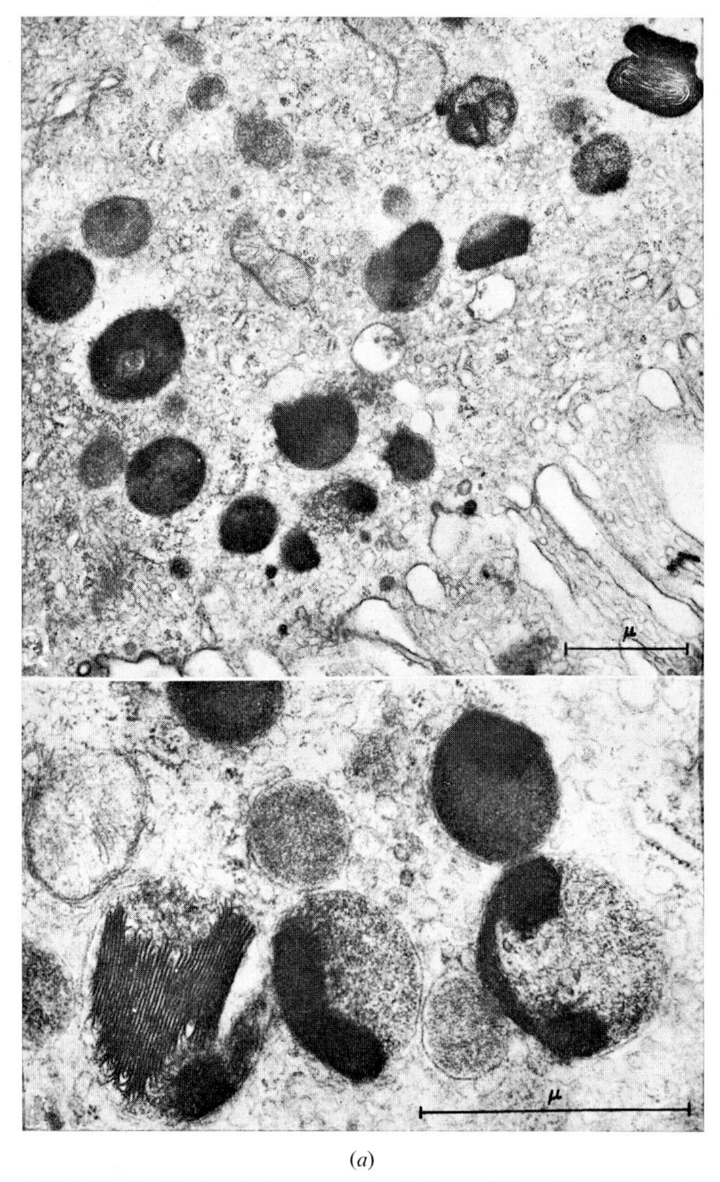

(*a*)

FIG. 15.29(*a*). Myeloid bodies in the pigment epithelium of the retina of the rat. (Dowling and Gibbons, 1962.

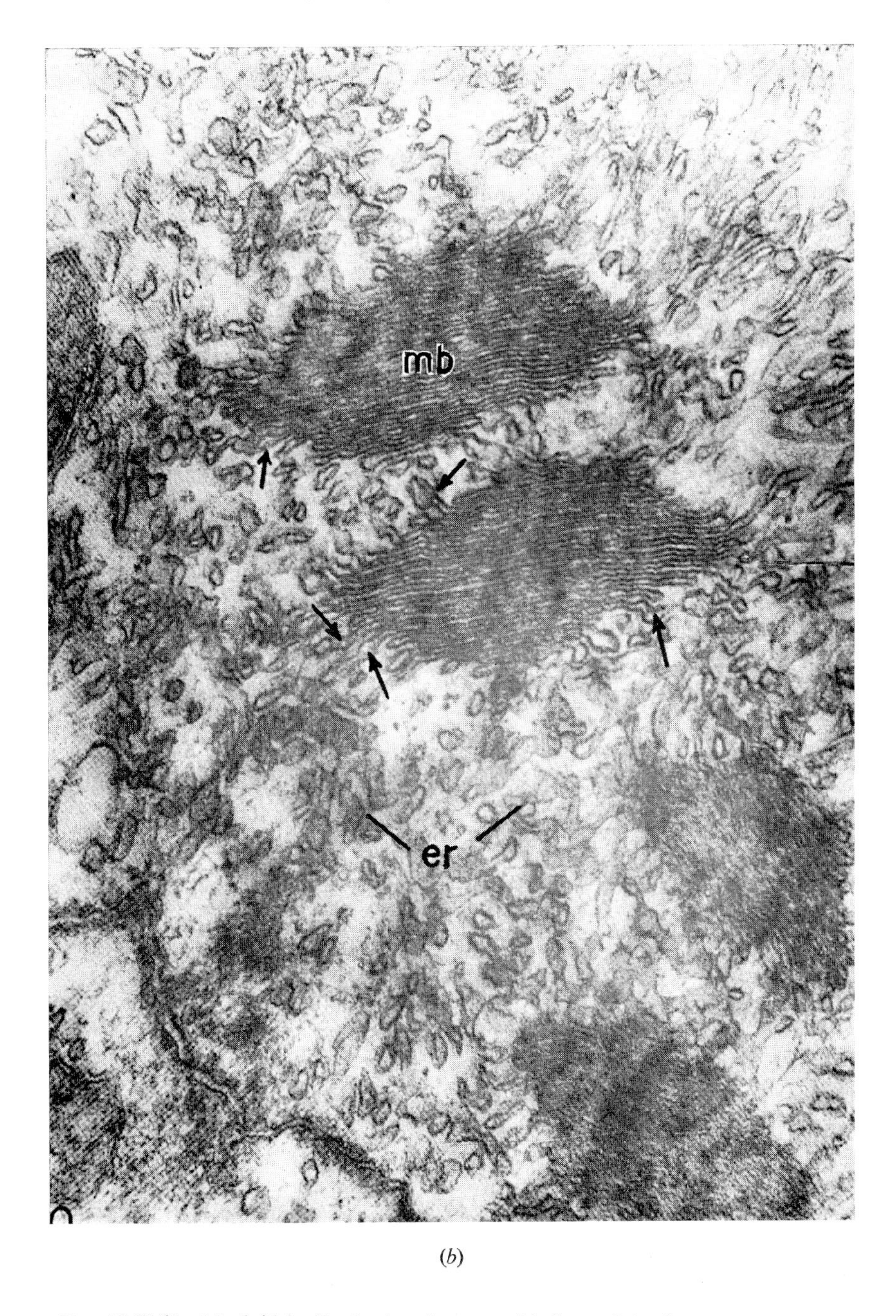

(b)

FIG. 15.29(b). Myeloid bodies in the pigment epithelium of the frog. er, Endoplasmic reticulum; mb, myeloid body; (Porter and Yamada, 1960.)

this difference has been dictated on purely functional grounds (i.e. providing sensitivity for rods by not absorbing light in their vicinity, and acuity for cones by preventing scatter and halation) or whether it reflects an inherent and physiological difference between the two groups of cells is not known. So-called myeloid bodies (in birds, frogs, and turtles) and lamellated bodies (in rats) (Fig. 15.29) are present in the cytoplasm of the pigment epithelium, and the structures of these bodies differ somewhat in the different species (Yamada, 1958; Dowling and Gibbons, 1962). Again, these differences could be related to the presence of rods or cones in the species mentioned, but their significance and function are not yet apparent.

A primary function of the pigment epithelium, which often escapes notice, is to transmit substances from the choroidal blood vessels as far as the rods and cones at least, and in some cases to the retina as a whole, for "retinal" blood vessels are virtually absent in some groups of animals, and, it may be remembered, external pressure on the eyeball, if sufficient to occlude the blood vessels, quickly brings temporary blindness. The pigment epithelium is thus far from inert. Moreover, it frequently contains stores of vitamin A (Greenberg and Popper, 1941). This occurs in the 11–*cis* form, the form in which the retinene (retinal, vitamin A aldehyde) is combined in rhodopsin, and it has been suggested that there is constant traffic in this commodity between these cells and the rods and cones (Dowling and Gibbons, 1962).

Beyond the margin of the retina, and over the ciliary processes, the pigment epithelium continues as a pigmented layer till it reaches the margin of the iris, where it turns back on itself and folds against the iris as a double-layered epithelium in which both layers are pigmented. When it again reaches the ciliary processes, it continues to cover the underlying pigment epithelium, but it loses its pigment and continues as a colourless columnar epithelium till it reaches the margin of the retina. Over the ciliary processes the cells of the inner and outer layers are very different in their microstructure (Fig. 15.30).

In the region of the iris and ciliary processes, the epithelium displays many interesting physiological features. Whatever the traffic may have been between the choroid vessels and the retina across the pigment epithelium proper, over the ciliary processes there is no doubt that a secretion of aqueous humour occurs into the eye, across the two layers of epithelium. This is not merely a filtration; the cells must be actively involved. It will be realized that the outer and pigmented layer, which has a basement membrane adjacent to the connective tissues of the choroid and sclera, secretes into the erstwhile cavity of the neural tube (i.e. the cavity that became occluded during the process of invagination of the optic cup), while the inner and unpigmented layer secretes from this potential space into the cavity of the bulb. Thus the inner and outer layers are, in fact, working in opposite directions with respect to the original cavity, but with a net transfer of fluid into the eye.

During the process of invagination of the optic vesicle, the cup so formed becomes invaded with mesenchyme cells from the surrounding tissues.

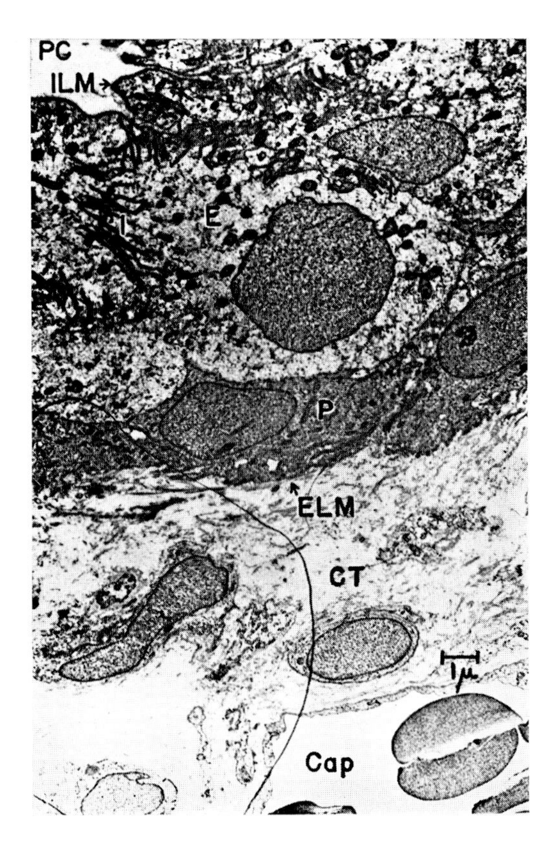

Fig. 15.30. Two classes of cells in the ciliary processes of the albino rabbit. Cap, Capillary; CT, connective tissue; E, epithelial cell next to posterior chamber; ELM, external limiting membrane; I, interdigitated cell margins; ILM, internal limiting membrane; P, epithelial cell (pigmented in normal rabbits); PC, posterior chamber. (Pappas and Smelser, 1961.)

Thus, when the pigment layer doubles back on itself at the inner margin of the iris, its basal surface maintains its original contact with mesenchymal cells and their products. It is not surprising, therefore, to find that the epithelium over the ciliary processes, in addition to having a basement membrane on the side next to the blood vessels, continues to be separated from the vitreous body and the contents of the eye by the formation of a basement membrane on that side also. For similar reasons the lens of the eye, which is again an epithelial structure that becomes buried in the mesenchyme, also becomes surrounded by a capsule which is essentially a basement membrane. Furthermore, the two structures, i.e. the lens capsule and the basement membrane over the ciliary processes, are connected by that important sling or hammock of fine collagen fibres, the zonula of Zinn. In it the tension is lessened by the contraction of the ciliary muscles, so that the lens fibres themselves (i.e. the cells composing the lens proper) can exert their elasticity and make the lens more spherical, thereby reducing its focal length and accommodating the eye for vision of near objects.

At this point a few words may be said about the lens itself. Its phylogenetic origin has already been adumbrated in considering the cephalic organ. In the vertebrates it is essentially a product of the ectoderm, and it generally develops directly or indirectly as the result of the approach of the optic vesicle to the under surface of this epithelium (see Huxley and de Beer, 1934; Spemann, 1938). In some animals, this or a similar inductive stimulus is necessary (e.g. in *Amblystoma*; Harrison, 1920); in others, induction is effective but not necessary (Spemann, 1912; Balinsky, 1951). Moreover, in some animals the optic cup can induce lens formation in an indifferent piece of ectoderm, e.g. a piece grafted from the ventral surface of the embryo. The nature of the inductive process has been discussed and investigated by numerous authors and by a variety of means. There is some evidence for the production of mucopolysaccarides between the two ectodermal layers; an initial contact between the layers appear to be necessary (McKeehan, 1951). The superficial ectoderm and the epithelium of the optic cup first come into close relationship with each other, though separated by criss-crossing fibres, and subsequently they become separated by elements from the connective tissue. There is evidence for the orientation of particles in the cytoplasm of the cells opposing each other (Weiss and Fitton Jackson, 1961). It has been suggested also that in many cases of induction there is a transfer of RNA (or more probably ribonucleoprotein) particles from the inducing agent to the induced (see Brachet, 1957), but the specific agent in this case has not been determined.

One point which does not seem to have been so fully investigated as it might have been is the local environment which may develop between these converging epithelia. Induction is still generally considered to be dependent on some specific and rather hormone-like influence. Though this may indeed often, or even always, be the case, it is equally possible that the aggregate of local changes in a cell's environment produced by the activities of neighbouring cells may also determine a cell's behaviour. Local differences in concentration, either up or down, in essential metabolites may be all that is necessary to remodel the activities of a cell. If this is so, it may turn out that specific inducing agents are less frequent than was at one time supposed.

At first, the optic vesicle enlarges and fills with fluid, but on touching the ectoderm it begins to invaginate and lose fluid from its inside. For a time the ectoderm cells overlying the vesicle are, as it were, drawn down with the invaginating cup, and eventually they form a little vesicle of their own which is closed off from the outside world (Fig. 15.31). This lens vesicle

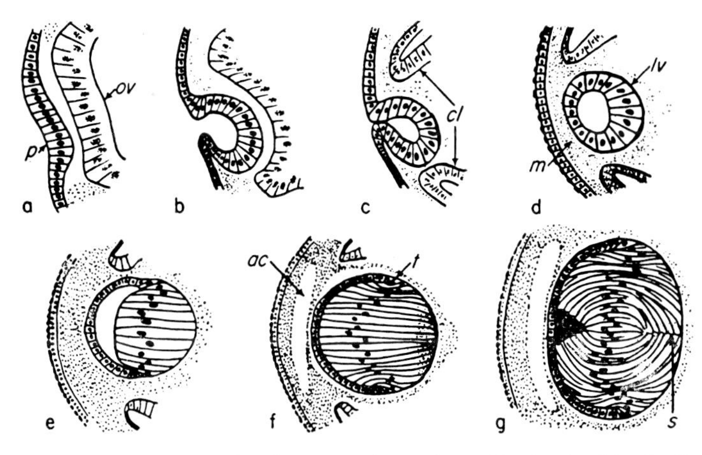

FIG. 15.31. Changes in the epithelium over the optic vesicle that lead to the formation of a lens. Note the blastula-like arrangement of the cells in d and e. ac, Anterior chamber; cl, lips of optic cup; lv, lens vesicle; m, mesenchyme; ov, optic vesicle; p, primordium of lens; s, suture; t, transitional zone. (From Walls, 1942, after Mann.)

is filled with fluid, and in some ways it resembles a little blastula with the same problems of stability and volume control (see p. 198). Interestingly enough, the cells of the lens then become visibly separable into two groups. The outer, more superficial cells remain as a low cuboidal epithelium; the inner cells elongate and begin to form the special lens proteins (crystallins). When this happens, the contained fluid disappears and the lens continues to grow as a solid body by the progressive addition of cells to both inner and outer layers resulting from cell divisions that mostly occur equatorially. The causes and consequences of these movements of fluid in the optic vesicle, lens, and eye chambers must be important to the full understanding of the behaviour of the cells in the various layers,

In *Xenopus* the lens can be regenerated from the inner layer of cells of the corneal epithelium under the influence of the eye-cup, and the behaviour of the differentiating cells is illuminating. An examination of the normal corneal epithelium from which regeneration can take place shows an even or random distribution of cells with one nucleolus (about 45%) and with two nucleoli (about 55%). However, in the particular localized area from which the lens will actually form, there is a great preponderance of cells with only one nucleolus. In the developing lens vesicles, some 78 cells have a single nucleolus to every 22 cells with more than one nucleolus, and there is a suggestion that most of the large cells that form the inner layer (i.e. those that form the lens fibres) possess only a single nucleolus (Freeman, 1963). This may simply indicate a high degree of protein synthesis on the part of the mononucleolar cells. It may also be pointed out, however, that the amoebocytic type of cell frequently has only a single nucleolus, whereas the mechanocytic type of cell usually has two or more. The change in the distribution of these two types in the corneal epithelium before lens formation actually begins is interesting, and it emphasizes possible changes in the local environments and a means of getting the two types of cells into the vesicle in proportions that lead to successful lens formation.

This brings us back again to the differentiation in the optic cup and to the retina itself. At the time when the cup is still expanding, the neuro-epithelial layer is a simple epithelium (neuroectoderm). During the invagination, when the inner layer is beginning to form the retina, however, cell divisions are going on rapidly and cells are being budded off into the deeper layers (e.g. in the direction facing the lens) in the same manner as nerve cells and neuroglia cells are budded off to form layers in the rest of the central nervous system. The orientation of the metaphase plates of the dividing cells is occasionally parallel to the surface, but far more frequently at right angles to it. This presumably determines the fate of the daughter cells: to stay in, or to stay attached to, the surface till a later division; to stay in the surface and differentiate there into rod or cone cells (the equivalent of ependymal cells in the rest of the nervous system); or to leave the surface and then become a ganglion cell, bipolar cell, or other cell of the inner layers of the retina (see Figs. 10.3, 11.4, 15.28).

When invagination of the optic vesicle begins, the fluid obviously leaves the cavity. As we have already seen in considering the epithelium behind the iris and over the ciliary processes, the inner and outer parts of the epithelium come closer and closer to each other until the layer of cells which will form the rods and cones of the retina lies in intimate contact with the pigment layer. All that is now left of the original cerebrospinal fluid between the two layers lies in the minute spaces between the rods and cones and the processes of the pigment cells. The composition of this fluid, which contains some acid mucopolysaccharides (Lillie, 1952; Sidman, 1958), is presumably determined by the interplay of the pigment cells and the retinal cells acting on the original cerebrospinal fluid, on materials available from the choroidal blood vessels on the one side, and on those from retinal vessels on the other. Visual cells therefore may well be concerned not only with vision but also, at least in their early stages, with ionic equilibria. Indeed, the strict preservation of this symbiosis between retinal cells and pigment cells could be just as important in determining retinal function as is the symbiosis between neurones and neuroglia in the formation and maintenance of the myelin sheath and in maintaining neural function in the rest of the central nervous system.

This ionic regulation may perhaps offer a clue to the real significance of rods and cones as distinct cell types. In some animals the structural and functional differences between rods and cones are obvious, and there can be no argument as to which is which (Fig. 15.32a). In others the

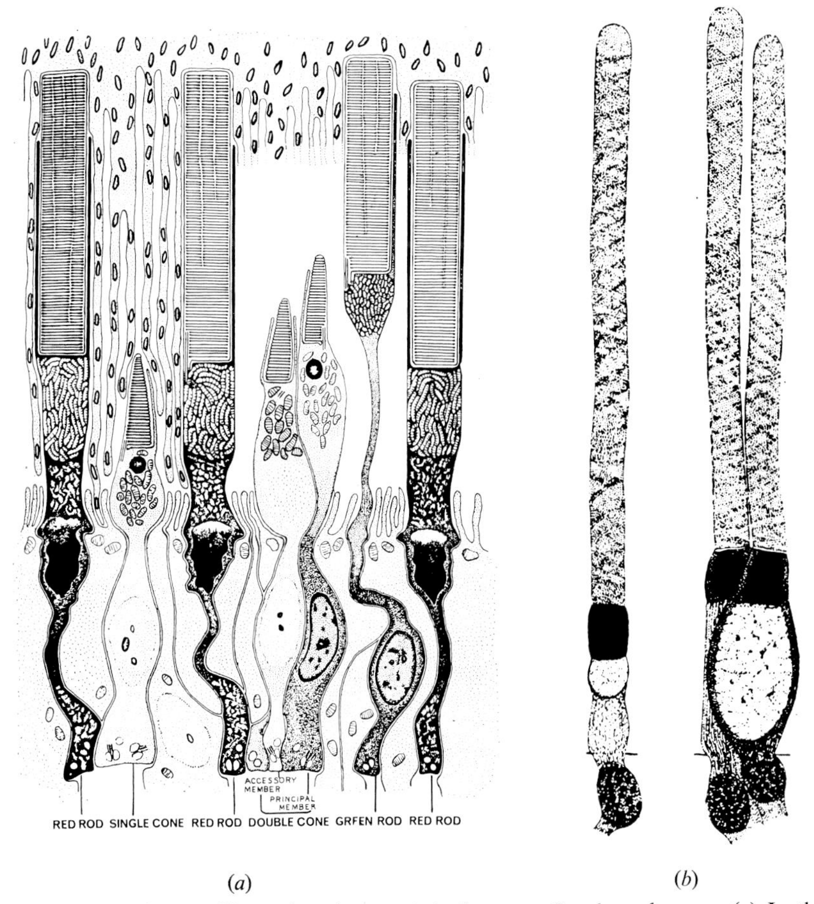

FIG. 15.32. Diagram illustrating characteristic features of rods and cones. (*a*) In the frog, *Rana pipiens* (Nilsson, 1964); (*b*) in the nocturnal gecko, *Coleonyx variegatus* (Walls, 1942).

distinction is not so clear, and often the classification has to be somewhat forced, e.g. in the rod-like cones of the nocturnal geckoes (Fig. 15.32b) (see Crescitelli, 1965; Dunn, 1966). If it is supposed that initially the neural ectoderm is, like most epithelia, a mosaic of two balancing or opposing types of cells (we have seen several indications that this is so), then, as the result of its folding to form the retina, there could be four groups of cells. All of them would be concerned with regulating the environment of the visual receptor processes, two in the pigment layer, and the rods and cones themselves, all of which could be considered the derivatives of the original two classes of cells (ecballic and emballic). Where rods and cones are clearly separable, the suggestion is strong that the cone is the more amoebocytic (emballic) and the rod the more mechanocytic (ecballic). The typical cone has a single nucleolus; it tends to store carbohydrate in the form of a paraboloid; it may also store lipids and carotenoids; it has a broad cone-pedicle. The typical rod generally has more than one nucleolus, no stored lipid, no paraboloid (though there are exceptions), and it has a simple synaptic "spherule". Evidence derived from the extraction of pigments from bovine and chick retinae and their recombination with proteins extracted from the same eyes indicates that retinal, by combination with different proteins, produces different visual pigments. On these grounds it would seem that the proteins (opsins) of cones differ from those of rods (Wald, 1958). The metabolism of rods is also evidently different from the metabolism of cones as judged by the selectively toxic action of sodium iodoacetate on the rods of mammals (Noell, 1960). A further suggestion of a difference in polarity comes from the observations of Young and Droz (1968). These authors have shown that when radioactive amino-acids are injected into frogs, autoradiographs of the retina show the labelling in the rods first in the cell body, whence it spreads to the base of the outer segment and then progressively moves up it. In the cones, however, the labelling occurs fairly evenly throughout the cell and its process. In both cell types the labelling progressively increases at the synaptic end. Other indications of "opposition" between rods and cones come from eyes in which both the rods and the cones have contractile myoids, e.g. those of the cat-fish (Fig. 15.33) and the crocodile (Detwiler, 1943). In such eyes, when the rods contract, the cones relax. Similarly, the actions of adrenalin in the eye are interesting: adrenalin causes the active contraction of the dilator pupillae muscle but does not affect the sphincter, though both are epitheliomuscular cells derived directly from the pigment epithelium. In amphibia, adrenalin can causes expansion of the pigment in the pigment layer together with contraction of the cone myoids. Finally, when tissue cultures are made from suspensions of retinal cells behaviour are produced. Frequently the cells aggregate into clusters and form hollow vesicles, tubules, and rosettes. At other times, or in other places, solid clusters

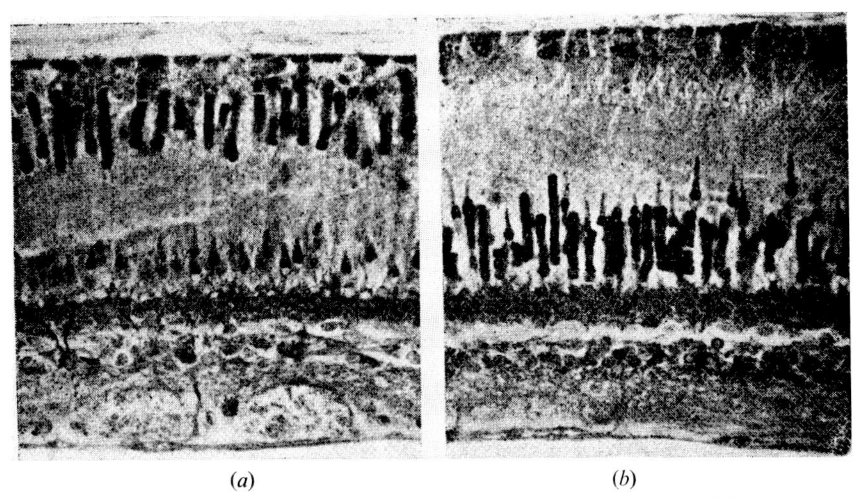

FIG. 15.33. Oppositely orientated movement of rods and cones in the cat-fish (*Ameiurus*). (*a*) Light-adapted; rods elongated; cones contracted. (*b*) Dark-adapted; rods contracted; cones elongated. (Detwiler, 1943.)

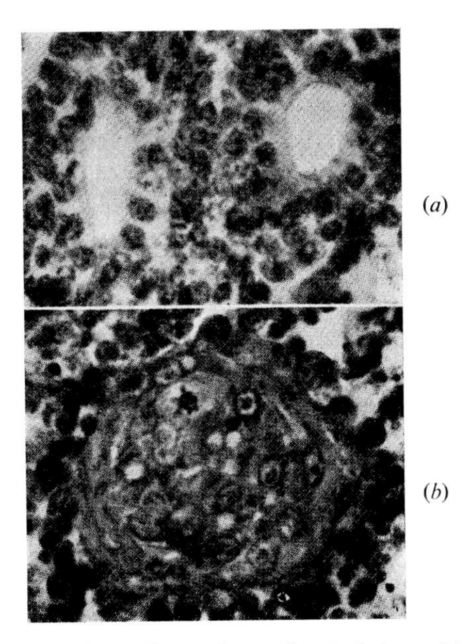

FIG. 15.34. Cell aggregates from dispersed neural retinal tissue. (*a*) Rosette-shaped cell aggregates, 2 days in suspension culture; (*b*) lentoid formation, 4 days on a plasma clot. Note that the cells in (*a*) have two or more nucleoli, whereas those in (*b*) mostly have one. (Moscona, 1957.)

of larger and more cytoplasmic units are formed as "lentoids", and the two groups may co-exist side by side (Tansley, 1933; Pomerat and Littlejohn, 1956; Moscona, 1957) (Fig. 15.34). These formations are reminiscent of the behaviour in culture of the proximal and distal tubules of the avian kidney (see p. 198) and speak rather strongly of orientated cells of two types with respect to the passage of fluid through them. They are also interesting in that the "lentoid" behaviour may be pertinent to the problem of lens regeneration from the dorsal parts of the retina in certain animals (see p. 382). Exactly what the cells in rosettes and lentoids are, in terms of functional retinal cells, is uncertain, but there is certainly an *a priori* case for considering the retina as being initially like any other epithelium and containing basically two types of cell. Exactly how these cells have to function in the photoreceptive layer of any given species must depend on the equilibria which they must maintain with their immediate surroundings and also with the environment of the various conducting and integrating cells in the rest of the retina. This equilibrium with the surroundings must depend not only on the cells of the rod and cone layer itself, but also on the activity of the cells of the pigment layer and on the extent to which the composition of the blood in the neighbourhood of these cell groups is controlled and suitable without further adjustment. In most of the higher vertebrates the blood composition as a whole is probably very well controlled, but exactly what the composition of the tissue fluid is in the immediate vicinity of the rods and cones may well vary in different species. In one species, rods may have to be very different from cones in order to maintain an ionic equilibrium which might otherwise be ill regulated and widely varying. In another species, the internal environment may already be so well regulated that the rods and cones have only to act as fine adjustments, pushing or pulling on either side of the equilibrium point. To do this, both may have to act more or less emballically if the concentrations tend to be high, or ecballically if they tend to be on the low side. Both units therefore might appear cone-like, or both might look like rods.

To summarize: it seems possible that the basic fact of retinal structure and function, like that of any other epithelium, is that the retina is composed of two groups of cells which work symbiotically and probably more or less antagonistically to maintain a particular traffic across their membranes. In this function the retinal cells are aided by the cells of the pigment epithelium, which are basically similar cells and probably are organized in a similar way. In such a system there are thus four primary units to be taken into account, two in the pigment layer and two in the rod and cone layer. Because theoretically only two are necessary, all sorts of combinations are possible provided that there are at least two acting in opposition to each other. Thus the argument about the nomenclature of rods and cones, i.e. whether a given cell should be called a rod or a cone, becomes less relevant than whether the cells are acting emballically or ecballically. In general, it would appear from electron-microscopic studies made so far that, in the vertebrates, visual receptors can develop from cells which are strongly flagellate and also from cells with rather feeble flagella and with numerous microvilli. The forms of these cells probably depend on the requirements for the stability of the localized environment. In the old terminology, rods correspond more to the flagellate end of the series, and cones to the microvillous end. But the opponent "rods" and "cones" could both be derivatives of the flagellate type, or, as in the amphibian mud-puppy, both could be derivatives of the microvillous type, provided that in each case the partners in the pigment layer-retinal complex have an adequately balanced system for controlling their immediate environment. Attention is called again to the contrasting behaviour of choroid plexus and ependymal cells when they are placed in tissue culture (see p. 403).

One point which will have occurred to many readers is that in the frog, for example, there are at least two sorts of cones and two sorts of rods (see Fig. 15.32). In the human eye there are cones with certainly two and possibly three different pigments (Willmer, 1955; Rushton, 1955; Marks *et al.*, 1964; Wald, 1964; Baker and Rushton, 1965). This probably means there are two (or three) different proteins in the containing cells. It may therefore be asked how these observations fit in with the idea of a primary duality. The probable answer is that there is plenty of room for variation within each group, just as in the sponge embryo there are cells that grade from flagellate type to amoeboid type, and in the adult sponge there are many forms of amoebocyte. In the retina of the frog also, one cone stores carbohydrate and its neighbour may store an oil globule. These seem to be variations on one theme, like the thesocytes and fuchsinophil cells in sponges which are both variants of the amoebocytes. Presumably something similar may be at the bottom of the difference between erythrolabe- and chlorolabe-containing cones in the human eye. If an eye is to remain as a functional system, it must adapt itself to the particular requirements of the animal to which it belongs. Thus, if sensitivity to light is based on absorption by carotenoid pigments, then any one pigment can only make the eye effective over a relatively narrow range of wave-lengths. Two pigments whose spectral sensitivity curves partly overlap could cover a wider range, and three would do even better. Moreover, such events as the migration of a population of animals from the relatively blue sea-water to the relatively brown fresh water may encourage the development of pigments with different spectral sensitivity (e.g. red-sensitive pigments) in the fresh-water population. If it happened that the new and old pigments developed in separate sensory cells, their effects could, with suitable integration, be used for the discrimination of wave-length as well as for widening the range of the visible spectrum. By similar means, eyes may have adapted themselves to fulfil the many and different special needs of animals in a variety of situations, and one development has led to another.

The cephalic organ of nemertines already has a very complex ciliated tube with several variations in the arrangement and organization of the ciliated cells (see Fig. 15.8) that suggest functional variations. Furthermore, the ciliated cells are already integrated into a functional neural mechanism in which the cellular pattern is in several ways comparable with the organization of the retina of vertebrates. Thus it seems at least possible that variations of the cephalic organ may help us to understand the enigma of the vertebrate eye.

These thoughts are in line with the possible derivation of the eye from an organ that was initially concerned with salt metabolism. They are unorthodox in that they consider the visual function of the retina secondary and dependent on the proper discharge of the primary function.

Cilia and such phospholipid systems as those in myelin and the visual receptors are, in themselves, surprisingly stable. However, the stability of the absolute visual threshold and the extraordinarily high visual acuity and colour-sensitivity of the vertebrate retina necessitate an almost incredible precision and durability in the underlying mechanism. Thus the part which the various units of the retina play in maintaining this mechanism must inevitably be a primary consideration.

For the development of the rest of the retina, the cells of the inner layer of the optic cup are again responsible, and the bipolar cells, ganglion cells, horizontal cells, amacrine cells, and probably the Müller's cells too are all derivatives of the original epithelial cells. Thus, if there is a duality in the primary layer, there is likely to be a similar duality in the other layers as well. This does not necessarily follow, however, because one group may divide much more frequently than the other, or the large number of cell deaths, which can be such a conspicuous feature of the developing retina (Glücksmann, 1940, 1951), may be highly selective for a particular class of cells. Considerations of this sort make Allen's observations (1965) that retinal ganglion cells, and also cells of the pigment epithelium, sometimes possess cilia much more intelligible than they perhaps would otherwise have been. It is also interesting to note that, although little is known about the transmitter substances at the first synaptic layer (between receptors and bipolar cells), dopamine and some noradrenalin have been detected in the second synaptic layer (Häggendal and Malmfors, 1965), and this layer also contains cholinesterase (Francis, 1953). There are thus probably two different transmitters in this synaptic layer.

This potential duality of the retina (see Willmer, 1965) raises the same sort of problems in relation to neural interaction as were raised in Chapter

11 (see p. 244). What has become abundantly clear in recent years is that, in the retina, the amount of interaction between the various cells is far greater than was at one time supposed, and the connexions between the rods and cones and their associated horizontal and bipolar cells are far more complex than the early Golgi impregnations implied. Tight junctions have been found between the central processes of rods and cones, for example, in addition to the more elaborate and typical synapses in the rod spherules and cone pedicles (Nilsson, 1964; Dowling and Boycott, 1966). The interactions are not all additive, and the synapses are not all of one kind, nor are they simply positive connexions; it appears that inhibition and excitation are much more intimately linked than was at one time supposed. Moreover, many of the retinal cells, e.g. the horizontal cells in fish, are not merely transmitters of impulses; they selectively store information, by acquiring higher or lower potentials, about such things as the wave-length of the exciting light (Fig. 15.35) (Svaetichin and MacNichol,

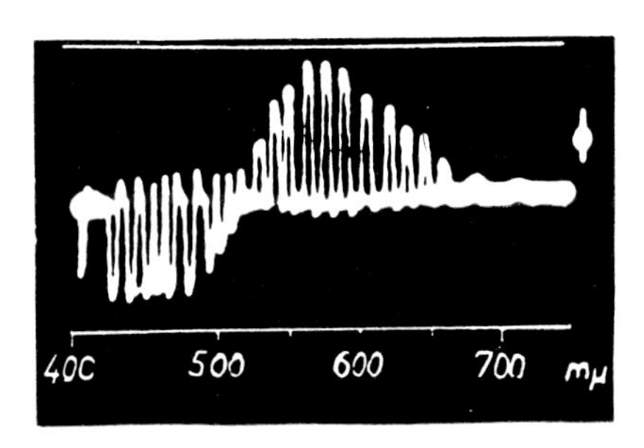

FIG. 15.35. Record showing the potentials of different size and sign developed in horizontal cells in the fish retina in response to lights of different wave-length as indicated on the abscissa. (Svaetichin and MacNichol, 1958.)

1958; Svaetichin *et al.*, 1965). Positive and negative interaction between cells is clearly rampant in the retina, as would be expected if the original cells from which the retina is derived were themselves divided into two opposing camps. However, what is not yet clear is whether activity in one group always acts as a stimulus to the other group to counteract the effects, or whether it evokes from the partner an inhibitory response which damps down its own activity. Both reactions seem to be possible, and the latter would have definite advantages in the economy of a nervous system, where any excited cell is told by its neighbours to be quiet and its message is passed on only if its immediate compensators cannot silence it. This would lead to economy in passing only the important information and would ensure the neglect of trivial signals in favour of the more important ones. As hinted above, in addition to the usual synaptic connexions in the retina there are places on the synaptic knobs and pedicles where adjacent cells come into direct contact (tight junctions) with others, whereas elsewhere they are completely insulated from each other by intervening cytoplasmic processes of Müller's cells. These tight junctions could be so-called electrical synapses. Moreover, in the frog's retina, where these things have been most fully investigated, the contacts are systematically arranged. For example, a single cone may have tight junctions with three red rods; and a red rod may make this sort of contact with a single cone, a principal cone of a double cone (or cone-pair) and another red rod. Green rods do not appear to make such contacts (Nilsson, 1964) (see Fig. 15.32(*a*)). What these special junctions mean functionally, if anything, is at present a mystery, though electrical synapses are possible and, since the freedom of intercellular communication between adjacent cells of epithelia has recently been stressed, it may be important here (Loewenstein *et al*,, 1965).

Recent electrical recording from retinal ganglion cells and optic nerve fibres make it clear that individual ganglion cells are specialized for handing on particular classes of information, not in the main from individual receptors but from groups of receptors, linked through the bipolar cells and the like. It is also clear that this information has already been sorted and coded in its passage through the retina, so that different ganglion cells hand on information about movement, shape, direction, colour, brightness, saturation only in so far as their receptive fields and intervening connexions allow. Furthermore, it is probable that in the process of evolution each species animal has its retina specially adapted to processing information particularly important to the well-being of the species (e.g. in relation to defence, offence, food detection) and that natural selection has brought about many variations of the simple theme of a retina basically composed of two balancing types of cells and their respective progenies. Mere inspection of the histology and cytology of the retina, though of cardinal importance, cannot pretend to elucidate all the coding and sifting that goes on within the visible structures.

In this chapter we have attempted to consider the evolution of the sense organs of the vertebrates from simpler beginnings in such creatures as the nemerteoids. We have shown how the pattern of cell behaviour already present at that stage could evolve in various ways to produce the main sense organs in the vertebrates. The only major sense organ that has not received much attention is the vestibular organ and the ear. The evolution of this sense organ takes place mostly within the vertebrates themselves, probably by direct specialization of the cells of the lateral-line canal system, which may itself have evolved from the cephalic grooves of the nemerteoids. The organs described as statocysts (Fig. 15.6) in certain nemertines are of dubious significance in this respect. Again, the presence in the vertebrate organs of "hair cells" with cilia and microvilli, supported by non-ciliated cells, emphasizes the same sort of duality and symbiosis as in the other organs discussed.

Proprioceptors for the greater efficiency of muscular co-ordination must also develop in relation to the evolution of somatic muscles and a skeleton. They, too, are therefore developed mainly within the vertebrate organization.

References

Allen, R. A. (1965). Isolated cilia in inner retinal neurons and in retinal pigment epithelium. *J. Ultrastruct. Res.* **12**, 730.

Andersson, B., and Jewell, P. A. (1957). The effect of long periods of continuous hydration on the neurosecretory material in the hypothalamus of the dog. *J. Endocrinol.* **15**, 332.

Assheton, R. (1896). Notes on the ciliation of the ectoderm of the amphibian embryo. *Quart. J. Microscop. Sci.* **38**, 465.

Bagnara, J. T. (1963). The pineal and the body lightening reaction of larval amphibians. *Gen. Comp. Endocrinol.* **3**, 86.

Baker, H. D., and Rushton, W. A. H. (1965). The red-sensitive pigment in normal cones. *J. Physiol.* (*London*) **176**, 56.

Balinsky, B. I. (1951). On the eye cup-lens correlation in some South African amphibians. *Experientia* **7**, 180.

Bernard, R. A., Halpern, B. P., and Kare, M. R. (1961). Effect of vitamin A deficiency on taste. *Proc. Soc. Exptl. Biol. Med.* **108**, 784.

Boycott, B. B., and Young, J. Z. (1956). The subpedunculate body and nerve and other organs associated with the optic tract of cephalopods. *In* "Bertil Hanström, Zoological Papers in Honour of His Sixty-Fifth Birthday" (K. G. Wingstrand, ed.). p. 76, Zoological Institute, Lund.

Brachet, J. (1957). "Biochemical Cytology". Academic Press, New York.

Brambell, F. W. R., and Cole, H. A. (1939). The pre-oral ciliary organ of the Enteropneusta: Its occurrence, structure and possible phylogenetic significance. *Proc. Zool. Soc. London* **B109**, 181.

Brown, P. K., Gibbons, I. R., and Wald, G. (1963). The visual cells and visual pigment of the mud puppy, *Necturus*. *J. Cell Biol.* **19**, 79.

Bürger, O. (1895). Die Nemertinen des Golfes von Neapel. *Fauna und Flora des Golfes von Neapel Monograph* **22**, 1.

Cameron, G. (1953). Secretory activity of the chorioid plexus in tissue culture. *Anat. Record* **117**, 115.

Charlton, H. M. (1966). The pineal gland and colour changes in *Xenopus laevis* Daudin. *Gen. Comp. Endocrinol.* **7**, 384.

Coe, W. R. (1927). The nervous system of pelagic nemerteans. *Biol. Bull.* **53**, 123.

Crescitelli, F. (1965). The spectral sensitivity and visual pigment content of the retina of *Gekko gekko*. *Ciba. Found. Symp. Colour Vision, Physiol. Exptl. Psychol.* p. 301.

Dendy, A. (1907). On the parietal sense-organs and associated structures in the New Zealand lamprey (*Geotria australis*). *Quart. J. Microscop. Sci.* **51**, 1.

de Robertis, E. (1960). Some observations on the ultrastructure and morphogenesis of photoreceptors. *J. Gen. Physiol.* **43**, Suppl. 2, 1.

Detwiler, S. R. (1943). "Vertebrate Photoreceptors". Macmillan, New York.

Dingle, J. T., and Lucy, J. A. (1965). Vitamin A, carotenoids and cell function. *Biol. Rev. Cambridge Phil. Soc.* **40**, 422.

Dodt, E. (1964). Physiologie des Pinealorgans Anurer Amphibien. *Vision Res.* **4**, 23.
Dodt, E., and Heerd, E. (1962). Mode of action of pineal nerves in frogs. *J. Neurophysiol.* **25**, 405.
Dowling, J. E., and Boycott, B. B. (1966). Organization of the primate retina: Electron microscopy. *Proc. Roy. Soc.* (*London*) **B166**, 80.
Dowling, J. E., and Gibbons, I. R. (1962). The fine structure of the pigment epithelium in the albino rat. *J. Cell Biol.* **14**, 459.
Dücker, M. (1924). Über die Augen der Zyklostomen. *Jena Z. Med. Naturw.* **60**, 471.
Dunn, R. F. (1966). Studies on the retina of the gecko, *Coleonyx variegatus*. 1. The visual cell classification. *J. Ultrastruct. Res.* **16**, 651.
Eakin, R. M. (1964). The effect of vitamin A deficiency on photoreceptors in the lizard *Sceloporus occidentalis*. *Vision Res.* **4**, 17.
Eakin, R. M., and Westfall, J. A. (1959). Fine structure of the retina in the reptilian third eye. *J. Biophys. Biochem. Cytol.* **6**, 133.
Eakin, R. M., and Westfall, J. A. (1960). Further observations on the fine structure of the parietal eye of lizards. *J. Biophys. Biochem. Cytol.* **8**, 483.
Eakin, R. M., Quay, W. B., and Westfall, J. A. (1961). Cytochemical and cytological studies of the parietal eye of the lizard, *Sceloporus occidentalis*. *Z. Zellforsch. Mikroskop. Anat.* **53**, 449.
Farrell, G. L. (1959). Glomerulotropic activity of an acetone extract of pineal tissue. *Endocrinology* **65**, 239.
Flock, Å., and Wersäll, I. (1962a). Synaptic structures in the lateral line canal organ of the teleost fish *Lota vulgaris*. *J. Cell Biol.* **13**, 337.
Flock, Å., and Wersäll, I. (1962b). A study of the orientation of the sensory hairs of the receptor cells in the lateral line organ of fish, with special reference to the function of the receptors. *J. Cell Biol.* **15**, 19.
Francis, C. M. (1953). Cholinesterase in the retina. *J. Physiol.* (*London*) **120**, 435.
Freeman, G. (1963). Lens regeneration from the cornea in *Xenopus laevis*. *J. Exptl. Zool.* **154**, 39.
Glücksmann, A. (1940). Development and differentiation of the tadpole eye. *Brit. J. Ophthalmol.* **24**, 154.
Glücksmann, A. (1951). Cell deaths in normal vertebrate ontogeny. *Biol. Rev. Cambridge Phil. Soc.* **26**, 59.
Gontcharoff, M. (1953). Le phototropisme chex *Lineus ruber* et *Lineus sanguineus* au cours de la régénération des yeux. *Ann. Sci. Nat. Zool. Biol. Animale* **15**, 369.
Gontcharoff, M. (1956). Le phototropisme chez *Lineus ruber* et *Lineus sanguineus* au cours de la régénération des yeux. *Proc. Intern. Congr. Zool. Copenhagen* **14**, 208.
Gontcharoff, M., and Lechenault, H. (1958). Sur le déterminisme de la ponte chez *Lineus lacteus*. *Compt. Rend.* **246**, 1929.
Goodrich, E. S. (1909). Cyclostomes and fishes. *In* "A Treatise on Zoology" (E. R. Lankester, ed.), Vol. 9. Black, London.
Greenberg, R., and Popper, H. (1941). Demonstration of vitamin A in the retina by fluorescence microscopy. *Am. J. Physiol.* **134**, 114.
Häggendal, J., and Malmfors, T. (1965). Identification and cellular localization of the catecholamines in the retina and the choroid of the rabbit. *Acta Physiol. Scand.* **64**, 58.
Harrison, R. G. (1920). Experiments on the lens in Amblystoma. *Proc. Soc. Exptl. Biol. Med.* **17**, 199.
Herrick, C. J. (1901). The cranial nerves and cutaneous sense organs of the North American siluroid fishes. *J. Comp. Neurol.* **11**, 177.
Hesse, R. (1899). Untersuchungen über die Organe der Lichtempfindung bei niederen Thieren. V. Die Augen der Polychäten Anneliden. *Z. Wiss. Zool.* **65**, 446.

Holtfreter, J. (1947). Changes of structure and the kinetics of differentiating embryonic cells. *J. Morphol.* **80**, 57.

Huxley, J. S., and de Beer, G. R. (1934). "The Elements of Experimental Embryology". Cambridge Univ. Press, London and New York.

Jensen, D. D. (1960). Hoplonemertines, myxinoids, and deuterostome origins. *Nature* **187**, 645.

Jensen, D. D. (1963). Hoplonemertines, myxinoids and vertebrate origins. *In* "The Lower Metazoa: Comparative Biology and Phylogeny" (E. C. Dougherty *et al.*, eds). p. 113. Univ. Calif. Press., Berkeley and Los Angeles.

Kappers, J. A. (1965). Survey of the innervation of the epiphysis cerebri and the accessory pineal organs of vertebrates. *Progr. Brain Res.* **10**, 87.

Kelly, D. E. (1962). Pineal organs, photoreceptors, secretion and development. *Am. Scientist* **50**, 597.

Knight-Jones, E. W. (1952). On the nervous system of *Saccoglossus cambrensis* (Enteropneusta). *Phil. Trans. Roy. Soc.* (*London*) **B236**, 315.

Lechenault, H. (1963). Sur l'existence de cellules neurosécrétrices chez les Hoplonemertes. Caractéristiques histochimiques de la neurosécrétion chez les Nemertes. *Compt. Rend.* **256**, 3201.

Lewin, R. A. (1953). Studies on the flagella of Algae. (ii) Formation of flagella by *Chlamydomonas* in light and darkness. *Ann. N.Y. Acad. Sci.* **56**, 1091.

Lillie, R. D. (1952). Histochemical studies on the retina. *Anat. Record* **112**, 477.

Ling, E. A. (1969). The structure and function of the cephalic organ af a nemertine, *Lineus ruber*. *Tissue and Cell.* **1,** 503.

Loewenstein, W. R., Socolar, S. J., Higashino, S., Kanno, Y., and Davidson, N. (1965). Intercellular communication, renal urinary bladder, sensory and salivary gland cells. *Science* **149**, 195.

Löwenstein, O., and Wersäll, I. (1959). A functional interpretation of the electronmicroscopic structure of the sensory hairs in the cristae of the elasmobranch, *Raja clavata*, in terms of directional sensitivity. *Nature* **184**, 1807.

McCord, C. P., and Allen, F. P. (1917). Evidences associating pineal gland function with alterations in pigmentation. *J. Exptl. Zool.* **23**, 207.

McIsaac, W. M., Farrell, G., Taborsky, R. G., and Taylor, A. N. (1965). Indole compounds: Isolation from pineal tissue. *Science* **148**, 102.

McKeehan, M. S. (1951). Cytological aspects of embryonic lens induction in the chick. *J. Exptl. Zool.* **117**, 31.

Marks, W. B., Dobelle, W. H., and MacNichol, E. .F, Jr. (1964). Visual pigments of single primate cones. *Science* **143**, 1181.

Maxwell, D. S., and Pease, D. C. (1956). The electron microscopy of the choroid plexus. *J. Biophys. Biochem. Cytol.* **2**, 467.

Merker, von E. (1934). Die Sichtbarkeit, ultravioletten Lichtes. *Biol. Rev. Cambridge Phil. Soc.* **9**, 49.

Millen, J. W., and Rogers, G. E. (1956). An electron microscope study of the chorioid plexus of the rabbit. *J. Biophys. Biochem. Cytol.* **2**, 407.

Moscona, A. (1957). Formation of lentoids by dissociated retinal cells of the chick embryo. *Science* **125**, 598.

Nawitzki, W. (1931). *Procarinina remanei.* Eine neue Paläonemertine der Kieler Förde. *Zool. Jahrb. Abt. Anat. Ontog. Tiere* **54**, 160.

Newth, D. R., and Ross, D. M. (1955). On the reaction to light of *Myxine glutinosa* L. *J. Exptl. Biol.* **32**, 41.

Nilsson, S. E. G. (1964). Interreceptor contacts in the retina of the frog. *J. Ultrastruct. Res.* **11**, 147.

Noel, R., and Jullien, A. (1933). Recherches histologiques sur le corps blanc des Céphalopoda. *Arch. Zool. Exptl. Gen.* **75**, 485.

Noell, W. K. (1960). The impairment of visual cell structure by iodoacetate. *J. Cellular Comp. Physiol.* **40**, 25.

Okano, M., Weber, A. F., and Frommes, S. P. (1967). Electron microscopic studies of the distal border of the canine olfactory epithelium. *J. Ultrastruct. Res.* **17**, 487.

Oksche, A. (1965). Survey of the development and comparative morphology of the pineal organ. *Progr. Brain Res.* **10**, 3.

Oksche, A., and Harnack M. (1963). Electronmicroscopische Untersuchungen am Stirnorgan von Anuren (sur Fragen der Lichtrezeptoren). *Z. Zellforsch. Mikroskop. Anat.* **59**, 239.

Oztan, N., and Gorbman, A. (1960). Responsiveness of the neurosecretory system of larval lampreys (*Petromyzon marinus*) to light. *Nature* **186**, 167.

Pappas, G. D., and Smelser, G. K. (1961). The fine structure of the ciliary epithelium in relation to aqueous humour secretion. *In* "The Structure of the Eye" (G. K. Smelser, ed.), p. 453. Academic Press, New York.

Plack, P. A., Kon, S. K., and Thompson, S. Y. (1959). Vitamin A_1 aldehyde in the eggs of the herring (*Clupea harengus*. L) and other marine teleosts. *Biochem. J.* **71**, 467.

Pomerat, C. M., and Littlejohn, L., Jr. (1956). Observations on tissue culture of the human eye. *Southern Med. J.* **49**, 230.

Porter, K. R., and Yamada, E. (1960). Studies on the endoplasmic reticulum. V. Its form and differentiation in pigment epithelial cells of the frog retina. *J. Biophys. Biochem. Cytol.* **8**, 181.

Press, N. (1959). Electron microscope study of the distal portion of a planarian retinular cell. *Biol. Bull.* **117**, 511.

Quay, W. B. (1963). Circadian rhythm in rat pineal serotonin and its modification by estrous cycle and photoperiod. *Gen. Comp. Endocrinol.* **3**, 473.

Quay, W. B. (1965). Histological structure and cytology of the pineal organ in birds and mammals. *Progr. Brain Res.* **10**, 49.

Quay, W. B. (1966). Rhythmic and light-induced changes in levels of pineal 5-hydroxyindoles in the pigeon (*Columba livia*). *Gen. Comp. Endocrinol.* **6**, 371.

Ritchie, A. (1960). A new interpretation of *Jamoytius kerwoodi* (White). *Nature* **188**, 647.

Röhlich, P., and Török, L. J. (1961). Elektronmikroskopische Untersuchungen des Auges von Planarien. *Z. Zellforsch. Mikroskop. Anat.* **54**, 361.

Rushton, W. A. H. (1955). Foveal photopigments in normal and colourblind. *J. Physiol. (London)* **129**, 41P.

Sager, R., and Granick, S. (1954). Nutritional control of sexuality in *Chlamydomonas reinhardi*. *J. Gen. Physiol.* **37**, 729.

Salensky, W. (1896). Bau und Metamorphose des Pilidium. *Z. Wiss. Zool.* **43,** 481.

Scharrer, E. (1941). The nucleus preopticus of *Fundulus heteroclitus*. *J. Comp. Neurol.* **74**, 81.

Schröder, O. (1918). Beitrage zur Kenntnis von *Geonemertes palaensis* (Semper). *Abhandl. Senckenberg. Naturforsch. Ges.* **35**, 155.

Sidman, R. L. (1958). Histochemical studies on photoreceptor cells. *Ann. N.Y. Acad. Sci.* **74**, 182.

Sjöstrand, F. S. (1949). An electron microscope study of the retinal rods of the guinea-pig eye. *J. Cellular Comp. Physiol.* **33**, 382.

Sjöstrand, F. S. (1961). Electron microscopy of the retina. *In* "The Structure of the Eye" (G. K. Smelser, ed.), p. 1. Academic Press, New York.

Spemann, H. (1912). Zur Entwicklung des Wirbeltierauges. *Zool. Jahrb. Abt. Allgem. Zool. Physiol.* **32**, 1.

Spemann, H. (1938). "Embryonic Development and Induction". Yale Univ. Press, New Haven, Connecticut.
Steven, D. M. (1955). Experiments on the light sense of the hag, *Myxine glutinosa* L. *J. Exptl. Biol.* **32**, 22.
Steyn, W. (1959). Ultrastructure of pineal eye sensory cells. *Nature* **183**, 764.
Stone, L. S. (1967). An investigation recording all salamanders which can and cannot regenerate a lens from the dorsal iris. *J. Exptl. Zool.* **164**, 87.
Svaetichin, G., and MacNichol, E. F. Jr, (1958). Retinal mechanisms for chromatic and achromatic vision. *Ann. N.Y. Acad. Sci.* **74**, 385.
Svaetichin, G., Negishi, K., and Fatehchand, R. (1965). Cellular mechanisms of a Young-Hering visual system. *Ciba Found. Symp. Colour Vision, Physiol. Exptl. Psychol.* p. 178.
Tansley, K. (1933). The formation of rosettes in the rat retina. *Brit. J. Ophthalmol.* **17**, 321.
Tokuyasu, K., and Yamada, E. (1959). The fine structure of the retina studied with the electron microscope. IV. Morphogenesis of outer segments of retinal rods. *J. Biophys. Biochem. Cytol.* **6**, 225.
Twitty, V. (1956). Eye. *In* "Analysis of Development" (B. H. Willier, P. Weiss, and V. Hamburger, eds.), p. 403. Saunders, Philadelphia, Pennsylvania.
Van de Kamer, J. C. (1965). Histological structure and cytology of the pineal complex in fishes, amphibians and reptiles. *Progr. Brain Res.* **10**, 30.
Villegas, G. M. (1960). Electronmicroscopic study of the vertebrate retina. *J. Gen. Physiol.* **43**, Suppl. 2, 15.
Wald, G. (1958). Retinal chemistry and the physiology of vision. *Natl. Phys. Lab. Gt. Brit. Proc. Symp.* **8**, 7.
Wald, G. (1964). The receptors of human colour vision. *Science* **145**, 1007.
Wald, G., and Krainin, J. M. (1963). The median eye of *Limulus*: An ultraviolet receptor. *Proc. Natl. Acad. Sci. U.S.* **50**, 1011.
Walls, G. L. (1942). "The Vertebrate Eye". Cranbrook Press, Bloomfield Hills, Mich.
Weiss, P., and Fitton Jackson, S. (1961). Fine-structural changes associated with lens determination in the avian embryo. *Develop. Biol.* **3**, 532.
Wells, M. J. (1964). Hormonal control of sexual maturity in cephalopods. *Bull. Natl. Inst. Sci. India* **27**, 63.
Wells, M. J., and Wells, J. (1959). Hormonal control of sexual maturity in *Octopus*. *J. Exptl. Biol.* **36**, 1.
White E. I. (1946). *Jamoytius kerwoodi*, a new chordate from the Silurian of Lanarkshire. *Geol. Mag.* **83**, 89.
Willmer, E. N. (1955). A physiological basis for human colour in the central fovea. *Doc. Ophthalmol.* **9**, 235.
Willmer, E. N. (1965). Duality in the retina. *Ciba Found. Symp. Colour Vision, Physiol. Exptl. Psychol.* p. 89.
Wilson, C. B. (1900). The habits and early development of *Cerebratulus lacteus*. Verill. *Quart. J. Microscop. Sci.* **43**, 97.
Wolken, J. J. (1958a). Retinal structure of mollusc cephalopods; *Octopus*, *Sepia*. *J. Biophys. Biochem. Cytol.* **4**, 835.
Wolken, J. J. (1958b). Studies on photoreceptor structures. *Ann. N.Y. Acad. Sci.* **75**, 161.
Wurtman, R. J., Axelrod, J., and Chu, E. W. (1963a). Melatonin, a pineal substance. Effect on the rat ovary. *Science* **141**, 277.
Wurtman, R. J., Axelrod, J., and Phillips, L. S. (1963b). Melatonin synthesis in the pineal gland: control by light. *Science* **142**, 1071.
Wurtman, R. J., Axelrod, J., and Fischer, J. E. (1964). Melatonin synthesis in the pineal gland: Effect of light mediated by the sympathetic nervous system. *Science* **143**, 1328.

Yamada, E. (1958). A peculiar lamellated body observed in the cells of the pigment epithelium of the retina of the bat. *J. Biophys. Biochem. Cytol.* **4**, 329.

Young, R. W., and Droz, B. (1968). The renewal of protein in retinal rods and cones. *J. Cell Biol.* **39**, 169.

Young, J. Z. (1935). The photoreceptors of lampreys. II. The functions of the pineal complex. *J. Exptl. Biol.* **12**, 254.

Young, J. Z. (1950). "The Life of Vertebrates." Oxford Univ. Press (Clarendon), London and New York.

Zonana, H. V. (1961). Fine structure of the squid retina. *Bull. John Hopkins Hosp.* **109**, 185.

CHAPTER 16

FROM NEMERTEOIDS TO VERTEBRATES: 4. THE VASCULAR SYSTEM, COELOM, AND URINOGENITAL SYSTEM

The Vascular System

The nemertines of to-day have a closed vascular system which displays many interesting features (see Chapter 12) in addition to great variability. Basically, it consists of lateral vessels joined fore and aft, and sometimes possessing a secondary loop in the cephalic region (see Fig. 12.10). Normally there is nothing in the way of a heart, though some of the vessels may be contractile. In view of the greater frequency of cross-connexions in the region of the intestine, the vascular system does not seem to be primarily respiratory in function, though it may be of some benefit in that way. It seems likely that its development assisted in the distribution of food materials, and perhaps of hormones, to the rest of the body and facilitated the carriage of waste products. It may also assist bodily movements by its hydrodynamic effects. As shown later, the vessels often have a close relationship with the excretory system; in some instances the vascular system and the rhynchocoel are intimately related (see Figs. 14.4, 14.5), at least in so far as the pressures of their contained fluids are concerned. The fluid in the vessels is presumably moved in a fairly thorough, if random, fashion by the general peristaltic, antiperistaltic, lengthening, and shortening movements of the body. Any localized movement of the body's circular muscles, for example, must displace the blood from the region of the contraction to other parts of the body. Thus one function of the blood system may be to allow such movements to occur more easily. In species that burrow, the local turgor produced in regions of relaxation of the circular muscles may serve to anchor the worm in its burrow, and thus facilitate movement of the rest of its body by means of extension and contraction in the longitudinal axis. It is, however, interesting to notice that *Cerebratulus*, by the periodic intake and expulsion of fluid into and out of its pharynx (Wilson, 1900), can make fresh surfaces available for receiving oxygen and eliminating CO_2, and these surfaces are well vascularized. Perhaps the large size of *Cerebratulus*, or some other condition of its life, e.g. swimming, necessitates greater supplies of oxygen.

It appears likely that the carriage or storage of oxygen may play some part in the function of the blood system because many species of nemertines have coloured corpuscles, though not apparently in great numbers, and, at least in some instances, the pigment has been identified as haemoglobin (Lankester, 1872; Hubrecht, 1883). This, in itself, is interesting as a pointer toward vertebrate organization, though it is by no means diagnostic. Haemoglobin, after all, is widely distributed in Nature from bacteria to man, but that it should be housed in relatively non-motile corpuscles in the blood is a point of some relevance in considering the evolution of the blood system. Haemoglobin is also said to occur in the nervous system of several nemertines, though for what purpose is unknown. A store of oxygen near the important neural tissue, however, could presumably always be useful in times of low oxygen content, e.g. when the worm is in its burrow. Large blood spaces in the immediate vicinity of the cephalic ganglia and the cephalic organ are a conspicuous feature of some nemertines. What is more suggestive and pertinent to the evolution of vertebrates, however, is the presence of several other classes of corpuscles in the blood of some nemertines (see Fig. 12.11). For example, Ohuye (1942) has described in one species (*Lineus fuscoviridis*), in addition to the oval, flattened nucleated and non-motile red blood corpuscles, large "lymphocytes", basophil cells, and eosinophil cells. The last two are granular corpuscles, and it is the granules that stain selectively. Among the eosinophils are those with large granules and those with small granules. Moreover, these granular corpuscles, like the granular corpuscles of vertebrates, give positive reactions to the test for the copper-containing peroxidase system (benzidine and H_2O_2) and also to the Nadi reaction. The lymphocytes do not. This nemertine therefore has virtually the same classes of blood cells as are found in the vertebrates in various forms, with the exception that the small lymphocyte is apparently missing. Spindle-shaped cells were also described as being present (see Fig. 18.8); they could be endothelial cells, or perhaps they are akin to the small lymphocytes of higher forms. The absence of the small lymphocyte as such is of considerable interest, because, as shown later, it appears to be a rather characteristically vertebrate cell and its origin, together with that of lymphoid tissue and immunity reactions in general, needs further discussion (see p. 525).

No information seems to be available on the place in the nemertine body where the various blood cells arise. They are presumably mesenchymal (mesohyl) cells derived from the walls of the blood vessels, but there is no certainty of this. The walls of some of the blood vessels frequently appear rather "tatty" in fixed preparations; this may be an indication that cells are leaving them and floating away freely in the contained fluid.

The smaller blood vessels are normally lined with endothelial cells which, when the vessel is not distended, bulge towards the centre of the lumen (Böhmig, 1897, 1898). In *Lineus*, some of the blood spaces near the cephalic organ have features suggesting a secretory function (Fig. 16.1). The endothelium normally rests on what appears to be an amorphous basement

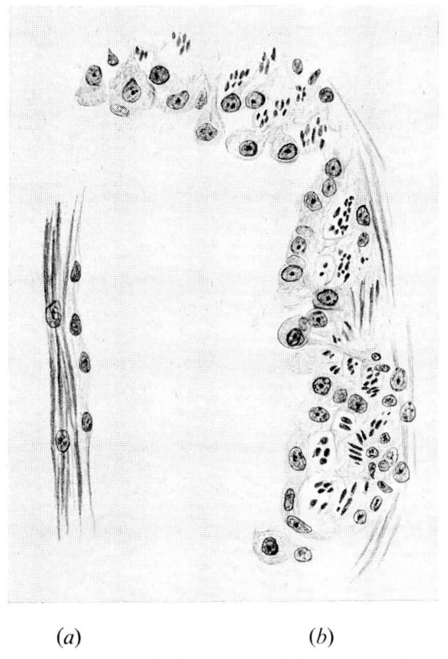

(*a*) (*b*)

FIG. 16.1. Epithelia in blood vessels in *Lineus*. (*a*) Endothelium of the lateral blood vessel; lumen on the right. (*b*) The lining cells of the wall of the rhynchodaeal blood vessel; lumen on the left.

membrane which may be surrounded by circular muscle fibres and an outer collagenous coat (Riepen, 1933). There is thus the basis for the development of blood vessels along the lines found in vertebrates, with the notable exception that there is no elastic tissue, at least as detected by ordinary staining methods. On the other hand, in the absence of a developed heart, the necessity for elasticity in the walls of the vessels had presumably not yet arisen. It may of course be argued that, in order to be efficient, a conducting path for blood must have certain properties and that therefore the similarity between the vessels of nemertines and the smaller vessels of vertebrates is of no evolutionary significance. Indeed, this argument receives strong support from the fact that the arteries of cephalopod molluscs are built on exactly the same general plan as are the arteries of vertebrates (Fig. 16.2). For this there would seem to be three possible explanations. First, this is the way that arteries need to be constructed to make them efficient. Second, the blood vessels of molluscs and vertebrates are both evolved from a common stock

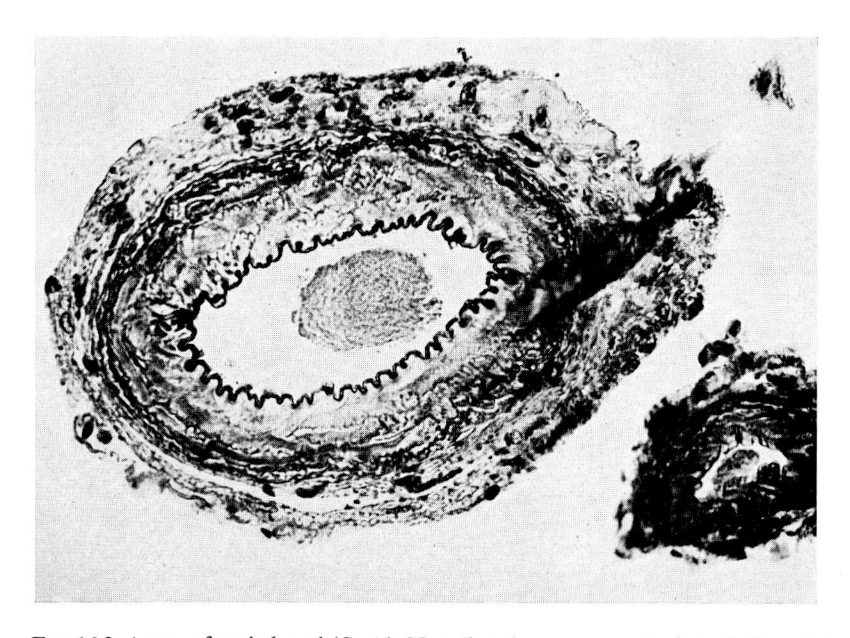

FIG. 16.2. Artery of cephalopod (*Sepia*). Note that the arrangement of elastic tissue into a thick internal lamina and into subsidiary laminae in the tunica media is similar to the arrangement of elastic fibres in the arteries of vertebrates. The tunica media, containing muscle fibres, and the tunica adventitia, composed largely of collagen fibres, are also arranged along vertebrate lines.

and are adaptations of similar materials to serve similar purposes. Third, the similarity is merely coincidental. At first sight, the second and third explanations appear to be unlikely, but, if the arguments used on p. 385 concerning the possible developments of the cephalic organ and the eye have any weight, the second one should not be dismissed immediately.

A peculiar and, at present, inexplicable feature of the vascular system of some nemertines is the presence of leucine-aminopeptidase in the endothelium of many of the vessels (Gibson and Jennings, 1967). In relation to certain

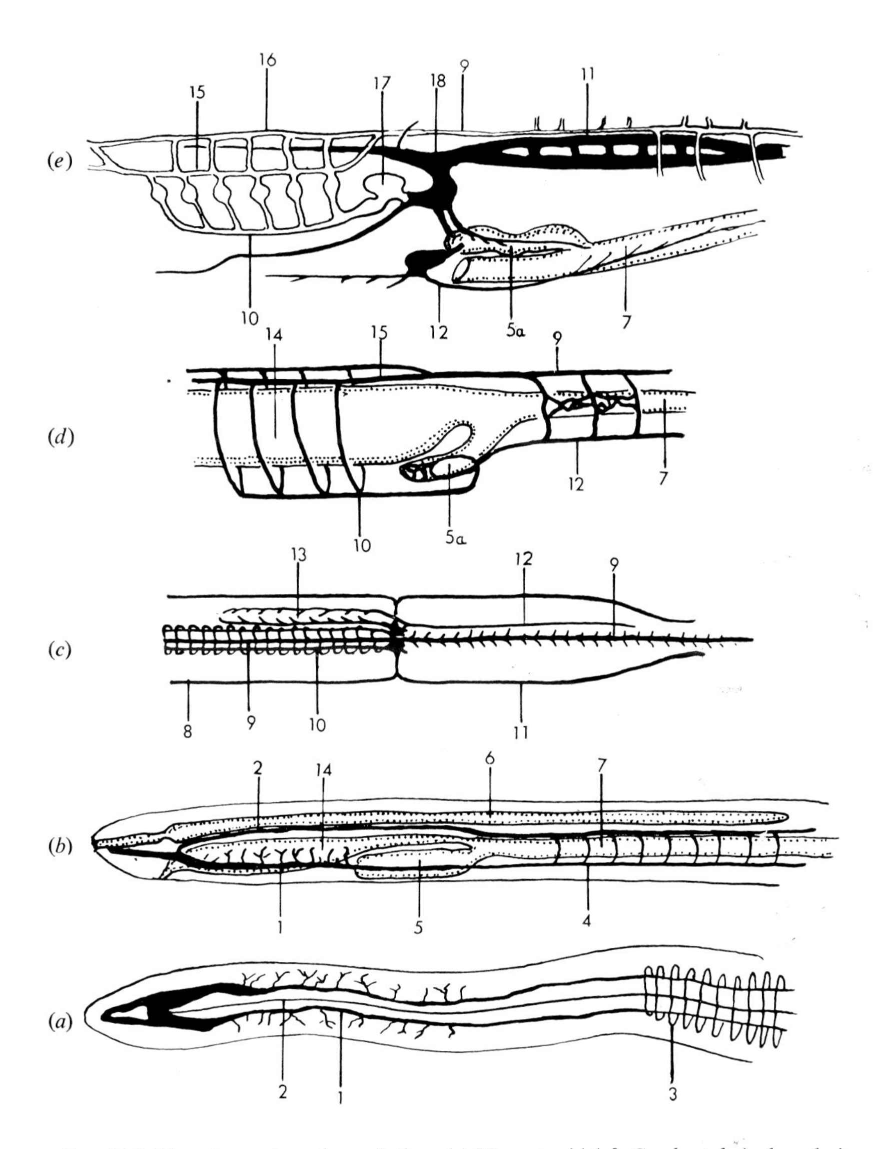

FIG. 16.3. Vascular systems in evolution. (*a*) Nemerteoid (cf. *Cerebratulus*), dorsal view; (*b*) nemerteoid, lateral view; (*c*) *Amphioxus*, dorsal view (after Young, 1950); (*d*) *Amphioxus*, lateral view (after Parker and Haswell, 1940); (*e*) *Bdellostoma* (after Goodrich, 1909). 1, Lateral vessel in pharyngeal region with numerous branches to pharyngeal wall; 2, dorsal vessel; 3, links between dorsal vessel and lateral vessel in intestinal region; 4, lateral vessel; 5, diverticulum; 5a, liver; 6, rhynchocoel; 7, intestine; 8, anterior cardinal vein; 9, dorsal aorta; 10, ventral aorta; 11, posterior cardinal vein; 12, sub-intestinal vein; 13, vessels to and from liver diverticulum; 14, pharynx; 15, paired anterior aorta; 16, median anterior aorta, 17, atrium; 18, sinus venosus.

later discussions (see p. 454) it may be significant that the enzyme is sometimes absent from the flattened endothelium of the lacunae.

The presence of an elementary vascular system in the nemertines opens the way to increased movement, more efficient excretion, hormonal control and also to the development of an oxygen-carrying system from any suitable respiratory surface. Thus the correlation between the richness of the supply of blood vessels to the pharynx of *Cerebratulus*, the worm which alternately fills and empties its pharynx (Wilson, 1900), is indicative of the possibilities of opening up a gill system in that area. Whilst the pharynx remained as a relatively closed system, as in *Cerebratulus* itself, the flow of fluid in the vessels would probably depend on the alternating suction and compression used in drawing fluid into and expelling it from the main cavity. The perforation of gill-slits and the movement of fluid through the pharynx by ciliary action, or by the animal maintaining a fixed position relative to a current of water, or by swimming through the water would, however, put an end to this suction-pumping action on the blood. Any circulation that was to continue would have to depend on active propulsion of the blood by some other means. Although this might still depend on bodily movements elsewhere, the development of rhythmically contractile vessels to drive blood through the vessels of the gill arches would be biologically advantageous. In this manner, it seems reasonable to account for the development of a heart in the form and position in which it is found in tunicate larvae, in *Amphioxus*, and in the vertebrates (Fig. 16.3). The heart muscle would presumably develop from the local mesenchymal cells, and tissue culture of vertebrate heart muscle shows it to be of mechanocyte origin.

Another consequence of the adoption of a filter-feeding mechanism and the type of movement associated with it is the change in what might be called the hydrostatic skeleton. In nemertines there are three main reservoirs of fluid: the alimentary canal, containing essentially external fluid; the extensive rhynchocoel cavity; and the more limited blood vessels. In some species, e.g. in *Lineus*, the cavities of the blood vessels and the rhynchocoel may be in communication, with regard to pressure if not actually, near the base of the proboscis, for at this point a conical tube projects from the rhynchocoel into

---

FIG. 16.4. Two transverse sections in the cephalic region of *Lineus*. Each section shows the proboscis in its rhynchocoel and the blood vessel just ventral to the latter. In section (*a*) the floor of the rhynchocoel shows the opening of a tube which becomes entirely surrounded by and enclosed within the blood vessel in section (*b*), which is a few sections posterior to (*a*). Note the almost columnar arrangement of the endothelial cells of the blood vessel. In section (*b*) the blood vessel is beginning to subdivide into two dorsal lateral vessels and a ventral vessel. These two vessels surround the cephalic organ. Note also that some of the nerve cells in the ganglia on either side of the proboscis are neurosecretory.

See facing page→

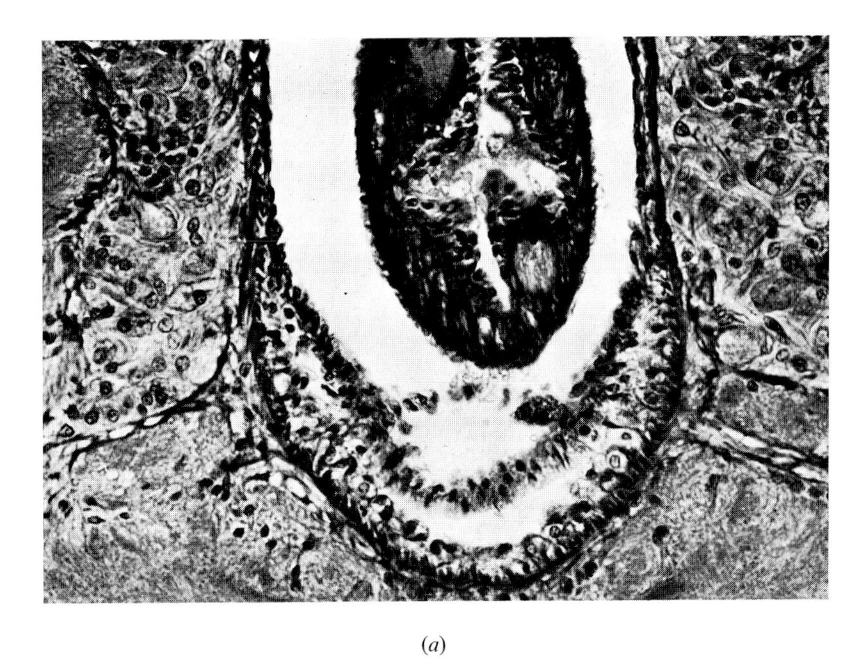

(a)

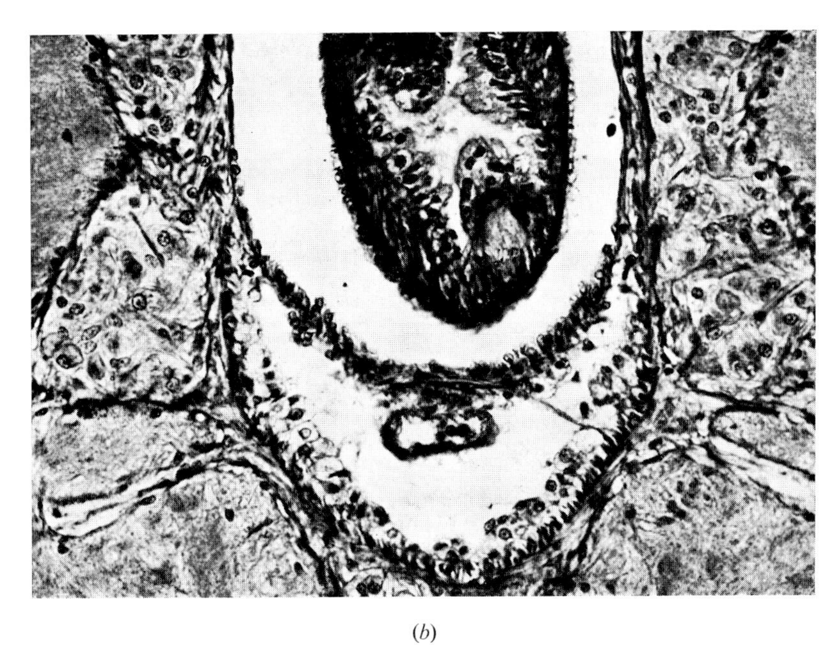

(b)

the dorsal blood vessel (Figs. 16.4 and, 14.5, 14.6). In addition to these three main cavities there are the gonocoel cavities, particularly in the male, and also the nephridial tubules which may contain significant quantities of fliud. There is no main coelomic or body cavity which could be regarded as the obvious homologue of that in the vertebrates.

The Nature of Body Cavities

Although names have been given to the various cavities present in the nemertines, partly on the basis of analogy with other organisms, the distinctions between them may not really be very sound. Unfortunately, little is known about the compositions of the fluids in the various cavities. In *Lineus*, for example, the rhynchocoel fluid and the blood both seem to have rather similar corpuscles. The rhynchocoel has been described as originating from the primary blastocoel cavity (Hubrecht, 1886) and as being derived by a split occurring between two muscle coats in the mesenchyme. In either case it would presumably be lined with mesenchymal cells. If its derivation is by the latter method, it is comparable to the described origin of blood vessels and of gonocoels. In the first instance, therefore, there may be little difference in structure or function between these variously named cavities, and they have not differentiated until later in response to changed conditions; the gonocoel for the more efficient management of the germ cells; the blood vessels for the circulation of an oxygenating as well as a nutrient and unifying fluid; and the rhynchocoel for the hydraulics of proboscis ejection and retraction. A fluid-filled space around the rhynchodaeum is also a conspicuous feature of some species.

A more searching study of the nature of the cells surrounding the various body cavities should be made, however. For example, the gonocoels are said to arise from splits in the mesenchyme, but the developing gonocoels of *Lineus* are surrounded by cells that have the qualities of epitheliocytes and not of secondarily flattened mesenchymal cells (Fig. 16.5). In vertebrates also, the cells of the coelomic cavity often display the properties and character of epitheliocytes, e.g. in the female amphibian the coelomic cavity may at times be lined with a ciliated epithelium (Rugh, 1935).

As evolution has proceeded, whether towards annelids, arthropods, or vertebrates, more specific body-cavities have been developed. In the case of the annelids the likely explanation is that, in order to satisfy the hydrodynamic requirements of a particular form of bodily movement, the gonocoel cavities developed prematurely, enlarged (Goodrich, 1945), and became secondarily connected with each other by a valvular system which allows the slow or damped transport of fluid from one cavity to the next (Clark, 1964).

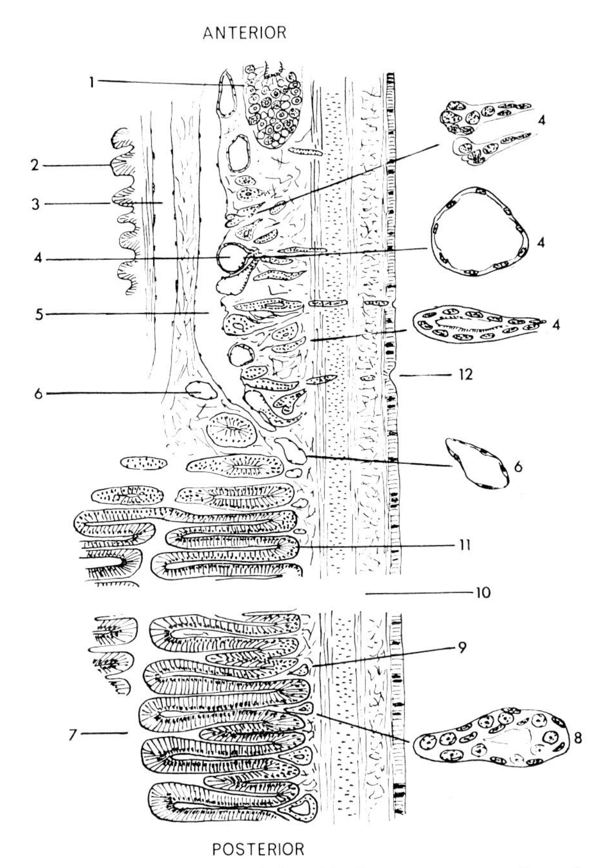

FIG. 16.5. Diagram to show the relationship between gonocoels and nephridia in a longitudinal section of *Lineus*. 1, Pharynx; 2, proboscis; 3, rhynchocoel; 4, nephridial tubes; 5, lateral blood vessel; 6, Branch of lateral blood vessel; 7, lumen of intestine; 8, gonocoel; 9, most anterior gonocoel; 10, space equivalent to about eight diverticula; 11, intestinal diverticulum; 12, nephridiopore opening on epidermis.

In the vertebrates the development of the main body-cavity is a more obscure process. It will be remembered that, embryologically, the main coelom of vertebrates is normally considered to be divisible into, or derived from, the myocoel, nephrocoel, and splanchnocoel cavities. It tends to develop, in part, as the myocoels in the cavities of the somites, but the main coelomic cavity itself arises from a hollowing-out of the lateral-plate mesoderm. How far this is primitive and how far it is a specialization is, however, undetermined. In one or two cases, e.g. in *Lampetra fluviatilis*, the lateral plate cavity is formed by the fusion of segmental cavities (Damas, 1944). If the possible origin for the myocoels discussed on an earlier page (p. 299) were accepted, the problem would resolve itself into the origin of the nephrocoel and splanchnocoel and the relationship of the gonocoels and nephridia or urinary tubules to these cavities. Along with the appearance of a main coelomic cavity in the chordates, the fate of the rhynchocoel of the nemerteoids and the development of the axial skeletal rudiments of the vertebrates must also be considered. The whole problem must be viewed in relation to the changes necessary in the fluid mechanics of the body in order to allow the creeping worm-like movements to be converted into those characteristic of a free-swimming fish-like creature.

The gonads of nemerteoids, which are arranged serially in the posterior part of the animal, are said to develop from the mesenchyme, simply as cavities lined with an epithelium (Fig. 16.5), though the nature of this epithelium is obscure. These cavities house the germ cells, and, when the animals are mature, usually send out-growths towards the exterior, the outer part of the out-growth being lined with a ciliated epithelium. Occasionally the body wall simply ruptures to provide the exit, a method of shedding the germ cells not unlike that in *Amphioxus*, where special cicatrices open temporarily for the purpose. The cells lining the gonocoels presumably must control the fluid contained in the cavity, and, when the tube opens to the outside, the ciliated cells lining the duct may also assist in this function. In animals inhabiting waters of varying salinity, such control may be specially important. One possibility for the development of the coelom, therefore, is that the gonocoels should develop prematurely and then enlarge. This is very likely what has occurred in the annelids. In the vertebrates the coelomic cavity could similarly be the result of the fusion of such serially arranged and enlarged gonocoels, the enlargement having perhaps occurred as a means of compensating, hydrodynamically, for the failure of a rhynchocoel to develop after the suppression of the original function of the proboscis. According to this view, the coelomic epithelium would be a secondary epithelium having been formed, like the lining of the blood vessels, by the flattening and cohesion of mesenchyme cells, though the caveat entered in relation to the nature of the gonocoels in the rhabdocoels (see p. 251) also applies here. The epithelium would, of course, continue to house the germ cells, at least in some part of it.

A second possibility is that the coelomic cavity of vertebrates is an essentially new structure formed, like the rhynchocoel before it, by a split in the mesenchyme in which fluid collected and which again became lined with a secondary epithelium formed of coherent mesenchyme cells and into which the gonads later projected. In view of the rather watery and gelatinous character of the mesenchyme in some nemertines, this does not seem to be a very unlikely event. It might well be encouraged in situations where the salt concentration in the external environment is low, as in estuaries, for under such conditions the water content of the animal visibly increases and all the cavities of the body enlarge. In 50% sea-water, for example, the pharynx of *Lineus* may for a time become almost completely surrounded by fluid-filled spaces crossed here and there by strands of connective tissues. A similar situation has been described as normal in *Procarinina* (Nawitzki, 1931) (Fig. 16.6.).

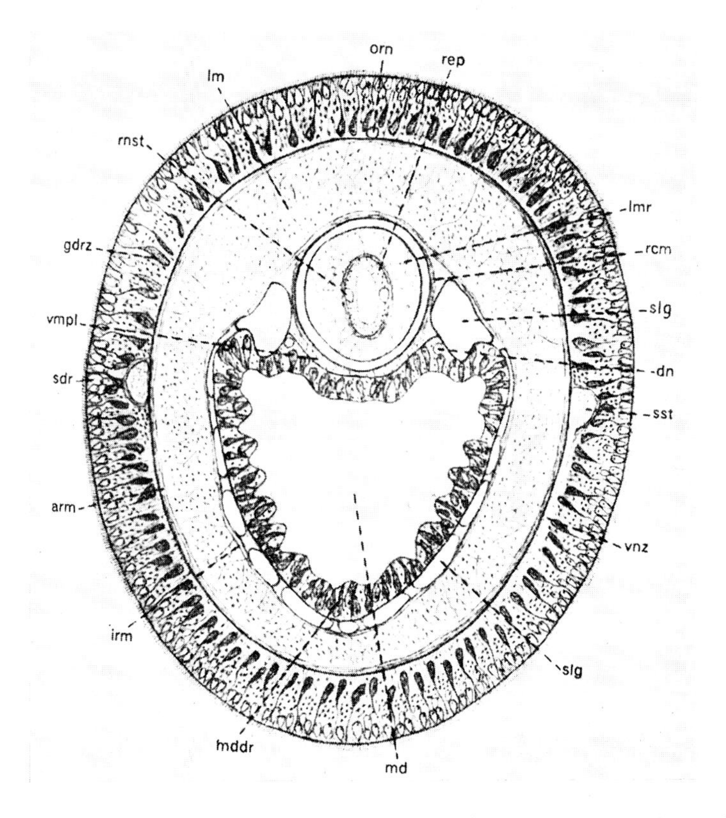

FIG. 16.6. Transverse section of pharyngeal region of *Procarinina*. arm, Outer circular muscle; dn, pharyngeal nerve; gdrz, large gland cells; irm, inner circular muscle; lm, longitudinal muscle; lmr, longitudinal muscle of proboscis; md, cavity of pharynx; mddr, pharyngeal glands; orn, dorsal nerve; rcm, rhynchocoel muscle; rnst, proboscis nerve; sdr, mucous glands; slg, blood vessels; sst, lateral nerve; vmpl, ventral muscle plate; vnz, ventral nerve. (Nawitzki, 1931.)

A third possibility, and the one usually accepted in vertebrate embryology, is that the coelomic cavity is formed by the downward extension of the cavities of the somites and the fusion of these extensions. These extensions would therefore, on the hypothesis outlined for the origin of somites (see

p. 299), correspond exactly to the diverticula of the gut and would be lined with primary epithelial cells. This certainly fits the facts in that the coelomic epithelium, though usually appearing as a typical mesothelium, may sometimes, as in amphibia, become ciliated and may also give rise to epithelial structures like the sex cords and their derivatives in the gonads. In turtles the coelomic epithelium can behave like a phagocytic membrane (Delaney, 1929). In opposition to this idea, however, it must be emphasized that gonocoels and sex-tubes exist in nemertines, side by side and alternating with the diverticula of the gut.

Urinogenital Tubules in the Invertebrates

These three possibilities for the development of the body cavity of the vertebrates have to be studied also in the light of the arrangements of the so-called segmental coelomoducts. These coelomoducts are a prominent feature of the cyclostomes and the more primitive gnathostomes in which they take the form of open pronephric funnels, joining an archinephric duct and then a series of mesonephric tubes (Fig. 16.7). In higher forms the latter series may be augmented by a series of metanephric tubules which, in mammals, constitute the definitive kidney and are dependent on the archinephric duct for their development. The nature of these various tubules and the cellular characters of the coelomic epithelium itself are the key problems, not only with regard to the origin of the vertebrate coelom, but also with regard to the nature of the kidney tubules, the cords of cells constituting the adrenal cortex, and the supporting tissues for the germ cells in the gonads themselves. It is therefore pertinent to examine the nemertines again with these problems in mind. Nemertines have two sets of tubules which are, or can be, serially arranged along the length of the body, namely the nephridia and the gonoducts. As already mentioned, the most common type of nephridium in these worms is a tube ending blindly in a flame cell or solenocyte. Usually these tubes join (or developmentally branch) to form multiple units (see Figs. 12.8, 12.9). They may open directly to the exterior, or they may join a longitudinally running duct which may have several openings or only a single common nephridiopore. Although nephridia are often distributed widely throughout the body, it is interesting that in the nemertines themselves they are seldom regular in their arrangement. In annelids, on the other hand, with the development of the coelomic septa, the nephridia may become strictly segmental.

Nephridia are essentially invaginations from the ectoderm, and their branching habit is strongly developed. Initially, it must be supposed, the activities of the cells composing them were variations of the activities of the

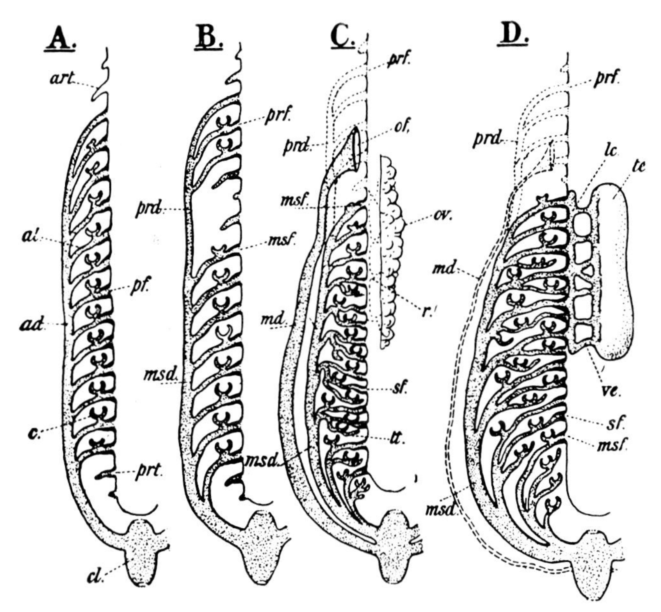

FIG. 16.7. Diagrams showing possible development of the urinogenital system in the Craniata. (*a*) Hypothetical ancestral stage with continuous archinephros; (*b*) cyclostome with anterior pronephros; (*c*) female gnathostome; (*d*) male gnathostome. ad, archinephric duct; art, anterior vestigial tube; at, archinephric tube; c, Malpighian capsule; cl, cloaca; lc, longitudinal canal; md, Müllerian duct; msd, mesonephric duct; msf, mesonephric funnel; of, coelomic funnel; ov, ovary; pf, coelomostome; prd, pronephric duct; prf, pronephric funnel; prt, posterior vestigial tube; r, vasa efferentia, vestigial; sf, secondary funnel; te, testis; tt, tertiary tubule; ve, vas efferens. The vestigial oviduct and the embryonic pronephros are represented by dotted lines in (*c*) and (*d*), (Goodrich, 1945.)

cells of the ectoderm from which they are derived. Thus, the cells composing the nephridia are likely to be variants of the initial surface-epithelial cells, i.e. potentially either ciliated or goblet cells, ecballic or emballic blastomeres with the usual mutually opposing characteristics, though they would, of course, not necessarily show the special characters of the epidermis of the nemertines themselves. In many cases the solenocytes, or flagellated flame-cells, are situated in close proximity to a blood vessel, e.g. in *Carinella*, *Geonemertes*, *Cephalothrix* (Hubrecht, 1885), and they may project into the lumen of the vessel. In some species, e.g. *Carinina* and *Procarinina*, the tubules may actually open into the lumen itself (Oudemans, 1885; Nawitzki, 1931) (Fig. 16.8). This is of particular interest because, in the cyclostomes *Bdellostoma* (Conel, 1917; Price, 1910) and *Myxine* (Holmgren, 1950), the relationship of the pronephric tubules to the adjacent veins is similar. There is one type of nephridium in the nemertine *Cephalothrix* (see p. 270 and Fig. 12.9) which has particular significance in relation to the present discussion

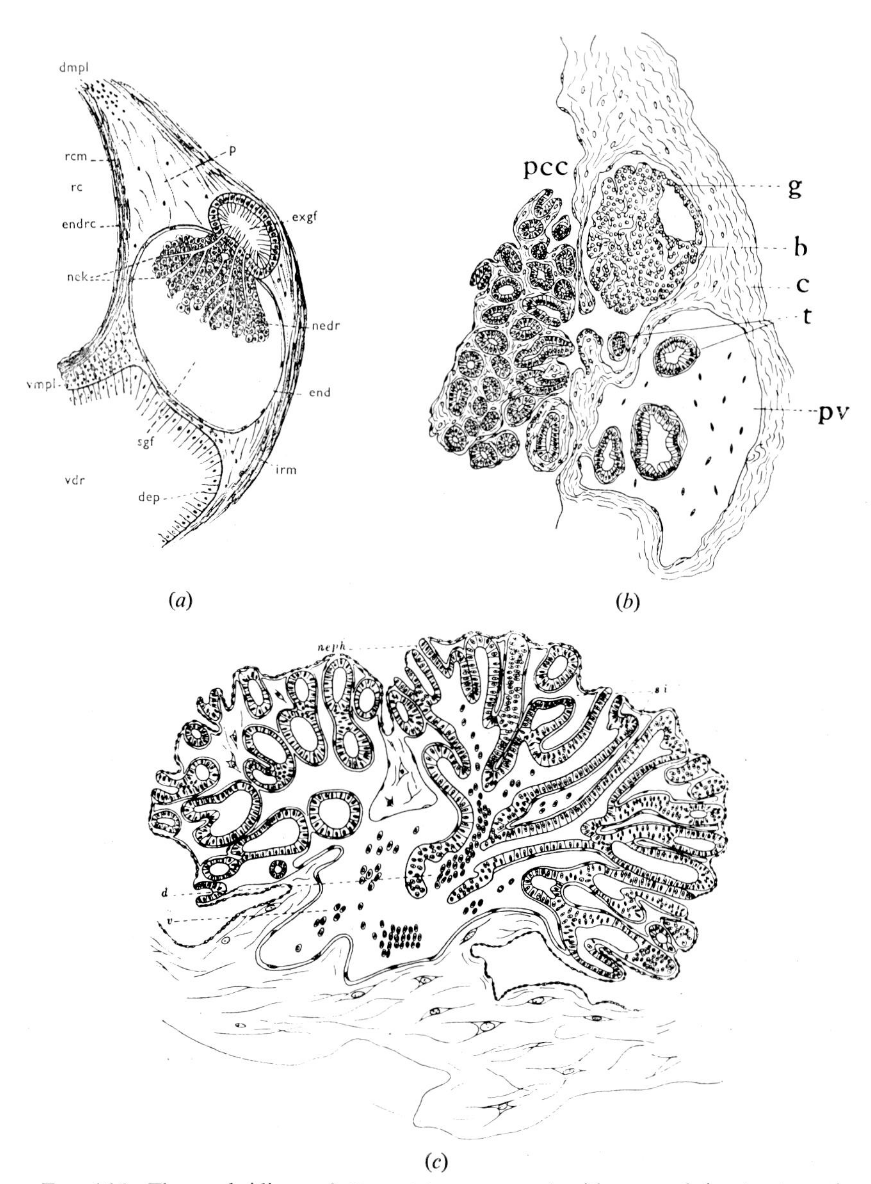

FIG. 16.8. The nephridium of *Procarinina* compared with pronephric structures in myxinoids. (*a*) Nephridium. dep, Gut epithelium; dmpl, dorsal muscle plate; end, endothelium; endrc, rhynchocoel endothelium; exgf, excretory canal; irm, inner circular muscle; nedr, nephridial tubes; p, parenchyma; rc, rhynchocoel; rcm, rhynchocoel muscle; sgf, lateral blood vessel; nek, nephridial openings; vdr, fore-gut; vmpl, ventral muscle plate. (*b*) Pronephros of *Bdellostoma*. b, Pocket of pericardial cavity; c, connective tissue; g, glomerulus; pcc, pericardial cavity; pv, pronephric vein; t, tubule. (*c*) Pronephros of *Bdellostoma*. d, Tubule opening into vein; neph, nephrostome; si, sinusoids; v, vein. (Nawitzki, 1931; Conel, 1917; Price, 1910.)

because it has a relatively large closed vesicle at its head instead of the usual solenocyte (Coe, 1930). The vesicle is placed in close contact with a blood vessel; its floor has ciliated cells while the rest of the wall of the vesicle is formed from a flat, pavement epithelium. The whole structure thus begins to resemble the capsule of Bowman at the head of the nephron in the kidneys of the higher vertebrates. This is especially significant when it is remembered that, in vertebrates, the neck tube and capsule may occasionally be, or may abnormally become, ciliated and that sex hormones may bring this about (see p. 509) because, curiously enough, in *Cephalothrix* this form of nephridium is apparently confined to the female sex. It is pertinent also to notice that the ciliated part of the nephridial tube of *Geonemertes* (a terrestrial nemertine) is rich in alkaline phosphatase (Danielli and Pantin, 1950), which is, of course, a conspicuous enzyme in the proximal tubules of vertebrates. Convergence, of course, on functional grounds is equally possible.

Another variation in nephridial organization, which occurs in *Baseodiscus* and *Taeniosoma*, is that some nephridial tubes open into the fore-gut (Hyman, 1951; Coe, 1906) rather than on to the external surface, an event that happens frequently in the annelids (Fig. 16.9) (Bahl, 1945) and could also be relevant to the development of Malpighian tubes in insects and possibly to certain structures in the vertebrates (see p. 543).

The nephridia, it will be recalled, are essentially excretory and homoiostatic organs and thus provide means, in addition to the epithelia of the skin and the gut, whereby the internal environment of the body may be controlled. An increase in the water-load in the tissues of *Geonemertes* has been shown (Pantin, 1947) to increase the rate of beat of the flagella of the solenocytes. This is a point that is important not only in relation to the function of nephridia, but also because it affords another example, like *Naegleria* and the pharyngeal epithelium of *Lineus*, of the connection between water-load and ciliary or flagellar activity.

Apart from the leakage of important metabolites, what leaves the nephridium and the composition of the fluid contained within its lumen are of no importance to the organism. What is left behind in the tissues is, however, important; it is presumably this that must determine the activity of the cells of the nephridium as a whole, i.e. of the solenocytes and the secretory or absorptive cells.

It is important to appreciate that, because the nephridia act as homoiostatic organs, they are not necessarily the only tissues that are concerned with water and salt balance; the skin and intestine may still be much involved. Regulation in different species of nemertines may be differently organized, and terrrestial species like *Geonemertes* have problems to solve which are different from those of their littoral and marine relatives. The number of nephridia in nemertines and in annelids varies widely, from one or two separate structures

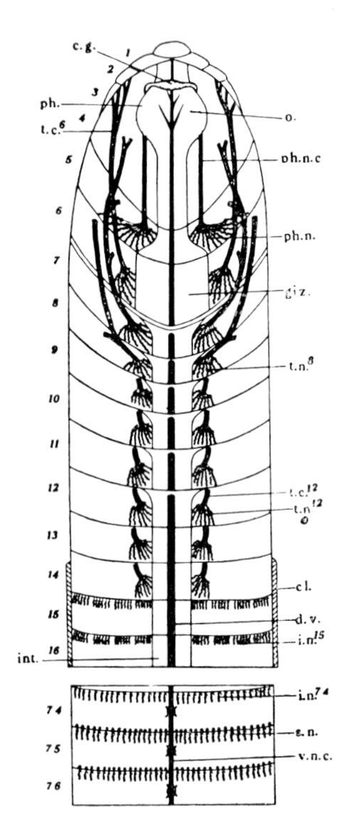

FIG. 16.9. Plan of part of the nephridial system of *Megascolex cochinensis*. c.g., Cerebral ganglia; cl, clitellum; dv, dorsal vessel; giz, gizzard; in, integumentary nephridia; int, intestine; o, concealed opening of pharyngeal duct into the lumen of the pharynx; ph, pharynx, ph.n., pharyngeal meronephridia; ph.n.c., bundles of ductules; s.n., septal meronephridia; t.c., ductules of exonephric tufted meronephridia. (From Goodrich, 1945, after Bahl, 1944).

to many thousands on either side. Their capacity for reduplication and branching thus seems to be immense (see Fig. 16.9).

There is little doubt that the nephridia of the annelids are the direct derivatives of those in rhabdocoeloid or nemerteoid organisms. It is thus perhaps illuminating to examine these animals to see the variety of ways in which nephridia can behave (see Goodrich, 1945). First, they can take on a strictly segmental arrangement, each nephridium opening separately on to the surface of the worm (e.g. in Nereidae). Second, they can unite to a common, backward-running duct and open into the gut in the region of the anus (Fig. 16.10) (e.g. *Hoplochaetella bifoveata*). Third, instead of ending blindly in solenocytes as in most nemertines and in the Phyllodocidae, for example, they may have open nephrostomes, i.e. they may have ciliated funnels leading directly out of the coelomic cavity (Fig. 16.11). In passing, it is interesting to note again that an effect of increased water-load on the beat of the cilia (similar to that seen in *Geonemertes*) has also been observed in the ciliated cells in the nephridia of the more highly organized annelids, *Lumbricus terrestris* and *Allolobophora chlorotica* (Roots, 1955). Fourth, the nephridia may, like those of *Baseodiscus*, open into the alimentary canal in the mouth region or at the beginning of the intestine, instead of on to the outer surface of the body (e.g. *Megascolex cochinensis*). This observation should be borne

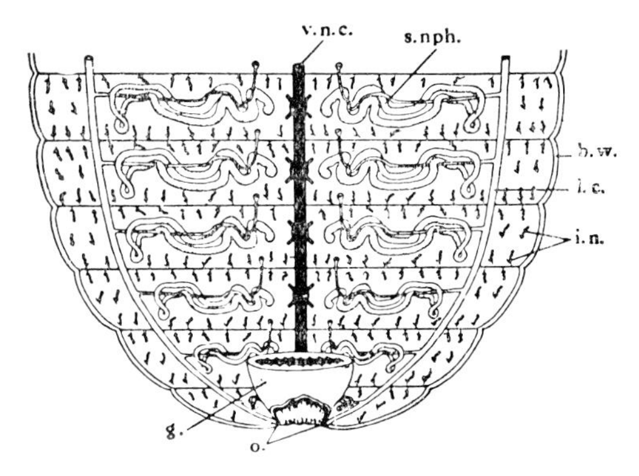

FIG. 16.10. Diagram of the nephridial system in the posterior segments of *Hoplochaetella bifoveata*, dorsal view. b.w., body wall cut open and spread out; g, hind end of gut, the rest cut away; in, integumentary meronephridia; l.c., longitudinal lateral canal; o, opening of excretory canal; s.nph., septal nephridium connecting with lateral canal; v.n.c., ventral nerve cord. (From Goodrich, 1945, after Bahl, 1942.)

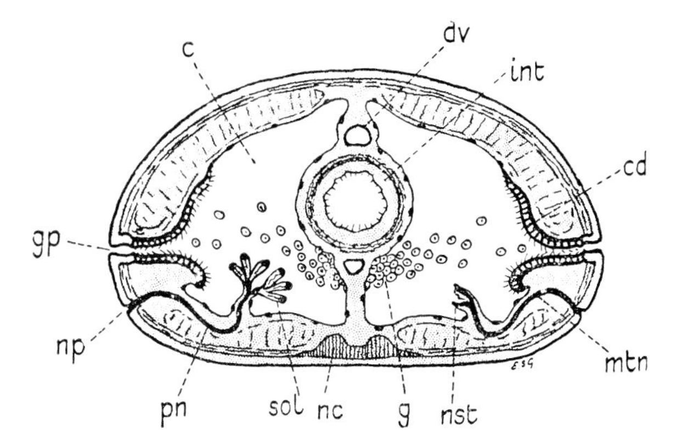

FIG. 16.11. Diagram showing two methods by which nephridia come into close contact with (left) or open into (right) the coelomic cavity in annelids. c, Coelomic cavity; cd, coelomoduct; dv, dorsal vessel; g, gonad; gp, gonopore; int, intestine; mtn, metanephridium; nc, nerve cord; np, nephridiopore; nst, nephridiostome; pn, protonephridium; sol, solenocyte. (Goodrich, 1945.)

in mind when the origins of the salivary glands (see p. 524) and pancreatic ducts (see p. 525) are being considered. For example, the action of aldosterone in altering the composition of saliva as well as that of urine could perhaps find an explanation along these lines. Fifth, nephridia may be reduplicated many times in each segment as in *Megascolex cochinensis*, a large Indian earthworm which has a vast number of small nephridia on each segment in addition to various other types (see Fig. 16.9). This potentiality for increasing the numbers of small nephridia is again worth remembering in considering how and why the terrestrial vertebrates produce sweat glands, which are

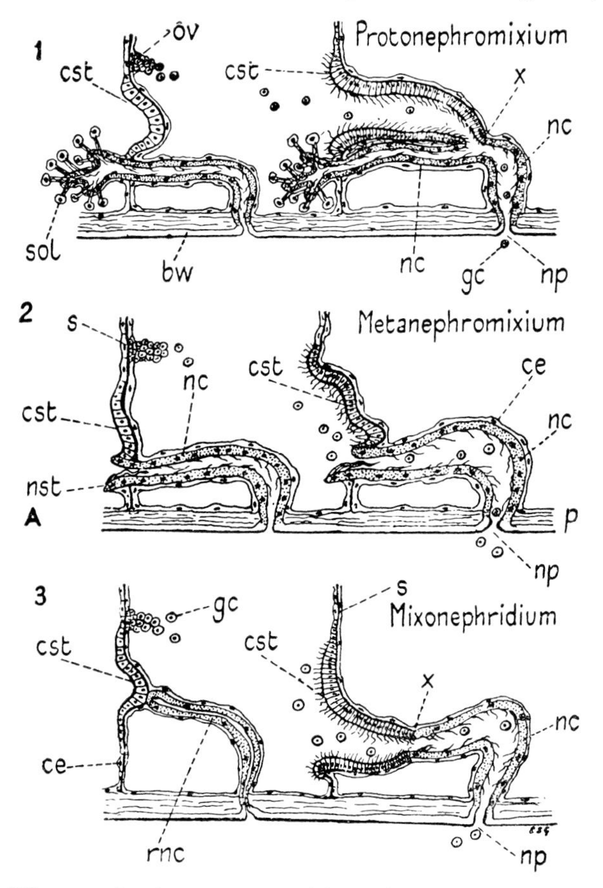

Fig. 16.12. Diagrams showing structure and formation of nephromixia by combination of coelomoduct with nephridium in three different ways. On the left is an early stage, and on the right the combination is complete. bw, Body wall; ce, coelomic epithelium; cst, coelomostome; ge, germ cell; nc, nephridial canal; np, nephridiopore; nst, nephridiostome; ov, ovary; rnc, rudiment of nephridium; s, intersegmental septum; sol, solenocyte; x, point of junction. (Goodrich, 1945.)

tubes produced in essentially the same way as nephridia, i.e. by invagination of the epithelial surface. Sixth, it is common to find that nephridia and gonoducts may unite, end to end, and then open in a single pore. When this happens union is always in the order gonoduct to nephridium, and the opening is strictly a nephridiopore (Fig. 16.12). This behaviour, as described by Goodrich (1945) in *Phyllodoce paretti* and other worms, may be compared with the union of sex cords and mesonephric tubules in the gonads of vertebrates (see Fig. 17.1). A similar arrangement is seen also in *Priapulus* (Fig. 16.13).

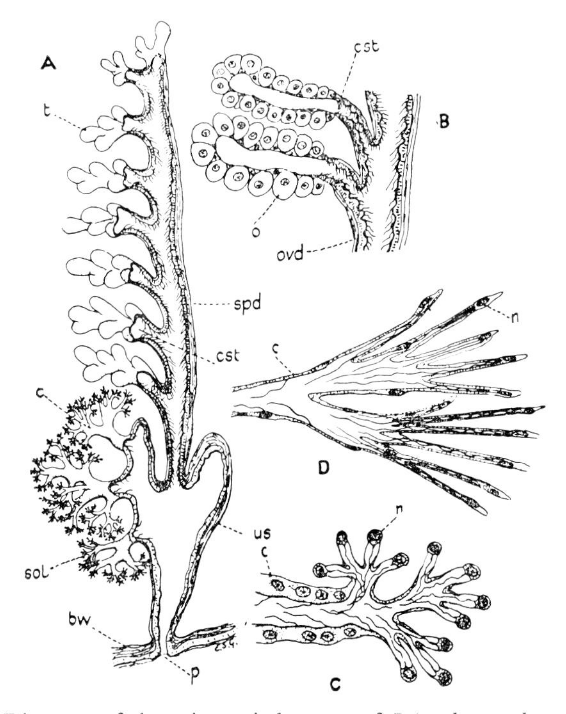

FIG. 16.13. Diagrams of the urinogenital organs of *Priapulus caudatus*. (*A*) Male; (*B*) female; (*C*) part of protonephridial canal with normal solenocytes; (*D*) part of protonephridial canal with elongated solenocytes. bw, Body wall; c, ciliated tube; cst, coelomostome; n, nucleus; o, ova; ovd, oviduct; p, urinogenital pore; sol, solenocytes; spd, sperm duct; t, testes. (Goodrich, 1945.)

As mentioned on p. 439, the contents of the nephridial tube are of no importance to the animal, but the nephridium must assist in maintaining the constant composition of the tissue fluid. On the other hand, the germ cells presumably require that their environment be to some extent regulated, so the contents of the gonocoel cavities and gonoducts must be controlled, together with any influx or efflux of fluids through the gonopore. Thus the two systems, nephridia and gonocoels, are both concerned with the regulation of the composition of fluids, but the activities of the cells are probably orientated in opposite directions. Gonocoels and gonoducts control the fluid within themselves, nephridia control the contents of the surrounding tissues. In some worms the external fluid of the nephridia is the internal fluid of the gonocoel (see Fig. 16.11). It is obvious therefore that, if the two tubular and regulatory systems are to unite, the requirements of the animal as a whole and of its germ cells in particular can only be met satisfactorily if the junction is made in the order indicated, namely gonoduct leading to nephridium. And so it always is. Goodrich (1945) has described three forms of such unions in annelids; they are the protonephromixium, the metanephromixium, and the mixonephridium, and in each case the order is the same (Fig. 16.12).

These observations, though referring primarily to the otherwise irrelevant annelids, give some insight into the manner in which nephridia and, to a lesser extent, gonoducts can behave, but they do not, of course, tell what has actually happened in the evolution of vertebrates. Moreover, the actual problems set to the excretory system as a whole must naturally vary with the water and salt relationships between the environment and the animal concerned.

Urinogenital Tubules: Transition to the Vertebrate Pattern

The problem of the transition from the nemerteoid to the vertebrate urinogenital system may also be approached from the other direction, namely from a study of the urinogenital systems in the primitive chordates and vertebrates.

In the Hemichordata, the "coelomic cavities" (see Fig. 13.6) of the proboscis and collar are connected to the exterior by ciliated tubes of unknown significance. They have been regarded by Goodrich as coelomoducts, and they have also been suggested to be organs regulating the turgidity of these cavities. If the interpretation of the proboscis and collar put forward on p. 285 is correct, these tubes are presumably secondary developments. Alternatively, they could be derived from nephridia penetrating into the rhynchocoel and "vascular" cavities, in much the same way as they penetrate the coelomic cavities in annelids and the blood vessels in nemertines. The proboscis pore itself could be a modification of the rhynchodaeum. The Hemichordata have clear gonoducts in the posterior part of the body, organized in a manner similar to that seen in the nemertines.

The Urochordata, in the relevant larval stages, have nothing in the way of nephridia. In the Cephalochordata, e.g. *Amphioxus*, there are well developed nephridia in the branchial region, with solenocytes projecting into the blood cavities (Fig. 16.14). These solenocytes are slightly different in their organization from those in annelids (at least from those in *Glycera*); the "rods" surrounding the flagellum are not united by a membrane. They are interesting, however, in another respect: they rest on the basement membrane by means of numerous branching and palisade-like feet. Attention has been called to the similarity in this respect between these cells and the podocytes of the Bowman's capsules in higher vertebrates (Brandenburg

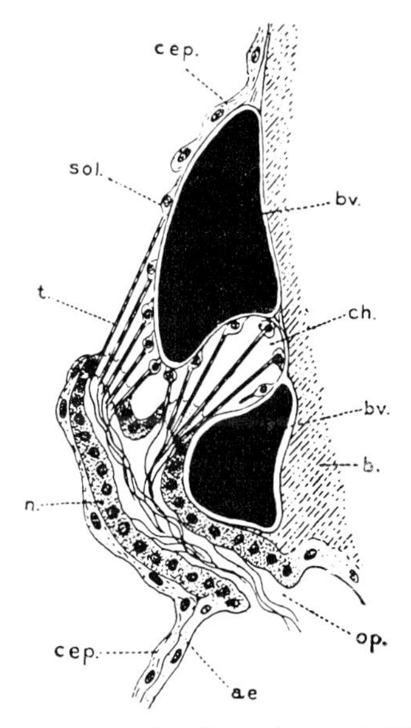

Fig. 16.14. Diagram of transverse section through a nephridium of *Amphioxus* showing the relationship between the solenocytes and the blood vessels. ae, Atrial epithelium; b, top of secondary gill bar; br, blood vessel; cep, coelomic epithelium; ch, chamber with solenocytes; n, wall of nephridial canal; op, nephridiopore; sol, solenocyte; t, solenocyte tube with flagellum. (Goodrich, 1945.)

and Kümmel, 1961). Is this necessitated by the permeability requirements, or are podocytes related to solenocytes?

There are no permanent gonoducts in *Amphioxus*; the gonocoels open temporarily through the body wall into the atrium by way of the cicatrices already mentioned, which may well be vestigial coelomoducts or nephridiopores. The gonads themselves (see Fig. 16.23) are formed as special entities which project into the coelomic cavities, into which they discharge their contents before these are liberated through the cicatrices. This arrangement does not suggest that the coelomic cavity in *Amphioxus* is an enlarged gonocoel.

The nature and relationships of the primary gonadal cavities and cicatrices in these creatures are the fundamental problems relevant to the present discussion. The gonadal cells first appear, embryologically, as small groups of cells in the floor of the myocoel, i.e. more or less in the same place as the nephrotome develops in the craniates. The gonads are segmentally arranged from the ninth or tenth segment to the thirty-fifth or so.

In the cyclostomes there are no nephridia as such. There are, however, serially arranged tubules which connect the coelomic cavity to the cloaca via a common duct as shown in Fig. 16.7. The tubules are subdivided into pronephric tubules and opisthonephric tubules, and, for a few segments in between, the tubules are generally rudimentary or absent. The pronephric funnels are intimately connected with blood vessels and have other interesting features discussed elsewhere (p. 526). Similar and serially arranged tubules occur in the elasmobranchs, particularly in their embryonic stages. Most of these tubules appear to have a primarily excretory function. Those in the pronephric region may be modified as open funnels and conduct the female genital products to the exterior; those in the mesonephric (or anterior opisthonephric) zone may be joined by connexions from the spaces or tubules in the male gonads and thus provide a direct means of exit for the male genital products.

In higher vertebrates, i.e. the amniotes, the posterior tubules (metanephros) finally take on all the excretory functions, for which purpose they become extremely numerous by systematic branching. The pronephros survives only in the form of the oviduct (though there may be other derivatives; see p. 531), and the mesonephros only in the form of the Wolffian duct, rete testis, epididymis, etc.

The situation in such nemertines as *Cerebratulus* and *Lineus* is interesting by way of comparison with the organization of the urinogenital system of *Amphioxus* and the lower vertebrates, e.g. *Myxine*. The nephridia are concentrated into a pair of large groups at about the level of the middle or lower part of the pharynx (Figs. 16.5, 16.15), and they have a very close connexion with the lateral blood vessels in that region. In *Lineus* they have two or three openings directly to the exterior. Just posterior to this, at about the level of the tenth intestinal diverticulum, the gonocoels begin to appear as regular epithelium-lined cavities lying between the diverticula (Fig. 16.5). The whole arrangement has an intriguing similarity to the pronephros and segmental tubules of vertebrates. For example, if the nephridial ducts continued posteriorly instead of opening at once, a sort of archinephric duct would appear, and this could then be joined by the serial gonoducts. Secondary branching and development of nephridia from the posterior part of the archinephric duct could then provide the meso- and metanephroi, and the whole system would resemble that of the primitive vertebrates. It is unlikely

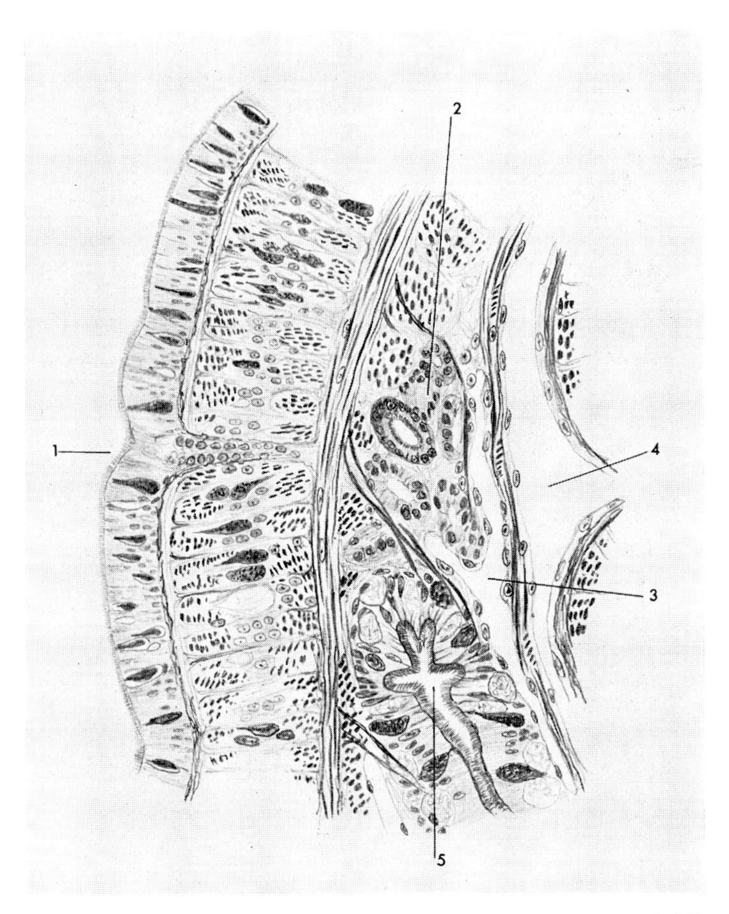

FIG. 16.15. Transverse section of the nephridial region in *Lineus*. 1, Nephridiopore; 2, nephridium (note that some tubules have a "brush border"); 3, lateral blood vessel; 4, rhynchocoel; 5, cavity of pharynx.

that *Lineus* itself has anything to do with the origin of the vertebrates or their urinogenital system. However, the necessary modifications that would have to be made are such that it is not unreasonable to suggest that a creature whose excretory and gonadal systems were similar to those now found in *Lineus* could provide the elements from which the vertebrate system has developed.

In the vertebrates the main questions connected with the origin of the urinogenital system are somewhat as follows: What is the nature of the serially arranged tubules (the "coelomoducts" of Goodrich) that are so characteristic of the embryos, particularly of the more primitive groups?

Have the vertebrates inherited anything corresponding to nephridia, and, if not, why not? (Among the chordates, nephridia as such are found only in *Amphioxus* and its relatives.) Is the coelom the result of the fusion of more primitive gonocoels, or is it a new cavity? Sex cords develop within the genital ridge when it differentiates; in the male these cords become patent as the seminiferous tubules and join with the mesonephric tubules; what is the nature of sex cords, are they remnants of some earlier tubular system, and why do they occur in both sexes?

These questions have never been satisfactorily answered. One school of thought regards the serial tubes as modified nephridia, another as coelomoducts derived essentially like the gonoducts of invertebrates from out-growths of the gonocoel cavities, and yet another considers them some combination of nephridial tubes and gonoducts, rather in the manner that is well established in the annelids, i.e. as nephromixia. Without penetrating too deeply into this controversial question, we may profitably review the problem briefly in relation to the hypothesis being discussed, namely that the vertebrates have evolved more or less directly from some nemerteoid-like creatures. On this assumption, the urinogenital apparatus of the vertebrates is likely to have arisen by modification of the cellular activities of the corresponding tissues in the nemerteoids.

Although both gonoducts and nephridia are initially concerned with ionic regulation, water balance, etc., they are, as has already been stated, differently orientated with reference to the fluid whose composition is being regulated. *A priori*, therefore, it would seem odd that the tubes in the vertebrates should be the equivalent of gonoducts only, because there would then be no organs, other than the skin and alimentary canal, that were primarily concerned with regulating the composition of the tissue fluids. Of course, we know this to be untrue, because the main function of the kidney in the higher vertebrates is undoubtedly homoiostasis. In addition, the complete disappearance of such an important system, as the nephridial system is known to be in most of these invertebrates and even in the cephalochordates, would seem in itself to be a very remarkable evolutionary phenomenon. Traces of this extremely primitive and widely distributed system would almost inevitably be present. There are indeed, as will be shown later (p. 541), indications of an earlier tubular system, possibly of a nephridial nature, in relation to the gill arches of vertebrates, and of course *Amphioxus* has its functional nephridia with well developed solenocytes in that region. It is also obvious that many aspects of kidney organization and function are comparable with those of nephridia. The problem is to decide whether this has resulted from phylogenetic relationship or from convergent evolution brought about by the similarity of homoiostatic requirements.

Although in the extant nemertines the nephridia are almost always closed tubes (*Carinina* excepted), the fact that open funnels exist in vertebrates does not necessarily argue in favour of the gonoduct origin of the pronephric funnels because, as we have seen (p. 441), the nephridia in many annelids have acquired open nephrostomes, and quite independently of coelomostomes. Furthermore, in the mesonephros of vertebrates and more particularly in the metanephros, the nephrons may never open to the coelom, or indeed come anywhere near it. However, their Bowman's capsules develop (Fig. 16.16)

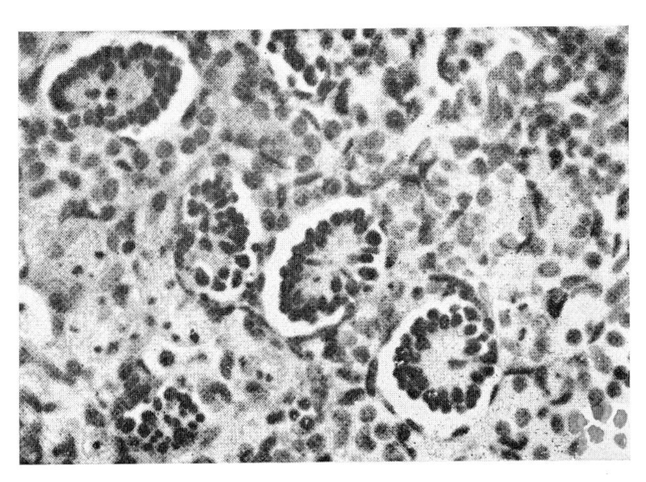

FIG. 16.16. Bowman's capsules as they appear in organ cultures of mammalian kidney. These structures should be compared with the nephridial capsules of *Cephalothrix* (see Fig. 12.9). (Trowell, 1959.)

by enlarging the blind ends of tubules which at their other end join the branches of the ingrowing system of collecting tubes derived from the main nephric duct. In so doing, they produce structures not unlike those found in the nemertine, *Cephalothrix* (see Fig. 12.9).

The essentially serial arrangement of the tubules in the primitive vertebrates does not really argue more in one direction than another. In nemertines the serial arrangement of the gonads is the more pronounced, the nephridia often being less regularly distributed. On the other hand, in annelids the arrangement of nephridia can be very rigidly segmental.

Although it does not apply to the pronephros (except perhaps in *Myxine*) and only to a limited extent to the mesonephros, the enormous multiplication of units in the metanephros is a character more in keeping with nephridial than with gonoduct behaviour. Of course, there are often numerous tubules in the testis, but how far this could be explained in terms of concentration and collection of units within one region rather than by reduplication is debatable.

The association of the nephric elements with blood vessels, particularly in the formation of glomeruli, is more intelligible on the hypothesis that the capsules related to the glomeruli are nephridial rather than of gonoduct origin. There is already a close association in the nemerteoids between blood vessels and nephridia. Since the primary function of the nephridia is to maintain the composition of the tissue fluids, this association has presumably come about because homoiostasis of the tissues can most easily be achieved by maintaining the composition of the blood.

Some light may be thrown on these problems by considering the functions of the cells in various parts of the tubules, the orientation of their activities, and the cytological features of the cells, though, once more, convergence may bedevil the interpretations. Moreover, it is extremely important to remember that each animal species, by inhabiting a particular ecological niche, sets different problems for its excretory cells to perform in order to maintain its body fluids. Thus marine animals, estuarine animals, fresh-water animals, terrestrial animals, desert animals, aerial animals, carnivorous animals, herbivorous animals, etc., all set different problems for their kidney tubules to perform.

In the mesonephros of amphibia, the fluid in the proximal tubules preserves the constancy of its initial osmotic pressure, chloride content, and ammonia content (Walker *et al.*, 1936) in spite of the complete removal of glucose (Walker and Hudson, 1936a) and of extensive movements of other substances, e.g. phosphate and water (Walker and Hudson, 1936c). In the distal tubule the urine is acidified (Montgomery and Pierce, 1936), the urea content rises, and most substances change rapidly in concentration (Figs. 16.17, 16.18) (Walker and Hudson, 1936b; Smith, 1951). The removal of glucose is an interesting point because glucose is completely absent from the coelomic fluid of earthworms (Bahl, 1946), and this fluid is hypothetically the fluid contained within enlarged gonocoels. The observation of Voglmayr *et al.* (1966) that there are only traces of glucose in the fluid in the seminiferous tubules of rams also fits into the same pattern. All these points together could indicate that the proximal tubule in these kidneys (i.e. the amphibian mesonephros) is really a modified coelomoduct and that, as such, it controls its own contents, rather than those of the surrounding tissues, while the distal tubule behaves more in the manner of a nephridium. This might suggest that an end-to-end fusion of coelomoduct and nephridium has occurred in the formation of the mesonephros. The metanephros of mammals may, of course, be different in kind from the mesonephros of the amphibia in that it develops in a different manner and has a different arrangement of its vascular supply, there being a renal portal supply to the amphibian mesonephros. There is less constancy about the contents of the proximal tubules in mammals (Walker *et al.*, 1941) than in amphibia and the bulk of the fluid is absorbed

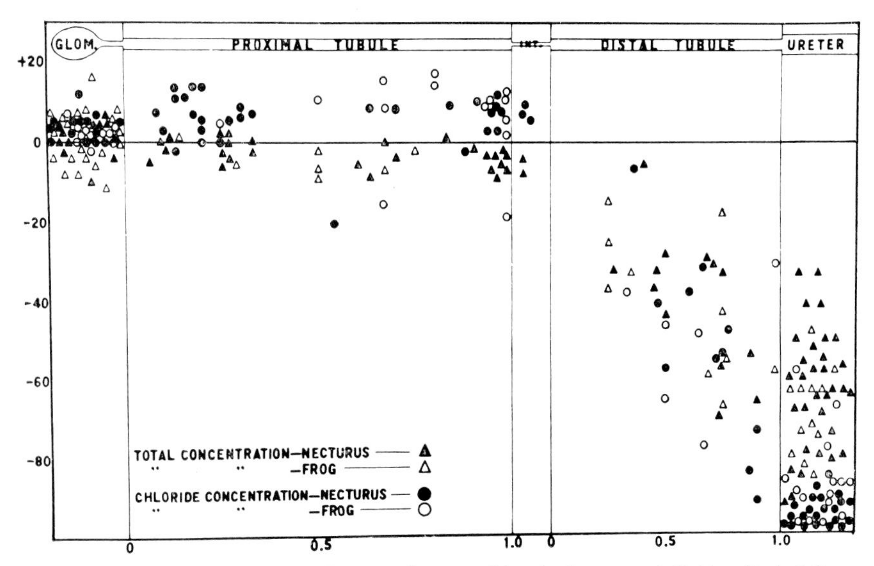

FIG. 16.17. Chart showing differences between blood plasma and fluid collected from various levels of the renal tubules with respect to total concentration and chloride concentration. Abscissae: distances along the proximal and distal tubules as fractions of the total length of each. Ordinates: Zero = plasma concentration; points above or below represent percentage increases or decreases. (Smith, 1951.)

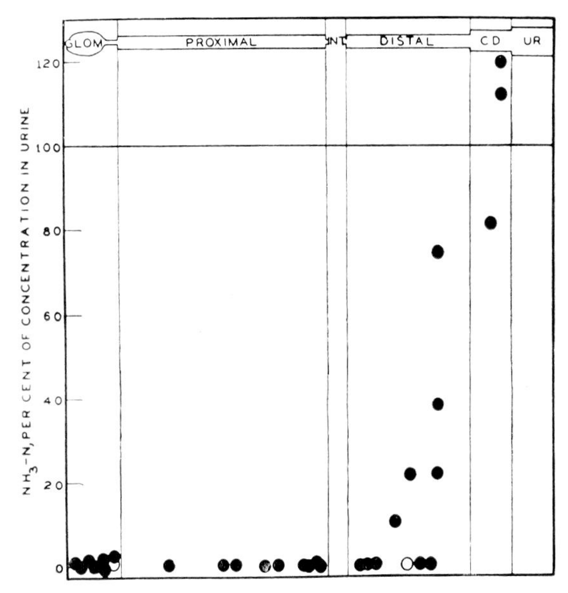

FIG. 16.18. Ammonia concentrations in the fluid in different parts of the mesonephric tubules as compared with the ammonia concentration in the urine. (Walker *et al.*, 1936.)

in the proximal tubules, but it is doubtful whether the differences are significant. Teleologically it may be argued that, if and when the nephridia became the main organs of homoiostasis for higher animals, they must reabsorb any glucose that enters their tubules, simply on account of its value as a foodstuff.

Arguments such as those above on the exact homologies of morphological structures are necessarily somewhat unsatisfactory unless all the physiological activities of the cells concerned can be appreciated at the same time. The supposition is that the kidneys of the vertebrates have developed by elaboration of existing tubules, i.e. nephridia, gonoducts, or mixonephridia. It is likely that the epithelia of both gonoducts and nephridia initially contained cells of opposing activities arranged on a gradient system, and the gonoducts controlled the fluid in the gonocoels while the nephridia controlled the tissue fluid around them. Thus the activities of the cells in the two classes of tubes were guided by events at opposite ends of the cells, the luminal end in the cells of the gonoduct and the tissue-fluid end in the nephridial cells. The gradients along the tubes might be very steep, delimiting two clearly marked groups of cells (e.g. flame-cells and granular nephrocytes), or there could be a much more gradual transition. Furthermore, it is almost inevitable that gradients of activity should develop in tubular systems acting like kidneys, because nearly every activity of the tubule creates its own gradient. For example, if a solution containing glucose is filtered from the blood through Bowman's capsule and the glucose is reabsorbed by the walls of the proximal tubule, then, unless there are compensating changes in the vascular system, the cells further down the tube from the capsule are faced with a much larger adverse concentration gradient of glucose across them than are the cells situated nearer the capsule. Similarly, in an animal loaded with a particulate dye like trypan blue, the smaller, and thus presumably more easily phagocytosed, particles of trypan blue find their way into the cells near the glomeruli, while the cells further along the tube are the ones to deal with the larger particles. Similarly, gradients of iron-containing pigment have been found in the proximal tubules of rats (Grafflin, 1942); this has additional interest because haemochromogens are conspicuous in the nephridial cells of certain worms (Bahl, 1945). Morphologically, such gradients are often evident in other ways, as by the fuchsinophilia of the same cells as have the iron pigment. In some lizards and other reptiles the proximal tubule may have a ciliated neck, a brush border for most of its length, and may end in a segment containing mucous cells, the last being, for some reason, particularly developed during the breeding season and differently developed in the two sexes (Regaud and Policard, 1903a, b; Cordier, 1928) (Fig. 16.19).

Somewhere in the evolutionary interval between nemerteoids and vertebrates (*Myxine* excepted), the body fluids have changed in composition from those in equilibrium with sea-water to the typical and more dilute body

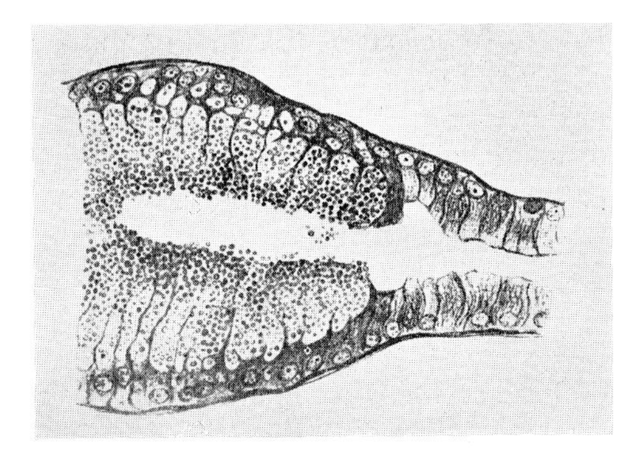

FIG. 16.19. The junction between the distal tubule and the "sex-segment" in the kidney of *Phrynosoma cornutum*. (Cordier, 1928.)

fluids of the vertebrates. These more dilute fluids may have arisen in response to a change to a fresh-water environment, or they may reflect the composition of the sea-water at the time when animals acquired control over the internal environment. The two events may, of course, have been simultaneous.

The nephridial tubules of a pelagic nemertine presumably are concerned mainly with excreting those substances that are in too high a concentration in the tissue fluid or blood and with maintaining the internal concentrations of those necessary metabolites which might otherwise escape into the sea-water. Ionic problems are minimal. The tubules of a fresh-water or estuarine animal are presumably also concerned with maintaining adequate concentrations of metabolites, but in addition they have the necessary concentrations of salts to maintain. The tubules as a whole must therefore be working very differently in the two cases, at least with respect to ions. The "kidneys" of sea-water nemerteoids and of *Myxine* may have similar functions to perform, but the "kidneys" of the terrestrial nemerteoids and of the migratory and land-locked fresh-water lampreys face very different problems. However, all are likely to carry out these functions by modifications of the same theme of nephridial and gonoduct activities. The ion-controlling functions of the skin and alimentary canal also have to be taken into account. They are directly important in aquatic creatures, but may become less so for animals in other environments when the skin tends to become impermeable.

Origin of the Coelom of Vertebrates

In the evolution of vertebrates along the lines indicated above, namely by the acquisition of a swimming habit in relation to filter-feeding, the invasion of new habitats (e.g. estuaries, rivers) would pose not only new problems of homoiostasis, but also problems connected with different types of food, different temperatures, and the like. Many of the morphological and physiological changes which occurred must have been immediately related to these changes of environment. In addition, the bodily movement would necessitate specialization of the vascular system as a circulatory system, which would need to become relatively independent of fluid movements caused by the primary movements of the body. In other words, the development of a large coelom or body cavity would have great advantages in isolating the movements of the contents of the gut and of the blood from the movements of the animal as a whole. If the coelom were to develop as a new schizocoel (and this is embryologically what the splanchnocoel often appears to do), the gonads could simply project into it but still have their own gonocoel cavities (as in *Amphioxus* and *Myxine*). These gonocoels might or might not then maintain or develop their own gonoducts, to open directly or via nephridial tubes to the outside. The shedding of the ova into the new coelom, as in cyclostomes and in many vertebrates, could be a secondary phenomenon. This concept visualizes the whole coelomic cavity as a space in the mesenchyme, and the epithelium lining it as a secondary epithelium, akin to the endothelium of blood vessels, i.e. formed of flattened mesenchyme cells.

An extension of this idea, and one that does not seem to have been much considered, is that the "blood spaces' of the nemerteoid could enlarge and subdivide into two groups. In an animal transitional between nemerteoid and vertebrate, two separate demands would be made on the vascular system First, it would be required to convey oxygen and possibly foodstuffs more quickly than before, at least to the active tissues. Second, in the absence of a rhynchocoel, it could assist with accommodating the fluid movements of the animal, i.e. providing the necessary flexibility so that the gut contents were not always moved whenever the animal as a whole moved. To achieve these ends it would seem possible that the vascular system actually became subdivided. A heart developed near the gills and established a true circulatory system initially for the most active tissues, with an increased number of corpuscles and greater oxygen capacity, while some of the other vascular spaces, which are very distensible (as, for example, the lateral vessels when the worm is placed in hypotonic media), separated off, became enlarged, and developed as coelomic cavities particularly surrounding the gut (see Fig. 16.6), and also surrounding the rhynchodaeum. The former could then become the main coelomic cavity; the latter might be represented in vertebrates by the head cavities (see p. 340).

If, on the other hand, the coelom were to be considered the result of the fusion of a series of enlarged gonocoels, then it could well be advantageous to remove the germ cells from the general cavity, as it grew larger, by letting them enter the gonoduct part of the gonocoels at an early stage. This might be the explanation for the origin of the sex cords in the embryonic gonads of the vertebrates. These cords could represent the remnants of the original gonoducts, which remain functional and hollow in the male. Thus, by uniting with the nephridial excretory system (the Wolffian body), they allow the escape of the ripe sperm. In the female they cease to do this (perhaps because the ova with their stored yolk, etc., become too large to pass easily through the general nephridial system). However, the cells of the sex cords provide the necessary environment for the development of the ova, which are then shed into the general coelomic cavity to be voided by one part of the nephridial system specially adapted for the purpose, i.e. the pronephric (Müllerian) duct, or by a special coelomic pore.

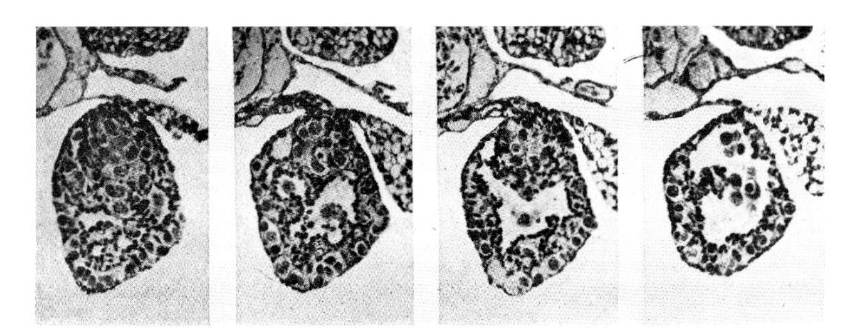

FIG. 16.20. Sections in antero-posterior sequence through the testes of a salamander, joined in parabiosis with a much larger female of another species. Four stages are shown in the conversion of the testis into an ovary by the degeneration of the medulla and development of the cortex. (Burns, 1956.)

This is an attractive hypothesis in that the germ cells would be housed in the gonoducts, i.e. in tubes whose primary function was initially regulative. Moreover, it is not unreasonable to suppose that the internal environment within the different parts of the tube would be different, so that, by lodging the germ cells at the appropriate point, they could find the conditions required for developing as either ova or sperm. It is well known (Witschi, 1929; Burns, 1956) that in amphibia, if the cortical part of the indifferent gonad is caused to develop by suitable hormonal treatment, then the germ cells contained therein develop as ova; if the medullary part is encouraged to develop, the germ cells therein develop into sperm (Fig. 16.20). This is important in two directions: first, the germ cells are evidently still plastic at this stage and, like *Naegleria*, can be caused either to become flagellate or to pinocytose and store yolk, etc.; second, the local external conditions as in *Naegleria*, can determine which course shall be taken.

This difference of environment for the developing germ cells in the male and female is a point of considerable cytological interest. It has already been

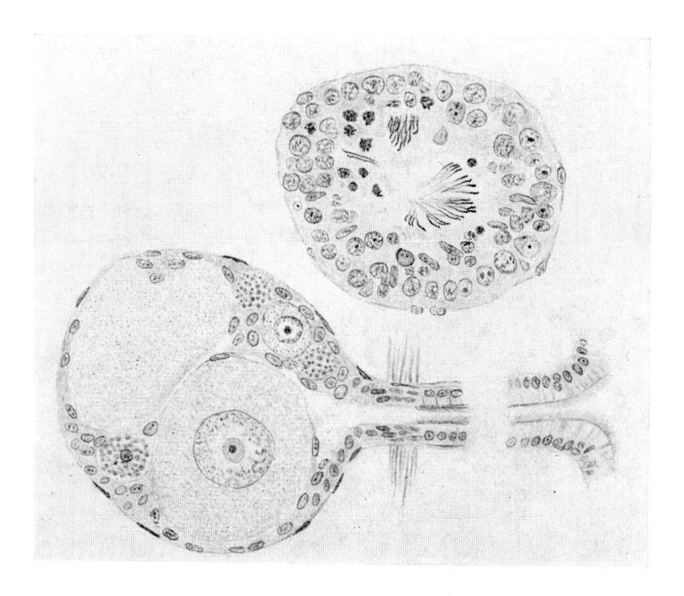

FIG. 16.21. The environment of germ cells in the nemertine. Male (*above*) and female gonocoels in *Lineus*. In the male the sperm develop in the cavity of the sac. In the female the oocytes are encased in follicle cells; the gonoduct is also shown.

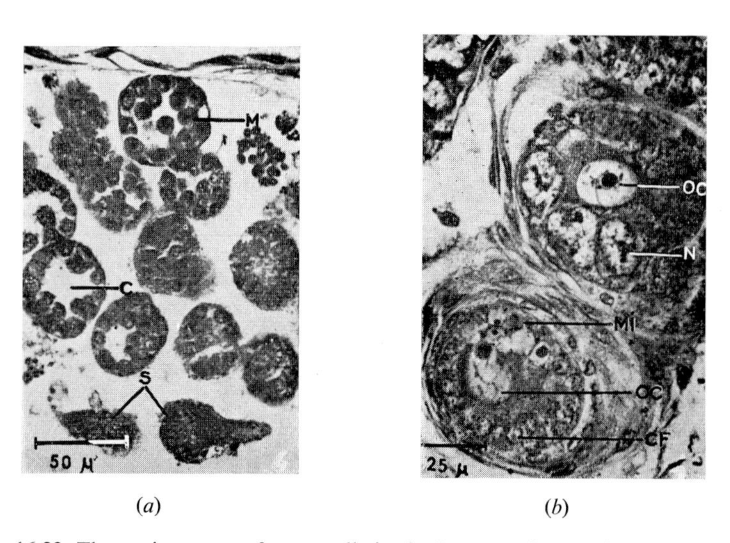

(*a*) (*b*)

FIG. 16.22. The environment of germ cells in the Insecta. Tissue cultures of gonads of *Galleria* showing (*a*) hollow follicles in the male, and (*b*) oocytes entirely surrounded by follicle cells in the female. C, Cavity; CF, follicle cell; M, meiosis; MI, Mitosis; N, nurse cell; OC, oocyte; S, spermatozoa. (Lender and Duveau-Hagège, 1962.)

noted as occurring at the rhabdocoeloid stage (p. 252 and Fig. 11.8). In nemerteoids the gonadal sacs (see Figs. 16.5, 16.21) of the male first appear as open follicles lined with germ cells and supporting cells. As the germ cells develop, they enter the cavity which is filled with fluid, presumably produced by the supporting cells which now line the sac. In the female there is a smaller cavity initially, but each ovum as it develops tends to become entirely surrounded by the supporting cells so that it is not in such direct contact with the fluid in the cavity; and the cavity may become almost completely obliterated. The same sort of arrangement is found in the gonads of insects (Fig. 16.22), of *Amphioxus*, and also in *Myxine* (Fig. 16.23). In all higher vertebrates too, the ovum develops when enclosed by follicular cells, and the sperm develops in the cavity of a tube lined with supporting cells (see Fig. 17.5).

If the coelom developed by lateral and ventral extensions of the rhynchocoel on either side of the gut by an elaboration of the process in evidence in

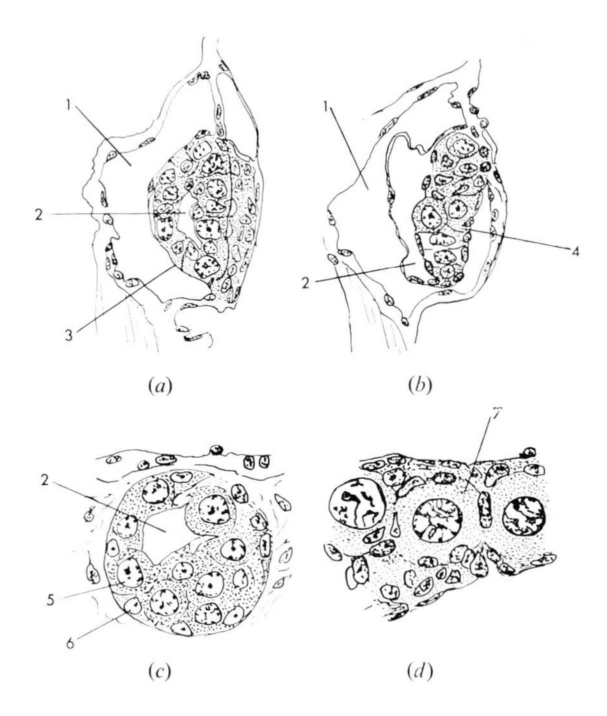

FIG. 16.23. The environment of the germ cells. Gonads of *Amphioxus* and *Myxine*. (*a*) Testis of *Amphioxus*; (*b*) ovary of *Amphioxus*; (*c*) testis of *Myxine*; (*d*) ovary of *Myxine*. 1, Coelom; 2, gonadal cavity; 3, germinal epithelium with germ cells mingling with supporting cells; 4, germ cells enclosed within a capsule of follicle cells; 5, germ cells; 6, Sertoli cells; 7, oocyte surrounded by follicle cells. (After Zarnik, 1904; and Schreiner, 1955.)

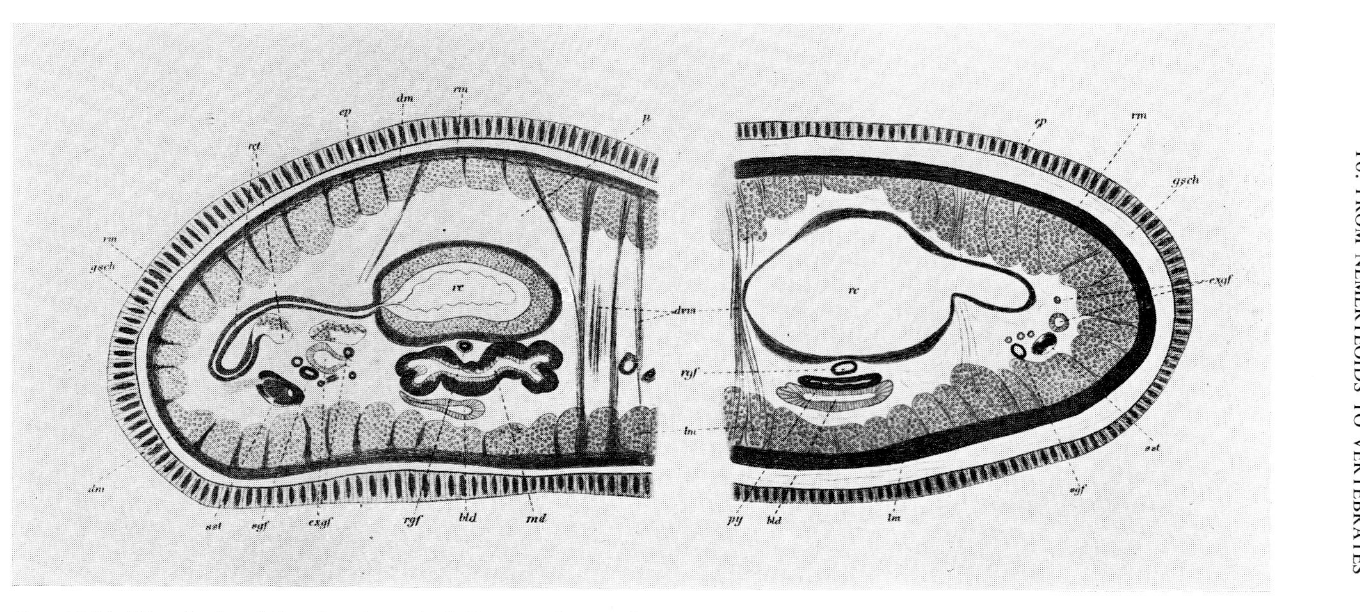

FIG. 16.24. Diverticula of the rhynchocoel of *Drepanophorus albolineatus* (left) and of *Amphiporus stanniusi* (right) bld, caecum; dm, diagonal muscle; dvm, dorso-ventral muscle; ep, epidermis; exgf, excretory vessel; gsch, basement membrane; lm, longitudinal muscle; md, mid-gut; p, parenchyma; py, pyloric tube; ret, diverticulum of rhynchocoel; rgf, dorsal vessel; rm, circular muscle; sgf, lateral vessel; sst, lateral nerve. (Bürger, 1895.)

Drepanophorus and *Amphiporus*, nemertines in which the rhynchocoel possesses diverticula (Fig. 16.24), the results would not differ fundamentally from those envisaged by the development of an entirely new schizocoel. The myotomes would presumably derive from the walls of the rhynchocoel, which already contains muscles. The archenteric pouches, as originator of the myotomes (cf. *Amphioxus*), would be difficult to integrate into this general picture. Indeed, there is very little, embryologically speaking, to suggest the probability of this course, nor are there any traces of coelomic cavities surrounding the notochord. The formation of the mesentery, however, would follow.

The coelomic cavity could have formed from the diverticula of the gut exactly as has been envisaged for the myocoels. Thus, while the upper parts of the diverticula have been nipped off as myocoels, the lower parts have remained connected (or become secondarily connected) to form the general body cavity and allow the free flow of fluid during bodily movement. As in the case of the schizocoel, the gonads, originally interdiverticular, would remain as separate gonocoels which would project into the coelom but not necessarily open into it. This hypothesis has several points in its favour. The coelomic epithelium would be a primary epithelium and not a secondary one. Thus, it would be more easy to understand some of its idiosyncrasies, like the formation of cilia or coelomic funnels, though less easy to understand others, like the mechanocytic behaviour of peritoneal cells in tissue culture. The germ cells, which are thought to originate in the wall of the archenteron, could be considered primary and natural inhabitants of the coelomic epithelium which later would invade the underlying mesenchyme and there proceed as they do in the mesenchyme of the nemerteoid, i.e. become enclosed in gonocoels. The embryological connexion of the myocoel, nephrocoel, and splanchnocoel cavities would thus have a phylogenetic explanation.

The fact that all these hypotheses can be put foward with some measure of probability emphasizes that a great gap exists in our knowledge and that the true nature of the coelomic epithelium remains obscure.

Whatever the correct explanation for the origin of the coelom in the early vertebrates, and whatever the relationship between the segmental tubules of primitive vertebrates and earlier nephridia or coelomoducts, the behaviour of these tubules is worth examining in more detail for the light which such an examination may throw on the physiology of the organs derived from them; this is the topic of the next chapter. Meanwhile, it may be concluded that an organism like the nemerteoid, with its pattern of nephridia on the one hand and of gonocoels and gonoducts on the other, is worthy of further investigation in relation to the functions of kidneys, gonads, and associated tissues as they appear in the vertebrates.

The concept, which was developed earlier, of a gradient of cell activity from strongly "ciliated" to strongly "amoeboid" can be applied with profit to the problem of mesenchyme, mesoderm, and germ cells. The primitive germ cells must be uncommitted cells at least in species in which they may be caused to develop into sperm or ova at quite a late stage, as in amphibia for example. They obviously have all the DNA to develop one way or the other as their immediate environmental conditions dictate. The environment and the cytoplasm (using the term broadly to include the non-chromosomal part of the cell) must be the deciding factors. The primitive germ cells then are probably derived from cells already in equilibrium with the average or middle environment, i.e. they are initially cells from the middle of the "gradient" from ciliated to amoeboid, and therefore rather easily sent along one line or the other of further differentiation, like *Naegleria*, but not necessarily showing any great tendency either way. Primitive germ cells in sponges and probably in the majority of animals are in fact in the amoeboid form before they start to differentiate.

When this concept is applied to the mesenchyme, mesoderm and their derivatives in general, two points ought to be taken into account: the nature of the mesoderm, or mesenchyme-forming epithelium with respect to its position on the "ciliated-amoeboid" gradient; and how far the epithelium had already differentiated into cells of two or more types, e.g. by virtue of its cells being derived from ciliated or amoeboid cells and of the two sorts having mingled before it became involved in mesoderm or mesenchyme formation? For example, the head mesenchyme of amphibia is largely derived by direct dehiscence of cells from the head ectoderm, and at an early stage the head ectoderm is already a mosaic of ciliated and non-ciliated cells (Assheton, 1896). One would expect, therefore, that such mesenchyme would be well provided both with mechanocytes and amoebocytes, with perhaps a preponderance of the former. It has already been noted (p. 436) that the mesothelium itself is a plastic sort of tissue and may appear as ciliated cells, phagocytic cells, or as a flat, pavement epithelium.

Returning once more to the gonads, kidneys, etc., we may note with interest that the segmental tubules of the primitive vertebrates do not start at the anterior end of the animal. This is equally true of the somites. If it is supposed that the somites in the vertebrates are the equivalent of the intestinal pouches of the nemertoid, then it is significant that the gonocoels of nemertoids also start in the same region and interdigitate with intestinal pouches. Nephridia sometimes occur more anteriorly than the intestine proper, and the fact that *Amphioxus* has nephridia interdigitating with the gill-slits, which on this hypothesis are pharyngeal in origin, is perhaps significant. It emphasizes that the pronephric and opisthonephric tubules of vertebrates actually follow the distribution of the gonoducts rather than that of the last functional nephridia, i.e. those of *Amphioxus*. It should be remembered, however, that the pronephros differs in several important respects from the opisthonephros and has features closely akin to those of certain nemertine nephridia. Thus it is perfectly possible that nephridia may have played their part in the formation of the segmental tubes of the primitive vertebrates, but that the segmental arrangement was dictated by the gonoducts. It would seem likely that the blind-ended protonephridium may have been temporarily abandoned in favour of the open nephridio- or coelomostome, only to be redeveloped, in a form similar to that in *Cephalothrix*, in the higher vertebrates in order to form the units of the metanephros and perhaps also of the mesonephros, or at least that part of it which develops later. In other words, the pronephros and perhaps part of the mesonephros are mixonephridial in origin, and the rest of the mesonephros and the metanephros are purely nephridial. More data are required.

References

Assheton, R. (1896). Notes on the ciliation of the amphibian embryo. *Quart. J. Microscop. Sci.* **38**, 465.

Bahl, K. N. (1942). Studies on the structure, development, and physiology of the nephridia of Oligochaeta. 1. General introduction and the nephridia of the sub-family Octochaetinae. *Quart. J. Microscop. Sci.* **83,** 423.

Bahl, K. N. (1944). Studies on the structure, development, and physiology of the nephridia of the Oligochaeta. IV. The enteronephric system in *Megascolex cochinensis* with remarks on vestigial nephridia. *Quart. J. Microscop. Sci.* **84**, 18.

Bahl, K. N. (1945). Studies on the structure, development and physiology of the nephridia of Oligochaeta. (VI) The physiology of excretion and the significance of the enteronephric type of nephridial system in Indian earthworms. *Quart. J. Microscop. Sci.* **85**, 343.

Bahl, K. N. (1946). Studies on the structure, development, and physiology of the nephridia of the Oligochaeta. VIII. Biochemical estimations of nutritive and excretory substances in the blood and coelomic fluid of the earthworm and their bearing on the role of the two fluids in metabolism. *Quart. J. Microscop. Sci.* **87**, 357.

Böhmig, L. (1897). Excretory organs and blood vascular system of *Tetrastemma. Ann. Mag. Nat. Hist. Ser.* 6 **20**, 324.

Böhmig, L. (1898). Beiträge zur Anatomie und Histologie der Nemertinen *Stichostemma graecense* (Böhmig) und *Geonemertes chalicophora* (Graff). *Z. Wiss. Zool.* **64**, 479.

Brandenburg, J., and Kümmel, G. (1961). Die Feinstruktur der Solenocyten. *J. Ultrastruct. Res.* **5**, 437.

Bürger, O. (1895). Die Nemertinen des Golfes von Neapel. *Fauna und Flora des Golfes von Neapel Monograph* **22**, 1.

Burns, R. K. (1956). Urinogential system. *In* "Analysis of Development" (B. H. Willier, P. Weiss and V. Hamburger, eds.), p. 462. Saunders, Philadelphia, Pennsylvania.

Clark, R. B. (1964). "Dynamics in Metazoan Evolution". Oxford Univ. Press (Clarendon), London and New York.

Coe, W. R. (1906). A peculiar type of nephridia in nemerteans. *Biol. Bull.* **11**, 47.

Coe, W. R. (1930). Unusual type of nephridia found in nemerteans. *Biol. Bull.* **58**, 203.

Conel, J. Le R. (1917). The urinogenital system of myxinoids. *J. Morphol.* **29**, 75.

Cordier, R. (1928). Études histophysiologiques sur le tube urinaire des Reptiles. *Arch. Biol.* (*Liège*) **38**, 111.

Damas, H. (1944). Recherches sur le developpement de *Lampetra fluviatilis* L. Contribution à l'étude de la céphalogenèse des Vertébrés. *Arch. Biol.* (*Liège*) **55**, 1.

Danielli, J. F., and Pantin, C. F. A. (1950). Alkaline phosphatase in protonephridia of terrestrial nemertines and planarians. *Quart. J. Microscop. Sci.* **91**, 209.

Delaney, P. A. (1929). Phagocytic mesothelium of turtles. *Anat. Record* **43**, 65.

Gibson, R., and Jennings, J. B. (1967). "Leucine aminopeptidase" activity in the blood system of Rhynchocoelan worms. *Comp. Biochem. Physoil.* **23**, 645.

Goodrich, E. S. (1909). Cyclostomes and fishes. *In* "A Treatise on Zoology" (E. R. Lankester, ed.), Vol. 9. Black, London.

Goodrich, E. S. (1945). The study of nephridia and genital ducts since 1895. *Quart. J. Microscop. Sci.* **86**, 113.

Grafflin, A. L. (1942). The storage and distribution of iron-containing pigment and the problem of segmental differentiation in the proximal tubule of the rat nephron. *Am. J. Anat.* **70**, 399.

Holmgren, N. (1950). On the pronephros and the blood in *Myxine glutinosa. Acta Zool.* (*Stockholm*) **31**, 233.

Hubrecht, A. A. W. (1883). On the ancestral form of the chordata. *Quart. J. Microscop. Sci.* **23**, 349.

Hubrecht, A. A. W. (1885). Der excretorische Apparat der Nemertinen. *Zool. Anz.* **8**, 51.

Hubrecht, A. A. W. (1886). Contributions to the embryology of Nemertea. *Quart. J. Microscop. Sci.* **26**, 417.

Hyman, L. H. (1951). "The Invertebrates. Vol. II. Platyhelminthes and Rhynchocoela", p. 492. McGraw-Hill, New York.

Lankester, E. R. (1872). A contribution to the knowledge of haemoglobin. *Proc. Roy. Soc.* (*London*) **B21**, 70.

Lender, T., and Duveau-Hagège, J. (1962). Survie et différentiation des gonades larvaires de *Galleria mellonella* en culture organotypique. *Compt. Rend.* **254**, 2825.

Montgomery, H., and Pierce, J. A. (1936). The site of acidification of the urine within the renal tubule of amphibia. *Am. J. Physiol.* **118**, 144.

Nawitzki, W. (1931). *Procarinina remanei.* Eine neue Paläonemertine der Kieler Förde. *Zool. Jahrb. Abt. Anat. Ontog. Tiere* **54**, 159.

Ohuye, T. (1942). On the blood corpuscles and the hemopoiesis of a Nemertean, *Lineus fuscoviridis*, and of a Sipunculid, *Dendrostoma minor. Sci. Rept. Tohoku Imp. Univ. Fourth Ser.* **17**, 187.

Oudemans, A. C. (1885). The circulatory and nephridial apparatus of the Nemertea. *Quart J. Microscop. Sci.* **25**, Suppl., p. 1.

Pantin, C. F. A. (1947). The nephridia of *Geonemertes dendyi. Quart. J. Microscop. Sci.* **88**, 15.

Parker, T. J., and Haswell, W. A. (1940). "A Text-book of Zoology", 6th ed. Foster Cooper, London.

Price, G. C. (1910). The structure and function of the adult head kidney of *Bdellostoma stouti. J. Exptl. Zool.* **9**, 849.

Regaud, C., and Policard, A. (1903a). Variations sexuelles de structure dans le segment preterminal du tube urinifère de quelques ophidiens. *Compt. Rend. Soc. Biol.* **55**, 216.

Regaud, C. and Policard, A. (1903b). Sur les variations sexuelles de structure dans le rein des reptiles. *Compt. Rend. Soc. Biol.* **55**, 973.

Riepen, O. (1933). Anatomie und Histologie von *Malacobdella grossa. Z. Wiss. Zool.* **143**, 324.

Roots, B. I. (1955). The water relations of earthworms. I. The activity of the nephridiostome cilia of *Lumbricus terrestris* L. and *Allolobophora chlorotica* (Savigny) in relation to the concentration of the bathing medium. *J. Exptl. Biol.* **32**, 765.

Rugh, R. (1935). Ovulation in the frog. ii. Follicular rupture to fertilisation. *J. Exptl. Zool.* **71**, 163.

Schreiner, K. E. (1955). Studies on the gonad of *Myxine glutinosa* L. *Univ. Bergen Årbok Nat. Rek.* **8**, 1.

Smith, H. W. (1951). "The Kidney. Structure and Function in Health and Disease". Oxford Univ. Press, London and New York.

Trowell, O. A. (1959). The culture of mature organs in a synthetic medium. *Exptl. Cell Res.* **16**, 118.

Voglmayr, J. K., Waites, G. M. H., and Setchell, B. P. (1966). Studies in spermatozoa and fluid collected directly from the testis of the conscious ram. *Nature* **210**, 861.

Walker, A. M., and Hudson, C, L. (1936a). The reabsorption of glucose from the renal tubule in amphibia and the action of phlorizin on it. *Am. J. Physiol.* **118**, 130.

Walker, A. M., and Hudson, C. L. (1936b). The role of the tubule in the excretion of urea by the amphibian kidney. *Am. J. Physiol.* **118**, 153.

Walker, A. M., and Hudson, C. L. (1936c). The role of the tubule in the excretion of inorganic phosphates by the amphibian kidney. *Am. J. Physiol.* **118**, 167.

Walker, A. M., Hudson, C. L., Findley, T., and Richards, A. N. (1936). The total molecular concentration and the chloride concentration of fluid from different segments of the renal tubule of amphibia. The site of chloride reabsorption. *Am. J. Physiol.* **118**, 121.

Walker, A. M., Bott, P. A., Oliver, J., and MacDowell, M. C. (1941). The collection and analysis of fluid from single nephrons of the mammalian kidney. *Am. J. Physiol.* **134**, 580.

Wilson, C. B. (1900). The habits and early development of *Cerebratulus lacteus* (Verill). *Quart. J. Microscop. Sci.* **43**, 97.

Witschi, E. (1929). Studies on sex differentiation and sex determination in amphibians. 1. Development and sexual differentiation of the gonads of *Rana sylvatica*. 2. Sex reversal in female tadpoles of *Rana sylvatica* following the application of high temperature. *J. Exptl. Zool.* **52**, 235.

Young, J. Z. (1950). "The Life of Vertebrates". Oxford Univ. Press (Clarendon), London and New York.

Zarnik, B. (1904). Uber die Geschlechtsorgane von *Amphioxus*. *Zool. Jahrb. Abt. Anat. Ontog. Tiere* **21**, 253.

CHAPTER 17

WITHIN THE VERTEBRATES: I. URINOGENITAL SYSTEM

In the preceding chapters the cytology and organization of the nemerteoids has been discussed at some length, and possible modes of further evolution towards the vertebrates have been suggested. It is now appropriate to consider some other evolutionary steps that have occurred mostly within the vertebrates themselves. It has already been pointed out that a great gap separates any invertebrate ancestor from the most primitive of existing vertebrates, and it will be shown that certain features of this gap are of interest in relation to cellular evolution.

The Divergence of Cell Types

Two major differences appear at once between the tissues of nemertines and those of early vertebrates, such as the cyclostomes. The first is that in the cyclostomes the grouping of cells and the cell boundaries have become much more regular and more clearly defined. Hence it is usually easier to assign functions to particular cell groups and to delimit the tissues that are concerned with particular functions. Presumably this means that, during differentiation, more specialization of the cells takes place. This is probably related to the other main difference: in the vertebrate there is greater variety but less plasticity of the tissues. One has the impression that, in the nemertines, only a few distinct classes of cell behaviour have become firmly established and that cells showing these limited types of behaviour occur in different parts of the body and are only just beginning to establish new processes of differentiation in each situation. These new processes must thus be regarded as variations on original themes. For example, the skin of the nemertine (see Fig. 12.2) seems to differentiate into ciliated cells and two varieties of secretory cells; cells which become filled with a protein mass that stains heavily red with azan and also with Masson's trichrome stain, and cells that stain with basic dyes and take on a deep purple colour with the bromine-alcian blue-PAS stain, and which are presumably producers of mucopolysaccharides. Probably these cell types are elaborations of the two primary classes of epitheliocyte, corresponding to the flagellated and amoeboid cells of the primitive blastuloid, or perhaps more accurately to the three groups of cells of those blastuloids in which a group of intermediate cells had also developed. Sometimes, as mentioned on p. 262, the cells of the nemertine skin all lie within the epidermis proper, but sometimes some of them are removed below the basement membrane and extend to the surface only by means of a long, thin process. These three kinds of epithelial cells presumably carry out all the functions of the skin except the sensory and neural functions, which are probably carried out by certain cells, originally epithelial, which have already specialized for these purposes. An examination of the pharynx and upper segment of the alimentary canal (see Figs. 12.6, 14.9) similarly shows three classes of cells. These cells are similar to the three classes in the skin, but now they presumably perform new functions or are at least specialize different aspects of their behaviour. In the pharynx some cells produce acid, some store iodine, and, if one may judge by the grouping of cells in particular places, e.g. the dorsal surface of the mouth, others may show elementary hypophysial functions, salivary gland functions, and so on. On the one hand, the cell which reacts positively for carbonic anhydrase in the pharynx and also stains heavily with eosin is comparable with the eosinophil cells of the epidermis; on the other, it could be considered the probable origin for the chloride cells of fish gills and the oxyntic cells of the mammalian stomach. Similarly, in the proboscis one sees variations of the activities of the same three types as seen in the skin. Even in the cephalic organ (see p. 370), which has reached quite an advanced state of cellular differentiation, the epithelial cells of the tube could be considered direct developments of the ciliated cells of the surface, whereas the secretory cells and vesicular cells could be legitimate descendants of the mucoid types. In the intestine also there are usually two main classes of cells, those possessing cilia or microvilli and those producing mucoproteins. In addition to these two classes are also found in the anterior region of the intestine proper (see Fig. 12.7) cells with granules that give the same staining reactions as cells in the skin and in the proboscis. The functions of these three classes of cells are almost certainly different from those of similar cells in the pharynx, the proboscis, or the skin, but the differentiation has not proceeded so far as to cover up the inherent similarities.

Thus in the various tissues of the nemerteoids, and in the evolution of animals up to the nemerteoid stage, one can see the process of differentiation and specialization in progress. One can also see how the variations of the behaviour of two initially antagonistic but compensatory cell types can lead to the differentiation of epithelial membranes of elaborate complexity and quite distinct major functions.

The two primary forms of cell, as exemplified by the flagellate and amoeboid forms of *Naegleria* and by the two forms of cells in the elementary blastuloid which are the potential precursors of ecballic mechanocytes and emballic amoebocytes, may have many other, if secondary, metabolic features which could, like the ionic ones, work in opposition to each other. For example, the mechanocyte liberates proline and hexosamine into its surroundings. Such secondary characters may lie at the root of much of the further differentiation and have led to the production of whole families of ecballic cells on the one hand, and of emballic cells, on the other. During these specializations the primary features of ecballism and emballism may even have become subsidiary to other distinctions which the cells in the two groups have secondarily acquired during the course of their specializations in other directions. In other words the production, on a large scale, of something that was initially a by-product may now overshadow the original function of moving ions or water, just as motor cars ousted the original bicycles from the Morris garage. Osteoblasts and chondroblasts are very different cells, but they are probably both derivatives of the originally ecballic cells, and, in tissue cultures, they both behave like mechanocytes. In the well regulated internal environment of the higher vertebrate the need for ecballism and emballism may be lessened and the cells may be made freer to specialize on other features of their activity. Presumably these specializations must always be consistent with the fact that the cells have originated either from ecballic or emballic progenitors, whose influence may reappear at any time. By analogy, men and women may specialize in all manner of activities, many of which are identical, and their characters may be modified thereby. Nevertheless, they remain men and women.

By the time the nemerteoid stage had evolved, it seems probable that two main classes of epitheliocytes were developed, with intermediates, and that, in the various epithelia of the body, specializations of these types had begun to be quite extensive, though most epitheliocytes still belonged recognizably to one class or the other. In addition to the epitheliocytes, numerous groups of mesohyl cells had also developed. These, too, were derived from primarily ecballic or emballic epitheliocytes and probably inherited most of the main characteristics of these types. Similarly nerve cells and their supporting cells each fall into two major groups.

Thus the epitheliocytes of the body can be broadly grouped into emballic and ecballic cells, as can cells of other classes as well. In general, the cells from the two families tend to pair off, so that the activities of the one are to a greater or lesser extent compensated by the activities of the other. In the first instance the balancing may have involved mainly water and ions, but with greater differentiation the more intimate biochemical processes may have had to be curbed and controlled on account of the effects of their products on other cells, or even on their own well-being.

Urinogenital Tubes and Sex Cords

With these ideas in mind, it is now appropriate to consider how the urinogenital system, as it appears in the early vertebrates, has become further elaborated within the phylum, and to consider what sort of behaviour might be expected from the cell types and organization found.

We have seen (Fig. 16.7) that in the cyclostomes there is a system of segmentally arranged tubules which are considered to be primarily excretory. They are not found at the anterior end of the body, and they are divisible into a pronephric series and an opisthonephric series. They excrete by a common duct. The tubules open internally into the coelomic cavity through ciliated funnels as in the pronephros, or end in large Malpighian capsules. Externally they open at the cloaca through a common archinephric duct. Apart from the fact that the coelomic cavity is not subdivided into segments, the position is somewhat comparable with that seen in the annelid *Hoplochaetella* (see Fig. 16.10), where the segmental nephridia join a common duct which opens at the anus. These tubules in the cyclostomes are not involved in carrying the genital products to the exterior; they appear to be mostly concerned with homoiostasis, though exactly how they function is uncertain. The pronephros also has functions connected with blood formation and filtration (see p. 526). The genital products are produced unilaterally, though this is probably a secondary phenomenon since lampreys with paired gonads have been found occasionally. The gonads occur as a series of pouches, ovarian anteriorly and testicular posteriorly, and these pouches project into the coelom. They do not open by separate ducts to the outside but shed their contents, when ripe, into the main body cavity from which they escape by two coelomoducts or coelomic (genital) pores to the exterior near the anus.

The series of gonadal pouches are intelligible as direct derivatives of the gonadal system of nemerteoids. Similarly, the serial kidney tubules are intelligible as derivatives of the nephridia. The real problem arises in determining the origin of the coelomic cavity, and why it has two paired openings to the exterior. It may be helpful to recapitulate briefly the main suggestions for this (see Chapter 16). First, the myocoels are formed from gut diverticula; therefore the nephrocoel and splanchnocoel could also be formed by their fusion. The pores to the cloaca could then be the relic of the connexion with the gut. Second, the main coelom is formed, as Goodrich (1945) concluded, from enlarged and fused gonocoel cavities, and the gonads themselves are the result of the germ cells having been housed prematurely in their gonoducts. The abdominal pores would then be either residual gonoducts or extremely shortened Müllerian ducts (i.e. remnants of the archinephric duct and probably nephridial in origin). Third, the coelom is a newly developed schizocoel. Fourth, the coelom is an enlarged and altered rhynchocoel. Fifth, the coelom is derived from part of the vascular system of the nemerteoid that separated off when a pumped circulation started in connection with the respiratory demands of increased muscular movement.

The evidence from cyclostomes suggests that the original serial tubes, which are excretory, are probably nephridial in origin. If Goodrich's (1945) view of the coelom is correct, however, the tubes could, as he suggested, be gonoducts, though it is curious that they do not function as such in the cyclostomes. Alternatively, the tubes could be a composite of gonoducts and nephridial tubes, comparable with somewhat similar tubes in the annelids and elsewhere (see Fig. 16.12).

Gonadal sacs are present as such in the cyclostomes, though they open secondarily into the coelom to shed their contents. In fish and the higher vertebrates the gonads, originating in the genital ridge, develop in the form of sex cords which then unite secondarily with the coelomoduct or nephridial system. Even in man, sex cords develop in the genital ridge and grow to join the mesonephric tubes (Fig. 17.1). What significance can be attached to these sex cords?

Embryologically speaking, cords of cells are believed to occur either before their hollowing out into tubes, or to occur where, in an ancestral form, there has been at one time a tubular system which no longer becomes functional, as, for example, in the ducts to the islets of Langerhans in the pancreas. In the male gonads of vertebrates the cords do indeed hollow out into tubes, which make direct connexion with the mesonephric tubes. The situation is thus similar to that which occurs in some annelids, where the gonoducts or coelomostomes open directly into nephridial tubes. In the female a temporary connexion may be made between the sex cords and the mesonephros, only to degenerate later.

Clearly the important questions to answer here are: What were the tubules of which the sex cords and seminiferous tubules are now the representatives? Are they the original gonoducts, or are they derivatives of the "coelomoduct" tubes of the cyclostomes? In relation to the body segments, some coelomoducts are missing between the pronephros and the mesonephros. This could

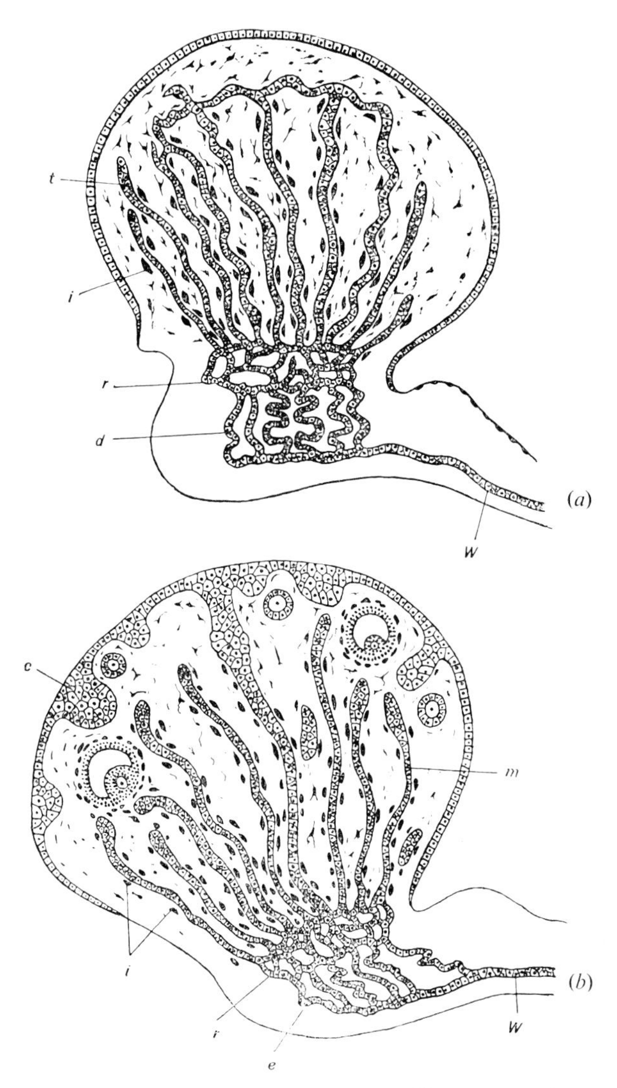

Fig. 17.1. Diagram of developing testis (*a*) and ovary (*b*). c, cortex; d, ductuli efferentes (urogenital junction); e, epoophoron (urogenital junction); i, interstitial cells; m, medullary cord; r, rete testis and rete ovarii; t, seminiferous tubule; W, Wolffian duct. (Maximow and Bloom, 1930.)

be significant in connexion with gonad function, though more or less developed Malpighian capsules may occur in this region suggesting that, for some reason, these intermediate tubes have simply atrophied.

It was shown earlier that the cells of both gonoducts and nephridia must initially have had fluid-regulating properties, though they are likely to have been orientated in reverse directions. Thus, even if it is not possible to be certain of the origin of the kidney tubes and sex cords of vertebrates, the two sets of tubes are likely to have much in common and may differ mainly in the orientation of their cells and the manner in which their activities are controlled. Moreover, if the ontogeny of the gonads is to be taken into account, the ontogeny and functions of the adrenal cortex must also be included, for this tissue is developed embryologically from the same genital ridge and has much in common developmentally and indeed physiologically with the gonadal tissues. It is therefore profitable to compare the cytological properties and cellular activities of these four tissues, testis, ovary, adrenal cortex, and kidney.

The first three, the testis, ovary, and adrenal cortex, all develop from the genital ridge, a thickening of mesenchymal cells on the dorsal wall of the coelomic cavity. Over the genital ridge the coelomic epithelium is peculiar in that it has no basement membrane (Gruenwald, 1942a, b) and it attracts the primary germ cells. In several groups of vertebrates there is good evidence (see Tyler, 1956) that the germ cells can first be localized as such elsewhere, i.e. among the endoderm cells (Everett, 1943) or in the yolk sac, and that they then migrate to their position on the genital ridge. In the frog they probably arise originally from the vegetal hemisphere of the ovum (Dennis Smith, 1966). The nature of the relationship between the germ cells and the so-called germinal epithelium of the genital ridge is, however, still quite obscure. While in some cases the germ cells are clearly separable from the other cells of the so-called germinal epithelium, in others it is possible that, like the germ cells themselves, the cells of the germinal epithelium as a whole are relatively "uncommitted" cells (like the archaeocytes of sponges for example).

During development the germinal epithelium over the genital ridge first thickens by cell multiplication. Then, in the formation both of the gonads and of the adrenal cortex, it invades the underlying mesenchyme with columns and, eventually, cords of cells which, interestingly enough, maintain or can maintain certain epithelial characteristics (Fig. 17.2) (Gruenwald, 1942a, b). In the gonads the germ cells are carried in with this invasion, and in the adrenal cortex it is pertinent to observe that certain large cells, which later degenerate, can also sometimes be seen among the invading cells (Fig. 17.3) (Hett, 1925; Uotila, 1940). Eventually these cords of cells differentiate into definitive sex cords in the gonads and into cell columns in the adrenal cortex, though, in the latter case the earliest cells to enter may degenerate and be replaced by a later invasion.

In the male gonad the process of ingrowth comes to a stop; mesenchyme cells and fibrous tissues intervene between the cord cells and the original

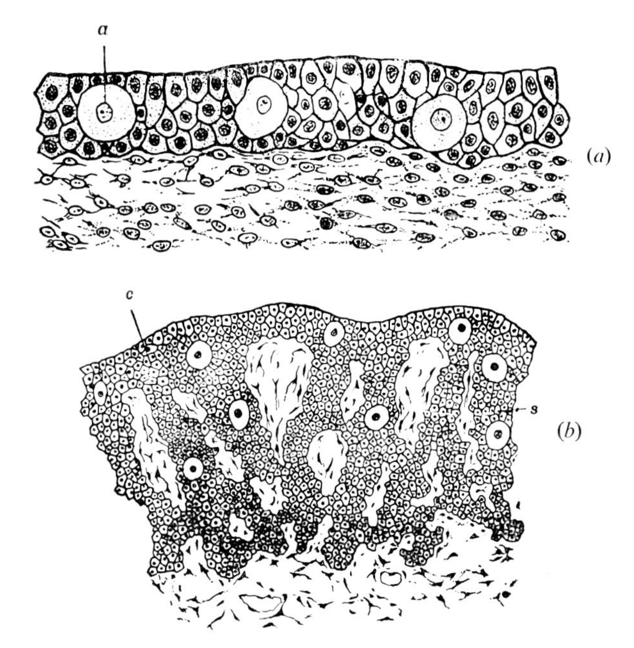

FIG. 17.2 Primordium of the gonad in the early indifferent stage. (*a*) Thickening of the germinal epithelum; (*b*) formation of the sex cords. a, Primitive sex cell; c, cortex; s, sex cords. (Maximow and Bloom, 1930.)

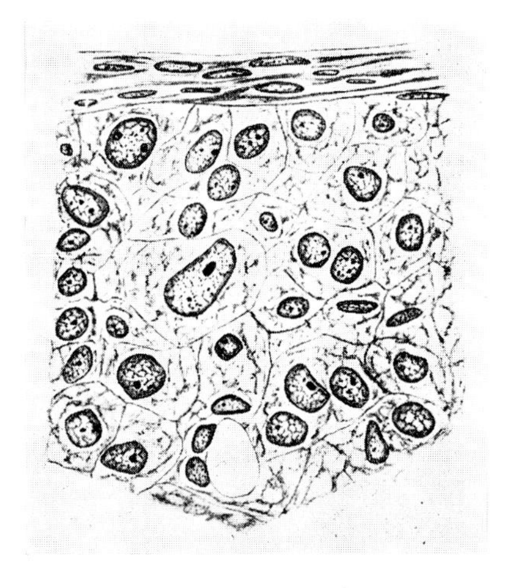

FIG. 17.3. Large cells in the outer adrenal cortex of 36 mm human embryo. (Hett, 1925.)

germinal epithelium, which then retrogresses to the usual, flattened, coelomic type. A basement membrane forms between the cells of the sex cords and the surrounding mesenchyme, though why this happens is obscure. There is a suggestion from observations of the salivary glands that basement membranes, or at least their mucopolysaccharide constituents, can be produced by the epithelial cells themselves (Kallman and Grobstein, 1966). It is notable that in the testis tubule and the ovarian follicles basement membranes start to develop or intensify when the tubules or follicles start to fill with fluid (Wimsatt, 1944). It may be that the polarity and the secretory activities of the epithelial cells are important in determining the formation and properties of basement membranes.

In the male gonad (Fig. 17.4a) the sex cords differentiate into definite testicular tubes which eventually make contact and connexion with the mesonephric tubes that have appeared in the walls of the nephrotome in association with the archinephric or Wolffian duct. Indeed, this connexion is probably of great importance to the process of differentiation because it allows the cells of the solid cords to form a hollow tubular arrangement with an outflow through the mesonephros. This outflow probably helps to establish in the cells of the tubules a permanent polarity favouring epitheliocyte behaviour and thus perhaps assisting in basement-membrane formation.

In the female gonad (Fig. 17.4b) the inflow of cells from the germinal epithelium is not permanently cut off by any intervening layer of mesenchyme cells, and the germinal epithelium persists in the form of a single layer of cells which are more cubical than those of the rest of the coelomic epithelium. The sex cords may make temporary contact with the mesonephric tubes, by way of the rete ovarii, but the connexion is not permanent and the sex cords do not form a basement membrane or hollow out. For the most part their cells become dispersed and enter the stroma of the ovary as groups of interstitial cells. The parts of the sex cords nearest the surface, the cortical cords, or new ingressions of germinal epithelial cells eventually enclose the invading germ cells and constitute the primary follicles. At a later stage these follicles begin to enlarge by growth in size of the ovum and by growth in number of the follicular cells. In each follicle the latter continue to multiply until two layers cover the ovum. Eventually the follicular cells, probably under the influence of the follicle-stimulating hormone of the pituitary, produce the follicular fluid which accumulates between the two layers (Fig. 17.5). At about that time the zona pellucida appears as a sort of basement membrane round the ovum, and a distinct basement membrane appears round the follicle as a whole, suggesting a connexion between cell polarity and the passage of fluid on the one hand and basement-membrane formation on the other. Furthermore, when the fluid begins to collect inside the follicle, the interstitial (sex-cord-derived) cells and blood vessels begin to congregate

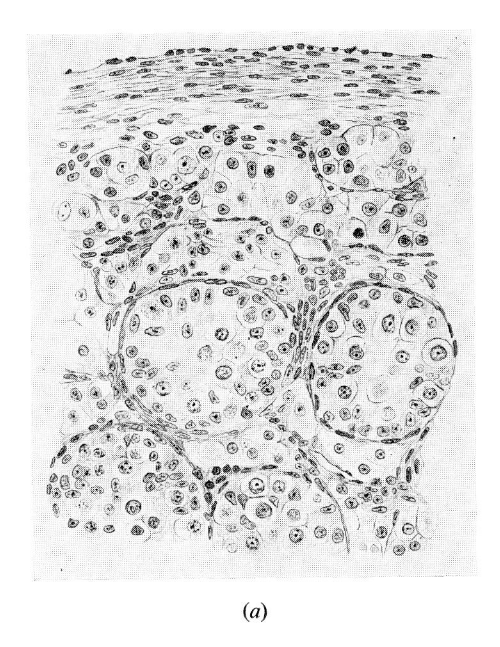

(*a*)

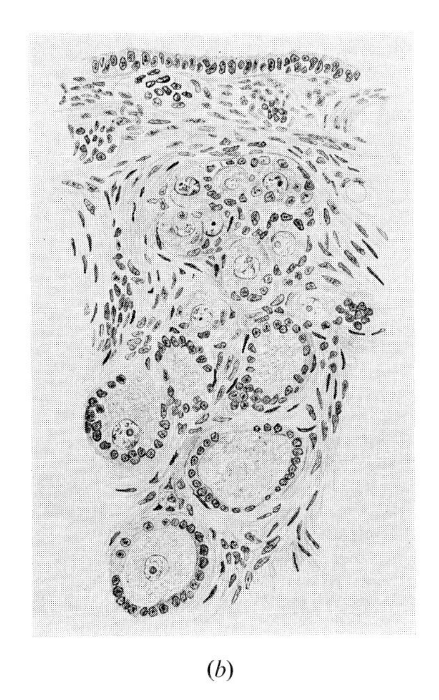

(*b*)

FIG. 17.4. (*a*) Developing testis. Note tubule formation centrally and the capsule with flattened epithelium; (*b*) developing ovary. Note follicle formation and persistent germinal epithelium.

outside the follicular membrane and constitute the theca interna. Outside the latter the mechanocytic mesenchyme cells again concentrate as the theca externa. The theca interna cells begin to enlarge as secretory "epithelioid" cells.

In the seminiferous tubules of the male the only cells present are the Sertoli cells and the germ cells. It must be presumed that the former provide the requisite environment for the latter (Fig. 17.5). Nevertheless there are also present in the testis the interstitial cells of Leydig whose development

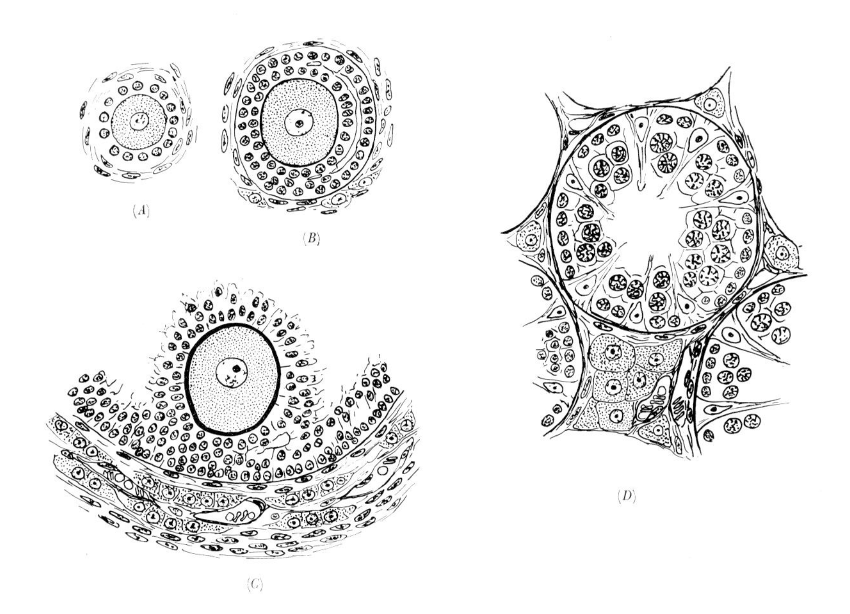

FIG. 17.5. Cellular arrangement in the ovary and testis. (*A*) Early follicle with single layer of nurse cells round the ovum; (*B*) beginning of accumulation of follicular fluid; (*C*) ovum in cumulus oophorus in mature follicle, showing zona pellucida and development of thecae interna and externa; (*D*) testis tubule showing the relationships of germ cells, Sertoli cells, and interstitial cells.

is somewhat obscure, though it apparently occurs in the mesenchyme. These cells produce the male hormones. When they are functional, they are clearly emballic (amoebocytic) cells with a vesicular and smooth endoplasmic reticulum (Fawcett and Burgos, 1960; Pudney, 1967), in contrast to the ecballic Sertoli cells which become functional at the same time and have a well developed, flat, membranous endoplasmic reticulum. There is thus a rather close parallel between what appears to be a functional interaction of the cells of the membrana granulosa with those of the theca interna, on the one hand, and the similar interaction between the Sertoli cells and the Leydig cells on the other.

All this makes an interesting enough story, but it also poses a great number of fundamental questions at the cytological level. An attempt at an analysis of some of these questions will not be out of place.

The Differentiation of Germ Cells

The first point, which has been made earlier (p. 455), is that it is necessary to provide the developing germ cells with the requisite local environments for their differentiation and development. The oocyte must be given a situation in which it will not divide more than twice, and then unequally, but in which it can accumulate masses of reserve storage materials. The spermatogonium must be encouraged to divide and must be given conditions in which each of its progeny will eventually get rid of most of its cytoplasm and develop a flagellum. The case has some parallelism with that of *Naegleria*, where conditions can be arranged in which the amoeboid form is preserved, while in other conditions flagella develop. It is, of course, also well known that many protozoa have a sexual phase in which cells become motile through the development of a flagellum, and this too is often determined by external conditions. In *Naegleria* the flagellum develops in a "dilute" environment in which the cell becomes ecballic (see pp. 169 and 175) and has to battle to preserve its necessary ions and not become flooded with water or lose too many metabolites. In a more concentrated environment the amoeboid (emballic) form may have to battle to maintain its water and to keep out excessive ions. The actual conditions that determine the behaviour of the germ cells are largely unknown, but it is perhaps of interest that testicular fluid, obtained from the tubules themselves (Voglmayr *et al.*, 1966), has a lower sodium content and a much higher potassium content than plasma (Table 17.1) (Voglmayr *et al.*, 1967). Setchell (1967) has shown that, after the concentration of various ions in the plasma has been raised, these ions are passed into the testicular fluid slowly and selectively. The fluid in the epididymis of the bull (Sørensen and Andersen, 1956) has very low sodium and a very high ratio of K^+/Na^+ which increases as the fluid passes down the tube (Crabo and Gustaffson, 1964). These observations are consistent with the idea that the spermatozoa develop in surroundings demanding ecballic activity. On the other hand, the follicular fluid of the cow, obtained after treatment of the latter with pregnant mare serum, also has a low sodium and high potassium content (Lutwak-Mann, 1954), and Smith (1937) found that the total osmotic pressure of the follicular fluid of the rabbit was equivalent to about 0.55% NaCl; the fluid is thus very hypotonic to plasma.

TABLE 17.1[a]

Sodium and Potassium Content of the Environments of Germ Cells

	Na (meq/l)	K (meq/l)	Osmotic pressures (m osm/l)
Plasma	144	4.18	298
Testicular fluid	133	12.7	283
			Glucose (mg/100 ml)
Plasma	142	4.8	(36–57)[b]
Follicular fluid	110	10.2	43

[a] From Voglmayr *et al.* (1967) and Lutwak-Mann (1954).
[b] Normal limits for cattle.

At first sight, therefore, these observations seem to argue very strongly against the amoeboid form of the ovum being determined by local conditions. The situation should perhaps be examined more closely, however.

In both gonads the fluids are probably secreted, the one in sufficient quantity to carry the sperm along the length of the seminiferous tubule, and the other in sufficient quantity to enlarge the follicle to a considerable size. In the case of the follicle, the cells which do this secretion are almost certainly the cells of the membrana granulosa derived from the original nurse cells of the primary follicle. In the case of the seminiferous tubule they are the cells of Sertoli, unless the spermatogonia themselves produce the fluid. A cell that can stand a dilute environment (using the word dilute with the wide connotation given to it in the present discussion) must be ecballic and capable of ejecting water at least as fast as it takes it in. If a cell is to produce a more dilute fluid at one end than it takes in at the other, it must be able to eject water forcibly from its distal end and, probably, eject ions at its proximal end. This is the situation for both the Sertoli cell and the granulosa cell (Fig. 17.5), if each is creating its respective dilute fluid; the production of a dilute fluid at one pole probably means a concentrated fluid at the other. In terms of what has been discussed earlier, both the granulosa cells and the Sertoli cells are probably ecballic towards the lumen, though they are not very strongly or permanently in this state. In tissue cultures the cells of the membrana granulosa and of the cumulus oophorous grow first as a continuous sheet (as epitheliocytes), but the sheet readily splits up and the cells then behave like typical mechanocytes with an occasional amoebocyte amongst them (Smith, 1952; Blandau and Rumery, 1962; Short, 1967). Esaki (1928) has reported similar behaviour for the Sertoli cells, which tend to grow as mechanocytes. Since these cells in the body are organized in a polarized epithelial manner, they must create different local environments at their opposite ends; dilute fluid, relatively rich in potassium, fills the tube or follicle, and relatively concentrated (or K^+ poor) fluid must tend to collect at the basal end. Dilute fluid collects within the testis tubule, therefore, and the sperm develop in this fluid. The cells which collect round the tubule, i.e. the cells of Leydig, are subjected to a relatively more concentrated environment, become emballic, and thus develop many of the characters of the amoebocytic type of cell, e.g. pinocytosis, lysosomes, vesicular endoplasmic reticulum, pigment, and other inclusions (Fawcett and Burgos, 1960). Thus the Sertoli cell and the Leydig cell form the same sort of symbiotic pair as the flagellate and amoeboid cells do in the blastula of the sponge. Similarly, the cells of the membrana granulosa pass the dilute fluid inwards into the ovarian follicle and tend to create around the outside of the follicle a more concentrated environment, which in its turn collects or produces emballic cells in the form of the theca interna cells. The sperm cell develops in the fluid at the dilute end of the Sertoli cell, but the peculiar distribution of the granulosa cells in the form of a cumulus oophorus means that the ovum develops in the concentrated environment at the basal end of these cells. The former thus becomes flagellate, the latter remains as an emballic amoebocyte whose surface becomes covered with microvilli, as it is found to be, for example, in the frog (Kemp, 1956) (Fig. 77.6) and rat (Sotelo and Porter, 1959).

If the granulosa and Sertoli cells are behaving ecballically (it has been mentioned that in culture they can both behave like mechanocytes), they might be expected to produce other products characteristic of such cells, e.g. mucopolysaccharides and collagen precursors. Follicular fluid does indeed contain mucopolysaccharides (Caravaglios and Cilotti, 1957; Wislocki *et al.*, 1947; Halmi and Davies, 1953), and its hexosamine content has been estimated as 36 mg per 100 ml (Lutwak-Mann, 1954). Testis fluid apparently does not. On the other hand, in the testis there is a high concentration of hyaluronidase which may function to break up the polysaccharides that might otherwise tend to form. Similarly, the mature follicle shows a sudden production of hyaluronidase (Zachariae and Jensen, 1958) which has been thought to be responsible for increasing the osmotic pressure of the contents just before ovulation (Smith, 1937). Lowering the adhesive properties of cells is important, both in the follicle just before rupture and in the seminiferous tubules all the time, and it is possible that the hyaluronidase may be involved in that way also.

With regard to the expected precursors of collagen, testicular fluid has a very high content of glutamic acid and of glycine (Setchell, 1967), though there is no evidence of any large proline or hydroxyproline content. The situation in the testicular tubule is, of course, complicated by the continuous breakdown of the cytoplasm of the developing sperm. Interestingly enough, the primary spermatocytes are said to imbibe a large quantity of water just as they begin to develop.

One important difference between follicular fluid and testis fluid is the presence of glucose in the former but not in the latter. It is, of course, possible that both fluids initially contain glucose but that, by the time the testicular fluid has reached the place where it can be collected, all the glucose has been used by the rapidly metabolizing sperm. Alternatively, the glucose could be removed from the seminiferous tubule in much the same way as from the proximal tubule of the kidney. The coelomic fluid of the earthworm does not contain any glucose (Bahl, 1947); if the coelomic sacs of the worm are considered to be enlarged gonocoels, the absence of glucose in the mammalian seminiferous tubules falls into line, but its presence in follicular fluid is unexplained.

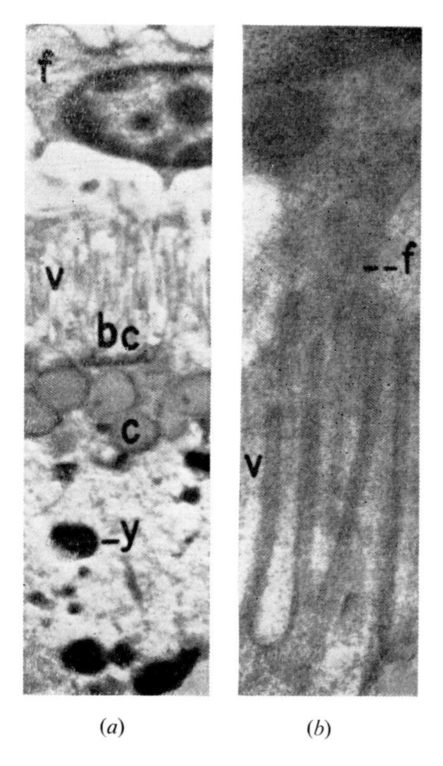

FIG. 17.6. (*a*) Microvilli in the surface of the ovum of the frog as seen with the electron-microscope. bc, Basal cortical cytoplasm; c, cortical granules; f, follicle cell; v, microvilli; y, yolk. (*b*) The villi at higher magnification. (Kemp, 1956.)

Interesting support for the concept that the local environment is responsible for the differentiation of the germ cells comes from an entirely different quarter. In the testis of the amphipod crustacean, *Orchestia*, the ripening sperm are enclosed in a capsule of mucus-producing cells, and the nurse cells of the ovary are of the flattened mechanocytic type of cell which usually surrounds developing ova (Fig. 17.7) (Meusy, 1963). Thus in this creature

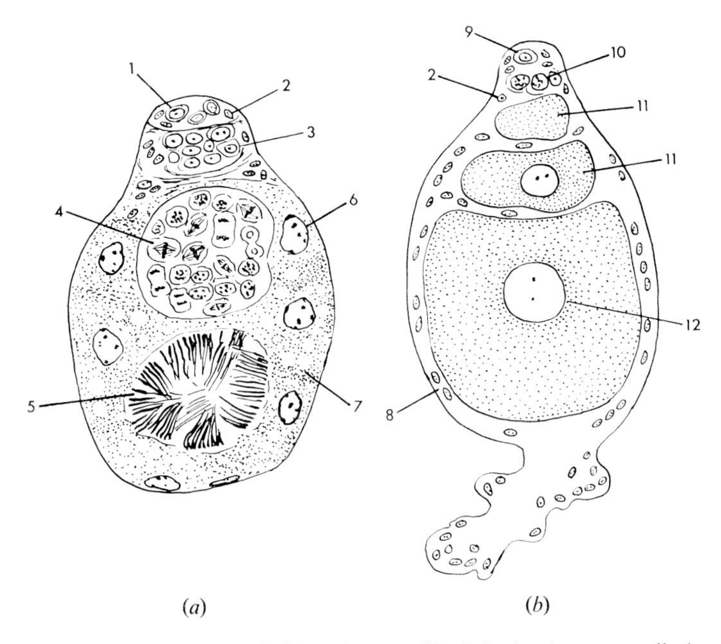

FIG. 17.7. Diagrams of the testis (*a*) and ovary (*b*) of *Orchestia gammarella* (amphipod crustacean). Note the large mucoid cells surrounding the developing sperm and the thinner follicle cells surrounding the oocytes. 1, Spermatogonia I; 2, mesenchyme cell; 3, spermatogonia II; 4, dividing spermatocytes; 5, spermatozoa; 6, giant nucleus of marginal cells; 7, foamy cytoplasm of marginal cells; 8, nuclei of follicle cells; 9, oogonia I; 10, oogonia II; 11, oocyte; 12, nucleus of oocyte. (After Meusy, 1963.)

the different environments are obtained by surrounding the developing germ cells with either mucoid (emballic?) cells or mechanocytic (ecballic?) cells. In the vertebrates the environments are apparently produced by making use of the polarity of the mechanocytic type of cell, as they probably are also in rhabdocoels (Fig. 11.8), nemertines (Fig. 16.21), and *Amphioxus* (Fig. 16.23). This is also true in cyclostomes (Fig. 16.23), many of which are hermaphrodite. In them the sperm develop in follicles filled with fluid and lined by a layer of "Sertoli" cells with large nucleoli, and each ovum is tightly enclosed in a capsule of follicle cells without any intervening fluid (Schreiner, 1955).

How these two sets of conditions are established within the one animal is not known, but they seem to provide what otherwise appears to be an uncommitted germ cell with an environment that determines its future development as male or female gamete. Tissue-culture studies of the gonads of insects (*Galleria*) (Lender and Duveau-Hagège, 1962) also show a similar state of affairs in that the testis tubules fill with fluid, like the corresponding structures of the developing nemertine, while the ova are tightly surrounded by their nurse cells, with no accumulation of fluid (Fig. 16.22).

It is interesting to notice that in *Orchestia*, where the gonads are essentially tubular, the specific properties of the supporting cells of the testis develop at the part of the tube where the maturing sex cells have reached the spermatocyte stage. This has a parallel in vertebrates, where it has been shown (Ohno and Smith, 1964) that, if the fetal oocyte fails to be enclosed in a complete follicle, then meiosis is not arrested and the oocyte enters mitosis and degenerates. Similarly, if at a later stage the follicle surrounding the oocyte is ruptured, further progress to the mature ovum is not interrupted as it normally is in the undisturbed follicle, but the maturation division occurs (Edwards, 1962). The surrounding cells are very important in determining the behaviour of the germ cells in both sexes.

One point of cardinal importance has so far been neglected in this discussion: the influence of the chromosomal sex differences on the development of the gonads and thus on the development of the germ cells. The different complement of DNA in the two sexes would, in general, be sufficient to account for the development of different local environments in the tissues of the two sexes, and particularly in the gonads. In them, the particular equilibrium set up would determine the manner in which the supporting cells differentiated and functioned. Thus, to take a simple case, it is easy to imagine that, in a tubular system like the gonocoel of a nemertine, the gradients across the walls and down the length could be sufficiently different in the two sexes to determine quite different local environments in the tubes of the two sexes so that the germ cells enclosed in them would develop quite differently. Much the same must apply in the more sophisticated gonads of the vertebrates.

There is, however, one more point. Among those animals in which sex determination is mainly chromosomal, there is the possibility of the germ cells themselves modifying the gonads specifically. For instance, in an individual in which all the cells are XY, the germ cells are, until they undergo meiosis, also XY. Thus, in the gonads, XY germ cells react with XY supporting cells, and the resulting environment and subsequent differentiation could well be very different from those determined by XX cells reacting with XX supporting cells, especially when it is remembered that, after meiosis, X and Y cells could behave quite differently. In fact, attempts to differentiate between X- and Y-bearing sperm have failed (Edwards and Gardner, 1967). This, however, is really only a special case of the original point that the chromosome content determines or affects the exact local environments in all tissues. Nevertheless, it does emphasize a fact which has also been somewhat neglected in the earlier discussion, namely that, in determining the local environment in which the germ cells develop, the germ cells themselves also play their part in establishing the *modus vivendi* with their supporting cells. There is always mutual interaction.

It should probably be pointed out that there is some danger of arguing in circles on all these questions. Thus, it has been suggested that, in the ovary, secretory activity of the membrana granulosa cells causes amoebocytic behaviour on the part of the theca interna cells, It could be that the theca interna cells, by becoming actively emballic, stimulate the secretory and ecballic activity of the membrana granulosa cells. The number of theca interna cells present at the time when the pituitary hormone initiates the follicular enlargement argues against this idea. Another point of uncertainty about this stage also arises from the exact meaning of the terms ecballic and emballic as applied to the cells in each particular situation. It is well to remember that, in the metaplasia of *Naegleria*, numerous factors can be involved in causing or preventing the assumption of the flagellate form, e.g. positive ions, negative ions, steroids. Certain concentrations of progesterone, for example, can actually encourage the amoebae to become flagellate; lactate and bicarbonate ions can have a similar effect. The effective agents in bringing about a particular pattern of behaviour in a metazoan cell may thus be numerous and diverse, and may involve the resultant of several competing actions.

Naegleria can change its activity in accordance with its surroundings. In this argument we are assuming that the germ cells retain this property and that perhaps other cells of the germinal epithelium do so too. Thus, in initiating the differentiation of the gonads, the manner in which the first cells differentiate, or the susceptibility of one particular group of cells to a sex hormone, may initiate a whole train of events. It is probably relevant to the discussion to observe that, in cultures, Esaki (1928) found that the spermatogonia were apt to produce giant cells (an amoebocytic property); in cultures from cryptorchid testes, without germ cells, the interstitial cell (and/or Sertoli cells) behaved like mechanocytes. It has already been noted that the germ cells are probably initially amoebocytic or emballic, as they were seen to be in the sponges.

The essence of most of these cases seems to be much the same. If cells lining a tube act ecballically towards the lumen, they are likely to induce emballic activity in cells lying near their basal surfaces and ecballic activity in those near their apices.

Emballic cells have been defined (p. 175) as cells which need to conserve water and to throw out ions. At first sight, therefore, it is difficult to see how a covering of emballic cells could, as in the amphipod crustacean, *Orchestia*, provide the necessary dilute environment for the sperm. The nature of the ions was purposely not specified, however, nor was any consideration given to polarity or to the specific metabolites of the emballic form of metabolism. There seem to be at least three possibilities to account for events in *Orchestia*. First, the supporting cells in this species surround but do not, like the Sertoli cells of other animals, intermingle with the germ cells, and the ions could be ejected towards the basement membrane so that the supply of them in the fluid of the cavity would be reduced. Second, the sperm do not so much require a dilute solution (though the activity of sperm can be prolonged by such; Weisel, 1948), but they do need a relatively high concentration of a particular ion, e.g. potassium. The mucous cells of *Orchestia* could be specifically ejecting this ion. Third, crustacean sperm is often anomalous in its behaviour and differentiation. Several observations on other animals argue in favour of the second concept. First, the motility of sperm is augmented by a high potassium content of the surrounding fluid (Schlenk and Kahmann, 1938). In the bull, this is certainly provided for the sperm by the fluid in the epididymis (Sørensen and Andersen, 1956) (see also Table 17.1). Second, in cases of abnormally increased numbers of goblet cells in the intestine, the contents of the intestine become exceptionally rich in potassium (Little, 1964), perhaps indicating that these emballic cells send potassium ions into the lumen.

However this may be, the cellular arrangements of the gonads in the vertebrates are such as to suggest that a more searching investigation into the local environments created for the germ cells could well lead to a better understanding of sexual physiology. When the adrenal cortex is under consideration, the importance of the Na/K ratio in the follicle and in the testis will be seen to have interesting implications.

The Adrenal Cortex

In the development of the adrenal cortex the cords of cells at first produce a pattern of cells similar to that of the sex cords in the gonads (Fig. 17.8). The zona glomerulosa corresponds to the cortical zone, and it is interesting to note that, in cattle (Fig. 17.9), the cells of this zone may actually be arranged in vesicles and separated from the rest of the cortex by a layer of connective tissue. They are thus comparable with the follicular (cortical) part of the ovary (Fig. 17.1b and 17.9c) (Weber, *et al.*, 1950). Even in man, the zona glomerulosa may contain vesicles at a certain stage of embryonic develop-

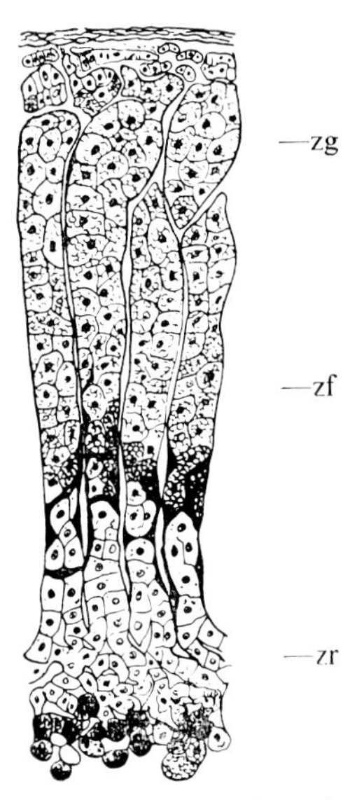

FIG. 17.8. The cellular arrangement in the mammalian adrenal cortex. zg, Zona glomerulosa; zf, zona fasciculata; zr, zona reticulata. (Cowdry, 1938.)

ment (Fig. 17.10) (Hett, 1925). In many species, cells continue to enter the zona glomerulosa from the coelomic epithelium or the capsular tissue long after the first development of cell cords and even into adult life, rather in the same way as cells continue to penetrate into the ovary from the germinal epithelium (Gruenwald and Konikov, 1944; Baxter, 1946; Gruenwald, 1942b, 1946).

The zona fasciculata corresponds similarly to the main sex cords, while presumably the zona reticulata is equivalent to the rete testis or ovarii, or perhaps to the connexions with the mesonephric elements. Just as the deeper parts of the sex cords in the female degenerate or become dispersed, the innermost parts of the cortex, the X-zone of mice and/or the foetal cortex of other species, also degenerate under the influence of sex hormones. In the male mouse the X-zone disappears at puberty. In the female it starts to retrogress periodically in the virgin, but disappears completely when the animal becomes pregnant (Fig. 17.11) (Deanesley, 1928).

In the adrenal cortex the tubular or cord-like arrangement of the cells is still evident, particularly in the mammals; moreover there is some evidence for a gradient along the cord. First, the occasional appearance of vesicles in the zona glomerulosa indicates that an inward passage of fluid tends to occur into this part of the cord, reminding us that the cord was once a tube. Second, in electron-microscopic sections across the cords of the zona fasciculata, the place where several cells meet is marked by a minute tubular space

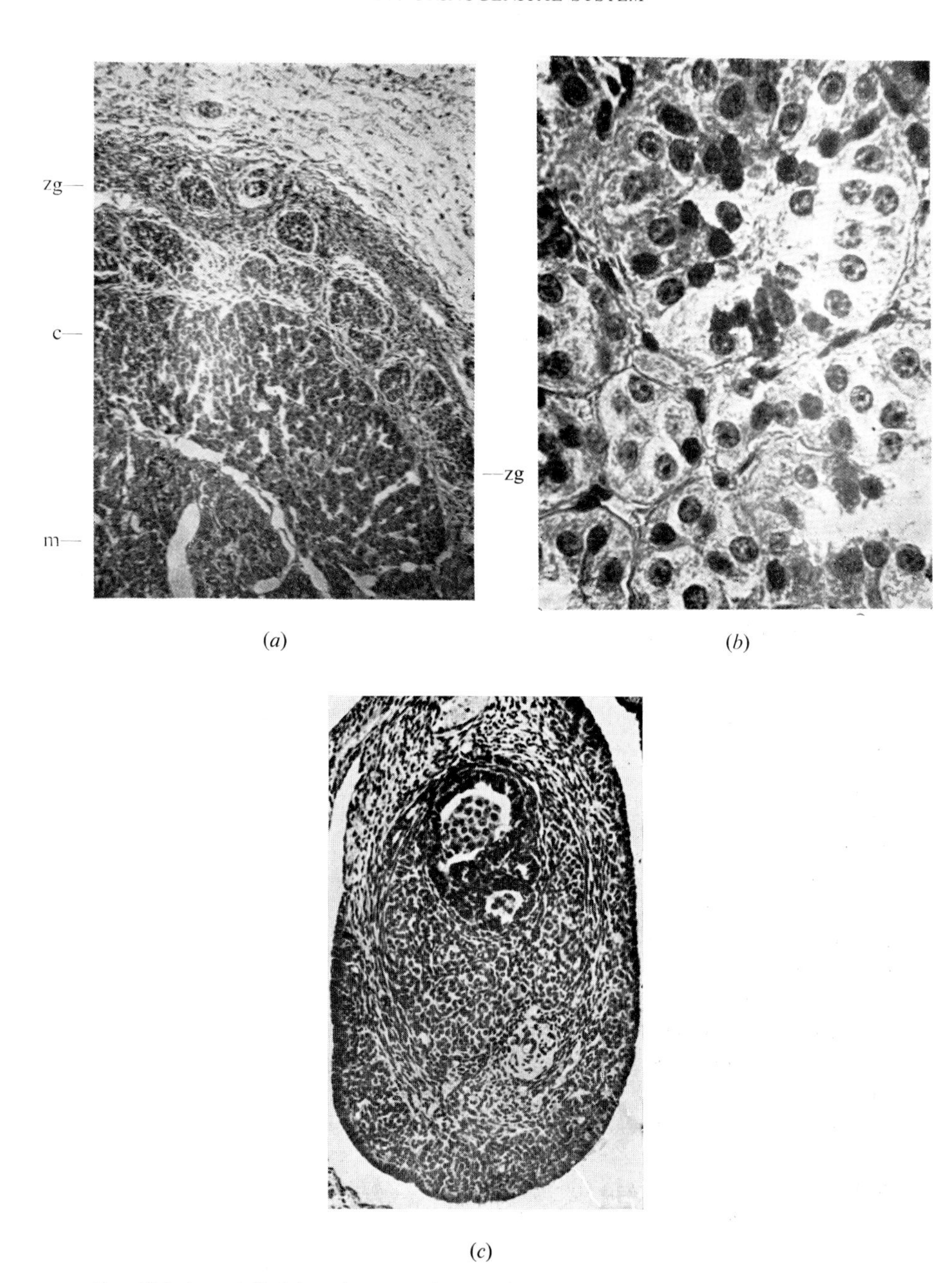

FIG. 17.9. (*a* and *b*) Adrenal cortex of cow. (*a*) General pattern. c, Cortex; m, medulla; zg, zona glomerulosa. (*b*) Vesicles in zona glomerulosa. (Weber *et al.*, 1950.) (*c*) Ovotestis of the opossum. Compare (*a*) with (*c*) (Burns, 1956.)

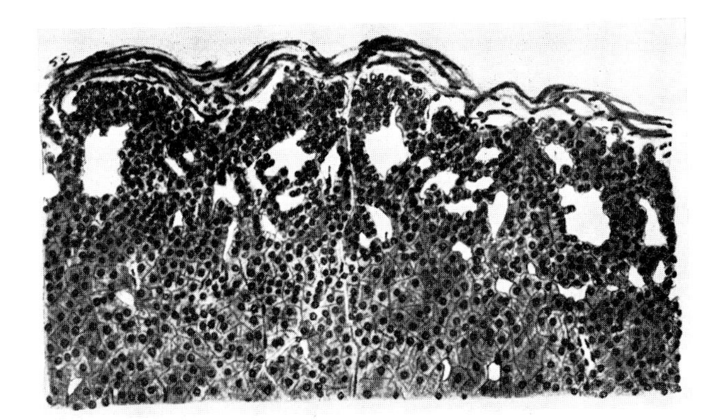

FIG. 17.10. Vesicles in adrenal cortex of 170 mm human embryo. (Hett, 1925.)

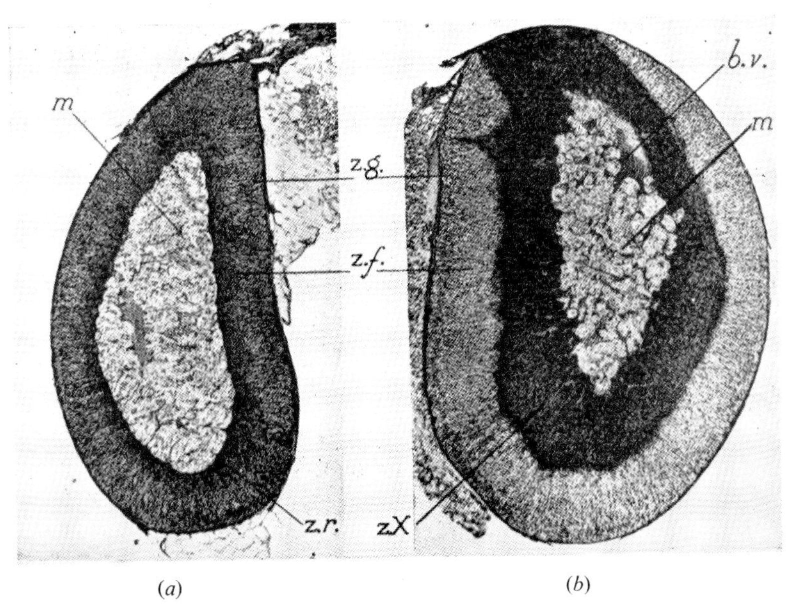

FIG. 17.11. Adrenal of mouse. (*a*) Male, 12 weeks old, X-zone absent; (*b*) unmated female, 12 weeks old; large X-zone. bv, Blood vessel; m, medulla; zf, zona fasciculata; zg, zona glomerulosa; zr, zona reticulata; zX, X-zone. (Deanesley, 1928.)

which has microvilli projecting into its lumen (Carr, 1958). Third, phospholipids tend to accumulate in the cells of the zona glomerulosa, and steroids and cholesterol accumulate in those of the zona fasciculata. Between the two zones there is a sudanophobic zone which is developed to a greater or lesser extent in different species. Castration can fill the zona fasciculata cells with lipids, and treatment with androgens can almost completely denude them of lipid inclusions, but in all this the lipids of the zona glomerulosa are almost unaffected (Fig. 17.12). Fourth, trypan blue is picked up as small granules by the cells of the zona glomerulosa, and as coarse granules in the deeper layers, in the same way as it accumulates in the tubules of the kidney (Baxter, 1946) (see p. 452).

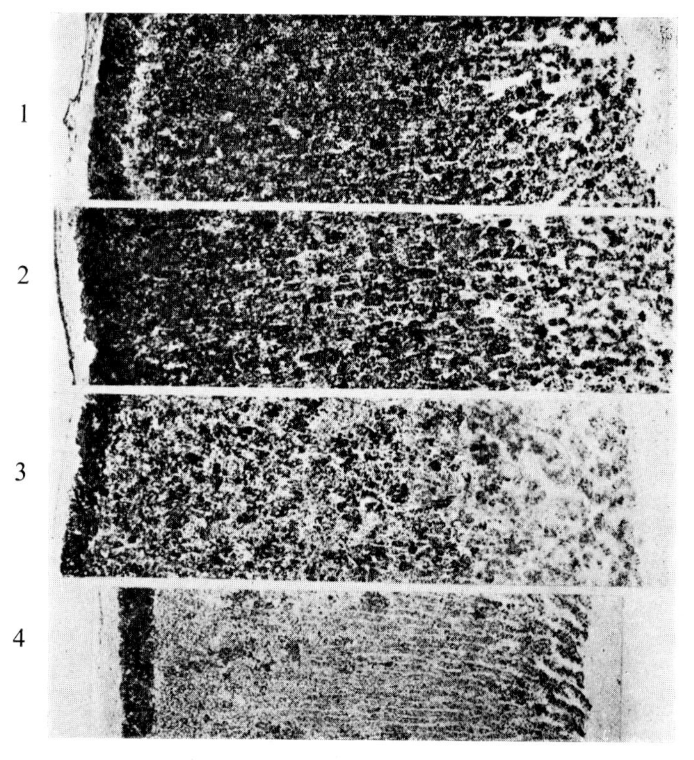

FIG. 17.12. The effects of castration and of male hormones on the adrenal cortex of the rat. 1, Control. Note the sudanophobic zone separating the zona glomerulosa from the zona fasciculata. 2, Castrated rat. Note the high content of lipids and relative absence of the sudanophobic zone. 3, Castrated rat, treated with 0.033 mg testosterone daily for 23 days. 4, Castrated rat, treated with 0.9 mg androsterone daily for 47 days. Note the reduction in lipids, except in the zona glomerulosa. (Hall and Korenchevsky, 1938.)

These observations are consistent with the view that the inward passage of fluid into the vesicles of the zona glomerulosa by ecballic cells working a polarized active transport system may change to an attempted outward passage of fluid by emballic and pinocytosing cells in the zona fasciculata. In terms of ions, the cells of the zona glomerulosa could be conserving ions and tending to increase their concentration on the side of the blood (assuming a dilute fluid in the vesicles like that in the follicles of the ovary). It would be interesting to know if there is a differential movement of sodium or potassium as there is in the gonads. When the blood from the zona glomerulosa flows past the cells of the zona fasciculata, the cells would need to keep out the excess ions and hence become emballic. Aldosterone, the hormone produced by the cells of the zona glomerulosa, is certainly concerned with Na^+ and K^+ concentrations.

When tissue cultures are made from the zona glomerulosa and zona fasciculata, those from the former have a tendency to become surrounded by pools of fluid, but the cells from the zona fasciculata do not, suggesting that the cells in the two zones differ in their relationships to the surrounding medium. Similarly, when grafts of cortical tissues are made into chambers in rabbit's ears, those from the zona glomerulosa become much more readily vascularized than those from the zona fasciculata, which undergo progressive atrophy. When both tissues are present, a "portal" vascular supply may lead from the glomerulosa tissues to those of the fasciculata (Williams, 1945). These vascular developments could well be dependent on the fluid relationships of the two tissues.

A comparison can probably be made legitimately between the cords of cells in the adrenal cortex and the coelomoducts and nephridia of more primitive creatures in which ecballic ciliated cells or flame-cells give place further down the tube to emballic cells or nephrocytes. In the nephridia the control may be exercised on the cells from the side of the tissue fluid; incloeomoducts it may be exercised from within. In the case of the adrenal cortex, though it is not known exactly how the zona glomerulosa is controlled, the zona fasciculata certainly responds to hormones circulating in the blood and thus, surprisingly perhaps, resembles the nephridia.

The Problem of the Steroids

One of the most interesting and challenging characters of the whole urinogenital system is its concern with steroids. Of all the tissues of the body, with the possible exception of the liver, those now under discussion specialize in steroid metabolism more extensively than any others. It is pertinent to ask why this should be so. The answers that the sex hormones are steroids and therefore they must be present, and that the kidney excretes unwanted steroids, do not take us any further. Why are the sex hormones steroids, and how do they act? If the kidney is concerned only with excreting steroids, why do steroids, and particularly oestrogens, accumulate in the kidneys of the carnivores only? Why do they do this only at certain times and under certain conditions and particularly when the secretions of pituitary gonado-trophins are favourable (Hewer *et al.*, 1948; Lobban, 1955, 1956)?

Before any suggestions are made along these lines, it is desirable to examine and consider more closely the occurrence of steroids in the gonadal, adrenal, and related tissues. Although the presence of steroids may result in the modification of the physiology and morphology of a wide variety of tissues in the body, they seem to be particularly associated with tissues in which the transport of ions, glucose, or water is specially important. For example, oestrogens accumulate in the ovarian follicles and the granulosa cells; testosterone is sometimes found in the seminiferous tubule, though the Sertoli cells themselves may contain oestrogens, particularly when they have become malignant (Huggins and Moulder, 1945); aldosterone affects kidney and salivary gland function. Corticoids affect the differentiation and functional activity of the intestinal epithelium (Moog, 1959) and the uptake of glucose by tissues; oestrogen and progesterone alter the secretory properties of the endometrium and the ionic equilibrium of both the endometrium and the myometrium; oestrogens collect in the tubules of the cat's kidney. Testosterone causes the cells of Bowman's capsule in the kidney of mice to become cubical and acquire a brush border (Selye, 1939); injections of stilboestrol produce tumours of ciliated cells in the kidney cortex of the hamster (Mannweiler and Bernhard, 1957); and there is often sexual dimorphism in the kidneys of reptiles (Cordier, 1928). All of these observations suggest a close relationship between kidney-tubule activity and steroid balance. In all these situations the transport of ions or fluids of one sort or another is going on, but we must beware of deciding too much about the extent of their action from the mere occurrence of the steroids in large or small amounts. If extracts of the adrenal cortex are made, the amount of aldosterone extracted is usually small; the amounts of other steroids may be much larger. Nevertheless, comparatively large quantities of aldosterone leave the cortex in the course of time and enter the blood. It is as though aldosterone is produced on demand and there is little storage of it, whereas other steroids, together with cholesterol, can be stored in relatively larger amounts as in the cells of the zona fasciculata. The fact that treatment of mice with deoxycorticosterone causes atrophy of the zona glomerulosa cells perhaps indicates the necessity for these cells to get rid of their aldosterone-like secretions. The fact that raising the Na/K ratio does the same thing (Miller, 1950) stresses the relationship between steroids and ionic balance.

The Mode of Action of Steroids

How steroids act is still an open question. There is evidence that aldosterone may accumulate in the nucleus of the target cell and act only after some delay (about 2 hours). From this it has been forcefully argued that steroids are able to act directly on some aspect of chromosome function and thus eventually to cause the production of new proteins and new enzymes. The moulting hormone of insects (ecdysone), which is also a steroid, has been thought to act in much the same way on its target cell. This hormone has been shown to cause puffs to appear at certain points on the giant chromosomes of the salivary gland of certain diptera and to determine the production of new proteins and enzymes (Karlson, 1967). Incidentally, ecdysone may be noted as an example of a steroid from another phylum that acts on cells with an active transport mechanism. It is more relevant to this discussion, however, and of considerable importance in general, that similar puffs appear on the giant chromosomes of the salivary glands, as the direct result of changes in the Na/K ratio of the target cells and, probably more particularly, on the ionic distribution within the nucleus, and also after treatment with Zn^{2+} ions (Kroeger, 1963, 1966, 1967).

Another view is that steroids directly affect the activity of enzymes, and numerous enzymes have been shown to be affected by the level of steroids. However, such is the complexity of the situation that in all cases it has been difficult to separate the direct and primary action of the steroid from the indirect consequences of such action.

Steroids, though extremely active biologically, are not very reactive chemically. They are, however, molecules with very definite shapes, and the precise shapes are known to be extremely important to the biological activity of the compounds in some cases. Steroids are so much more soluble in lipids than in water that it is difficult to visualize how sufficient concentrations of the more physiologically active ones, when they are used in minimal effective doses, could occur anywhere in cells in sufficient quantity except, in the lipid compartments or in the hydrophobic portions of proteins. Steroids have been shown to affect the permeability of artificial lipid membranes. For example, membranes of lecithin-cholesterol-cetyl phosphoric acid mixtures change their permeability to sodium and potassium in response to certain steroids in much the same manner and on much the same scale as cellular membranes would respond under similar conditions (Bangham *et al.*, 1965). From such observations it could be argued that the primary seat of action of the steroids is at lipid surfaces and interfaces. Sometimes the affected surface could be the cell surface, e.g. the action of progesterone on *Naegleria* (Pearson and Willmer, 1963); sometimes it could be the surface of mitochondria (Gallagher, 1960); sometimes the surface of the endoplasmic reticulum (thus affecting enzymic activity indirectly) (Villee and Engel, 1961); sometimes the lysosomal surface (e.g., cortisol; see Dingle and Lucy, 1965); finally, it could be the surfaces of the Golgi body or indeed any of the so-called unit membranes in the cell. Which surface is affected by which steroid must depend on a number of factors, among which may be mentioned accessibility of the surface, the chemical composition of the surface, the molecular configuration of the surface, and the requisite steric form of the steroid itself. Accessibility may be determined by the presence or absence of covering layers of protein or polysaccharide, by the presence of other structures (e.g. fat droplets) capable of absorbing the steroid, or by the ability of the steroid itself to penetrate barrier surfaces.

Since the steroid probably has to fit into the lipid surface in order to exert its specific action there, the nature of the lipids already in the layer must be important (Willmer, 1960). The packing and the arrangement of the molecules in a layer depend not only on the chemical nature of the constituents but also on their numerical relationships (Dervichian, 1958). Each of the unit membranes mentioned above may well differ from the others both chemically and sterically. From the opposite aspect, it is well known that only certain isomers and configurations of the steroids are active. For many of the steroids to be

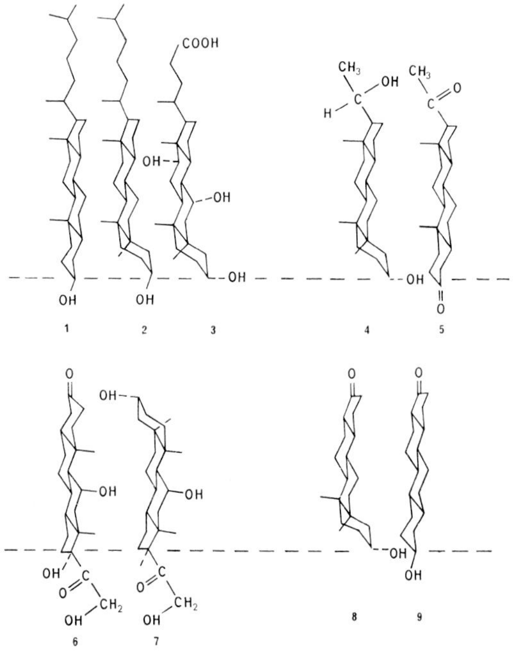

FIG. 17.13. Diagram to illustrate the molecular shape of inactive or excretory steroids compared with those of the corresponding physiologically active steroids. The steroids whose names are in italics are active in cells. 1, *Cholesterol*; 2, coprostanol; 3, cholic acid; 4, pregnanediol; 5, *progesterone*; 6, *cortisol*; 7, tetrahydrocortisol; 8, aetiocholanolone; 9, *androsterone*. (Willmer, 1960.)

biological active, one face of the molecule has to be flat. Similarly, of the fifty-odd steroids classed as adrenocorticoids, twenty are 5β compounds in which the first ring is bent away from the plane of the rest of the molecule. These steroids are all found in urine and are not reported to have any physiolocical action, while of the twenty-four that have physiological activity none are 5β compounds. Other 5β compounds also occur as excretory products or inactive isomers of other steroids (Fig. 17.13).

If the biologically active classes of steroids are set out in steric form, e.g. as androgens, oestrogens, etc, it is difficult at first sight to see any reason for the grouping. If, however, the assumption is made that the molecule, in order to be active, has to penetrate and pack into a lipid layer, then the obvious method of packing is with the long axis of the molecule parallel to the lipid chains of the phospholipid molecules. Thus the readiness of the molecule to enter such an hydrophobic layer must be determined at least partly by the hydrophobic properties of the molecule, and especially of the groups at the ends of the molecule. If one end is more hydrophobic than the other, that end is more likely to enter the lipid layer. When, therefore, the active steroids are arranged with these points considered (Fig. 17.14), it is of some interest to note that the terminal group which projects towards the water (the more hydrophilic terminal group) broadly determines the type of physiological action, though not the potency. Thus the oestrogens expose phenolic groups; progestagens, ketonic groups; and so on.

Some idea of the possible complexity of the lipid part of the unit membrane, on the assumption that it is basically a bimolecular leaflet, is illustrated in Fig. 17.15 where various biological phospholipids and steroids are incorporated in a theoretical model. It should be emphasized that, in this model, phospholipids are many and various, though those of biological importance have a certain homogeneity. Nevertheless, those which occur in biological systems differ in the length of their aliphatic chains, in the saturation of the chains and consequently in their isomeric form, and in their reactive hydrophilic groups, e.g. choline, ethanolamine, inositol, serine, etc. All these variations affect the packing and other properties that are important for the formation, stability, and biological properties of monolayers and unit membranes. When to this model are added the proteins and carbohydrates that become more or less securely linked to the phospholipid layers, partly by solution in the hydrophobic portion and partly by combination with the hydrophilic portion, some idea of the extraordinary complexity, variability, and (most important) specificity of biological membranes can be obtained.

Recent work (Dingle and Lucy, 1965) has suggested that yet another variant is possible in some membranes. The lipids may not form a continuous bimolecular leaflet like those assumed so far, but may exist in micellar form, stabilized and bound with protein (Fig. 17.16). Although superficially and probably functionally very different, these systems could be just as sensitive to steroids as the bimolecular leaflets, and the activities of the steroids would depend on similar principles. The existence of these two possible patterns of membranes, with the added possibility of rapid change from one to the other,

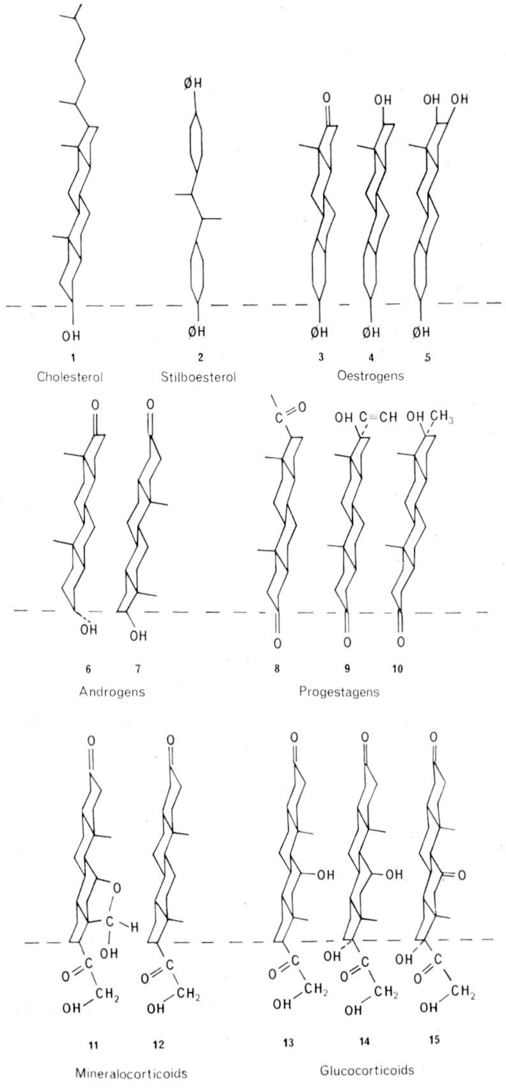

FIG. 17.14. Diagram to show how the physiologically active steroids might pack into a phospholipid monolayer. The dotted line separates the lipid layer above from the aqueous phase below. For explanation see text, p. 491. (Willmer, 1960.)

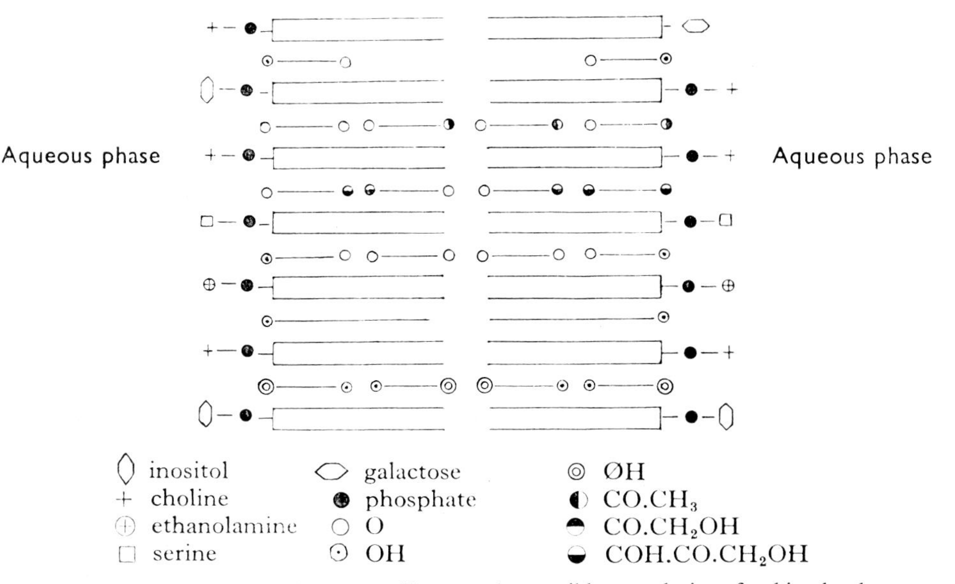

FIG. 17.15. Simplified diagram to illustrate the possible complexity of a bimolecular leaflet, assuming that the various naturally occurring phospholipids and steroids could pack together in the manner outlined on p. 491. The full complexity can be realized only when the various proteins and polysaccharides in the aqueous phase are visualized as competing for the polar groups at the lipid-water boundary and the mutual interactions between the lipids themselves are appreciated. (Willmer, 1960.)

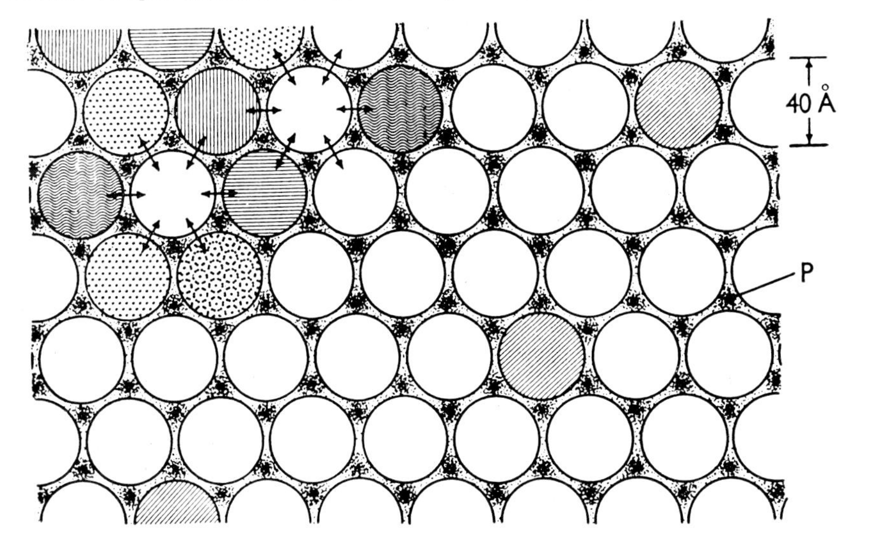

FIG. 17.16. Surface view of the lipid part of a membrane organized in the form of micelles of lipids (unshaded circles). The shaded units are globular proteins of various types. The spaces between the micelles could have water-filled pores (P). (Dingle and Lucy, 1965.)

open up intriguing ideas in relation to differentiation and functional changes of permeability and the like. It must be remembered, however, that there is as yet little finality concerning the exact nature of cell membranes and that, provocatively, some of the biological properties of membranes remain unaltered after most of the lipid has been extracted. However, something along the lines suggested above, involving proteins locking into orientated lipid systems and being themselves stabilized in particular steric configurations, probably gives as close a picture as it is possible to obtain at the present time. Carbohydrates and glycoproteins probably also enter the complex and add their quota to the specificity of the steric pattern, a pattern that is in a perpetual state of flux as the various molecules jockey for position, alter their configurations, combine with each other and generally react rogether.

How, then, is this general picture of cell membranes and the possible part that steroids play in their formation to be reconciled with other ideas that have been put forward for the mode of action of steroid hormones? In the first place it is well to remember that the site of accumulation of the active substance is not necessarily the site of its functional activity. Water may accumulate in a pool below a power station, but it is in passing through the power station that it does its work. Thus it could be that, in passing through the various membranes of the cell, the steroids have their effect. The sort of effects they are likely to produce would almost certainly include alteration in the packing and distribution of the molecules. This would mean an alteration in the distribution of the polar groups exposed on the hydrophilic surface of the lipid layers. This alteration would lead, in turn, to changes in the proteins attached to the layer, perhaps changes in enzymic activity. In general, such an alteration of cell equilibrium could ensue that it could easily result in altered gene action and the production of new proteins more appropriate to the altered state of the cell and necessary for the preservation of its equilibrium.

With reference to aldosterone again, although it may accumulate in the nuclear area (Edelman *et al.*, 1963), it has been shown to act in extremely small quantities (minimum dose 3.3×10^{-10} M); treatment of the cells for a period of only 5 minutes is all that is required to produce a response; and the suggestions are strong that it occupies a particular binding site (Sharp and Leaf, 1966). That its effect is not observable for 2 hours or so after it has been applied, and can be blocked by agents preventing protein synthesis, at first sight indicates that the binding site is perhaps nuclear. However, this does not necessarily follow, and the example of *Naegleria* is relevant (see p. 171). When *Naegleria* is put into distilled water, it begins to alter its form. After a delay of about 2 hours, during which time it has made new proteins and built a flagellum or two, it begins to deal with sodium ions and glucose in a different manner (see Fig. 8.14). Does the water, we may ask, act directly and specifically on the nucleus, or does it alter the whole balance of the cell in such a way that there is necessarily a redistribution of lipids and proteins? The latter seems to be the more likely, and it can be observed to happen. Furthermore, when deoxycorticosterone or NaCl is added to *Naegleria* which is in the flagellate form, the change to the amoeboid form is almost immediate (Fig. 17.17). It occurs within minutes, and apparently it does not depend

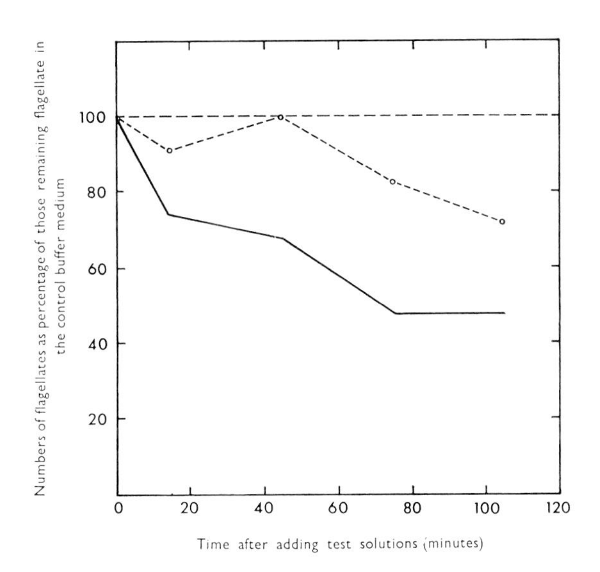

FIG. 17.17. Immediate effect of adding 10^{-5} *M* deoxycorticosterone to *Naegleria* in the flagellate form in the presence of 10^{-3} *M* KCl or 10^{-3} NaCl. The numbers of swimming flagellates are compared with the numbers treated with buffer solution alone. —— KCl; O - - - O NaCl.

directly on the formation of any new protein, though the surface of the cell becomes noticeably different.

Although not a steroid, vitamin A (retinol) has a molecule which would be expected to enter phospholipid layers in a manner similar to that suggested for steroids. This substance has been shown to increase the permeability of red cells and also that of lysosomes and mitochondria. It is pertinent to this discussion to notice that the effect on lysosomes is inhibited by hydrocortisone (Dingle and Lucy, 1965), which also stabilizes them.

Although it is extremely likely that steroids associate freely with phospholipid layers in the manner outlined above, other associations are possible. Many proteins contain hydrophobic regions which may have very specific shapes and patterns. Thus the association of particular steroids with particular proteins certainly occurs, and such associations could well lead to alterations in the configuration of the proteins and thus affect their biological or enzymatic properties. The requirement that certain steroids, e.g. aldosterone, act by combination of the steroid at some binding site (Sharp and Leaf, 1966) in the cell could be met either by its direct association with a protein or by its entry into the lipid of some lipo-protein complex like a cell membrane whose structure it may then change or disrupt. The recent and important observations on the distribution of aldosterone in the various subcellular fractions of kidney cortex indicate that this steroid becomes bound to a nuclear structure from which it can be specifically liberated by proteolytic enzymes and in which the binding is reduced by the competitor 9α-fluorocortisol (Fanestil and Edelman, 1966). However, in a system of this sort many other interpretations of the observations are possible in addition to the obvious one that the physiological action depends on this association with the nuclear protein. Moreover, the distinction between nuclear protein and nucleoprotein is important.

It is clear from this discussion that the systems within the cell are so closely linked that it is almost impossible to change one without affecting the others; thus, secondary effects become readily confused with primary ones. At this stage it is safe to say only that the mode of action of steroids on cells is not understood, but their association with the hydrophobic parts of phospholipids and of proteins is likely to be important. However, there are several other features of steroid activity which have a bearing on this discussion of their presence in the urinogenital system.

First, although steroids are most concentrated in the tissues of the urinogenital system in the vertebrates, their action is by no means confined to them. As just indicated, steroids, in doses which may be considered physiological, have even been shown to have considerable influence on the behaviour of *Naegleria* (Pearson and Willmer, 1963). Here the actions are rapid, and they certainly influence the surface properties of the organism to increase or decrease the stability of the phase, or even to change the phase. Moreover, the action of the steroid, e.g. deoxycorticosterone or progesterone, is not necessarily the same when it is applied to the two different phases; and it is greatly influenced by the concentration (Fig. 17.18). At high enough concentrations all the steroids tested favour the amoeboid form of activity, but at lower doses several of them encourage the flagellate form (e.g. progesterone and deoxycorticosterone). In these experiments the steroid was applied in pure form (apart from the ethanol used as a carrier) in an extremely dilute and simple buffer solution; in some cases, changes in the state of the amoeba can be observed within 15 minutes and probably less. Since the two phases of *Naegleria* have quite different relationships with the medium so far as ions and water are concerned, it is quite clear that steroids influence these relationships either directly or indirectly. It is not surprising therefore that, if the steroids act similarly on other cells, they are so important to the particular tissues of the vertebrate body that are concerned with the balance of the organism with

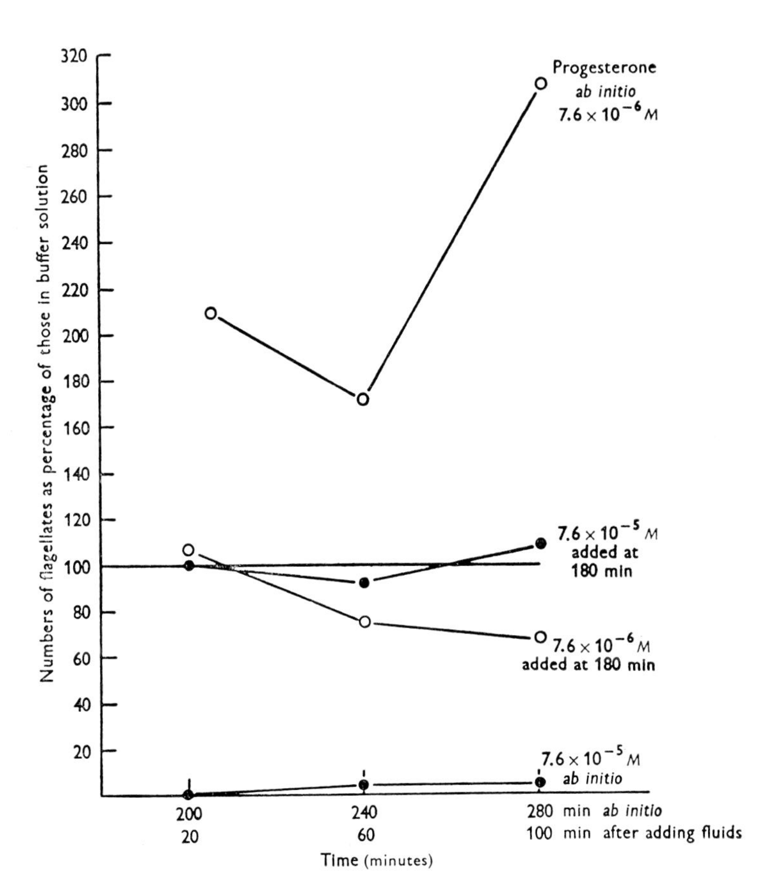

FIG. 17.18. The effects of progesterone at two different concentrations when applied to *Naegleria* in the amoeboid phase (*ab initio*) and in the flagellate form (i.e. 180 minutes after the suspension of *Naegleria* is placed in dilute buffer solution to encourage the acquisition of flagella). (Pearson and Willmer, 1963.)

its ionic environment. Moreover, that steroids seem to be produced in these urinogenital tissues suggests that they may be active participants in the actual movements of ions, water, and glucose. In some instances the steroid apparently accumulates within the cell (e.g. cortisone in the cells of the zona fasciculata); in other cases the steroid is passed out into the blood without any apparent intracellular inclusion (e.g. aldosterone from the cells of the zona glomerulosa). Similarly, oestrogens appear in follicular fluid, whereas progesterone is actually located within the cells of the corpus luteum. If the steroids are made on enzymes within the cytoplasm, perhaps some emerge through the membranes easily while others have to be helped out, or excreted in bulk. All this raises the question of how steroids can, in fact, escape from their position of equilibrium in the hydrophobic parts of proteins, or in solution in the lipids of the various membranes and other structures, in order to enter the water phases of the body in which they are normally so insoluble. In other words, how can a hormonal steroid cease to be active?

It has long been known that steroids are excreted via the kidney either as sulphates or as glucuronides, presumably because these compounds are more soluble in water than they are in lipids. They are among the few steroid compounds of which this is true (apart from some of the salts with carboxylic acids). The traffic in, and mobility of, steroids are thus likely to be influenced by the presence of sulphates and uronic acids in their vicinity. Conversely, the presence or absence of these metabolites may be related to the steroid metabolism of the tissues concerned. This point is also worthy of consideration in relation to the different activities of the mechanocyte and the amoebocyte. Sulphate and uronic acid are two very important constituents of the ground-substances of connective tissues. One may ask whether or not this means that they are in some way essential metabolites of the mechanocyte, perhaps concerned with the movement of ions. Ballantyne (1967) has recently called attention to the presence of β-glucuronidase in various tissues notable for their capacity for sodium transport, and has suggested that the enzyme may be concerned with liberating free steroid for some purpose. It may also be recalled that sea-urchin eggs are "animalized" in the absence of sulphate ions (see Needham, 1942; Runnström *et al.*, 1964), and that *Naegleria* is held in the amoeboid form when sulphate ions are present (Willmer, 1956). All these observations seem to suggest that sulphate ions are necessary also for the proper working of amoebocytes.

Ecballism, Emballism, and Renal Function

As has been repeatedly emphasized, mechanocytes and amoebocytes are complementary to each other. If the problem is reduced to its simplest terms, the membranes of the cells in the two groups could be thought of as being oppositely polarized, i.e. ecballic and emballic. A more detailed study of the activities of these two groups of cells, in the light of some of the information about steroids, may be informative with reference not only to the mechanocytes and amoebocytes themselves, but also to ecballic and emballic cells in general and to the homoiostatic cells of the urinogenital system in particular. The concentrations of steroids in the circulating fluids at different times presumably can act, as they do in *Naegleria*, to increase or decrease the ecballic or emballic activities of the cells with which they come into contact. Thus in a gradient system, as in the excretory tubes, or in a balanced system as in blastulae and many epithelia, the position of equilibrium may be altered by the steroids present. Although the distinction between ecballic and emballic cells is probably a fundamental one, it is at the same time a relative one.

Mechanocytes are postulated to be more active in a dilute ionic medium; as a class, they tend to feed their surroundings with hexosamines, uronic acids, sulphates, glutamic acid, proline, glycine, and combinations thereof. In the presence of ascorbic acid, collagen can be formed extracellularly by mechanocytes and its formation can be inhibited by cortisone. Furthermore, an analogue of ascorbic acid, hexenolactone, inhibits the activity of mechanocytes (Medawar *et al.*, 1943) (see p. 44).

In the neighbourhood of the mechanocytic or ecballic cells in the urinogenital system and adrenal cortex, oestrogens appear near the cells of the membrana granulosa and in connexion with Sertoli cells, and aldosterone escapes from the cells of the zona glomerulosa.

On the other hand, the more amoebocytic (emballic) cells prefer more concentrated media (cf. the behaviour of the pharyngeal cells in the nemertines where the ciliated cells survive well in dilute sea-water and the mucoid cells flourish when the sea-water is strengthened with extra salt), and their cytoplasm tends to fill with mucopolysaccharides (uronic acid and hexosamine derivates which are sometimes sulphated) and sometimes with ascorbic acid. In the urinogenital system the emballic or amoebocytic cells also store steroids. Thus the cells of the zona fasciculata store hydrocortisone and other glucocorticoids, the Leydig cells of the testis store testosterone, and the cells of the corpus luteum store progesterone. Also, all three tissues selectively accumulate cholesterol and ascorbic acid, and they tend to accumulate vitamin A and other carotenoids. Whether the actual cell membranes of the two groups of cells are simply polarized in opposite directions is of course dubious, though there is no doubt that they are structurally and functionally different.

Some of the important features of ecballic and emballic cells are illustrated in Fig. 17.19. In interpreting this figure in terms of any particular cell it is important to take into account the position and the polarity of the cell in question and the nature of the surfaces in the different parts of the cell. For example, in epithelial cells the luminal (apical) surface may differ greatly from the vascular (basal) surface, and both may differ from the surfaces which the cell exposes to its neighbours. An isolated mechanocyte or an amoebocyte probably has membranes all around it which are relatively uniform and which show few and small local variations. On the other hand, an epithelial cell may have on one side a membrane that behaves like that of an amoebocyte and, on the other side, a membrane that behaves like that of a mechanocyte. In studying epithelia, therefore, the important criteria are the cell surfaces exposed to the lumen on the one side and to the blood vessels on the other. For example, the behaviour of the cells of the adrenal cortex is probably rather closely related to the events in the blood vessels and dominated by the nature of the surfaces on that side. As far as we know, the "lumen" side of the cells of the adrenal cortex is now of secondary importance. Admittedly, we do now know much, but the cells are presumably activated by conditions in the blood and respond by pouring hormones into it. In the kidney, on the contrary, the cells separate

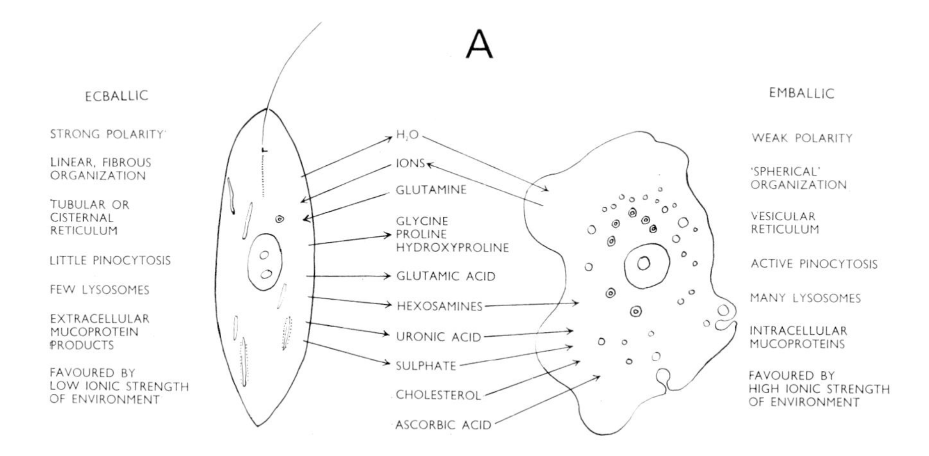

CELLS IN AN UNIFORM ENVIRONMENT

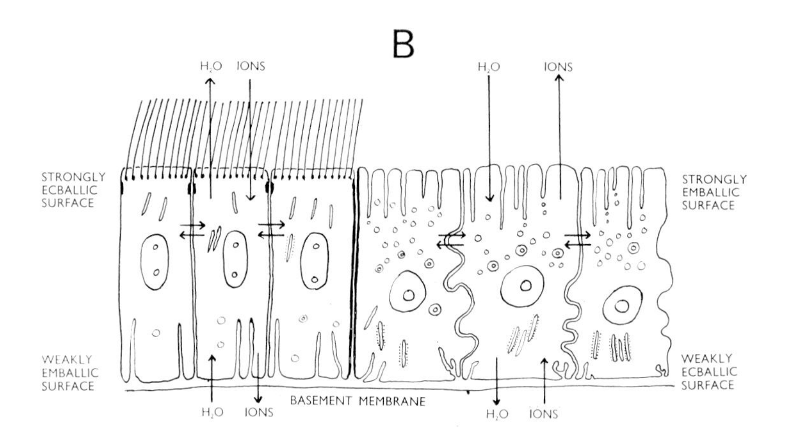

CELLS IN AN EPITHELIUM SEPARATING TWO DIFFERENT ENVIRONMENTS

the fluid in the lumen of the tubule from the blood in the vessels, and there are certainly two functional membranes to be considered, It seems likely, that in some parts at least, intake of sodium ions across one membrane is balanced by its extrusion through another. They need not, of course, both be active processes. Moreover, as pointed out in connexion with nephridia, it is likely that the control was initially related to conditions on the side of the blood rather than on the luminal side. The morphology of the cell membranes on the two sides of nephric cells can be extremely different. The luminal side of the proximal tubules is covered with microvilli, while the basal surface is heavily plicated and indented with folds of the membrane, a situation that is rather closely paralleled in the cells forming the Malpighian tubules of grasshoppers (Beams *et al.*, 1955) and other insects (Fig. 17.20) (Smith and Littau, 1960).

In the kidneys of vertebrates there is considerable variation from species to species in the extent of the different regions of the nephron and in the manner in which these parts function, because the requirements of the animals may differ greatly according to their diet and environmental conditions. Nevertheless, as a generalization, the proximal tubules in mammals are specifically concerned with the reabsorption of glucose, whereas the thick part of the loop of Henle and the distal tubule deal with the absorption of Na, the excretion of K, the regulation of blood pH, and, indirectly, with the regulation of the water balance of the body. The kidney tubule thus forms an

---

FIG. 17.19. Ecballic and emballic systems compared. Diagram to illustrate some of the main concepts suggested by the behaviour of *Naegleria*, mechanocytes, amoebocytes, blastulae, and epithelia, and the general principles of ecballism and emballism as applied to cell surfaces.

In (*A*) the cells are considered to have uniform surfaces and be subjected to an uniform environment. In (*B*) the surfaces of the cells vary from place to place and in relation to different environments.

The unlabelled arrows indicate the free passage of water and ions between adjacent cells. The thick line separating the two groups of cells indicates that the passage of water and ions may occur less readily.

The nature of the ions is not specified. *Naegleria* apparently makes little distinction between Na^+ and K^+, for example, but for most cells the specific ions are very important, and ecballism for Na^+ may mean emballism for K^+, and so on. The specificities of the surfaces may become very great.

In *Naegleria*, ecballism or emballism is imposed by the environment. In other cells, the two forms of activity are assumed to be determined at least partly from within the cell.

In *Naegleria*, ecballism is favoured by: deoxycorticosterone, 10^{-5} *M*; progesterone, 10^{-6} *M*; HCO_3, < 10 *mM*; lactate, < 10 *mM*; phosphate, < 10 *mM*; NaCl, < 1 *mM*; Emballism is favoured by: many steroids, $> 10^{-5}$ *M*; NaCl, > 1 *mM* (compulsory above 50 *mM*); Li^+, > 1 *mM* (compulsory above 5 *mM*); Mg^{2+}, > 1 *mM*; methonium compounds, $> 10^{-4}$ *M*.

←*See facing page*

interesting parallel with the unit of the adrenal cortex. The latter is composed of glomerulosa, sudanophobic, fasciculata, and reticulata regions. As suggested earlier, phylogenetic and embryological evidence points to many common features, both cytological and functional, in the two units.

It is important to notice that the salt-regulating steroids are released from the zona glomerulosa and probably function on the distal tubules (including the thick loop of Henle), while the glucocorticoids are produced from the

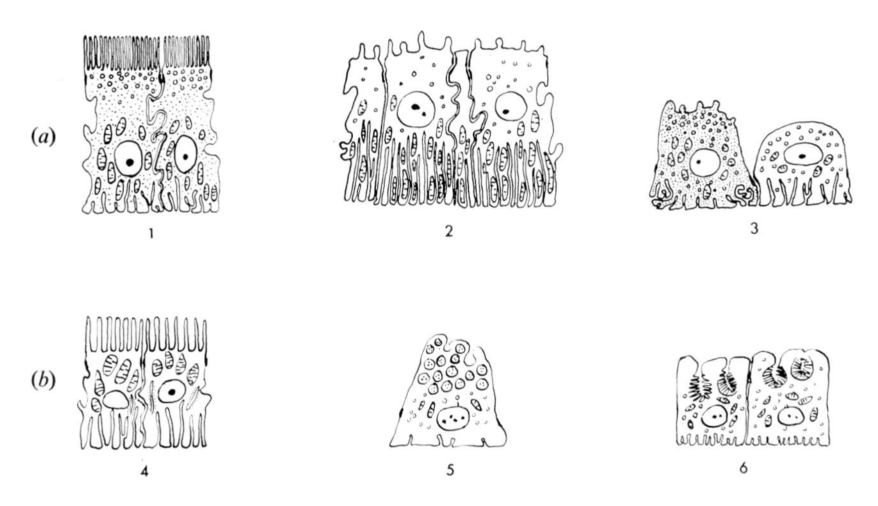

FIG. 17.20. Cells of mammalian kidney tubules (*a*) and cells of Malpighian tubules (*b*) of an homopteran insect compared. 1, Proximal tubule; 2, distal tubule; 3, collecting tubule; 4, segment 1; 5, segment 3; 6, segments 2 and 4. (Malpighian tubes after Smith and Littau 1960.)

zona fasciculata and probably act on the proximal tubules. The idea is not unreasonable that the adrenal cortex has been set aside primarily as a controller of kidney function. Its liberation of steroids, dependent directly or indirectly on the composition and volume of the blood, controls the activity of the kidney, and incidentally of other tissues too, in their handling of glucose, Na and K, and also of water. All this may well depend on the nature of the cell membranes, and, as in *Naegleria*, the steroids can probably determine this. It is even possible that they actually take part in the transfer mechanisms themselves, and that is why they are sometimes liberated into the surroundings, like aldosterone, or accumulate within cells, like cortisone and hydrocortisone. Although most of the cells of the body are concerned to a greater or lesser extent with movements of salt, water, and glucose, the excretory mechanisms would naturally be the most active in this respect. This may have been the cause of the specially active steroid metabolism associated with these tissues, a feature which has been elaborated gradually for a host of other specific purposes. The steroids in the kidneys of the carnivores are thus possibly a remnant of this initial regulatory activity and are perhaps related to the high intake of sodium and relatively low intake of potassium by these creatures, a consequence of their carnivorous diet.

It may be pointed out that steroids, though normally very lipophilic and hydrophobic, may, by virtue of their hydrophilic groups, afford a means of introducing limited amounts of water into otherwise purely lipid membranes. This property could be important in regulating water permeability and ion transfer through the various membranes of the cell. A similar role has recently been assigned to certain cyclic polypeptides, e.g. valinomycin (Mueller and Rudin, 1967), which expose an exterior surface of hydrophobic groups but have a hollow hydrophilic centre. Such molecules could presumably act as ion or water carriers through the lipid films. Is it too wild to suggest that the glycine-proline containing polypeptide precursors of collagen could be involved in this sort of function in the ecballic mechanocyte?

Two other points about the kidney are particularly relevant at this point in the discussion. First, there is an hereditary disorder of kidney function in which glycine, proline, and hydroxyproline appear in the urine, and a raised proline concentration in the blood leads to more glycine and hydroxyproline in the urine (Scriver *et al.*, 1961). Since these amino acids are characteristic products of mechanocytic or ecballic activity, the disease probably represents some abnormality of mechanocyte-like cells either in the kidney or elsewhere. If the disease is renal in origin, the location of the effect could be very illuminating with respect to the character of the cells concerned. The proximal tubule normally contains some mucosubstances in the form of "uromucoid" (Keutel, 1965), but whether this is anything more than that normally present among the microvilli of the brush border is obscure.

Second, there is the interesting structure in the mammalian kidney known as the juxta-glomerular apparatus which is responsible for the production of renin and, directly or indirectly, for the secretion of aldosterone from the adrenal cortex. The ascending loop of Henle comes into close contact with the glomerulus before it convolutes away as the distal tubule. At this point the tube itself has a group of specialized cells that are closely packed in its wall and form the macula densa. These cells lie, without any intevening basement membrane, adjacent to some large, roundish cells which constitute part of the wall of the afferent glomerular artery; only the endothelium separates them from the lumen of this artery (Fig. 17.21). These large, apparently endocrine, cells have been implicated in the production of renin, and PAS-positive granules appear in them when the animal is deficient in sodium and when the adrenals are removed. The granules disappear when the sodium level is raised (see Goormaghtigh, 1939, 1945; Brown *et al.*,

1963; Hartroft and Hartroft, 1955; Dunihue, 1949; Dunihue and Robertson, 1957).

The structure and arrangement of the cells forming the macula densa give the impression (Fig. 17.21) that they are epitheliocytes which are ecballic in character; they are situated at the head of the ascending loop of Henle. This is the part of the nephron primarily involved in the adjustment of water

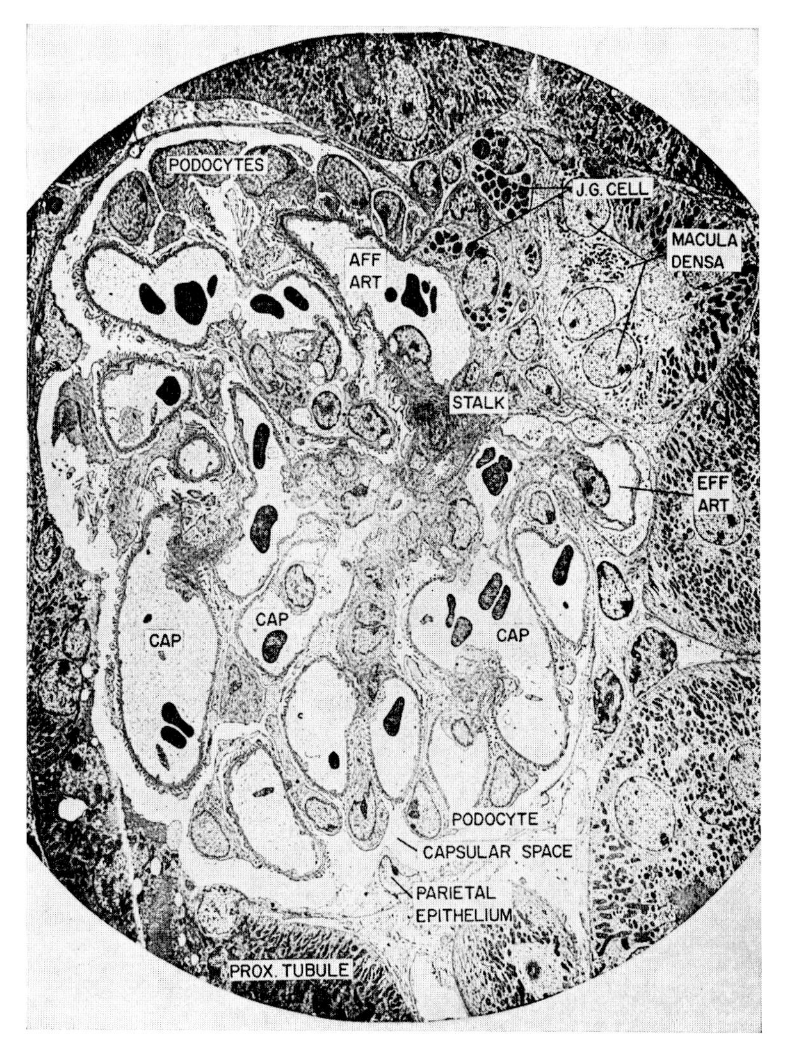

FIG. 17.21. The glomerulus and the juxta-glomerular apparatus. (Ham and Leeson, 1961.)

and salt balance, to which it contributes actively by pumping sodium from the lumen into the surrounding tissues creating the countercurrent, concentrating system of the medulla. It is likely, therefore, that when the fluid in the lumen of the ascending limb reaches the macula densa, it is sodium-deficient. Thus by analogy with *Naegleria*, it would favour strongly ecballic activity on the part of the cells of the macula densa. The lower the salt content, the higher this activity would be; thus the neighbouring juxtaglomerular cells, which structurally give the impression of being amoebocytic and emballic, may need to work harder to compensate the imbalance set up. Renin production may then be the result of this activity. Thus the situation could be directly comparable with that existing between the two groups of cells in the elementary blastuloid (see p. 198), and with that suggested for the onset of activity of the theca interna cells of the ovary (p. 477) or of the Leydig cells of the testis (p. 477). The placing of the juxtaglomerular apparatus at the end of the sodium-reabsorbing portion of the loop of Henle would seem to be very suitable from the point of view of monitoring the effectiveness of the salt-absorbing system and detecting salt deficiency (Vander and Miller, 1964; Peart, 1965). Why the afferent arterial wall is the place where the renin is released is not so clear, if the hormone or its derivative, angiotensin, eventually acts only on the zona glomerulosa of the adrenal cortex. If, however, the renin also acts by constricting the efferent artery of the glomerulus, it may so raise the pressure in the glomerulus that more fluid is filtered and more water is lost from the body (Gross *et al.*, 1965; Peart, 1965). The actual level of sodium in the lumen of the tubule at the level of the macula densa must depend on a number of factors, e.g. the rate of passage of fluid through the nephron, the initial concentration, and the rate of pumping of sodium in the ascending limb. Furthermore, the kidney is concerned not only with regulating the total water content of the body, but also with the sodium content, the potassium content, and in fact the general constancy of the composition of the blood.

The Pituitary and Control Systems

A common feature of the adreno-urinogenital system which has not received much attention in this discussion is its control by the pituitary gland, a control exercised by means of hormones that are either protein (many of them are mucoproteins) or polypeptides.

In the discussion on p. 343 it was considered probable that the anterior pituitary gland had its origin in the roof of the mouth of the nemerteoids. Since this membrane meets the full impact of the environmental variations in salt concentration, etc., it must immediately react in such a way as to limit the effects of the impact on the rest of the organism. If it were possible to make simultaneous, immediate, and appropriate alterations in the activity of the excretory and regulatory systems as soon as the buccal membrane was assailed, enormous advantages would be gained. This probably applies equally to the cephalic organ which, as we have seen, seems to have a sensory mechanism that responds to the salinity of the external medium; and, as outlined on p. 386, during evolution the neural mechanism of this is likely to have become incorporated into the nuclei of the hypothalamic area. The impacts on the two mechanisms are slightly different: the cephalic organ is a sensory mechanism; the buccal cavity is directly concerned with opposing the excessive entry of water or salts. The build-up of muco-substances in the deeper cells of the buccal cavity in response to high salt in the ambient fluid and their run-down in more aqueous media would appear to be relevant to anterior pituitary function. In the pituitary, however, the primitive direct response to the buccal attack has been eliminated and, in the course of evolution, replaced by a sensory and monitoring system based on the composition of the body fluids rather than that of the external medium. The cells of the anterior pituitary are supplied by blood which has passed through the neural tissues. Thus these cells are in a strong position to respond to neurosecretory substances and to monitor the fluid which has just bathed the dominating cells of the body. The various cells of the anterior pituitary, it will be remembered, have much the same gamut of staining reactions, dependent largely on mucopolysaccharides, as do the buccal cells of the nemerteoid (see Fig. 14.3). At present we do not know what the build-up of mucosubstances in the buccal cavity of the nemertines or in the cells of the anterior pituitary really signifies. However, it does seem at least possible that the cells of the pituitary liberate particular proteins which help to stabilize the cells of the excretory and gonadal systems in such a way that their secretory and regulating activities are controlled. It was suggested earlier that the steroid-phospholipid pattern of the surfaces and membranes of cells must both determine, and be determined by, the proteins attached to them, and *vice versa*. Thus, to control functions involving steroids it may be appropriate to use protein hormones to encourage and stabilize particular steroid-phospholipid patterns in the relevant cells. The pituitary hormones certainly fulfil these requirements; it is interesting, too, that angiotensin, which affects the zona glomerulosa, is also a polypeptide. The reverse process, which could be used in feed-back controlling mechanisms based on steroid hormones, is equally possible. The steroids could be used to determine the patterns of cell membranes and, in consequence, the amount and nature of the proteins associated with them.

Not all the cells of the anterior pituitary build up mucoproteins, and indeed some are ciliated. It will be recalled that the ciliated cells of the buccal cavity of the nemerteoid are highly active in a dilute sea-water medium. By-products of the activities of these cells may again be used to modify the activities of cells elsewhere and thus become hormones. Prolactin, or something like it, is known to alter the activity of the kidney in fish and is especially active in fresh-water fish (Potts *et al.*, 1967). Moreover, the prolactin-producing cells frequently have cilia in the mammalian pituitary (Barnes, 1961). Some other actions of prolactin are also interesting at the cellular level if prolactin is one of the products of activity of the ecballic ciliated cells, First, there is its stimulation of the production of milk by causing the cells of the mammary gland to indulge in what would appear to be essentially emballic activity because the cells become engorged with fat and protein. Second, it induces similar changes in the cells of the pigeon's crop which lead to the shedding of engorged cells and the production of pigeon's milk. Third, in some species, e.g. the rat, it activites the cells of the corpus luteum in their production of progesterone.

Thus, on this hypothesis, the hormones of the pituitary gland are probably metabolites which the cells of the buccal cavity produced and passed into the tissue fluids in their responses to changes in the fluids entering the mouth. The action of these metabolites then, in their turn, and especially when a vascular system developed, activated or repressed the cells of other homoiostatic tissues, e.g. the skin, the alimentary canal, the nephridia, and the gonocoels. How these metabolites will be found actually to produce their effects is, of course, not easy to forecast, since they may either enhance or reduce activity elsewhere. For example, a cell which is to excrete sodium has to pick up sodium on one side and eliminate it on the other; either or both of these processes could be "active", and either or both could be subject to activation or to blocking. On the other hand, cells like those in the buccal cavity of nemertines, which have to keep excess sodium out of the body, may have no traffic of sodium across either surface. Thus, in deciding how hormonal control occurs, it is necessary to take into account both stimulatory and inhibitory effects and also the physiology of the actual membranes and the exact sites of action where the hormones are involved.

In the case of the kidney, which regulates both the salt and water balance of the body, it is not immediately obvious how any one cell in a tubule "knows how hard it has to work" in order to preserve the ionic equilibrium of the body as a whole. It is important that the composition of the blood must at some point or points be monitored (as the intraluminal fluid in the nephron appears to be monitored by the macula densa), and a kidney-controlling factor must be produced in response to deviations from the norm. It is also clear that this factor, or these factors, must raise or lower the level of activity of all the relevant kidney cells with respect to the concentrations in question. The cells of the hypothalamus and the pituitary clearly act in this way with respect to water in their production of vasopressin in mammals and of prolactin in fish.

The adrenal cortex is a tissue which is directly concerned with the regulation of at least one aspect of renal activity, which it carries out by liberating mineralocorticoids from the zona glomerulosa into the blood. It liberates these compounds in response to angiotensin, but whether there are other effective excitants is not yet known. From general observations, and if the behaviour of *Naegleria* is any guide, the mineralocorticoids cause certain membranes to discriminate more effectively between sodium and potassium in their transport (see Fig. 17.17). If the cells of the zona glomerulosa are homologous with the membrana granulosa cells of the ovary as suggested on p. 482, the observation that these cells discriminate between sodium and potassium (follicular fluid having low sodium and high potassium) is relevant. It also indicates that the activities in one group of cells may in some way be utilized to control similar activities elsewhere. In other words, the Na/K equilibrium set up in the zona glomerulosa and the effects of angiotensin on it should probably be considered the chief monitor for salt balance. Meanwhile, the hypothalamus-pituitary monitoring system is responsible for water balance. In these systems the main onus of control is on the renal or output side. In the control of the glucose level, on the other hand, the onus is on the storage side; but, in so far as the input and output mechanisms are controlled, the adrenal cortex again acts as a monitor and produces glucocorticoids in the zona fasciculata. Here it is influenced by the pituitary product ACTH just as the zona glomerulosa is influenced by angiotensin and probably other factors. Attention is again called to the fact that mineralocorticoids produced from ecballic cells of the zona glomerulosa act on more emballic cells of the distal tubules of the kidney, whereas the glucocorticoids produced from the emballic cells of the zona fasciculata act primarily on the ecballic proximal tubules. The experiments of Chambers and Kempton (1933) on tissue culture of the chick kidney (see p. 198) should also be borne in mind here with regard to the direction of movement of fluid in the two parts of the tubules particularly since adrenal cortical compounds affect the degree of swelling of the closed vesicles (Chambers and Cameron, 1944). Furthermore, whereas there is glucose in the follicular fluid in the ovary (produced by the cortical cells of the gonad), the seminiferous tubules (originating from the medullary part of the gonad) contain fluid that is lacking in glucose (see p. 450), perhaps indicating a glucose-reabsorbing mechanism like that of the proximal tubules of the kidney.

The suggestion that is really being made here is that the adrenal cortex, kidney, ovary, and testis are all basically derived from a system of regulatory tubes. Like their precursors, the nephridia or the gonoducts or both, they probably displayed gradients along their lengths akin to the primary gradient of the blastuloid. The zona glomerulosa, membrana granulosa, Sertoli cells of the seminiferous tubules, glomerular and proximal tubule parts of the kidney correspond roughly to the ecballic part; the zona fasciculata, theca interna, Leydig cells, and distal tubules are more emballic. Among these cells it is interesting to notice that cilia have been occasionally observed on glomerulosa cells (Propst and Müller, 1966), follicle cells (Adams and Hertig, 1964), and in the kidney where they are of most frequent occurrence in the parietal layer of Bowman's capsule and in the neck tube, though they may also occur in other parts of the nephron as well. In all cases the positions are relative; the gradients may be sharp or gradual; and the point of change-over from the ecballic to emballic must necessarily be variable. Thus, in the adrenal cortex there may be a sudden transition between the ecballic glomerulosa cells and the emballic fasciculata cells in some species (just like the sudden change from flagellate cells to amoeboid cells in the blastula of the sponge, *Clathrina blanca*). In other species this transition may be gradual (as in *Leucoselenia*) and may entail the development of cells with intermediate activities, i.e. a more or less well developed sudanophobic zone (Fig. 17.22). The essence of such a gradient system, with reversal of activities, is that, by opposing actions, the cells of one part of the tube so balance those of the other that an equilibrium point can be securely established. On either side of that point the cells are working in opposite directions with respect to the activity that is being controlled. Just as in the other situations discussed in relation to other epithelia and the retina, the degree of ecballism or emballism in any system depends partly on the limits of variability of the commodity to be controlled. In this system the width of the zona glomerulosa of the rat can be increased (i.e. the sudanophobic zone is reduced) by low sodium, by high potassium, and by hypophysectomy. It can be decreased by desoxycorticosterone and by low potassium. These observations (Deane *et al.*, 1948) suggests that the position of the equilibrium point in the gradient system, i.e. the sudanophobic zone, does actually move with changed conditions, though the immediate stimulus (or stimuli) to which the cells respond remains somewhat obscure.

Two other points observed in the behaviour of more primitive creatures are relevant to these systems for controlling concentrations and establishing equilibria. First, when *Lineus* is treated with abnormal concentrations of sea-water, there are changes in the histological appearance of the pharynx within an hour or less, and the normal equilibria are thrown off balance (see p. 240). However, if the worms are adapted for a matter of days to the new saline conditions, the histological appearance is largely restored to normal. Presumably the rate of working of the various groups of cells has been adjusted, and a new equilibrium established, but exactly how this is achieved is not immediately obvious.

Second, in *Naegleria* there is a wide range of sodium concentrations in

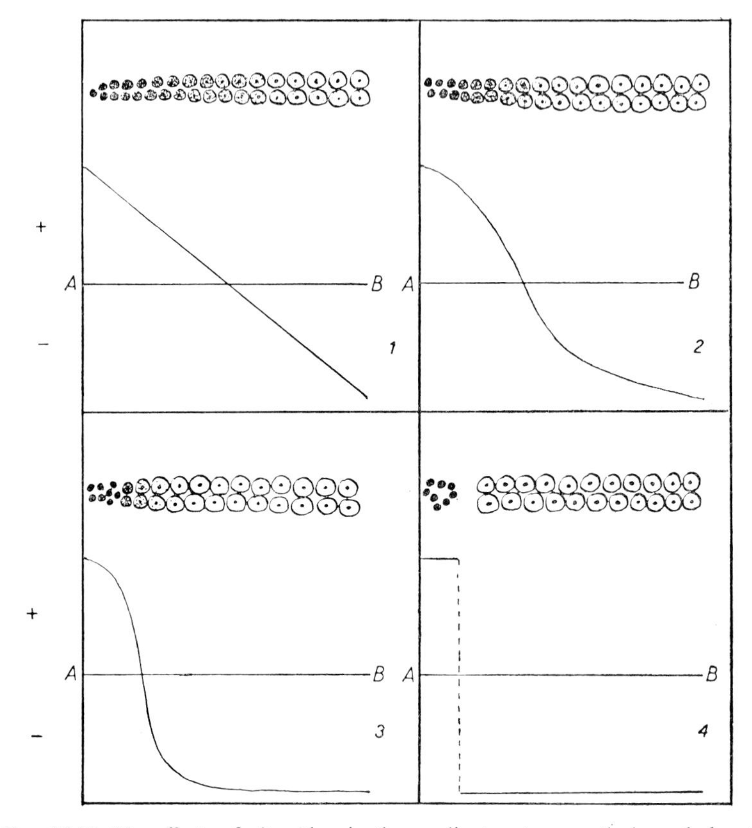

FIG. 17.22. The effects of alteration in the gradient system on the morphology of the adrenal cortex. The line AB in each case represents the change-over from ecballic to emballic activity. ⊛ Intermediate cell. ● Glomerulosa (ecballic) cell. ⊙ Fasciculata (emballic) cell.

which both flagellates and amoebae exist together (see Fig. 8.11). There is, however, for each amoeba at any one time probably one concentration which is critical for that amoeba. In other words, all *Naegleriae* do not have exactly the same equilibrium point. Selection could therefore act either to make all specimens have the same equilibrium point or to preserve a wide range. It is worth considering the implications of these two alternatives for the kinds of tubular systems with gradients which have been considered in this chapter. In a tubular system where all the units have the same equilibrium point, there must be a sharp transition between one phase and the other, i.e. a steep gradient. The greater the variation in the equilibrium point, the less steep is the gradient, and probably the more flexible the system. A permanent shift in the concentration of the fluid to be regulated could lead either to a shift in the proportions of the cells of the two types or to an alteration in the rates at which they both work.

Ontogenetically, phylogenetically, and physiologically, all these effects seem to be possible in relation to the local environments of tissue cells in the organs under discussion. Instead of thinking of kidney cells or adrenal cortical cells as a whole, it might be much more profitable to consider the cells and their immediate environments in the various parts of these organs with respect to their positions in regulatory and other gradients, bearing in mind that it is not only sodium that can determine important gradients.

References

Adams, E. C., and Hertig, A. T. (1964). Studies on guinea-pig oocytes. 1. Electron- microscopic observations on the development of cytoplasmic organelles in oocytes of primordial and primary follicles. *J. Cell Biol.* **21**, 397.

Bahl, K. N. (1947). Excretion in the Oligochaeta. *Biol. Rev. Cambridge Phil. Soc.* **22**, 109.

Ballantyne, B. (1967). Histochemical correlates of β-glucuronidase and sodium ion transport. *J. Physiol. (London)* **192**, 13P.

Bangham, A. D., Standish, M. M., and Weissman, G. (1965). The action of steroids and streptolysin S on the permeability of phospholipid structures to cations. *J. Mol. Biol.* **13**, 253.

Barnes, B. G. (1961). Ciliated secretory cells in the pars distalis of the mouse hypophysis. *J. Ultrastruct. Res.* **5**, 453.

Baxter, J. S. (1946). The growth cycle of the cells of the adrenal cortex in the adult rat. *J. Anat.* **80**, 139.

Beams, H. W., Tahmisian, T. N., and Devine, R. L. (1955). Electron microscope studies on the cells of the malpighian tubules of the grasshopper (Orthoptera, Acrididae). *J. Biophys. Biochem. Cytol.* **1**, 197.

Blandau, R. J., and Rumery, R. (1962). Observations on cultured granulosa cells from ovarian follicles and ovulated ova of the rat. *Fertility Sterility* **13**, 335.

Brown J. J., Davies, D. L., Lever, A. F., Parker, R. A., and Robertson, J. I. S. (1963). Assay of renin in single glomeruli. Renin distribution in the normal rabbit kidney. *Lancet* **ii**, 668.

Burns, R. K. (1956). Urinogenital system in "Analysis of Development." (B. H. Willier, P. A. Weiss and V. Hamburger, eds.) p. 462. Saunders, Philadelphia and London.

Caravaglios, R., and Cilotti, R. (1957). A study of the proteins in the follicular fluid of the cow. *J. Endocrinol.* **15**, 273.

Carr, I. A. (1958). Microvilli of the cells of the human adrenal cortex. *Nature* **182**, 607.

Chambers, R., and Cameron, G. (1944). Action of adrenal cortical compounds and l-ascorbic acid on secreting kidney tubules in tissue culture. *Am. J. Physiol.* **141**, 138.

Chambers, R., and Kempton, R. T. (1933). Indications of function of the chick mesonephros in tissue culture with phenol red. *J. Cellular Comp. Physiol.* **3**, 131.

Cordier, R. (1928). Études histophysiologiques sur le tube urinaire des Reptiles. *Arch. Biol. (Liège)* **38**, 111.

Cowdry, E. V. (1938). "Text-book of Histology". Kimpton, London.

Crabo, B., and Gustaffson, B. (1964). Distribution of sodium and potassium and its relation to sperm concentration in the epididymal plasma of the bull. *J. Reprod. Fertility* **7**, 337.

Deane, H. W., Shaw, J. H., and Greep, R. O. (1948). The effect of altered sodium or potassium intake on the width and cytochemistry of the zona glomerulosa of the rat adrenal cortex. *Endocrinology* **43**, 133.

Deanesley, R. (1928). A study of the adrenal cortex in the mouse and its relations to the gonads. *Proc. Roy. Soc.* (*London*) **B103**, 523.

Dennis Smith, L. (1966). The role of the germinal plasm in the formation of primordial germ cells in *Rana pipiens*. *Develop. Biol.* **14**, 330.

Dervichian, D. G. (1958). The existence and significance of molecular associations in monolayers. *In* "Surface Phenomena in Chemistry and Biology" (J. F. Danielli, K. G. A. Pankhurst, and A. C. Riddiford, eds.). p. 70. (Pergamon) Macmillan, New York.

Dingle, J. T., and Lucy, J. A. (1965). Vitamin A, carotenoids and cell function. *Biol. Rev. Cambridge Phil. Soc.* **40**, 422.

Dunihue, F. W. (1949). The effect of adrenal insufficiency and desoxycorticosterone acetate on the juxtaglomerular apparatus *Anat. Record* **104,** 442.

Dunihue, F. W., and Robertson, Van B. (1957). The effect of desoxycorticosterone acetate and of sodium on the juxtaglomerular apparatus. *Endocrinology* **61**, 293.

Edelman, I. S., Bogoroch, R., and Porter, G. A. (1963). On the mechanism of action of aldosterone on sodium transport. The role of protein synthesis. *Proc. Natl. Acad. Sci. U.S.* **50**, 1169.

Edwards, R. G. (1962). Meiosis in ovarian oocytes of adult mammals. *Nature* **196**, 446.

Edwards, R. G., and Gardner, R. L. (1967). Sexing of live rabbit blastocysts. *Nature* **214**, 576.

Esaki, S. (1928). Über Kulturen des Hodengewebes der Säugetiere und über die Natur des interstitiellen Hoden-gewebes und der Zwischenzellen. *Z. Mikroskop. Anat. Forsch.* **15**, 368.

Everett, N. B. (1943). Observational and experimental evidences relating to the origin and differentiation of the definitive germ cells in mice. *J. Exptl. Zool.* **92**, 49.

Fanestil, D. D., and Edelman, I. S. (1966). Characteristics of the renal nuclear receptors for aldosterone. *Proc. Natl. Acad. Sci. U.S.* **56**, 872.

Fawcett, D. W., and Burgos, M. H. (1960). Studies on the fine structure of the mammalian testis. II. The human interstitial tissue. *Am. J. Anat.* **107**, 245.

Gallagher, C. H. (1960). The mechanism of action of hydrocortisone on mitochondrial metabolism. *Biochem. J.* **74**, 38.

Goodrich, E. S. (1945) The study of nephridia and genital ducts since 1895. *Quart. J. Microscop. Sci.* **86**, 115.

Goormaghtigh, N. (1939). Existence of an endocrine gland in the media of the renal arterioles. *Proc. Soc. Exptl. Biol. Med.* **42**, 688.

Goormaghtigh, N. (1945). Facts in favour of an endocrine function of the renal arterioles. *J. Pathol. Bacteriol.* **57**, 392.

Gross, F., Brunner, H., and Ziegler, M. (1965). Renin-angiotensin system, aldosterone and sodium balance. *Recent Progr. Hormone Res.* **21**, 119.

Gruenwald, P. (1942a). The development of the sex cords in the gonads of men and mammals. *Am. J. Anat.* **70**, 359.

Gruenwald, P. (1942b). Common traits in development and structure of the organs originating from the coelomic wall. *J. Morphol.* **70**, 353.

Gruenwald, P. (1946). Embryonic and postnatal development of the adrenal cortex, particularly the zona glomerulosa and accessory nodules. *Anat. Record* **95**, 391.

Gruenwald, P., and Konikov, W. M. (1944). Cell replacement and its relation to the zona glomerulosa in the adrenal cortex of mammals. *Anat. Record* **89**, 1.

Hall, K., and Korenchevsky, V. (1938). Effects of castration and of sexual hormones on the adrenals of male rats. *J. Physiol.* (*London*) **91**, 365.

Halmi, N. S., and Davies, J. (1953). Comparison of aldehyde fuchsin staining, metachromasia and periodic acid-Schiff reactivity of various tissues. *J. Histochem.* **1**, 447.

Ham, A. W., and Leeson, T. S. (1961). "Histology". Pitman, New York.

Hartroft, P. M., and Hartroft, W. S. (1955). Studies on renal juxtaglomerular cells. Correlation of the degree of granulation of juxtaglomerular cells with width of the zona glomerulosa of the adrenal cortex. *J. Exptl. Med.* **102**, 205.

Hett, J. (1925). Ein Beitrag zur Histogenese der menschlichen Nebenniere. *Z. Mikroskop. Anat. Forsch.* **3**, 179.

Hewer, T. F., Matthews, L. H., and Malkin, T. (1948). Lipuria in tigers. *Proc. Zool. Soc. London* **118**, 924.

Huggins, C., and Moulder, P. V. (1945). Estrogen production by Sertoli cell tumours of the testis. *Cancer Res.* **5**, 510.

Kallman, F., and Grobstein, C. (1966). Localization of glucosamine-incorporating material at epithelial surfaces during salivary epithelio-mesenchymal interaction *in vitro*. *Develop. Biol.* **14**, 52.

Karlson, P. (1967). The effects of ecdysone on giant chromosomes, RNA metabolism and enzyme induction. *Mem. Soc. Endocrinol.* **15**, 67.

Kemp, N. E. (1956). Electronmicroscopy of growing oocytes of *Rana pipiens*. *J. Biophys. Biochem. Cytol.* **2**, 281.

Keutel, H. J. (1965). Localization of uromucoid in human kidney and in sections of human kidney stone with fluorescent antibody techniques. *J. Histochem. Cytochem.* **13**, 155.

Kroeger, H. (1963). Chemical nature of the system controlling gene activities in insect cells. *Nature* **200**, 1234.

Kroeger, H. (1966). Potentialdifferenz und Puff-muster. Elektrophysiologische und cytologische Untersuchungen an den Speicheldrusen von *Chironomus thummi*. *Exptl. Cell Res.* **41**, 64.

Kroeger, H. (1967). Hormones, ion balances and gene activity in dipteran chromosomes. *Mem. Soc. Endocrinol.* **15**, 55.

Lender, T., and Duveau-Hagège, J. (1962). Survie et différenciation des gonades larvaires de *Galleria mellonella* en culture organotypique. *Compt. Rend.* **254**, 2825.

Little, J. M. (1964). Potassium imbalance and rectosigmoid neoplasia. *Lancet* **i**, 302.

Lobban, M. C. (1955). Some observations on the intracellular lipid in the kidney of the cat. *J. Anat.* **89**, 92.

Lobban, M. C. (1956). The role of gonadotrophins in the production of cytological changes in the kidney and adrenal cortex of the male cat. *J. Physiol.* (*London*) **135**, 11P.

Lutwak-Mann, C. (1954). Note on the chemical composition of bovine follicular fluid. *J. Agr. Sci.* **44**, 477.

Mannweiler, K. I., and Bernhard, W. (1957). Recherches ultrastructurales sur une tumeur rénale expérimentale du Hamster. *J. Ultrastruct. Res.* **1**, 158.

Maximow, A. A., and Bloom, W. (1930). "Text-book of Histology". Saunders, Philadelphia, Pennsylvania.

Medawar, P. B., Robinson, G. M., and Robinson, R. (1943). A synthetic differential growth inhibitor. *Nature* **151**, 195.

Meusy, J. (1963). La gamétogenèse *d'Orchestia gammarella* Pallas, Crustacé amphipode. *Bull. Soc. Zool.* **88**, 197.

Miller, R. A. (1950). Cytological phenomena associated with experimental alterations of secretory activity in the adrenal cortex of mice. *Am. J. Anat.* **86**, 405.

Moog, F. (1959). The adaptations of alkaline and acid phosphatases in development. *In*

"Cell, Organism and Milieu" (D. Rudnick, ed.). p. 121. Ronald Press, New York.

Mueller, P., and Rudin, D. O. (1967). Development of K^{+}-Na^{+} discrimination in experimental bimolecular lipid membranes by macrocyclic antibiotics. *Biochem. Biophys. Res. Commun.* **26**, 398.

Needham, J. (1942). "Biochemistry and Morphogenesis," p. 486. Cambridge Univ. Press, London and New York.

Ohno, S., and Smith, J. B. (1964). Role of fetal follicular cells in meiosis of mammalian oocytes. *Cytogenetics* **3**, 324.

Pearson, J. L., and Willmer, E. N. (1963). Some observations on the actions of steroids on the metaplasia of the amoeba, *Naegleria gruberi*. *J. Exptl. Biol.* **40**, 493.

Peart, W. S. (1965). The functions for renin and angiotensin. *Recent Progr. Hormone Res.* **21**, 73.

Potts, W. T. W., Foster, M. A., Rudy, P. P., and Howells, G. P. (1967). Sodium and water balance in the cichlid teleost, *Tilapia mossambica*. *J. Exptl. Biol.* **47**, 461.

Propst, A., and Müller, O. (1966). Die Zonen der Nebennierenrinde der Ratte. Elektronmikroskopische Untersuchung. *Z. Zellforsch. Mikroskop. Anat.* **75**, 404.

Pudney, J. (1967). Ultrastructural studies on the testicular interstitial tissue of grey squirrels (*Sciurus carolinensis* Gmelin). *Communication to Soc. for Exptl. Biol.*, Jan 6th.

Runnström, J., Hörstadius, S., Immers, J., and Fudge-Mastrengelo, M. (1964). An analysis of the role of sulphate in the embryonic differentiation of the sea-urchin (*Paracentrotus lividus*). *Rev. Suisse Zool.* **71**, 21.

Schlenk, W., Jr., and Kahmann, H. (1938). Die chemische Zusammensetzung des Spermaliquors und ihre physiologische Bedeutung. Untersuchung am Forellensperma. *Biochem. Z.* **295**, 283.

Schreiner, K. E. (1955). Studies on the gonad of *Myxine glutinosa* L. *Univ. Bergen Årbok Naturvitenskap. Rekke* **8**, 1.

Scriver, C. R., Schafer, I. A., and Efron, M. L. (1961). New renal tubular amino-acid transport system and a new hereditary disorder of amino-acid transport system. *Nature* **192**, 672.

Selye, H. (1939). The effect of testosterone on the kidney. *J. Urol.* **42**, 637.

Setchell, B. P. (1967). The blood-testicular fluid barrier in sheep. *J. Physiol.* (*London*) **189**, 63P.

Sharp, G. W. G., and Leaf, A. (1966). Mechanism of action of aldosterone. *Physiol. Rev.* **46**, 593.

Short, R. V. (1967). Personal communication.

Smith, A. U. (1952). Cultivation of ovarian granulosa cells after cooling to very low temperatures. *Exptl. Cell Res.* **3**, 574.

Smith, D. S., and Littau, V. C. (1960). Cellular specialisation in the excretory epithelia of an insect (*Macrosteles fascifrons*) (Homoptera). *J. Biophys. Biochem. Cytol.* **8**, 103.

Smith, I. T. (1937). *Am. J. Obstet. Gynecol.* **33**, 820. Quoted in Zachariae and Jensen (1958).

Sørensen, E., and Andersen, S. (1956). The influence of sodium and potassium ions upon the motility of sperm cells. *Intern. Congr. Animal Reprod., Cambridge.*

Sotelo, J. R., and Porter, K. R. (1959). An electronmicroscope study of the rat ovum. *J. Biophys. Biochem. Cytol.* **5**, 327.

Tyler, A. (1956). Gametogenesis, fertilization and parthenogenesis. *In* "Analysis of Development" (B. H. Willier, P. Weiss, and V. Hamburger, eds.), p. 170. Saunders, Philadelphia, Pennsylvania.

Uotila, U. U. (1940). The early embryonic development of the foetal and permanent adrenal cortex in man. *Anat. Record* **76**, 183.

Vander, A. J., and Miller, R. (1964). Control of renin secretion in the anaesthetized dog. *Am. J. Physiol.* **207**, 537.

Villee, C. A., and Engel, L. L. (1961). "Mechanism of Action of Steroid Hormones". Macmillan (Pergamon), New York.

Voglmayr, J. K., Waites, G. M. H., and Setchell, B. P. (1966). Studies on spermatozoa and fluid collected directly from the testis of the conscious ram. *Nature* **210**, 861.

Volgmayr, J. K., Scott, T. W., Setchell, B. P., and Waites, G. M. H. (1967). Metabolism of testicular spermatozoa and characteristics of testicular fluid collected from conscious rams. *J. Reprod. Fertility* **14**, 87.

Weber, A. F., McNutt, S. H., and Morgan, B. B. (1950). Structure and arrangement of zona glomerulosa cells in the bovine adrenal. *J. Morphol.* **87**, 393.

Weisel, G. F. (1948). Relation of salinity to the activity of the spermatozoa of *Gillichthys*, a marine teleost. *Physiol. Zool.* **21**, 40.

Williams, R. G. (1945). The characteristics and behaviour of living cells in autogenous grafts of adrenal cortex in rabbits. *Am. J. Anat.* **77**, 53.

Willmer, E. N. (1956). Factors which influence the acquisition of flagella by the amoeba, *Naegleria gruberi*. *J. Exptl. Biol.* **33**, 583.

Willmer, E. N. (1960). Steroids and cell surfaces. *Biol. Rev. Cambridge Phil. Soc.* **36**, 368.

Wimsatt, W. A. (1944). Growth of the ovarian follicle and ovulation in *Myotis lucifugus lucifugus*. *Am. J. Anat.* **74**, 128–174.

Wislocki, G. B., Bunting, H., and Dempsey, E. W. (1947). Metachromasia in mammalian tissues and its relationship to mucopolysaccharides. *Am. J. Anat.* **81**, 1.

Zachariae, F., and Jensen, C. E. (1958). Studies on the mechanism of ovulation. Histochemical and physicochemical investigations on genuine follicular fluids. *Acta Endocrinol.* **27**, 343.

CHAPTER 18

WITHIN THE VERTEBRATES: 2. THE ALIMENTARY CANAL, BLOOD CELLS, BONES, AND TEETH

The Alimentary Canal

The cytology of an alimentary canal is necessarily so adapted to the diet of the animal that consideration of its evolution in any detail is not either possible or profitable. The alimentary canal of every animal is, in a sense, a special case. On the other hand, if, as we have supposed, the vertebrates have descended from some nemerteoid stock, then the main features of the vertebrate alimentary canal must have had their origin in the system already developed in the nemerteoids. The relevant problems are therefore to examine the canal, as it exists in nemertines, in search of features that could be adapted along the necessary lines and to bear in mind, all the time, the functions of an alimentary canal. These functions are: to make food substances available, by digestion; to absorb the required digested products; and in many ways to act as a protective barrier between the blood and tissues on the one hand and the variable contents of the canal and the outside world on the other.

As already described, the mouth opens into a fore-gut which has an epithelium of considerable complexity and which responds both rapidly and cytologically to changes in ionic concentration of the medium. This, of course, is interesting in itself. It also provides a basis for considerable variation in activity and for the correct response to be made when a sea-water-adapted animal invades a fresh-water environment or *vice versa*, or becomes terrestrial, or feeds on micro-organisms or on something more solid. The ciliated cells and the various forms of mucoprotein-producing and secretory cells provide a good basis for the development of glands (e.g. salivary glands of various sorts) and, as we have seen, for providing endocrine systems responding to environmental changes (e.g. the thyroid and the pituitary), and for providing the special covering of gills. It will be recalled that some of the cells of the buccal cavity have staining reactions akin to those of chloride-secreting cells and oxyntic cells (see p. 350). Thus both of these rather special cell types could arise as direct derivatives of the acidophil, carbonic anhydrase-containing, and probably acid-producing cells akin to those found in existing nemertines (Jennings, 1962). In nemertines in which digestive processes have been investigated, the fore-gut does not appear to contribute much to the actual digestion. No true peptic digestion has been found in the fore-gut. This, however, is not very surprising, as peptic digestion seems to be a process that is confined to the vertebrates and is absent from both *Amphioxus* and *Balanoglossus* (Barrington, 1942b). On the other hand, digestion goes on rapidly in the lumen of the upper part of the intestine proper at a pH which has been colorimetrically estimated as 5 to 5.5. After the main protein structure of the ingested food has been broken down by cathepsin C from the gland cells of the intestine, the particles are phagocytosed and further broken down within the chief cells with the help of leucine amino-peptidase (Jennings, 1962).

The manner in which the fore-gut is subdivided into segments (see Fig. 13.4), labelled oesophagus, stomach, and pylorus, varies from species to species among the nemertines and, except in one or two cases, not much is known about the cytology and functions of these different morphological parts. That such divisions do occur, however, is in itself interesting in that it may foreshadow the corresponding subdivisions in the vertebrates, i.e. into mouth, pharynx, oesophagus, and various special portions of the stomach, as they occur in adaptation to special forms of diet.

In the cyclostomes, e.g. in the ammocoete larva of *Petromyzon*, small groups of cells are budded off the epithelium of the fore-gut just before it joins the mid-gut and are often associated with the bile duct and extreme anterior end of the mid-gut. These nests of cells sink into the underlying connective tissue as the follicles of Langerhans. On the grounds that their destruction leads to raised levels of blood sugar and that injection of glucose leads to their vacuolation (Barrington, 1942a), it has been suggested that these follicles are the precursors of the islets of Langerhans of the pancreas (Fig. 18.1) There are cells with different staining reactions in these follicles, but it is not yet clear how, if at all, they relate to α and β cells of the islets of higher animals. Nothing in the way of islet tissue has so far been reported from the corresponding region of nemertines, but further investigations are necessary before such structures can be definitely ruled out. It is interesting that the staining reactions of the islet cells in elasmobranchs (Thomas, 1940) are similar to those of the pharyngeal mucosa of the nemertine (see also p. 342).

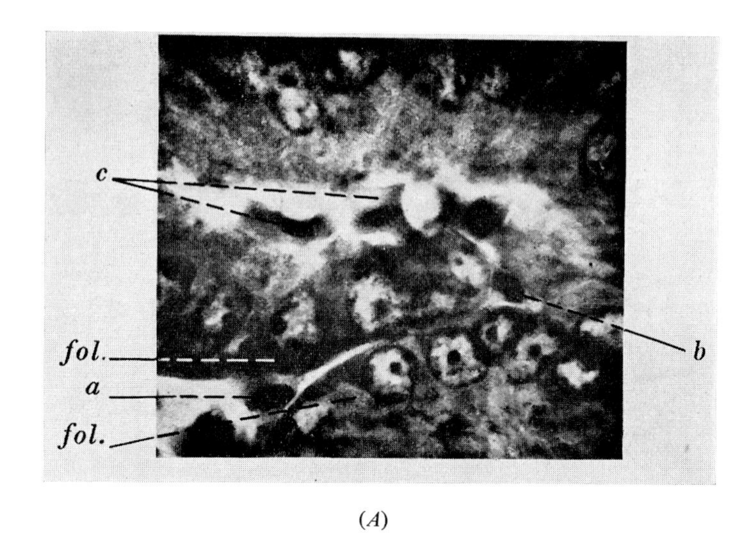

(*A*)

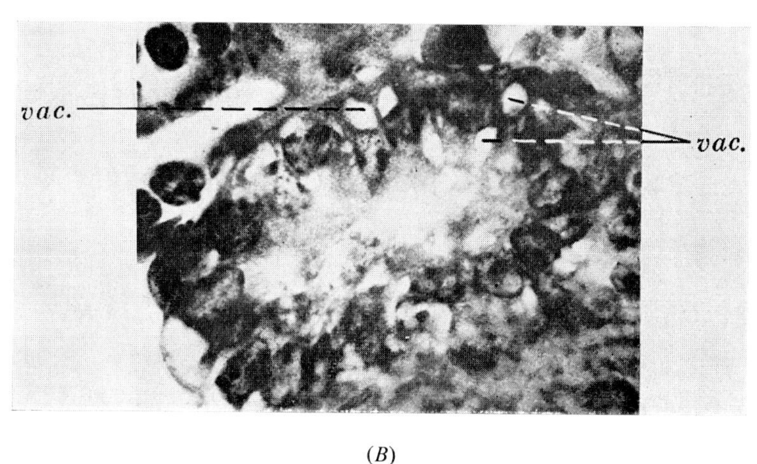

(*B*)

FIG. 18.1. Follicles of Langerhans in the ammocoete larva. (*A*) normal follicle; (*B*) after four injections of glucose (0.05 cc of 20% glucose). a, b, c, Blood corpuscles; fol., follicles; vac., vacuoles. (Barrington, 1942a.)

In the upper part of the intestine of the cyclostomes there are, in addition to the chief cells, others that are filled with azanophilic granules and which have large single nucleoli in their basally situated nuclei (Fig. 18.2). Similar cells are present in the nemerteoid intestine in the same place (see Fig. 12.7). These cells are rich in cathepsin C (endopeptidase), and it has been suggested by Barrington (1936, 1945) and others that those in the cyclostomes could be the precursors of the acinar cells of the pancreas. These suggestions about

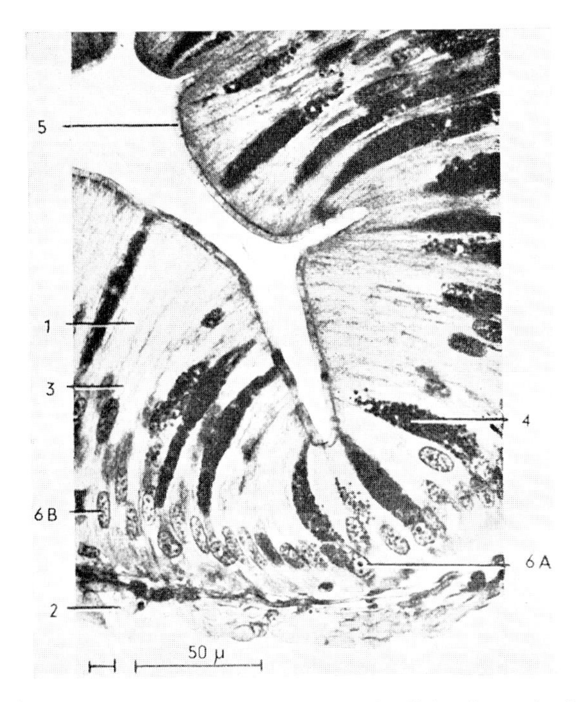

FIG. 18.2. Intestinal epithelium of *Myxine*. 1, Epithelial cells; 2, lamina propria; 3, cylinder cells; 4, zymogen granule cells; 5, brush-border zone with terminal web and terminal bars; 6A, B, nuclei of zymogen cell (with nucleolus) and of cylinder cell, respectively. (Adam, 1963.)

the follicles and the peptidase-containing cells are not unreasonable as a working hypothesis concerning the origin of the pancreas, an organ which appears for the first time as a separate entity in the fishes. In *Myxine*, an "islet organ" associated with the entry of the bile duct is considered homologous with the islet tissue of the pancreas, and it has α and β cells with characteristic staining reactions (Schirner, 1963). The exocrine "pancreas", on the other hand, is diffuse and separate from the islet tissue

In the pancreas of elasmobranchs, islet cells may appear to enclose, or to be incorporated into, the walls of the ducts, though the lumen of the latter is always surrounded by simple duct cells (Thomas, 1940). Alternatively, they may appear as isolated cords of cells or even as separate islets. Similarly, cells very like the zymogen cells of *Myxine* and of the other cyclostomes may be incorporated into the walls of the ducts.

These observations give some indication of how the pancreas may have come into existence, but they leave several matters unexplained, especially the origin of the duct tissue and the relationship of the islet tissue to it. The significance of the cell cords and their eventual organization into islets within the acinar tissue need clarification.

However this may be with regard to the pancreas, there is a strong cytological resemblance between the beginning of the intestine in *Lineus ruber* and that of the corresponding region of *Myxine glutinosa* (cf. Figs. 12.7 and 18.2) which again argues in favour of a relationship between these groups of animals, especially as the intestine of *Myxine* is in many ways simpler than that of higher vertebrates. The epithelium in *Lineus* is composed of chief cells and cells with secretory granules just like that of *Myxine*. The chief cells of the intestine of *Lineus* are rich in alkaline phosphatases (Jennings, 1962), and this is a characteristic feature of the corresponding cells in the alimentary canal of the vertebrates. It would be interesting to know if the development of these phosphatases in *Lineus* is in any way dependent on cortisone or cortisol as it is in higher animals (Moog *et al.*, 1954). In the vertebrates the appearance of the enzyme in the intestine, though not in the kidney (Junqueira, 1952), is related to the state of differentiation of the cell, and cortisone, even in cultures *in vitro*, accelerates its appearance (Moog, 1959). The difference between the action of corticoids on the intestine and the kidney phosphatases is interesting and peculiar, since in both tissues the phosphatase is associated with microvilli and the presence of PAS-positive material.

In spite of the similarity between the intestinal epithelium of *Lineus* and that of the cyclostomes it must at the same time be noted that the intestine of *Myxine* and of all vertebrates is backed by connective tissue and muscle coats, while that of most nemerteoids has no muscle of its own and depends on the movement of the body as a whole for the mixing and onward passage of food.

The intestinal epithelium of nemertines varies considerably in its appearance according to its functional activity. In the fasting animal it may be covered with long cilia or microvilli, and the ciliated cells may appear empty and elongate. In the well-fed animal the cilia seem to lose their definition and to cohere, while the whole of the cytoplasm may be filled with phagocytotic and pinocytotic vacuoles. Intermediately, the epithelium has a fairly typical "brush border" which surmounts an area of cytoplasm that may contain mucoid material in the form of droplets, sometimes small and numerous, at other times larger and fewer. Attention was called earlier (p. 311) to these vacuoles in connexion with similar vacuolation in the epithelial cells of myotomes.

Myxine and *Lineus* therefore show a similar arrangement in their chief cells, and among the chief cells are found the granular cells already mentioned. Whereas goblet cells are a conspicuous feature of the whole of the intestinal epithelium of the higher vertebrates, they do not appear as such in the intestine of either *Myxine* or the nemertines, though mucous-secreting cells are present. This raises the question of what contribution goblet cells make to the economy of the intestine. The simple and popular view is that, as the faecal matter becomes more solid by the absorption of the fluid contents, the production and outpouring of mucus preserves its mobility by a simple process of lubrication. This view, of course, does not really explain the existence of goblet cells in the duodenum, where the contents are largely fluid, nor in several other situations where goblet cells are present, e.g. the trachea, although the increase in their number in the lower parts of the intestine would seem to support the idea. In the trachea the trapping of foreign particles in the moving coat of mucus may, of course, be regarded as sufficient reason for the presence of goblet cells. Nevertheless, an interesting observation, to which reference has already been made in another connexion (see p. 482), may have some significance with regard to goblet cell function. It is that in some abnormal cases where there is a great increase in goblet cells in the human rectum, caused by a benign tumour, the contents of the organ were found to be exceptionally rich in potassium and the sodium concentration was normal (Little, 1964). It is thus possible that water absorption and potassium excretion are part of the stock-in-trade of goblet cells and that the presence or absence of these cells may be determined partly by ionic gradients across the intestinal wall. In other words, the chief cells and the goblet cells in the intestines of higher vertebrates between them preserve the ionic composition of the intestinal blood in the face of the wide variations that may occur in that of the intestinal contents. This does not, however, throw any light on why the cellular composition of the intestines of nemertines and *Myxine* is so uniform and so similar. It may nevertheless be pertinent to note that the ionic composition on both sides of the intestine in these creatures is not far from that of sea-water, so that the need for ionic regulation is much less than it would be in fresh-water or terrestrial creatures.

The development of a separate pancreas and of the submucosal glands (Brunner's glands) in the vertebrates may have been initiated by the development of a true peptic digestion in a highly acid medium in the stomach and by the necessity for the neutralization of the acid before the more complete tryptic digestion could be initiated in the neutral medium of the intestine. As soon as these two separate phases of protein digestion developed, and this may have depended on the development of the necessary sphincter muscle to separate them, they would necessitate the simultaneous development of a coordinating mechanism to turn one process off when the other started. Thus the development of the intestinal hormones, secretin, pancreazymin, gastrin, and enterogastrone, became progressively necessary. However, exactly which cells became responsible for their production and how they did so is still debatable. In some cases the necessary stimulus remains to be determined also.

Another feature that has developed among the vertebrates and is not present in nemertines or cyclostomes is the intestinal villus. In the vertebrates such villi become extremely numerous and important. Their distribution on the surface of the intestine is by no means random but, in some species at least, conforms to a very definite pattern. Not only is the pattern evident as villi and crypts alternating in orderly sequence characteristic of the position in the intestine, but also the villi develop according to a precise pattern of folding of the intestinal mucosa which gives the impression of being initiated by mechanical factors in the early development of the intestine (Clarke, 1967) (Fig. 18.3). Moreover, in intestines in which villi are conspicuous, so are the numerous cell divisions in the intestinal epithelium and especially in the crypts, where they provide replacements for the cells lost from the tips of the villi. The evolution of this whole mechanism of cell replacement and cell migration is still somewhat mysterious.

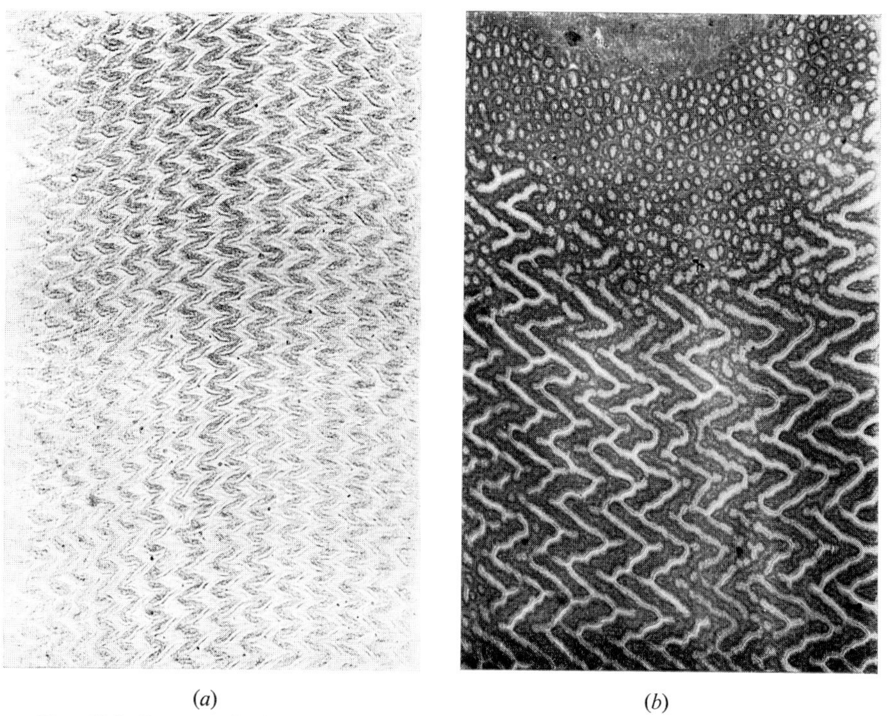

FIG. 18.3. Pattern of villi in the duodenum of the bird. (*a*) Whole mount of opened duodenum of 16-day embryo; (*b*) oblique horizontal section of opened duodenum of 2-day-old-chick. (Clarke, 1967.)

Whereas in nemertines much of the later part of the digestion of a meal is carried out intracellularly, the succus entericus of the vertebrate gut contains proteolytic enzymes akin to the intracellular lysosomal enzymes. There is evidence that numerous epithelial cells break down at the tips of the villi and presumably liberate their enzymes. A similar loss of cells into the gut occurs in *Balanoglossus* (Barrington, 1940). The high growth rate of the cells in the crypts must be seen as part of the digestive mechanism which depends partly on holocrine secretion of the enzymes by the destruction of whole cells. However, the intestine is not only involved in digestion; it also absorbs, and absorption takes place on the villi. Thus it seems that the intestinal epithelium has developed in a thoroughly economic way. The cells in the crypts multiply; they move up the villi and act as mechanisms for absorbing the digested products; when they reach the top, they are released into the lumen where they disintegrate and liberate their enzymes to provide more products of digestion for the rising generation of cells to absorb. This rapid turnover of cells with their "purposeful" destruction is parallel to the turnover of red blood corpuscles in the higher vertebrates, in which these cells are produced in enormous numbers to function with great efficiency for a limited time and then are discarded. In both cases, some of the breakdown products are used again. An intriguing feature of the intestinal system is that, in order to produce the orderly sequence of events on the very regular intestinal villi, a well determined pattern of cell division must also have been evolved (Clarke, 1967).

The comparatively simple picture we have been considering of the events going on in the intestine is in fact complicated by the contributions which the pancreas and liver make to the overall pattern. It seems probable that the first signs of the liver of the vertebrates are to be seen in the forward-running diverticulum of the gut as, for example, in *Amphioxus* (see Fig. 16.3), though in *Myxine* and the lampreys it has already become a well developed organ with bile ducts and gall bladder. There is no doubt that the liver develops embryologically as a diverticulum of the gut, but it is not immediately obvious how or why, phylogenetically, such a diverticulum should become converted into a liver. The highly vascular diverticulum in *Amphioxus* (see Fig. 16.3) is in the right place; it probably indicates that, in the creatures that were actually the forerunners of the vertebrates, there was a similar diverticulum which could be so converted. It is therefore legitimate to ask what functions the diverticula in these creatures performed. Apart from any information that can be gleaned from *Amphioxus* itself, the question is likely to remain unanswered. Again, however, it is possible that the nemertines may come to the rescue. It will be remembered that some nemertines have a forward-running diverticulum or paired diverticula of the gut originating from the junction between the "pharynx" and intestine (see Figs. 16.3 and 13.4). An investigation of such structures could well throw light on the beginnings of the various liver functions, e.g. production of a digestive juice, storage of products of digestion and absorption, metabolism of food products, detoxication and excretion, and initiation of blood formation. These beginnings may occur in the nemerteoids. The fact that the diverticula were originally paired is interesting. Perhaps the relationship between the bile ducts and the pancreatic ducts may offer a clue here, though, as we have seen, the cellular origin of the pancreas seems likely to have been somewhat different.

A topic touched on before may now be mentioned again. The alimentary canal of vertebrates receives into its lumen the secretions of several glands, in particular salivary glands, pancreas, liver, and, in elasmobranchs, rectal glands. In addition to these actual glands there are certain patches of lymphoid tissue, e.g. tonsils, appendix, Peyer's patches, bursa of Fabricius,

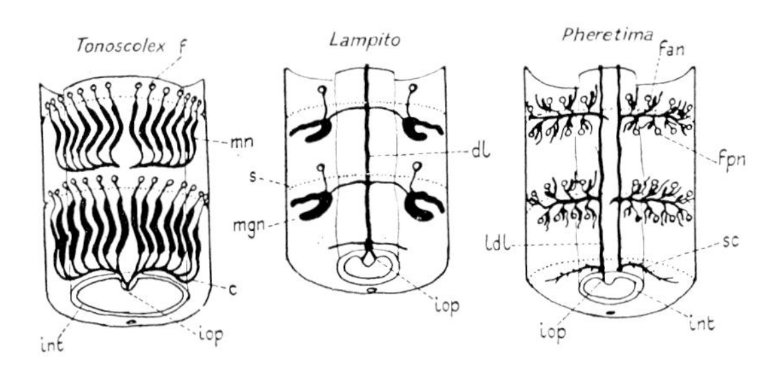

FIG. 18.4 Diagrams of enteronephric systems in *Tonoscolex*, *Lampito* (*Megascolex*) *trilobata*, and *Pheretima*. c, Terminal canals which join to a terminal duct; dl, median longitudinal duct; f, nephridiostome; fan, fpn, nephridiostomes of anterior rows and posterior rows of septal meronephridia; int, intestine; iop, opening into intestine; ldl, left dorsal longitudinal duct; mgn, meronephridium; mn, septal meronephridium; s, position of intersegmental septum; sc, septal canal. (Goodrich, 1945.)

which turn up in set places and for which no evolutionary explanation is forthcoming. As we have seen, suitable cells and cellular arrangements which could initiate the development of salivary glands, liver, and pancreas are present in the appropriate regions of the gut of nemerteoids. Nevertheless, it is just possible that another observation should be taken into consideration. In the annelids, in which nephridia became important organs, widely distributed throughout the body (presumably because the segmentation limited the unification of the body by means of a common tissue fluid), it is not unusual to find some of the nephridia opening into the alimentary canal rather than on to the surface of the body (Bahl, 1947) (Fig. 18.4). Curiously enough, these openings do not occur just anywhere. The mouth, the anterior part of the intestine proper, and the anal region are the three regions favoured in this respect. They are, of course, also the places where the main glands open into the canal of vertebrates. This may be a coincidence, but, in view of the actions of mineralocorticoids on the activities of the salivary glands as well as on the kidneys and of the relationship between glucocorticoids and the islet tissue of the pancreas, it is perhaps worth bearing in mind. The sexual differences between the salivary glands of rodents might also find their explanation along these lines. One wonders, also, if the follicles of Langerhans are vestigial nephridia? In the nemertine, *Taeniosoma cingulatum*, Coe (1906) records the presence of a pair of nephridia opening into the alimentary canal at the lower end of the pharynx; thus they are in the same relative position as the follicles of Langerhans in the cyclostomes (see p. 517).

The Cells of the Blood

It is well established that the liver originates embryologically as a diverticulum of the gut. For obvious reasons the diverticula of *Amphioxus* and the nemerteoids have been suggested as relevant progenitors. Nevertheless, there is much to be explained in the conversion of a relatively simple diverticulum into a complicated system of bile ducts leading to a series of tubular or plate-like structures which, during development, are separated by tissues which become specialized as primary sources of erythrocytes and granulocytes. The latter arise, ontogenetically, from the mesenchymal elements, and three questions are at once raised thereby. First, why do the cells of the mesenchyme in this situation do something similar to what the cells of the extra-embryonic mesenchyme covering the yolk sac also do? Second, why are mainly erythrocytes, granulocytes, and megacaryocytes formed here, as in the bone marrow, while lymphocytes are not formed, or at least not extensively? Third, why does the endothelium which finally lines the sinusoids between the liver cells assume phagocytic properties, whereas elsewhere, except in the bone marrow and adrenal cortex, the vascular endothelium is not normally phagocytic? These three questions raise fundamental issues concerning the nature of the cells that circulate in the blood, and this topic is now taken up in some detail.

First, let it be stated that it is probable that the study of blood until recently almost exclusively by means of smears and Romanowsky stains has to some extent restricted our knowledge of some of the essential features of blood and its formation. For example, we now know that several distinct classes of cells masquerade under the heading of lymphocyte. Only through more searching tests and investigations will it be possible to appreciate the physiology of the various cells that may circulate in the blood stream, lymphatics, and tissue spaces and which may appear in the various haemopoietic organs and tissues. It is a salutary observation that an embryonic heart fibroblast, when treated with trypsin, may readily assume a shape and form that could easily be confused with that of a large lymphocyte if both cells were stained with Leishman's stain.

Next, it appears that erythrocytes and granulocytes habitually arise together in certain situations (e.g. the liver and the bone marrow), whereas lymphocytes and monocytes, though they may occur and indeed multiply in bone marrow, are essentially the cells of the lymph nodes. Lymphocyte-like cells (thymocytes) dominate the thymus, in which they coexist with a "reticulum" of epithelial origin. "Lymphocytes" also occur together with macrophages in the white pulp of the spleen, the tonsils, the lymph nodules,

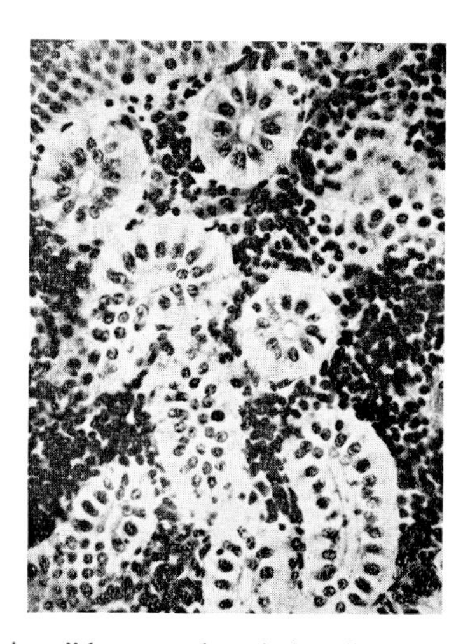

FIG. 18.5. "Lymphoid tissue" between the tubules of the aglomerular kidney of *Opsanus tau*. (Smith, 1951.)

and Peyer's patches in the intestine, the appendix, the bursa of Fabricius, and, abnormally, in areas of chronic infection. They may also appear in the zona glomerulosa of the adrenal cortex, the parotid gland, the thyroid, and kidneys (Kingsbury, 1945), particularly those of aglomerular fish (Fig. 18.5) (Smith, 1951). In the pronephros of cyclostomes, periods of formation of "lymphocytes" and monocytes may alternate with periods of formation of erythrocytes, granulocytes, and megacaryocytes (Fig. 18.6) (Holmgren, 1950).

FIG. 18.6. Blood formation in the pronephros of *Myxine*. (*a*) Pronephros of a 9.6 cm specimen with haemocytoblasts and proerythroblasts in the central mass and formative cells in the epithelium of the right tubule. (*b*) Pronephros of a 28 cm specimen. Myelopoiesis with megacaryocytes on the right; "fibre cells" and lymphocytes coming from the tubules in the upper left. (Holmgren, 1950.)

See facing page→

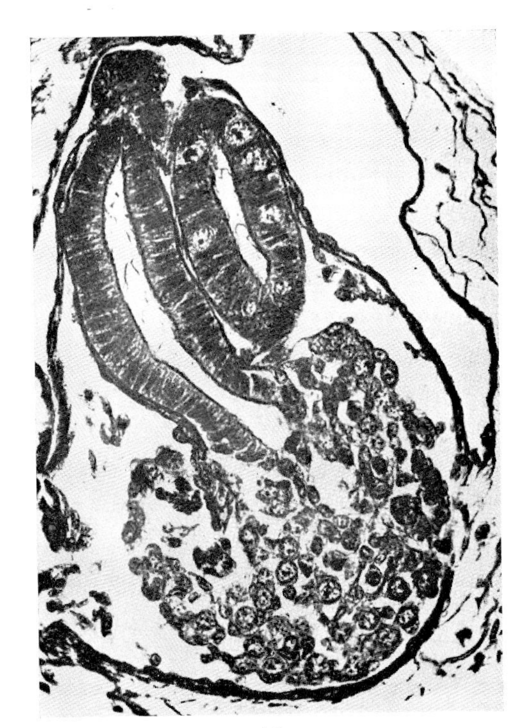

(*a*)

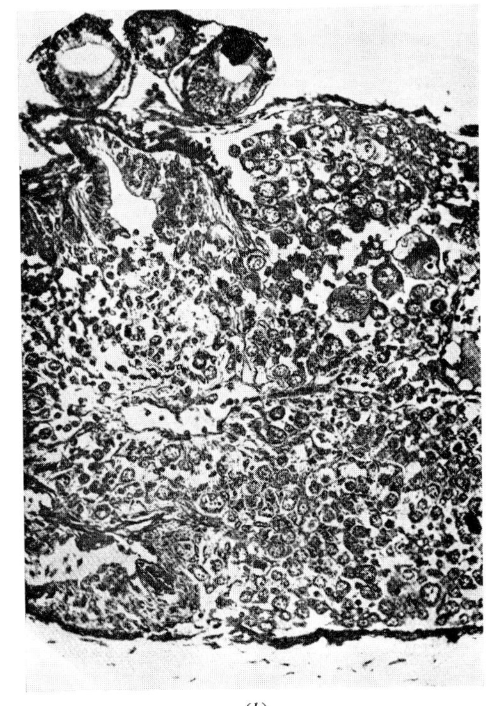

(*b*)

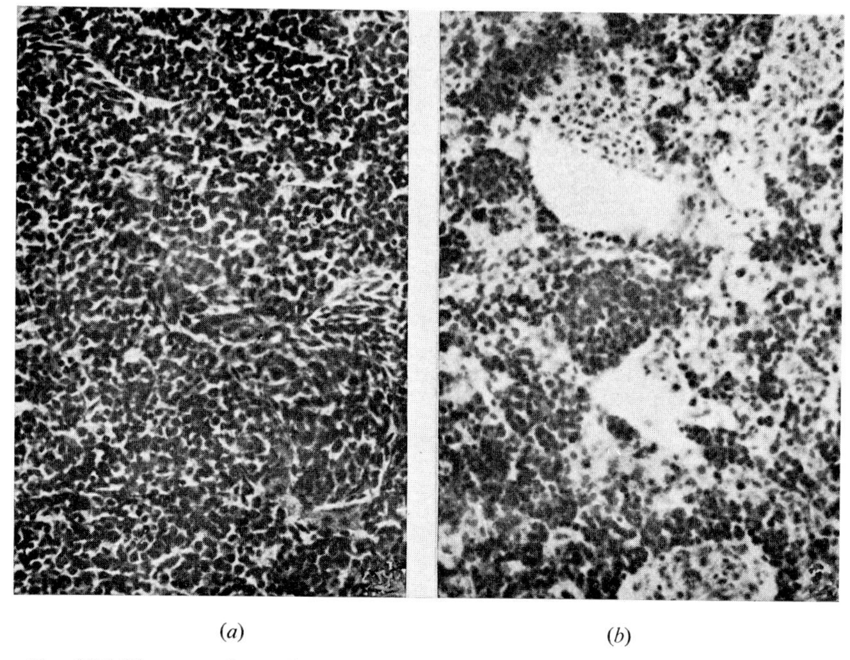

(*a*) (*b*)

FIG. 18.7. The pronephros of *Astyanax*. (*a*) In the normal animal, showing a high content of lymphocytes. (*b*) In an animal injected with adrenocorticotrophic hormone. Note the reduction in lymphocytes; macrophages were simultaneously increased. (Rasquin, 1951.)

The head kidneys of other fish (e.g. *Astyanax*) may also contain masses of lymphocytes (Fig. 18.7) (Rasquin, 1951). There thus seems to be some connexion between nephridial or excretory tissue with blood cell formation and particularly that of lymphocytes. Nevertheless the occurrence of lymphocytes in a tissue is not necessarily evidence of their origin in that tissue. Among the invertebrates, lymphoid tissue is apparently very sparse, but, in crustacea (e.g. the crayfish), something like it occurs on the dorsal surface of the stomach, probably as a modification of excretory tissues (Fischer-Piette, 1931). Lymphocyte-like cells are present in myxinoids, but these animals lack immunoglobulins and show no delayed hypersensitivity. In lampreys, on the other hand, hypersensitivity is present together with homograft immunity, and typical lymphocytes occur. Plasma cells make their first appearance in the chondrosteans. Thus it seems that such immunity phenomena came in with the vertebrates and were developed stage by stage (Good *et al.*, 1966).

In the liver, in the blood islands of the blastoderm, and in the bone marrow, the blood cells derive directly from the mesenchyme; in the head kidneys of fish and cyclostomes, and probably in the thymus, the epithelial cells directly undergo these conversions (see Figs. 18.6b and 18.7). The situation is similar to that discussed in relation to the origins of muscle. It appears that several types of cell can act as precursors. In all cases the original tissues contain the two classes of cells which we have classed as ecballic, mechanocytic cells and emballic, amoebocytic cells. The question that then arises is which types of cells in the blood can these two original groups produce. The problem is a difficult one, since all the cells are mobile (except the adult erythrocyte); most populations are mixed; cell divisions as well as differentiations are in progress; and the certain labelling of cells at different stages in their differentiation is not always easy. It may be significant that, in Murray's (1932) analysis of erythrocyte production in tissue cultures of the chick blastoderm, erythropoiesis occurred only from the cells of the posterior three-quarters of the blastoderm. Thus (comparing the chick blastoderm with a flattened blastula) the cells of the vegetal three-quarters would be concerned primarily; this suggests that erythropoiesis occurs among the amoebocytes. Another question in need of an answer is why lymphocytes and monocytes tend to occur together in some haemopoietic tissues while erythrocytes and granulocytes develop together in others. Is this a feature of the "stem" cells or of the situations in which development is occurring? On grounds of such processes as protoplasmic movement, pinocytosis, storage of polysaccharides, numbers of lysosomes, type of endoplasmic reticulum, numbers of mitochondria, and occurrence of clasmatosis, i.e. the shedding of portions of cytoplasm, the monocytes and granulocytes have affinities with amoebocytes, whereas the erythrocytes and lymphocytes do not.

When lymphoid tissue appears in the adrenal cortex, or in aglomerular kidneys, it appears in place of the zona glomerulosa or of the glomerular end of the nephron. Treatment of mammals with adrenocorticotrophic hormone (ACTH) not only causes the zona fasciculata of the adrenal cortex to develop and secrete, but also causes the zona glomerulosa to atrophy. Moreover, probably because of the corticoids released, the macrophages in the lymph nodes enlarge, become actively phagocytic, and fill with brown pigment, while the lymphocytes atrophy (Dougherty and White, 1945). At the same time the lymphoid tissue becomes oedematous, an interesting point in view of the likelihood of the zona glomerulosa, when grown *in vitro*, to become floating in fluid, and of the involvement of the upper part of the nephric tube with the fluid balance of the body. Furthermore, because lymphoid tissue is first encountered in quantity in phylogeny in the pronephros of fishes, it is relevant that, when the teleost, *Astyanax*, is treated with ACTH, the lymphocytes in its pronephroi disappear and large numbers of macrophage-like cells appear instead (Rasquin, 1951) (see Fig. 18.7). Further afield, in the gephyrean, *Phoronis*, at a certain stage in development the solenocytes of the nephridia break down and float away in the blood stream as small lymphocyte-like cells (Goodrich, 1903, 1945). In nemerteoids the blood cells are thought to be derived from the walls of the vessels and adjacent mesenchyme cells amongst which there are mechanocytes and amoebocytes. In *Lineus* there are no small lymphocytes (large lymphocytes are present; see p. 272), but in the blood and in the rhynchocoel fluid there are spindle-shaped cells resembling fibroblasts (Fig. 18.8). This is interesting on three counts: first, perhaps it emphasizes the relationship between lymphocytes and the mechanocytic (ecballic) cells; second, it brings up again the relationship of

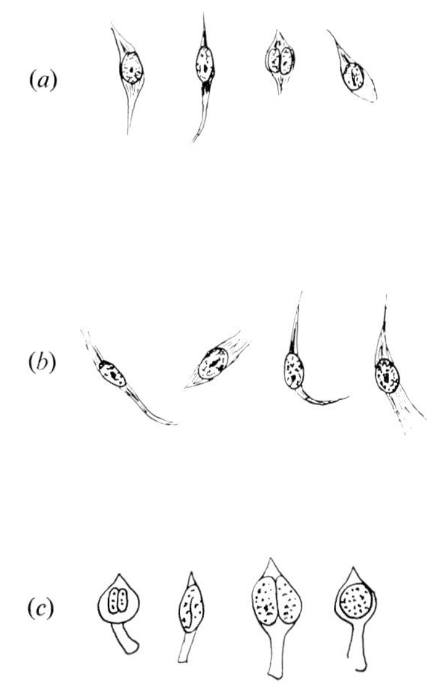

FIG. 18.8. Spindle cells in the blood (*a*) and the rhynchocoel fluid (*b*) of *Lineus*, and in the blood of *Myxine* (*c*). (*Myxine* after Holmgren, 1950.)

the coelomic fluid, the rhynchocoel fluid, and blood; and, third, it points to another similarity to *Myxine*.

In connexion with the last two points, if the idea that the coelomic cavity of vertebrates is really a subdivision of the haemocoel of the nemerteoids can be substantiated (see p. 454), the state of affairs in the pronephros of the myxinoids and ammocoetes perhaps becomes explicable. As stated earlier, the pronephric tubes in these creatures make direct and open connexions between the coelomic cavity and the blood vessels (see Fig. 16.8). It seems possible that these organs act like valves regulating the fluid in the two regions, for there is evidence of the passage of fluid from the coelomic cavity, through the pronephric funnels and the "lymphoid" tissue, into the blood of the pronephric vein. Presumably, such passage of fluid, either initially depended on, or became dependent on, the presence of a heart exerting sufficient hydrostatic pressure on the blood to force fluid through the glomeruli into the coelom, thus providing the necessary head of pressure. It may also be that the main source of blood cells for the haemocoel and blood system initially became concentrated in the pronephric area. The "filtration" of the coelomic fluid, which is initially discharged into the coelomic cavity through the large glomerular structures (Fig. 18.9) projecting

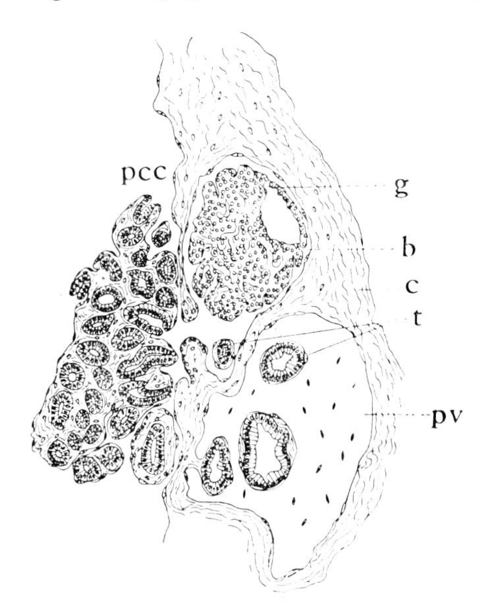

FIG. 18.9. Diagram showing the relationship of the glomerulus (g) to the pericardial cavity (pcc), the pronephric tubules (t), and the phonephric vein (pv) in *Bdellostoma stouti*. c = connective tissue. (Conel, 1917.)

into the coelom before it returned to the blood, is certainly a function similar to that of the more sophisticated lymph nodes when they "filter" the tissue fluid before it returns to the blood. In ammocoetes, athrocytosis occurs in the "blood-forming" tissue of the pronephros just as it does in lymph nodes (Gérard, 1933). In both cases, cells are added to the fluid in its passage towards the blood. In *Myxine* itself the situation is peculiarly interesting because the pronephric tubules are periodically and alternately converted into fibrous tissue with lymphocyte-like cells and into typical erythropoetic tissue (see Fig. 18.6). A detailed cytological and functional study of these versatile pronephric tissues could be very illuminating in many respects. Which cells of the pronephric tubules give rise to which cells in the blood? What are the necessary conditions? At certain stages, definite "formative" cells are visible among the epithelial cells of the tubules (see Fig. 18.6a), but their subsequent fate has not yet been adequately followed. Presumably, some local conditions determine events.

The observation of Trowell (1966) that, for the successful cultivation of lymph nodes in organ culture, it is advantageous to lower the concentration of sodium chloride in the medium below normal (0.8 %), and even to 0.4 %, seems to indicate that in this respect the lymphocyte resembles the flagellate form of *Naegleria*, i.e. it belongs to the ecballic (mechanocytic) group and is adapted to a "dilute" environment. The observation that a herbivore (rabbit) has many more lymphocytes than carnivores (cat and dog) (Ehrich, 1946) could be relevant here also, since the diet of the rabbit is much richer in K^+ than that of the carnivores, which tends to be high in Na^+. The intestinal flora is, of course, very different too.

Direct and cinematographic studies of lymphocytes in culture, and in observation chambers in the ears of rabbits, have shown that the movement of lymphocytes, although rapid and peculiar to lymphocytes, is generally well polarized and has more affinity with the much slower gliding movement of fibroblasts than with the relatively random and labile movement of macrophages (de Bruyn, 1946; Harris, 1953; Lewis, 1931, 1933). Although lymphocytes, like macrophages and monocytes, undergo phases of movement and phases of apparent inactivity, they show no tendency towards fusion or giant-cell formation. On the other hand, they show no contact inhibition, nor do they form desmosomes. In fact, the lymphocytes remain very firmly themselves even when in intimate contact with other cells, e.g. during emperipolesis with macrophages and other cells (Trowell, 1966; Shelton and Dalton, 1959) (Fig. 18.10). In this untouchability, lymphocytes have something in common with erythrocytes. Neither cell adheres to glass. During the so-called resting phases, lymphocytes may put out and retract long fibrous or pointed pseudopodia and in this way show more likeness to mechanocytes than to macrophages. There is no tendency for storage of materials in lymphocytes, whereas there is evidence that they or their derivatives actively produce RNA and liberate specific proteins, namely antibodies. However, they do this only under certain conditions, e.g. when they have been specifically activated, as by the intervention of sensitized macrophages (Ford *et al.*, 1966). In peripolesis and emperipolesis, small lymphocytes enter into active association with, or even within, macrophages and possibly other cells, sometimes for a matter of hours; apparently they benefit from this temporary association. These phenomena make it probable that the small lymphocytes receive substances or "information" of some sort during the process. There is no visible sign of transfer of material from one cell to the other (rhopheocytosis) as there is, for example, when erythroblasts receive iron-containing compounds from Kupffer cells during erythropoiesis in the embryonic liver or from reticulum cells in the bone marrow (Policard and Bessis, 1958; Sorenson, 1960). Nevertheless, this relationship is specially interesting because for many years it has been thought that macrophages may hand on activating or nutrient materials to mechanocytes, and the case of the lymphocytes seems to offer a parallel. The activated lymphocytes are thought to "home" into lymphoid tissue and then become active antibody producers as plasma cells (Gowans and Knight, 1964).

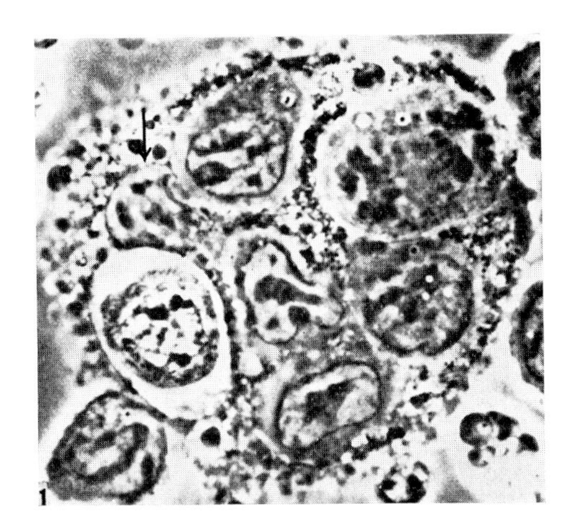

FIG. 18.10. Emperipolesis. Several lymphocytes (from an ascites tumour), one of which is in mitosis, lie within the cytoplasm of a macrophage. The arrow indicates the nucleus of the macrophage. (Shelton and Dalton, 1959.)

Another line of evidence suggesting that lymphocytes are related to fibroblasts comes from the actions of corticosteroids on these cells. Cortisone and hydrocortisone are extremely toxic to lymphocytes (Trowell, 1958, 1966) and inhibitory to fibroblasts (Grossfeld, 1959; Biggers, 1966). In lymph nodes of mammals treated with ACTH, and thus probably under the influence of corticosteroids, the macrophages remain well and healthy, whereas the lymphocytes are reduced and disintegrate (Dougherty and White, 1945), just as in the pronephros of *Astyanax* (see p. 528).

All these points make it likely that the small lymphocyte should not be classed in the same group of cells as the monocytes, but that it should be regarded as having closer affinities to the fibroblast or mechanocyte. Lymphocytes and monocytes (macrophages) are thus likely to be another example of a symbiotic pair of cells, ecballic and emballic perhaps, and comparable with pairs like the ciliated and goblet epithelial cells and the rest. These postulates, of course, assume that the lymphocyte is a definite cell type and not a phase of activity of a convertible cell. There are many claims in the literature that, under some conditions, small lymphocytes change into monocytes and macrophages, as well as claims that macrophages change into fibroblasts and *vice versa*. When small lymphocytes are treated with phytohaemagglutinin or with some antigens, they acquire cytoplasm, begin to resemble monocytes, and start to divide. It is thus possible that the lymphocytes under these special conditions may, like *Naegleria*, be able to change phase. But, when they function normally as lymphocytes within the body, they are probably ecballic cells balancing the emballic activities of macrophages. *Mutatis mutandis*, much the same may apply equally to fibroblasts and macrophages.

If then, in lymphoid tissue, macrophages and lymphocytes act as symbiotic pairs, what is the situation in the myeloid tissues? Why do some cells give rise to erythrocytes and others to granulocytes?

If this question is considered broadly at the level of the higher vertebrates, then in both lymphoid and erythropoietic tissue there is initially present a three-dimensional reticulum. This reticulum is composed (except in the thymus) of reticulin threads on which, and within the meshes of which, the contained population of cells, based on the so-called "reticulum cell", both maintains the structure of the reticulum and undergoes development and differentiation into the characteristic haemopoietic cells. There is no *a priori* reason for believing that all the "reticulum cells" are the same. Indeed, it seems likely that amongst them there must be some fibroblasts (mechanocytes), since reticulin is essentially a collagen-like material, and cells of the amoebocyte class do not usually form collagen. Questions are thus raised about whether the cells from which lymphocytes, erythrocytes, and myelocytes arise are all the same and are the embryonic "reticulum" cells which then differentiate; whether there are special stem cells for each type; or whether the mechanocytes and amoebocytes of the reticulum are the stem cells. If the last is true, do the mechanocytes give rise to some types, and the amoebocytes to others? For example, it would be reasonable to think that in the lymph nodes the mechanocyte type of cell of the reticulum could undergo multiplication and differentiation to produce the small lymphocytes, and that the amoebocyte cell on the reticulum might continue to produce macrophages and monocytes. There would then be some explanation for the mechanocytic properties of lymphocytes.

While macrophages (amoebocytes) are certainly present on the reticulum of the red pulp of the mammalian spleen, some of the cells on the reticulum have also been found to have cilia (Roberts and Latta, 1964; Abdel and Sorenson, 1965) thus indicating their probable relationship to mechanocytes. This again favours the view that two classes of cells are present on the reticulum and that the term "reticulum cell" may be misleading.

There is little doubt that in the bone marrow the more mature granulocytes have some of the essential characteristics of amoebocytes and could logically be derived from similar elements of the reticulum. The early erythroblast, on the other hand, when studied by means of the electron-microscope, has some of the features of the mechanocyte group and, like the lymphocyte, could possibly be derived from the fibroblastic elements in the reticulum. The means of locomotion of these early forms could be of assistance, but it is difficult to interpret owing to the difficulty of identifying the living cells with their fixed and stained counterparts. Cells which migrate from cultures of bone marrow and which have been classed as "myelocytes" have been observed to migrate in a polarized manner, like the lymphocyte, but with a corkscrew motion peculiar to themselves (Rich *et al.*, 1939) and not seen in lymphocytes. The exact nature of this cell, however, is in some doubt.

In the liver of the fetal rabbit, where the blood cells arise from the mesenchyme, as they do in the bone marrow, there are almost certainly both classes of mesenchyme cell. The electron-microscopic appearance of the early cells of the erythroblast series (Sorenson, 1960), in which the cytoplasm is devoid of vacuoles, granules, etc., but rich in ribosomes and a few elements of parallel side endoplasmic reticulum, is consistent with the idea that these cells are mechanocytic. Furthermore, the appearance of the myeloblasts, with their more vacuolated and inclusion-containing cytoplasm, suggests that they are probably amoebocytes (Fig. 18.11). Admittedly, this is not good evidence, since cells vary greatly in microstructure according to their functional activity and local conditions, but at least it does not point in the opposite direction. When embryonic liver is grown *in vitro*, it displays active emigration of both amoebocytes and mechanocytes (Ephrussi and Lacassagne, 1933). This growth of mechanocytes from the embryonic liver is interesting because cultures of liver from older animals display growth predominantly of the liver epithelium and of macrophages (Garvey, 1961; Bang and Warwick, 1965). Adult liver is not normally erythropoietic. Does this again indicate a connexion between fibroblasts and erythropoiesis? Although organ cultures of liver sometimes show haemopoiesis, the conditions under which it occurs have not been established. It is possible that both in liver and in bone marrow the "stem" cells are seeded there from other sources, e.g. yolk sac, thymus.

During the process of maturation of the erythrocytes, ferritin is transferred to them from the hepatocytes and Kupffer cells in the fetal liver and from macrophages in the bone marrow by micropinocytosis and, probably, by a process of rhopheocytosis. The latter process might be considered to be parallel to the emperipolesis of lymphocytes (see p. 532) whereby material is thought to be transferred to the lymphocytes from the macrophages. Electron-microscopic studies show that some ferritin is picked up by adhesion to preferred spots on the cell membrane followed by a process of micro-

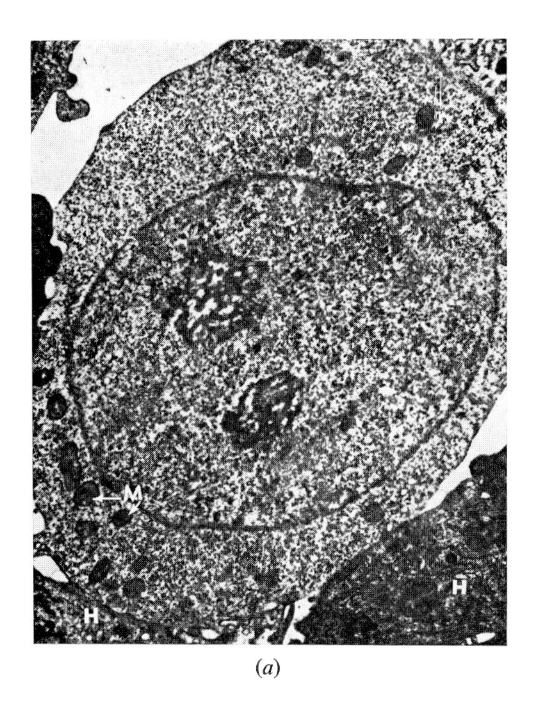

(*a*)

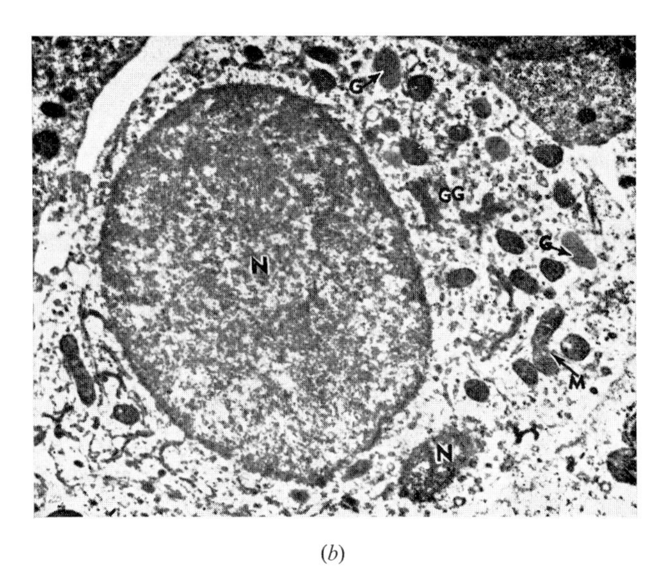

(*b*)

FIG. 18.11. (*a*) Haematopoietic stem cell in fetal liver. (*b*) Heterophil myelocyte in fetal liver. G, Granule; GG, Golgi zone; H, hepatocyte; M, Mitochondria; N, Nucleus. (Sorenson, 1960.)

pinocytosis (Fawcett, 1964, 1965; Sorenson, 1960). This process of micro-pinocytosis is akin to that which occurs in the cells of the ordinary capillary endothelium, cells which have many of the properties of mechanocytes (see p. 56).

To add to the complexity of the situation, the bone marrow normally contains large numbers of "lymphocytes". In view of this and of the enormous numbers of lymphocytes which apparently enter the blood stream daily from the thoracic duct, Yoffey (1950) at one time suggested that the lymphocytes entering the bone marrow could provide the necessary stem cells for the erythrocytes. There are now good reasons for rejecting this view. It is important to remember, however, that there is direct electron-microscopic evidence that lymphocytes do cross the walls of sinusoids in the bone marrow (Hudson and Yoffey, 1966), though the direction in which these cells move has not been established. Accumulations of lymphocytes can also occur in the bone marrow, and the sinusoids are often loaded with them (Yoffey, 1962), but the interpretation of these findings is not clear. Difficulties undoubtedly arise in distinguishing between small lymphocytes, transitional cells, lymphoblasts, and erythroblasts (Yoffey *et al.*, 1965a) and, when phytohaemagglutinin is applied to small lymphocytes, several types of transitional forms are obtained (Yoffey *et al.*, 1965b). Any dogmatic statements on the situation must obviously be regarded with scepticism until better methods are established for identifying lymphocyte-like cells with more certainty.

In mice which have been lethally irradiated, it has been found that bone marrow cells from another compatible donor, and marked by the presence of a chromosome defect, can: recolonize the bone marrow to produce erythropoiesis and granulopoiesis; recolonize the spleen both haemopoietically and lymphopoietically; recolonize the thymus to produce thymocytes; and recolonize the lymph glands, producing mostly lymphopoiesis but some medullary granulopoiesis. On the other hand, cells from lymph glands cannot recolonize the bone marrow nor the thymus (Micklem *et al.*, 1966), though under other conditions small lymphocytes certainly enter the thymus. Cells from lymph glands can temporarily recolonize the lymph glands of lethally irradiated animals, but only cells from the bone marrow have been found to effect a permanent recolonization. This observation seems to indicate that the lymphocytes from lymphoid tissue are different from the lymphocytes or lymphocyte-forming cells of the bone marrow.

Two other points from these experiments with irradiated mice are important in relation to these problems of cell type and cell differentiation in the haemopoietic systems. First, injection of thymus cells induces granulopoiesis in the bone marrow. It is not clear exactly what the cells from the thymus really are. These so-called thymocytes, small lymphocyte-like cells, may be a mixed collection, for it is known that cells from the bone marrow can colonize the thymus and also that thymocytes can differentiate directly from the epithelial reticulum of the thymus itself (Murray, 1947; Auerbach, 1960). Nevertheless, it is odd that the thymus cells which do not produce granulopoiesis in the thymus itself can induce it in the bone marrow. Second, granulopoiesis sometimes occurs in the medulla of lymph nodes injected with marrow cells, though apparently not in the cortical nodules. Both of these observations, and indeed many others, suggest that the local conditions in which certain "embryonic" cells find themselves can be very important in determining the type of differentiation the cells can undergo. They also indicate that "embryonic" cells from different sources can have different potentialities.

In investigating the effects of local conditions, it may be pointed out that the reticulum of the bone marrow, of the medulla of the lymph nodes, and of the red pulp of the spleen (at least in the mouse, which is the relevant animal in this case) is loose and coarse and has a preponderance of phagocytic cells attached to it, or in its vicinity. The reticulum of the germinal centres and lymph cords in the lymph nodes and also of the Malpighian corpuscles in the spleen is generally denser; phagocytic cells are rarer. The proximity of the blood in the two groups is also different. A denser reticulum with little phagocytosis suggests a preponderance of fibroblasts over amoebocytes; conversely, a loose reticulum and greater phagocytosis indicate greater numbers of amoebocytes and fewer fibroblasts.

While the situation with regard to the various cells of the vascular and lymphatic systems is anything but clear, all the points discussed above indicate that there is some basis for the hypothesis that the blood cells also conform to the general plan of a balance between mechanocytic and amoebocytic cells. In the lymph glands the cells of the germinal centres, which proliferate through the stage of large lymphocyte towards the small lymphocyte, may belong essentially to the mechanocytes while the amoebocytes produce the monocytes and the histiocytes. In the bone marrow the mechanocytic cells may become the erythroblasts or may emigrate as lymphocytes, while the amoebocytes proliferate as myelocytes and megacaryocytes. In other words, it is tentatively suggested that lymphocytes and erythrocytes are the end-products of two forms of differentiation of the mechanocytic, ecballic cell, while the myelocytes, monocytes, and macrophages are the end-products of the amoebocytic, emballic, type of cell. As elsewhere, the particular form of differentiation followed is determined by the local conditions. This hypothesis concerning an age-old problem is in all probability a gross oversimplification. Nevertheless, it does emphasize the need to investigate the local conditions in yolk sac, liver, bone marrow, spleen, lymph node, and thymus, and the influence which they may have on the behaviour of more standardized "blast" cells.

The hypothesis runs counter to the conclusions of Trowell (1966) who, in reviewing the subject of "lymphocytes", was convinced by the alternative hypothesis that the large lymphocyte derives from phagocytic cells, partly because the greenish pigment and pycnotic remnants of small lymphocytes are frequently found in cells looking like large lymphocytes in tissue cultures. Such phagocytic cells would, however, be expected to occur in cultures of lymph nodes and could simply be monocytes, or monocytes in the making. The many records of the apparent transformation of small lymphocytes into monocytes and macrophages also support Trowell's contentions, in which, however, there is often doubt about the correct identification of the initial and final cells. "Small lymphocytes" are undoubtedly a "mixed bag". For example, it seems fairly clear from the work of Gowans and Knight (1964) and of Marchesi and Gowans (1964) that there are probably at least two functional sorts of small lymphocyte, or perhaps several phases of lymphocyte life represented in the "lymphocytes" of the thoracic duct of the rat. One group of lymphocytes probably lasts for less than 2 weeks, another survives for more than 8 weeks. There are small lymphocytes (about 35%) which do not label with tritiated adenosine, and about 65% which do. In addition to the small cells, there are large ones that label with tritiated thymidine; if they are infused into another animal, they may be represented later by a population of small lymphocytes. Electron-micrographs show three types of cell in the thoracic duct lymph, small lymphocytes, large lymphocytes with extensive endoplasmic reticulum, and monocytes (?) with numerous small vesicles and a prominent Golgi apparatus. Cinefilms of the cells in the thoracic duct of the dog show mostly small lymphocytes, among which are a few larger ones, and some 5% of monocytes whose locomotion is entirely different from that of the lymphocytes. The small lymphocytes of the thoracic duct probably contain a minority of cells that have come recently from the germinal centres of lymph glands and a majority of cells that have circulated many times between blood and lymph. A free passage of cells from blood to lymph has been shown to occur via special venules found near the lymph cords in the medullary portions of lymph glands (Fig. 18.12). The phylogeny of these peculiar vessels would be an interesting study. The recirculating lymphocytes may have been conditioned by contact with antigens in the tissues, or with macrophages that have been affected by antigens, and these may then differentiate into plasma cells and produce antibodies. Curiously enough, the recirculating lymphocytes apparently do not circulate to the thymus, though the thymus can be recolonized by cells from the bone marrow with cells that produce "thymocytes". The thymic cells thus seem to be more related to the bone marrow cells. On the other hand, it has been suggested that "lymphocytes" from the thymus can provide stem cells for the lymphocytes in lymph nodes (Miller, 1961), though some of the data can also be interpreted in terms

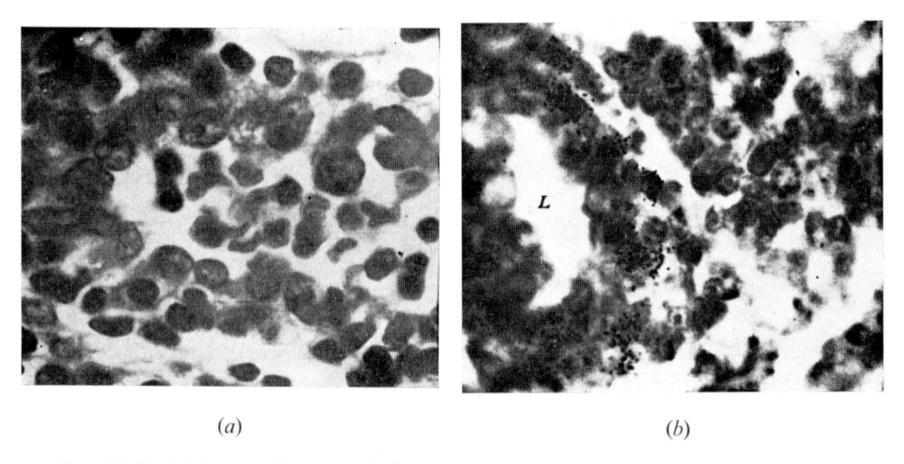

(*a*) (*b*)

FIG. 18.12. (*a*) Postcapillary venule in a cervical lymph node of a normal rat. Note the large, pale nuclei of the endothelium. (*b*) Similar venule in a mesenteric lymph node after an infusion of labelled thoracic duct cells into the blood. L marks the lumen, and lymphocytes can be seen to have penetrated the wall. (Gowans and Knight, 1964.)

of stimulation of division of lymphocytes by a hormone from the thymus.

Some of these puzzles would perhaps come nearer to solution if there were more knowledge of the comparative aspects of the blood-forming and lymphoid tissues. The suggestion has been made that the myeloid and erythroid cells develop from the vascular endothelium, as they probably do in nemertines and certainly do in blood islands of the chick blastoderm, in embryonic liver, kidney, and spleen, and presumably, but with less certainty, in bone marrow, where the status of the reticulum cell is doubtful. If this suggestion holds, we are left in doubt about what determines whether a given cell leaving the vascular endothelium should become an erythrocyte or follow one of the myelocyte lines. Local conditions could be the determining factor. Alternatively, since it has been shown that, whereas most endothelia grow *in vitro* as mechanocytes, there are also some endothelia in the body (e.g. myeloid, hepatic, and adrenocortical) that are phagocytic. There are, therefore, probably two sorts of endothelial cell (ecballic and emballic?). It is thus possible that erythrocytes could develop from one sort of cell, and myelocytes from the other. The manner of colonization of the red pulp of the spleen of irradiated animals by cells from the bone marrow tends to support the idea of the introduction of two different types of stem cells. First, only a limited number of colonies develop; second, some of them are myeloid, others erythroid. Perhaps, however, this view merely betrays our ignorance of the inherent complexity of the red pulp of the spleen and of the number of special local environments that are present in it.

It is perhaps relevant that red-cell formation depends on the presence in the blood of a glycoprotein, erythropoietin, the amount of which is related to the condition of oxygenation of the animal (White *et al.*, 1960). It was stressed earlier that the mechanocytes depend for their growth on rather special conditions which can be partly satisfied by the activities of leucocytes or by the provision of "embryo extract". Thus it would seem possible that erythropoietin acts somewhat in the manner of embryo extract on mechanocytes when it stimulates the growth of the erythroblast. In a spleen that is being recolonized by injected marrow cells, oxygen lack and erythropoietin activate erythroid colonies but leave the myeloid ones relatively unaltered (Feldman *et al.*, 1967), thus indicating the different requirements of the myeloid and erythroid cells.

Just to add to the confusion of the whole scene, Boak *et al.*, (1968) have recently produced strong evidence that under certain conditions of intense stimulation of the reticulo-endothelial system, the macrophages (Kupffer cells) of the liver can be derived from almost pure populations of small lymphocytes obtained from the thoracic duct.

As already remarked, the problem of the origin of the lymphocytes is essentially one of vertebrate phylogeny and ontogeny. That both lymphocytes and macrophages are such motile cells means, of course, that lymphoid tissues could finally locate themselves almost anywhere where they found favourable conditions for their development, and that the distribution of lymphoid tissues may mean nothing more than that the tissue occurs in those particular situations in which it can do a useful job. Nevertheless the affinity between lymphoid tissue, nephridia, pronephroi, kidneys, adrenal cortex, and gonads does lend support to the idea that, in some way, lymphocytes are related to the ecballic cells of these tissues. It will be recalled, for example, that *Amphioxus* has functional nephridia on the gill arches, while higher vertebrates have a number of rather obscure tissues associated with these arches, e.g. tonsils, parathyroids, epithelioid bodies, carotid bodies, and particularly the thymus. The thymus, with its epithelial reticulum and the peculiar nature of its lymphocytes (thymocytes), seems to be a particularly strong candidate for consideration in this way. In its embryonic development the sac-like invaginations which become cords of epithelial cells strongly suggest an originally tubular structure (Bell, 1906). The cells of the parathyroid, though not lymphoid, could well be derived similarly from the ecballic nephridial cells which have specialized in a different direction, one that is more clearly related to the original nephridial function, namely assisting in maintaining the level of calcium in the blood. It is interesting that among the actions of the parathyroid hormone must be included that of increasing the activity of the osteoclasts. Parathyroid cells are presumably ecballic, osteoclasts are emballic; whether there is also a depressant action of parathormone on osteoblasts in relation to matrix-formation is not clear, but there is much to suggest it. It is interesting too that thyrocalcitonin, which opposes the action of the parathyroid, is secreted from the C cells in the thyroid gland cells that are argyrophilic and have many properties suggesting that they are emballic, including a highly vesiculated cytoplasm (Fig. 19.13) (Pearse, 1966). It is interesting, too, that

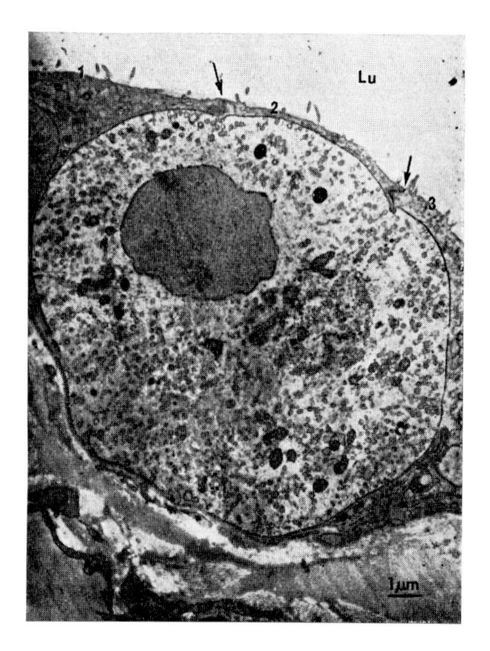

FIG. 18.13. A C-cell (thyrocalcitonin-producing?) in the thyroid gland. Note that it is separated from the lumen (Lu) by portions of A cells, which have microvilli and are held together by desmosomes (arrows). (Pearse, 1966.)

parathyroid hormone also affects the excretion of phosphate by the kidney, again perhaps indicating a different aspect of the originally nephridial origin and function of parathyroid cells. The thyroid C cells, like the rest of the thyroid (see p. 353), presumably originate from the cells of the pharyngeal floor whose actions would initially be more concerned with regulation of entry than with excretion of substances. The parathyroid cells could also be direct derivatives from the ciliated cells of this epithelium rather than of nephridia. In several species of mammal they have been observed to possess an occasional cilium (Munger and Roth, 1963; Stoeckel and Porte, 1966).

It has been mentioned several times that there are examples of nephridia opening into the alimentary canal instead of on to the surface of the body.

If this method of opening was used by the ancestors of the vertebrates, either embryological or functional traces of it might still be discernible in some of the living vertebrates. Thus possible remnants of ancient nephridia in structures associated with the gut in vertebrates might include the following, in addition to the tonsils, thymus, and parathyroids, already noted as opening on to the gill apertures, and in addition to the salivary glands, the pancreatic islet, and duct tissue already discussed (p. 524): the spleen, the bursa of Fabricius in birds, and the rectal salt glands of elasmobranchs. The spleen is perhaps the organ of most interest in this connexion, because its origin is completely mysterious. It occurs as a definite structure in close proximity to the intestine, generally in the neighbourhood of the pancreas or stomach, in all vertebrates from the fish upwards. Yet it is represented in myxinoids only by some nodules of well vascularized lymphoid tissue associated with some large fat-filled cells in the sub-mucosa of the intestine (Mawas, 1922). It seems possible that these nodules, like the similar pancreatic follicles, could be vestiges of earlier nephridia. In the case of the spleen, this would bring the white pulp into line with other lymphoid tissue as originating from nephridial tissues. These nodules may be relatively insignificant in the myxinoids because of the peculiar lymphoid and erythroid activities of the pronephros in these animals. Lampreys have a definite spleen (Jordan and Speidel, 1930).

These ideas, however, do not indicate at all how the bone marrow, which is obviously secondary and unlikely to have anything to do with nephridia, has come to take over the dominating position which the experiments outlined above indicate that it now has with regard to erythroid and myeloid development. It probably also has a dominant role with regard to the lymphoid tissue in the adult, for, in lethally irradiated mice, cells from the bone marrow, and incidentally a very small fraction of them indeed, can give rise to colonies in the spleen; these colonies may be lymphoid in the white pulp and either myeloid or erythroid in the red pulp. Such colonies are in all probability, clones initiated from single cells. Thus two questions must be asked. First, is there one stem cell in the marrow which can develop in any one of these three ways according to where it comes to lodge in the spleen and where local condition then determine further development? Second, are there three different stem cells, one for each type of development, and does each "know where to go and what to do"? These are the two extreme positions, and any combination of them is possible. However, the data at present available are simply not adequate to make further discussion profitable, and the riddle of the bone marrow has, at present, something in common with the riddle of the sphinx.

Bones and Teeth

As an appropriate addenum to this chapter, a short discussion of bones and teeth is not out of place.

A conspicuous feature of the alimentary canal of the vertebrates, and one which has been widely used in the more orthodox studies of the evolution of vertebrates and particularly of mammals, is the development of teeth. These structures are interesting not only because of their permanence and value in the fossil record and because of the diagnostic importance of their form and arrangement, but also in relation to the evolution of their cytology.

In the most primitive living vertebrates the teeth are purely ectodermal structures of keratin (Fig. 18.14). Whether or not these "teeth" are directly

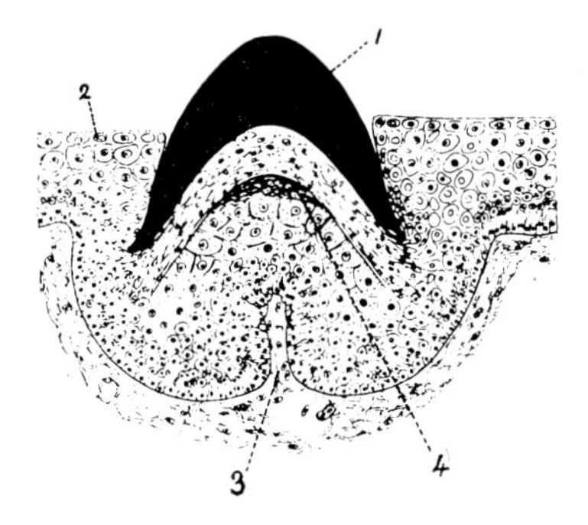

FIG. 18.14. Section of a developing tooth of *Petromyzon marinus* L. 1, Functional epidermal tooth; 2, epidermis; 3, dermal papilla; 4, successional tooth beginning to cornify. (Goodrich, 1909.)

connected with the more widespread composite teeth, which have both ectodermal and mesenchymal components, is uncertain. However, these early horny teeth are interesting in that, almost for the first time, they exploit the properties of keratin, the resistant scleroprotein specially associated with epitheliocytes, and the protein which becomes of cardinal importance in the integument of all the higher vertebrates, and, in particular, of those that have left the water and adopted a terrestrial or aerial habitat. Keratin is, of course, also of great importance in the enamel of the composite tooth of higher animals, and the modifications of epithelial behaviour which lead to the development of enamel are very relevant to the general theme of cytological evolution.

It seems probable that teeth, as such, are modifications of a more general and initially protective coating of denticles or scales, or both, which preceded the main development of the bony, protective, and supporting axial skeleton. The formation and evolution of denticles is thus of cardinal importance to the physiology of teeth and that of bone. Most of the scales and denticles of present-day fishes are dual structures formed by the co-operation of mesenchymal elements with those of the superficial epidermis. Just as the teeth of cyclostomes are purely epithelial, however, so there are denticles and dermal scales that are mainly, if not entirely, mesenchymal structures, and in this way they resemble dentine and bone (Fig. 18.15).

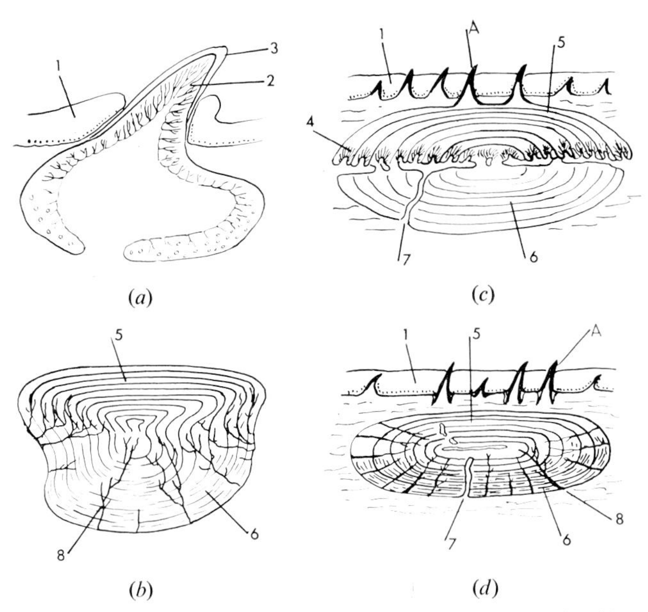

Fig. 18.15. Denticles and scales. (*a*) Denticle; (*b*) acanthodian scale; (*c*) palaeoniscoid scale; (*d*) lepidosteoid scale. 1, Epidermis; 2, dentine; 3, enamel; 4, cosmine; 5, ganoine; 6, bone; 7, vascular canals; 8, canaliculi. (After Goodrich, 1909.)

In many elasmobranchs a succession of denticles is formed just within the mouth cavity. As each denticle develops, it moves up on to the jaw and then outwards and finally moves away from the mouth to take its place in the general scaly covering of the skin (Fig. 18.16). As the denticles pass over the summit of the jaw, they function as teeth for the capture and holding of prey. On the mouth side of the functional teeth are those that are developing to take their place, and on the external side the worn-down denticles move away in a continuous slow stream. It seems probable therefore that this continual and apparently unlimited succession, or another very like it (several types are found in fishes), is at the root of the much more limited dental succession as it is seen in mammals and man. The dental lamina of man seems to resemble, in a more restricted way, the corresponding epithelial groove in elasmobranchs.

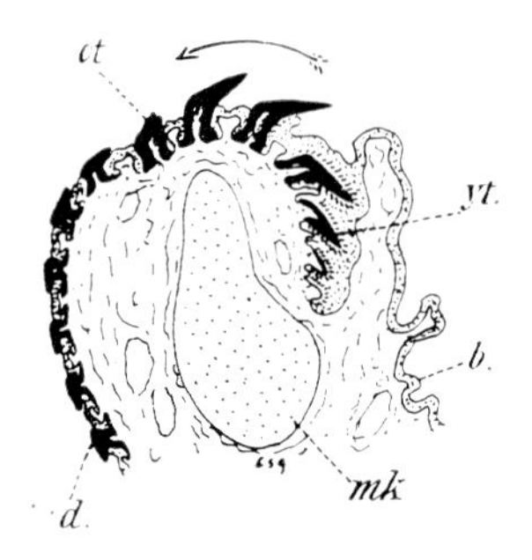

FIG. 18.16. Diagram showing the succession of teeth in an elasmobranch. b, Mucous membrane of mouth; d, denticle in skin of outer surface; mk, Meckel's cartilage; ot, old teeth; yt, young teeth. (Goodrich, 1909.)

When such a composite denticle or tooth starts to develop, two things happen almost simultaneously, and it is not yet certain which one is the leader. The epithelium locally thickens, and the underlying mesenchyme "condenses". In other words, each in their own way, the epithelial cells and the mesenchymal cells, cluster together, and each group influences the other. The epithelium begins to form the enamel organ (Fig. 18.17), a curious tissue consisting of ameloblasts and stellate reticulum, while the mesenchyme produces the odontoblasts and the dental papilla (see Fig. 2.24).

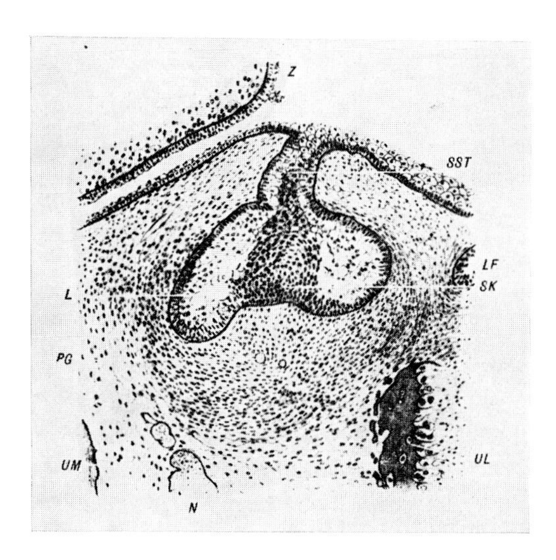

FIG. 18.17. Early stage in tooth formation. L, Beginning of stellate reticulum; LF, lingual end of lip furrow; N, nerve; PG, blood vessels in papilla; SK, enamel knot; SST, enamel cord; UL, UM, bone; Z, tongue. (Maximow and Bloom, 1930.)

The causes and consequences of this clustering of cells have not been adequately analysed at the cellular level in spite of many investigations (Glasstone, 1965; Dryburgh, 1966). Similar events occur also in the production of the limb buds and of feather germs in birds. In the latter, cell division is probably not responsible for the initial clustering but occurs freely later (Wessells, 1964). Whether this applies in other situations remains to be seen. We need to know the answers to such questions as the following. Are all the cells in a condensation of the same type? Why do they aggregate? Is contact inhibition increased and, if so, how? Pycnotic and dying cells are sometimes observed, but which cells are thus dying? When the cells aggregate, what effects does this have on the local environments of the cells? For example, when pre-cartilage cells aggregate, as in a limb-bone rudiment, the aggregation is quickly followed by the production of chondroitin sulphate and the like. Is this related to the effects of cell crowding on the abundance or scarcity of some necessary nutrient or metabolite? These and many similar questions will have to be answered before the glib phrase "condensation of mesenchyme" can have much physiological meaning.

Just as in other situations (e.g. the interaction of epidermis and gastric mesenchyme; McLoughlin, 1961), where the character of the mesenchyme determines the behaviour of epithelial cells, whatever goes on in the mesenchyme of the tooth germ certainly affects what goes on in the enamel organ, where events of considerable interest take place. The epithelial cells of the dental lamina at first form a thickened stratified layer in which the basal cells are polarized in the usual way with their apex towards the cavity of the mouth. Where the basal cells come into contact with the condensation ofmesenchyme cells, they become changed. First they become much more columnar than the cells of the normal basal layer; second, it seems probable that they reverse their polarity, for the Golgi apparatus moves to the side of the nucleus nearer the basement membrane (Fig. 18.18) (Beams and King,

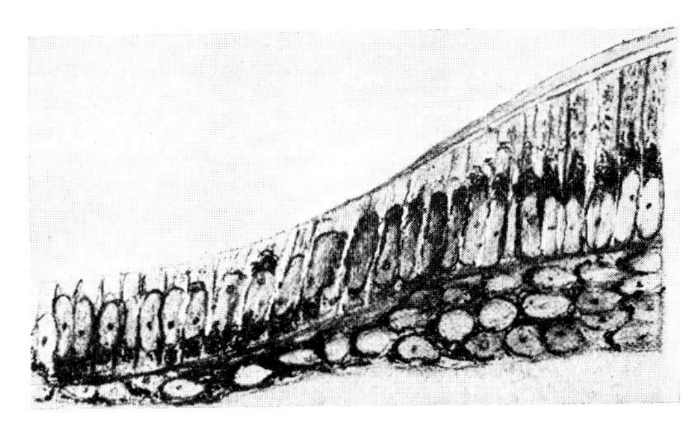

FIG. 18.18. Reversal of polarity of the basal layer of epidermal cells as they assume the character of ameloblasts over the dental papilla (right). (Beams and King, 1933.)

1933). The reason for this change of polarity is obscure. The mesenchymal condensation probably has its part to play, but how it does this is not known. As cause or effect of this change of polarity, probably the latter, the cells in the epithelium itself, i.e. those between the basal layer and the epithelial surface, also change character in a significant way. Those immediately adjacent to the basal layer become flattened and extended in a plane parallel to the basement membrane and acquire a curiously fibroblast-like appearance for epithelial cells. Outside this flattened layer, the cells form a loose network, with branching processes connected by desmosomes and containing tonofibrils (Pannese, 1960), the spaces of which are filled with a mucoprotein-containing fluid. Microvilli project from the cell surfaces. These cells of the "stellate reticulum" are quite unlike any other epithelial cells in the vertebrate body, and they superficially resemble mesenchyme cells in appearance. They do not, however, appear to form collagen.

If the cells in the basal layer of the enamel organ over the dermal papilla do, in fact, change their polarity in a physiological sense, the other cells in the epithelium of the enamel organ (i.e. those corresponding to the prickle cells) must be subjected to those products and by-products of the basal cells which (other things being equal) would normally be directed towards the mesenchyme and *vice versa*, substances must now be extracted by the ameloblasts from the other epithelial cells which would normally be extracted from the mesenchyme cells. Thus it seems possible that the changes in morphology of the epithelial cells in the deeper layers of the enamel organ that produce the stellate cells may be direct consequences of the primary change of polarity of the ameloblasts. In addition, because the deeper cells are now subjected to an environment much more like that of the mesenchyme cell than that of the normal epithelial cell, they change their character in the direction of the latter, When the enamel organ becomes more mature, these stellate epithelial cells are much less numerous and have shorter processes. Then they lie immediately adjacent to the ameloblasts, with which, it would be reasonable to suppose, they form some sort of partnership. At this stage, blood vessels press inwards between the groups or papillae of stellate cells, and the boundary between epithelium and mesenchyme become irregular and difficult to define, though a thin basement membrane does actually separate the two (Kallenbach, 1966). In cases of vitamin A deficiency or magnesium deficiency the epithelial component becomes clearly separable and loses its characteristic organization (Irving, 1940) (Fig. 18.19) It would be interesting to know more about the nature of these changes.

Meanwhile, in the normal organ the ameloblasts, instead of combining with the mesenchyme to produce a normal basement membrane, react quite differently to the dermal papilla. Each builds a prism of keratin (Fig. 18.20) on that part of its surface membrane which is in contact with the mesenchyme

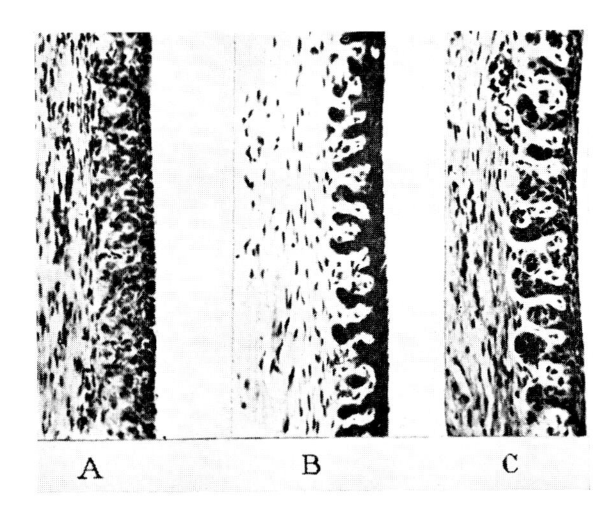

FIG. 18.19. The enamel organ in the rat. (*A*) The normal organ. (*B*) Deficient in Mg^{2+}. (*C*) Deficient in vitamin A. (Irving, 1940.)

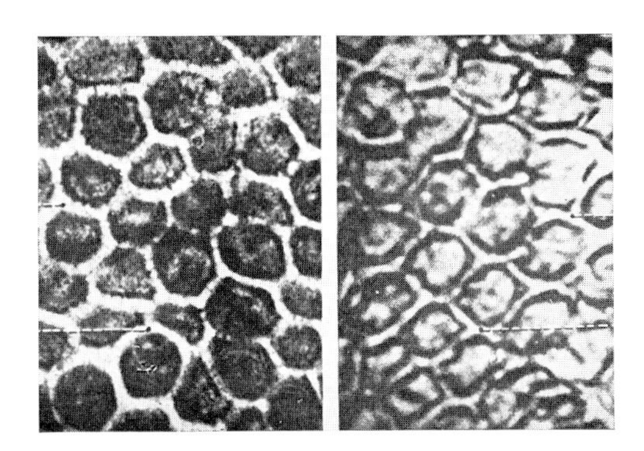

FIG. 18.20. Transverse sections of enamel prisms, with interprismatic substance. (From Ham, 1961).

and, with the help of an alkaline phosphatase, impregnates the prism with apatite-like crystals whose long axes are orientated parallel to the long axis of the prism. The prisms are held together by a continuous keratinous matrix which is also calcified and in which the crystals are at right angles to those in the prisms (Fig. 18.21) (Glimcher *et al.*, 1965). This too is a peculiar type of behaviour for epithelial cells, and it has few, if any, parallels elsewhere. If the epithelial cells occupied fixed positions relative to each other, the prisms would be regular and straight and would exactly follow the spacing of the underlying cells, A sort of micro-Giant's Causeway would

FIG. 18.21. Orientation of the apatite crystals parallel to the long axis of the enamel prisms, and in other directions in the interprismatic spaces. (Glimcher *et al.*, 1965.)

be formed. Sections of enamel, however, show that the prisms are not usually straight but are woven together in a gently curving manner (Fig. 18.22). This, though possibly inevitable since the tooth is a growing system, is probably of mechanical and functional importance because it must reduce the risks of cleavage occurring along the lines of junction of the crystals. It is also of interest in showing that, just as in tissue cultures, cells are seldom really static; even in a closely packed epithelium, they may be constantly moving round each other. In this case, since the moving cells continue to secrete the enamel from their now apical (but originally basal) surfaces, the enamel prisms are woven together as the cells slowly withdraw. In contrast to the cells of the enamel organ, however, the epidermal epitheliocytes of an insect, when producing the cuticle, apparently remain more nearly stationary

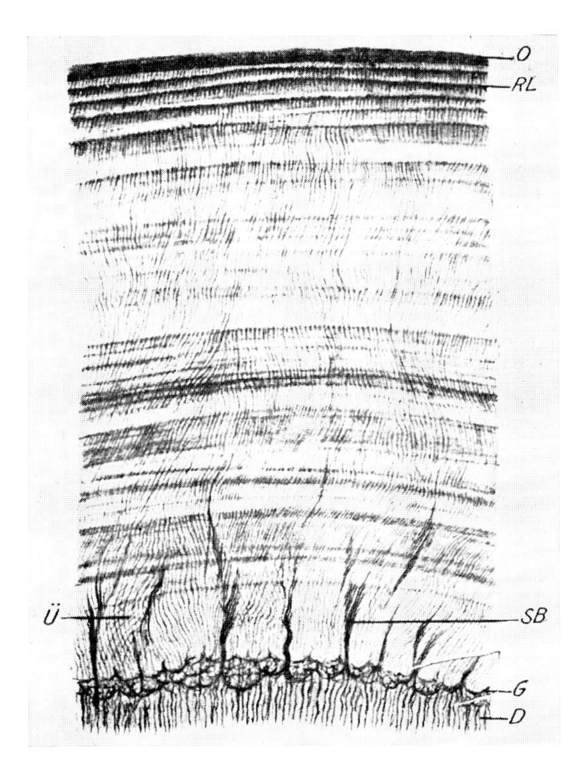

FIG. 18.22. Ground section through the crown of a human cuspid tooth. D, Dentine; G, wavy boundary between dentine and enamel; O, surface of tooth; RL, parallel stripes of Retzius indicating periods of altered calcification; SB, enamel tuft; Ü, crossing of enamel prisms. (Maximow and Bloom, 1930.)

with respect to each other. Thus each situation must be studied separately. A third interesting feature of the enamel organ is that the actual ameloblasts are apparently all identical cells, i.e. unlike most epithelia, the layer of ameloblasts is not composed of more than one type of cell. It must be remembered, however, that the ameloblasts are directly related to and in communication with the cells of the fibrous and stellate layers of the enamel organ, for these cells were initially epithelial cells of the dental lamina. The ameloblasts, rich in alkaline phosphatase, are probably ecballic cells. It may be that, because only one class of cell is present in the actual enamel-forming layer and the activity of the ameloblasts is therefore not immediately opposed by other cells, the enamel goes on being deposited.

The last comment may apply also to the specialized mesenchyme cells which originally lie along the inner surface of the enamel organ, but which gradually become separated from it by the intervention of the enamel prisms and by the opposing lamina of dentine that the layer of mesenchymal odontoblasts themselves produce. Again, this layer of odontoblasts is peculiar. First, these mechanocyte-like cells, though lining a surface, do not flatten themselves against it, as in an endothelium, but, more in the manner of osteoblasts and synovioblasts, they arrange themselves in a much more columnar pattern (see Fig. 2.24). Second, the cells in the odontoblast layer are again apparently all the same, i.e. modified mechanocytes that are producing dentine, which is a collagenous matrix, calcified with the assistance of alkaline phosphatase. They, like the ameloblasts, are all orientated and polarized in the same direction, but in a direction opposite to that of the ameloblasts. The special relationship between the epitheliocytic ameloblasts and the mechanocytic odontoblasts is clearly the key to the situation. Normally the junction between epitheliocytes and mechanocytes leads to the production of a basement membrane of some sort, but here enamel and dentine result. Both types of cell are probably ecballic, at least at their adjacent surfaces, but there is still much that is mysterious in their relationship. The collagen fibres of the dentine (and they are true collagen) are all laid down at one pole of the cells, the pole nearest to the enamel, and not all around the cell as in fibrocytes and osteocytes. Perhaps the word "pole" is slightly misleading, because that end of the cell is not simply a single surface, like that of the ameloblast. It sends out long cytoplasmic processes between the collagen fibres that are being formed, and these processes continue to penetrate the dentinal mass even when it has thickened, solidified, and become impregnated, as in enamel and bone, with apatite crystals. Whereas enamel is a completely "dead" tissue and the amelobasts retreat from the product that they have laid down, dentine, like bone, remains penetrated by canals containing cytoplasmic processes (Fig. 18.23) and is often capable of further modification, renewal, and revitalization. Moreover, it seems very probable that the fine processes of the odontoblasts are the agents responsible for the extreme tactile, thermal, and osmotic sensitivity of the enamel-dentine junction. In other words, these mechanocytes not only produce the dentine but, like some of their earlier ancestors (see p. 244), are capable of acting as sensory nerve endings and fibres. Again, since the odontoblasts are all apparently identical, dentine formation goes on unchecked, rather in the same way that the vesicles formed by the proximal tubules in tissue cultures of kidney cortex (see p. 198) continue to fill with fluid, apparently unchecked.

It is well known that teeth are amongst the hardest and strongest of organic products. They are indeed very "strong solids". It is thus somewhat humiliating to human pride to realize that the most recent metallurgical researches have only just become aware that, if crystalline solids, which, though very hard in themselves, may be brittle, are impregnated with criss-crossing fibres of great tensile strength they acquire the properties of very "strong solids" indeed. Yet enamel, dentine, and bone, which are all variations of this essential structure, are well recognized as the hardest constituents of man's body and indeed, because of their strength and durability, have been used for making his tools since time immemorial.

These are some aspects of tooth formation which are brought into special prominence by the sort of cytological approach that attempts to follow the evolution of cell function. As such, they can be profitably compared and contrasted with corresponding events in bone formation.

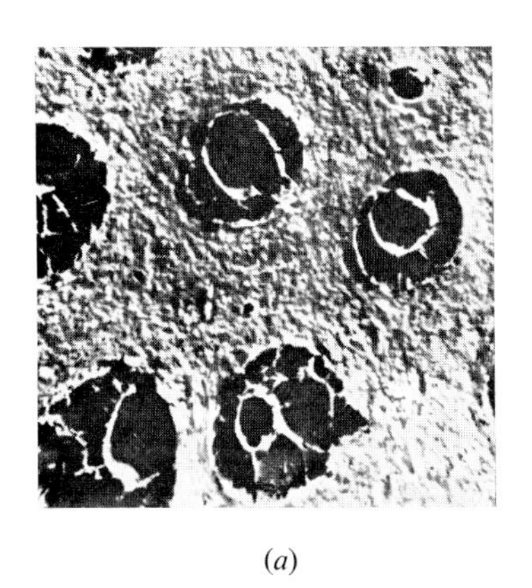

(*a*)

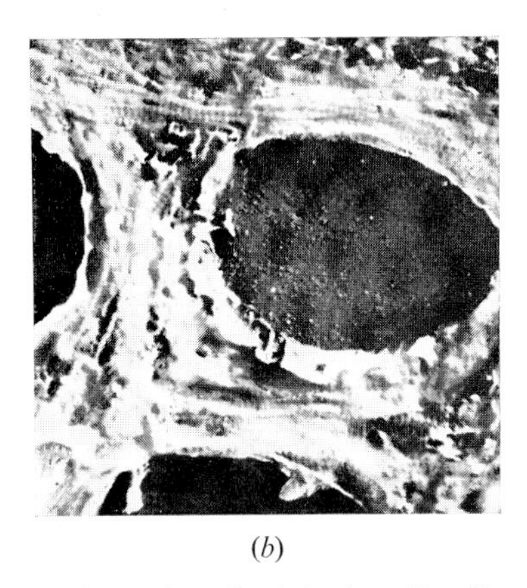

(*b*)

Fig. 18.23. (*a*) Thin-walled odontoblastic processes in demineralized dentine; (*b*) collagen fibrils in demineralized dentine. (Scott, 1954.)

Bone, which as a tissue is characterized by the presence of alternating laminae of orientated collagen fibres that are then systematically calcified, probably originated as scales in the dermis. In such scales, collagenous fibres are laid down in laminae and then, under the influence of alkaline phosphatase and some rather special local environmental conditions, become impregnated with apatite in a regular and semicrystalline manner (see Fig. 18.15) It is an interesting phenomenon that collagen fibres in the dermis, particularly in the immediate neighbourhood of the epidermis, i.e. in the basement lamella, tend to be arranged in layers and that in each layer the fibres run at right angles to those in the adjacent layer (Fig. 18,24) (Weiss and Ferris, 1956). If the hypothesis concerning the origin of vertebrates from nemerteoids or the like is substantiated, this arrangement could be a relic of the geodetic arrangements of collagenous fibres in the relatively cylindrical worms, where it was necessary to permit considerable changes of shape without corresponding changes in the volume of the animal as a whole (Fig. 18.25). It is a striking phenomenon that, when a wound in amphibian skin begins to heal, the first collagen fibres are laid down in a more or less random

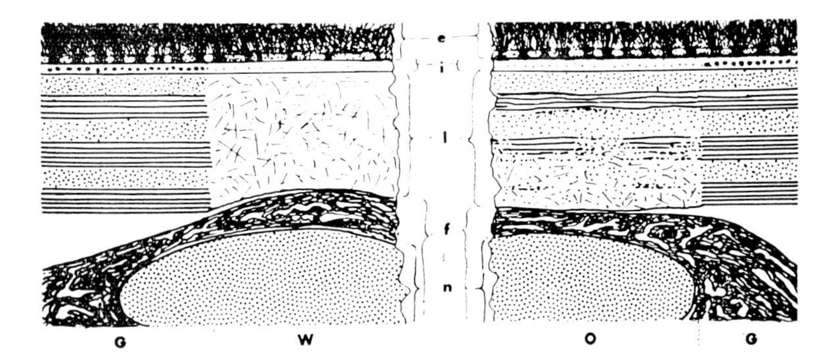

FIG. 18.24. Two stages in the repair of a wound in amphibian skin. G, Normal arrangement of fibres in undamaged membrane. W, The collagen fibrils in the wound being laid down with random orientation, in contrast to the laminated pattern in the normal skin. O, At a later stage, the alternating lamination proceeding from the epidermis downwards, replacing the random pattern. e, Base of epidermal cells; i, "cement" film; l, basement membrane; f, fibroblast; n, nucleus of fibroblast. (After Weiss, 1956.)

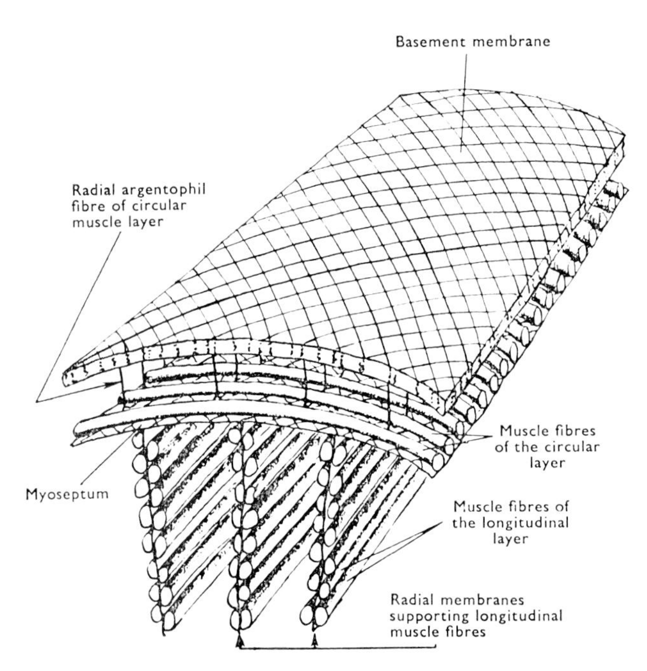

FIG. 18.25. Orientation of collagen fibres and muscle fibres in the body wall of a nemertine (*Amphiporus lactifloreus*). (Cowey, 1952.)

fashion, but within a day or two they have begun to be arranged in criss-crossing layers (Fig. 18.24). The mechanism of this transformation is a mystery. At first, one might suspect that the cell processes of the fibroblasts cause a continuous turn-over of the collagen fibres and that the cells have become adapted, perhaps by local conditions of tension, etc., to organizing the fibres in this special way to meet the requirements of the animal as a whole. Indeed, this may well be so, but there is another situation wherein a similar event is enacted but in which the actors are quite different. The cuticle of earthworms, which lies on the outside of the epidermis, is peculiar in that it is formed of a sort of collagen, which, though chemically not quite identical with most mesenchymal collagens because it has an exceptionally high hydroxyproline content (Mathews, 1967), is laid down in a similar arrangement of alternating layers of fibres running at right angles to each other. It is true that the cuticle needs to have properties, in relation to the size and movements of the animal, similar to those that the basement membrane of the epidermis of other worms must have, but it is of quite different origin. Nevertheless, this cuticle provides another interesting example of epitheliocytes actually forming a collagen, thus fulfilling the suggestion made earlier (see p. 215) that collagen precursors are among the by-products of ecballic epitheliocytes which could potentially be turned to advantage in various ways (e.g. for nematocysts, etc.). In this case the collagen fibrils actually lie among the microvilli of the epidermal cells. (Coggeshall, 1966) (Fig. 18.26). This organization and layering of the collagen fibres is characteristic of dermal connective tissue, of dentine, cosmine (the substance of cosmoid fish-scales) and bone. In dentine the fibres typically run in the direction of the long axis of the tooth, but they are arranged in laminae which are radial near the odontoblasts and curve away progressively so that they become tangential to the dentino-enamel junction. This arrangement is best seen in such regular cylindrical teeth as the tusks of elephants and carnivores (Fig. 18.27). In ganoine and the bony scales in the dermis, the laminae are close-packed and parallel to the surface; later they may become organized in relation to blood vessels, as in the Haversian system of bone. An examination of the scales in various fish suggests that, if the "ossification" process in the dermis comes under the direct influence of the epidermis, then dentine and tooth-like structures can be produced at the junctional surface; if it occurs in the deeper parts of the dermis, a more bone-like structure results. Intermediately, a superficial dentinal structure can readily attach itself to and fuse with more deeply situated and true osseous tissues. This can happen in the production of various types of scale in fish (see Fig. 18.15) or, on a somewhat more sophisticated basis, between mammalian teeth, which are essentially modified denticles, and their bony sockets. In the last example the junction is not a solid fusion of the matrices; it is made by a zone of

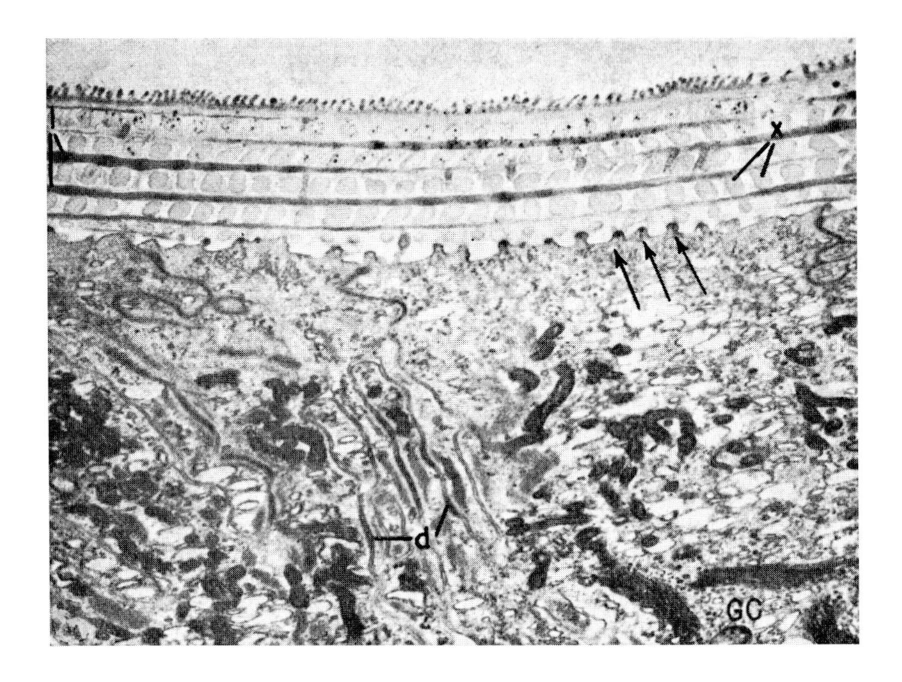

FIG. 18.26. Orientation of collagen fibrils in the cuticle of the earthworm. Arrows show half-desmosome-like structures joining the cuticle to the surface of the epithelial cells. Microvilli can be seen in places between the collagen fibres. d, Desmosomes between adjacent cells; GC, Golgi complex; l, longtitudinal fibres in cuticle; x, transverse fibres in cuticle. (Coggeshall, 1966.)

non-calcifying collagen fibres running at right angles (in the third plane of space) to the fibres in both the dentine and the bone.

Whereas the pulp of denticles, cosmoid scales, and teeth is well vascularized, dentine itself is entirely avascular. Similarly, bone has an enormously rich blood supply, but the laminae themselves, though they are penetrated by innumerable fine canaliculi, containing the processes of the osteoblasts, tend not to be penetrated by blood vessels as such but to be arranged around them. The osteoblast has less regular processes than the odontoblast and, unlike the latter, becomes entirely enclosed within the matrix that it produces, except for the fine canaliculi. Dentine tends to stay where it is put, though in some animals it is capable of being continuously augmented; bone, on the other hand, can be here today and gone tomorrow. It is always in a state of flux, adjusting its position and shape, in some ill-understood manner, to the mechanical stresses of the moment. The initial pattern of a developing limb-bone seems to be determined by factors within the limb rudiment, but, once that basic pattern is laid down, external influences begin to exert

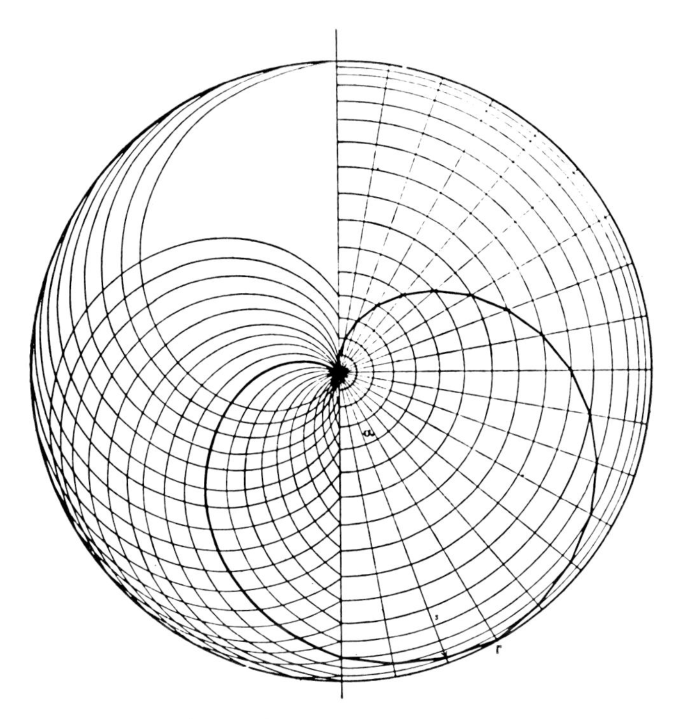

Fig. 18.27. Orientation of fibre laminae in a tusk. Idealized transverse section through the tusk. The lines on the left indicate the curving of layers or sheets of fibres. The fibres themselves run longitudinally and therefore would be cut transversely in such a section. The geometry of the design is indicated in the right half of the diagram. (From Picken, 1960; after Gebhardt.)

some control. In this connexion it is interesting that among the osteoblasts (which are mechanocytes) there are always osteoclasts (which are amoebocytes). Therefore the constantly adjusting bone is probably the result of balanced activities of these two groups of cells. It is, however, probably too naïve a view to hold osteoblasts responsible for bone formation, and osteoclasts for bone destruction. Osteoblasts or, more correctly, osteocytes can move about within the bone matrix, e.g. they move to the outside of the explant when pieces of solid bone are cultured *in vitro*. Presumably they can move in this way only by a process of removal of bone on one side and its deposit on the other, much in the way that cartilage cells must be able to move within their matrix when small-celled cartilage converts progressively into hyaline and hypertrophic cartilage. Again, it seems to be a reasonable suggestion that unbalanced osteoblasts tend to lay down bone excessively, but the presence of osteoclasts can balance this action or even reverse it, so that, in the extreme, no matrix results.

It should therefore be considered that osteoblasts and osteoclasts act symbiotically, as other mechanocytes and amoebocytes do elsewhere, and that whether or not bone substance appears depends on the degree of imbalance. The presence of the bone marrow, with a high proportion of amoebocytes in its reticulum, and of the periosteum, with a high proportion of mechanocytes (osteoblasts and fibroblasts), perhaps lends support to this idea. The hollowing-out of bones to house the marrow is a physiological requirement as functionally important as the growth of bones by their increase in girth. Both may depend on the balanced activities of mechanocytes and amoebocytes.

This does not, of course, deny the fact that osteoclasts, as good amoebocytes, do under certain conditions (e.g. after treatment of a bone culture with parathyroid hormone) actively phagocytose bone (Gaillard, 1955a, b). Bone matrix can, however, also come and go without the apparent intervention of visible osteoclasts (Fell and Mellanby, 1952), presumably because of local conditions affecting the activity of the osteoblasts. The word "apparent" is used advisedly because the multinucleate cell called the osteoclast is formed by the fusion of mononucleate cells (Hancox, 1965), which could easily escape recognition unless they were being specifically sought. Nevertheless, there is reason to suppose that they mainly function as phagocytic agents in the multinucleate state. The observations on the effects of vitamin A on cartilage and bone matrices (Fell and Mellanby, 1952; Lucy *et al.*, 1961; Dingle and Lucy, 1965) suggest that the liberation of enzymes from lysosomes in the bone and cartilage cells is an important factor in the dissolution of the matrix.

From their original position in the dermis, bones gradually invade and replace the softer cartilaginous skeleton which forms in the connective tissues of vertebrates, especially in relation to the development of the tubular nervous system and all the changes that that brought with it. Cartilage itself, unlike bone, is not a new development in the vertebrates. Similar substances appear in the porifera, polychaetes, and arachnids, and material almost identical with that in vertebrates is present in the molluscs, as, for example, round the eyes of cephalopods (Fig. 18.28) (Person and Mathews, 1967; Person and Philpott, 1963). In nemertines there are also cells lying in an amorphous matrix which could be the precursors of the great development of cartilage in some of the lower vertebrates. Although the differences between bone and cartilage are clear, osteoblasts and chondroblasts are probably closely related mechanocytes. It is even possible that they are interchangeable, i.e. one may be a modulation of the other; both are modulations of the mechanocyte (Fell, 1933) (see p. 49). Nevertheless, it is not known what determines whether a mechanocyte differentiates to produce chondroitin sulphate in a big way and collagen in a small way, or to produce collagen in

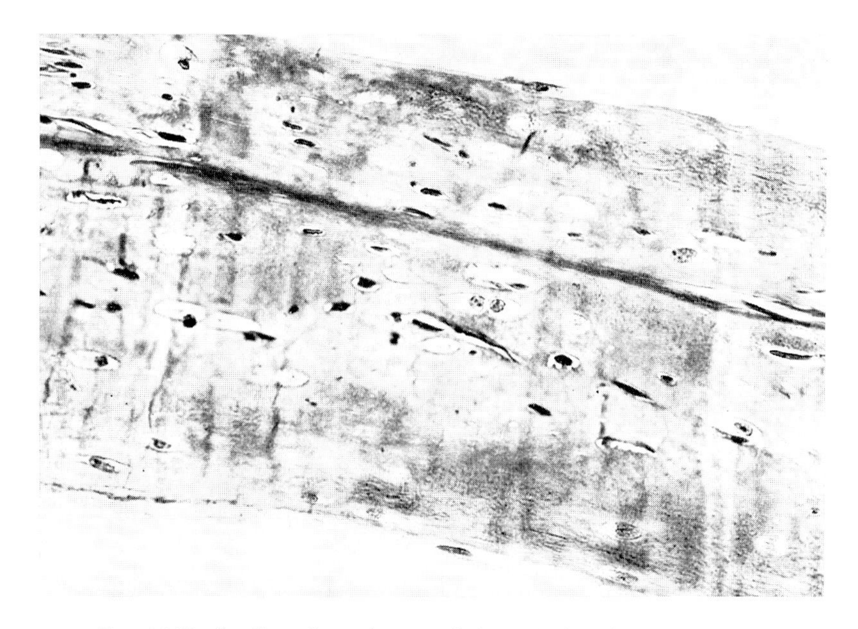

FIG. 18.28. Cartilage from the eye of the cephalopod mollusc, *Sepia*.

a big way and chondroitin-like substances in minimal amount. This question would seem to lead us back again to the beginning of our story, and to reopen the question of the factors involved in the differentiation of cells and how they can be investigated by the methods of tissue culture.

REFERENCES

Abdel, B. W., and Sorenson, G. D. (1965). Ciliated Cells in the spleen of adult rats. *Anat. Record.* **152**, 481.

Adam, H. (1963). Structure and histochemistry of the alimentary canal. *In* "The Biology of Myxine" (A. Brodal and R. Fänge, eds.). p. 256. Universitetsforlaget, Oslo.

Auerbach, R. (1960). Experimental analysis of the origin of cell types in the development of the mouse thymus. *Develop. Biol.* **3**, 336.

Bahl, K. N. (1947). Excretion in the Oligochaeta. *Biol. Rev. Cambridge Phil. Soc.* **32**, 109.

Bang, F. B., and Warwick, A. C. (1965). Liver. *In* "Cells and Tissues in Culture" (E. N. Willmer, ed.), Vol. 2, p. 620. Academic Press, New York.

Barrington, E. J. W. (1936). Proteolytic digestion and the problem of the pancreas in the ammocoete larva of *Lampetra planeri*. *Proc. Roy. Soc.* (*London*) **B121**, 221.

Barrington, E. J. W. (1940). Observations on feeding and digestion in *Glossobalanus*. *Quart. J. Microscop. Sci.* **82**, 227.

Barrington, E. J. W. (1942a). Blood sugar and the follicles of Langerhans in the ammocoete larva. *J. Exptl. Biol.* **19**, 45.

Barrington, E. J. W. (1942b). Gastric digestion in the lower vertebrates. *Biol. Rev. Cambridge Phil. Soc.* **17**, 1.

Barrington, E. J. W. (1945). The supposed pancreatic organs of *Petromyzon fluviatilis* and *Myxine glutinosa*. *Quart. J. Microscop. Sci.* **85**, 391.

Beams, H. W., and King, R. L. (1933). The Golgi apparatus in the developing tooth with special reference to polarity. *Anat. Record* **57**, 29.

Bell, E. T. (1906). The development of the thymus. *Am. J. Anat.* **5**, 29.

Biggers, J. D. (1966). Cartilage and bone. *In* "Cells and Tissues in Culture" (E. N. Willmer, ed.), Vol. 2, p. 197. Academic Press, New York.

Boak, J. L., Christie, G. H., Ford, W. L., and Howard, J. G. (1968). Pathways in the development of liver macrophages: alternative precursors contained in populations of lymphocytes and bone-marrow cells. *Proc. Roy. Soc.* (*London*) **B169,** 307.

Clarke, R. (1967). On the constancy of the number of villi in the duodenum of the post-embryonic domestic fowl. *J. Embryol. Exptl. Morphol.* **17**, 131.

Coe, W. R. (1906). A peculiar type of nephridia in nemerteans. *Biol. Bull.* **11**, 47.

Coggeshall, R. E. (1966). A fine-structural analysis of the epidermis of the earthworm, *Lumbricus terrestris* L. *J. Cell Biol.* **28**, 95.

Conel, J. Le R. (1917). The urogenital system of myxinoids. *J. Morphol.* **29,** 75.

Cowey, J. B. (1952). The structure and function of the basement membrane muscle system in *Amphiporus lacteus* (Nemertea). *Quart. J. Microscop. Sci.* **93**, 1.

de Bruyn, P. P. H. (1946). The amoeboid movement of the mammalian leucocytes in tissue culture. *Anat. Record* **95**, 177.

Dingle, J. T., and Lucy, J. A. (1965). Vitamin A, carotenoids and cell function. *Biol. Rev. Cambridge Phil. Soc.* **40**, 422.

Dougherty, T. F., and White, A. (1945). Functional alterations in lymphoid tissue induced by adrenal cortical secretion. *Am. J. Anat.* **77**, 81.

Dryburgh, L. (1966). Personal communication.

Ehrich, W. E. (1946). The role of the lymphocyte in the circulation of lymph. *Ann. N.Y. Acad. Sci.* **46**, 823.

Ephrussi, B., and Lacassagne, A. (1933). Essais de culture comparative de tissus hépatique et rénal de lapins embryonnaires, nouveau-nés, adultes et vieux. *Compt. Rend. Soc. Biol.* **113**, 976.

Fawcett, D. W. (1964). Local specialization of the plasmalemma in micropinocytosis vesicles of erythroblasts. *Anat. Record* **148**, 369.

Fawcett, D. W. (1965). Surface specializations of absorbing cells. *J. Histochem. Cytochem.* **13**, 75.

Feldman, M., Bleiberg, J., and Liron, M. (1967). The regulation of intrasplenic formation of erythroid clones. *Ann. N.Y. Acad. Sci.* **129**, 864.

Fell, H. B. (1933). Chondrogenesis in cultures of endosteum. *Proc. Roy. Soc.* (*London*) **B112**, 417

Fell, H. B., and Mellanby, E. (1952). The effect of hypervitaminosis A on embryonic limb-bones cultivated *in vitro*. *J. Physiol.* (*London*) **116**, 320.

Fischer-Piette, E. (1931). Culture de tissus de crustacés. La glande lymphatique du Homard. *Arch. Zool. Exptl. Gen.* **74**, 33.

Ford, W. L., Gowans, J. L., and McCullagh, P. J. (1966). The origin and function of lymphocytes. *Ciba Found. Symp. Thymus*: *Exptl. Clin. Studies*, p. 58.

Gaillard, P. J. (1955a). Parathyroid gland tissue and bone *in vitro*. I. *Exptl. Cell Res. Suppl.* **3**, 154.

Gaillard, P. J. (1955b). Parathyroid gland tissue and bone *in vitro*. II. *Koninkl. Ned. Akad. Wetenschap., Proc.* **C58**, 279.

Garvey, J. S. (1961). Separation and *in vitro* culture of cells from liver tissue. *Nature* **191**, 972.

Gérard, P. (1933). Sur le système athrophagocytaire chez l'Ammocoete de la *Lampetra planeri* (Block). *Arch. Biol.* (*Liège*) **44**, 327.
Glasstone, S. (1965). The development of tooth germs in tissue culture. *In* "Cells and Tissues in Culture" (E. N. Willmer, ed.), Vol. 2, p. 273. Academic Press, New York.
Glimcher, M. J., Daniel, E. J., Travis, D. F., and Kamhi, S. (1965). Electron optical and X-ray diffraction studies of the organisation of the organic crystals in embryonic bovine enamel. *J. Ultrastruct. Res. Suppl.* **7**, 14.
Good, R. A., Gabrielsen, A. E., Peterson, R. D. A., Finstad, J., and Cooper, M. D. (1966). The development of the central and peripheral lymphoid tissue. Ontogenetic and phylogenetic considerations. *Ciba Found. Symp.Thymus*: *Exptl. Clin. Studies*, p. 181.
Goodrich, E. S. (1903). On the body cavities and nephridia of the Actinotrocha larva. *Quart. J. Microscop. Sci.* **47**, 103.
Goodrich, E. S. (1909). Cyclostomes and Fishes. *In* "Treatise on Zoology" (E. R. Lankester, ed.), Vol. 9. Black, London.
Goodrich, E. S. (1945). The study of nephridia and genital ducts since 1895. *Quart. J. Microscop. Sci.* **86**, 303.
Gowans, J. L., and Knight, E. J. (1964). The route of re-circulation of lymphocytes in the rat. *Proc. Roy. Soc.* (*London*) **B159**, 257.
Grossfeld, H. (1959). Action of adrenal cortical steroids on cultured cells. *Endocrinology* **65**, 776.
Ham, A. W. (1961). "Histology." Pitman, New York.
Hancox, N. M. (1965). The Osteoclast. *In* "Cells and Tissues in Culture" (E. N. Willmer, ed.), Vol. 2, p. 261. Academic Press, New York.
Harris, H. (1953). The movement of lymphocytes. *Brit. J. Exptl. Pathol.* **34**, 599.
Holmgren, N. (1950). On the pronephros and the blood in *Myxine glutinosa*. *Acta Zool.* (*Stockholm*) **31**, 234.
Hudson, G., and Yoffey, J. M. (1966). The passage of lymphocytes through the sinusoidal endothelium of guniea-pig bone marrow. *Proc. Roy. Soc.* (*London*) **B165**, 486.
Irving, J. T. (1940). The influence of diets low in magnesium upon the histological appearance of the incisor tooth of the rat. *J. Physiol.* (*London*) **99**, 8.
Jennings, J. B. (1962). A histochemical study of digestion and digestive enzymes in the rhynchocoelan, *Lineus ruber* (O. F. Müller). *Biol. Bull.* **122**, 63.
Jordan, H. E., and Speidel C. C. (1930). Blood formation in cyclostomes. *Am. J. Anat.* **46**, 355.
Junqueira, L. C. U. (1952). Phosphomonoesterase content and localization in the mesonephros and metanephros of the chick embryo. *Quart. J. Microscop. Sci.* **93**, 247.
Kallenbach, E. (1966). Electron microscopy of the papillary layer of rat incisor enamel organ during enamel maturation. *J. Ultrastruct. Res.* **14**, 518.
Kingsbury, B. F. (1945). Lymphatic tissue and regressive structure, with particular reference to degeneration of glands. *Am. J. Anat.* **77**, 159.
Lewis, W. H. (1931). Locomotion of lymphocytes. *Bull. Johns Hopkins Hosp.* **49**, 29.
Lewis, W. H. (1933). Locomotion of rat lymphocytes in tissue cultures. *Bull Johns Hopkins Hosp.* **53**, 147.
Little, J. M. (1964). Potassium imbalance and rectosigmoid neoplasia. *Lancet* **i**, 302.
Lucy, J. A., Dingle, J. T., and Fell, H. B. (1961). Studies on the mode of action of excess of vitamin A. 2. A possible role of intracellular proteases in the degradation of cartilage matrix. *Biochem. J.* **79**, 500.
McLoughlin, C. B. (1961). The importance of mesenchymal factors in the differentiation of chick epidermis. II. Modification of epidermal differentiation by contact with different types of mesenchyme. *J. Embryol. Exptl. Morphol.* **9**, 385.

Marchesi, V. T., and Gowans, J. L. (1964). The migration of lymphocytes through the endothelium of venules in lymph nodes: an electron microscope study. *Proc. Roy. Soc.* (*London*) **B159**, 283.
Mathews, M. B. (1967). Macromolecular evolution of connective tissue. *Biol. Rev. Cambridge Phil. Soc.* **42**, 499.
Mawas, J. (1922). Sur le tissu lymphoïde de l'intestin moyen des Myxinoïdes et sur sa signification morphologique. *Compt. Rend.* **174**, 889.
Maximow, A. A., and Bloom, W. (1930). "Text-book of Histology". Saunders, Philadelphia, Pennsylvania.
Micklem, H. S., Ford, C. E., Evans, E. P., and Gray, J. (1966). Interrelationships of myeloid and lymphoid cells: Studies with chromosome-marked cells transfused into lethally irradiated mice. *Proc. Roy. Soc.* (*London*) **B165**, 78.
Miller, J. F. A. P. (1961). Immunological function of the thymus. *Lancet* **ii**, 748.
Moog, F. (1959). The adaptations of alkaline and acid phosphatases in development. *In* "Cell, Organism and Milieu" (D. Rudnick ed.), p. 121. Ronald Press, New York.
Moog, F., Bennett, D., and Dean, C. (1954). Growth and cytochemistry of the adrenal gland of the mouse from birth to maturity. *Anat. Record* **120,** 873.
Munger, B. L., and Roth, S. I. (1963). The cytology of the normal parathyroid glands of man and virginia deer. *J. Cell Biol.* **16,** 379.
Murray, P. D. F. (1932). The development *in vitro* of the blood of the early chick embryo. *Proc. Roy. Soc.* (*London*) **B111**, 497.
Murray, R. G. (1947). Pure cultures of rabbit thymus epithelium. *Am. J. Anat.* **81**, 369.
Pannese, E. (1960). Observations on the structure of the enamel organ. 1. Stellate reticulum and stratum intermedium. *J. Ultrastruct. Res.* **4**, 372.
Pearse, A. G. E. (1966). The cytochemistry of the thyroid C cells and their relationship to calcitonin. *Proc. Roy. Soc.* (*London*) **B164**, 478.
Person, P., and Mathews, M. B. (1967). Endoskeletal cartilage in a marine polychaete, *Eudistylia polymorpha. Biol. Bull.* **132**, 244.
Person, P., and Philpott, D. E. (1963). Invertebrate cartilages. *Ann. N.Y. Acad. Sci.* **109**, 113.
Picken, L. E. R. (1960). "The Organization of Cells and Other Organisms". Oxford Univ. Press, London and New York.
Policard, A., and Bessis, M. (1958). Sur un mode d'incorporation des macromolécules par la cellule visible au microscope électronique: La rhophéocytose. *Compt. Rend.* **246,** 3194.
Rasquin, P. (1951). Effects of carp pituitary and mammalian ACTH on the endocrine and lymphoid systems of the Teleost, *Astyanax mexicanus. J. Exptl. Zool.* **117**, 317.
Rich, A. R., Wintrobe, M. M., and Lewis, M. R. (1939). The differentiation of myeloblasts from lymphoblasts by their manner of locomotion. *Bull. Johns Hopkins Hosp.* **65**, 291.
Roberts, D. K., and Latta, J. S. (1964). Electron-microscopic studies on the red pulp of the rabbit spleen. *Anat. Record.* **148,** 81.
Schirner, H. (1963). The pancreas. *In* "The Biology of Myxine" (A. Brodal and R. Fänge, eds.), p. 481. Universitetsforlaget, Oslo.
Scott, D. B. (1954). The electronmicroscopy of enamel and dentine. *Ann. N.Y. Acad. Sci.* **60**, 575.
Shelton, E., and Dalton, A. J. (1959). Electron microscopy of emperipolesis. *J. Biophys. Biochem. Cytol.* **6**, 513.
Smith, H. W. (1951). "The Kidney. Structure and Function in Health and Disease". Oxford Univ. Press, London and New York.
Sorenson, G. D. (1960). An electron microscopic study of haematopoiesis in the liver of the fetal rabbit. *Am. J. Anat.* **106**, 27.

Stoeckel, M. E., and Porte, A. (1966). Observations ultrastructurales sur la parathyroide de souris. *Z. Zellforsch. Mikroskop. Anat.* **73,** 488.
Thomas, T. B. (1940). Islet tissue in the pancreas of the elasmobranchii. *Anat. Record* **76**, 1.
Trowell, O. A. (1958). The lymphocyte. *Intern. Rev. Cytol.* **7**, 236.
Trowell, O. A. (1966). The lymphocyte. *In* "Cells and Tissues in Culture" (E. N. Willmer, ed.), Vol. 2, p. 96. Academic Press, New York.
Weiss, P. (1956). Cellular components of the wound. *In* "Wound Healing and Tissue Repair" (W. B. Patterson, ed.), p. 16. Univ. of Chicago Press, Chicago, Illinois.
Weiss, P., and Ferris, W. (1956). The basement lamella of amphibian skin: its reconstruction after wounding. *J. Biophys. Biochem. Cytol.* **2**, Suppl., 275.
Wessels, N. K. (1964). Tissue interactions and cytodifferentiation. *J. Exptl. Zool.* **157**, 139.
White, W. F., Gurney, C. W., Goldwasser, E., and Jacobson, L. O. (1960). Studies on erythropoietin. *Recent Progr. Hormone Res.* **16**, 219.
Yoffey, J. M. (1950). The mammalian lymphocyte. *Biol. Rev. Cambridge Phil. Soc.* **25**, 314.
Yoffey, J. M. (1962). The present status of the lymphocyte problem. *Lancet* **i**, 206.
Yoffey, J. M., Hudson, G., and Osmond, D. G. (1965a). The lymphocyte in guinea-pig bone marrow. *J. Anat.* **99**, 841.
Yoffey, J. M., Winter, G. C. B., Osmond, D. G., and Meek, E. S. (1965b). Morphological studies in the culture of human leucocytes with phytohaemagglutinin. *Brit. J. Haematol.* **11**, 488.

CHAPTER 19

THE CONCEPT OF EVOLUTIONARY CYTOLOGY

In the preceding chapters the various types of cells found in the tissues of vertebrates have been examined and classified according to their behaviour in tissue and cell cultures. Then the outline of a possible scheme has been drawn up which might account for how, stage by stage during the long process of evolution, these various cells and tissues have come to occupy the positions and to discharge the functions that they now do in the vertebrate organism. An account has been given of some of the various modifications of pre-existing structures and functions which seem likely to have occurred in the long and complicated pre-history of the vertebrates. This search has been partly based on, and has partly accentuated, the importance of several basic principles which could perhaps be regarded as the pillars of cytological evolution and the buttresses of vertebrate cytology.

The Principle of Perpetual Adaptation

Any living cell at any given instant of time exhibits an intricate pattern of complex molecules with a variety of properties and a very special temporal and spatial distribution. As the instant is extended in duration, the molecules move around, changing their positions, their relative numbers, and even their form and character, but in so doing they must preserve in the cell a certain stability of pattern and even of behaviour. If something should occur to upset the pattern beyond certain limits, the whole complex system may have to remodel itself in a new and dynamically stable way, or else break down, become inert, and die. This living pattern of the cell is the result of at least four components.

First, the cell is endowed with a set of chromosomes and other genetic material, containing, in its complement of DNA, all the basic information for the construction of the proteins of the cell. There is not only information for those proteins which it actually possesses at one given instant in time, but probably also for the vast majority of the proteins which any cell belonging to any individual of that species might ever have been able to produce at any time, and perhaps even for others that have been produced by some ancestral individuals in the past which are not now normally produced. We may broadly refer to this wealth of inherited information as the "genome", remembering that only part of it is functional at any one time, the rest being suppressed. The part of this genome which regulates the cell at any one time could conveniently be called the "hegemon" (ἡγεμονία = leadership). This is the second component.

Thirdly, the cell with which we are concerned was not suddenly made *de novo*. It may have been produced by division from another cell, almost completely identical with itself except in size; it may have been modified from its parent during that division; or it may have become modified since that division. However that may be, the cell in addition to its hegemon is now provided with a particular "econome", (οἰκονομία = house management). This term means the particular pattern of the cell's cytoplasmic and nuclear membranes, organelles, and the like. This pattern has resulted from the pattern of the parental cell, and from any modification of the parental pattern that occurred during the cell division by which the cell in question came into being, together with any subsequent modifications of it that may have occurred since the cell began its separate existence.

Fourth, a cell in a multicellular organism does not exist *in vacuo*, but in a very specially organized local "environment" which happens to be suitable for its present existence and activity. This organized meeting place of the cell with its neighbours and the surrounding tissue fluids may be called the "agoranome" (ἀγορά meeting place; νόμος, law).

At any given instant, the genome, the hegemon, the econome, and the agoranome may be considered to be in a state of suspended animation. At the next instant, however, all may have changed. Moreover, such changes may, and usually do, affect all four components, since all four systems are inextricably linked with each other. When a cell is instantaneously fixed (this is the ideal aimed at in preparing tissues for electron-microscopy), animation is suspended, and, at best, the result bears about as much resemblance to the living reality as the information presented by an "instantaneous" photograph of events at one fence of the Grand National does to the concept of the whole race. In addition, the process of fixation is never quite instantaneous; inevitably it constitutes a very radical change in the environment which may indeed cause notable alterations to the econome before the latter becomes stabilized or "petrified" by the fixative. In the living cell, any change in the agoranome inevitably starts a chain of responses in the econome, and some of them may then modify the relationship between the econome and the hegemon. Altered protein distribution in the econome, for example, may lead to altered activity of the hegemon so that different parts of the genome are called into action, new proteins are produced, and a more appropriate pattern of cellular activity is developed. Similarly, a reaction going on in the econome may bring about changes both in the hegemon and in the agoranome. Whether the genome itself can initiate changes is perhaps of purely academic interest, because, owing to the vital character of the econome and the improbability of the environment of any cell remaining absolutely constant, even momentarily, the genome is unlikely ever to be in a completely stable situation. All this makes it quite clear that the behaviour of a cell, including both its development and its function, is the resultant of perpetual interactions between genome, hegemon, econome, and agoranome.

One uncertainty about the genome and its responsibility for the production of numerous proteins is whether all proteins that are produced fulfil only the one function for which their DNA code was selected, or whether it is possible that the same protein may fulfil several functions in the cell. Could a protein perform more than one function either by occurring in different places or by having its tertiary structure modified by local conditions in such ways that it catalyses different reactions, stabilizes different lipids, forms different membranes, is activated by a different prosthetic group, or fulfils any other role? Similar questions are: Is it possible for a protein to act in one way in one cell and quite differently in another? Can two different economes make use of the same protein in different ways?

The Principle of the Constancy of the Specific Environment: the Agoranome

Most natural environments for animals are variable, and many of them become incompatible with active life at times. Claude Bernard showed that much of the physiology of higher animals could be considered as the integration of those processes by which the cells, tissues, and organs of the body work together to provide the constituent cells with as constant an "internal environment" as possible, so that greater changes in the external environment can be tolerated by the animal as a whole than would otherwise be possible. The organs and tissues achieve this constancy very largely by regulating the volume, composition, and temperature of the unifying fluid, namely the blood. As long as this circulating fluid is kept constant, the cells and tissues bathed by it can adapt themselves to work within specified narrow limits; thus their activities can be far more certainly determined. This concept of the constancy of the internal environment of the whole animal is, of course, of the utmost importance and well recognized, but it is by no means the whole story.

From the cytological point of view, the environment that matters to each cell is not the blood as a whole, but the agoranome of the cell. This includes the particular tissue fluid in contact with each part of the cell surface, the nature of any neighbouring cell surface, the fibres and mucoproteins in the vicinity, or any basement membrane with which the cell in question may be in contact. In other words, cytologically speaking, the local environment is of far more immediate importance than the total internal environment, from which it may, in fact, differ extensively. Furthermore, this orderly local environment (the agoranome) of each cell is usually a complex of several different environments, any one of which may at any time react with the econome, both to alter the latter and, very likely, to be altered by it. Tissue-culture work has shown this principle of local and micro-environments emphatically.

In organ cultures, where the physiologically normal local environments are relatively well preserved, cell function is likely to continue much as it does in the body, because most of the inner cells of the explant retain something resembling their natural surroundings. When cells are separated by trypsinization and the like, and then suspended in a uniform physiological(!) fluid, however, the diverse and special local environments, the agoranomes, all disappear as such and the cells are immersed on all sides in an almost uniform milieu. Local specializations of surface, gradients, polarities, and the like all tend to disappear unless internally maintained by the econome. Even so, the econome adapts to the new and uniform environment and, being thus altered, modifies the hegemon and, if the cell is to survive, calls forth more appropriate proteins. As soon as they are produced, these proteins again alter the econome, and new metabolites escape into the environment, changing the latter. This changed agoranome further modifies the econome and hence the hegemon; and so the process goes on and on. In due time, most of the cells in suspension cultures, and indeed in the more usual cell cultures, tend to produce populations of "tissue-culture cells" (see Figs. 2.15, 3.10) which are presumably the cells that are best adapted to life in that particular medium and under those particular conditions. It matters comparatively little from what organ of the animal the initial cells were derived. Even the differences between the cells of different species would be expected to be reduced under such conditions, though it is remarkable that their antigenic properties in fact appear to be very stable. Even in such apparently uniform suspensions, the cells may be quite different at the end of the growth phase from what they were at the beginning, the environment having changed meanwhile as the result of the cellular activity within it. The morphology of the cells in sparsely populated cultures is also quite different from that of crowded cultures.

A cell can perform its physiological function—the manner in which it normally functions in the body—only if the local environments of the cell are so constituted that they undergo only such changes as they normally do in relation to that cell in the body. A cell isolated from its normal neighbours soon ceases to be, and may never be able to return to being its normal self. During the process of adaptation there is a remodelling of the econome; cell membranes may change their constitution (see p. 491); new biochemical cycles and patterns may be necessary to ensure the continuing stability of the cell. The cell either modifies successfully or breaks down. The econome, by modifying the hegemon, may call on the genome for more or less production of this protein or of that one, and the genome may be able to supply a variety of proteins which the cell would never produce in the ordinary course of events. A cell cannot, of course, produce an endless number of proteins, but only those determined by the hegemon, i.e. those for which there is the necessary information in that part of the DNA which can function in that particular econome. However, on account of the long history of successful survival in a wide range of environments, any given cell probably has available the necessary DNA for an extensive repertoire of proteins, provided that the econome allows the necessary parts of the genome to be activated or freed from repression, i.e. determines the appropriate hegemon.

It is interesting that the neoblasts of planarian worms when cultured *in vitro* have, at first, features similar to those of cell cultures of vertebrate tissues, i.e. they are undifferentiated and one cell looks much like another. However, when these cells aggregate into masses, as they very soon begin to, differentiation begins to occur (Betchaku, 1967)*. Presumably this differentiation is the result of the development of local environments, i.e. of specific agoranomes in place of the uniform tissue-culture medium. The aggregation of cells from suspensions of tissue cells of higher animals sometimes leads to similar phenomena, though there is not much information about how long cell cultures can retain the ability to redifferentiate when eventually allowed to aggregate.

For the development of the full physiological function of a tissue it is necessary to build up the requisite special environments for all the various reacting cells concerned. Of course, this is largely what embryology is about: every time an animal develops, it produces the proper local environments for all its cells and thereby evokes the appropriate behaviour from them.

J. Exptl. Zool **164**, 407 (1967).

It has learned to do this during its phylogenetic history, and this is largely what evolutionary cytology is about.

In the past it has been the fashion to consider differentiation in terms of inducing agents eliciting specific responses from cells, rather in the manner of hormones or neural transmitters. Such inducing agents undoubtedly occur, but this concept of differentiation is probably a gross over-simplification. In most cases, induction is more likely to be the result of a particular agoranome having been set up for each econome and of every cell having its own particular niche in which numerous and diverse factors may be involved. It may be necessary, for example, to establish a particular gradient across a cell for the movement of particular ions or of some special metabolite. This could sometimes be brought about by an inducing agent or evocator adding something positive or causing an excess of some metabolite; it might equally well be set up by exactly the opposite procedure, namely by the so-called inducing agent or evocator creating a local shortage of some important metabolite(s). The evocator may take the form of a single substance or of several substances; it could be a cell or part of a cell, or even a particular arrangement of cells. It is, in fact, the agoranome.

When due consideration is given, for example, to the possible variations in the lipid constituents of a cell membrane, to say nothing of its proteins, glycoproteins, and other constituents, it becomes obvious that differentiation, although possibly triggered occasionally by some one stimulus, is, in general, a far more subtle and elaborate process than most of us have been prepared to admit.

In this context the agoranome has appeared to be the major contributor to the organization of the econome. The hegemon must, however, also contribute its share, since the production of any protein may by its presence modify its own production or that of others. Thus the econome is the resultant of influences from both sides, i.e. not only from the agoranome but also from the hegemon and, ultimately, the genome.

The Principle of Balanced Activity

In general physiology there are numerous examples of a steady state of activity being maintained by the balance of two, or sometimes more, opposing or differing actions. The frequency of the heart-beat is under the influence of an accelerator in the sympathetic innervation and a retarding agent in the parasympathetic supply. Cholinergic sphincter pupillae and adrenergic dilator pupillae control the amount of light that enters the eye. Parathyroid hormone raises the calcium in the blood; thyrocalcitonin lowers it. Many other examples will occur to the reader.

This principle is probably far more widespread than was at one time supposed, and in cytology there appear to be numerous examples in which cells work together in pairs or small groups, either for their mutual benefit and control or for the production of a functional system benefitting the whole body. Mention has been made of nerve cells and their supporting neuroglia, Schwann cells, satellite cells, etc. The germ cells and their supporting cells afford another well known example. Evolutionary cytology has indicated that such symbiotic pairs may, in fact, permeate the whole of physiology, because of the early necessity to control the contents of the elementary blastuloid organism. This control, it was suggested, was achieved partly by the cells establishing their own equilibria as in the protozoa, but mainly by evolving two sorts of cells in the walls of the blastuloid: one ecballic and one emballic; one attuned to dilute environments and one attuned to those that are more concentrated, out-pushers and in-pushers. By balancing the activities of one sort against those of the other an equilibrium state could most easily be obtained, and any drift away from the normal could be prevented. Because of the different mechanisms involved in these two opposing processes, the metabolism of the two groups of cells was necessarily very different. In consequence, ecballic (mechanocytic) and emballic (amoebocytic) cells have become firmly established as complementary pairs in the majority of the tissues of the body.

This idea of a simple duality is probably too simple. Equilibria may sometimes be attained more efficiently by balancing more than two opposing actions. In many blastulae there are more than the two basic types of cells. A third class of cell, perhaps intermediate, perhaps merely different, is frequently present and, indeed, seems to pervade many of the epithelia in the more primitive groups of animals and even in the respiratory epithelium of man.

Although the duality was originally postulated in terms of ionic balance as suggested by the observations on *Naegleria*, where ions are clearly important, it is probable that other aspects of metabolism are stabilized in a similar manner.

Epithelia can now be comprehended not only as layers of cells covering surfaces, providing protection, or producing secretions, but can also be understood as providing active barriers between two different environments, e.g. the outside world and the tissue fluid, the cerebrospinal fluid and the nerve cells, the blood and the aqueous humour in the eye. These barriers control or encourage the passage of particular substances from one environment to the other. In exercising this control, the constituent cells seem likely to be oppositely orientated, so that excessive activity on the part of one class may be counteracted by the opposing efforts of the other. A continuous flow of fluid towards one side, or the accumulation of secretory products on one side or the other, tends to indicate that there is some degree of imbalance in the system and that, temporarily at least, one class of cells is being allowed to have its own way more or less unchecked.

A more intimate study of epithelia, combined with a more searching analysis of this two-way controlled traffic across them, should lead to a better understanding of the functions of particular epithelia. Many studies have been made on the permeability of the frog's skin, especially to sodium ions; and theories of membrane actions, pumping mechanisms, and so on have been based on the results. The preparations of skin that have sometimes been used for such analyses have contained at least the following elements: superficial epithelial cells; basal epithelial cells; mucous glands; a basement membrane under the epithelial cells that may or may not differ from that under the mucous gland cells; connective tissue with fibroblasts, and in some cases plain muscle cells; another basement membrane and the endothelium of the lymph space. This complexity may not matter when only the net result is in question, but, when the mechanism of transfer is under investigation, the less said about the interpretations the better. The naïve view of the function of the small intestine is that it absorbs the digested food products, and that in this process the chief cells do the work while the goblet cells lubricate the "bolus" with their mucous secretions. The problems which the intestinal epithelium has solved in keeping the blood, into which the digested products are being passed, from being contaminated and altered by all sorts of fluctuations in ionic concentration, including pH changes, products of digestion, and the generally unhygienic and variable contents of the gut, often pass unnoticed. Events in the intestinal epithelium during the digestion and absorption of a meal must be exceedingly complex, and it is likely that both classes of cells contribute in more ways than one. Moreover, the goblet cells and the chief cells probably are another example of symbiotic activity. In such situations it must be emphasized again (see p. 500) that, since the conditions on the external or luminal side are very different from those on the internal or vascular surface, cells which are essentially ecballic may be capable of acting ecballically at one end or surface and emballically at the other, or *vice versa*. It must be remembered, too, that not only can *Naegleria* be flagellate or amoeboid as a whole, but also under certain conditions can take an intermediate stance and produce pseudopodia at one end and flagella at the other. Epithelial cells are probably often in a comparable state, and, as in *Naegleria*, their degree of emballism or ecballism may be variable from time to time.

Not only does the idea of a primary duality and subsequent symbiosis, which arises in simple epithelia, remain applicable to epithelia themselves, but also it probably extends to the descendents of epithelia. We have stressed the symbiosis between the ecballic mechanocytes and the emballic amoebocytes as it seems to go on in the various connective tissues of the body. A proper appreciation and investigation of this relationship should materially assist in creating an understanding of the means by which bones, cartilage, tendons, and the various connective tissues achieve their special properties of elasticity, strength, rigidity, and the like, without which their functions would be impaired. Such an understanding should assist in providing means for alleviating or repairing the ravages of some of the more intractable and crippling of human diseases, rheumatism, arthritis, fibrositis, arteriosclerosis, osteitis, and many more.

In the nervous system, too, duality is probably widespread; it certainly seems to be present in the retina. On first principles, it would seem that to oppose the action of one group of cells by that of another would be an economical way to deal with the vast input of information which must certainly flow in from the sense organs and be, for the most part, quite unimportant to the animal. With such a dual system, in which it is possible to oppose every action by some sort of counter mechanism, only such stimuli as are strong enough not to be damped out by opposing forces would be transmitted to the next stage in the neural hierarchy. Sometimes this damping can be performed by like cells, as happens with the rods in the eye, but advantages may also be gained by using unlike cells. From the retina, for instance, the information that would go forward to the brain if the surrounding rods inhibited the impulses inaugurated by the rods at the centre of the receptive field of a ganglion cell would be different from that which would go forward if the same central rods were inhibited by cones, because the cones would be activated by wave-lengths of light different from those that activated the rods.

Neurones and neuroglia cells also offer examples of mutual interaction, probably established in the interest of sustained neural action. "Neural action" is intended to imply more than the simple transmission of impulses, for that is probably but one aspect of neural activity. By analogy with similar organisms, it may be assumed that the cell membranes of *Naegleria* are excitable and capable of passing that excitation to other parts of the cell. Moreover, it is probable that this is true for both forms of the organism. One is therefore entitled to wonder which form of cell membrane approximates more closely the squid axon or the sciatic nerve of the frog.

In many situations the balance of activity of a cell system can be achieved by adjacent cells or by forming suitably mixed populations. At other times, as in the elementary blastuloid and in excretory tubules, the opposing cells may be arranged in groups or in a gradient system. The gradients may be steep or gentle; this observation means that the cells in the two groups need not be "black" or "white" in their actions. Intermediate shades of grey are permissible, the greyness being a measure of the efficiency with which the positive or negative action is carried out by the cell in question. In more concrete and specific terms, some cells may be able to eject Na^+ ions, for example, very efficiently, and others much less so. It is unlikely that the cells of higher animals are simply divisible into two classes in this respect, e.g those that eject Na^+ ions and those that do not. It is much more likely that they vary in their efficiency and that this efficiency is partly dependent on external concentrations. Some cells, like *Naegleria*, that normally eject Na^+ may even be able to switch to an intake of Na^+ should the conditions favour such a change. The gradients across epithelial membranes, as for example in the kidney, and the effects of hormones on cell membranes in such situations both indicate that too rigid an interpretation of ecballism and emballism is likely to be fallacious.

The Principle of the Permanence of the Genome

It was pointed out earlier that, by the time the Protozoa, as we know them, appeared on the scene, practically the whole gamut of biochemical processes, cycles, and syntheses had probably been achieved. With one or two comparatively minor exceptions, the evolution of the Metazoa did not depend on the development of new biochemical pathways or syntheses. It depended on parcelling out and ordering what was already there into new and more appropriate arrangements to meet each new situation as it developed. It is true that new and species-specific proteins have appeared during the course of evolution, but many of them probably differ only in some minor and secondary way, at least in so far as their main biological actions are concerned; the main enzymatic or essential physiological activity is necessarily preserved. The example of the haemoglobins has already been given; the insulins and posterior pituitary hormones are others. From the physiological point of view, the extraordinarily sensitive antibody system probably exaggerates the diversity of proteins.

However, the presence of this enormous body of biochemical competence in even the simplest organisms emphasizes what may be called the principle of permanence. Genetic information exists and must be preserved so that what has already been achieved shall not be lost. It seems that Nature has done this by incorporating all the required information for making the right proteins into organized DNA threads and then devising a mechanism whereby much of the information is rendered temporarily inoperative by a process of suppression. In any given cell, only certain parts of the genome or genetic code are available for the purpose of producing enzymes, proteins, etc., under the particular set of conditions reigning at the time. These available pieces, as we have said, constitute the hegemon.

As an embryo develops, the information in the genome of the fertilized ovum is progressively and systematically unlocked and relocked as the daughter cells of each division progressively move into slightly different environments which they themselves are helping to create. In other words, the hegemon of each cell differs from that of its neighbours because the process of cleavage systematically alters the agoranomes of the cells, and each cell acquires a specific econome. This process is the result of an orderly chain of events in development that is characteristic of each species. Because of its complexity, when differences occur in the chain they tend to be harmful more often than beneficial. Thus animals showing such differences tend to be eliminated by natural selection, and the standard developmental sequences are thereby preserved very conservatively. For these reasons embryological processes and events tend to be broadly similar in large groups of animals. A study of developmental patterns shows that they often have great significance in establishing or disproving genetic relationships, though it must also be remembered that modifications in response to special environments also occur as, for example, in the production of larval forms.

Evolutionary processes tend to act by modifying the uses of structures that are already developed, or partially developed, rather than by producing entirely new features. On earlier pages we have suggested how the proboscis of early nemerteoids may have been modified to fulfil a variety of functions, from that of an external gut (as in the Pogonophora) to a notochord (as in the chordates). Similarly, the nephridial tube may be the fore-runner of the kidneys, of various glands, and perhaps of lymphoid tissue.

Thus it would appear that not only is the actual DNA-coded information (the genome) permanently stored, but also that the system by which the relevant parts are released from time to time is rather rigidly controlled, i.e. the hegemon has some stability. Undoubtedly this depends mainly on the econome. The econome of the germ cells of each species is unique to that species, and it is achieved largely by an orderly system of development which ensures that the germ cells meet essentially the same environmental conditions, i.e. the same agoranome is always developed around them. The relationships among the agoranome, econome, and hegemon of the fertilized ovum are such that the correct parts of the genome become available in the correct places at the correct times during subsequent development. Thus, not only the genome, by virtue of its actual DNA pattern, and the hegemon, or the parts of the pattern that are available, but also the econome, by varying the hegemon, and the agoranome, by varying the econome, all determine the course of development of the ovum. As cleavage occurs, new agoranomes induce new economes, and hence new hegemons for the emerging cells.

For all these reasons there arises a degree of permanence about cell activity in the fully developed animal with its controlled internal environment. The agoranome, the econome, and the hegemon are so integrated for each cell that the cell continues to function in a particular way for as long as reasonably stable conditions prevail. This fact has led to the erroneous idea that liver cells, kidney cells, and so on are inherently specific and relatively permanent types of cells. In the body they are certainly well characterized and permanent, but, when they are taken out of their particular environment, they start to change and to adapt to the new conditions. Observations on tissue cultures suggest that the adaptation goes on slowly and continuously, but that it is accelerated when the cells are encouraged to divide. Parker's experiments, in which tissues were maintained as differentiated tissues for years by feeding them only with serum or heparinized plasma (see p. 32), strongly contrast with the almost complete lack of tissue characteristics among cells that have been stimulated to grow rapidly as cell cultures or cell suspensions. It is probable that in the latter case both the relative uniformity of the environment and the fact that the cells are frequently dividing have speeded the adaptation. Numerous examples from the field of malignant growth indicate that cells adapt to new situations more quickly when they are dividing. Whether this is due to actual mutations, i.e. changes in the base sequences of DNA, or to the fact that changes in the parts of the genome that are active or repressed occur more easily during mitosis, remains to be elucidated. The hegemon is probably more directly available to the econome during mitosis, when the nuclear membrane has disappeared and the chromosomes are more directly in contact with cytoplasmic components than at any other time. This may be important in determining modifications in the repressor system and the like. If, as seems probable, mitosis is the main period during which modifications and adaptations occur, this would help to accentuate the relative constancy and stability of cellular performance in those differentiated tissues in which cell division has virtually ceased. At the same time it would help to explain why the cells in rapidly growing systems, in cultures and in malignant growths for example, so often adapt quickly to such conditions and produce uniform populations of generalized cells that can continue to multiply rapidly. The experiment described on p. 40 in which a dormant tissue culture was treated with two successive doses of growth-stimulating embryo extract is relevant here. It was shown that a second dose of extract was not effective in causing mitoses, unless it was applied during or after the mitoses induced by the first dose. This observation clearly supports the view that the nuclear apparatus is more open to external influences during or immediately after division. On the other hand, it must also be emphasized that the agoranome can certainly produce changes in the hegemon, presumably via the econome, at times other than at mitosis.

The arguments developed in this book suggest that the study of cytology should not be made *in vacuo* as it were, but that the cells, like people, are better understood when considered in relation to their environment and their social history. The pattern of activities seen in the cells of an animal are just as elaborate and diverse as the pattern of activities of the people of a large city like London. Any given cell is, in many ways, comparable with any given Londoner. It has its own individuality and, by virtue of its inherited qualities and its upbringing, it fits into a particular niche in the economy of the body and it performs a definite job. Its character, the job that it does, and the general organization of its activities are dictated by its own properties in relation to the organization of the body as a whole. Both the cell and the body to which it belongs are the product of their respective evolutionary histories, just as Londoners and London's organization have their own characters and are the products of their evolutionary histories. Thus evolution is as important in the study of cytology as human history is in the study of the sociology of London.

There are, of course, many ways of studying the sociology of London, just as there are many ways of approaching the study of cytology. As extreme examples of different points of view towards London and towards cytology, we may compare those of two birds. The ubiquitous starling, like the majority of scientists of the present day, is an assiduous and determined bird. Flocks of them swoop down on a particular field, or perhaps a heap of old rubbish, and then chattering, jostling, and competing with each other, these "eager beavers" pick up whatever crumbs they can find, till the whole area is systematically cleared of all that is edible, and the flock moves on. The starlings are indubitably successful and efficient birds, but their vision is limited except in the leaders of the flocks. Aloft a solitary eagle may be soaring freely on the up-currents of hot air rising from below, its head in the clouds for some of the time, and enjoying, with a fine disregard of gravity the wide sweep of town and country that unfolds below him. Every now and then he swoops headlong downwards in mad pursuit of some flight of fancy or to chase some attractive but elusive hare. Numerically, he may not be so successful as the starling, but his life is exciting and romantic. His food may be limited to that for which his talons and beak are adapted, but at least he has the joys of the chase. He is an individualist, not to be seen in such clacking hordes as are starlings in their roosts or scientists at their international conferences. He gets an enjoyment quite unknown to the gregarious and mundane starling. He may even assist the latter from time to time by providing half-eaten carcases to be demolished.

May the pages of this book, penned by a sparrow on an envious flight into the dizzy heights of the eagle's world, persuade a few starlings to soar aloft occasionally to take a wider view, and may the half-picked carcases herein provide a few tit-bits to be wrangled over, enjoyed, or firmly rejected as inedible.

AUTHOR INDEX

Numbers in italics refer to the pages on which the complete references are listed.

Abdel, B. W., 534, *559*
Abercrombie, M., 24, 43, *59*, 114, *119*, 126, 138
Adam, H., 349, *360*, 519, *559*
Adams, E. C., 509, *511*
Adelmann, H. B., 340, *360*
Åkesson, B., 328, *334*
Algire, G. H., 17, *18*, 91, *103*
Allen, A., 43, *61*
Allen, C. R., 245, *258*
Allen, F. P., 396, *421*
Allen, R. A., 416, *419*
Allen, R. D., 192, *203*
Allgower, M., 220, *235*
Ambrose, E. J., 23, *59*
Andersen, S., 475, 482, *514*
Anderson, H. K., 116, *119*
Andersson, B., 386, 389, *419*
Assheton, R., 227, *234*, 246, *257*, 402, *419*, 460, *461*
Auclair, W., 240, *257*
Auerbach, R., 538, *559*
Augustinsson, K. B., 246, *257*
Axelrod, J., 396, 397, 398, *423*

Bacq, Z. M., 250, *258*, 263, 265, *279*
Baggerman, B., 356, 357, *360*
Bagnara, J. T., 395, *419*
Bahl, K. N., 439, 440, 441, 450, 452, *461*, 478, *511*, 524, *559*
Baker, H. D., 415, *419*
Baker, L. E., 42, 43, *59*, 90, *103*
Baker, P. C., 227, *234*
Balamuth, W., 156, 169, *176*
Balfour, F. M., 142, *154*, 326, *334*
Balfour, W. E., 354, 355, *360*
Balinsky, B. I., 165, *176*, 188, 189, *203*, 228, *234*, 408, *419*
Ballantyne, B., 498, *511*
Band, N., 165, *176*
Bang, F. B., 79, *85*, 100, *104*, 535, *559*
Bangham, A. D., 489, *511*
Barnes, B. G., 343, *360*, 507, *511*
Barr, L., 188, *204*
Barraclough Fell, H., 283, *334*
Barrington, E. J. W., 285, 286, *334*, 352, 353, 354, 356, *360*, *361*, 517, 518, 523, *559*, *560*
Barta, E., 96, *103*
Baxter, J. S., 483, 486, *511*
Bayer, G., 246, *258*
Beadle, L. C., 199, *203*, 254, *258*
Beams, H. W., 501, *511*, 547, *560*
Bell, E. T., 541, *560*
Bennett, D., 520, *562*
Berg, O., *361*
Berger, A. J., 316, 321, *335*
Bernard, R. A., 380, *419*
Bernhard, W., 4, *5*, 488, *513*
Bessis, M., 533, *562*
Best, C. H., 99, *103*
Biggers, J. D., 214, 221, *234*, 533, *560*
Blandau, R. J., 324, *336*, 476, *511*
Blanquet, R., 216, *234*
Bleiberg, J., 541, *560*
Bloom, W., 25, 53, 57, *59*, *61*, 469, 471, *513*, 546, 551, *562*
Boak, J. L., 541, *559*
Böhmig, L., 271, *279*, 427, *461*
Boell, E. J., 246, *258*
Bogoroch, R., 494, *512*
Bone, Q., 307, 310, 311, *334*

Booth, F. A., 199, *203*
Bornstein, M. B., 79, *84*
Borojevic, R., 143, 151, 152, *154*
Bott, P. A., 450, *463*
Bouillon, J., 141, *154*
Boycott, B. B., 381, 386, 387, 388, 417, *419*, *420*
Boyd, I. A., 318, 319, *334*
Brachet, J., 408, *419*
Bradley, C. F., 114, 115, *120*
Brambell, F. W. R , 289, 290, *334*, 367, *419*
Brandenburg, J., 445, *461*
Briggs, R., 44, *59*, 130, *139*, 193, *204*
Brodal, A., 349, *361*
Brown, J. J., 503, *511*
Brown, P. K., 386, 392, *419*
Brumbaugh, J. E., 255, *258*
Brunner, H., 505, *512*
Bryant, J. C., 8, 12, *18*, 46, *60*, 127, *138*
Bülbring, E., 244, 245, 246, *258*
Bürger, O., 260, 271, 272, 274, *279*, 287, *334*, 369, 385, *419*, 458, *461*
Bullock, T. H., 290, *334*
Bunting, H., 477, *515*
Bunting, M., 156, 157, 165, *176*
Burdon-Jones, C., 286, *334*
Burgos, M. H., 474, 477, *512*
Burn, J. H., 244, 245, *258*
Burnett, A. L., 217, *236*
Burns, J., 352, *361*
Burns, R. K., 455, *461*, 484, *511*
Butt, F. H., 328, *336*

Cameron, G., 71, *84*, 403, *419*, 508, *511*
Cantero, A., 215, *236*
Canti, R. G., 8, *18*, *19*
Caravaglios, R., 477, *511*
Cardini, C. E., 221, *235*
Carr, I A., 485, *511*
Carrel, A., 8, *18*, 42, 43, *59*, 88, 90, 92, 94, 99, *103*, 124, 125, *138*
Cash, J. R., 249, *258*
Castiaux, P., 141, *154*
Chambers, R., 65, 71, *84*, 124 *138*, 189, 198, 199, *203*, 508, *511*
Chang, H. C., 245, *258*, *259*
Chapman-Andresen, C., 100, *103*, 175, *176*
Charlton, H. M., 397, *419*
Chèvremont, M., 42, *59*, 99, 100, *103*, 320, *334*
Chèvremont-Comhaire, S., 100, *103*
Christensen, H. N., 221, *234*
Christie, G. H., 541, *560*
Chu, E. W., 396 *423*
Cieciura S J , 46, *62*, 72, 80, *86*
Cilotti, R., 477, *511*
Clark, E. L., 17, *18*, 91, *103*
Clark, E. R., 17, *18*, 91, *103*
Clark R. B., 250, *258*, 263, *279*, 282, *334*, 432, *461*
Clarke, R , 522, 523, *560*
Clermont, Y., 209, *234*
Cobb, J. L. S., 232 *234*
Coe, W. R., 269 270, 274, 275, 276, *279*, 297, 326, 329, *334*, *335*, 345 *361*, 367, *419*, 439, *461*, 525, *560*
Coggeshall, R. E., 555, 556, *560*
Cole, H. A., 289, 290, *334*, 367, *419*
Collier, H. D. J., 220, *234*
Conel, J Le R., 437, 438, *461* 531, *560*
Conklin, E. G , 301, *335*
Coombs, R R. A., 134, *138*
Cooper, M. D., 528, *561*
Copeland, D. E., 352, *361*
Cordier, R., 452, 453, *462*, 488, *511*
Corssen, G., 245, *258*
Costero, I., 108 109 117, *119*
Cowdry, E. V., 483, *511*
Cowey, J. B., 264, *279*, 302, *335*, 554, *560*
Crabo, B., 475, *511*
Creaser, C. W., 356, *361*
Crescitelli, F., 412, *419*

Daems, W. T., 213, *236*
Dakin, W. J., 292, *335*
Dalton, A. J., 532, 533, *562*
Damas, H., 433, *462*
Dan, M., 184, *203*
Daniel, E. J., 549, 550, *561*
Danielli, J. F., 439, *462*
Darke, S. J., 99, *104*
Davidson, J. N., 41, 43, *59*
Davidson, N., 189, *204*, 418, *421*
Davies, D. L., 503, *511*
Davies, J. 477, *513*
Davydoff, C. 186, *203*
Dean, C , 520, *562*
Deane, H W., 509, *512*
Deanesley, R., 485, *512*
de Beer, G. R., 408, *421*

de Bruyn, P. P. H., 24, *59*, 532, *560*
de Graaf, A. R , 180, *204*
Deitch, A. D , 112, *119*
Delaney, P A., 436, *462*
Dempsey, E. W., 477, *515*
Dendy, A., 391, 392, *419*
Dennis Smith, L., 470, *512*
de Rényi, G. S., 29, 52, *59*
de Robertis, E., 380, *419*
Dervichian, D G., 490, *512*
Detwiler, S. R., 412, 413, *419*
Devine, R. L., 501, *511*
Devis, R., 188, *204*, 228, *234*
Dewey, M. M., 188, *204*
Dingle, A. D., 159, 160, 161, *176*
Dingle, J. T., 52, *59*, 70, *84*, 380, *419*, 489, 491, 493, 495, *512*, 558, *560*, *561*
Dobelle, W. H., 415, *421*
Dodt, E., 396, 397, *420*
Doljanski, L., 43, *60*, 72 73, 75, *84*, *85*
Dorey, A. E., 219, 229, 230, 232, *234*
Dorfman, A. 222, *234*
Dougherty T. F , 529, 533, *560*
Dowling J E., 404, 406, 417, *420*
Drew, A. H., 77, *85*
Droz, B., 412, *424*
Dryburgh L. 547, *560*
Duboscq, O., 141 *154*, 190, 191, *204*
Dücker, M. 398, 399, *420*
Dunihue, F W., 504, *512*
Dunn, R. F., 412, *420*
Duveau-Hagège, J., 456, *462*, 480, *513*

Eagle, H., 47, 48, *60*, 73, *85*, 221, *234*
Eakin, R. M., 299, *335* 391, 392, 394, 397, *420*
Earle, W. R., 5, *6*, 8, 12, 13, *18*, 36, 38, 46, *60*, *61*, *62*, 81, *85*, 127, *138*
Ebeling, A. H., 74, 76, 77, *86*, 88, 90, 92, 94, *103*, 124, *138*
Ebert, R., H., 17, *18*, 91, *104*
Eccles, J. C., 188, *204*, 321, *335*
Eccles R. M, 321, *335*
Edelman, I. S., 494, 496, *512*
Edwards, R. G. 480, 481, *512*
Efron, M. L., 503, *514*
Eggleston, L. V., 221, *235*, *237*
Ehrich, W. E., 532, *560*
Ehrmann, R. L., 79, *85*
Elsden, D. F., 55, *62*
Ely, J. O., 214, *234*
Engel, L. L., 489, *515*
Ephrussi, B., 535, *560*
Ernyei, S., 282, *335*
Esaki, S., 476, 481, *512*
Evans, E. P., 537, *562*
Evans, V. J., 8, 12, 13, *18*, 46, *60*, 81, *85*, 127, *138*
Everett, N. B., 470, *512*
Ewer, D. W., 180, *204*

Fänge, R., 349 *361*
Fanestil, D. D., 496, *512*
Farquhar M. G., 188, *204*
Farrell, G. L., 396, *420*, *421*
Fatehchand, R., 417, *423*
Fauré-Fremiet, E., *176*, 220, *236*
Fauré-Fremiet, M. E., 143, 144, 148, 149, *154*, 159
Fawcett, D. W., 90, 91, 95, 96, *105*, 220, *234*, 474, 477, *512*, 537, *560*
Feldberg, W., 118 *119*
Feldman, D., 112, *119*
Feldman, M., 541, *560*
Fell H. B., 31, 33, 34, 49, 51, 52, *60*, 69, 70, 82, *85*, 97, 98, *104*, 124, *138*, 558, *560*, *561*
Ferris, W., 553, *563*
Findley, T., 450, 451, *463*
Finstad J. 528, *561*
Fioramonti, M. C., 127, *138*
Firket, H., 39, *60*
Fischer, A., 8, *18*, 21, 22, 28, 31, 43, 48, *60*, 65, 67, 68, 76, 77, *85*, 94, *104*, 123, 124, 125, *138*
Fischer, J. E., 397, *423*
Fischer-Piette, E., 528, *560*
Fisher, D. C., 8, *18*, 37, 42, 47, *61*
Fisher, H. W., 36, 46, *62*, 72, 80, *86*
Fitton Jackson S., 213, *234*, 408, *423*
Fleischman, R., 48, *60*
Flesch, P., 70, *85*
Flock, Å., 367, 368, *420*
Flood, P. R., 232, *234*, 307, 310, 312, 316, 329, *335*
Florey, H. W., 17, *18*, 91, 98, *104*
Fontaine, M., 357, *361*
Ford, C. E., 537, *562*
Ford, W. L., 532, 541, *560*
Fordham, M. G. C., 292, *335*

Foshee, A. M., 255, *258*
Foster, M. A., 507, *514*
Fourman, J., 246, *258*
Fowden, L., 356, *361*
Franchi, L. L., 353, 354, *360*
Francis, C. M., 114, *119*, 416, *420*
Franks, D. 134, *138*
Frédéric, J., 42, *59*
Freeman, A. E., 47, *60*
Freeman, G., 410, *420*
Fries, N., 214, *235*
Frommes, S. P., 363, 365, *422*
Fudge-Mastrengelo, M., 498, *514*
Fulton, C., 159, 160, 161, *176*

Gabe, M., 330, *335*
Gabrielsen, A. E., 528, *561*
Gaillard, P. J. 98, *104*, 558, *560*
Gallagher, C. H., 489, *512*
Gardner, R. L., 481, *512*
Garvey, J. S., 535, *560*
Gauchery, M., 220, *236*
Gauthier, G. F., 316, *335*
Geiger, R. S., 108, 111, 112, *119*
George, J. C., 316 321 *335*
Gérard P., 531, *561*
Geren, B. B., 118, *119*
Gey, G. O., 12, *19*, 79, *85*, 100, *104*
Gey, M. K., 12, *19*, 100, *104*
Gibbons, I. R., 386, 392, 404, 406 *419*, *420*
Gibson, R., 428, *462*
Gill, P. M., 73, *85*
Glasstone, S. 547, *561*
Glees, P., 118, *119*
Glimcher, M. J., 549, 550, *561*
Glücksmann, A., 51, *60*, 416, *420*
Godman, G. C., 213, *235*, *236*
Goldberg, B., 213, *235*
Goldman, H. H., 246, *258*
Goldwasser, E., 541, *563*
Gontcharoff, M., 379, 388, 394, *420*
Good, R. A., 528, *561*
Goodrich, E. S., 390, *420*, 429, 432, 437. 440 441, 442, 443, 444, 445, *461*, 468, *512*, 524, 529, 544, 545, 546, *561*
Goormaghtigh, N., 503, *512*
Gorbman, A., 353, 356, *361*, 387, *422*
Gowans, J. L., 532, 533, 539, 540, *560*, *561*, *562*
Grafflin, A. L., 452, *462*
Granick, S., 380, *422*
Gray, J., 299, *335*, 537, *562*
Green, H., 213, *235*
Green, K., 356, *361*
Greenberg, R., 406, *420*
Greenlee, T. K., Jr., 124, *139*, 188, *205*, 228, *236*
Greep, R. O., 509, *512*
Grobstein, C., 74, 75, 77, 78, 79, *85*, 124, *138*, 472, *513*
Gross, F., 505, *512*
Gross, J., 54, *60*, 71, *85*, 146, 147, *154*
Grossfeld, H., 213, *235*, 533, *561*
Gruenwald, P., 470, 483, *512*
Gurdon, J. B., 130, 131, *138*, 180, 193, *204*
Gurner, B. S., 134, *138*
Gurney, C. W., 541, *563*
Gustaffson, B., 475, *511*
Gustafson, T., 113, *119*, 246, *257*

Hadži, J., 135, *138*, 210, *235*
Häggendal, J., 416, *420*
Härde, S. 189 *204*, 221, *236*
Hagey, P. W., 324, *336*
Hagiwara, A., 184, *203*
Hall, K., 486, *513*
Hallauer, C., 125, *138*
Halmi, N. S., 477, *513*
Halpern, B. P., 380, *419*
Ham, A. W., 504, *513*, 549, *561*
Ham, R. G., 215, *235*
Hancox, N. M., 97, 98, *104*, 558, *560*
Hanson, E. D., 231, *235*
Hanson, J., 314, 323, *335*
Hanström, B., 243, 247, *258*
Harnack, M., 394, *422*
Harris, H., 131, *138*, 532, *561*
Harris, J. E., 303, *335*
Harris, M., 43, *60*
Harrison, R. G., 8, 10, *18*, 408, *420*
Hartline, E. J. H., 162, 163, *176*
Hartroft, P. M., 504, *513*
Hartroft, W. S., 504, *513*
Haswell, W. A., 429, *462*
Hauschka, S. D., 79, *85*, 250, *258*
Hauschka, T., 44, *61*
Hawkins, N. M., 81, *85*
Healy, G. M., 8, *18*, 37, 42, 47, *61*, 214, *234*
Heaton, J. B., 44, *61*
Heaysman, J. E. M., 24, *59*, 126, *138*

Heerd, E., 396, *420*
Herrick, C. J., 367, *420*
Hertig, A. T., 509, *511*
Hess, A , 312, 313, *335*
Hesse, R., 399, 400, *420*
Hett, J., 470, 471, 483, 485, *513*
Hewer, T. F., 488, *513*
Hickman, C. P., 357, *361*
Higashino, S , 189, *204*, 418, *421*
Highberger, J. H., 54, *60*
Hillis, W. D., 79, *85*
Hiscoe, H. B., 107, *120*
Hoar, W. S., 357, *361*
Hörstadius, S., 136, *138*, 200, *204*, 498, *514*
Hoffman, R. S., 43, *60*
Hofmann, H. J., 302, *335*
Hogue, M. J., 29, 52, *59*, 111, 117, *119*, 215, *236*
Hollande, A., 165, *176*
Holmgren, N., 437, *462*, 526, 530, *561*
Holtfreter, J., 71, 72, *85*, 126, *138*, 174, *176*, 401, 402, *421*
Horton, C. L., 48, *60*
Houlihan, R. K., 244, 245, 246, *259*
Howard, A., 39, *61*
Howard, J. G , 541, *560*
Howells, G. P., 507, *514*
Hubrecht, A. A. W., 273, *279*, 296, 326, *335*, 426, 432, 437, *462*
Hudson, C. L., 450, 451, *463*
Hudson, G., 537, *561*, *563*
Hueper, W. C , 43, *61*, 95, *104*
Huggins, C., 488, *513*
Hughes, A., 106, 107, *119*
Hulliger, L., 99, *104*, 220, *235*
Huntsman, R. G., 184, *204*
Huxley, J. S., 146, *154*, 408, *421*
Hyman, L. H., 251, 253, *258*, 439, *462*

Immers, J., 44, *61*, 202, *204*, 221, 223, *235*, 498, *514*
Irving, J. T., 548, 549, *561*
Ivanov, A. V., 291, 293, *335*
Iwata, F., 328, *335*

Jackson, S. F., 213, 220, *235*, *237*
Jacobson, L. O., 541, *563*
Jacoby, F., 8, *18*, 30, 39, 40, 41, 43, 44, *61*, *63*, 91, 92, 93, 99, *104*, 124, *138*
Jaffe, L., 192, *204*
Jakus, M. A., 218, *235*
James, D. W., 188, *204*, 220, 228, *234*, *235*
Jennings, J. B., 268, 269, *279*, 351, 352, *361*, 428, *462*, 517, 520, *561*
Jennings, M. A., 98, *104*
Jensen, C. E., 477, *515*
Jensen, D. D., 296, *335*, *336*, 370, *421*
Jewell, P. A., 386, 389, *419*
Johannsen, O. A., 328, *336*
Johnson, B. C., 70, *86*
Johnson, F. B., 216, *235*
Johnson, M. L., 114, *119*
Johnson, W. H., 255, *258*
Jones, B. M., 137, *138*
Jordan, H. E., 543, *561*
Josephson, R. K., 243, *258*
Jullien, A., 389, *422*
Junqueira, L. C. U., 520, *561*
Jurand, A., 200, *204*

Kahmann, H., 482, *514*
Kallenbach, E., 548, *561*
Kallman, F., 472, *513*
Kamhi, S., 549, 550, *561*
Kanno, Y., 189, *204*, 418, *421*
Kappers, J. A., 394, 397, *421*
Kare, M. R., 380, *419*
Karlson, P., 489, *513*
Katsuma Dan, 188, *204*
Keilin, D., 183, 184, *204*
Kelly, D. E., 188, *204*, 394, *421*
Kemp, N. E., 477, 478, *513*
Kempton, R. T., 198, 199, *203*, 508, *511*
Kendal, L. P., 43, *63*
Kent, P. W., 98, *104*
Kepner, W. A., 249, 255, *258*
Kerr, J. G., 309, 311, *336*
Keutel, H. J., 503, *513*
Keys, A. B., 352, *361*
King, E. S. J., 56, *61*
King, R. L., 547, *560*
King, T. J., 130, *139*, 193, *204*
Kingsbury, B. F., 526, *561*
Klein, N., 249, *258*
Knight, E. J., 533, 539, 540, *561*
Knight-Jones, E. W., 286, 288, 289, *336*, 367, *421*
Koblick, D. C., 246, *258*
Koch, H. J., 246, *258*
Koenig, H., 112, *119*

Kon, S. K., 380, *422*
Konigsberg, I. R., 79, *85*, 250, *258*, 312, 321, 322, 323, *336*
Konikov, W. M., 483, *512*
Kordik, P., 244, 245, *258*
Korenchevsky, V., 486, *513*
Kozak, W., 321, *335*
Krahl M. E., 179, *204*
Krainin, J. M., 399, *423*
Krebs, H. A., 221, *235*, *237*
Kroeger, H., 489, *513*
Krüger, P., 316, *336*
Kümmel, G., 445, *461*
Kuffler, S. W., 316, 331, *336*
Kulonen, E., 223, *235*
Kutsky R. J., 43, *60*

Lacassagne, A., 535, *560*
Lännergren J., 317, *336*, *337*
Lamont, D. M., 43, *59*
Lamport, D. T. A., 214, *235*
Langley, J. N., 116, *119*
Lankester, E. R., 426, *462*
Lasnitski, I., 43, *61*
Latta, J. S., 534, *562*
Leaf, A., 494, 496, *514*
Leblond, C. P., 209, *234*, *235*
Lechenault, H., 330, *336*, 371, 388, *420*, *421*
Leeson, T. S., 504, *513*
Legallais, F. Y., 17, *18*
Lehmann, H., 184, *204*
Leloir, L. F., 221, *235*
Leloup, J., 354, *361*
Lender, T., 249, *258*, 456, *461*, 480, *513*
Lenhoff, H. M., 216, *234*, *235*
Lenicque, P., 113, *119*
Lentz, T. L., 149, *154*, 217, 229, *235*, 240, 247, *258*
Leslie, I., 41, *59*
Lever, A. F., 503, *511*
Levi G., 113, *119*
Levi, G. M., 213, *235*
Levintow, L., 47, *60*, 221, *234*
Levy, M., 47, 48, *60*, 73, *85*
Lewin, R. A., 159, *176*, 380, *421*
Lewis, M. R., 24, 25, 26, 29, 30, 52, 56, *61*, 319, 320, *336*, 535, *562*
Lewis, W. H., 23, 24, 25, 26, 29, 30, 52, 56, *61*, 95, 97, 100, *104*, 319, 320, *336*, 532, *561*
Liebman, E., 137, *139*
Likely, G. D., 5, *6*, 36, 46, *61*, *62*
Lillie, R. D., 410, *421*
Lindahl, P. E., 202, *204*
Ling, E. A., *265*, *280*, 347, 353, *361*, 370, 375, 379, 401, *421*
Linker, A., 213, *235*
Liron, M., 541, *560*
Littau, V. C., 501, 502, *514*
Little, J. M., 482, *513*, 521, *561*
Littlejohn, L., Jr., 90, *104*, 414, *422*
Liu, C. K., 352, *361*
Lobban, M. C., 488, *513*
Locke, M., 188, *204*
Loeb, L., 8, *18*, 123, *139*
Loewenstein, W. R., 189, *204*, 418, *421*
Lourie, E. M., 246, *258*
Löwenstein, O., 367, *421*
Lowther, D. A., 221, *235*
Lowy, J., 314, 323, *335*
Lucy, J. A., 52, *59*, 70, *84*, 380, *419*, 489, 491, 493, 495, *512*, 558, *560*, *561*
Lumsden, C. E., 118, *119*
Luse, S. A., 118, *120*
Lutwak-Mann, C., 475, 476, 477, *513*

McCord, C. P., 396, *421*
McCullagh, P. J., 532, *560*
MacDowell, M. C., 450, *463*
McIssac, W. M., 396, *421*
Mackay, B., 306, 307, 309, *336*
McKeehan, M. S., 408, *421*
Mackie, G. O., 244, *259*
McKinney, R. L., 213, *235*
Macklin M., 243, *258*
McLoughlin, C. B., 79, *85*, 547, *561*
MacNichol, E. F., Jr., 415, 417, *421*, *423*
McNutt, S. H., 482, 484, *515*
McQuilkin, W. T., 127, *138*
Malkin, T., 488, *513*
Malmfors, T., 416, *420*
Mannweiler, K. I., 4, *5*, 488, *513*
Manton, S. M., 293, *336*
Mapes, M. O., 180, *205*
Marchesi, V. T., 539, *562*
Marcus, P. I., 46, *62*
Marks, W. B., 415, *421*
Marques-Pereira, J. P., 209, *235*
Marriott, C. H., 255, *259*
Martinovic, P. M., 108, 111, *120*

Mathews, M. B., 53, *61*, 216, 222, *236*, 555, 558, *562*
Matthews, L. H., 488, *513*
Matty, A. J., 356, *361*
Mauro, A., 332, *336*
Mawas, J., 543, *562*
Maximow, A. A. 57, *61*, 469, 471, *513*, 546, 551, *562*
Maxwell, D. S., 402, *421*
Mears, K., 180, *205*
Medawar, P. B., 8, *18*, 44, 45, *61*, 73, 77, *86*, 125, *139*, 499, *513*
Meek, E. S., 537, *563*
Meisel, E., 317 *337*
Mellanby, E., 52, *60*, 69, 70, 82, *85*, 97, 98, *104*, 558, *560*
Merker, von E., 391, 397, *421*
Metschnikoff, E. 198, *204* 283, 326, *336*
Meusy, J., 479, *513*
Meyer, H., 113, *119*
Meyer, K., 213, *235*
Micklem, H. S., 537, 562
Mihálik, P. V., 242, *259*
Millen, J. W., 402, *421*
Miller, C. A., 255, *258*
Miller, J. F. A. P., 539, *562*
Miller, R., 505, *514*
Miller, R. A., 488, *513*
Minchin, E. A., 142, 145, 146, 150, *154*, 166, *176*, 209, *236*
Miszurski, B., 68, *86*
Mitchell, J. F., 265, *280*, 344, *361*
Mitropolitanskaya, R. L., 246, *259*
Monné, L., 189, *204*, 221, *236*
Montgomery, H., 450, *462*
Moog, F., 488, *513*, 520, *562*
Morgan, B. B., 482, 484, *515*
Morgan, J. F., 8, *19*, 48, *62*, 214, *236*
Morris, C. C., 213, *236*
Morris R., 355, *361*
Morton, H. J., 8 *19*, 48, *62*, 214, *236*
Moscona, A., 13, *19*, 71, *86*, 125, *139*, 221, *236*, 413, 414, *421*
Moscona, H., 125, *139*
Moulder, P. V., 488, *513*
Müller, O., 509, *514*
Mueller, P., 503, *514*
Mugard, H., 159, *176*
Munger, B. L., 542, *562*
Munshi, J. S. D., 352, *361*
Murnaghan, D. P., 113, *120*
Murray, M. R., 108, 110, 112, 114, 115, 118, *119*, *120*
Murray, P. D. F., 529, *562*
Murray, R. G., 132, *139*, 538, *562*

Nachmias B. T., 317, *336*
Nawitzki, W., 273, *280*, 369, *421*, 435, 437, 438, *462*
Needham, J., 202, *204*, 223, *236*, 498, *514*
Negishi, K., 417, *423*
New, D. A. T., 200, 201, *204*
Newth, D. R., 398, *421*
Nilsson, S. E. G., 411, 417, 418, *421*
Noel, R., 389, *422*
Noell, W. K., 412, *422*
Nonidez, J. F., 114, *120*
Nordmann, M. 82, *86*
Northcote, D. H., 214, *235*

Ohno, S., 480, *514*
Ohuye, T., 272, *280*, 426, *462*
Okano, M., 363, 365, *422*
Oksche, A., 394, 398, *422*
Oliver, J., 450, *463*
Olivo, O. M., 106, *120*
Osmond, D. G., 537, *563*
Oudemans, A. C., 273, *280*, 437, *462*
Outka, D. E., 156, 172, *176*
Overton, J., 188, 189, *204*
Owens, O. von H., 12, *19*
Oyama, V. I., 47, 48, *60*, 73, *85*
Oztan, N., 387, *422*

Pace, N., 246, *258*
Padykula, H. A., 316, 317, *335*, *336*, *337*
Page, S. G., 312, *336*
Palade, G. E., 100, *104*, 188, *204*
Pannese, E., 548, *562*
Pantin, C. F. A., 439, *462*
Pappas, G. D., 407, *422*
Pardoe, A. V., 246, *258*
Parker, G. H., 197. *205*, 229, *236*
Parker R. A., 503, *511*
Parker, R. C., 8, 17, *18*, *19*, 30, 31, 32, 37, 39, 42, 47, *60*, *61*, *62*, 88, 89, 99, *104*, 117, *120*, 124, *139*, 214 *234*
Parker. T. J., 429, *462*
Parshley, M. S., 71, *86*
Partridge, S. M., 55, *62*

Pasieka, A. E., 48, *62*
Patchett, A. A., 214, *237*
Patel, N. G., 217, *236*
Paterson, M. C., 200, *205*
Paton, S., 303, *336*
Paul, J., 99, *104*
Peachey, L., 307, *336*
Pearse, A. G. E., 357, *361*, 542, *562*
Pearson, J. L., 489, 496, 497, *514*
Peart, W. S., 505, *514*
Pease, D. C., 402, *421*
Pedersen, K. T., 232, *236*
Pelc, S. R., 39, *61*, 70, *85*
Penn, R. D., 189, *205*
Person, P., 558, *562*
Peters, A., 306, 307, 309, *336*
Peterson, E. R., 114, *120*
Peterson, R. D. A., 528, *561*
Phillips, J. H., 216, *236*
Phillips, L. S., 396, 397, 398, *423*
Philpott, D. E., 558, *562*
Picken, L. E. R., 216, *236*, 557, *562*
Piddington, R., 221, *236*
Pierce, J. A., 450, *462*
Pittam, M. D., 157, *176*
Plack, P. A., 380, *422*
Platt, M., 43, *61*
Policard, A., 452, *462*, 533, *562*
Pollack, H., 189, *203*
Pollard, J. K., 214, *236*, *237*
Pomerat, C. M., 43, 44, *62*, 90, *104*, 108, 109, 117, 118, *119*, *120*, 414, *422*
Popper, H., 406, *420*
Porte, A., 542, *563*
Porter, G. A., 494, *512*
Porter, K. R., 4, *6*, 220, *234*, *236*, 405, *422*, 477, *514*
Potts, W. T. W., 507, *514*
Press, N., 255, *259*
Price, G. C., 437, 438, *462*
Propst, A., 509, *514*
Ptashna, M., 133, *139*
Puck, T. T., 36, 46, *62*, 72, 80, *86*
Pudney, J., 474, *514*

Quastel, J. H., 215, *236*
Quay, W. B., 391, 398, *420*, *422*

Randall, J. T., 55, *62*, 213, *234*
Rasmont, R., 141, *154*
Rasquin, P., 528, 529, *562*
Rawles, M. E., 312, *337*
Regaud, C., 452, *462*
Reutter, K., *344*
Rex, R. O., 91, *103*
Rich, A. R., 535, *562*
Richards, A. N. 450, 451, *463*
Richardson, A. P., 240, *259*
Riepen, O., 271, *280*, 325, *336*, 427, *462*
Riggs, T. R., 221, *234*
Rinaldini, L. M. J., 13, *19*, 25, 27, *62*, 71, *86*, 125, *139*
Ritchie, A., 390, *422*
Roberts, D. K., 534, *562*
Robertson, J. I. S., 503, *511*
Robertson, Van B., 504, *512*
Robinow, C., 125, *139*
Robinson, G. M., 44, *61*, 125, *139*, 499, *513*
Robinson, R., 44, *61*, 125, *139*, 499, *513*
Robison, R., 31, 51, *60*
Röhlich, P., 255, 256, *259*, 380, 383, *422*
Rogers, G. E., 402, *421*
Rogers, H. J., 221, *235*
Roots, B. I., 441, *463*
Rose, G. G., 100, 101, *104*, *105*
Rosetti, F., 114, *120*
Ross, D. M., 398, *421*
Ross, R., 124, *139*, 188, *205*, 220, 228, *236*
Roth, S. I., 542, *562*
Rougvie, M., 146, 147, *154*
Rouiller, C., 220, *236*
Royle, J., 44, *62*
Rubin, A., 215, *236*
Rudin, D. O., 503, *514*
Rudy, P. P., 507, *514*
Rugh, R., 432, *463*
Rumery, R. E., 324, *336*, 476, *511*
Runnström, J., 202, *204*, 223, *235*, 498, *514*
Rushton, W. A. H., 415, *419*, *422*,
Russell, M., 43, *61*
Russell, M. A., 95, *104*
Ruud, J. T., 180, *205*
Ryley, J. F., 183, *204*

Sage, M., 353, 356, *360*
Sager, R., 380, *422*
Salensky, W., 328, *336*, 384, *422*
Sanders, A. G., 17, *18*, 91, *104*
Sandison, J. C., 17, *19*, 91, *105*

Sanford, K. K., 5, *6*, 8, 13, *18*, 36, 46, *61*, *62*, 127, *138*
Sauer, F. C., 208, *236*, 244, *259*
Schafer, I. A., 503, *514*
Schanen, J. M., 214, *234*
Schardinger, F., 156, *176*
Scharrer, B., 330, *337*
Scharrer, E., 389, *422*
Scherft, J. P., 213, *236*
Schilling, E. L., 12, *18*, 46, *60*
Schirner, H., 519, *562*
Schlenk, W., Jr., 482, *514*
Schmidt, G. A., 326, 327, 328, 329, *337*
Schmitt, F. O., 54, *60*
Schreiner, K. E., 457, *463*, 479, *514*
Schröder, O., 383, 391, *422*
Schuster, F., 159, 167, *176*
Scott, D. B., 553, *562*
Scott, T. W., 475, 476, *515*
Scriver, C. R., 503, *514*
Seaman, G. R., 244, 245, 246, *259*
Sedar, A. W., 4, *6*, 220, *236*
Seed, J., 41, *62*
Selye, H., 488, *514*
Setchell, B. P., 450, *463*, 475, 476, 477, *514*, *515*
Shaffer, B. M., 31, *62*
Shannon, J. E., 8, 13, *18*
Sharp, G. W. G., 494, 496, *514*
Shaw, J. H., 509, *512*
Shelley, H. J., 245, *258*
Shelton, E., 532, 533, *562*
Shen, S. C., 246, *258*
Short, R. V., 476, *514*
Shostak, S. 217, *236*
Sidman, R. L., 410, *422*
Siegel, B. W., 240, *257*
Simms, H. S., 9, *19*, 43, *62*, 71, *86*
Sjöstrand, F. S., 380, 392, *422*
Skaer, R. J., 211, 216, 218, *236*, *237*, 240, 242, *259*
Sleigh, M. A., 193, *205*
Smelser, G. K., 407, *422*
Smith, A. U., 476, *514*
Smith, D. S., 501, 502, *514*
Smith, H. W., 450, 451, *463*, 526, *562*
Smith, I. T., 475, 477, *514*
Smith, J., 180, *205*
Smith, J. B., 480, *514*
Smith, J. E., 326, *337*
Smith, R. H., 220, *235*, *237*
Smith, R. S., 317, *336*, *337*
Socolar, S. J., 189, *204*, 418, *421*
Sørensen, E., 475, 482, *514*
Sokal, Z., 146, 147, *154*
Sorenson, G. D., 533, 534, 535, 536, 537, *559*, *562*
Sotelo, J. R., 477, *514*
Speidel, C. C., 113, *120*, 543, *561*
Spemann, H., 408, *422*, *423*
Springer, G. F., 215, *236*
Standish, M. M., 489, *511*
Stein, J. M., 316, *337*
Stephenson, E. M., 43, *59*
Steven, D. M., 398, *423*
Stevenson, R., 134, *138*
Steward, F. C., 180, *205*, 214, *236*, *237*
Steyn, W., 391, *423*
Stillman, N. P., 9, *19*, 43, *62*
Stoeckel, M. E., 542, *563*
Stone, L. S., 382, *423*
Stordahl, A., 202, *204*
Stout, A. P., 108, 110, 114, 115, *120*
Strangeways, T. S. P., 8, *19*
Straus, W. L., 312, *337*
Svaetichin, G., 417, *423*

Taborsky, R. G., 396, *421*
Tahmisian, T. N., 501, *511*
Taliaferro, W. H., 249, 255, *258*, *259*
Tansley, K., 414, *423*
Tardent, R., 215, *237*
Taylor, A. N., 396, *421*
Taylor, N. B., 99, *103*
Terner, C., 221, *237*
Thomas, J., 55, *62*
Thomas, J. A., 100, *105*, 125, *139*
Thomas, T. B., 517, 519, *563*
Thompson, J. F., 214, *237*
Thompson, S. Y., 380, *422*
Thorpe, A., 356, *360*, *361*
Tibbs, J., 246, *259*
Törö, E., 74, 76, *86*
Török, L. J., 255, 256, *259*, 380, 383, *422*
Tokuyasu, K., 380, *423*
Travis, D. F., 549, 550, *561*
Trowell, O. A., 31, 40, 41, 43, *61*, *62*, 449, *463*, 532, 533, 539, *563*
Tuft, P., 200, 201, *204*, *205*
Tull, F. A., 214, *234*

Turner, R. S., 243, *259*
Tuzet, O., 141, *154*, 190, 191, *204*
Twitty, V. C., 246, *259*, 382, *423*
Tyler, A., 470, *514*
Tyzzer, E. E., 155, 156, *176*

Ude, J., 249, *259*
Uotila, U. U. 470, *514*

Van de Kamer, J. C., 395, *423*
Vander, A. J., 505, *514*
Vandermersche, G., 141, *154*
Varandani, P. T., 70, *86*
Vaubel, E., 51, 56, *62*
Vaughan Williams, E. M., 316, *336*
Vickers, T., 167, *176*, 352, *361*
Villee, C. A., 489, *515*
Villegas, G. M., *423*
Voglmayr, J. K., 450, *463*, 475, 476, *515*
Vogt, M., 118, *119*

Wachstein, M., 317, *337*
Waddington, C. H., 78, *86*
Waites, G. M. H., 450, *463*, 475, 476, *515*
Wald, G., 386, 392, 399, 412, 415, *419*, *423*
Walker, A. M., 450, 451, *463*
Walker, P. M. B., 42, *63*
Walls, G. L., 409, 411, *423*
Waltz, H. K., 8, 13, *18*
Wang, H., 113, 116, *120*
Wang, Y. L., 184 *204*
Warwick, A. C., 535, *559*
Watterson, R. L., 208, *237*
Waymouth, C., 8, *19*, 41, 43, 47, *59*, *63*, 221, *237*
Webb, M., 214, *234*
Weber, A. F., 363, 365, *422*, 482, 484, *515*
Weisel, G. F., 482, *515*
Weiss, L. P., 90 91, 95, 96, *105*
Weiss, P., 13, 14, 15, *19*, 72, *86*, 107, 113, 114, 115, 116, *120*, 408, *423*, 553, 554, *563*
Weissman, G., 489, *511*
Wells, J., 386, *423*
Wells, M. J., 386, *423*
Wen, I. C., 245, *259*
Wenrich, D. H., 155, *176*
Wense, T., 246, *258*
Wersäll, I., 367, 368, *420*, *421*
Wessels, N. K., 547, *563*
Westfall, B. B., 81, *85*
Westfall, J. A., 299, *335*, 391, 392, 394, 397, *420*
White, A., 529, 533, *560*
White, E. I., 390, *423*
White, P. R., 37, 47, *63*
White, W. F., 541, *563*
Whitehouse, M. W., 98, *104*
Whiting, H. P., 303, *335*
Williams, R. G., 487, *515*
Willmer, E. N., 8, 17, *18*, *19*, 30, 36, 37, 39, 40, 41, 43, 44, *61*, *62*, *63*, 158, 169, 170, 171, 172, *176*, 202, *205*, 265, *280*, 344, 352, 354, 355, *360*, *361*, 403, 415, 416, *423*, 489, 490, 492, 493, 496, 497, 498, *514*, *515*
Wilson, C. B., 287, *337*, 367, *423*, 426, 429, *463*
Wilson, H. V., 146, *154*
Wimsatt, W. A. 472, *515*
Winter, G. C. B., 537, *563*
Wintrebert, P., 303, *337*
Wintrobe, M. M., 535, *562*
Wislocki, G. B., 477 *515*
Witkop B., 214, *237*
Witschi, E., 455, *463*
Wolf, G., 70, *86*
Wolken, J. J., 380, 385, *423*
Wong, A., 245, *258*, *259*
Wood, R. L., 188, 189, *205*, 228, *237*
Woodward, G., 43, *61*
Wulfert, J., 225, *237*
Wurtman, R. J., 396, 397, 398, *423*

Yamada, E., 380, 405, 406, *422*, *423*, *424*
Yates, H. B., 42, *63*
Yoffey, J. M., 537, *561*, *563*
Young, J. Z., 107, *120*, 282, 312, *335*, 337, 367, 381, 383, 386, 387, 388 394, 395, 396, *419*, *424*, 429, *463*
Young, R. W., 412, *424*

Zachariae, F., 477, *515*
Zarnik, B., 457, *463*
Zeleny, C., 200, *205*
Ziegler, B., 343, *361*
Ziegler, M., 505, *512*
Zonana, H. V., 385, *424*

INDEX OF ANIMAL AND PLANT NAMES

Specific names in brackets after the generic name indicate that the particular species is not always specified.

Groups above genera will be found in the Subject Index.

Alciopa, 399
Allolobophora chlorotica, 441
Amblystoma, 382, 408
Ameiurus, 413
Amoeba (*proteus*), 4, 100, 175
Amphioxus, 3, 232, 294-5, 299-301, 304-8, 310, 312, 315, 329-30, 332, 353, 363, 383, 429-30, 434, 444-6, 448, 454, 457, 459-60, 479, 517, 523, 525, 541
Amphiporus (*lactifloreus*), 269, 271, 554
stanniusi, 458-9
Arbacia punctulata, 137
Ascaris, 216, 323
Astyanax, 528, 529, 533

Balanoglossus, 285, 517, 523
Baseodiscus, 439, 441
Bdellostoma (stouti), 429, 437-8, 531
Bdelloura candida, 243
Bean, 214
Birkenia elegans, 390

Carinella, 437
Carinina, 437, 449
Carinoma, 271
Carrot, 180, 214
Cat, 319, 488, 532
Cat fish, 412-3
Cephalothrix major, 269-70, 437, 439, 449, 461
Cerebratulus (*lacteus*), 200, 271, 287, 294, 298, 359, 370, 398, 426, 429-30, 446
Chaenocephalus aceratus, 180
Chaos diffluens, 175
Chlamydomonas moewusii, 159, 380
reinhardi, 380
Clathrina blanca, 509
coriacea, 145, 150-1
Codonosiga botrytis, 165-6
Coleonyx variegatus, 411
Cordylophora lacustris, 199
Cormorant, 398
Corynactis viridis, 216
Cow (cattle), 316, 412, 475-6, 482, 484
Crayfish, 528
Crocodile, 412
Crown-gall, 214
Cuneonemertes, 367

Dendrocoelum lacteum, 255
Dog, 98, 363, 365, 389, 532, 539
Drepanophorus (*albolineatus*), 274, 370, 383, 385, 458-9
Duck, 398

Earthworm (see also under specific names), 478
Echinus, 202
Eel, 180-181, 299, 352
Elephant, 555
Eriocheir sinensis, 246
Espejoia mucicola, 159
Eunemertes, 274
Eupolia, 274

Ficulina ficus, 144, 148-9

Flounder, 357
Fowl (chick), 8, 11, 21, 23-7, 29, 33, 39, 40, 55, 65, 69, 70, 82, 91-3, 125, 155, 188-9, 198-201, 312, 340, 412, 522, 540
Frog, 9-10, 91, 180, 227, 245, 332, 395, 399, 401, 405, 411-2, 477
Fucus, 192
Fundulus heteroclitus, 389

Galleria, 456, 480
Garter Snake, 313
Gasterosteus aculeatus, 357
Geonemertes (*palaensis*), 269, 383, 391, 394, 397, 399, 437, 439, 441
Geotria australis, 391-2
Glossobalanus minutus, 286
Glycera, 445
Gonium pectorale, 165-6, 187
Gonothyraea loveni, 225, 229
Gorgonorhynchus, 292-3
Grantia compressa, 141, 190
Grasshopper, 501
Guinea-pig, 92-3, 392
Gyratrix hermaphroditus, 251-2, 260, 276

Hag-fish, 389
Hamster, 215, 488
Histomonas meleagridis, 155-6
Hoplochaetella bifoveata, 440-1, 467
Hubrechtia, 271
Hydra, 189, 217, 228, 244

Jamoytius kerwoodii, 390

Lambellisabella zachsi, 291
Lampetra fluviatilis, 433
  planeri, 394
Lampito (*Megascolex*) *trilobata*, 524
Lamprey, (see also Subject Index), 294-5, 306, 308-9, 354, 391-2, 396
Lepidosiren, 309, 311
Leucoselenia, 150, 509
Limulus (*polyphemus*), 123, 137, 399
Lineus fuscoviridis, 426
  lacteus, 388
  ruber, 240-1, 264, 267, 272-3, 307, 329, 331, 340-2, 345-7, 349-51, 353-6, 363, 370-1, 379, 388, 394, 398, 427, 430, 432-3, 435, 439, 446-7, 456, 502, 509, 530.
Lingula, 185
Loligo, 387
Lumbricus terrestris, 441

Megascolex cochinensis, 440-2
Microstomum caudatum, 248-9
Micrura, 274
Monkey, 209
Monocelis, 369
Mouse, 17, 38, 81, 91, 127, 282, 483, 485, 488, 537
Mud-puppy, 386, 392-3, 415
Mycale contarenii, 152
Mytilus edulis, 245
Myxine, 349, 398-9, 437, 446, 449, 452-4, 457, 519-21, 523, 526-7, 530-1

Naegleria gruberi, 156-76, 177, 184-5, 187, 190, 195-7, 202, 220, 226, 231, 240, 350, 380, 439, 455, 460, 466, 475, 481, 489, 494-8, 501, 505, 508-10, 532, 534, 570-3
Nectonemertes mirabilis, 297
Necturus, 451
Nemertopsis, 287
Neuronemertes, 275

Octopus, 381, 387, 389
Opossum, 484
Opsanus tau, 526
Orchestia gammarella, 479-80, 482
Ototyphlonemertes, 369

Paramecium, 4, 218, 246
Petromyzon, 353, 363-4, 517, 544
Pheasant, 155
Pheretima, 524
Phoronis, 529
Phrynosoma cornutum, 453
Phyllodoce paretti, 443
Pigeon, 398, 507
Polycelis tenuis, 217
Potato, 214
Priapulus caudatus, 443
Procarinina, 269, 369, 435, 437-8
Prorhynchus, 253
Prosorhochmus, 287, 294
Prostoma, 269

Rabbit, 17, 91, 245, 407, 532, 535
Rana pipiens, 411
Rat, 209, 341-2, 396, 398, 404, 406, 452, 477, 486, 507, 539-40, 549

Saccoglossus horsti, 285, 286, 356
Sepia, 388, 428, 559
Sheep, 450
Shrew, 316
Siboglinum caulleryi, 291, 293
Sparrow, 398
Sphenodon, 397
Squid, 385, 572
Stenostoma, 248-9
Stickleback, 356-7
Sycamore, 214
Sycandra raphanus, 142
Sycon (*raphanus*), 149, 190-1

Taeniosoma (*cingulatum*), 439, 525
Tetrahymena, 246
Tetramitus rostratus, 155, 157, 159, 162, 164, 165, 172, 187
Tetrastemma, 274
Toad, 317, 356
Tonoscolex, 524
Triturus, 382
Tubulanus, 269
Turkeys, 155
Turtle, 436

Valencinia, 271, 274
Vanadis formosa, 399, 400
Volvox, 187

Whiting, 299

Xenopus laevis, 180, 200, 201, 317, 409

SUBJECT INDEX

Abdominal pores, 467-8
Absorption
  intestinal, 523
Acetaldehyde
  and mechanocytes, 44
Acetylcholine
  and ciliated cells, 173, 244-6
  and muscles of *Lineus*, 264, 304, 307, 344
  and *Naegleria*, 173
  and placenta, 246
  and proboscis, 265, 344
  synthesis of, 118
  as transmitter, 244, 246
Acetylcholinesterase (see also Cholinesterase) and bipolar cells, 149, 247
Acetylthiocholine
  method for nerve endings, 308-9
Acoel(a) (Acoeloid stage), 206
  characters of, 229-34
  derivation of, 231
  sagittocysts in, 217
Actin (see also Actomyosin)
  in insect muscle, 314
  and myosin, 3, 132, 315
Actinomycin, 8
Actomyosin
  distribution of, 325
  myoblasts and, 53
  organization of, 314-5, 323
Adaptation
  of cells in culture, 4, 5, 23, 37, 47-8, 128, 133-4, 567, 575
  of cells to media, 4, 37, 59, 567
  and econome, 567-8
  and mitosis, 133, 575
  of *Naegleria*, 162
  perpetual, 564-6
Adenohypophysis (see also Pituitary)
  of *Myxine*, 349
Adenosine
  tritiated, and lymphocytes, 539
Adenosine triphosphate (ATP)
  and collagen formation, 54
Adhesion (see also Cohesion)
  of cells, 71, 146, 175, 203, 208
  hyaluronidase and, 477
Adrenal cortex, 482-7
  control of cells in, 500
  cords of cells in, 436, 470, 482-3
  ecballism and emballism in, 499, 508
  endothelium in, 56, 95, 525
  foetal, 483
  and genital ridge, 470
  gradients in, 483, 509-10
  grafts of, 487
  and kidney, 501-2, 508
  large cells in, 470-1
  and Na/K ratio, 482
  pineal and, 396
  steroids and, 485, 488
  sudanophobic zone of, 485-6, 509
  trypan blue and, 486
Adrenalin(e)
  in eye, 412
  and nemertine muscle, 264-5, 307
  and retractor of proboscis, 344
  in sponges, 149, 247
Adrenal medulla, 2, 484-5
Adrenocorticoids, 491
Adrenocorticotrophic hormone (ACTH)
  and adrenal cortex, 508, 529
  and lymph nodes, 533

and pronephros, 528-9, 533
Aetiocholanolone, 490
Agar
and epithelial growth, 66, 163
Age
of chick, for extract, 39-40
of chondroblasts, 53
of cultures, 37
of tissue cultured, 51, 107
Aggregation
of cells, 71-2, 165, 547, 568
of retinal cells, 413-4
of sponge cells, 71, 146
Agoranome, 565-9, 574
Alanine
and growth, 48
and nematocysts, 216
Alar plate, 329
Albumen
amoebocytes and, 99, 125
culture on, 200-1
flotation, 90
serum, 99
Alcian blue
and cephalic organ, 374
and mucous cells, 263, 269
and nemertine skin, 263
staining of buccal cavity, 341
Aldehydes
and mechanocytes, 44
Aldosterone
and adrenal cortex, 487-8
and kidney, 488, 496
mode of action of, 488, 492, 494, 496, 502
renin and, 503
and saliva, 442, 488
secretion of, 488, 497, 502
and zona glomerulosa, 487-8, 499
Algae
iodine and, 356
Alimentary system (see also Gut, Intestine)
and ionic balance, 448
muscle and, 268
in nemerteoids, 267-8, 279, 430, 516
and proboscis, 262, 277, 293
in rhabdocoeloids, 250-1
in vertebrates, 516-25
Amacrine cells
in retina, 403, 416
Ameloblasts, 57, 546-51
Amino acids
and bradykinin, 220
and collagen, 48, 214
D and L forms, 47-8
and growth, 42-3, 47-8, 73
and ionic balance, 220-1
of keratin, 68
pool of, 47
specific sequences of, 179, 184
uptake of, in rods and cones, 412
Amino sugars, 148, 221
Ammocoete larva
chloride cells in, 352
iodine in, 353-6
and light, 394-5
and origin of pancreas, 517-8
pronephros in, 531
Ammonia (NH_3)
and amino sugars, 221
as excretory product, 254
in kidney tubules, 450-1
Ammonium
chloride and *Naegleria*, 174, 380
nitrate and flagella, 380
quaternary, and macrophages, 100, 125
quaternary, and *Naegleria*, 172-4
Amnion
in nemertine embryogeny, 326, 329
Amoeba(e)
epithelioid behaviour of, 163-4
flagellate forms of, 155-76, 509-10
"limax" movement of, 126
pinocytosis in, 100, 175
polarity of, 192
Amoebocytes
as cell family, 103, 134
clones of, 127
definition of, 20
emballic, 194, 210, 220, 223, 225, 466, 498-9, 505, 529, 538
and flagellate cells, 187, 415
of fowl, 92
and germ cells, 147, 151
and giant cells, 96-8, 148
growth of, 87-103
of guinea-pig, 92-3
and haemopoiesis, 529, 534-5, 538
of king-crab, 123, 137
and mechanocytes, 21, 92-3, 220, 223-4, 231, 498, 529, 538, 557

from membrana granulosa, 476
in mesenchyme, 460
monocytes and, 89-90, 538
movement of, 91-3, 147
and muscle, 307, 312, 314, 321-2
from neural crest, 136
and neurones, 230, 247, 331
and nucleoli, 147, 410
origin of, 151-3, 155, 212
osteoclasts and, 97, 557-8
and phagocytosis, 89, 91, 99, 124, 126, 224-5
and planuloid stage, 210, 213, 225
properties of, 123-4
pseudopodia of, 87, 106, 124
pure cultures of, 98, 224
Schwann cells and, 115-6
of sponges, 137-8, 142, 144-51, 415
and sulphate, 498
surface properties of, 247, 307, 332
and totipotence, 150, 152, 210
and transport system, 213
and trypan blue, 147-8
Amphibia
animal and vegetal poles of, 200
coelomic cavity of, 432, 436
culture of tissues from, 8, 20, 22, 71-2
gastrulation in, 227
isolated cells from, 126
lens in, 382
Mauthner cells in, 114
melanophores in, 395-6
mesonephros in, 450
photoreceptors in, 384, 411-2
skin of, 553-4
Amphiblastula, 141, 190
Amphiporin, 263
Ampulla
of cephalic organ, 370, 372, 379-80, 384-5
of lateral line, 363
of nephridium, 269
of otocyst, 370
Anaemia
sickle-celled, 179, 184
Anaerobic conditions
and glycogen, 73
Anaphase
of neuroblast, 208
Anaspida
lateral line of, 367
position of, 390
Androgens
and adrenal cortex, 486
structure of, 492
Androsterone
and adrenal cortex, 486
and *Naegleria*, 173
structure of, 490
Angiotensin
and zona glomerulosa, 505-6, 508
Animalization
of embryos, 202
sulphate and, 223, 498
Animal pole
of amphibian embryo, 200, 202
blastulae from, 200
cells composing, 192, 197
in sponges, 196
Annelids
coelomic cavities in, 432, 434
eyes in, 399
gonoducts in, 468
metamerism of, 281, 298, 449
muscles of, 312, 314, 323
nephridia in, 282, 436, 439-43, 445, 467-8, 524
nerve cord in, 289
neurosecretion in, 399-400
phylogeny of, 279, 282, 284
Antero-posterior axis
establishment of, 213, 232
Antibiotics
and growth of skin, 77
and tissue cultures, 7-8
Antibodies, 532-3, 573
Antigens
and lymphocytes, 534, 539
Anura (see also Frog)
pineal eyes of, 394
Anus
absence of, in rhabdocoeloids, 251
in nemerteoids, 260-1, 268, 279, 294-5
and nephridia, 440
Aorta
development of, 429
endothelium of, 55
Aortic body
origin of, 359, 363
Apatite, 549-53
Apopyle, 142
Appendix, 524, 526

Aqueous humour
  secretion of, 406
Arachnids
  cartilage in, 558
Archaeocytes, 148
  derivatives of, 152
  and germ cells, 150-2, 470
  and trypan blue, 147-8
Archenteron
  diverticula from, 301-2, 304, 312, 314-5, 459
  of enteropneusts, 286
  fluid relationships of, 201
  and germ cells, 459
  meaning of, 226
Archinephric duct
  derivatives from, 468, 472
  origin of, 436-7, 446, 467
Archinephros, 437
Argentaffin cells, 84
Arginine
  in bradykinin, 220
  and echinoderms, 283
  and growth, 73
  and rhabdites, 218
Argyrophil cells, 82, 84
Argyrophil fibres (see also Reticulin)
  mechanocytes and, 25, 32, 36, 51, 53
  vitamin C and, 54
Armature
  of nemertine proboscis, 260-1, 277, 338
Arsenious oxide
  and cells in culture, 125
Arteries
  in cephalopods and vertebrates, 428
Arteriosclerosis, 572
Arthritis
  and cell metabolism, 224, 572
Arthropods
  eyes of, 399
  muscles of, 312
  nerve cord in, 289
  neurosecretion in, 399
  photoreceptors in, 380, 396
  phylogeny of, 279, 282, 284, 360, 367-8, 399
  segmentation of, 281
Ascidian
  blastula of, 186
Ascites tumour cells
  glycine and, 220
Ascorbic acid (see also Vitamin C)
  and collagen formation, 71, 499
  and epithelial cells, 71
  storage of 499-500
Asepsis, 7-8
Aspartic acid
  and growth, 48
  in nematocyst, 216
Astrocytes
  in culture, 116-7
  and myelination, 118
  and nerve cells, 331
Athrocytosis, 531
Atrium
  of *Amphioxus*, 445
  origin of cardiac, 429
  in pineal eye, 392
  in sponge, 191
Atrophy
  of disuse, 32, 181-2
  of muscle, 32
Atropine
  and nemertine muscle, 264
Auditory organ (see also Ear, Otocyst, etc.)
  cell types in, 84
  origin of, 362, 369
Autolysis
  and production of trephones, 99
  and selection of cells, 27
Autonomic nervous system
  fibres in, 116, 333
  origin of, 330, 333
Autoradiograph
  and iodine, 355
  and retina, 412
  and uptake of sulphur, 70
Autotomy
  in nemertines and enteropneusts, 285
Axial skeleton, 544
  myotomes and, 284, 310
Axons
  to muscle spindles, 318
  myelination of, 118
  of nerve cells in culture, 107, 112-4
  of squid, 572

Bacilli
  and giant cells, 96

Bacteria
adaptation of, 5, 37
defence against, 7-8, 242
as food for *Naegleria*, 156
haemoglobin and, 426
population size of, 13
and sex, 226
suspensions of, 17
Bacteriophage
repression of DNA in, 133
Balance (see also Symbiosis)
of cellular activities, 189, 198, 569-73
Basal granule (body)
of choanocytes, 141, 143
and cilia, 219, 365
of *Naegleria*, 159, 165, 167, 175
of photoreceptors, 393
Basal plate, 329
Basement membrane
in acoeloid, 230, 233
and agoranome, 567
in blastulae, 189
in blood vessels, 427
and cephalic organ, 374
in ciliary processes, 406-8
and differentiation, 12, 78-9
and ependyma, 330
and epidermis, 233, 239-40, 242, 262-3, 363, 548
in gonads, 470, 472
in lateral line organ, 368
and macula densa, 503
in nemerteoids, 262-3, 458
in nephridia, 445
and nerve cords, 275
and position of nuclei, 262, 330, 374
of proboscis, 276
in relation to muscle, 242, 307
in rhabdocoeloid, 238-9, 257
in salivary glands, 472
in teeth, 548, 552
Basophil cells
in nemerteoids, 267, 271-2, 350, 358, 426
Basophilia
of cartilage matrix, 52
Behaviour patterns, see Patterns
Benzidine
and peroxidase, 426
Bicarbonate (HCO_3^-)
and culture media, 8
and ecballism, 501
and hyaluronic acid, 223
and *Naegleria*, 170-1, 481
and salt balance, 353
Bile duct, 517, 519, 523-5
Bipolar cells
in cephalic organ, 375-7, 384, 401
in elasmobranchs, 333
in frontal organ, 363
in retina, 332, 376-7, 384, 403, 416-8
in sponges, 149-50
Birds
culture of tissues from, 20, 22
duodenum of, 522
light sensitivity in, 233
navigation by, 398
pigment epithelium in, 406
pineal in, 398
prechordal plate in, 340
"Blackhead"
in turkeys, 155
Blastocoel
fluid in, 189, 200, 220-1
meaning of, 226-7
and rhynchocoel, 432
Blastoderm
and blastula, 529
blood islands in, 529, 540
fluid relationships of, 200-1
Blastomeres
ecballic and emballic, 210, 213, 226
genetic information in, 180
specific proteins in, 180
of sponge, 151
Blastopore
occlusion of, 212-3
Blastula(e)
and animal pole cells, 200
and blastoderm, 529
buoyancy of, 202
and Ca^{2+}, 71
cell cohesion in, 165, 185
and cell polarity, 164
cell types in, 153, 166
culture of, 46
in different phyla, 186
of echinoderm, 46, 202
gastraea and, 224
gradient in, 192, 196, 509
ionic content in, 198, 201

and *Naegleria*, 165, 185
origin of, 184-5
osmotic regulation in, 201
of sponges, 140, 150, 165-6, 175, 185, 187 209
symbiosis in, 226, 477, 570
of *Xenopus*, 200
Blastuloid stage
cell classes in, 465
collagen fibres and, 230
developments from, 206-7, 230
gastraea and, 206, 224
gradients in, 190, 192, 254, 509, 572
and internal environment, 198-203
nephridia and, 254
properties of, 177-203
symbiosis in, 203, 332, 570
Blood
buffy coat of, 90, 98
cell formation in, (see also Haemopoiesis etc.) 47, 132, 427, 467, 525-43
corpuscles of, 179, 184, 271-2, 426, 454
ecballism and emballism in, 529
fibrinogen in, 42-3
and homoiostasis, 273, 502, 507, 566
islands, 528
movement of, 454
in nemerteoids, 271, 530
sugar, 517
Blood vessels (spaces)
and blood cell formation, 427
in bone, 556
and cephalic organ, 371, 388, 426-7
choroidal, 403, 406, 410
and coelom, 289, 435, 454, 530
contractile, 271, 425, 430
in culture, 28, 323
and enamel organ, 548
endothelia of, 55, 66, 71, 271, 454
of molluscs, 428
and movement, 425-6, 454
in nemerteoids, 269-71, 279, 425-33,454
and nephridia, 269, 272, 438-9, 444-5, 450
oxytocin and, 344
of Pogonophora, 293
portal, 349
and pronephros, 446, 530
retinal, 406
and respiration, 454
and rhynchocoel, 346-7, 349, 388, 425, 429-32
and rhynchodaeum, 277, 289, 340, 427
Bodian stain
and nerve cells, 111, 374
Bone
blood vessels in, 556
and cartilage, 49-52
culture of, 28, 33-6, 40, 97-8, 557-8
evolution of, 296
neural crest and, 136
parathormone and, 558
in primitive fish, 367-8
properties of, 553
remodelling of, 556-8
as strong solid, 552-3, 555
tension and, 51
vitamin A and, 52, 70, 97-8, 558
vitamin C and, 71
Bone marrow
and bone, 558
cells, injection of, 537, 539
differentiation in, 132
endothelium in, 56, 95, 525
and erythropoiesis, 525, 528, 533, 537, 540, 543
granulocytes in, 534, 537-8
lymphocytes in, 537
macrophages in, 535
origin of, 543
recolonization of, 537
Boutons terminaux
in pineal receptors, 393
in cord of *Amphioxus*, 308, 310, 329-30
Bowman's capsule
cilia in, 439, 509
and glucose, 452
nephridia and, 439, 445, 449
testosterone and, 488
Brachiopod
blastula of, 186
evolution of, 185
Bradykinin
origin of, 220
Brain
in acoeloids, 229-30
amoebocytes in, 90
culture of, 90, 106, 108, 111
growth-promoting extracts from, 43
in nemerteoids, 273-4, 289
in rhabdocoeloids, 238, 242-3

Branchial arches (see also Gill arches)
muscles of, 304
Brunner's glands, 84, 521
Brush-border (striated border) (See also Microvilli)
of intestinal cell, 83, 194-5, 519-20
of kidney tubule, 194-5, 198, 452, 488, 502
in nephridia, 447
in sensory cells, 255
Bryozoon
blastula of, 186
Buccal cavity (see also Mouth)
and anterior pituitary, 340-3, 388
derivatives of, 517
and salinity, 506
"Buffy coat"
of blood, 90, 98
Buoyancy
of blastulae, 202
in nemerteoids, 298
Bursa of Fabricius, 524, 526, 543
Butyrylcholinesterase
and myotomal muscle, 307
and sodium transport, 246

Caecum (see also Diverticulum)
intestinal, 287, 339, 458
Calcification, 551
Calcium (Ca^{2+})
and culture media, 8, 51, 71
and dissociation of cells, 13, 27, 71, 125
and goitre, 357
and *Naegleria*, 168-9
and nemertine, 357
parathyroid and, 541
precipitation in cultures, 35, 51
and sponge spicules, 146
and thyrocalcitonin, 357
Callus
growth of cells from, 180
Canaliculi
osteoblasts and, 556
in scales, 545
Canals
dentinal, 553
Haversian, *q.v.*
semicircular, 370
vascular, in scales, 545
Capillaries
amoebocytes and, 95
Carbohydrates
association with protein, 179
and membranes, 494
in retinal cells, 412
Carbon dioxide (CO_2) (see also Bicarbonate)
excretion of, 254, 426
and hanging-drop cultures, 9
and survival of eggs, 46
Carbonic anhydrase
inhibition of, 8
in nemertine pharynx, 268, 465, 517
Carcinoma
cell strains from, 131
epitheliocytes and, 124
Cardiac muscle, see Heart, Muscle
Carmine
and choanocytes, 146
Carnivores
kidneys of, 487-8, 503
lymphocytes of, 532
teeth of, 555
Carotene
and vitamin A, 182
Carotenoid pigments
in nemerteoids, 263
occurrence of, 233, 255, 499
and vision, 233, 379, 412, 415
Carotid body
origin of, 359, 363, 541
Carrel flasks, 17, 30
Cartilage (see also Chondroitin)
cells and matrix, 52, 557
and condensation of mesenchyme, 547
development from bone, 49, 57
differentiation of, 49, 50-1
growth-promoting extracts of, 43
hypertrophy of, 51, 557
in invertebrates, 558-9
lysosomes in, 52, 70, 558
Meckel's, 546
and neural crest, 136
and origin of vertebrates, 296
and tubular nervous system, 558
and vitamin A, 52, 70, 558
Castration
and adrenal cortex, 485-6
Catechol amines, 344
Cathepsin
in intestinal cells, 268-9, 351, 517-8

Caves
  animals in, 181
Cell contacts (see also Desmosomes, Tight-junctions, etc.)
  in blastulae, 189
  with fibroblasts, 125
  with glass, *q.v.*
  types of, 23, 65, 188, 245, 418
Cell cycle, 39, 41
Cell deaths, 68, 330, 416, 547
Cell division (see also Mitosis)
  and culture medium, 30
  and dehiscence, 208, 244
  and differentiation, 35, 132-3, 209, 565, 575
  in epidermal wounds, 64
  inhibition of, 8
  in intestinal epithelium, 522-3
  in macrophages, 92
  in mechanocytes, 30, 36, 40-1
  and mesenchymal condensation, 547
  and nerve cells, 108, 110-2, 244-5
  in retina, 410
  in vitro, 8
Cell lineage
  in sponges, 150-4, 175
Cell membrane (see also Membranes; Cells, surface of)
  carbohydrates and, 69, 221-2
  of cell types, 124, 224
  and differentiation, 123
  and flagella, 4, 374
  interdigitations of, 125, 188
  micropinocytosis and, 535, 537
  permeability of, 47
  properties of, 123, 244, 331, 500-1
  undulating, 116
  uridine triphosphate and, 221
  variations in, 124, 187, 500-1
Cells
  bipolar, *q.v.*
  breeding true, 21-2, 42, 80
  degeneration of, 330
  density of, 37-8, 568
  division of, see Cell division
  "fibre", 526
  membranes of, 9, 12-3, 23, 33, 36-7, 39, 43-4, 87, 214, 227, 550
  neurosecretory, *q.v.*
  orientation of, 13-6, 187
  petaloid, 137
  pigmented, see Pigment cells
  polarity of, 126, 150, 187, 331, 500
  races of, 42, 46, 103, 128, 134, 150
  replacement of, 211, 240, 242, 257, 522-3
  size of, 14
  strains of, 8, 36, 46-7, 81
  surface of, 23, 123, 125-6, 130, 167, 172, 187, 192-3, 197, 208, 222-3, 312, 331, 500
  suspensions of, 12-3, 17, 27, 46-7, 146, 148, 567, 575
  tissue culture and, 4 *et seq.*
  types of, 21-2, 83, 134, 465
  visual, (see also Rods, Cones, etc.) 410-1
Cellulose, 222
  acetate rafts, 31
  polymerization of, 222
Cell wall
  in plants, 214-5
Cement substance
  between cells, 65, 69, 71
Centriole
  in fibroblast, 58
  in *Naegleria*, 165
  in photoreceptors, 393
Centrosphere
  in amoebocytes, 95-6
  in mechanocytes, 25
Cephalaspida
  lateral line in, 367
Cephalic canal (tube) 371-2, 375
Cephalic (cerebral) ganglion
  blood spaces and, 426
  in evolution, 330
  in nemerteoids, 274-5, 379
  in rhabdocoels, 249
Cephalic glands
  in nemertines, 287
Cephalic grooves (see also Ciliated grooves)
  flagellated cells in, 193, 362, 366
  and lateral line, 418
Cephalic lobe
  of pogonophoran, 291
Cephalic (cerebral) organ
  and blood spaces, 388, 426-7
  cell classes in, 465
  and cephalopod eye, 385, 399, 428
  development of, 273, 327, 384
  and hypothalamus, 386-90

and light, 373, 379, 384
melanin in, 385
in nemerteoids, 260-1, 273-4, 279, 288, 362, 370-9, 386-7, 391
and otocyst, 370
and salt balance, 373, 387, 506
secretory cells in, 349, 373, 386-7, 400
and vertebrate eye, 384, 394, 398-9, 403, 416
Cephalic pits
microvilli in, 386
of nemertines, 366
nuclei in, 330
in rhabdocoeloids, 248-9, 257, 380
Cephalization (see also Head), 333-4
Cephalochordates, 360
and iodine uptake, 353, 356
and nephridia, 444, 448
Cerebellum
culture of, 109, 111
oligodendroglia in, 118
Cerebral cortex
culture of, 111
Cerebrospinal fluid
composition of, 402
culture of nerve in, 108
origin of, 328
and retina, 410
Cestodes
phylogeny of, 278, 281-2
Cetacea
filter-feeding in, 296
Chaetognathe
blastula of, 186
Chemoreceptors (see also Frontal organ, etc.)
in nemerteoids, 273
in rhabdocoeloids, 249
Chief cells
of intestine, 268-9, 306
of optic gland, 387
Chitin
cuticle of, 254
as epithelial product, 69, 221
and molluscs, 282
occurrence of, 222, 281-2
Chloride (Cl^-)
and culture media, 8
and iodine uptake, 355
in kidney tubules, 450-1
and *Naegleria*, 169-71
and salt balance, 353
-secreting cells, 167, 195, 350-3, 465, 517
Chlorolabe, 415
Choanocytes, 141-6, 151, 165
aggregation of, 71
and cell lineage, 150-2
and development of embryo, 190-1
neurohumours and, 150
Cholesterol
and *Naegleria*, 172
storage of, 488, 499-500
structure of 490, 492
and zona fasciculata, 485
Cholic acid, 490
Choline
and amoebocytes, 100, 125
and *Naegleria*, 173-4
and phospholipids, 491, 493
Cholinesterase
and cilia, 244, 246, 265
and ecballism, 246
in fish sperm, 246
lithium salts and, 246
and myotomal muscle, 307
in nemertines, 265
in platyhelminthes, 250
in retina, 416
in sea-urchin embryo, 246
and sodium transport, 246
and trypanosomes, 246
Chondroblasts
and cilia, 213
in culture, 28, 51-3
and embryo juice. 55
as myxoblasts, 57
and osteoblasts, 49-50, 466, 558
Chondroitin (sulphates)
and ecballic cells, 221-2
formation of, 53, 547
mechanocytes and, 51, 53, 69, 558-9
and nematocyst, 216
and sulphate metabolism, 223
vitamin A and, 70
Chondrosteans, 528
Chordates, 294-8
urinogenital system of, 444-5, 448
Chorionic villi
and acetylcholine, 246

Choroid
  blood vessels of, 403, 406, 410
Choroid plexus
  cell types in, 84, 194, 402-3
  in culture, 403
  and ependyma, 403, 415
  fluid transfer in, 348, 402
Chromatophores
  migration of, 257
  and neural crest, 136
Chrome-alum haematoxylin
  and neurosecretion, 374
Chromosomes
  and cytoplasm, 575
  and differentiation, 42, 128
  elimination of, 132
  fragmentation of, 42
  and genetic information, 178, 564-5
  giant, 489
  marking by, 537
  in nerve cells, 112
  steroids and, 489
  X and Y, 480-1
Cicatrices
  of *Amphioxus*, 434, 445-6
Cilia (see also Ciliated cells)
  and cell polarity, 126, 174
  in cephalic organ, etc., 366-7, 370 374-5, 386
  cholinesterase and, 244-6, 265
  in choroid plexus, 402-3
  on cnidoblasts, 216
  on coelomic epithelium, 432, 436, 459
  currents created by, 228-9
  and filter-feeding, 294
  on hair cells, 419
  in kidney, 509
  on mechanocytes, 213
  metachronal rhythm of, 197, 229
  movement of, 11, 193
  in Naeglerioid, 175
  and nerve cells, 330
  in olfactory organ, 365
  in parathyroid cells, 542
  on pigment epithelium, 394, 416
  and polysaccharide production, 69
  on reticulum cells, 534
  on retinal ganglion cells, 416
  and sense cells, 365-8, 385, 416, 419
  structure of, 3-4, 193, 219
  and trichocysts, 219
  in tumour, 4
  water load and, 441
  in zona glomerulosa, 509
Ciliary muscle, 408
Ciliary processes,
  blood vessels of, 403
  cells of, 194, 406-7
  and pigment epithelium, 403, 406
Ciliate (Protozoa)
  acoels and, 231
Ciliated bands
  in enteropneusts, 288
Ciliated cells (see also Flagellated cells, Cilia, Flagella, etc.)
  and acetylcholine, 244
  in anterior pituitary, 341, 506
  in cell gradient, 193-4, 459-60
  in cephalic pits, 363, 416
  cholinesterase and, 244, 246, 265
  in choroid plexus, 84
  and dilute media, 240-1, 273, 350, 353, 499
  in echinoderm embryos, 202, 240
  in ectoderm of gastrula, 401
  in endostyle, 353, 355-6
  in enteropneusts, 286
  in epidermis, 227, 229, 231, 233, 238-9, 242, 257, 262-3, 330, 465
  in gonoducts, 84
  and halogens, 355-6
  in intestine, 268, 520
  in kidney, 488, 509
  and mechanocytes, 213-4
  metabolites from, 216-7, 221
  microvilli on, 193
  mosaic of, 227, 246, 401, 460
  in nemertine pharynx, 240-1, 267-8, 273, 350, 353, 499, 516
  in nephridia, 437, 439, 441
  in nervous system, 402
  in ovarian follicles, 509
  PAS stain and, 69, 214
  root fibres in, 219
  in sense organs, 363, 394
  and vitamin A, 82-3, 394
  in zona glomerulosa, 509
Ciliated funnels, 440, 467
Ciliated groove, 365-70
  and lateral line, 370

in nemerteoids 260-1, 273, 279, 289, 362, 371
and vestibule, 370
Ciliated pits (see also Cephalic pits)
in nemerteoids, 273, 288-9, 362-3, 370
in rhabdocoeloids, 248-9, 257, 273, 370
Cinephotography
of amoebocytes, 87, 100
of bone resorption, 97-8
of cultured cells, 8, 11, 121
of epitheliocytes, 67, 73
of lymphocytes, 532, 539
of mechanocytes, 23, 36
of muscle contraction, 266
of nerve cells, 113
of neurites, 106
of neuroglia, 118
of sponge cells, 147
Cisternae
of reticulum, 124
Clasmatosis, 529
Cleavage
and new economes, 574
spiral, 296, 326
Clitellum, 440
Cloaca, 437, 446, 467
Clones
of amoebocytes, 99, 127
of epithelioid cells, 80
of 'fibroblasts' (mechanocytes), 36-8, 46, 72, 80, 127, 323
of myoblasts, 323
in spleen, 543
variations within, 5
Clot (Coagulum)
and differentiation, 78
epitheliocytes and, 66-7
fibrinogen and, 43
liquefaction of, 10, 40, 66
penetration of, by cells, 10, 27, 93-5, 124
properties of, 8
shape of cells in, 12-4
Cnidaria
blastulae of, 186
cnidoblasts of, 215-7
collagen in, 216, 231
desmosomes in, 189
gastraea of, 206
and iodine, 353
light-sensitivity in, 233
muscle in, 303
phylogeny of, 233, 281
spike-potentials in cells of, 244
Cnidoblasts
cilia on, 216
and connective tissue, 217
and interstitial cells, 199, 215
Cnidocil
barbs on, 278
Coagulum, see clot
Cochlea
and cephalic organ, 370
Coconut milk
and growth of plant cells, 214
Coelenterate, see Cnidaria
Coelom (Coelomic cavity)
absence from rhabdocoels, 251
in *Amphioxus*, 445, 457
definition of, 226
of enteropneusts, 285-6, 290, 444
epithelium of, 292, 432 (see also Coelomic epithelium)
and gonocoel, 434, 448, 457, 459, 468, 478
homologues of, 289, 467, 530
nephridia and, 440-1
origin of, 291, 296, 432-6, 453-61, 467-8
and ova, 454
of Pogonophora, 293
of proboscis, 285, 289-90
and rhynchocoel, 271, 289, 291, 457
Coelomic epithelium, 292
and basement membrane, 470
ciliated, 432, 436, 459
and germ cells, 459
phagocytic, 436
and zona glomerulosa, 483
Coelomic fluid, 271
and blood, 530
formation of, 531
glucose and, 450, 478
and rhynchocoel fluid, 530
Coelomic funnels, 437
Coelomic pore, 455, 467
Coelomoducts (see also Gonoducts)
and adrenal cortical cords, 487
in *Amphioxus*, 445
in annelids, 282
arrangement of, 436, 468
of cyclostomes, 436, 467-8
of Hemichordata, 444

mesonephric tubules and, 450
nature of, 447-8
and nephridia, 282, 442, 448
Coelomostome, 437, 442-3, 449, 461, 468
Coenocyte(s)
definition of, 319
muscle fibres as, 306, 314, 319, 325
Cohesion (see also Adhesion)
in acoels, 232
in blastulae, 165, 185, 187-9, 207
in epithelia, 187, 207
in nemerteoids, 278
Collagen (fibres)
in acoels, 230-1
alternating laminae of, 553-5
amino acids and, 48, 68, 215, 555
and ascorbic acid, 71, 499
banding of, 54-5, 147
in basement membrane, 239, 257, 555
in blood vessels, 427-8
and bradykinin, 220
in Cnidaria, 217
cortisone and, 499
in dentine, 552-3, 555-7
of earthworm cuticle, 555-6
and ecballism, 213, 555
and enamel organ, 548
and epithelial tissues, 53, 216, 555
and eye, 408
mechanocytes and, 25, 36, 53, 58, 68-9, 80, 99, 125, 213-4, 231, 477, 499, 558
and myelinating cells, 282
in nematocysts, 216
in nemerteoids, 264, 278, 553-4
and organization of muscle, 53, 249-50, 264
in planuloid stage, 210
produced in culture, 25, 31-2, 51, 99
and reticulum, 534
and root fibres, 219
in sponges, 144, 146-8, 213, 230-1
as substrate for cultures, 79
and trichocysts, 218
turnover of, 555
vitamin A and, 52
Collar
of enteropneusts, 285-6, 289, 444
Collar cells, see Choanocytes
Collencytes
in calcareous sponges, 144, 146-7
in siliceous sponges, 152
"Colloid"
of thyroid, 358
Colonies
of amoebocytes, 92, 94, 99
behaviour of, 124, 127
blastuloid, 202
formation of, 165, 184, 197
haemopoietic, 537-41
Commissures (neural)
in nemerteoids, 273, 275, 277, 287, 294-5, 339, 345, 371
"Common origin"
meaning of, 183
Conduction
in muscle, 321
in nerve fibres, 114, 331
Cones
emballism of, 412, 414
and flagella, 384
functions of, 411
nomenclature of, 414-5
nutrition of, 410-1
and pigment epithelium, 393, 403
pigments in, 412, 415
and pineal eye, 391, 393
polarity in, 412-3
properties of, 393
and rods, 391, 393, 399-418, 572
single and double, 418
of vertebrate retina, 84, 383, 393-4
Conjunctiva
growth of, 73, 81
keratinization of, 68-9
Connective tissue (cells)
culture of, 10, 28, 36
development of, *in vitro*, 32
from heart tissue, 23, 25, 27
products of, 25, 69
of sponge, 141, 148
symbiosis between cells in, 224, 571
Consanguinity
meaning of, 183
Contact inhibition (see also Cell contacts) 24, 44, 124-5, 532
Contractile vacuole
of *Espejoia*, 159
of *Gonium*, 166
of *Naegleria*, 156-8, 167, 172
of *Tetramitus*, 164

Convergence (Convergent evolution), 178, 233, 239, 382, 428
Coprostanol, 490
Cornea
  and lens regeneration, 409-10
  and pineal, 392
Corpus allatum
  origin of, 399
Corpus luteum
  and progesterone, 497, 499, 507
  prolactin and, 507
Corticoids (Corticosteroids) (see also specific hormones) and intestine, 488
  and lymphocytes, 533
  and macrophages, 529
Cortisol (see also Hydrocortisone)
  and alkaline phosphatase, 520
  and fibroblasts, 533
  and lymphocytes, 533
  and lysosomes, 489
  and *Naegleria*, 173
  structure of, 490
Cortisone
  accumulation of, 497, 502
  and alkaline phosphatase, 520
  and collagen formation, 499
  and fibroblasts, 533
  and lymphocytes, 533
  and zona fasciculata, 497
Cosmine, 545, 555
Creatine phosphate
  and echinoderms, 283
Crustacea
  blastula of, 186
  and cilia, 3
  gonad of, 479, 482
  light sensitivity in, 233
  lymphoid tissue and, 528
Crystallins
  and lens formation, 409
Culture chambers
  in eyes, 17
  in mice, 17, 91
  in rabbits, 17, 91, 487, 532
  size of, 13
Cumulus oophorus
  culture of, 476
  and ovum, 474, 477
Cuticle
  of *Ascaris*, 216
  chitinous, 254
  of earthworm, 555-6
  of epithelial cells, 69, 550
Cyclostomes (see also Lampreys, Myxinoids, etc.)
  cell boundaries in, 464
  chloride cells in, 352
  hypophysial organ in, 294-5, 339-40
  intestine of, 518
  lateral line in, 367
  myotomes of, 306, 308-10
  and origin of pancreas, 517
  otocysts in, 370
  phylogenetic position of, 284
  skin of, 242
  teeth of, 545
  urinogenital system in, 436-8, 446, 454, 467-8, 526, 528
Cysteine
  and hexenolactone, 44
Cystine
  and growth, 48, 70, 73
  and sulphur, 70
Cysts
  of *Naegleria*, 163
Cytochrome
  oxidase in muscle, 315
  systems, 182
Cytoplasm
  influence of, on nucleus, 130-1
Cytostome
  of *Espejoia*, 159

Dark-field illumination, 11
Day
  length of, 255, 395
Decamethonium
  on *Naegleria*, 172-4
Dedifferentiation
  of epithelia, 74
  among mechanocytes, 23, 27, 56
Defence systems, 359
Degeneration (see also Cell deaths)
  of cells in development, 330-1
Dehiscence
  and gastrula formation, 206-7
  from germ layers, 209
  methods of, 207-8
  and planula formation, 210, 212
  significance of, 211

Demilune cells, 82
Dendrites
  of nerve cells in culture, 107, 112-3
  of rods and cones, 393
Dendritic cells
  and neural crest, 136
  in skin, 82, 84
Dental lamina, 545-7, 551
Dental papilla, 546, 548
Denticles
  formation of, 545
  pulp of, 556
  succession of, 544-6
  types of, 545
Dentine, 551, 555-7
  mesenchyme and, 545
  odontoblasts and, 51, 57, 552-3
  as "strong solid", 552-3
Deoxycorticosterone
  and ecballism, 501
  and *Naegleria*, 172-3, 495-6
  and zona glomerulosa, 488
Deoxyribonucleoprotein (see also DNA), 39
Dermis
  collagen fibres in, 553-5
  culture of, 64, 77
Desmosomes
  in blastulae, 188
  and dehiscence, 207-8
  and earthworm cuticle, 556
  in enamel organ, 548
  in epithelia, 65, 124-5, 165, 188
  in fibroblasts, 124, 228
  in heart, 320
  lymphocytes and, 532
  and neuroid transmission, 197
  septate, 188-9, 228
  in thyroid cells, 542
Desor larva, 326-7
Diaphragm
  muscles of, 316
Diaphysis, 50
Diet
  and alimentary canal, 516
  and kidney, 503
  and lymphocytes, 532
Differentiation
  of bone and cartilage, 33-5, 48-9
  and cell division, 132-3, 209, 575
  and cell membranes, 569
  and closure of blastopore, 212-3
  in cultures, 30, 32, 48
  DNA and, 42, 132-3
  of epithelia, 73-84
  of gonads, 469-71, 481
  local environment and, 51, 569
  of muscle fibres, 31-2, 311, 321
  nature of, 123-34, 464
  of nerve cells, 106, 109, 245
  of optic cup, 403
  phylogenetic, 5
  stages in, 465-6
  and synthetic media, 47
  temporal, 175
  time sequences in, 51
Digestion
  in acoeloid, 232
  and intestinal function, 351, 516, 571
  intracellular, 228, 232, 251, 351, 523
  in nemerteoids, 351-2, 516
  peptic, 517, 521
  by proboscis, 293
  in rhabdocoeloid, 251
  tryptic, 521
Di-iodotyrosine
  in urochordates, 356
Dilator pupillae, 403
  and adrenaline, 412, 569
Diverticulum (a)
  of archenteron, and muscle, 301-7, 312, 459
  of intestine, 260-1, 268, 278-9, 307, 436, 446, 459, 468, 523, 525
  of nemertine mouth, 359
  of pharynx, 289, 350
  of rhynchocoel, 346-7, 458-9
  and segmentation, 333
DNA
  and cell division, 39, 41-2
  changes in (mutations), 178-9, 575
  and different haemoglobins, 184
  and differentiation, 42, 132-3, 197
  and genes, 179, 182
  in germ cells, 460
  and gonads, 480
  and hegemon, 573
  in mitochondria, 42
  and *Naegleria*, 196-7
  and Protozoa, 167
  in somatic cells, 42

as store of information, 565-6
suppression of, 42, 132-3, 197, 573
"Dolly cells"
in sponges, 191
Dopamine
in retina, 416
Dorsal nerve roots, 282, 312
Dorso-ventral axis
development of, 209, 213, 232
Duality
in epithelia, 83-4
in neural tissues, 136, 572
in otic tissues, 419
in retina, 415-6, 572
in tissues, 136
Ductuli efferentes, 469
Duodenum
goblet cells in, 521
villi and crypts in, 522

Ear
culture chamber in, 17, 91, 487, 532
human, 2
origin of, 418-9
Earthworm (see also under specific names)
coelomic fluid in, 450, 478
cuticle of, 555-6
Ecballism (Hydrecballism)
and acetyl choline, 246
in adrenal cortex, 486, 499, 508-10
and ameloblasts, 551-2
and blastomeres, 220, 225, 437, 570
of blood cells, 529
and collagen, 555
and corticoids, 508
definition of, 194
and endothelia, 540
and erythroblasts, 538
factors favouring, 481-2, 499-501
features of, 221, 499-501
and flagellate cells, 194, 227, 246
in gonads, 474, 477, 479, 481, 509
and internal environment, 466
and invagination, 225
of lymphocytes, 532-4, 538
in macula densa, 505
and mechanocytes, 224, 466, 534
and muscle fibres, 312
of *Naegleria*, 466, 475, 501, 532
in nephridia, 437
of nerve cells, 330
and odontoblasts, 552
origin of, 175, 570
in parathyroid, 541
and photoreceptors, 386, 412, 414
and renal function, 498-505, 508-9
variability of, 571-3
Ecdysone, 489
Echinoderm
blastulae of, 46, 165, 186, 189, 202, 240, 283, 326
eggs of, 46
gastrula of, 207, 283
leucocytes of, 137
muscle processes of, 232
relationships of, 279, 283-4
Econome
definition of, 565
of germ cells, 574
interrelationships of, 566-9
and mitosis, 575
specificity of, 574
Ectoderm
aggregation of cells from, 71-2
and animal pole, 227
and ecballic cells, 227
epitheliocytes from, 67, 135
fluid relationships of, 199-201
folding of, 325-6
of gastrula, 401
in hydrozoa, 225
and interstitial cells, 215
larval (nemertine), 327-9
and lens formation, 408
meaning of, 211, 226-7
mesenchymal cells from, 460
nephridia and, 436
neural, 382, 400
vesicles from, 199
Ectomesenchyme
and muscle, 315
Elasmobranchs
bipolar nerve cells in, 333
islet tissue in, 517, 519
muscular movement in, 303
myotomes in, 307, 309-10
notochord of, 299-300
rectal glands in, 69, 524, 543
teeth in, 545-6
urinogenital system in, 446

Elastase
and cell suspensions, 27, 125
Elastic(n) fibres
and blood vessels, 427-8
and lysine, 55
mechanocytes and, 25, 282
Schwann cells and, 282
Electrical stimulation
of cells in culture, 11
Electron micrographs
interpretation of, 121
Emballism (Hydremballism)
in adrenal cortex, 486, 508-10
and amoebocytes, 194, 220, 223, 227, 466, 498, 505, 529, 538
and blastomeres, 220, 225, 227, 437, 570
of blood cells, 529
and corticoids, 508
definition of, 194, 482
and endothelia, 540
factors, favouring, 481-2
features of, 499-501
in gonads, 474, 477, 479, 481, 509
and internal environment, 466
and invagination, 225
of juxta-glomerular cells, 505
and megacaryocytes, 538
and metabolism, 482, 490, 500
and monocytes, 533-4, 538
and myelocytes, 538
and myoblasts, 312
of *Naegleria*, 466, 475, 501
and nephridia, 437
and nerve cells, 330
origin of, 175, 570
in osteoclasts, 541
and photoreceptors, 386, 412, 414
and potassium, 482
and renal function, 498-505, 508-9
in thyroid, 542
variability of, 571-3
Embryo(s)
culture of chick, 8
specializations of, 207
subdivision of amphibian, 200
Embryo juice (extract)
and amoebocytes, 88, 92, 100, 124
and cell form, 11, 92
and culture of bone, 33
and culture of nerve, 108
and epitheliocytes, 72-3
first use of, 8, 43
and glucose, 43
heated, 73
and mechanocytes, 23, 31-2, 36-40, 43, 46-7, 55, 73, 92, 124, 133, 541
and mitosis, 36-41, 133, 575
proteoses and, 42-3
Emperipolesis, 532-3, 535
Enamel (see also Ameloblasts)
formation of, 57
keratin-like protein and, 68, 544, 549
organ, 57, 68, 546-50
prisms, 548-51
in scales, etc., 545
as "strong solid", 552-3
Endoderm
in *Amphioxus*, 301
and emballic cells, 227
epitheliocytes from, 67, 135, 314
fluid relationships of, 199-201
and germ cells, 470
in hydrozoa, 199, 215
and intestine, 225
meaning of, 211, 226-7
and neutral red, 199
nuclei in, 130
in planula, 225
and somites, 314
and vegetal pole, 227
Endometrium, 488
Endopeptidase, 518
Endothelial cells
amoebocytes and, 95
of blood vessels, 55, 66, 71, 271, 426-31, 454
and calcium, 71
in cell gradient, 194
contractility of, 271
culture of, 66, 540
from heart, 23, 25, 55-6
of liver, 56, 95, 525
and mechanocytes, 56, 65, 80, 537, 540
in nemerteoids, 426-31
and silver nitrate, 65
sponge cells and, 144, 146
tension striae in, 55
transformation of, 56
two types of, 56, 540
Endothelium (see also Endothelial cells)
of adrenal cortex, 56, 95

of bone marrow, 56, 95
ecballic, 540
emballic, 540
erythroid cells from, 540
of heart, 25-6, 55
and leucine-aminopeptidase, 428, 430
micropinocytosis in, 537
myeloid cells from, 540
phagocytic, 56, 95, 525, 540
in post-capillary venules, 540
Endoplasmic reticulum
of amoebocytes, 102, 124, 529
consistency of 121
of epitheliocytes, 124
of erythroblasts, 535
in granulocytes, 529
in large lymphocytes, 539
in Leydig cells, 474, 477
of mechanocytes, 24, 58, 124
in monocytes, 539
in muscle, 313, 315
in naeglerioid, 162, 167, 175
pattern of, 121, 130, 500
in pigment epithelium, 405
and polysaccharides, 222
in Sertoli cells, 474
and steroids, 489
Endosteal bone, 33
and cartilage, 49
Endostyle
of ammocoetes, 354-6
of *Amphioxus*, 3, 353
and thyroid, 3, 353
End-plates
"en grappe et en plaque," 307, 312, 319
in myotomal muscles, 308-10, 314-6
and transmitter, 322
Enterogastrone, 521
Enteron
definition of, 226
Enteropneusta
ciliated grooves in, 288, 367
and iodine, 356
and nemertines compared, 285-92
phylogeny of, 285, 360
Environment, external
and alimentary canal, 516
of blastuloid, 187, 189
and evolution, 178, 453
and excretion, 450, 501
of *Naegleria*, 167-8, 173, 196, 226, 350
in ontogeny, 212
of protozoa, 184
Environment, internal
blastopore and creation of, 212-3
of blastuloid, 187, 189
constancy of, 131, 190, 198-203, 566-9
and contractile vacuoles, 167
development of, 213, 453
ecballism, emballism and, 466
epithelia and creation of, 189, 240
in planuloid, 206
regulation of, 273, 359, 439
Environment, ionic
of blastuloid, 198, 202
constancy of internal, 453
and differentiation, 220, 350
of *Naegleria* etc., 156-61, 167-76
Environments, local
and agoranome, 565
and amnia, 328
of cells in culture, 17, 21-2, 46
on cell surfaces, 197
epithelia and creation of, 189, 400
and epithelial differentiation, 78-9, 132, 209
formation of, 188
on gene action, 130-1, 197
of germ cells, 226, 252, 443, 455, 457, 460, 475-80
and gradients, 511
and haemopoiesis, 538-40
and internal environment, 567
in lens formation, 408, 410
mechanocytes and, 78
for neural tissues, 107, 402
on visual cells, 412, 414-5
Enzymes (see also under specific names)
formation of, in cultures, 95
lysosomal, 523, 558
in muscle, 315-7
in naeglerioid, 167
proteolytic, 496, 523
steroids and, 489, 496
Eosinophil cells
and ions 350, 353, 355
in nemerteoids, 267, 271-2, 350, 426, 465
in vertebrates, 272
Ependymal cells
and choroid plexus, 403, 415

cilia, flagella and, 330, 402
in culture, 106, 403, 415
photosensitivity of, 396
and retina, 221, 403, 410
terminal bars of, 244
Epidermis (see also Skin)
in acoels, 229-31, 233
of amphibia, 126
basement membrane of, 239-40, 242, 330, 363, 465, 555
cell boundaries in, 125
and cell recognition, 79
cell replacement in, 211, 240, 242, 257
and collagen, 555-6
and denticles, etc., 544
and dermis, 77, 553-4
of enteropneusts, 285
and gastric mesenchyme, 547
of insect, 550
keratin and, 68, 254
of nemertine, 262, 465
and nerve cells, 243, 329-30, 333
position of nuclei in, 262, 329-30, 465
of proboscis, 276-8, 286, 465
respiratory, 239-40
in rhabdocoels, 238-42
in wound repair, 64, 554
Epididymis
fluid in, 475
origin of, 446
and potassium, 482
and stereocilia, 193-4
Epitheliocytes (Epithelial cells)
as cell family, 20, 79-80, 134
and chitin, 221
ciliated vacuoles in, 242
and collagen, 555
in culture, 12, 64-86
in cultures of brain, 106
differentiation of, 73-84
duality of, 83-4, 135, 190, 245, 412, 466
ecballism, emballism and, 194, 466, 504, 555
of gonocoels, 432
and germ layers, 134-5
and hexenolactone, 44-5, 80
and ionic permeability, 188-9, 418
and keratin, 68-70, 80, 83, 544, 548-9
and mechanocytes, 22, 66, 80-1, 123, 187
from membrana granulosa, 476
movement of, 65-7, 550-1
and mucoprotein, 123
and muscle, 302-3, 315, 321-2
nuclei of, 128-9
polarity of, 146, 190-1, 499, 547
properties of, 68, 123-4, 134
and spike-potentials, 244
versatility of, 76, 499
Epithelium(a)
as barrier, 189
in blastuloid, 203
dehiscence from, 206-10
in gastraea, 207
glandular, 67, 76
gradients in, 193-7, 460
Naegleria and, 163
of nemertine intestine, 268, 277-8
of nemertine pharynx, 240-1, 267
of nemertine proboscis, 277-8
of nemertine skin, 262-3
in planuloid, 206-10
primary, 146, 150-1
respiratory, 239-40, 242, 257, 351, 570
secondary, 144, 146, 150-1, 434, 454
of sponges, 137-8, 141-2, 144-5, 150-1, 210
stratified, 67, 547
Epiphysis
of bone, 50
and pineal eyes, 391, 394-5, 397
Epithelioid bodies, 541
Epithelioid cells
in cultures of amoebocytes, 95-6, 137
in cultures of heart, 25, 27, 55-6
and endothelia, 66, 95
Epitheliomuscular cells
in acoeloid, 233
evolution from, 249
in iris, 403, 412
in planuloid stage, 210
in sponges, 141
Epoophoron, 469
Erythroblasts
and erythropoietin, 541
rhopheocytosis and, 532
structure of, 535-7
Erythrocytes (Red corpuscles)
and amoebocytes, 529
in cultures, 10, 90
ferritin and, 535
of frog, 272
and granulocytes, 525-6, 529

haemoglobin and, 180, 272
life-span of, 132
and lymphocytes, 534, 537
nuclei of avian, 130
origin of, 525 534, 540
and pronephros, 526
in sickle-celled anaemia, 179, 184
vitamin A and, 380, 495
Erythrolabe, 415
Erythrolitmin, 90
Erythropoiesis
and blastoderm, 529, 540
in bone marrow, 540-1
and fibroblasts, 535
in *Myxine*, 531
rhopheocytosis in, 532
in spleen, 540-1
Erythropoietin, 541
Eserine and ciliary activity, 245
Ethanolamine, 491, 493
Evocation
nature of, 78, 569
Evolution
convergent, 178, 239, 382
of metazoa, 185
rates of, 381
Excitation
in muscle, 321
in nerve, 114
in retina, 417
Excretion
and environment, 450, 453, 501
by inner mass cells, 207, 214, 232
in rhabdocoels, 254
and vascular system, 425
Exoskeleton
of arthropods, 284, 369
Eye(s)
of acoels, 230
of annelids, 399
carotenoids and, 233
cephalopod, 381-2, 385-6, 399, 559
as culture chamber, 17
cup (see also Optic cup), 256, 260-1
diurnal rhythms and, 255, 395-7
effects of disuse of, 181-2
and enteropneusts, 292
fluids in, 406, 409-10, 414
functions of, 255, 396
of *Jamoytius*, 390
of lamprey, 360
of *Limulus*, 399
melanin and, 257, 385, 396, 404
muscles of, 304
of *Myxine*, 398-9
of nemerteoids, 362, 379, 383, 397
parietal, 398
and pineal eyes, 390-99
receptors in, 374, 382, 385-6, 392-4
in rhabdocoeloids, 238, 256, 383
-spots, 255-7, 273, 334, 379, 382-3, 385, 391, 394, 397, 399
vertebrate, origin of, 255, 284, 360, 362, 381-2, 384-5, 399

Fallopian tube
cell types in, 84
Families of cells, 52-3, 79, 103, 124-5
Fat (see also Lipids)
droplets in cells, 23-4, 316
in muscle, 312, 315-7
Feather-germs, 547
Feeder cells for mechanocytes, 46
Feldenstruktur, 315-6
Ferritin, 535
Fertilization membrane
impermeability of, 202
Fibres (see also Collagen, Elastin, Muscle Nerve, etc.)
adrenergic nerve, 116
cholinergic nerve, 116
ciliary, 3-4
lens, 408
myelinate and amyelinate, 115
in sponges, 146-8
Fibrillenstruktur, 315-6
Fibrin
digests of, 42, 47
Fibrinogen
amoebocytes and, 125
function of, 42
origin of, 99
Fibroblasts (see also mechanocytes)
amino acids and, 42, 47-8, 73, 214
clones of, 37-8, 72, 127, 323
constancy of, 42
corticosteroids and, 533
in culture, 21-2, 41, 52
definition of, 20, 25, 28
differentiation of, 50-1, 117

endothelia and, 56
and epithelia, 76-7
and epithelioid cells, 27, 72
and erythropoiesis, 535
fine structure of, 58
and hydroxyproline, 220
ionic balance and, 220
in latent life, 31-2, 117
movement of, 43-4, 532
and muscle cells, 26, 321, 323
as myxoblasts, 53, 57
nucleus of, 24, 58, 128
origin of, 43
and phagocytosis, 123
products of, 25, 36, 51, 53
on reticulum, 534-5, 538
and Schwann cells, 116
strain L, 48, 73, 127
and trephones, 99
trypsin and, 525-6
in wound-healing, 554
Fibrocytes
in culture, 28
and fibre production, 25, 552
of periosteum, 50
Fibrositis, 572
Filter-feeding
consequences of, 349, 360, 430
origin of, 292, 294
and swimming, 296-7, 453-4
Fins
of *Birkenia*, 390
of *Jamoytius*, 390
on nemerteoids, 263, 296-7
pectoral, 3
and swimming, 263, 299
Fish
chloride-secreting cells in, 167, 195
culture of tissues from, 20-2
eggs of, 380
gonads of, 468
and haemoglobin, 180-1
horizontal cells in, 417
lateral line organ of, 2, 367
light-sensitivity in, 233
pancreas in, 519
prolactin and, 507
scales of, 13, 15, 544-5
siluroid, 367
skin of, 242
sperm of, 246
thyroid activity in, 357
Fixation
events during, 121-2, 565
Flagellum (a)
activity of, 155
in cephalic organ, 370, 374-5, 384-5, 400
in cephalic pits, 288
of choanocytes, 141, 143, 146
cholinesterase and, 246
development of, 158-61, 455, 475, 494
with dilated membranes, 374-5, 400
and ependymal cells, 330
and gametes, 246, 380, 475
ions and, 158-9, 169-70, 380
movement of, 169-70, 193
of naeglerioid, 167, 169-70, 175, 475
orientation of, 187, 190
and photosensitivity, 380, 382, 396
and pineal eye, 391
and polarity, 190
rate of beat of, 439
of solenocytes, 439, 445
steroids and, 494
structure of, 3, 193
Flagellated cells (see also Ciliated cells, Flagella)
aggregation of, 146
in cephalic organ, 374-6, 379, 400
cohesion of, 165, 228
derivation of, 151
and ecballism, 246
in ependyma, 402
and frontal organ, 364
and mechanocytes, 213-4
microvilli on, 191, 193
and nephridia, 270
on neural folds, 246
occurrence of, 3, 202
relation of, to amoeboid cells, 153, 155-76, 187, 193-4
and retina, 221, 394, 415
of sponge embryo, 142, 150, 190-2, 227
structure of, 219, 366
Flagellated chamber, 142-4
Flame cells (see also solenocytes)
ecballic, 487
in nemerteoids, 269, 436
in nephridia, 254, 437, 452, 487
in rhabdocoeloids, 238, 257

Fluid skeleton
in nemerteoids, 263-4
in rhabdocoeloids, 249-50
9α-fluorocortisol
and aldosterone, 496
Follicle(s)
cells, 252-3, 456, 472, 478-9
fluid in, 457, 472, 474, 476
hyaluronidase in, 477
of Langerhans, 517-8, 525, 543
oestrogens and, 488
ovarian, 252-3, 456-7, 472-4
primary, 472-3, 476
rupture of, 477, 480
and zona glomerulosa, 482
Follicle-stimulating hormone (FSH), 472
Follicular fluid
formation of, 472, 474, 477
glucose in, 476, 478, 508
mucopolysaccharides in, 477
oestrogens in, 497
properties of, 475-6
sodium and potassium in, 475-6, 508
Fore-gut (see also pharynx)
digestion in, 517
in nemerteoids, 267, 287, 359, 439
Foreign bodies
and giant cells, 96
Formaldehyde
and mechanocytes, 44
Fossils
myotomes in, 302, 390
Fowl-pest virus
and epitheliocytes, etc., 125
Frenulum
of pogonophoran, 291
Frog
gastrula of, 401
kidney tubules of, 451
neurula of, 227, 401-2
ovum of, 477-8
pigment epithelium of, 406
retina of, 411-2, 415, 418
sciatic nerve of, 572
skin of, 571
trachea of, 240, 245
Frontal organ
of acoels, 230, 232
evolution of, 294-5, 363-5
of nemerteoids, 260-1, 273, 287, 339, 362
of rhabdocoeloids, 238
Fuchsinophil cells
in nephric tissue, 452
of sponges, 144, 150
Fungichromin
and pigment cells, 255

Galactosamine
in matrix, 69, 222
in nematocyst, 216
Galactose
and phospholipins, 493
and keratosulphate, 222
Gall bladder, 523
Gametes (see also Germ cells)
flagellated, 155, 175
induction of, 380
Ganglion (cells)
of annelids, 399-400
blocking of, 173
cephalic, 257, 274, 285, 330, 388, 391
in cephalic organ, 370-1, 375-6
and cephalic pit, 248-9
cerebral, 371, 376, 379, 388, 440
with cilia, 416
in enteropneusts, 285
migration of, 230
in nemerteoids, 260-1, 273-5, 287, 331, 347
parasympathetic, 114, 332
of pineal eye, 392, 394
in planula, 225
and retina, 113-4, 332, 382, 384-5, 403, 410, 416, 418
in rhabdocoeloids, 243, 257
and satellite cells, 332
spinal (dorsal root), 112-3, 115, 136, 282, 332-3
superior cervical, 394, 397
sympathetic, 2, 108, 110-1, 114, 136, 282, 332-3
Ganoine, 545, 555
Gastraea
formation of, 206
and planuloid, 224
of Scyphozoa, 226
Gastral cavity
of sponge, 142
Gastrin, 521
Gastrula
ectoderm of, 174, 401

formation of, 202, 206-7, 209, 283
and formation of planula, 212
invagination of, 227
transference of nuclei of, 130, 193
Geckoes
visual cells of, 411-2
Genes
absence of, 182-3
activation of, 128, 132, 494
cytoplasm and, 131
and DNA, 179, 182
environment and, 131
and functional eyes, 181
ineffective, 183
and physiological characters, 178-9
and proteins, 130, 179, 494
and RNA, 131
Genetic characters
and cells in culture, 22
and cellular organization, 130
and collagen, 231
Genetic code
activation of, 133
and DNA, 179, 184
and haemoglobins, 180-2, 184
and *Naegleria*, 185
suppression of, 181
Genetic information
and appearance of characters, 80, 239
in nuclei, 130
in ova, 130, 180
suppression of, 180-1
use and storage of, 178-84
Genital apparatus (genitalia)
specialization of, 253
Genital pore (see also Gonopore)
in Pogonophora, 293
Genital ridge
and adrenal cortex, 470
and germ cells, 470
sex cords and, 448, 468
Genome (see also Genes, etc.)
in development, 573-4
and hegemon, 565, 573
mitosis and, 575
mutation in, 181
permanence of, 573-6
and proteins, 131, 181, 565-6
repression of, 565, 573-4
in somatic cells, 180
Germ cells (see also Gametes, Ova, etc.)
of acoels, 230
agoranome of, 574
and amoebocytes, 147, 210, 226, 460, 481
and archaeocytes, 150-1
chromosomes and, 480
differentiation of, 226, 475-82
econome of, 574
emballic, 481
environment of, 443, 455, 457, 474-80, 482, 570
extrusion of, 278, 434, 467
and gonadal cavities, 434, 442, 454-7, 459 480
origin of, 459, 470
plasticity of, 455, 460
of rhabdocoels, 251
in sponges, 460, 481
Germinal epithelium, 470-3, 481, 483
Germinal centres, 538
Germ layers
and cell types, 67, 134-8, 226-7
Giant cells (see also Multinucleate cells)
formation of, 95-8, 108, 110, 532
and muscle, 306-7, 319-20
of *Naegleria*, 162-3
from spermatogonia, 481
in sponges, 147-8
Gill arches
derivatives from, 359, 541, 543
effects of, 359
nephridia and, 448, 460, 541, 543
Gill-plates
of mussel, 245
Gills
chloride cells and, 167, 195, 351-2, 465
consequences of, 296, 454
of crab, 246
of enteropneusts, 286, 292
of *Jamoytius*, 390
origin of, 284, 287, 296, 430, 517
Gizzard
keratin in, 69
in *Megascolex*, 440
Gland(s)
formation of, in epithelia, 66
optic, 386-7
submandibular, 78-9, 82
Gland cells
in buccal epithelium, 359

in cephalic organ, 371-9
in cephalic pits, 249
in culture, 76-7
in endostyle, 353-4, 356
in epidermis, 240
in olfactory organs, 363-4
Glass
and cell migration, 9, 23, 66, 90-1, 282
effects of, on cells, 12, 72, 78-9, 127
effects of grooves on, 16
lymphocytes and, 532
Schwann cells and, 282
Globin
production of, 182
properties of, 183-4
Glucocorticoids
and pancreas, 525
structure of, 492
and zona fasciculata, 502, 508
Glucosamine
acetyl-, 221
in epithelial polysaccharides, 69
and hyaluronic acid, 69, 221
mechanocytes and, 69-70, 214
and sarcoma cells, 214-5
and sulphur, 70
and uridine diphosphate, 221
Glucose
and coelomic fluid, 450, 478
control of, 508
and culture media, 8, 73
and echinoderm embryo, 202
and follicles of Langerhans, 517-8
in follicular fluid, 476, 478, 508
in gonocoels, 450, 478
in kidney tubules, 450, 452, 478
and *Naegleria*, 168-71, 494
in seminiferous tubules, 478, 508
and steroids, 488
uptake by cells, 43
Glucuronic acid
and hyaluronic acid, 221
in nematocyst, 216
and steroids, 498
β-Glucuronidase
and sodium transport, 498
Glomerulus(i) (Glomerular body)
blood vessels in, 450, 504-5
and coelomic fluid, 531
in enteropneusts, 285, 289-90
and lymphoid tissue, 529
and nephridia, 450
of pronephros, 438
Glutamic acid (Glutamate)
ecballism and, 500
and growth, 48, 214
and haemoglobin, 179, 184
and hexenolactone, 44
and ionic balance, 220-1, 500
mechanocytes and, 499
and *Naegleria*, 170-1
in nematocyst, 216
in testicular fluid, 477
Glutamine
and ecballism, 500
and growth, 48, 73, 221
synthetase, 221
Glyceraldehyde
and mechanocytes, 44
Glycine
in bradykinin, 220
and collagen, 48, 53
and growth, 48
and ionic balance, 220-1, 500, 503
mechanocytes and, 220, 499, 503
in nematocysts, 216
in testicular fluid, 477
in urine, 503
Glycogen
in cultures of liver, 73, 76
in muscle fibres, 307, 312, 315
Glycolysis
aerobic, 43
and cell differentiation, 198
in muscles, 321
Glycoprotein
and cell membranes, 494, 569
and collagen formation, 54
and erythropoiesis, 541
Gnathostomes
urinogenital system in, 436-7
Goblet cells
in cell gradient, 194
and concentrated media, 240
in cultures, 69, 82
in epidermis, 229, 231, 233, 257, 330
functions of, 571
in intestine 82, 84, 482, 520-1, 571
and nephridia, 437
and potassium, 482, 521

of trachea, 83
Goitre, 357
Gold chloride
and nerve cells, 111
Golgi complex (apparatus, body)
in ameloblasts, 547
in amoebocytes, 102
in epidermis, 547
in fibroblast, 58
in macrophage, 102
in monocyte, 539
in myelocyte, 536
in *Naegleria*, 167
in neurosecretory cells, 374, 376
pattern of, 130
in pineal receptor, 393
and polarity, 547
steroids and, 489
Gomori technique
for neurosecretory material, 347
Gonadotrophin
and fat in kidney, 487-8
Gonads (Sex glands)
in *Amphioxus*, 446, 454, 457
in annelids, 441
chromosomes and, 480
and coelom, 436, 454, 467-8
cortex and medulla or, 455, 471, 508
of cyclostomes, 467-8
differentiation of, 469-71, 481
in elasmobranchs, 446
of enteropneusts, 289
in insects, 456-7, 480
in *Myxine*, 454, 457
in nemerteoids, 278-9, 289, 302, 434, 457, 467
optic gland and, 386
of Pogonophora, 294
in rhabdocoeloids, 238, 252, 457
serial arrangement of, 278-9, 434, 467
steroids and, 488
Gonocoel (Gonadal cavity)
in *Amphioxus*, 445, 454, 457
and coelom, 434, 448, 450, 454, 459, 468
distribution of, 446, 460
gradients in, 480
of *Lineus*, 433, 446
in *Myxine*, 454, 457
in nemerteoids, 260-1, 278, 432-3, 436, 480
and nephridia, 433-4, 443-4, 454
in rhabdocoeloids, 251, 257
and seminiferous tube, 448
Gonocytes (see also Germ cells)
of sponges, 191
Gonoducts
Amphioxus and, 445, 454
germ cells in, 468
in Hemichordata, 444
and homoiostasis, 443
and kidneys, 452, 460-1
in nemerteoids, 278, 436, 456
and nephridia, 443, 452, 468
and sex cords, 448, 455
Gonopore, 260-1, 441, 443-4
Gradients, 192-7
in adrenal cortex, 508-11
animal-vegetal, 192, 196-7, 203, 209
axial, 192, 196-7, 203
and balance, 572
in blastuloid, 190, 192, 196, 254, 415, 509, 572
and cell differentiation, 132, 197, 227
of concentrations, 79
and differentiation of germ cells, 460, 480
ecballic to emballic, 194, 227, 459-60, 572-3
electrical, 192
of epithelial cells, 193-6, 460, 573
in excretory tubes, 254, 452, 498, 508, 511, 573
ionic, 569
in planula, 225
Graeff's organ, 400
Granules (see also Basal granule)
in cultured cells, 23-4
in leucocytes, 271-2
neurosecretory, 376
in sponge cells, 147
Granulocytes
and amoebocytes, 529
in culture, 90
origin of, 272, 525-6, 534, 537
and pronephros, 526
properties of, 534-5
Granulopoiesis, 537-8
Grey crescent
and invagination, 227
Ground substance (see also Matrix)
amoebocytes and, 98, 231
ascorbic acid and, 71

and differentiation, 34, 78
glucosamine and, 69
mechanocytes and, 230-1
in rhabdocoels, 251
of sponge connective tissue, 143, 148, 213
sulphate and, 498
uronic acid and, 498
Growth
and cell movement, 43-4
and differentiation, 35, 47-8, 51, 133
embryo extract and, 8, 37-8, 43, 575
environmental changes during, 575
of epitheliocytes, 72-3
hydroxyproline and, 214
inhibition of, 44-5, 214
of intestinal epithelium, 523
malignant, 575
of mechanocytes, 36-45
organized, 18, 49, 122
patterns of, 24
rates of, 75-6, 107
stimulants of, 7, 43-4
and synthetic media, 47
unorganized, 18, 48, 73-4, 106, 122
Guinea-pig
amoebocytes from, 92-3
Gustatory organ
cell types in, 84, 367, 369
Gut (see also Intestine, Alimentary canal)
contents of, 454
diverticula of, 278, 285, 302, 305, 314-5
of enteropneusts, 285-6
innervation of, 332
movement of, 454
of nemerteoids, 262, 294-5, 302
of Pogonophora, 293
regulatory function of, 254, 439

Haem
distribution of, 182
and globin, 183
Haematein test (Baker's)
and phospholipins, 125
Haemochromogens
in nephridia, 452
Haemocoel
and coelom, 530-1
of nemertines, 285, 530-1
Haemocytoblasts, 526-7
Haemoglobin
distribution of, 180-4
and DNA, 182
in fish, 180
foetal, 184
formation of, 132
in nemerteoids, 271-2, 279, 426
origin of, 183
in Protozoa, 183
in sickle-celled anaemia, 179, 184
ova and, 130
Haemopoiesis, 534-41
Hair cell
in auditory organ, 84, 419
in gustatory organ, 84
in lateral line organ, 368, 419
in olfactory organ, 84
Hair follicles, 68
Halophytes
and iodine, 356
"Hand-mirror"
form of lymphocyte, 23
Hanging-drop culture
of "buffy coat", 90
cell-form in, 17, 20, 23
critique of, 9, 11
of echinoderm eggs, 46
of heart, 23
of tissues kept in "latent life", 32
Haversian systems
blood vessels and, 33
collagen fibres and, 555
in cultures of bone, 33
Head
cavities, 340, 454
and collar of enteropneusts, 285
first appearance of, 273
of *Jamoytius*, 390
and spawning, 388
Heart
and coelomic fluid, 531
culture of muscle of, 13, 23-7, 321
and elastic tissue, 25, 427
of enteropneusts, 290
frequency of beat, 569
glutamine and, 48
growth-promoting properties of, 43
mechanocytes from, 23-27, 33, 35-6, 40, 48, 52, 55, 133, 315, 321, 430
muscle contraction, 25-7

origin of, 270, 284, 425, 430, 454
in Pogonophora, 293
as syncytium, 319
and trypsin, 27, 35, 525-6
Hegemon
changing specificity of, 574
definition of, 565, 573
interrelationships of, 566, 569, 574
and mitosis, 574-5
stability of, 574
HeLa cells
with avian erythrocyte nuclei, 131
pinocytosis in, 100-1
Hemichordata
urinogenital system, etc. in, 444
Heparin
use of in tissue culture, 31, 72, 108
Hepatocytes
and embryo juice, 72
and ferritin, 535
Herbivores
and lymphocytes, 532
Heterozygosity, 183
Hexamethonium, 172, 174
Hexenolactone
and cells, 44, 80, 125, 499
Hexosamines (see also Glucosamine, etc.)
ecballic cells and, 221, 500
in follicular fluid, 477
mechanocytes and, 214, 220, 466, 499
and mucopolysaccharides, 499
in sponges, 146
utilization of, 216, 223
Hexuronic acid
and uridine diphosphate, 221
Histidine
and growth, 73
Histiocytes, see Macrophages
Histones
and differentiation, 42
Homograft immunity, 528
Homoiostasis
buccal cavity and, 507
in cyclostomes, 467
ecballism, emballism and, 498
nephridia and, 439, 450, 452, 507
organs regulating, 448, 452, 507
and origin of vertebrates, 454
pharynx and, 359
Homology
interpretation of, 3, 5, 183
Hoplonemertines
pharynx in, 359
Horizontal cells
in retina, 403, 416-7
Hormones (see also under specific names)
acting on genes, 131
and cell membranes, 573
and control of muscle, 264
and induction, 569
intestinal, 521
and *Naegleria*, 172
moulting, 489
pituitary, 505-7
sites of action of, 487-98
from thymus, 540
vascular system and, 425
Hyaluronic acid
ecballic cells and, 221-2
and ion-binding, 223
mechanocytes and, 51, 69
vitreous body and, 53
Hyaluronidase
in ovary and testis, 477
Hydranth
regeneration of, 199
Hydrecballism
definition of, 175
Hydremballism
definition of, 175
Hydrocortisone (see also Cortisol)
accumulation of, 502
and fibroblasts, 533
and lymphocytes, 533
and lysosomes, 495
and zona fasciculata, 499
Hydroxyindole-O-methyl transferase
and pineal, 396
Hydroxyproline
and collagen, 48, 53, 215, 477, 555
in cuticle, 555
and growth, 48, 214
and ionic balance, 220-1, 500
mechanocytes and, 220
and nematocysts, 216
and plant-cell growth, 214
and testicular fluid, 477
in urine, 503

5-Hydroxytryptamine
in pineal, 398
and retractor of proboscis, 344
in sponges, 149, 247
Hypersensitivity, 528
Hypertonicity (see also Salinity)
and nerve cells, 401-2
Hypophysectomy
and zona glomerulosa, 509
Hypophysial organ, 339, 363-4
Hypophysial tube, 294-5, 340, 363
Hypophysis (see also Pituitary)
anterior lobe, 342
Hypothalamus
and kidney function, 507-8
neurosecretion in, 374
origin of, 330, 349, 386-90, 506
and salinity, 506

Iliofibularis muscle, 317
Image
formation of, 381
Immunity reactions, 426, 528
Immunoglobulins, 528
Immunology
of cells in culture, 134
in cyclostomes, 528
Impulses
and neural activity, 114, 331-2, 572
Inclusion body
in optic gland, 387
Indole
and diurnal rhythms, 396
Induction
and agoranomes, 569
of lens, 408
nature of, 78
Information (see also Genetic information)
and antibodies, 532
sifting of, in retina, 417-8, 572
Inhibition
of cells in culture, 8, 44-5, 214
contact, 43-4
in elementary neural systems, 246-7
in hormonal control, 507
and nerves, 114
in retina, 417, 572
Inner mass (see also Mesohyl)
and nerve cells, 243
neurohumours in cells of, 149-50
of planuloid, 209
of sponge, 146, 151
Inophragmata
in heart, 319-20
Inositol
and phospholipins, 491, 493
Insects
chromosomes in, 132, 489
culture of tissues from, 22, 137
folding of epithelia in, 328
gonads in, 456-7, 480
hormones in, 489
light-sensitivity in, 233
Malpighian tubes in, 439, 501-2
mimicry in, 382
multinucleate fibres in, 314, 322
Insulin
species differences in, 179
Integument, see Skin
Intercalated discs
in heart, 319-20
Internal environment, see Environment
Interstitial cells (see also Leydig cells)
cnidoblasts and, 199, 215
ectoderm and, 199
of gonads, 469, 472, 474
as mechanocytes, 481
Intestine (see also Gut, Alimentary canal)
in acoeloid, 232-3
brush border in, 84, 194-5
cathepsin and, 351
cell types in, 82, 84, 194, 306, 465, 571
corticoids and, 488, 520
culture of, 28, 64, 67, 81, 135
of cyclostomes, 518
development of, 326
differentiation of, 74, 76, 78
diverticula (pouches) from, 260-1, 268, 278-9, 285, 302, 304-7, 309, 312, 314-5, 333, 436, 460
and endoderm, 67, 135, 225
of enteropneusts, 286, 289
epithelium of, 64, 72-3, 78, 132, 257, 263, 268-9, 518-9, 522
functions of, 571
goblet cells in, 83-4, 482, 520-1, 571
and homoiostasis, 254, 439, 453
hormones of, 521
and iodine, 353-4
and muscle, 305-6, 520

in nemerteoids, 260-1, 268, 279, 302, 339, 433, 518, 520
nephridia and, 440-1, 524-5, 542
oxytocin and, 344
phosphatases in, 520
potassium in, 482, 521
in rhabdocoeloid, 238, 255, 257
and spleen, 543
and vascular system, 425, 429
villi and crypts of, 302, 522-3
Invagination
of buccal epithelium, 359
centre of, 227
emballic cells and, 225, 227-8
and gastrula formation, 202, 206-7, 209
and nervous tissue, 285, 287
of optic cup, 409-10
and planuloid formation, 207, 210, 225
and proboscis formation, 262, 275, 293
stomodaeal, 294
Iodine, 353-7, 465
Iodotyrosine
in land plants, 356
Ionic balance
amoebocytes, mechanocytes and, 220, 224, 466
and blastula, 185, 189, 192, 198, 201, 220, 467
and cephalic organ, 380, 386
and collagen formation, 214, 220
gonoducts and, 448
hyaluronic acid and, 223
and metaplasia of *Naegleria*, 168-75, 220, 466, 475, 570
nephridia and, 448
in optic vesicle, 411, 414
and polarity, 174, 189
in protozoa, 167
in rhabdocoeloids, 242
and specific metabolism, 130, 220-1, 223-4
steroids and, 488, 503
and *Tetramitus*, 172
Ionic pump
in *Naegleria*, 172
Ionic regulation
gills and, 359
intestine and, 254, 448, 521
in marine animals, 201, 453
nephridia and, 254, 269, 448
in neural tube, 402-3
pinocytosis and, 100
skin and, 242, 254, 448
visual cells and, 380, 411, 414
Ions
and adrenal cortex, 486-7
and buoyancy, 201-2
and cell types, 500
hyaluronic acid and, 223
and mucous secretion, 242
and *Naegleria*, 168-72, 202, 481, 496
passage from cell to cell, 188, 418, 500
and steroids, 496-7
sulphate and, 498
uronic acid and, 498
Iris
in cephalopod, 381
culture of, 65, 67-9, 76-7
epithelium of, 67-9, 76-7, 406-7
muscle of, 303, 315
pigment formation in, 72, 75
regeneration from, 382
Iron
and erythroblasts, 532
in excretory tubes, 452
Irradiation
and inhibition of growth, 46
lethal, 537, 540, 543
by ultraviolet light, 396-7
Islets of Langerhans
α and β cells in, 517, 519
ducts to, 468
origin of, 517
Isoleucine
and growth, 73
Isorhiza
of *Corynactis*, 216

Jacobson's organ
origin of, 362
Juxta-glomerular apparatus, 503-5

Kalyptorhynchia
chitin in, 282
proboscis in, 251, 257, 277
Keratin
formation of, by epitheliocytes, 68-9, 80, 83-4, 125

SH groups and, 68
and teeth, 544, 548-9
and water-proofing, 254
Keratosulphate
osteoblasts and, 53
units of, 222
Ketoglutarate, 171
Kidney (see also Pronephros, etc.)
and adrenal cortex, 470, 501-2
aggregation of cells from, 71
aglomerular, 526, 529
aldosterone and, 488, 496
brush-border in, 194-5, 452
of carnivores, 450, 487-8
cell membranes in, 501-2
cell types in, 82, 84, 194
ciliated cells in, 452, 488
control of, 502, 507
culture of, 64, 67, 78, 198-9, 414, 449, 508, 552
differentiation of, 74, 76-8, 436
and embryo juice, 72, 133
extracts of, 44
and glucose, 452, 478, 501
and gonadotrophins, 488
and homoiostasis, 273, 448, 450
lymphocytes in, 526, 529
and mesoderm, 135
oestrogens and, 487-8
origin of, 284, 467
and osmoregulation, 273, 450, 453
phosphatase in, 520
prolactin and, 507
and proline, 503
of reptiles, 452-3, 488
and sodium transport, 246
specificity of cells in, 79, 575
steroids and, 439, 488, 498, 502, 520
trypan blue and, 452, 486
tubules of, 82, 84, 194, 198-9, 414, 450-1, 501-2, 508-9, 552
vesicles in, 198, 414, 552
King crab (*Limulus*)
amoebocytes of, 123
ocelli of, 399
Kinocilia
in lateral line cells, 195, 367-8
Kölliker's pit, 294-5, 363
Kupffer cells, 84, 95
and erythropoiesis, 532, 535
from lymphocytes, 541
serum proteins and, 99

Labyrinth
ciliated grooves and, 370
stereocilia in, 195
and swimming, 370
Lactate
and *Naegleria*, 170, 481, 501
Lactic dehydrogenase
in muscle, 315
Lamellated bodies, 406
Lampreys (see also Cyclostomes, Petromyzon, etc.)
eyes of, 360
gonads of, 467
and hypophysis, 363-4
immunity and, 528
iodine and, 353-4
ionic regulation in, 453
liver in, 523
melanophores in, 395
muscular movement in, 303
myotomes in, 306-9
neuropil in, 333
neurosecretory cells in, 387
notochord of, 294-5, 308
photosensitivity of, 383, 396
pineal eye of, 391-2, 394-5
spleen in, 543
Larval forms
as specializations, 283, 574
Latent life
of cultures in serum, 31-2
Latent period
and mitosis, 39-40, 92
Lateral line organ
cilia in, 195
derivatives from 2, 418
origin of, 249, 362, 367-9
Lateral pit
in nemertine, 326
Lateral plate, 433
and muscle fibres, 312, 332-3
Leishman's stain, 526
Lens
capsule of, 76, 408
of cephalopod eye, 381-2, 385
culture of, 67

in eye-spots, 383, 391, 394
induction of, 408-9
origin of, 284, 384-5, 408
of pineal eye, 193, 391-2, 394
regeneration of, 382, 384, 409-10, 414
stereocilia and, 193
and ultraviolet light, 391
of vertebrate eye, 382, 408
vesicle of, 409
Lentoids
in culture of retina, 413-4
Leptocephalus larva
haemoglobin and, 180-1
Leucaemia
and amoebocytes, 124
Leucine
aminopeptidase, 428, 517
and growth, 73
Leucocytes (White cells) (see also Granulocytes, Lymphocytes, etc.)
in culture, 12, 90, 94-5, 99
of echinoderms, 137
life-span of, 132
of nemerteoids, 271-2, 279
nuclei of, 128-9
and trephones, 99
in vertebrates, 525-43
Leydig cells
emballism of, 474-5, 477, 505, 509
and testosterone, 499
Light
and cephalic organ, 373, 379-80
and eye-spots, 255
and formation of gametes, 380
and iodine metabolism, 356
and melanophores, 395
and neurosecretion, 249, 387
and photoreceptors, 382
and pineal, 395-8
reactions to 394-5
and sexual activity, 387
Light-sensitivity, see Photosensitivity
Limb-bud
differentiation of, 51, 547
Lipid (granules) (see also Fat)
in adrenal cortex, 485-6
in cell membranes, 168, 490-4, 569
in muscle, 312, 315-7
in photoreceptors, 393, 412
and proteins, 179, 566
Lipofuscin
in eyes, 404
Lipoprotein(s)
in embryo juice, 43
specificity of, 491
and steroids, 496
and vision, 380
Lithium (Li^+)
and cholinesterase, 246
and emballism, 501
and *Naegleria*, 169, 501
and vegetalization, 202, 246
Lithium carmine
amoebocytes and, 90, 95
Liver
aggregation of cells of, 71-2
and blood cells, 525, 528, 535-6
blood vessels and, 429
cell nuclei, 128-9
cell types in, 84
culture of, 64, 67, 72-3, 81, 133, 535
extracts of, 44
functions of, 99, 524
glycogen in, 73, 76
origin of, 284, 523-4
phagocytosis in, 525
regeneration in, 131-2
rhopheocytosis in, 532, 535
sinusoids in, 56, 95, 525
specificity of cells of, 79, 575
steroids and, 487
Lizard
kidney tutules in, 452-3
pineal eye of, 193, 391-2, 394, 397-8
London
sociology of, 576
Loop of Henle, 84
and macula densa, 503-5
sodium transport in, 246, 501, 503-5
L-strain (of "fibroblasts'), 48, 73, 127
Lung cells
aggregation of, 72
Lycopodium
and giant cells, 96
Lymph
cells from, 90, 99, 539, 541
as culture medium, 8-10
fibrinogen in, 42-3
Lymph cords
reticulum of, 538

venules of, 539
Lymph nodes (glands)
and ACTH, 529, 533
amoebocytes and, 90
cells of, 526, 539
culture of, 532, 539
germinal centres of, 538-9
granulopoiesis in, 538
lymph cords in, 539
medulla of, 538
origin of, 531
recolonization of, 537, 539
recolonization from, 537
venules in, 539-40
Lymphoblasts, 537
Lymphocytes
ACTH and, 528, 533
and antibodies, 532-3
and antigens, 534, 539
and amoebocytes, 529, 534, 539
in bone marrow, 537-8
classes of, 525, 537, 539
corticosteroids and, 529, 533
in culture, 90
and diet, 532
ecballism of, 530, 532-3, 538
and emperipolesis, 532-3, 535
and erythrocytes, 529, 532, 534, 537
in haemopoietic tissues, 525, 537
in pronephros, 526-9, 531
and Kupffer cells, 541
large, 526, 530, 538
and mechanocytes, 530, 532-4, 538
and monocytes, 526, 529, 533, 539
movement of, 23, 90, 532, 535
and NaCl, 532
in nemerteoids, 272, 426, 530
nucleus of, 128-9
numbers of, 537
origin of, 426, 534, 541
and plasma cells, 533
recirculation of, 539-40
and RNA, 532
small, 426, 530, 533-4, 537-9
and thymus, 537, 540-1
transformation of, 534, 539
Lymphoid tissue (see also Lymph nodes)
and ACTH, 529
in invertebrates, 528
and kidney, 526, 529-30, 541
lymphocytes and, 533-4
oedema in, 529
origin of, 426, 524, 541, 543
reticulum of, 534
and spleen, 543
Lymphopoiesis, 537
Lysine
and bradykinin, 220
and elastin fibres, 55, 220
and growth, 48, 73
myoblasts and, 48
osteoblasts and, 48, 54
Lysosomes
acid phosphatase in, 95
and amoebocytes, 224, 529
and cell types, 500
enzymes of, 523, 558
and granulocytes, 529
in Leydig cells, 477
microkinetospheres and, 101
and monocytes, 529
steroids and, 489
vitamin A and, 52, 70, 380, 495

Macronucleus
of *Espejoia*, 159
Macrophages (see also Amoebocytes)
and ACTH, 529, 533
and antibodies, 532
in bone marrow, 535
chromosomes in, 42
and corticoids, 529
culture of, 88-9, 92
in cultures of muscle, 99-100, 320
in echinoderms, 137
and emperipolesis, 532-3
and ferritin, 535
free and fixed, 91
from heart tissue, 23, 25
in liver, 535, 541
from lymphocytes, 541
and mechanocytes, 99-100, 533
and monocytes, 88, 90-1, 533, 539
movement of, 91, 532, 541
and phagocytosis, 99,
and pinocytosis, 100, 103
and pronephros, 528
and Schwann cells, 116
in spleen, 526
structure of, 102

transformation of, 99-100, 534
and trypan blue, 25, 90-1
in wound-healing, 99, 102
Macula
adherens, 188
densa, 503-5, 507
of otocysts, 370
Magnesium (Mg^{2+}.)
and culture media, 8
and dissociation of cells, 13, 27, 71, 125
and emballism, 501
and enamel organ, 548-9
and *Naegleria*, 169
and polymerization, 222
and polysaccharides, 222, 231
and vegetalization, 202
Maize
extracts of, 45
Malaria
and sickle cells, 179, 184
Malignancy (Malignant cells)
and adaptation, 575
development of, 5, 73
keratin and, 68
and synthetic media, 47
Malpighian capsule, 437, 467, 469
Malpighian corpuscle, 538
Malpighian tubes
in insects, 439, 501-2
Malt extracts
effects on cells, 44-5, 73, 80
Mammals
culture of tissues from, 20, 22, 92
Mammary gland
cell types in, 84
prolactin in, 507
Mastigophores
of *Corynactis*, 216
Matrix (see also Ground substance)
calcification of, 541-2, 549, 552
collagenous, 552
epitheliocytes and, 549
keratinous, 549
mechanocytes and, 69, 216, 231, 557
production of, 34, 69, 213, 217, 221, 231, 358, 542
relation of cells to, 35, 224, 557-8
vitamin A and, 52, 97-8
Mauthner cells
and thyroxin, 114
Mechanocytes (see also Fibroblasts)
and amino acids, 48, 214, 220, 499
and aldehydes, 44
and amoebocytes, 21, 87, 99-100, 137, 220, 224, 498, 557
and argyrophil fibres, 25, 32, 36, 53
and ascorbic acid, 499
from bone, 33-5, 48, 51
as cell family, 40, 52, 55-9, 134
chromosomes in, 42
and ciliated cells, 213-4
and collagen, 51, 68-9, 231, 552
colonies of, 92, 100
cortisone and, 499
definition of, 20, 25
desmosomes in, 228
ecballic, 213, 220, 224, 226, 466
electrical (surface) properties of, 247, 332
and endothelia, 65, 80
and epitheliocytes, 22, 66, 74, 77-8, 80-1, 187, 212, 552
as evocators, 78
fine structure of, 58
and germ layers, 135
and glutamic acid, 48, 214, 220, 499
and glycine, 48, 220, 499
and ground substance, 52, 69, 78, 230-1
growth of, 36-46, 541
and haemopoiesis, 532, 535
from heart, 24-7, 33, 35, 55, 220, 315
and hexenolactone, 44-5, 80, 499
and hexosamines, 69, 214, 220, 466, 499
and intestinal cells, 64, 76, 78
and liver, 535
and low ionic strength, 499
and lymphocytes, 530-4, 538
from membrana granulosa, 476-7
in mesenchyme, 460
and mesohyl, 210-3
and mucoprotein, 123
and muscle, 52, 55, 64, 99, 307, 312, 314, 319, 322
and myelination, 282
nature of, 27-35, 134, 213-24
and neural crest, 136
and neurones, 230, 247, 331
and nucleoli, 24
odontoblasts and, 51, 53, 56, 552
origin of, 151-3, 155, 210, 212-3, 231
pinocytosis in, 100, 124

and plain muscle, 52, 323
in planuloid stage, 210, 213
and proline, 48, 214, 220, 466, 499
properties of, 46, 123-5, 134, 213-24
pure strains of, 36, 46-7, 80, 127, 224
races of, 40, 46, 48-52
and reticulin, 25, 213
and sarcoma cells, 214-5
from Sertoli cells, 476-7, 481
sources of, 30, 36, 39
in sponges, 137-8, 147-51
and sulphate, 70, 223, 498-9
transformation of, 35, 80, 100
and uronic acid, 220, 223, 498-9
and vitamin A, 52, 69-70
Media
adaptation to, 37-8
and cell form, 12-4, 25, 30, 51, 67, 80, 82, 126
chemically-defined, 38
and growth of single cells, 46, 99
growth-promoting, 33, 36-7, 51
synthetic, 8, 36-7, 47-8, 99, 127
for tissue culture, 7-10, 23
Medusae
umbrella of, 217
Megacaryocyte
as amoebocyte, 538
nucleus of, 128-9
origin of, 525
and pronephros, 526
Meiosis
arrest of, 480
establishment of, 175, 226
in insects, 456
in Naeglerioid, 167
Melanin
in eye-spots, 255-7, 334, 385
in nemertines, 385, 396
and neural crest, 334
in planarians, 255-7
and vertebrate eye, 385, 404
Melanophores
melatonin and, 396-7
and neural crest, 136, 334
and pigment cells in planaria, 255, 257
and pituitary, 397
rhythm in, 395
Melatonin
in pineal, 396-8
Membrana granulosa
culture of, 476-7
and follicular fluid, 476-7
and oestrogens, 488-9
and theca interna, 474, 481
and zona glomerulosa, 508
Membrane(s) (see also Cell membranes)
and cell polarity, 168, 187, 190, 499
ecballism, emballism and, 499-501
external limiting, 330, 363, 374, 407
fertilization, 202
of flagella, 374
glycoproteins and, 494, 569
hormones and, 573
internal limiting, 407
lamelliform (undulating), 87-8, 116-7, 147, 167
lipid, 70, 131, 380, 489-91, 496, 498, 503, 506, 566, 569
of *Naegleria*, 157, 167
and neural action, 572
nuclear, 130, 167, 575
olfactory, 380
of ovum, 192, 478
proteins and, 131, 496, 498, 506, 566, 569
specificity of, 123
steroids and, 489, 494-6, 498, 502-3, 506
unit, 167, 490-4
vitamin A and, 70, 495
β-Mercaptoethanol
and vegetal pole cells, 200-1
Mercury droplets
for isolating cells, 77
Meronephridia, 440-1, 524
Mesenchyme (cells)
and blood cell formation, 427, 525, 528 530, 535
and blood vessels, 432, 434
and buoyancy, 298
and coelom, 434, 454
condensation of, 546-7
differentiation of, 50, 53, 80, 135
and epithelia, 75-80, 212, 454, 547-8
in eye, 407-8
and gonocoels, 432
and head cavities, 340
meaning of, 226, 460
mechanocytes and amoebocytes in, 135, 460
and muscle, 304, 314-5, 430

neurohumours in, 149
in ovary, 212
and rhynchocoel, 432
and teeth, 544-5
Mesentery
of enteropneusts, 286, 289
origin of, 459
Mesoblast, 314
Mesoderm
in *Amphioxus*, 301
epitheliocytes from, 67, 135
gradient of cells in, 460
lateral plate, 433
meaning of, 226
and mesohyl, 135 210-1
Mesoglea
in Hydrozoa, 215, 217
Mesohyl
in acoeloid, 231-3
and blood cells, 427
cnidoblasts and, 216
definition of, 135
dependence on epitheliocytes, 211, 227, 466
development of, 283
and epidermis, 211, 233, 240
and muscle formation, 267, 301, 303
from neural crest, 136
and neurohumours, 229
and neurones, 243
in planuloid, 209-10, 213, 225
in rhabdocoeloid, 240, 243
waste products and, 214
Mesonephros
of amphibia, 450-1
and blood supply, 450
cultures of, 198-9, 449
development of, 437
and metanephros, 450, 461
mixonephridial, 461
and sex cords, 443, 448, 468, 472, 483
tubules in, 199, 436, 446, 449, 472
urine formation in, 450-1
Mesosoma
of pogonophoran, 291, 293
Mesothelium
ciliated or phagocytic, 436, 460
of coelom, 436
and endothelium, 66
from heart, 25
of rhynchocoel, 276
as secondary epithelium, 144
Metabolic pool
in blastulae, 189
dilution of, 47, 189
Metabolism
and ionic balance, 130, 220-4
Metabolites
disposal of, 214
Metachromasia
of matrix, 52
Metachronal rhythm
in body segments, 298
of cilia, 197, 229
Metamerism
of annelids, 281, 449
and nemertines, 296
of Pogonophora,
in vertebrates, 304
Metamorphosis
of sponge larva, 141
Metanephridium, 441
Metanephromixium, 442, 444
Metanephros
differentiation of, 77, 79, 436
and mesonephros, 449-50, 461
multiplication of units in, 446, 449
and nephridia, 461
Metaphase
direction of plate, 208, 244-5, 410
of neuroblast, 208, 244
of neurone, 111
Metaplasia
of epithelia, 98
of *Naegleria*, (*q.v.*), 481
Metasoma
of pogonophoran, 291, 293
Metazoa
from protozoa, 177
Methionine
and growth, 48, 73
and sulphur, 70
Methonium compounds
and *Naegleria*, 172-5, 501
Methylene blue
and nerve cells, 112-4
Micelles
in biological membranes, 491-3
and neurofibrils, 113
and orientation of cells, 13

Microdissection, 11
Microglia
as amoebocytes, 90, 119
Microinjection, 11
Microkinetospheres
and pinocytosis, 101
Micropinocytosis, 535, 537
Microtubules
in naeglerioid, 175
and rhabdites, 217
Microvilli
alkaline phosphatase and, 520
and cell gradients, 194
in cephalic organ, 366-7, 370, 374, 386
of choanocytes, 143, 151
in choroid plexus, 194, 402-3
and collagen, 555-6
in enamel organ, 548
and epididymis, 193, 195
on intestinal cells, 194-5, 268, 465
in kidney tubules, 194, 198, 502-3
in lateral line organ, 368, 419
and lens cells, 193, 195
in naeglerioid, 167, 175
on ovum, 477-8
and photosensitivity, 380, 384-5, 394, 396
and polysaccharide secretion, 69, 502
in sensory cells, 255-6, 363, 365, 368, 380, 382, 394, 415, 419
of sponge embryo, 190-1
in thyroid cells, 542
Migration
of cells, *q.v.*
and mitosis, 39, 112
Milk
oxytocin and ejection of, 344
pigeon's, 507
prolactin and, 507
Millipore filters
in tissue cultures, 75
Mimicry
in insects, 382
Mineralocorticoids
and salivary glands, 525
structure of, 492
and zona glomerulosa, 502, 508
Mitochondria
in amoebocytes, 102, 529
in blood cells, 529, 536
DNA and, 42
and fibre formation, 113
in mechanocytes, 24, 58
membranes of, 131
at mitosis, 128
in muscle cells, 315-6
in naeglerioid, 167, 175
in nerve cells, 113
pattern of, 130
in sensory cells, 256, 393
steroids and, 172, 489
vitamin A and, 495
Mitosis
and cell movement, 36-7, 39, 43-4
in cultures, 36, 39-40, 92
and differentiation, 133, 575
embryo extract and, 36, 39-40, 575
mitochondria in, 128
in naeglerioid, 167, 175
in nerve cells, 106, 110-1, 114
oocyte and, 456, 480
and repression, 575
separation of cells after, 65, 112
spindle in, 113
thyroxin and, 114
Mitotic index
in various media, 37, 39, 41, 92
Mitotic rates
of amoebocytes, 92
of epitheliocytes, 72
of mechanocytes, 36, 39-40
of myxoblasts and myoblasts, 55
Mixonephridium, 442, 444, 452, 461
Modulation
of cells in sponges, 150
of osteoblasts and chondroblasts, 558
in relation to RNA, 133
Molluscs
blastulae of, 186
blood vessels in, 428
cartilage in, 558-9
cephalopod, 380-2, 396, 399, 428, 558-9
light-sensitivity in, 233, 380, 396
muscles of, 314, 323
phylogeny of, 279, 282, 284, 360, 367-8
tissue cultures of, 22, 137
Monkey
retina of, 376-7, 384
Monoamine oxidase
in sponges, 149, 247

Monocytes
as amoebocytes, 88-9, 96, 529, 533, 538
in blood, 88, 90-1, 272
colonies of, 94
in cultures of lymph nodes, 539
in haemopoietic tissues, 529
and macrophages, 90-1, 539
movement of, 90, 539
nuclei of, 128
origin of, 526
in pronephros, 526
and surfaces, 90
in thoracic duct lymph, 90, 539
Mono-iodotyrosine
in enteropneusts, 356
Morphogenetic movements, 309
Morphology
as criterion of differentiation, 35
Morula(e)
and early metazoa, 185
from vegetal pole cells, 200
Motor nerves
development of, 307, 325
Moulds, 7
Mouth
of enteropneusts, 290
of *Jamoytius*, 390
in nemerteoids, 267, 273, 277, 279, 287, 289-90
and nephridia, 441
and Rathke's pouch, 340
in rhabdocoeloids, 251
and rhynchodaeum, 289, 338
Mucin (mucus) (see also Mucoprotein, etc.)
and epithelial cells, 69, 82
and *Espejoia*, 159
goblet cells and, 521
iodinated, 353
and myxoblasts, 53
from *Naegleria*, 156, 175
of nemerteoids, 263, 276
Mucoitin sulphuric acid
and epithelial cells, 69
Mucopolysaccharides
of anterior pituitary, 506
of basement membrane, 472
in blastocoel, 221
and cartilage, 70
and collagen, 54, 217
and epitheliocytes, 69, 125
formulae of, 222
and gonadal fluids, 477
and induction, 408
and mechanocytes, 125, 477
in nemertine pharynx, 499, 506
in nemertine skin, 465
and notochord, 299
and retina, 410
and surface of amoebocytes, 125, 223
Mucoprotein (including mucosubstances)
and agoranome, 567
and basement membrane, 239
in blastocoel, 189
and cartilage matrix, 51-2
cell types and, 123, 125, 465, 500
in enamel organ, 548
and epithelia, 69, 80, 516
and ground substance, 251
and growth, 43
mechanocytes and, 80, 123, 213, 231
myxoblasts and, 53
in nematocysts, 216
and notochord, 296
and trichocysts, 219
and vitamin A, 69-70
Mucous glands (cells) (see also Goblet cells)
and acetylcholine, 245
and concentrated media, 240-1, 499
differentiation of, 76-7, 82-3
in enteropneusts, 286
in epidermis, 82-3, 239, 262
and frontal organ, 363-4
in gonads, 479
in intestine, 269, 306, 520-1
and iodine binding, 353-4
and ions, 242
in kidney tubules, 452-3
in nemertine pharynx, 240-1, 267-8
and potassium, 482
in rhabdocoeloids, 238, 242
and sulphur, 98
Müllerian duct, 437, 455, 468
Müller's cells
in retina, 374, 416, 418
Multinucleate cells
from bone, 96-8, 558
ciliate, 231
and muscle, 30, 312, 314, 319-21
from nerve, 108, 110
in sponges, 147-8

Muscle(s)
of acoels, 229-30, 232-3
adrenergic, 264
and amoebocytes, 315, 320-2
in blood vessels, 271, 323, 428
buds, 29-30, 319
cardiac, 23-7, 43, 52, 430
cholinergic, 265, 304
and collagen, 53, 250, 554
contractile system of, 3, 315
culture of, 28, 31-2, 35, 48, 55, 319, 321-3
derivation of, 231, 303
development of, 247, 249-50, 301-2, 305, 311
ecballic and emballic, 312
as epitheliocytes, 321-2
fast and slow, 312-3, 316-7, 321
fat in, 312, 315-7
fine structure of, 315-6
heterogeneity of, 315-25
innervation of, 305, 307-10, 318-9, 321
of insects, 314, 322
intestinal, 28, 268
and mechanocytes, 315, 319, 321-2
multinucleate, 30, 312, 314, 319-21
myoglobin in, 180, 315
in nemerteoids, 262-4, 344, 554
plain (smooth), 28, 52, 128-9, 250, 257, 264, 315, 323, 332
and planuloid stage, 228
of proboscis, 251, 265-7, 276-7, 344
process as nerve, 232, 307, 310
red and white, 310, 312, 316
in rhabdocoels, 238, 249-50
skeletal, 28-30, 52, 128-9, 314-5, 332
spindles, 318-9, 363
spiral, 323
striation of, 52, 302, 322, 324
and swimming, 299
symbiosis and, 332
Mutation
and adaptation, 575
effects of, 179, 181, 183
and haemoglobin, 181
Myelination
in annelids and arthropods, 282
astrocytes and, 118
development of, 284
mechanism of, 113-4, 333
and nemertines, 333
oligodendroglia and, 118
phospholipins and, 416
Myelin figures
and swellings on nerves, 113
Myelin sheath
neuroglia and, 118, 411
Myeloblasts
character of, 535
Myelocytes
migration of, 535
origin of, 534, 538, 540
structure of, 536
Myeloid
bodies, 404-6
cells, 540
colonies, 540
Myelopoiesis (see also Granulopoiesis), 526-7, 534, 540-1, 543
Myoblasts
amino acids and, 48
bipolarity of, 321
cardiac, 57, 324
classes of, 52-9, 311
in culture, 29, 321-3
ecballic and emballic, 312
and muscle-fibre formation, 132, 311, 322
and satellite cells, 332
union between, 322
Myocoel
and gonad, 446
origin of, 305, 459, 468
and splanchnocoel, 315, 433, 459
Myocommata, 304
Myoepithelial cells
and planuloid, 229
Myofibrils (myofibrillae)
in cultured cells, 25, 52, 324
in developing muscle, 306-9, 311, 314-6
Myoglobin
and haemoglobin, 184
in muscles, 180, 312, 315
Myoids
of rods and cones, 412
Myoinositol
and stabilizing action of serum, 47
Myometrium
steroids and, 488
Myosin, 3, 132, 315
Myotomes (see also Somites)
of *Amphioxus*, 304, 306-8, 310, 312, 329, 332

contraction in, 303
in early fossils, 302
of elasmobranchs, 303, 307, 310
and epithelial cells, 303-4
and eye muscles, 304
innervation of, 304-5, 307-10, 325, 332
of *Jamoytius*, 390
of lampreys, 303, 306-10
of *Lepidosiren*, 309-11
origin of, 284, 314-5, 459
vacuolation in, 311, 520
Myxinoids (see also *Myxine*)
hypophysis in, 294-5, 349
and immunoglobulins, 528
intestine in, 519-20
nasal sac in, 364
pancreas in, 519-20
pronephros in, 438, 527, 531, 543
spleen in, 543
Myxoblasts
classification of, 52-9

Nadi reaction
in nemerteoids, 426
Naeglerioid stage (see also *Naegleria*)
leading to blastuloid, 155-76, 185
Nasal mucosa, 82, 240
Nasal sac, 339, 364
Navigation
by birds, 398
Necrosis (see also Cell death)
in cultures, 27
Nemathelminth
blastula of, 186
Nematocyst
in *Gonothyraea*, 225
cnidocil of, 278
composition of, 216-7, 555
structure of, 215
and trichocysts, 218
Nematode(s)
cuticle of, 216
muscles of, 323
phylogeny of, 278, 281
Nemerteoid stage, 260-79
off-shoots from, 281
Nemertines (Nemertea), 260-79
blastula of, 186, 200
blood spaces in, 340, 425-6
cell boundaries in, 278
collagen fibres in, 553-4
embryogenesis in, 294-5, 325-9
fate of proboscis of, 290-6
gastrulation in, 283, 326
iodine binding in, 354-5
littoral, 370
matrix in, 558
movement of, 263, 285, 325
muscle development in, 312, 325
nephridia (*q.v.*) of, 269-71, 448, 461
pelagic, 263, 296-8, 329, 345, 370, 453
transmitters in, 264
and vertebrates, 5, 281 *et seq.*
"Nemertine"
and nicotine, 263
Neoblasts
of planaria, 568
Neoteny
and notochord, 299
and origin of species, 181
Nephridia
and adrenal cortex, 487
in annelids, 282, 436, 439-43, 445, 448, 467-8, 524
and blood vessels, 269-72, 279, 437, 446
and body cavities, 432, 441, 444
capsule of, 270, 279
and carotid bodies, 541
of *Cephalothrix*, 270, 437, 439, 461
in chordates, 444-5, 448, 460, 541
and distal tubules, 450
distribution of, 460
ecballism and emballism in, 487
functions of, 254, 439, 443
of *Geonemertes*, 437, 439, 441
and gonocoels, etc., 434, 442-4, 455
and homoiostasis, 439, 443-4, 448
and lymphocytes, 528, 541
in nemerteoids, 260-1, 269-70, 279, 433, 436-7, 439-40, 446-7, 461
opening into gut (enteronephric), 439-41, 524-5, 542
and pancreas, 442, 524-5, 543
and parathyroid, 541
pigments in, 452
of Pogonophora, 293
in *Priapulus*, 443
and pronephros, 437-8, 449, 467
reduplication of, 440, 442
in rhabdocoeloids, 238, 254, 257, 440

and thymus, 541
and tonsils, 541
in vertebrates, 448, 452
Nephridiopore
in *Amphioxus*, 445
in annelids, 441-3
in nemerteoids, 260-1, 269-70, 433, 436 447
Nephridiostome (see also Nephrostome), 441-2, 461, 524
Nephrocoel, 433-4, 459, 468
Nephrocytes
emballic, 487
granular, 452
Nephromixia, 442, 448
Nephrostome (see also Nephridiostome), 438, 440, 449
Nephrotome, 446, 472
Nereidae
nephridia of, 440
Nerve cells (see also Neurones)
in acoels, 229-30, 232-3
and amoebocytes, 231, 331
in cephalic organ, 384
and ciliated cells, 244-5, 330
in culture, 20, 106-19
differentiation of, 106, 109, 135, 245, 403
ecballic and emballic, 245, 330, 466
giant, 110, 287
glutamate and, 221
hypertonicity and, 401-2
longevity of, 132
and mechanocytes, 231, 331
mitosis in, 106, 108, 110-2
movement of, 111-2, 401
in nemerteoids, 279, 346-7
and neuroglia, 331, 570
nuclei of, 113, 128-9
in planuloid stage, 210, 229
in rhabdocoels, 243
types of, 109-11, 114
unipolar, 243, 333
Nerve cord(s)
in *Amphioxus*, 301, 308
dorsal, 274-5, 286, 289, 296, 329
in enteropneusts, 286-7, 289-92
ganglionation of, 274-5, 296, 329
and myotomes, 308
in nemerteoids, 260-1, 274-5, 296, 329
in rhabdocoels, 238, 242-3, 257
Nerve endings, 307, 310, 312, 319
Nerve fibres (see also Axons)
adrenergic and cholinergic, 114, 116
amoeboid processes of, 9-10, 232
degeneration of, 114
fluid transport in, 106-7
growth of, 8-10
myelination of, 113-2, 282, 333
and myotomes, 307-9, 315
and nemertine epidermis, 333
and neuroglia, 114-6, 275, 332
to proboscis, 276, 341
specificity of, 116, 250
swellings on, 113
Nerve impulses
and classes of nerve cells, 114, 331-2
Nerve plexus
in rhabdocoeloids, 242-3
Nerve ring
in enteropneusts, 285-6
in Pogonophora, 293
Nervous system
in acoels, 230, 233
in annelids, 282
in arthropods, 282
and blastuloid, 197
and control of muscle, 303
duality of 332
and haemoglobin, 426
in nemerteoids, 273-5, 279, 326
order of differentiation in, 332
in rhabdocoels, 242-3, 257
tubular, 243, 296, 325-34, 558
in vertebrates, 243, 303
Neural action
nature of, 114, 572
Neural crest
derivatives of, 135-6
pigment cells and, 136, 334
and sympathetic ganglia, 110
Neural ectoderm, 244, 401-2, 412
Neural folds
cells in, 328, 401-2
and cholinesterase, 246
and neural crest, 135, 334
Neural plate
cells in, 401
and myelination, 333
and nerve cells, 135, 329

Neural tissue
  and acoeloid, 229-30
  and blastuloid, 197
  and collagen, 53
  and differentiation, 79, 244
  in rhabdocoeloids, 242
  and sponges, 150
Neural tube
  and differentiation, 79, 111, 114, 135-6, 333, 403
  environment within, 328, 330
  and myotomes, 329
  and optic vesicles, 384, 386, 402, 406
  origin of, 284-5, 287, 328-30
  polarity of cells in, 208, 333
Neurites
  pinocytosis in, 106-7
Neuroblasts
  in culture, 106
  in developing neural tissue, 208, 333, 401
  formation of, 208
  movement of, 333, 401
  multiplication of, 244-5
Neurocord
  of enteropneusts, 285-7, 290
Neuro-epithelium, 233
  of optic cup, 410
  and vertebrate nervous system, 230, 330
Neurofibrillae
  in nerve cells in culture, 112-3
Neuroglia
  and acetylcholine, 118
  in culture, 106-19
  and myelin sheath, 118, 333
  in nemerteoids, 273, 275, 282, 333, 376
  and nerve cells, 113, 275, 331-2, 570, 572
  nuclei of, 128-9
  origin of, 135, 330, 403
  in pineal nerve, 394
  in polyclads, 243
  in rhabdocoeloids, 257
Neurohumours (Transmitters)
  in retina, 416
  in sponge cells, 149, 229
Neurohypophysis (see also Pituitary), 349
Neuroid transmission, 197, 203, 229
Neuroma
  and posterior pituitary, 341
Neuromuscular junctions (see also Nerve endings), 308-9
Neuromuscular system
  origin of, 250
Neurone(s) (see also Nerve cells)
  chain of, 244, 247
  differentiation of, 106-19
  from mesohyl, 247
  and neuroglia, 411, 572
  theory, 9
  varieties of, 247
Neuropil
  in lampreys, 333
  in nemerteoids, 273-4, 371
Neurosecretion
  and anterior pituitary, 506
  in cephalic ganglia, 371, 430-1
  in cephalic organ, 349, 376, 386
  and hypothalamus, 330, 349, 386
  and light, 387
  in *Myxine*, 399
  in nemerteoids, 274, 330-1, 346-8, 399
  in rhabdocoeloids, 249, 257
  and salinity, 387
Neurula
  hypertonicity and, 401-2
  mosaic of cells in, 227, 401
Neutral red
  amoebocytes and, 90
  and cells from heart, 24, 27
  and endoderm, 199
  and mechanocytes, 24
  and nerve cells, 113
Neutrophil leucocytes, 272
Nexus (see also Tight junction), 188
Nissl's substance
  in nerve cells in culture, 112
Nitrogen requirements
  of epithelia, 73
Noradrenaline
  in retina, 416
  in sponges, 149
Notochord
  of *Amphioxus*, 294-5, 299-301, 304, 308
  of chick, 340
  of elasmobranchs, 299-300
  of enteropneusts, 289-90
  and iodine, 354
  of *Jamoytius*, 390
  of lamprey, 308, 354
  origin of, 284, 296, 299-315, 338, 356, 459
  and pituitary, 338

of tunicates, 299
Nuchal skeleton
in enteropneusts, 290
Nuclear bag fibre, 318
Nuclear chain fibre, 318
Nucleoli
in amoebocytes, 410
in corneal epithelium, 410
in cultured cells, 24, 128, 413
of fibroblasts, 24, 58
in intestinal cells, 269, 518-9
and naeglerioid stage, 167
in neurosecretory cells, 376, 389
in parolfactory cells, 388
in pharyngeal cells, 267
of rods and cones, 393, 412
in vesicular cells, 378
Nucleus(i)
of brain stem, 330
in cephalic organ, 374-6
controlled by cytoplasm, 130, 192-3, 197
DNA and, 42
identity of, 128, 130
in muscle, 316, 318-9
and naeglerioid form, 167
of photoreceptors, 393
position of, in epidermis, 262, 329-30
preoptic, 374
puncture of, 65
rotation of, 113
steroids and, 489, 494, 496
transference of, 130, 180, 193
types of, 24, 128-9, 479
Nurse cells
and ova, 253, 456, 474, 476, 480
Nutrition
of blastuloid, 198
of inner-mass cells, 207, 232

Ocelli
of arthropods, 399
Odontoblasts
in culture, 51, 53
development of, 546
ecballism and, 552
as mechanocytes, 51, 53, 552
as myxoblasts, 56-7
and neural crest, 136
and osteoblasts, 51, 552, 556
polarity of, 552, 555
Oesophagus
development of, 326, 359, 517
maturation of cells in, 209
Oestradiol
and *Naegleria*, 173
Oestrogens
and kidneys, 487-8
and *Naegleria*, 172
and ovarian follicles, 488, 497, 499
and Sertoli cells, 488, 499
structure of, 491-2
and uterus, 488
Oil
in retinal cells, 415
Oleic acid (oleate)
and cell surface, 123
and pigment cells, 255
Olfactory organ
cell types in, 84, 363-5
origin of, 362, 369
retinol and, 380
Oligodendrocytes (Oligodendroglia), 117-8, 331
Olynthus, 141-3
Oocytes (see also Ovum)
of crustacean, 479
enclosure of, within follicle, 456-7, 480
environment and, 456, 474-5
of insects, 456
and meiosis, 456, 480
and mitosis, 456, 480
of sponges, 191
Oogonia
of *Orchestia*, 479
Opisthonephros, 446, 460-1, 467
Opsin
and rods and cones, 412
and visual pigments, 182, 379
Optic cup, 408, 410, 416
Optic gland
of cephalopods, 386-7
Optic nerve
in nemertines, 388
recordings from, 418
Optic vesicle
cells in, 402-3
invagination of, 407, 409
ionic regulation in, 403
and lens formation, 408
origin of, 400

Organ culture
advantages of, 17-8, 122, 134, 567
of kidney, 449
of lymph nodes, 532
and synthetic media, 47
Organization
in embryonic development, 78
molecular, 130
serial, 257, 270, 279, 282, 284
into tissues, 278
Orientation (see also Polarity)
of cells, 13-6, 187, 570
cephalic organ and, 370
of collagen fibres, 553-7
in epithelia, 358, 570
of flagella, 187, 374
and induction, 78, 408
in membrana granulosa, 477
of metaphase plate, 208, 244
of micelles, etc., 13, 113, 322, 408
of nephridia and gonoducts, 443-4, 448, 450, 452, 470
of nerve cells, 244-5
of rods and cones, 413-4
Osculum, 141-2
Osmoregulation
in nemerteoids, 273
pharynx and, 357
Osmotic pressure
and blood volume, 272
and cell organization, 130
in gonads, 475-7
in kidney tubules, 450
and marine organisms, 201
and *Naegleria*, 168, 175
and teeth, 552
Ossification
and epidermis, 555
in limb bud, 51
Osteitis, 572
Osteoblasts
and chondroblasts, 49-51, 466, 558
and cilia, 213
in culture, 28, 33, 35, 39, 48, 53, 98
cystine and, 48
and embryo juice, 55
lysine and, 48, 54
and matrix, 33-5, 53, 542, 557
as mechanocytes, 28, 51-2, 97, 557-8
as myxoblasts, 53, 57
and odontoblasts, 51, 56, 552, 556
and osteoclasts, 97-8, 557-8
parathormone and, 541-2
processes of, 556
Osteoclasts
as amoebocytes, 97, 557-8
in culture, 33, 97-8
emballism of, 541
and matrix, 97-8
and osteoblasts, 557-8
and parathyroid, 98, 541, 558
and vitamin A, 97-8
Osteocytes
culture of, 33, 35
and matrix, 35, 552, 557
Osteoid tissue
production of, in culture, 34, 51
Ostium, 142
Otocyst (statocyst)
in acoels, 230, 369
and cephalic organ, 370
Outgrowth
from bone, 34-5
from cerebellum, 109
from chick heart, 21, 24-5
and embryo jjuice, 31, 38
mitosis in, 36-7, 110-1
from muscle, 31
from perichondrium, 11
zone of, 12, 17, 28, 31, 34
Ovary
of *Amphioxus*, 457
basement membrane in, 472
cortex and medulla of, 471
development of, 437, 469-74
ecballism and emballism in, 481
and germinal epithelium, 472
of insect, 456
of *Myxine*, 457
in nemerteoids, 278, 456
of *Orchestia*, 479-80
in rhabdocoeloids, 253
and testis, 455, 474
in vertebrates, 212, 481
Oviduct
origin of, 446
in *Priapulus*, 443
Ovotestis
of opossum, 484

Ovum(a) (see also oocyte)
characters of, 226
and coelom, 454-5
emballism of, 477
environment of. 252, 455-7, 472-4, 476, 480
genetic information in, 180
gradient in, 192, 197
and haemoglolin, 180
nucleus of, 128-9
in *Priapulus*, 443
in rhabdocoeloids, 252
spiral cleavage of, 296-7
of sponges, 147, 190, 192
surface of, 192, 477-8
transference of nuclei to, 130, 180
Oxygen
and cell mobility, 232
and erythropoiesis, 541
and giant cells, 96
and growth of amoebocytes, 94
and growth of epithelia, 73
and growth of mechanocytes, 94
in hanging-drop cultures, 9, 13
and vascular system, 426, 454
Oxyntic cells, 84
and nemerteoid pharynx, 350-3, 465, 517
surface of, 195
Oxytocin
and nemertine muscle, 265-6
and retractor of proboscis, 265-6, 344-5

Pancreas
cell nuclei, 128-9
cell types in, 84
culture of, 64, 67
glucocorticoids and, 525
islets in, 517-9, 525, 543
and nephridia, 442, 524-5, 543
origin of, 442, 517-21, 524, 543
Pancreazymin, 521
Paneth cells, 82, 84
Papillae
on Pogonophora, 291
on proboscis, 251-2, 276-7
Parabiosis
and differentiation of gonad, 455
Paraboloid
in photoreceptors, 393, 412
Paraldehyde-fuchsin
and "neurosecretion", 249, 347, 374
Parapodia
and swimming, 298
Parathormone, 542, 558, 569
Parathyroid
cell types in, 84
cilia in, 542
and cultures of bone, 98
ecballism in, 541
nephridia and, 541-3
origin of, 359, 541-2
and osteoclasts, 98, 541, 558
Parenchyma
of acoels, 232
and coelom, 291
of nemertine, 458
of rhabdocoels, 240, 251, 254
of sponge, 145
Parolfactory body
of cephalopods, 386, 388
Parotid gland, 526
Pars distalis, etc., see Pituitary
Patterns
of cell behaviour, 20, 24, 79, 123, 564, 566, 568
of cell organization, 130, 564-5, 568
of embryonic form, 150
Peniculus
of *Espejoia*, 159
Pentamethonium
on *Naegleria*, 172, 174
Pepsin
and digests of fibrin, 42
origin of, 521
Peptic cells, 84
Pericardium
mesothelium of, 66
and pronephros, 438, 531
Perichondrium, 11, 28, 51
Periodic acid—Schiff reaction (PAS)
and amoebocytes, 95
in cephalic organ, 374
and ciliated cells, 69, 214
in juxta-glomerular apparatus, 503
and microvilli, 520
and pharyngeal epithelium, 342, 358
in pineal eye, 397-8
Periosteum
and bone, 50-1

culture of, 28, 36, 40-1, 48
osteoblasts of, 33, 56, 558
Peripolesis, 532
Peristalsis
of nemertine body, 302, 425
Peritoneum
amoebocytes from, 92-3
mechanocytes from, 459
mesothelium of, 66, 194
Permeability
and cell surfaces, 126, 189, 380
of rhynchocoel, 345
Peroxidase systems
in blood cells, 272, 279, 426
Peyer's patches, 524, 526
pH (and Hydrogen ion concentration)
of blastocoel fluid, 189
and giant cells, 96
and intestine, 517, 571
and kidney tubes, 450, 501
and *Naegleria*, 168, 171
in nemertine fore-gut, 351, 353
and survival of eggs, 46
Phagocytosis
in acoeloid, 232-3
by amoebocytes, 89, 91, 99, 124, 126, 224-5
in bone, 558
and emballism, 225
of endothelia, 56, 95, 525, 540
by epitheliocytes, 153, 155
in haemopoietic tissues, 525, 538-9
in intestine, 520
by mesothelium, 460
in naeglerioid, 167, 175
and nephridia, 254
in nephron, 452
in planuloid, 228
stimulation of, 123, 529
Phagosomes, 224
Pharynx (see also Fore-gut)
of acoels, 230, 232-3
cell types in, 465, 499
derivatives from, 289, 338-360
of enteropneusts, 290
eversible, 251, 257, 260
iodine and, 353-7, 465
of *Lineus*, and concentration, 240-1, 273, 353, 435, 439, 509
of nemerteoids, 267, 279, 287, 290, 339, 357, 433, 499, 517
nephridia and, 440
pouches in, 287, 349
and respiration, 287, 426
of rhabdocoels, 238, 251, 257
vascularity of, 429-30, 435
Phase contrast, 11
Phenol red
and kidney tubules, 199
Phenylalanine
and bradykinin, 220
and growth, 73
Phosphatase
acid, 95-6
alkaline, 51, 439, 520, 549, 551-3
Phosphate (PO_4^{3-})
and culture media, 8
and ecballism, 501
and growth of epithelium, 71
in kidney tubules, 450
and *Naegleria*, 170-1, 501
parathyroid and, 542
Phospholipins
and membranes, 70, 125, 416, 491-4
and steroids, 491-4
vitamin A and, 495
in zona glomerulosa, 485
Photoreceptors (see also Eyes, Sense cells, etc.)
organization of, 233-4, 386, 393, 416
in rhabdocoels, 257, 369
in squids, 385
Photosensitivity, 255
of *Amphioxus*,, 383
and carotenoids, 182, 233, 415
in enteropneusts, 292
of lampreys, 383
of *Myxine*, 398
of pineal, 395-6
Photosynthesis
and blastuloids, 187
Phyllodocidae
nephridia in, 440
Phytohaemagglutinin, 534, 537
Pia mater
origin of, 136
Pigment
cells, 255, 401
in ectoderm, 401
in eye-spots, 255-7, 383
formation of, 72, 75

iron-containing, 452
in large lymphocytes, 539
in Leydig cells, 477
in macrophages, 529
migration of, 255
in nemerteoids, 263, 396
and neural crest, 136, 334
photosensitive, 233, 412, 415
in pineal eyes, 392-4
Pigment epithelium
cells of, 403-7, 412, 416
and cilia, 394, 416
culture of, 64, 67, 72
function of, 75, 406
keratin and, 69
and rods and cones, 393, 406, 410, 414
vitamin A and, 406
Pilidium larva, 326, 328
Pineal body (see also Pineal eye)
origin of, 284, 391
vesicles in, 398
Pineal eye, 390-9
and diurnal rhythm, 395
lens of, 193
origin of, 362
receptors of, 392-3
vitamin A and, 394
Pinnules
of Pogonophora, 291
Pinocytosis
by amoebocytes, 100-1, 124, 126, 224, 529
and cell types, 500
by epithelia, 124, 153, 155
in intestine, 520
and Leydig cells, 477
by mechanocytes, 100, 124
and microvilli, 195
by monocytes, 529
in naeglerioid, 167, 175
by nerve cells, 106-8
and ova, 455
and salt concentration, 100, 175
in zona fasciculata, 486
Pituitary
ACTH and, 508
anterior, 84, 338-43, 357-8, 388-9, 505-6
cell types in, 84, 342-3
ciliated cells in, 341, 506
FSH and, 472, 481
and melanophores, 397
origin of, 284, 294-5, 338-49, 505, 517
pars distalis, 294-5, 340-2
pars intermedia, 294-5, 340-1
pars nervosa, 294-5, 341
pars tuberalis, 294-5, 340
posterior, 341, 344-9, 388-9
prolactin and, 507
and thyroid, 357-8
and urinogenital system, 505-11
Placenta
acetylcholine and, 246
in sponge development, 190
Placodes
in nemertine development, 326-9, 384, 399-400
Planaria
epidermis of, 211
neoblasts of, 568
photoreceptors in, 255-6, 380, 383
regeneration in, 180
and ultraviolet light, 391, 397
Planula
formation of, 207
of *Gonothyraea*, 225
of sponges, 224
Planuloid stage
characters of, 206-29
and collagen, 210, 214
special features of, 224
Plasma
and amoebocytes, 88, 90, 93-4
and cell form, 11, 25, 31-2
and cell orientation, 13-4
cells, 528, 533, 539
and cell suspensions, 27
and culture of bone, 51
and culture of nerve, 108
and differentiation, 31, 76
and epitheliocytes, 66, 72
and gonadal fluids, 475-6
and growth of mechanocytes, 36-7, 39, 43, 46
protective action of, 46
and tissue culture, 8-11
and urine, 451
Platelets, 272
Platyhelminthes
balancing organ in, 369
chitin in, 282

cholinesterase in, 250
Podocytes, 445
Pogonophora, 291, 293, 360, 574
and *Amphioxus*, 301
Poison
in nemertines, 276-7
Polar groups
on cell membranes, 493
Polarity
and basement membrane, 472
in blastuloid, 187, 190-1, 203
of cells, 124, 126, 190, 499-501
in dental lamina, 547-8
of epithelia, 146, 472
and local environments, 479
of lymphocytes, 532
and movement, 164, 168
of myelocytes, 535
of *Naegleria*, 156-8, 164, 168, 174-5, 190
of nerve cells, 111, 244, 247, 331
of neural tube cells, 333, 403
of neuroglia cells, 118
reversal of, 164, 174, 191, 547-8
of rods and cones, 412
Polarized light, 11
Pole (see also Animal pole, etc. and Polarity) 190, 192, 196-7, 552
Polychaetes
blastula of, 186
cartilage of, 558
eyes of, 399-400
swimming of, 298
Polypeptides
of bradykinin, 220
ecballic cells and, 221, 503
in haemoglobin, 183-4
as hormones, 505-6
in insulins, 179
mechanocytes and, 220
and permeability, 345, 503
Polyploidy
in ganglion cells, 110
Polypoid cells
of choroid plexus, 84, 194, 402
Polysaccharides
and amoebocytes, 223, 231
and cilia, 214
and ground substances, 70-1, 146, 148, 214
magnesium and, 222, 231
in nematocysts, 216
and phospholipins, 493
sulphated, and pseudopodia, 44
and uridine triphosphate, 221-2
Pore(s)
in biological membranes, 493
in nuclear membrane, 167
of sponge, 141-6
Porifera (see also Sponges)
calcareous, 140
cartilage in, 558
collagen fibres and, 231
muscle in, 303
and origin of cell types, 137-8, 140-54
phylogeny of, 231, 233, 281
Porocyte (pore cell)
contractility of, 144-5, 150-1
derivation of, 146, 210
Porphyropsin, 379
Potassium (K^+)
and adrenal cortex, 482, 487
and carnivores, 503, 532
and culture media, 8
and follicular fluid, 475-7, 508
glutamate and, 220-1
glycine and loss of, 220-1
and goblet cells, 482, 521
hyaluronic acid and, 223
and ionic balance, 223, 353
and kidney tubules, 501, 505
and membranes, 489
mineralocorticoids and, 508
and *Naegleria*, 168-9, 172, 175, 495, 501
and sodium ratio, 231, 475, 482, 488-9, 508
and spermatozoa, 475-6, 482
in testicular fluid, 475-7
Preamoebocyte, 210
Prechordal plate
and rhynchodaeum, 340
Pregnanediol, 490
Premechanocyte, 210
Preoptic-neurohypophysial tract, 349, 386, 389
Proboscis
and acetycholine, etc., 265, 344
armature of, 260-1, 277, 338
cell types in, 465
development of, 275, 296, 326-7, 329
of enteropneusts, 285-6, 290, 292, 356, 444
eversion of, 276-7, 287, 289, 339, 341, 364
evolution of, 277, 294-6, 574

functions of, 276
and gut, 262, 574
hinge of, 341, 345
and iodine, 354
in Kalyptorhynchia, 251-2, 257, 277
in nemerteoids, 251, 260, 262, 273, 277, 279, 287, 290, 292, 362, 433
nerves to, 275-6, 290, 341, 344-5
and notochord, 290, 296, 299-300, 304, 574
origin of, 251
in Pogonophora, 293, 574
pores, 444
retractor of, 251, 260, 265-7, 290, 341, 344
and rhynchocoel, 271, 279, 287, 344-5, 432-4
significance of, 277, 284, 338
structure of, 275-7, 300, 435
Proerythroblasts, 526-7
Progestagens, 491-2
Progesterone
and corpus luteum, 497, 499, 507
and ecballism, 501
and *Naegleria*, 172-3, 481, 489, 496-7, 501
structure of, 490
and uterus, 488
Prolactin, 507-8
Proline
in bradykinin, 220
and collagen, 48. 53
and growth, 48, 214
and hamster cell-line, 215
and ionic blance, 220-1, 500, 503
mechanocytes and, 48, 214, 220, 466, 499, 503
and polypeptides, 220-1, 503
urine and, 503
Pronephros
ACTH and, 528-9, 533
athrocytosis in, 531
blood formation in, 467, 526-7, 531
in cyclostomes, 437, 446, 467, 531
funnels in, 436, 449, 530
lymphocytes in, 526, 528-9
macrophages in, 529
in myxinoids, 438, 526-7, 530-1, 543
and nephridia, 438, 446, 449, 455, 460-1
Prophase of neuroblast, 208
Proprioceptive mechanisms, 363, 419
Propylaldehyde
and mechanocytes, 44
Prosopyle, 142
Prosthetic group, 566
Protein
and carbohydrates, 179
digests of, 42
and DNA, 42, 179, 565-6, 568
and the econome, 566, 568
fibrous, 13
genes and, 179
iodinated, 353
and lipids, 179, 493, 566
and membranes, 131, 168, 506, 566
and nucleoli, 410
serum, 99
species specific, 183, 573
specific functions of, 130-1, 184, 566
and steroids, 489-90, 498, 506
synthesis of, 132-3
Proteoses
and growth, 42-3, 73
Protochordates, 285-94
Protonephridium, 441, 443, 461
Protonephromixium, 442, 444
Protosoma
of Pogonophora, 291, 293
Protozoa
biochemical processes and, 166-7, 177, 183
colony-forming, 165-6, 184, 187
and DNA, 167
and ionic content, 198
to metazoa, 177, 184-7, 231
and origin of cell types, 5, 138, 153, 155
photosensitivity of, 233
and sex, 167, 226, 475
trichocysts in, 218
Protractor muscle
of proboscis, 251
Pseudopodia (cell processes)
of amoebocytes, 87, 106, 124, 147
of cells in culture, 23, 81, 107, 124
in enamel organ, 548
filiform, 126, 158, 193
lamelliform, 167
lobose, 126, 156-8, 167-8, 187, 195
of lymphocytes, 532
of nerve cells, 9-10, 106-7, 111-3
of odontoblasts, 552-3
of Schwann cells, 115-18
and sulphated polysaccharides, 44
types of, 80, 175, 232

Pteraspida
lateral line in, 367
Pure strains
of amoebocytes, 91
in culture, 4
of epithelia, 76, 81
of mechanocytes (fibroblasts), 46-7, 76, 99-100, 127
Purkinje cells, 111
Puromycin, 8
Pylorus
keratin and, 69
in nemertine, 287, 458, 517

Rathke's pouch
derivatives from 357-8, 364
origin of, 294-5, 338, 340-1, 350, 357
Rayon
rafts of, 31
Receptive field
of ganglion cell, 418
Receptors (see also Sense cells, Photoreceptors, etc.)
in cephalopod retina, 385-6
in eye-spots, 255-6, 382-3
and movement, 325
in pineal eyes, 392-4
in vertebrate retina, 374, 382, 386, 392-4
Recessives
selection of, 183
Rectal gland
and mucin production, 69
and nephridia, 524, 543
Rectum
abnormal potassium in, 521
Red blood cells, see Erythrocytes
Reflexes
development of, 325
Regeneration
in planaria, 180, 240
Renin, 503, 505
Repression (see also Suppressors, Hegemon and Genome)
of genes, 132-3, 575
Reptiles
culture of tissues from, 22
kidney tubules in, 452, 488
Reserve cells
and differentiation, 49
Respiration
in *Cerebratulus*, 287, 426
and differentiation, 198
gill slits and, 296, 359
of inner-mass cells, 207
in muscles, 321
and swimming, 296, 426
Rete ovarii, 469, 472, 483
Rete testis, 446, 469, 483
Reticulin
in erythropoietic tissue, 534
formation of, 25, 213
Reticulum (see also Endoplasmic reticulum)
cells, 533-4, 540
in erythropoietic tissue, 533-5, 538, 558
in lymphoid tissue, 534, 538
sarcoplasmic, 313
stellate, 546, 548, 551
of thymus, 526, 534, 538, 541
Retina
bipolar cells in, 332, 376-7, 384, 403, 410, 416-8
cell deaths in, 416
cell types in, 84, 90
of cephalopod eye, 381-2, 385
cholinesterase in, 416
differentiation in, 403, 410
direct, 382, 391, 394
duality in, 572
ecballism and emballism in, 412
ganglion cells of, 114, 382, 384
glutamate and, 221
glutamine-synthetase and, 221
information provided by, 397
interconnections in, 417-8
inverted, 284, 382-4, 394
of monkey, 376-7, 384
and optic vesicle, 400, 403
pigment epithelium of, 67
of pineal, 391-2
and regeneration of lens, 414
tissue culture of, 90, 412-3
transmitters in, 416
of vertebrate eye, 382, 416
Retinene (Retinal)
distribution of, 182, 380
and opsins, 379, 412
and pigment epithelium, 406
Retinula
in cephalopod eye, 382

Retractor muscle
of proboscis, 251, 260-1, 265-6, 277, 290, 343-4
Rhabdites
and bradykinin, 220
composition of, 218
function of, 242
in nemerteoids, 263, 276-7
in rhabdocoels, 217, 238-40, 257
Rhabdocoeloid stage, 238-57
off-shoots from, 399
Rhabdocoels
alimentary canal in, 250-1
cephalic pits in, 248-9
connective tissue in, 239-40
epidermis in, 239-42
and gonocoel, 251, 257, 434, 479
and molluscs, 282
muscle development in, 249-50
neural tissue in, 242-9
pigmentation in, 255-7
rhabdites in, 217, 239-40, 257
and salt, 380
and swimming, 298
urinogenital system in, 251-5
Rhabdom
in cephalopod eye, 382
Rheoceptors
evolution of, 369
in lateral line, 367-8
in nemerteoids, 273
in rhabdocoeloids, 249, 369
in semicircular canal, 370
Rheumatism, 224, 572
Rhodopsin
in *Dendrocoelum*, 255
ova and, 130
in relation to DNA, 182
retinene and, 182, 406
and vision, 379, 391
Rhopheocytosis, 532, 535
Rhynchocoel
and blood vessel, 346-7, 349, 388, 425, 429, 430-1
and cephalic organ, 371
coelom and, 276, 289, 434, 457, 468
corpuscles in, 530
diverticula of, 347, 458-9
fluid relationships of, 260, 430, 432, 530
homologue of, 289
nephridia and, 444
origin of, 276
pressure in, 341
and proboscis, 271, 277, 289-90, 294-5, 338-9, 344-7, 371, 430-1, 433, 435
villus body and, 285, 289-90, 347-8
volume of, 270-1, 277, 344-5
Rhynchocoela, 260-79
Rhynchodaeum
blood vessels and, 260-1, 285, 289-90, 340, 432, 454
and eversion of proboscis, 276-7, 287, 371
and frontal organ, 363-4
and mouth, 289, 338-40
and proboscis pore, 444
Rhythm
and cardiac muscle, 25-6
diurnal, 395-8
Ribonucleoproteins (see alss RNA)
and growth stimulation, 41, 43
and induction, 408
in nerve cells, 132
Ribosomes
in erythroblasts, 535
in mechanocytes, 25
in naeglerioid, 167, 175
suppression and, 197
Ribs
culture of, 51
RNA
and antibodies, 532
and cell division, 41, 133
and differentiation, 42
genome and, 130-2
and globin, 182
and induction, 408
messenger, 197
and repression, 133, 197
ribosomal, 197
soluble, 197
Rods
and cones, 391, 393, 403-19, 572
ecballism of, 412
and flagella, 384
functions of, 411, 572
nomenclature of, 411-2, 414
nutrition of, 406
and pigment epithelium, 393, 403, 411-2, 414
and pineal eye, 391, 393

polarity in, 412
properties of, 393
red and green, 411, 418
of solenocytes, 445
of vertebrate retina, 84, 383, 394
Roller tubes, 17, 108
Romanowsky stains, 525
Root fibres
in cephalic organ, 370
in cephalic pit, 288-9, 366-7
in flagellate cells, 193, 219, 288-9
and metachronal rhythm, 197, 229
in *Naegleria*, 159, 167
in olfactory organ, 365
Root nodules
haemoglobin and, 184
Rosettes
in cultures of retina, 412-3
Rotifer
blastula of, 186
Rous sarcoma
and amoebocytes, 125
Ruffles
and cell movement, 23
and pinocytosis, 100

Sagittocysts, 217
Salinity (see also Sodium chloride, Salt)
and body fluids, 434-5, 444
and cephalic organ, 370, 373, 379, 385, 416, 506
and cephalic pit, 249, 370, 380
and iodine, 355-7
and migration, 356-7
and nemerteoids, 240, 273, 350, 353, 357, 387, 505, 509
and photosensitivity, 380, 385, 387
preference, 356
response to, 171-5, 434-5, 505, 509
Saliva
aldosterone and, 442
and PAS, 95
Salivary gland
aldosterone and, 488
basement membranes and, 472
cell types in, 82-4
chromosomes of, 489
differentiation of, 74-5
mineralocorticoids and, 525
origin of, 359, 442, 465, 516, 524, 543
Salt (solutions) (see also Salinity, Sodium chloride, etc.)
and cell suspensions, 27
and chloride cells, 352
reaction of cells to, 240-1, 353, 386
and sexual activity, 387
for tissue cultures, 7, 9, 23
Sarcoma
and glucosamine, 214-5
and mechanocytes, 124-5
Satellite cells
in muscle, 332
and nerve cells, 282, 570
in sponges, 191
Scales (fish)
blood vessels and, 556
collagen fibres and, 553, 555
growth on, 13, 15
of *Jamoytius*, 390
and teeth, 544, 555
types of, 545, 555
Schizocoel
coelom as, 276, 454, 459, 468
meaning of, 226
rhynchocoel as, 276
Schwann cells
in culture, 114-9
and myelination, 113, 115-6, 118, 282
and nerve fibres, 570
origin of, 135-6
types of, 114-6, 136, 331
Scleroblasts, 144-8, 151, 210
Scyphozoa
gastraea of, 226
Sea urchin
animalization of, 223
blastula of, 221, 228
cholinesterase in embryos of, 246
Sea-water
and body-fluids, 434-5, 452-3, 521
and cephalic organ, 378
hypertonic, 378-9, 401-2
and iodine, 355-6
and nemertine muscle, 265, 344
and sponge cells, 146
and visual pigments, 415
Secretin, 521
Secretion
and basement membrane, 472
holocrine, 523

Secretory cells
in cephalic organ, 372-4
in frontal organ, 364
Segmentation
and metamerism, 282, 284, 304, 460
and swimming, 298
Segregation
of dyes, 89-90, 124, 126, 147-8
Self-recognition
by cells, 165
Seminiferous tubules
differentiation in, 132, 474, 476
ecballism in, 476, 509
fluid in, 476-7
and glucose, 450, 478
gonocoels and, 448
sex cords and, 448, 468-9
testosterone and, 488
Sense cells (Sensory cells)
in acoeloid, 232, 242
in cephalic organ, 372
and cilia, 363, 367-8
in eye-spots, 255-6, 382-3
in frontal organ, 363-4
in nemerteoids, 262, 288, 383
of pineal, 391-2
in rhabdocoeloids, 238, 242-3, 248-9, 257
in teeth, 552
and tubular nervous system, 328, 332
Sense organs
and exoskeleton, 369
in nemerteoids, 273, 345, 362, 367
Serialization
in nemerteoids, 270, 278-9
in rhabdocoels, 257
and segmentation, 282, 284
Serine
and bradykinin, 220
and growth, 48
and phospholipins, 491, 493
Serotonin
in pineal, 396, 398
Serous glands (cells)
differentiation of, 77, 82-3
in nemertine epidermis, 262-3
in nemertine pharynx, 267
in salivary glands, 84
Sertoli cells
in cyclostomes, 457, 479
ecballism of, 474-7, 481, 509
and oestrogens, 488, 499
polarity, etc., of, 477, 482
Serum
and cell types, 124
in culture media, 23, 37, 43, 73, 99, 108
and differentiation, 30-1, 575
and form of monocytes, 89
growth in, 37, 90, 92
and phagocytosis, 99
of pregnant mare, 475
protective action of, 44, 47, 99
proteins in, 99
and Schwann cells in culture, 115
Sex cords
and adrenal cortex, 470, 482-3
formation of, 212, 436, 469, 472
and genital ridge, 448, 468, 470-1
as gonoducts, 455
and mesonephric tubes, 443, 468, 472
Sex determination, 480
Sex glands, see gonads
Sex-hormone
and adrenal cortex, 483
and differentiation, 481
and kidney, 439, 488
steroid nature of, 487
Sex-segment
in kidney tubule, 452-3
Sexuality
development of, 167, 226, 386
Silver (nitrate) (staining)
in cephalic organ, 374
and epithelial boundaries, 65, 125, 189
of *Espejoia*, 159
of nerve cells, 111-2, 114
Sinusoids
in bone marrow, 95, 537
in liver, 95, 525
in pronephros, 438
Sinus venosus
origin of, 429
Sipunculids, 328
Skeleton
axial, *q.v.*
cartilaginous, 558
development of limb, 51
and effects on muscle, 322
exo-, 284, 322
fluid, *q.v.*
origin of, 284, 544

proboscis, 286
and proprioceptors, 419
Skin
and Ca^{2+}, 71
cell types in, 82, 84, 134, 242, 465
cholinesterase and, 246
culture chambers in, 17, 91
cultures of, 64, 67, 135
differentiation of, 77, 79
environment and, 263, 453
frog's, 571
function of, 242
and homoiostasis, 439
and ionic regulation, 254, 273, 453
and keratinization, 69
mucus-producing, 69-70, 82, 263, 465
of nemerteoids, 262-3, 465
photosensitivity of, 398
types of, 242
vitamin A and, 69-70, 82
Snake
muscle fibres in, 313
nerve endings in, 312
Sodium (Na^+)
and adrenal cortex, 487, 508
and carnivores, 503, 532
chloride, and cephalic organ, 373, 379, 386
chloride, and flagella, 157, 168-72, 495
chloride and iodine, 355-6
chloride and lymphocytes, 532
chloride and nemerteoids, 273
chloride and pinocytosis, 100
citrate, 27
and culture media, 8
ecballism, emballism and, 501, 573
and follicular fluid, 475-6, 508
glutamate and mechanocytes, 220
glycine and uptake of, 220
hyaluronic acid and, 223
iodoacetate and retina, 412
and ionic balance, 501
and juxta-glomerular apparatus, 503, 505
and kidney tubules, 501, 503, 505
and light, 380
mineralocorticoids and, 508
and *Naegleria*, 168-72, 494-5, 501, 508-9, 573
oleate, and cell surface, 123
and pharynx of nemertines, 240-1, 273, 353
and potassium ratio, 231, 475-6, 482, 488-9 508
in testicular fluid, 475-6
transport, 246, 489, 498, 501, 507-8, 573
Solenocytes (Flame-cells)
in *Amphioxus*, 445, 448
in annelids, 441-2
and lymphocytes, 529
in nemerteoids, 269, 436-7. 439-40
in *Phoronis*, 529
in *Priapulus*, 443
in rhabdocoeloids, 238, 254
Somatic cells
information in, 42, 180
Somatic motor system
development of, 315, 325
Somatopleure
and muscles, 312
Somites (see also Myotomes)
cavity of, 301, 305
and coelomic cavity, 435
development of muscle from, 304-6, 312, 314-5
distribution of, 303-4, 460
innervation of, 305, 332
in insects, 314
origin of, 299-315, 435-6, 460
transmitter for, 304, 344
Spawning
and pharyngeal epithelium, 357
and salinity, 388
Species characters
of cells in culture, 134
Specificity
of biological membranes, 491
Spectral sensitivity
of visual pigments, 415
Spermatid
nucleus of, 128-9
in rhabdocoeloids, 253
Spermatocyte
nucleus of, 128-9
of *Orchestia*, 479-80
in rhabdocoeloids, 253
and water, 478
Spermatogonia
division of 209, 475
and giant cells, 481
of *Orchestia*, 479
in rhabdocoeloids, 253

and testicular fluid, 476
Spermatozoa
characters of 226
cholinesterase in, 246
crustacean, 482
ecballism of, 475
and glucose, 478
in insects, 456
local environment of, 252-3, 455-6, 477
in nemerteoids, 456
of *Orchestia*, 479, 482
and potassium, 477, 482
release of, 455
of rhabdocoeloids, 252-3
X and Y-bearing, 480-1
Sphincter
pyloric, 521
Sphincter pupillae
and adrenalin, 412
cholinergic, 569
muscle, 403
Spicules
calcareous, 144-8, 152
siliceous, 152
Spike potentials
in cells of Hydra, 244
Spinal cord
of *Amphioxus*, 307-8, 310, 329
of frog, 9-10
Spindle cells
in blood of *Myxine*, 530
in blood of nemertines, 272, 426, 530
Splanchnocoel, 315, 433-4, 454, 459, 468
Splanchnopleure
and muscles, 312
Spleen
culture of, 90
endothelium of, 56
lymphocytes and, 526
origin of, 543
recolonization of, 537, 540-1, 543
red pulp of, 534, 538, 540, 543
white pulp of, 526, 543
Sponges
blastulae of, 140, 153, 187, 196, 207, 209-10
calcareous, 140, 152, 187, 196, 207, 210, 213
cell lineage in, 137-8, 150-4
cell suspensions from, 146
and collagen fibres, 213
fresh-water, 189
germ cells in, 151-2, 460
gradients in, 415
and iodine, 353
and *Naegleria*, 153
siliceous, 151-2
Spongin fibres, 144, 147
Spurs and metachronal rhythm, 197, 229
Squid
axon of, 572
photoreceptors of, 385
Statocyst
of acoels, 230
of nemertines, 362, 369, 418
Stem cells (formative cells)
in erythrocyte formation, 132, 529, 534-7, 543
for lymphocytes, 529, 534, 539, 543
Stereocilia (see also Microvilli)
and cell gradient, 193-4
in cephalic pit, 366-7
on choanocytes, 141, 143
in epididymis, 193-5
on intestinal cells, 268
and naeglerioid, 167
and photoreceptors, 384
in rheoceptors, 367-8
of sponge embryo, 190
Steroid(s), 487-98
and adrenal cortex, 485, 502
and cell surfaces, 131, 172, 490, 494, 496, 502, 506
and ecballism, etc., 499, 501
excretion of, 498
formation of, 497
and glucose, 488, 497
hydrophilic groups in, 491
and ions, 488-9, 496-8
metabolism of, 487-8, 502-3
mode of action of, 488-98, 502
and *Naegleria*, 172-3, 175, 481, 496, 501
and phospholipins, 490-3, 498, 506
and protein, 489-90, 498, 506
structure of, 490
and sulphate, 498
and uronic acid, 498
Stilboestrol
and kidney cortex, 483
structure of, 492
Stomach
cell types in, 84

epithelium of, 79
in nemertine, 287, 326, 359, 517
oxyntic cells in, 195, 351-2, 465
peptic digestion in, 521
and rhythms, 398
Stomatoblastula
orientation of cells in, 190
Stomochord
of enteropneusts, 289
Stomodaeum
invagination of, 290, 294-5
Stratum spinosum
in oesophagus, 209
Striations (in muscle)
in cultures, 25-6, 52
development of, 302, 324
and skeleton, 322
structure of, 308-9, 313
Stripes of Retzius, 551
Stroboscope, 11
"Strong solids," 552-3
Subpedunculate body
of cephalopods, 386
Substratum
locomotion and, 168, 207
relation of cells to, 13-6, 157, 168
Succinic dehydrogenase, 315, 317
Succus entericus, 523
Sucrose
and *Naegleria*, 168-9
Sudan Black
and phospholipins, 125
Sudanophobic zone
of adrenal, 485-6, 509
Sugar (see also glucose, etc.)
blood, 517
and glycogen in liver, 73
and neutral red vacuoles, 24
Sulphanilamide
and carbonic anhydrase, 8
Sulphate
and cell types, 500
and connective tissues, 498
mechanocytes and, 499
and mucus-secreting cells, 98
and *Naegleria*, 170, 498
and polysaccharides, 44, 222
and steroids, 498
and vegetalization, 202, 223, 498
Sulphonamide drugs
and cell division, 8
and mechanocytes, 44
Sulphur (-S, -SH)
and keratin formation, 68
and nematocysts, 216
and vitamin A, 70, 98
Suppressor(s)
in early development, 182
of genetic information, 42, 180-3
in *Naegleria*, 197
Surface (see also Cell surface)
and differentiation, 56, 76
and growth of cells, 9-14, 64, 67, 72, 87, 90
Sweat glands
cell types in, 84
and fluid regulation, 254
nephridia and, 442
Swimming
and origin of chordates, 263, 294, 296-9, 304, 385, 434, 453
and respiration, 426, 430
and sense organs, 325, 369-70, 385, 391
Symbiosis (see also Balance)
of cells in blastuloid, 192, 203, 220, 226, 332, 477, 570
between chief cells and goblet cells, 534, 571
between ecballic and emballic cells, 533, 570-1
between epitheliocytes, 357-8
between fibroblasts and macrophages, 224, 534, 571
between flagellate and amoeboid, 332, 477
between hair-cells and supporting cells, 419
between Leydig and Sertoli cells, 477
between lymphocytes and monocytes, 533-4
between neuroglia and nerve, 118-9, 411, 332, 570
between osteoblasts and osteoclasts, 558
in retinal cells, 411, 414
Symmetry
bilateral, 209, 213, 232
Synapses
cell polarity and, 245
and desmosomes, 188, 197
electrical, 197, 418
retinal, 412, 416-8

transmitters at, 245-6, 416
types of, 188, 308, 312
Synaptic vesicles, 188
in *Amphioxus*, 308, 310, 312, 330
and pinocytosis, 108
Syncytium
definition of, 319
muscle fibre as, 306
Synovioblasts
in culture, 50-3, 56
and odontoblasts, 56-7, 552

Tail
fins on, 296-7
origin of, 284
photosensitivity in, 383, 396
tadpole's, 91
Tantalum grid
for cultures, 31
Taste buds
and cephalic pits, 370
origin of, 249, 362, 367
Teeth, 544-57
ascorbic acid and, 71
collagen fibres in, 555
development of, 546
evolution of, 544
of keratin, 544
Mg^{2+} and, 548-9
odontoblasts of, 51, 56
pulp of, 556
sockets of, 555-6
as strong solids, 552-3
structure of, 551
succession of, 545-6
vitamin A and, 548-9
Teleosts
muscular movement in, 303
Temperature
and cell organization, 130
and cultures, 9
and evolution, 454
and formation of flagella, 170
and giant cells, 163
and iodine binding, 353, 357
Tendon, 55
Tension
and bone, 51
in cultured cells, 23, 66
effects of, on cells, 14, 555
-striae in cells, 52, 55
Tentacles
of Pogonophora, 291
Terminal bars (zonae occludentes)
in blastulae, 188-9
and dehiscence, 208
in development of muscle, 303, 306
in epithelia, 69, 125, 195, 211, 519
of neuroblasts, 208, 244
in olfactory organ, 365
Testicular fluid, 475-6
Testis
of *Amphioxus*, 457
basement membrane in, 472
cell division in, 209
development of, 437, 469-70, 473-4
fluid in, 475-7
in *Galleria*, 456, 480
Leydig cells in, 474-5, 477, 499, 505, 509
of *Myxine*, 457
of *Orchestia*, 479
in nemerteoids, 278, 456
in *Priapulus*, 443
in rhabdocoeloids, 253
of salamander, 455
tubules in, 449, 472
Testosterone
and adrenal cortex, 486
and kidney, 488
and Leydig cells, 499
and *Naegleria*, 172-3
and seminiferous tubules, 488
Tetraethylammonium iodide, 174
Tetrahydrocortisol, 490
Theca
externa, 474
interna, 474, 477, 481, 505, 509
Thesocytes, 150
Thiouracil
and endostyle, 353, 356
Thiourea
effect of, on fish, 356-7
Thoracic duct
cells in lymph from, 90, 539, 541
Threonine
and growth, 73
Thrombocytes
in fowl blood, 90-1
Thymidine
tritiated, and lymphocytes, 539

Thymocytes, 132, 526, 537-9
Thymus
  differentiation of cells in, 132, 528
  and growth promotion, 43
  hormone from, 540
  lymphocytes in, 537, 541
  origin of, 359, 541, 543
  recolonization by, 537
  recolonization of, 537
  stem cells in, 526, 535, 538
Thyrocalcitonin
  and calcium, 357, 569
  and thyroid, 357, 542
Thyroglobulin, 358
Thyroid
  cell types in, 84
  culture of, 64, 67, 74
  differentiation in, 74, 76, 358
  lymphocytes in, 526
  and migration, 357
  origin of, 3, 284, 353, 356-8, 517
  -pituitary relationship, 357-8,
  -stimulating hormone (TSH), 357-8
  and thyrocalcitonin, 357, 542
Thyronine
  in halophytes, 356
  and nemertines, 354-5
Thyrotrophic cells, 358
Thyroxin
  and amoebocytes, 89
  in chordates, 353, 356
  and fish, 357
  formation of, 358
  and Na^+ transport, 356
  and nerve cells, 114
Tibia
  culture of, 33
Tight junctions (nexus)
  in blastulae, 188-9
  and ecballic cells, 228
  in epithelia, 65, 125, 165
  between rods and cones, 417-8
Tissue culture
  adaptation in, 4, 37-8, 59, 567
  behaviour of cells in, 4, 5, 7, 12-8, 133, 153-4
  cells, 20, 47, 59, 73, 81, 133, 567
  of choroid plexus, 403, 415
  critique, of, 9, 122, 153
  of ependyma, 106, 403, 415
  and germ layers, 134-7
  of insect gonads, 456
  and local environments, 567
  methods of, 7-12
  of retinal cells, 412, 414
  of sponge tissues, 148-9
Tissue fluid, 12, 414
Tonicity (see also Osmotic pressure, Salinity)
  and neurosecretion, 249
Tonofibrils
  in enamel organ, 548
Tonsils, 524, 526, 541, 543
Totipotency
  of archaeocytes, 150, 152, 210
Toxicity
  and inhibition, 44
Trachea
  cell types in, 83-4
  ciliary beat in, 240, 245
  epitheliocytes, in, 194
  goblet cells in, 521
  keratinization of, 68-9
Tractella
  on *Naegleria*, 157, 164
Transformations
  of amoeboid and flagellate, 164
  between cell types, 27, 35, 80, 99-100
Transmitter(s) (neural)
  and induction, 569
  and muscle systems, 250, 264, 322
  oxytocin as, 344
  and polarity of cells, 244-6
  in retina, 416
  and synapses, 246, 257
  and types of nerve cells, 119
Trematodes, 278, 281
Trephocytes, 137
Trephones
  and fibroblasts, 99
Trichocysts
  in protozoa, 218-9
Tri-iodothyronine (T_3)
  endostyle and, 353
  and thyroid, 356, 358
Trimethyl-ethyl-ammonium chloride, 100
Tropocollagen
  and collagen fibres, 54
  production of, 220

Trypan blue
and adrenal cortex, 486
amoebocytes and, 25, 90-1, 147-8
and kidney tubule, 452, 486
mechanocytes and, 25
Trypanosomes
and cholinesterase, 246
Trypsin
and activation of tissues, 9, 43
and cell suspensions, 13, 27
and epithelial cells, 71, 79, 125
and heart fibroblasts, 27, 35, 525-6
and intestine, 521
Tryptophane
and growth, 73
T-system
of muscle, 312-3
Tubules
cytoplasmic, 167, 217
kidney, *q.v.*
Tumour (see also Carcinoma, Sarcoma, etc.)
cilia in renal, 4, 488
Schwann cells, 114
Tunica adventitia, 428
Tunica media, 428
Tunicates
heart of, 430
iodine and, 353
notochord of, 299
Turbellarians
eye-spots in, 379, 391
frontal organ in, 363
gastrulation in, 283
gut in, 302
nerve cells in, 243
and rhabdocoeloid stage, 238
Turtles
coelomic epithelium in, 436
myeloid bodies in, 406
Tusk, 555, 557
Typhlosole
homology of, 298
Tyrode's solution, 23, 27
and embryo extract, 39-40
and Schwann cells, 115
Tyrosine
and growth, 73

Ultraviolet light
on cultures, 77
and eyes, 391, 395-7
and *Limulus*, 399
and Nissl's substance, 112
and planaria, 391, 397
and rhabdites, 218
Umbrella
of medusae, 217
Unipolar cells, 243, 333
Urea
and *Naegleria*, 168, 174
in urine, 450
Uridine triphosphate
and cell membranes, 221
Urine
amino acids in, 503
formation of, 450-1
Urinogenital system
in annelids and arthropods, 282, 440-2
homoiostasis and, 498-505
in invertebrates, 436-444
in nemerteoids, 269-70
pituitary control of, 505-11
in *Priapulus*, 443
of primitive chordates, 444
in rhabdocoeloids, 251-4
and sex cords, 467-75
and steroids, 487-98
in vertebrates, 282, 437, 444-53
Urochordates
derivation of, 360
and iodine, 353, 356
and nephridia, 444
Uroid
of lymphocytes, 23
of *Naegleria*, 156
Uromucoid
in proximal tubules, 503
Uronic acid
in cnidoblasts, 216
and connective tissue, 498
ecballic cells and, 220-1, 499-500
emballic cells and, 499-500
and mucopolysaccharides, 499
and steroids, 498
and sulphate, 223
Uterus
cell types in, 84
culture of, 67
oxytocin and, 344
steroids and, 488

Vacuoles (see also Contractile vacuoles)
in cells of gut, 268, 520
ciliated, 242
in cultured cells, 23-4
in developing muscle, 311, 520
in follicles of Langerhans, 517-8
in myeloblasts, 535
in *Naegleria*, 162
in neurites, 106-7
phagocytotic, 102
pinocytotic, 100-1
Vagina
keratinization of, 68
Valine
and growth, 73
and haemoglobin, 179, 184
Valinomycin
and ion transfer, 503
Van Gieson's stain
and cartilage matrix, 52
Vascular system
and coelom, 454, 468
development of, 454
and hormones, 344, 507
in nemerteoids, 270-1, 429
Vas efferens
cell types in, 84
and coelomoducts, 437
Vasopressin
and control of kidney, 507-8
and rhynchocoel, 345
Vegetalization
of embryos, 202, 246
Vegetal pole
of amphibian embryo, 200, 202
cells composing, 192, 197
emballism and, 227
and lithium, 202, 246
mercaptoethanol and, 200
morulae from, 200
and origin of germ cells, 470
of sponge embryo, 196, 227
Ventral nerve roots
origin of, 310, 312, 315, 332
Venule
post capillary, 539-40
Vesicles
in adrenal cortex, 482, 484-5
in cephalic organ, 374, 376, 378
in cultures of ependyma, etc., 403
in cultures of kidney, 198-9, 508, 552
in cultures of retina, 412-3
from ectoderm, 199
formed by chick blastoderm, 201
Golgi, 167
lens, 409
in monocytes, 539
in olfactory cells, 365
optic, 384, 403, 409
in pineal, 398
in planarian eye-cup, 256
synaptic, 108, 188, 308, 310, 312, 330
thyroid, 74, 76, 358
Vesicular tissues
in cephalic organ, 371-2, 376, 378, 386, 388, 465
Vestibular organ
origin of, 249, 362, 369-70, 418
Vie libre
blastulae et la, 202
Villi
chorionic, 246
of choroid plexus, 402
of intestine, 302, 522-3
Villus body
in rhynchocoel, 285, 289-90, 347-8
Vision
photosensitivity and, 233, 397
Visual acuity, 406, 416
Visual pigments (see also Rhodopsin, etc.)
distribution of, 233, 379, 412, 415
Visual threshold, 416
Vital dyes
and amoebocytes, 95, 124
and mechanocytes, 25, 27, 124
and macrophages, 89-90
Vitamin A (retinol)
on bone and cartilage, 97-8, 558
and cell differentiation, 82-3
and enamel organ, 548-9
and keratinization, 68, 83
and lysosomes, 52, 70, 380, 495, 558
and membranes, 70, 380, 495
and mitochondria, 495
and mucoprotein, 69-70, 82
and pigment epithelium, 406
and pineal eyes, 394
and retinene, 182, 255, 379
and SH-groups, 70
storage of, 499

Vitamin C (see also Ascorbic acid)
and collagen, 54
and intercellular substances, 71
Vitelline membrane
cultures on, 200-1
Vitreous body
and ciliary processes, 408
hyaluronic acid in, 53

Wandering cells, see Amoebocytes
Watch-glass
culture in, 31, 201
Water
balance (see also ionic balance), 167, 174, 254, 448, 501
and beat of flagella, 439-41
and cell types, 500
and *Chlamydomonas*, 159
and *Espejoia*, 159
and hypothalamus, 389
and *Naegleria*, 156, 174, 494
and nephric tubule, 450, 501, 505
and nucleus, 494
and ultraviolet light, 391
Wave-length
discrimination of, 415, 417
White body
of cephalopod, 381, 386, 389
Wings, 3
Witte's peptone, 73
Wolffian duct (body), 446, 455, 469, 472
Wound
behaviour of cells in, 17, 64, 99, 134
collagen in repairing, 553-4
fibrin in, 42
vitamin C and, 71

X-ray diffraction
and collagen, 53, 68
and keratin, 68
X-zone
of adrenal cortex, 483, 485

Yolk
in myoblasts, 311
in rhabdocoeloids, 253
storage in ova, 455, 478
Yolk sac
and blood cells, 525, 535
and germ cells, 470

Zinc
on chromosomes, 489
Zonae adherentes, 188
Zonae occludentes, 188-9
Zona fasciculata, 483
and ACTH, 529
and castration, 485-6
cholesterol and, 488
and cortisol, 499
cultures of, 487
emballism in, 486, 509-10
and glucocorticoids, 499, 502, 508
microvilli in, 485
pinocytosis in, 486
and steroids, 485-6, 488, 497, 499
Zona glomerulosa, 483
and ACTH, 529
aldosterone and, 487, 497, 499
angiotensin and, 505-6, 508
atrophy of, 488, 509, 529
cilia in, 509
cultures of, 487, 529
deoxycorticosterone and, 488, 509
ecballism in, 486, 508-10
hypophysectomy and, 509
lymphocytes in, 526, 529
and membrana granulosa, 508-9
and mineralocorticoids, 502, 508
phospholipins and, 485-6
and pineal, 396
and sex cords, 482
and sodium and potassium, 509
vascularity of, 487
vesicles in 482, 484-6
Zona pellucida, 472, 474
Zona reticulata, 483, 485
Zonula of Zinn, 408
Zymogen granules, 519